Digital Filter Design
Handbook

ELECTRICAL ENGINEERING AND ELECTRONICS

A Series of Reference Books and Textbooks

Editors

Marlin O. Thurston
Department of Electrical
Engineering
The Ohio State University
Columbus, Ohio

William Middendorf
Department of Electrical
and Computer Engineering
University of Cincinnati
Cincinnati, Ohio

Other Volumes in Preparation

Digital Filter Design Handbook

FRED J. TAYLOR

University of Florida
Gainesville, Florida

MARCEL DEKKER, INC.   New York and Basel

Library of Congress Cataloging in Publication Data

Taylor, Fred J. , [date]
  Digital filter design handbook.

  (Electrical engineering and electronics; 18)
  Includes bibliographies and index.
  1. Signal processing--Digital techniques.
2. Electric filters, Digital. I. Title. II. Series.
TK5102.5.T338   1983   6 21.38'043   83-18940
ISBN 0-8247-1357-5

MARCEL DEKKER, INC.
270 Madison Avenue, New York, New York 10016

Current printing (last digit):
10 9 8 7 6 5 4 3 2

PRINTED IN THE UNITED STATES OF AMERICA

To the men in my life—
  Rusty, my son,
  and Merton and Ralph,
  who knew him all too briefly

Preface

Digital signal processing (DSP) is the study of systems and signals with
respect to the constraints and attributes imposed on them by digital computing machinery. This science has become an essential element of modern
technology and has achieved widespread acceptance in such areas as filtering, data communication, picture processing, biomedicine, speech, radar,
sonar, electronic countermeasures, and many more. The history of this
study has been essentially evolutionary, with a few scattered key revolutions.

The origins of DSP can be found in the computational algorithms of
the seventeenth- and eighteenth-century mathematicians Sir Isaac Newton
and Karl Friedrich Gauss. Contemporary digital signal processing began
to form its identity in the 1940s and 1950s. In this period, sample data
systems appeared. They were used to perform low-frequency control operations which impose unacceptable design constraints on pure analog systems.
However, because of the absence of affordable analog delay lines, sample
data systems remained essentially an extension of linear continuous system
theory.

In the late 1950s and early 1960s, the digital processing of signals
using general-purpose computers began to appear. Digital correlation
methods were developed for processing seismic and aerospace data. Vocoders were simulated digitally and techniques were proposed for performing power spectrum analysis during this period. These early works de-
manded large quantities of computer time and resources. In addition, these
early DSP systems were resident in software. It would be a decade before
dedicated hardware systems would become a practical reality.

In 1960, I. J. Good developed methods by which a discrete Fourier
transform (DFT) could be computed in terms of a sparse matrix transform.
Unfortunately, there was a reluctance to allocate the limited computer resources of the day to exploit Good's discovery. Five years later, when
interest and computing resources became more abundant, J. W. Cooley

and J. W. Tukey achieved a similar breakthrough with their celebrated fast Fourier transform (FFT) algorithm. This extremely popular result represents one of the early revolutions in DSP and was important for several reasons. First, the FFT was able to achieve transform rates that were unattainable using existing methods. In addition, efficient memory utilization was achieved. Second, and perhaps most important, the FFT ushered in the era of true digital signal processing. The FFT transforms a digital rather than a sampled data base into the frequency domain. In addition, the algorithm was developed with regard to physical digital computer constraints. From this time on, an understanding of the structure of digital machinery would be essential to the study of digital signal processing systems.

The FFT was instrumental in creating new theories and design methodologies. Linear system theoretic concepts (such as convolution, correlation, power spectra, transfer functions) have been interpreted in a DFT sense. Besides serving in many application areas, the FFT caused us to consider the effects of finite word length, digital architecture, and digital arithmetic.

During this transition phase, another revolution of sorts was witnessed when J. F. Kaiser popularized the use of several versions of the z transform for use as a digital filter design tool. It was soon discovered that filters, which emulated their classic analog counterparts, could be designed in a straightforward manner. It was found that digital filters could be configured to approximate, or simulate, analog filters over a wide range of frequencies. Digital filters were found to be particularly effective in processing low-frequency signals. Because of their digital nature, results were always repeatable and therefore superior to their analog cousins, whose performance is sensitive to parametric and environmental effects. Furthermore, filters of high precision and large dynamic range can be achieved using digital hardware. Finally, unlike analog filters, digital filters may be reconfigured under program control.

Various filter configurations, premised on minimizing arithmetic error or execution speed, have been developed and tested. However, these early filters were generally implemented in software. Although they provide excellent performance in some speech, sonar, seismic, and medical applications, the high cost of computer mainframes a decade ago precluded the development of dedicated filters. As long as digital signal processing systems were restricted to reside in software, their utility would be limited.

The technological explosion of the 1970s carried digital signal processing into new prominence. Theory was mated with LSI, VLSI, microprocessors, high-speed digital arithmetic units, bipolar high-density memory, charge transfer devices, and so forth, to produce a wealth of new signal processing systems. The science of digital signal processing has become a study unto itself without dependence on analog experience of modeling methods.

The design and analysis of digital filters has been enriched by theoretical contributions in error prediction and minimization, system invariances, homomorphic and orthogonal transforms, and so on. However, it has been technology that has been the primal force in establishing the potential of this field. The superior speed, cost, and packaging metrics associated with digital filters are obtained principally through ever-improving hardware.

In this book, the design and analysis of digital filters are investigated in depth. Design procedures, protocols, architectures, and hardware are explored. This study represents the thoughts of many contemporary signal processing scientists. The material presented spans a wide range of theoretical and technological attitudes and practices. The study begins with an overview of the classic filters.

The material presented is the result of many years of study, experimentation, and research. The author has attempted to provide a comprehensive perspective on DSP theory and design. This includes computer-aided design (CAD), analysis, synthesis, and technology dependence. It is hoped that this work will serve the reader well in the years to come.

The author owes a major debt of gratitude to all those who helped in the preparation of this book. Special appreciation goes to the National Science Foundation, the Engineering Foundation, and the Air Force Office of Scientific Research for their support. Most of the praise must be reserved for my highly motivated and competent graduate students. Much of the experimental work between these covers can be attributed to them. Where possible, I have tried to give them the credit that they so richly deserve.

Fred J. Taylor

Contents

1
Introduction to Signal Processing

1.1 INTRODUCTION

The study of digital filters and signal processing systems requires a blend of theory and practice. Signal processing theory, design practices, architectures, and realizations are developed in later chapters. In this section the attributes of a digital filter are reviewed. However, to appreciate the virtues of a digital filter, its competitor, the analog filter, needs to be more fully understood. Accordingly, classical analog filter filtering is reviewed to provide a foundation for comparing these two filter philosophies.

1.2 CLASSICAL FILTER DESIGN

For decades, the frequency domain manipulation of analog data has been the responsibility of active and passive filter design theorists. Active and passive filters are generally classified in terms of the filter function they perform. The basic functions are:

1. Lowpass
2. Highpass
3. Bandpass
4. Bandreject or bandstop

In graphical terms, these filters appear in Fig. 1.2.1. It can be seen that the filters are specified in terms of

1. Passband
2. Transition band
3. Stopband

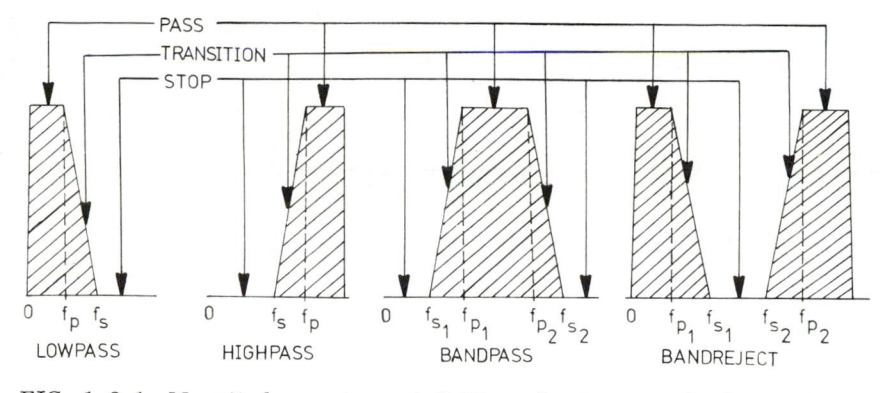

FIG. 1.2.1 Magnitude spectrum definitions for lowpass, highpass, band-pass, and bandreject filters.

The analytic properties of an active filter can be stated in terms of the filter's transfer function $H(s) = Z(s)/P(s)$. The zeros of the filter are found in $Z(s)$ while the poles come from $P(s)$. The analysis of an active filter concentrates on the study of the system pole-zero patterns. Various ladder and lattice filter architectures have been developed which simplify system design and analysis. The synthesis of an analog filter usually requires that the filter structure be prespecified. The desired frequency response is generally specified as well. In support of this design task, several classical approximation schemes have been developed which will now be studied (see Fig. 1.2.2).

1. Butterworth filter: The Butterworth filter provides a maximally flat passband (around zero frequency) and a monotonically decreasing gain in the stopband. The slope, or "skirt," of the filter's frequency response is less than the steep-skirt Chebyshev and elliptic filters.
2. Chebyshev filter: The Chebyshev filter provides an equiripple passband and a monotonically decreasing gain in the stopband.
3. Inverse Chebyshev filter: The inverse Chebyshev filter provides a maximally flat passband (around zero) and an equiripple stopband.
4. Elliptic filter: The elliptic filter provides an equiripple passband and an equiripple stopband.

The Butterworth approximation is named after its English author, while the Chebyshev approximation derives its name from the celebrated Russian mathematician. The elliptic approximation refers to a set of elliptic integral equations that define its structure. The elliptic filter is also known as the Cauer or Darlington filter. Each of these optimal filters satisfies a different criterion of "optimality." This should be remembered if and when one wishes to compare the performance of these classic filters.

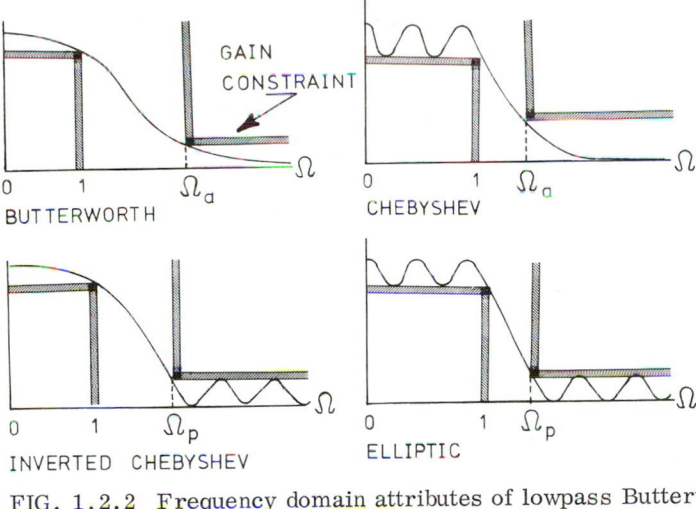

FIG. 1.2.2 Frequency domain attributes of lowpass Butterworth, Chebyshev, inverted Chebyshev, and elliptic filters.

Optimal filters, whether lowpass, bandpass, or other types, are generally referenced to a normalized lowpass model referred to as a <u>prototype</u> filter. For example, several important prototype optimal filters have the following magnitude-squared response (see Fig. 1.2.3).

$$H(j\omega)^2 = \frac{1}{1 + \omega^{2n}} \qquad \text{Butterworth}$$

A normalized lowpass filter can be mapped into a desired lowpass, bandpass, or other filter using the transformations listed in Table 1.2.1.

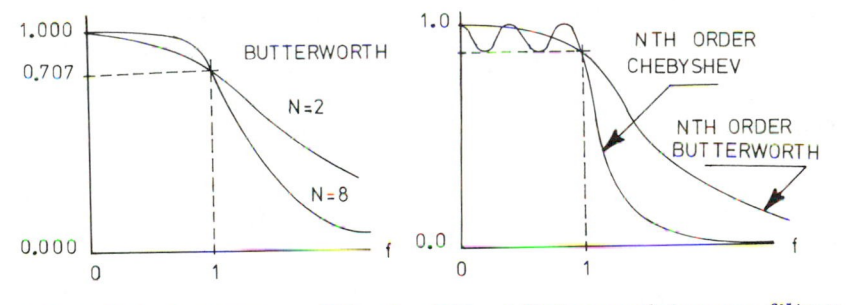

FIG. 1.2.3 Comparison of N = 2 and N = 8 Butterworth lowpass filters as well as a lowpass Butterworth versus Chebyshev performance as a function of filter order.

TABLE 1.2.1   Classical Frequency-to-Frequency Transforms for Translating a Normal Lowpass Filter to Low, High, Bandpass, or Bandreject Filters

PROTOTYPE LOWPASS TO:	TRANSFORM	CRITICAL FREQUENCIES
LP→LP(LOWPASS) $H(s) \rightarrow H(\tilde{s})$	$\tilde{s} = s/\omega_p$	ω_p
LP→HP(HIGHPASS) $H(s) \rightarrow H(\tilde{s})$	$\tilde{s} = \omega_p/s$	ω_p
LP→BP(BANDPASS) $H(s) \rightarrow H(\tilde{s})$	$\tilde{s} = (s^2 + \omega_{p_1}\omega_{p_2})/(s(\omega_{p_2} - \omega_{p_1}))$	$\omega_{p_1}, \omega_{p_2}$
LP→BS(BANDSTOP) $H(s) \rightarrow H(\tilde{s})$	$\tilde{s} = (s(\omega_{p_2} - \omega_{p_1}))/(s^2 + \omega_{p_1}\omega_{p_2})$	$\omega_{p_1}, \omega_{p_2}$

The design of optimal analog filters has become a highly computerized study. However, to understand this class of optimal filters more completely, and to lay the groundwork for their digital realizations, a conventional treatment of these basic filters will be pursued. In addition, for the purpose of completeness, a fourth filter, known as the Bessel filter, will be introduced.

Butterworth

A normalized nth-order Butterworth filter model can be expressed as $H(j\omega)$, where

$$|H(j\omega)|^2 = \frac{1}{1 + \epsilon^2 \omega^{2n}} \qquad (1.2.1)$$

having a gain at the normalized critical frequency of $\omega = 1$, equal to $|H(j\omega)|^2 = 1/(1 + \epsilon^2)$. In terms of a decibel metric, the filter gain at $\omega = 1$ is -3 dB if $\epsilon = 1.0$, and the resulting prototype filter is referred to as a "3-dB" filter. The poles of a 3-dB filter, for order less than four, are shown in Fig. 1.2.4.

Observe that the poles of an nth-order Butterworth filter lie on the periphery of a unit circle in the s domain and are separated from each other by π/n radians. A given 3-dB lowpass filter can be converted to a $-10 \log(1 + \epsilon^2)$-dB lowpass filter with a bandpass cutoff frequency of ω_c by replacing s by $s = \epsilon^{1/n}(s/\omega_c)$.

EXAMPLE 1.2.1 A first-order 1-dB Butterworth filter with a 1000-rad/sec bandpass cutoff frequency can be designed as follows. The lowpass prototype filter is given by

$$H(s) = \frac{1}{s+1}$$

Direct computation implies that –1 dB equates to $\epsilon = 0.51$. Therefore, the resulting desired lowpass filter satisfies

$$H(s)\Big|_{s \to 0.51s/10^3} = \frac{10^3/0.51}{s + 10^3/0.51} = \frac{(1.96)10^3}{s + (1.96)10^3}$$

Chebyshev

The magnitude-squared response of an nth-order Chebyshev filter is given by

$$|H(j\omega)|^2 = \frac{1}{1 + \epsilon^2 C_n(\omega)} \qquad (1.2.2)$$

where $C_n^2(\omega) = n \cosh^{-1}(\omega)$. The Chebyshev polynomials can be computed recursively as follows:

$$C_0(\omega) = 1$$

$$C_1(\omega) = \omega \qquad (1.2.3)$$

$$C_{n+1}(\omega) = 2\omega C_n(\omega) - C_{n-1}(\omega)$$

The gain parameter ϵ is chosen so as to satisfy a prespecified magnitude (gain) constraint established at a prespecified passband cutoff frequency. The poles of a few Chebyshev filters are shown in Fig. 1.2.5.

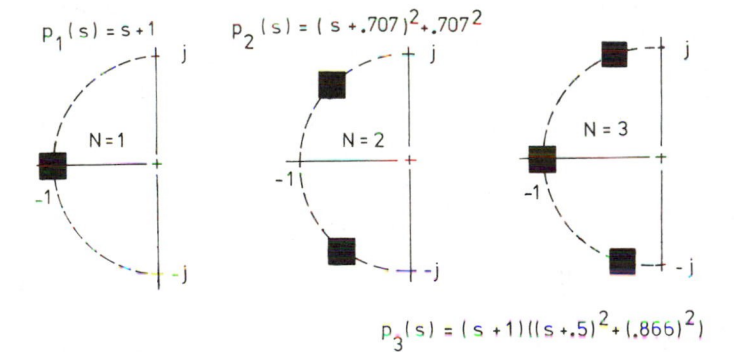

$$P_1(s) = s+1 \qquad P_2(s) = (s+.707)^2+.707^2$$

$$P_3(s) = (s+1)((s+.5)^2+(.866)^2)$$

FIG. 1.2.4 Pole locations of first-, second-, and third-order Butterworth filters in the s domain.

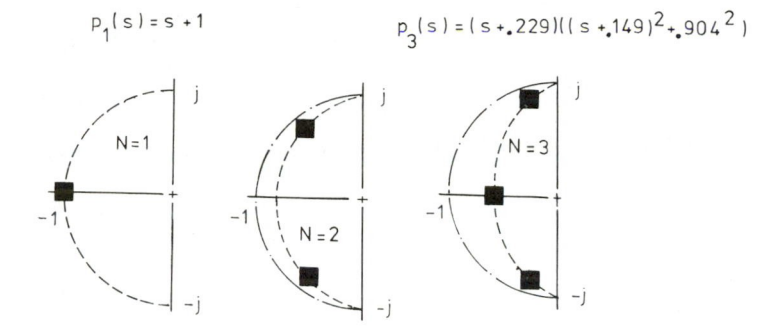

$$P_1(s) = s + 1 \qquad\qquad P_3(s) = (s + .229)((s + .149)^2 + .904^2)$$

$$P_2(s) = ((s + .325)^2 + .777^2)$$

FIG. 1.2.5 Pole locations of first-, second-, and third-order Chebyshev filters in the s- domain.

Observe that the poles are no longer located on the unit circle. In fact, the poles of the nth-order Chebyshev filter fall on an ellipse with real- and imaginary-axis intercepts given by

$$\pm \sinh\left[\frac{1}{n}\ \sinh^{-1}\left(\frac{1}{\epsilon}\right)\right], \quad \pm \cosh\left[\frac{1}{n}\ \sinh^{-1}\left(\frac{1}{\epsilon}\right)\right] \qquad (1.2.4)$$

Compared to the Butterworth filter, the Chebyshev filter leads to improved overall filter performance. The price paid is the added algebra required to compute the poles of an arbitrary Chebyshev filter.

Elliptic

Elliptic filters derive their name from the elliptic integral equations used to define this filter function. Elliptic filters are more efficient than Butterworth or Chebyshev filters in optimizing the transition from passband to stopband. Therefore, this class of filter is sometimes referred to as a steep-skirt filter. The details of this complex filter are developed in detail in a future section on discrete elliptic filters.

Bessel

In general, a filter's frequency response can be expressed as

$$H(\omega) = H(\omega)\ \exp\left[j\phi(\omega)\right] \qquad (1.2.5)$$

The first derivative of the phase function with respect to ϕ is called the group delay. A Bessel lowpass prototype filter has the property that the filter's group delay is maximally flat about $\omega = 0$.

The Bessel filter is defined in terms of Bessel polynomials. However, this filter has not gained much popularity in the digital arena. If phase linearity is a major concern, another class of digital filter (the nonrecursive filter) is known to be superior. The development of this class of filter is presented in a subsequent chapter in a group delay context.

1.3 LADDER STRUCTURES

Another important structure is the <u>ladder filter</u> of alternating series and parallel impedances. A block diagram of a ladder structure is shown in Fig. 1.3.1.

The impedance elements of this structure can generally be computed using long division over a real coefficient field. That is, the filter's impedance, say $Z_{in}(s)$, can be expressed as in Eq. (1.3.1).

$$Z_{in}(s) = Z_1(s) + \cfrac{1}{Y_2(s) + \cfrac{1}{Z_3(s) + \cdots + 1/Y_n(s)}} \qquad (1.3.1)$$

Finally, a filter's transfer function can, with some algebraic manipulation, be written in terms of the network impedance $Z_{in}(s)$.

EXAMPLE 1.3.1 The analysis of the following network is formed as follows:

$$Z_{in}(s) = \frac{3 + 4s + s^2}{2s + s^2}$$

$$= \frac{3}{2s} + \frac{1}{1/[(25/2s) + 1]}$$

$$= \frac{1}{sC_1} + \frac{1}{(1/R_1) + [1/(sC_2 + R_2)]}$$

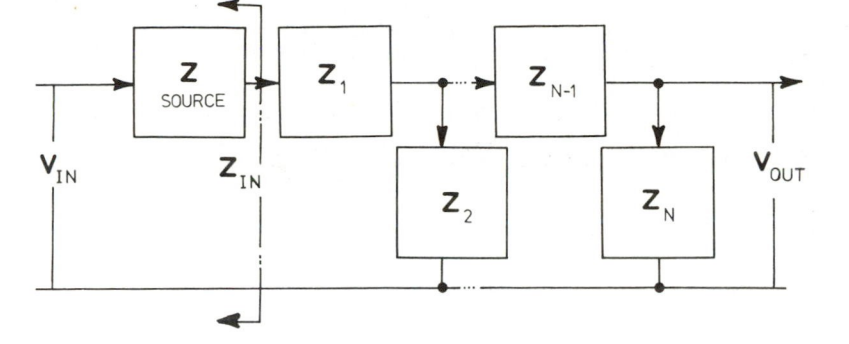

FIG. 1.3.1 Classical ladder structures, where Z_i denotes a general impedance, Z_{source} a source impedance, and Z_{in} an input impedance.

1.4 ANALOG FILTER PERFORMANCE

Active and passive filters represent a classic study. Such filters can be realized with the use of RC, LC, or RLC components. The theory and practice of this mature field is well documented and the interested reader should have no difficulty finding suitable reference material on this subject. However, some shortcomings to analog filtering are known to exist. For example, active and passive filters must be designed with realistic values for R, L, and C. However, because of the physical limitations of passive components, the frequency response of a realizable analog filter is limited. They are known to perform poorly at very low or high frequencies. Analog filters often require extensive and expensive alignment and adjustment. The maintenance of an analog filter is required throughout the entire life span of the filter. Linear phase behavior is difficult to achieve. The fact that analog filters cannot be reprogrammed limits their utility and design flexibility. The alternative technology, the digital filter, overcomes these limitations.

1.5 CRITICAL COMPARISON

Digital signal processing is a child of the solid-state electronics explosion of the 1970s. The fallout from this explosion is a wealth of high-performance, low-cost electronic widgets that are capable of manipulating signals in an algorithmic manner never before possible. Although it is true that the analog filter industry has felt the impact of integrated circuits, the degree of this impact can in no way compare to that caused in the digital semi-conductor industry.

   For example, active filters have for decades been based on operational amplifiers. The availability of these devices in high-density,

low-cost packages has made analog filters more economical but has otherwise left the design methodology of the field unchanged. This is not the
case in the digital area, however. Here, advances in technology, such as
the microprocessor metal-oxide semiconductor, have created a whole new
field known as digital signal processing. The ability to combine a $5 analog-
to-digital converter with a $10 microprocessor, $20 memory, and a $2
digital-to-analog converter has propelled this field into a position of prominence. Therefore, it is safe to say that digital signal processing has had a
most precocious youth.

In fairness, one should note that several new analog-type technologies
have recently evolved: the charge-coupled device and the switched capacitance device. Although restricted in their ability to work over a wide range
of frequencies, they do represent an innovation. Therefore, these two
units are given special consideration in this book.

From a theoretical viewpoint, digital filters provide a rich forum for
study. Because of its richer theoretical base, one should expect more
advanced digital signal processing (DSP) systems to be developed in the
future which are based on mathematical abstractions and thought. The reason that DSP has developed a higher level of mathematical consciousness is
attributable to the hardware in which these systems reside. Whereas analog systems are defined over a real coefficient field, digital filters are
defined over a finite coefficient field. The finite field is a result of the fact
that digital words have a finite precision. The mathematical skill levels
required to work in this area of algebra at an advanced level is higher than
that required of real algebra.

The digital versus analog filter controversy can be quantified in terms
of the following observations:

1. Digital filters can be fabricated in available high-density, low-cost
   digital hardware. This should continue to improve as digital signal
   processing "bootstraps" itself to the continued advancements in the
   digital electronics industry.
2. Guaranteed stability of certain classes of digital filters.
3. There are no impedance-matching problems.
4. Digital filters are able to work at extremely low frequencies that cannot be supported by analog filters.
5. Digital filters are programmable if realized with a programmable
   processor.
6. Digital filters efficiently support computer-aided data analysis [simple
   input-output (I/O) logging and transfer], and can often be interfaced to
   the power supplies and mechanical structure of existing digital processors.
7. Digital filters can work over a wide range of critical frequencies.
   This is a difficult task for analog filters.
8. Digital filters can be used in conjunction with data compression (i.e.,
   input and/or output) schemes.

9. Certain digital filters possess outstanding phase linearity.
10. High accuracy and precision can be achieved using digital filters. At best, the precision of an analog filter may be expected not to exceed 60 to 70 dB. The precision of a digital filter, however, can be extended simply by increasing word length. For example, 10 bits of resolution corresponds to the dynamic range 20 log(1024) to 60 dB.
11. Digital filters do not require periodic alignment as is the case for analog filters. Furthermore, digital filters do not "drift" due to parameter aging or environmental changes.
12. Because of their digital nature, digital filters are less sensitive to certain classes of noise which corrupt analog filters (e.g., line-frequency noise).

The disadvantages of the digital filter are few. They are as follows:

1. Digital filters are subject to quantization noise disturbances.
2. Digital filters may exhibit limit cycling.
3. Hardware development time may be longer.
4. The design and synthesis of a digital filter may be more algebraically difficult unless a general-purpose digital computer is used to support the design task.

Many of these disadvantages can be overcome through the use of good digital design practices, a familiarity with hardware, and solid analysis protocols and procedures. The intent of this book is to provide these skills.

BIBLIOGRAPHY

Budak, A. (1974), _Passive and Active Network Analysis and Synthesis_, Houghton Mifflin, Boston.

Cooley, J. W., and J. W. Tukey (1965), An Algorithm for the Machine Calculation of Complex Fourier Series, Math. Comput., 19.

Kaiser, J. F. (1966), _Digital Filters in System Analysis by Digital Computer_, Wiley, New York.

McClellan, J. H., and C. M. Rader (1979), _Number Theory in Digital Signal Processing_, Prentice-Hall, Englewood Cliffs, N.J.

Sedra, A. S., and P. O. Brackett (1978), _Filter Theory and Design: Active and Passive_, Matrix Publishers, Champaign, Ill.

Temes, G. C., and S. K. Mitra (1973), _Modern Filter Theory and Design_, Wiley, New York.

Zveriv, A. I. (1967), _Handbook of Filter Synthesis_, Wiley, New York.

2
Mathematical Preliminaries

2.1 INTRODUCTION

Signal processing concerns itself with the problem of manipulating and managing information. The information source—signals—may appear in one of several forms. A continuous signal process is one that is continuously resolved in both its independent and dependent axis. A discrete signal process is discretely resolved (quantized) in the dependent variable. A digital signal process is discretely resolved in both variables. For the moment our study will concentrate solely on discrete signal representation. Periodic (uniform) sampling is the most commonly used method of converting an analog signal to a discrete format. It will be assumed, unless otherwise stated, that periodic sampling is to be used to create a discrete data base. The mathematical connection between a time series and its parent analog signal was established by Shannon (1949) in his celebrated sampling theorem. This theorem equates the information content of a time series with that of the sampled analog signal.

Shannon's Theorem. Shannon stated that if $x(t)$ is a signal having f_{max} as its highest-frequency component, and $x(t)$ is periodically sampled so that the sample period T is less than or equal to $T \leq 1/2f_{max}$ (referred to as the Nyquist sample rate), then $x(t)$ can be interpolated from its time series under the following rule:

$$x(t) = \sum_{n=-\infty}^{\infty} \frac{x(nT) \sin [\omega_c(t - nT)]}{\omega_c(t - nT)} \qquad (2.1.1)$$

By weighting the sample values of $x(t)$ by a $(\sin x)/x$ type of window, the original analog signal can be reconstructed. Equation (2.1.1) represents an interpolation formula which is a theoretical milestone. Unfortunately, this window is difficult to implement in hardware. Instead, simpler windows are generally used. These interpolation schemes generally produce acceptable data reconstruction in limited hardware. Two of the more commonly used interpolation forms are the zero-order hold and the first-order hold, which are discussed next.

2.2 INTERPOLATION FORMS

The zero-order hold is the simplest of the two reconstruction algorithms. It can be modeled in the time domain in the manner suggested in Fig. 2.2.1. The impulse response of the zero-order hold is given by h(t), where

$$h(t) = \begin{cases} 1, & 0 \le t \le T \\ 0, & \text{otherwise} \end{cases}$$

$$H(s) = \frac{1 - \exp(-sT)}{s}$$

$$H(j\omega) = \frac{1 - \exp(-j\omega T)}{T}$$

$$= \frac{T \exp(-j\omega T) \sin(\omega T/2)}{\omega T/2}$$

(2.2.1)

Suppose that the sample values of an analog signal process x(t) are given by $\{x(n)\}$. Using the zero-order hold, a piecewise constant approximation of x(t), say x(t), can be generated in the manner outlined in Fig. 2.2.2. The electronic subsystems required to support the zero-order-hold operation are readily available in low-cost hardware.

The first-order hold is more complex mathematically and electronically and has the following interpretation:

$$x \simeq x(n) + \frac{x(n) - x(n-1)}{T}$$

(2.2.2)

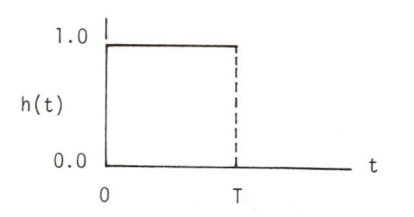

FIG. 2.2.1 Graphical interpretation of zero-order-hold transfer function.

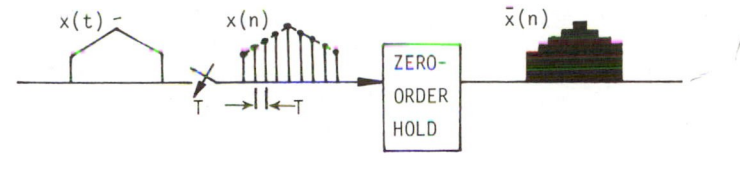

FIG. 2.2.2 Input–output signal traces for a zero-order-hold and signal re-construction network.

which induces the filter impulse response h(t) shown in Fig. 2.2.3.

2.3 TRANSFORM REPRESENTATIONS

In the study of discrete systems, interpolation formulas provide the user with the means to reconstruct an analog approximation of an original signal process. If the original signal were not continuous, however, f_{max} would be unbounded (i.e., infinite). According to Shannon, the reconstruction of the original signal from its sample values would be a futile exercise. There are times, however, when signal reconstruction is of little or no concern to the user. In particular, the signal processing scientist may only be interested in manipulating and analyzing a signal process solely on the basis of its sample values. Under this specific hypothesis, algebraic reconstruction is not required.

The analysis of a discrete system can be performed using transform methods. Just as transforms played a major role in linear system theory (i.e., Laplace, Fourier), transforms are also instrumental in the study of discrete systems. One of the popular discrete transforms is the z transform. The z transform belongs to a class of algebraic operations referred to as <u>impulse invariant transforms</u>. An impulse-invariant transform pre-serves an important system property that relates to the sample values of

$$h(t) = \begin{cases} (t+1)/T; & -T/2 \le t \le 0 \\ -(t+1)/T; & 0 \le t \le T/2 \end{cases}$$

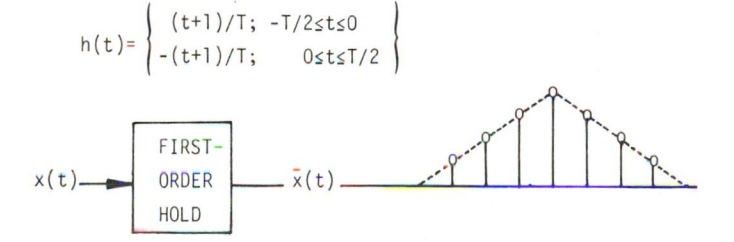

FIG. 2.2.3 First-order-hold and signal reconstruction network.

a transformed signal and system. Suppose that the output of a transformed analog system, having an impulse response h(t) and forcing function x(t), is y(t). Mathematically, y(t) is defined in terms of the convolution of x(t) and h(t), which is denoted $y(t) = x(t) * h(t)$. Furthermore, suppose that we wish to represent these system functions as a weighted infinite series of the form

$$z(t) = \sum_{i=-\infty}^{\infty} \overline{z}(i) \Phi_i(t) \qquad (2.3.1)$$

where $\overline{z}(i)$ is the sampled value of z(t) at $t = iT$, with T denoting the sample period. An impulse-invariant transform would guarantee that the output of the analog system and transformed system agree at the sample instances. That is, if y (i) is the output of the system based on an Eq. (2.3.1) representation at $t = iT$, then impulse invariance implies that $y(i) = y(iT)$. Oppenheim and Johnson (1972) have shown that if

$$\Phi_n(s) = [\Phi_1(s)]^n \qquad (2.3.2)$$

the resulting transform is impulse invariant. It is also known that a reconstructed signal derived using Shannon's interpolation formula will always agree with the original signal at the sample instances. This is true independent of the bandwidth restriction imposed on the original analog signal. Therefore, it can be observed from Eqs. (2.2.1) and (2.3.1) that $\Phi_i(t)$ satisfies

$$\Phi_i(t) = \frac{\sin(\pi/T)(t - nT)}{\pi(t - nT)} \qquad (2.3.3)$$

and it follows that

$$\Phi_i(s) = \begin{cases} \exp(snT) = [\exp(sT)]^n, & |\omega T| \leq \omega \\ 0, & \text{otherwise} \end{cases} \qquad (2.3.4)$$

The preceding statement may appear familiar to some readers since it represents the <u>standard z transform</u>, which can now be explicitly defined as follows:

$$z = \exp(sT) \qquad (2.3.5)$$

This specific transform is but one of many versions of the z transform found in contemporary use. However, the standard z transform is one of the most popular forms of the family of z transforms. In addition, it provides an excellent conceptual forum from which many of the salient features of such transforms can be discussed.

2.4 THE Z TRANSFORM

The z transform of a signal x(i) shares many attributes with the Laplace transform of x(t). Functionally, they both rely on similar methods of performing forward and inverse transforms. For example, both make extensive use of standard tables of transforms which allow the desired map to be determined through "table-look up" operations. More specifically, a given mathematical expression would be algebraically manipulated into a form containing terms found in a standard table of transforms. These recognized terms would then be collected mathematically to form the required transform. Some of the more useful transforms are summarized in Taylor (1975) or Smith and Cadzow (1974).

The fundamental algebraic and system theoretic properties of the z transform are listed below. There are many extensions to the data presented in this list and the interested reader should refer to a reference on discrete linear system theory if a deeper understanding of these properties is desired.

1. The z transform of $\{x(i)\}$ exists if and only if $\{x(i)\}$ is unique and bounded for all n.
2. Two z transforms $X(z)$ and $Y(z)$ are equal if and only if $\{x(i)\} = \{y(i)\}$.
3. The z transform of $x(t/T)$ is independent of the sampling period T.
4. The z transform is a <u>linear operator</u> in that if $x(n) \rightarrow X(z)$ and $y(n) \rightarrow Y(z)$, then if $z(n) = x(n) + y(n)$, it follows that $Z(z) = X(z) + Y(z)$.
5. The shifting property holds in that if $y(n) = x(n + m)$ and $x(n) \rightarrow X(n)$, then $Y(z) = z^m X(z)$.
6. The convolution property holds in that if $x(n) \rightarrow X(z)$ and $y(n) \rightarrow Y(z)$, then $z(n) = x(n) * y(n)$ implies that $Z(z) = X(z)Y(z)$. In terms of a more formal mathematical statement, one observes that

$$z(n) = \sum_{k=-\infty}^{\infty} x(k)y(n - k)$$

$$= \sum_{k=-\infty}^{\infty} x(n - k)y(k) \overset{Z}{\longleftrightarrow} X(z)Y(z) \qquad (2.4.1)$$

The system theoretic properties of discrete linear systems are explored in greater depth later in the chapter.

From the definition of z [i.e., $z = \exp(sT)$], one notes that the z transform will be unique only if the modulus of sT is bounded by $\pm\pi$ (i.e., $|sT| \leq \pi$).

Letting $s = j\omega = j2\pi f$, one immediately concludes that $|2\pi fT| \leq \pi$, which in turn implies that $|f| \leq 1/2T$. The parameter $1/2T$ holds a place of particular significance in signal processing and is referred to as the Nyquist frequency. The portion of the s plane which admits a unique mapping into the z plane is summarized in Fig. 2.4.1. Referring to this figure, the fundamental signal processing concept known as aliasing can be explained. An aliasing signal derives its name from the fact that it "impersonates" another signal. This is a result of the fact that the z transform, $z = \exp(sT)$, possesses cyclic behavior. It is immediately apparent that $s = \sigma + j\omega$ will be mapped into $z = \exp(\sigma T) \exp(j\omega T)$. Concentrating on the second term in this expression it can be noted that the value of z is repeated for all $\omega T = 2k\pi + \omega_0$. That is, the z transform of a pure harmonic oscillation at a frequency of ω_0 is identical to the transform applied to a signal having a frequency of $\omega = \omega_0 + 2k\pi/T$. For example, the sample values of $x_1(t) = 1$ and $x_2(t) = \cos(2\pi t/T)$ are identical if the sample period if T. The result of confusing $x_1(k)$ with its impersonator $x_2(k)$ is referred to as an aliasing error. Aliasing errors will occur only if the signal being sampled at a rate that is less than the Nyquist sample rate. To minimize the probability and effect of an aliasing error, systems are often outfitted with antialiasing filters. These filters are lowpass units having strong stopband attenuation (e.g., -80 dB) and a cutoff frequency equal to the Nyquist frequency. Such a filter will suppress high-frequency signal which would otherwise cause aliasing errors.

It is worthwhile to observe that the $j\omega$ axis in the s plane is mapped onto the periphery of the unit circle in the z plane. Also, it is well known that the stable region of the s plane, the left-half plane, is mapped into the unit circle in the z plane.[†] That is, poles in the s domain having a negative real part will be mapped into an interior point of the unit circle. The mapping constraints are summarized graphically in Fig. 2.4.1.

The inversion of a given X(z) is formally defined in terms of the following integral equation:

$$x(n) = Z^{-1}[X(z)] = \frac{1}{2\pi j} \oint_C \frac{X(z)z^n \, dz}{z} \tag{2.4.2}$$

where C is a restricted closed path found in the z plane. However, one rarely approaches the problem of inversion through the use of this integral equation. Instead, more simplified inversion methods have been developed which expedite the inversion process. The most popular of these methods are:

[†]Some authors graphically interpret the z transform in the z^{-1} space. Here the stable region is found outside the unit circle.

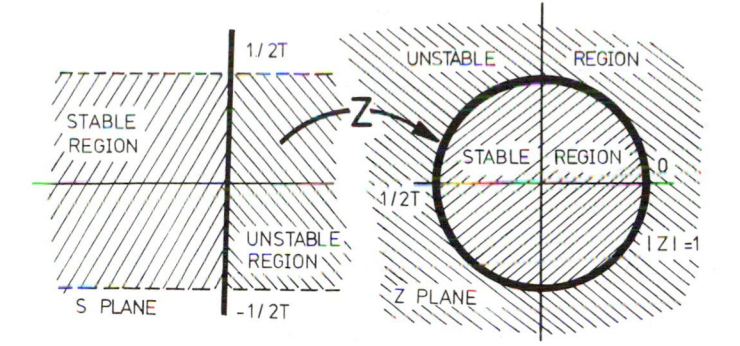

FIG. 2.4.1 Conformal mapping of the s plane into the z plane under the z transform.

1. Long division
2. Partial fraction expansion
3. Residue theorem

These three methods will now be explained algebraically.

Long Division

If $X(z)$ is the ratio of two rational polynomials, say $X(z) = N(z)/D(z)$, given by

$$X(z) = \frac{N(z)}{D(z)} = \frac{\displaystyle\prod_{i=0}^{M} a_i z^{-i}}{\displaystyle\prod_{i=0}^{N} b_i z^{-i}}$$

$$= \prod_{i=0}^{\infty} c_i z^{-i} \qquad\qquad (2.4.3)$$

then one concludes, in light of Eq. (2.4.3), that the coefficients c_i are the elements of the time series $\{x(n)\}$. This observation can be motivated in terms of the following examples.

EXAMPLE 2.4.1

$$X(t) = \begin{cases} 1, & t \geq 0 \\ 0, & \text{otherwise} \end{cases}$$

$$X(t) \overset{Z}{\longleftrightarrow} \frac{z}{z-1} = \frac{N(z)}{D(z)} = X(z)$$

EXAMPLE 2.4.2

$$\frac{N(z)}{D(z)} \Rightarrow z - 1 \sqrt{z} \; \frac{1 + z^{-1} + z^{-2} +}{\rule{3cm}{0.4pt}} \quad \cdots \Rightarrow X(z) = 1 + z^{-1} + z^{-2} + z^{-3} + \cdots$$

$$\underline{z - 1}$$
$$1$$
$$\underline{1 - z^{-2}}$$
$$z^{-2} \qquad \left\{ x(n)^Z \right\} = \left\{ 1, \ 1, \ 1, \ 1, \ \ldots \right\}$$

Partial Fraction Expansion (Heaviside's Method)

Consider again $X(z)$ defined as before. Suppose further that $D(z)$ is an nth-order polynomial which admits the following factorization:

$$D(z) = \prod_{i=1}^{L} (z - \lambda_i)^{n(i)}; \quad \sum_{i=1}^{L} n(i) = n \qquad (2.4.4)$$

where λ_i is a root (eigenvalue) of $D(z)$ and therefore is also a pole of $X(z)$. The parameter $n(i)$ represents the multiplicity of the pole λ_i. For the sake of clarity it will be assumed that $n(1) = 1.0$ and $n(2) = M$. Then

$$X(z) = \frac{N(z)}{D(z)}$$

$$= \frac{c_{11} z}{z - \lambda_1} + \left[\frac{c_{21} z}{z - \lambda_2} + \frac{c_{22} z}{(z - \lambda_2)^2} + \cdots \right.$$

$$= \left. \frac{c_{2M} z}{(z - \lambda_2)^M} \right] + \text{others} \qquad (2.4.5)$$

It is well known that the Heaviside coefficients c_{ij} are given by

$$c_{11} = \lim_{z \to \lambda_1} (z - \lambda_1) \frac{X(z)}{z}$$

$$c_{2M} = \lim_{z \to \lambda_2} (z - \lambda_2)^M \frac{X(z)}{z}$$

$$c_{2, M-1} = \lim_{z \to \lambda_2} \frac{d}{dz} (z - \lambda_2)^M \frac{X(z)}{z} \qquad (2.4.6)$$

$$c_{2, j} = \lim_{z \to \lambda_2} \frac{1}{j!} \frac{d^{M-j}}{dz^{M-j}} (z - \lambda_2)^M \frac{X(z)}{z}, \quad 1 \le j < M$$

which can be further motivated in terms of the following example.

EXAMPLE 2.4.3 For the sake of demonstration, define x(t) in the following manner:

$$x(t) = 1 + t$$

$$X(z) = \frac{z^2}{(z-1)^2}$$

Then

$$\frac{X(z)}{z} = \frac{c_{11}}{z-1} + \frac{c_{12}}{(z-1)^2}$$

$$c_{12} = \frac{(z-1)^2}{z} X(z)\Big|_{z=1} = z\Big|_{z=1} = 1$$

$$c_{11} = \frac{d}{dz} \frac{(z-1)^2 X(z)}{z}\Big|_{z=1} = \frac{dz}{dz}\Big|_{z=1} = 1$$

That is,

$$\frac{X(z)}{z} = \frac{1}{z-1} + \frac{1}{(z-1)^2}$$

or

$$X(z) = \frac{z}{z-1} + \frac{z}{(z-1)^2}$$

Referring to the table of z transforms, it becomes immediately apparent
that

$$\frac{z}{z-1} \longrightarrow 1; \qquad \frac{z}{(z-1)^2} \longrightarrow kT$$

or, in terms of an analog process, x(t) = 1 + t.

Residue Theorem

A function $X(z)$ is said to be analytic at $z = z_0$ if and only if it is single
valued and uniquely differentiable at z_0. If $X(z)$ is analytic at z_0, then
$X(z)$ admits a Taylor series expansion of the form

$$X(z) = \sum_{i=0}^{\infty} a_i(z - z_0)^i$$

$$a_0 = X(z_0)$$

(2.4.7)

$$a_n = \frac{1}{n!} \left. \frac{d^n X(z)}{dz^n} \right|_{z=z_0}$$

If $X(z)$ is not analytic at z_0, a Laurent series expansion is required and is given by

$$X(z) = \sum_{i=0}^{\infty} a_{-i}(z - z_0)^{-i}$$

(2.4.8)

$$a_{-i} = \frac{1}{2\pi j} \oint_C \frac{X(z)\ dz}{(z - z_0)^{i+1}}$$

The coefficient a_{-1}, called the <u>residue of X(z) at</u> z_0, holds an important place in z-transform mechanics. It is known that a_{-1} satisfies

$$a_{-1} = \frac{1}{2\pi j} \oint_C X(z)\ dz$$

(2.4.9)

where the closed integration path C contains no singularities (poles) of $X(z)$ except at z_0. If z_0 is not a pole of $X(z)$, then a_{-1} has a value of zero. Using this principle, the residue theorem states that

$$x(n) = \sum [\text{residues of } X(z)z^{n-1}]$$

(2.4.10)

which is further motivated in the following example.

EXAMPLE 2.4.4   Consider again $X(z) = z^2/(z - 1)^2$. It was previously shown that $x(n) = 1 + n$. Using a Heaviside expansion, one obtains

$$X(z)z^{n-1} = \frac{z^2(z^{n-1})}{(z - 1)^2} = \frac{z^{n+1}}{(z - 1)^2}$$

$$= \frac{c_{-2}}{(z - 1)^2} + \frac{c_{-1}}{z - 1}$$

where the residue c_{-1} is given by

$$c_{-1} = \lim_{z \to 1} \frac{d}{dz} (z-1)^2 X(z) z^{n-1}$$

$$= \frac{d}{dz} (z^{n+1}) \Big|_{z=1} = n+1$$

which can be recognized to be the correct answer.

EXAMPLE 2.4.5 Consider now an example of added complexity. Let $X(z)$ be given by $X(z) = (z+1)^3/(z+0.5)(z-0.5)^2$. There is a simple pole at $z = -0.5$ and a pole of multiplicity two at $z = +0.5$. It follows that

$$X(z)z^{n-1} = \frac{(z+1)^3 z^{n-1}}{(z+0.5)(z-0.5)^2} \triangleq Q(z)$$

Observe that for $n = 0$, $X(z)z^{n-1}$ becomes $(z+1)^3/z(z+0.5)(z-0.5) = Q_0(z)$, which can be expressed as

$$Q_0(z) = \frac{A}{z} + \frac{B}{(z+0.5)} + \frac{C}{(z-0.5)^2} + \frac{D}{(z-0.5)}$$

where A, B, and D are residues. There values are computed to be

$$A = zQ_0(z)\Big|_{z=0} = \frac{1}{(0.5)(-0.5)^2} = 8; \quad B = (z+0.5)Q_0(z)\Big|_{z=-0.5} = -0.25$$

$$D = (z-0.5)^2 Q_0(z)\Big|_{z=0.5} = -6.75 \quad \text{(C not a residue, not computed)}$$

For $n > 0$,

$$Q_n(z) = z^{n-1} X(z) = \frac{(z+1)^3 z^{n-1}}{(z+0.5)(z-0.5)^2}$$

or, in partial fraction form

$$Q_n(z) = \frac{A}{z+0.5} + \frac{B}{(z-0.5)^2} + \frac{C}{(z-0.5)}$$

where A and C are now residues. Their values are given by

$$A = \frac{z^{n-1}(z+1)^3}{(z-0.5)^2}\Bigg|_{z=-0.5} = \frac{(-0.5)^{n-1}(0.5)^3}{(-1)^2} = -0.25(-0.5)^n$$

$$C = \frac{d}{dz}\left[\frac{z^{n-1}(z+1)^3}{(z+0.5)}\right]\Bigg|_{z=0.5} = \frac{[(n-1)z^{n-2}(z+1)^3 + z^{n-1}(z+1)^2]}{(z+0.5)}$$

$$+ z = 0.5 - \frac{[z^{n-1}(z+1)^3]}{(z+0.5)^2}\Bigg|_{z=0.5}$$

$$= \frac{n(1.5)^3(0.5)^n}{(0.5)^2} - \frac{(1.5)^3(0.5)^n}{(0.5)^2} + \frac{3(1.5)^2(0.5)^n}{(0.5)} - \frac{(1.5)^3(0.5)^n}{(0.5)}$$

$$= 13.5n(0.5)^n - 6.75(0.5)^n$$

As a result, the resulting time series is given by

$$x(n) = \begin{cases} (A+B+D) = 8.0 - 0.25 - 6.75 = 1.0, & \text{for } n = 0 \\ 13.5n(0.5)^n - 6.75(0.5)^n, & \text{for } n > 0 \end{cases}$$

As a check, one could test the first several values of $x(n)$ using long division. Observe also that $x(\infty) = 0$.

2.5 STEADY-STATE ANALYSIS

The standard z transform for the most commonly encountered signal can be found in Cadzow (1974) and Taylor and Smith (1975). For the sake of discussion, consider the one-sided exponential $h(t) = \exp(-\lambda t)$ for $t > 0$. This function is interpreted graphically in Fig. 2.5.1. The transform of this signal can be deduced from the knowledge that $y(n) = a^n$ translates to a $Y(z) = z/(z-a)$. Matching the sample values of $h(n)$ and $y(n)$, one concludes

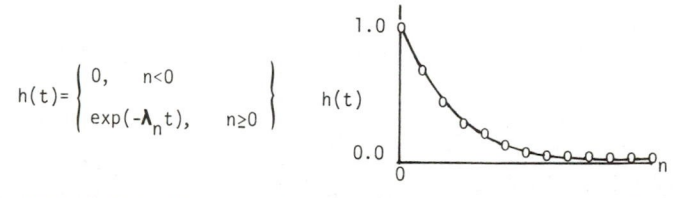

$$h(t) = \begin{cases} 0, & n < 0 \\ \exp(-\lambda_n t), & n \geq 0 \end{cases}$$

FIG. 2.5.1 Periodically sampled exponential signal to form an exponential time series.

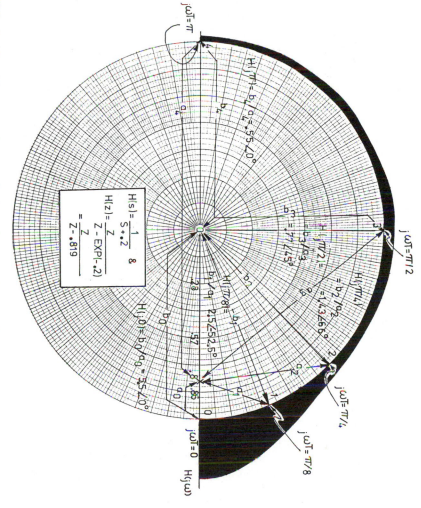

FIG. 2.5.2 Pole-zero diagram in the z domain of a simple transfer function H(z) with a graphically derived magnitude frequency response shown along the periphery of the unit circle.

that $a^n = \exp(-n\lambda T)$ and $H(z) = z/[z - \exp(-\lambda T)]$. One is often interested in quantifying the frequency response of this filter function [i.e., $H(z)$] in terms of its response to a sinusoidal input. In analog signal and system analysis, the steady-state frequency response of a filter was determined by evaluating $H(s = j\omega)$. In an analogous fashion, the steady-state frequency response of a discrete filter is defined in terms of $H(z)$ evaluated at $z = \exp(sT)$, where $s = j\omega$. Therefore, the steady-state frequency response is given by $H(z) = \exp(j\omega)$. In terms of the simple example problem, one concludes that $H[\exp(j\omega)] = \exp(j\omega T)/[\exp(j\omega T) - \exp(-\lambda T)]$. By evaluating this function over the baseband of frequencies residing from 0 to $1/2T$ Hz, the frequency spectrum of $x(n)$ can be quantified.

A sometimes useful steady-state analysis methodology is based on a graphical interpretation of the poles and zeros of $X(z)$. Based on traditional time-invariant linear system analysis methods, the magnitude and phase of $X[\exp(j\omega)]$ can be determined graphically. For example, the pole-zero diagram of the sample exponential $x(n) = \exp(-0.2n)$ is displayed in Fig. 2.5.2. The pole is located at $z = 0.819$, while the zero is found at $z = 0$. The steady-state frequency response at frequency ω_i is shown to be the ratio of the distances a_i and b_i found in this figure [i.e., $X[\exp(j\omega_i)] = a_i/b_i$]. The parameters a_i and b_i can be seen to be vectors with a magnitude and phase orientation, where the quotient a_i/b_i represents complex division. The vector-valued magnitude and phase parameters can be determined graphically using a ruler and protractor. Here $X[\exp(j\omega_i)]$ is determined for selected values of ω_i and the results plotted in an acceptable form. In Fig. 2.5.2 the magnitude data are plotted along the periphery of the unit circle which uniformly resolves the frequency interval $[0, 1/2T]$ hertz.

The steady-state response of a given $X(z)$ can be determined numerically using a digital computer. One can simply compute the ratio of rational polynomials using complex arithmetic or use a Fourier transform system library routine if one exists. The digital generation of a Fourier transform will be studied in Chapter 3.

EXAMPLE 2.5.2 In Chapter 7 a class of optimal discrete recursive filters will be studied. One of some importance is called the elliptic filter. The pole-zero diagram of a fourth-order elliptic filter having a magnitude response in Fig. 2.5.3 is shown in Fig. 2.5.4. Using graphical techniques and knowledge of the pole-zero pattern, the magnitude response of the discrete elliptic filter can be verified qualitatively.

2.6 OTHER DISCRETE TRANSFORMS

The standard z transform is by no means the only mapping of the s domain to the z domain found in practical use today. Many others have been considered, with the following enjoying a certain degree of popularity:

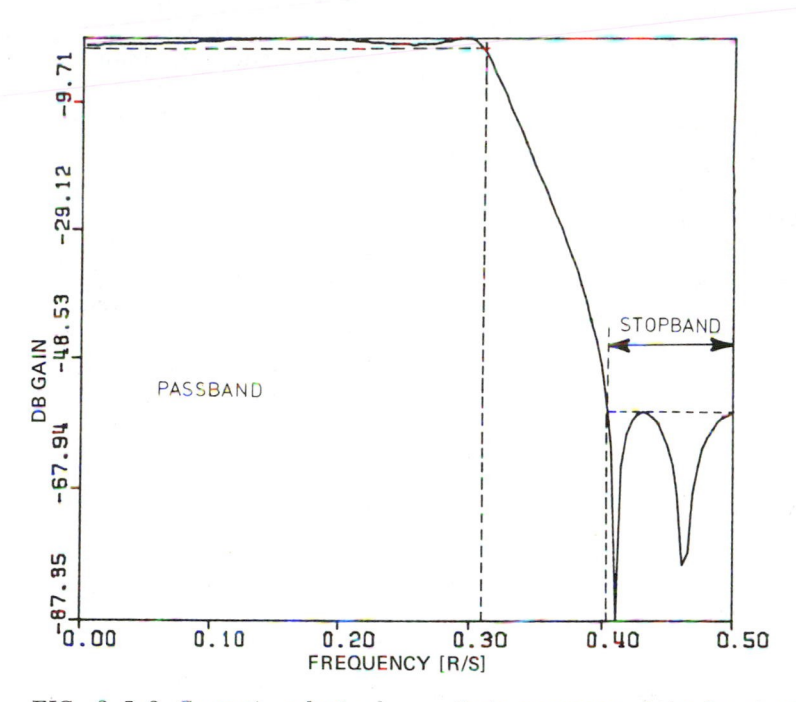

FIG. 2.5.3 Computer-derived magnitude response of the fourth-order elliptic filter. FS = 1.0.

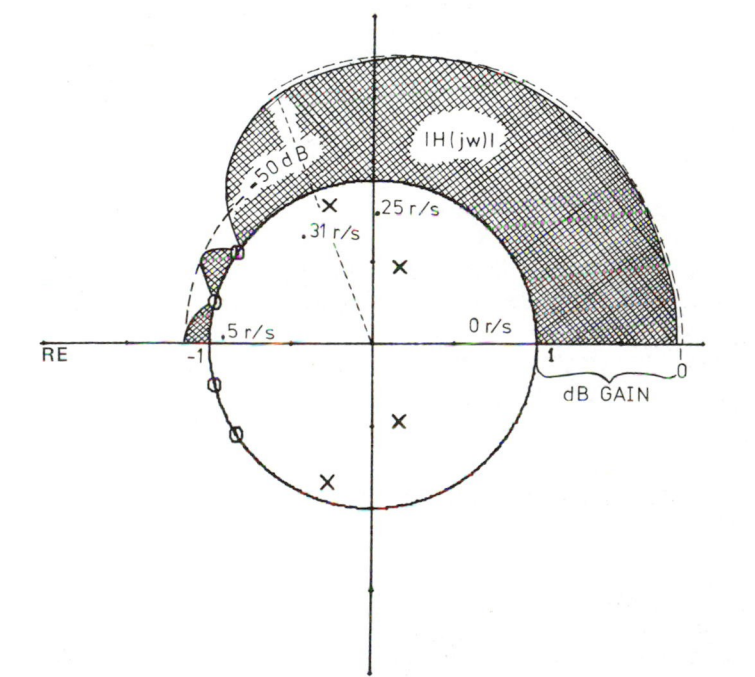

FIG. 2.5.4 Pole-zero diagram and magnitude response in the z domain for a fourth-order elliptic filter.

1. Trapezoidal transform
2. Right-rectangular integration transform
3. Bilinear transform
4. Matched z transform

The first two have limited popularity and are, in fact, only versions of the standard z transform. The standard z transform was modeled as a rectangular integrator (i.e., zero-order hold) whose sample values were found at the left-hand edge of the sample interval. The right-hand integration rule simply specifies the defining sample value to reside at the right-hand limit of the sample interval. The trapezoidal rule combines the right- and left-hand methods to form a sample value which equals their arithmetic sum. A simple one-pole lowpass filter, given by $H(s) = 1/(s - a)$, would yield the following discrete transforms:

H(s)	Standard z	Right-hand integration	Trapezoidal rule
$\dfrac{1}{s + a}$	$\dfrac{z}{z - \exp(aT)}$	$\dfrac{1}{z - \exp(aT)}$	$\dfrac{z + \exp(aT)}{z[z - \exp(aT)]}$

In general, the right-hand integration and trapezoidal transforms are of limited interest to the modern signal processing scientist. However, the remaining two transforms, the bilinear and matched, are extremely popular and possess attributes not found in the standard z transform. These two transforms are next.

2.7 BILINEAR Z TRANSFORM

In the introductory study of interpolation forms, the zero- and first-order holds were discussed. One can now interpret the zero-order hold, found in Eq. (2.2.1), to read

$$H(s) = \frac{1 - z^{-1}}{s} \qquad\qquad (2.7.1)$$

Consider now the first-order hold expressed in terms of the Riemannian difference equation.

$$\hat{x}(n + 1) \triangleq \hat{x}(n) + \frac{T}{2}[x(n) + x(n + 1)]$$

$$\simeq \int_a^b x(t)\ dt; \qquad a = nT, \quad b = (n + 1)T \qquad (2.7.2)$$

Formally, upon z transforming Eq. (2.7.2), one obtains

$$\hat{X}(z) = z^{-1}\hat{X}(z) + \frac{T}{2}[z^{-1}X(z) + X(z)] \qquad (2.7.3)$$

or

$$H(z) = \frac{\hat{X}(z)}{X(z)} = \frac{T(z+1)}{2(z-1)} \qquad (2.7.4)$$

Since the zero-order-hold circuit is designed to emulate a Riemann integrator, it should possess an s-domain equivalent model $H(s) = 1/s$. Therefore, upon equating these two results, the celebrated bilinear z transform is naturally defined as

$$s = \frac{2(z-1)}{T(z+1)} \qquad (2.7.5)$$

An obvious strength of this class of transform is its ability to map a given continuous signal, having a Laplace representation $H(s)$, into the z domain without the algebraically complex operations normally associated with the standard z transform. However, the bilinear z transform is not an impulse-invariant transform. For example, $H(s) = 1/s$ is realized as $H(z) = Tz/(z-1)$ using the standard z transform and $H(z) = T(z+1)/2(z-1)$ in the bilinear case. Using $T = 1$ and long division, one obtains

z Transform:

$$z^2 - 2z + 1\,\sqrt{\strut\,3}\quad \overset{z^{-1} + 2z^{-2} + 3z^{-3} + \cdots}{} \Longrightarrow \{x(n)\} = \{0,\ 1,\ 2,\ 3,\ \ldots\}$$

Bilinear z transform:

$$2z + 2\,\sqrt{\strut\,z-1}\quad \overset{0.5 + z^{-1} + z^{-2} + \cdots}{} \Longrightarrow \{x(n)\} = \{0.5,\ 1,\ 1,\ \ldots\}$$

Whereas the time series derived from the standard z transform agrees with the pure integrator's known impulse response at the sample instances, the bilinear transform imitates a "leaky" integrator. Thus, the utility of the bilinear transform is not in preserving sample value integrity in an impulse-invariant sense. Therefore, the strength of the bilinear transform must be found elsewhere. It is the interpolation power of the first-order hold, which is known to be superior to the zero-order hold, which holds the key. One should normally expect smoother, or more "analog-like" behavior from this class of transforms. This thesis will be developed later when it will be shown that the frequency response of a discrete filter is a function

of the type of z transform used. In general, the frequency response of
a bilinear transform system is superior to that of the standard z transformed system. Here the metric of comparison is assumed to be the similarity of the frequency response of an anlog filter H(s) and its discrete
model H(z). However, before the spectral properties of the bilinear transform can be discussed in detail, the concept of <u>warping</u> must be developed.

2.8 WARPING

The bilinear z transform possesses an interesting frequency domain image.
Solving the bilinear z transform defining equation for z, one obtains

$$z = \frac{(2/T) + s}{(2/T) - s} \qquad\qquad (2.8.1)$$

Suppose that we wish to emulate the frequency response of an analog continuous filter evaluated at $j\omega$ to be H(s = $j\omega$). Using a standard z transform,
the frequency response of the discrete filter version of H(s), namely H(z),
is found by evaluating H(z) at z = exp($j\phi$), where $0 \leq \phi \leq \pi$. Using the
bilinear transform, a discrete filter of the form H(z) is defined as follows:

$$H(z) = H(s)\big|_{s = (2/T)[(z-1)(z+1)]} \qquad\qquad (2.8.2)$$

Under this transform, the arc given by z = exp($j\phi$) can be seen to correspond
to the $j\omega$ axis in the s domain through the relationship

$$\exp(j\phi) = \frac{(T/2) + j\omega}{(T/2) - j\omega} \qquad\qquad (2.8.3)$$

where ϕ represents a frequency in the z plane corresponding to the analog
frequency ω in the s plane. After a modest amount of algebra, one can reduce this relationship to read

$$\omega = \frac{2}{T} \tan\left(\frac{\phi T}{2}\right) \qquad\qquad (2.8.4)$$

which is interpreted graphically in Fig. 2.8.1. For ϕ small (say ≤ 0.3T),
the discrete and analog filter frequency response will be in reasonably
good agreement since $\phi \approx \omega$. However, for larger values of ϕ, a significant differenct between the discrete and analog frequency locations may
exist. This effect is known as <u>warping</u>. The nonlinear relationship that
exists between the independent frequency axis in the discrete and continuous
space (i.e., z and s planes) must be understood if the bilinear z transform
is to be used as a design tool. One obvious consequence of Eq. (2.8.1) is

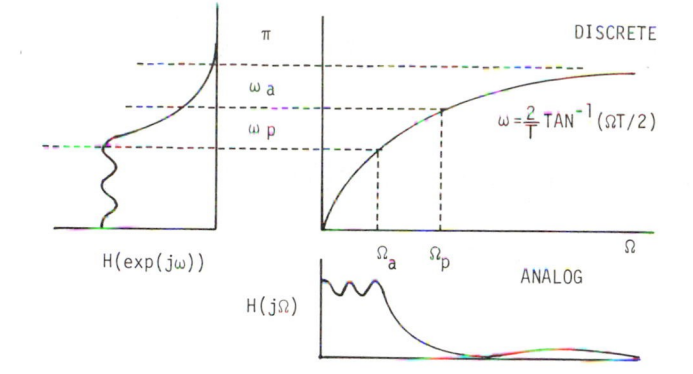

FIG. 2.8.1 Frequency warping map of the analog frequency axis into the discrete frequency axis under the bilinear transform.

that the bilinear z transform is free of "aliasing" errors. This can be argued in the following manner. Observe that under the bilinear z transform that z is single valued for all choices of s and therefore all $j\omega$. This is in sharp contrast with the constraint imposed on the standard z transform, where uniqueness could be guaranteed only if $j\omega$ was restricted to reside in the interval $[-\pi/T, \pi/T]$. The distortion introduced by the warping effect can be overcome using a technique called <u>prewarping</u>. Prewarping is a compensation technique which allows a discrete filter to be realized having prespecified critical frequencies. Simply stated, if ϕ_i is a critical frequency of an analog filter, the corresponding frequency of the realized discrete filter would have to be chosen to be

$$\phi_i = \frac{2}{T} \tan^{-1}\left(\frac{\omega_i T}{2}\right) \qquad\qquad (2.8.5)$$

so that no warping distortion would occur. However, one should note that it is only the critical frequency ω_i that is corrected for warping. Other frequencies will have to take their chances under Eq. (2.8.4).

EXAMPLE 2.8.1 Suppose that an analog prototype lowpass filter has a cutoff frequency of (1) 1 kHz ($2\pi \times 10^3$ rad/sec) or (2) 10 kHz ($2\pi \times 10^4$ rad/sec); then these critical frequencies under the bilinear transform, for an assumed sample rate of 10 kHz, are given by

$$\phi_1 = \frac{2}{10^{-4}}\ \tan^{-1}\left(\frac{2\pi \times 10^3 \times 10^{-4}}{2}\right) = 6.088 \text{ krad/sec}$$

$$\Rightarrow 969 \text{ Hz}$$

$$\phi_2 = \frac{2}{10^{-4}} \tan^{-1}\left(\frac{2\pi \times 10^4 \times 10^{-4}}{2}\right) = 25.253 \text{ krad/sec}$$

$$\Rightarrow 4.019 \text{ kHz}$$

As $\omega \to \infty$, the resulting discrete critical frequency can be seen to approach

$$\phi \to \frac{2}{T} \tan^{-1}(\infty) = \frac{2}{T}\frac{\pi}{2} = \frac{\pi}{T} \quad \text{rad/sec}$$

$$\Rightarrow \frac{1}{2T} \quad \text{Hz} \tag{2.8.6}$$

where $1/2T$ hertz is the Nyquist frequency. This limiting condition illustrates another difference that exists between the standard z transform and the bilinear transform. The principal angle requirement imposed on the standard z transform required that $|f_{max}| \le 1/2T$. Choosing to sample at or above the Nyquist sample rate ensured that the resulting transform is impulse invariant. However, in the bilinear case, there are no limitations on transformable analog frequencies. As a result, this class of transform is not impulse invariant.

In a typical application, a prototype analog filter, say H(s), would be hypothesized. Upon substituting $s = (2/T)(z - 1)/(z + 1)$, the bilinear form H(z) would result. Because of the simple algebraic structure of this map, it lends itself to computer-aided study. Using a general-purpose digital computer, the bilinear transform of a filter based on a cascaded or parallel architecture can be automated. Specifically, these two architectures would be defined as follows:

1. Cascade:

$$H(s) = K[H_1(s)H_2(s) \cdots H_n(s)] \tag{2.8.7}$$

2. Parallel:

$$H(s) = K[H_1(s) + H_2(s) + \cdots + H_n(S)] \tag{2.8.8}$$

where $H_i(s)$ is at most a second-order section having the following structure:

$$H_i(s) = \frac{A_i(s)}{B_i(s)} = \frac{A_{0i} + A_{1j}s + A_{2j}s^2}{B_{0i} + B_{1i}s + B_{2i}s^2}; \quad i = 1, 2, \ldots, n \tag{2.8.9}$$

It is possible that the roots (poles) of such a second-order section are complex numbers. Using a second-order model, however, will guarantee that all the filter coefficients are real due to the fact that complex roots always appear in complex conjugate pairs. The modeling of a discrete filter over a real coefficient field obviously has some merit. It means that in the hardware realization of the discrete filter, only real multipliers would be needed. For complex coefficients, a complex multiplication would necessitate the use of four real multipliers (i.e., real × real, real × imaginary, imaginary × real, and imaginary × imaginary) per complex multiply call.

The factors of $H_1(s)$ can be computed manually or through the use of a general-purpose digital computer. If the general-purpose computer includes root-finding subroutines, such as the library function POLRT in an IBM or DEC sense, the software analysis is greatly simplified. The mechanics of converting a given $H(s)$ to $H(z)$ under a bilinear z transform is accomplished as follows. Direct calculation shows that

$$
H(s) = \left.\frac{a_0 + a_1 s + a_2 s^2}{b_0 + b_1 s + b_2 s^2}\right|_{s=c(z-1)/(z+1)}
$$

$$
= \frac{(z_2/p_2)z^2 + (z_1/p_2)z^1 + (z_0/p_2)}{z^2 + (p_1/p_2)z + (p_0/p_2)} \tag{2.8.10}
$$

where

$$
\begin{aligned}
z_0 &= a_0 - a_1 c + a_2 c^2 \\
z_1 &= 2(a_0 - a_2 c^2) \\
z_2 &= a_0 + a_1 c + a_2 c^2 \\
p_0 &= b_0 - b_1 c + b_2 c^2 \\
p_1 &= 2(b_0 - b_2 c^2) \\
p_2 &= b_0 + b_1 c + b_2 c^2.
\end{aligned} \tag{2.8.11}
$$

It should be apparent that a first-order filter section would be modeled as a second-order section except that a_2 and b_2 would be assigned a value of zero. A flowchart of a typical bilinear z-transform source program is shown in Fig. 2.8.2.

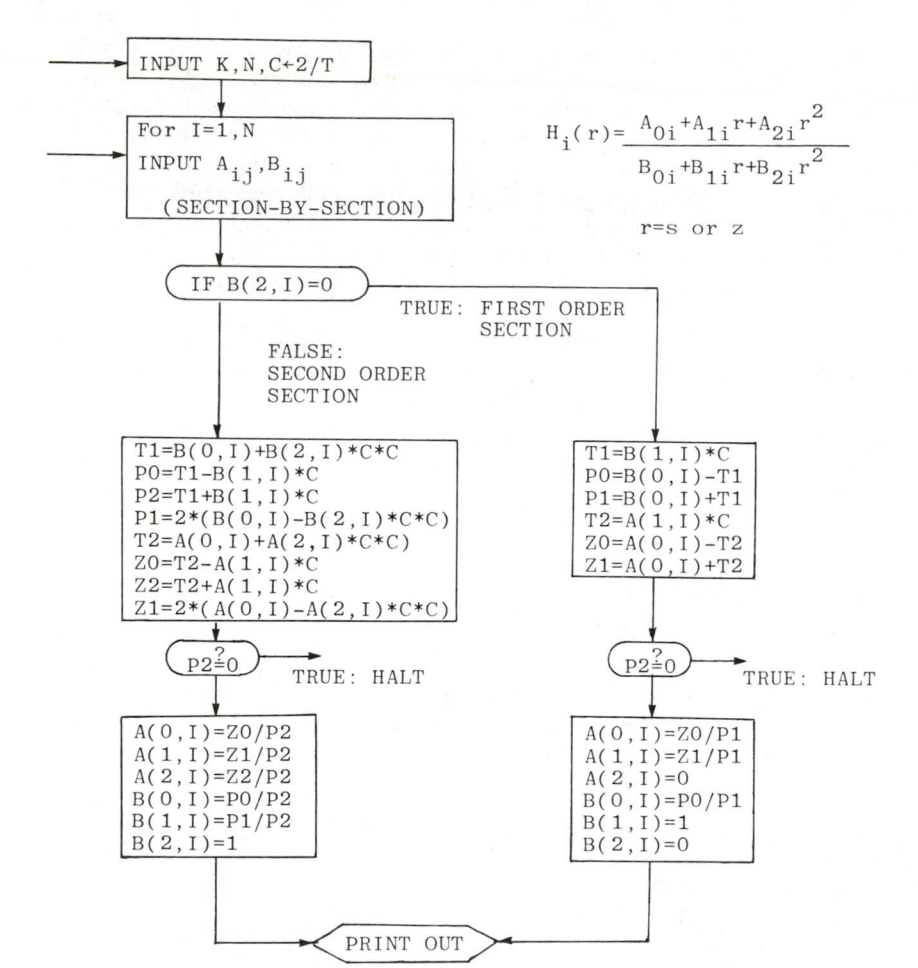

$$H_i(r) = \frac{A_{0i} + A_{1i}r + A_{2i}r^2}{B_{0i} + B_{1i}r + B_{2i}r^2}$$

$$r = s \text{ or } z$$

FIG. 2.8.2 FORTRAN source code for computing the bilinear z transform of an analog transfer function.

One can now pause and reflect on the z-transform methods developed to this point. One important difference is found in the formal definition of the bilinear z transform [see Eqs. (2.8.1) and (2.7.5)]. It can be noted that z can be explicitly defined in terms of s. Therefore, the conversion of H(s) to H(z) is simply a substitution of variables. Second, there exists in the definition of the bilinear z transform a denominator term equal to (z + 1). This means that the poles of H(s) will give rise to zeros of H(z) located at z = -1. However, z = -1 is located at the Nyquist frequency in the z plane and at $j = \pi/T$ in the s plane. As a result, the system gain will automatically be zero at z = -1. For example, consider the simple first-order filter H(s) = 1/(s + a) realized as H(z) using the z-transform methods discussed in this section. The result of each transform is summarized, together with the steady-state frequency response, in Fig. 2.8.3. A more ambitious second-order example produces the results shown in Fig. 2.8.4.

2.9 MATCHED Z TRANSFORM

The matched z transform is occasionally used in signal processing. It can be effective in modeling narrowband frequency-selective filters. However, it is ineffective in modeling ripple-producing filters such as Chebyshev and elliptic filters. Also, this transform is not recommended for use with wideband filters. Briefly stated, if an analog filter is given by H(s), where

$$H(s) = \frac{\prod\limits_{i=1}^{N} (s - a_i)}{\prod\limits_{i=1}^{N} (s - b_i)} \qquad (2.9.1)$$

TRANSFORM	FORM	POLE-ZERO DIAGRAM (MAGNITUDE)
STANDARD Z TRANSFORM	$\dfrac{z}{z-\exp(aT)}$	
RIGHT RECTANGULAR INTEGRATION	$\dfrac{1}{z-\exp(aT)}$	
TRAPEZOIDAL	$\dfrac{K(z+\exp(aT))}{(z-\exp(aT))}$	
BILINEAR Z TRANSFORM	$\dfrac{K'(z+1)}{(z+(a-2/T)/(a+2/T))}$	

FIG. 2.8.3 Example of standard, rectangular, trapezoidal, and bilinear z transforms of H(s) = 1/(s + a).

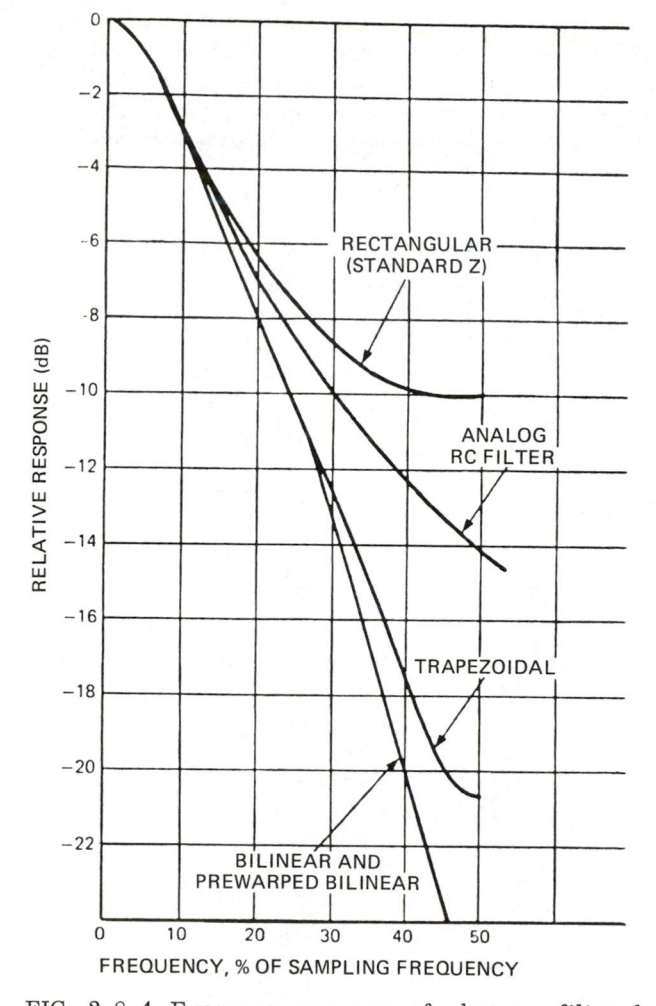

FIG. 2.8.4 Frequency response of a lowpass filter designed on the basis of standard, trapezoidal, and bilinear z transforms. (After Leon and Bass, 1974; reprinted with the permission of EDN.)

then the matched z transform of H(s) is given by

$$H(z) = \frac{\prod\limits_{i=1}^{N} [1 - \exp(a_i T)z^{-1}]}{\prod\limits_{i=1}^{N} [1 - \exp(b_i T)z^{-1}]} \qquad (2.9.2)$$

From the data found in the z-transform table, it can be noted that if $x(t) = \exp(-at)$, then $X(s) = 1/(s + a)$ and $X(z) = 1/[1 - \exp(-aT)z^{-1}]$. Therefore, the poles of the matched z-transformed filter $H(z)$ are those of the z-transformed version of $H(s)$. However, the zeros are found in different locations. In fact, using the z transform of $(s + a_j)$ cited previously, it is apparent that the zeros of $H(s)$ are mapped directly to the zeros of $H(z)$ under $[1 - \exp(a_jT)z^{-1}]$.

BIBLIOGRAPHY

Cadzow, J. A. (1974), Discrete Time Systems, Prentice-Hall, Englewood Cliffs, N.J.

Kuo, F. F., and J. F. Kaiser (1966), System Analysis by Digital Computer, Wiley, New York.

Leon, B. J., and S. C. Bass (1974), Designers' Guide to Digital Filters [six-part series], EDN, January-June.

Oppenheim, A. V., and D. H. Johnson (1972), Discrete Representation of Signals, Proc. IEEE, 60, June, pp. 681-691.

Rabiner, L. R., and B. Gold (1975), Theory and Application of Digital Signal Processing, Prentice-Hall, Englewood Cliffs, N.J.

Schwartz, M., and L. Shaw (1975), Signal Processing: Discrete Spectral Analysis, McGraw-Hill, New York.

Shannon, C. E. (1949), Communication in the Presence of Noise, Proc. IRE, 30, No. 1, pp. 30-41.

Taylor, F. J., and S. L. Smith (1975), Digital Signal Processing in FORTRAN, Lexington Books, Lexington, Mass.

Tretter, S. A. (1975), Introduction to Discrete Time Spectral Analysis, Wiley, New York.

3
Spectral Analysis

3.1 INTRODUCTION

In addition to discrete system transforms (i.e., z transforms), signal processing scientists study other classes of transforms as well. The most common of these transforms are known as the spectral transform algorithms. These transforms map a given time series into the frequency domain using the discrete form of the Fourier transform. Since spectral analysis is found in just about every engineering endeavor, the applications of such transforms are almost uncountable. Fortunately, an efficient computational technique exists which allows spectra to be generated using a general-purpose digital computer. This algorithm, known as the fast Fourier transform (FFT), is discussed in this chapter. In addition, other important spectral transform algorithms are developed. They are the chirp-z cosine, and Walsh transforms. Their mathematical description and signal processing utility are also discussed.

3.2 DISCRETE FOURIER TRANSFORM

The discrete Fourier transform (DFT) of a periodic time series $\{x(n)\}$ is a set of N-distinct harmonics $\{X(k)\}$. The transform pairs, referred to as the forward and reverse DFTs, are defined below:

1. Forward DFT:
$$X(k) = \sum_{n=0}^{N-1} x(n) W_N^{-kn} \qquad (3.2.1)$$

2. Inverse DFT: $x(n) = \dfrac{1}{N} \displaystyle\sum_{k=0}^{N-1} X(n)W_N^{kn}$ (3.2.2)

where $W_N = \exp(j2\pi/N) = \cos(2\pi/N) + j\,\sin(2\pi/N)$ and both $x(n)$ and $X(k)$ are complex numbers (i.e., mapping of the complex numbers into the complex numbers). The complex weighting coefficients W_N are obviously N-periodic (i.e., $W_N^{i+N} + W_N^i$; $W_N^{iN} = 1.0$. Observe that $X(0)$ is N times the value normally attributed to the dc value of the time series $\{x(n)\}$. If one desires to normalize the set of harmonics $X(k)$ by $1/N$, the set of derived Fourier coefficients would then be comparable to the Fourier coefficients of $x(t)$. The DFT can also be interpreted in the context of the z transform. The z transform of a time series $\{v(n)\}$, defined to be

$$v(n) = \begin{cases} x(n), & 0 \le n \le N-1 \\ 0, & \text{otherwise} \end{cases}$$ (3.2.3)

and is given by

$$V(z) = \sum_{n=0}^{\infty} v(n)z^{-n} = \sum_{n=0}^{N-1} x(n)z^{-n}$$ (3.2.4)

Observe that if one chooses to evaluate z along the periphery of the unit circle [i.e., $z = \exp(j2\pi/N)$], Eq. (3.2.4) can be seen to be equal to Eq. (3.2.1).

For T denoting the sample period, the ith DFT coefficient $X(i)$ is the ith harmonic of $x(n)$ located at the real frequency $f_i = 2\pi i/NT$. Other DFT frequency-resolution parameters are summarized as follows:

Sample size = N samples
Sample period = T seconds
Data record length = NT seconds
Frequency resolution = 1/NT hertz per harmonic
Nyquist frequency = 1/2T hertz

Through the judicious choice of N and T, a desired frequency resolution can be achieved. However, it will soon be noted that DFT throughput (computation speed) is inversely related to N. Therefore, execution speed is purchased at the expense of spectral resolution.

The frequency selectivity of the DFT can be determined by testing its response to a pure harmonic time series $\{x(n)\}$. That is, it will be assumed that $x(n) = \exp(j\omega_0 nT)$. The DFT of $\{x(n)\}$ is given by

$$X(k) = \sum_{n=0}^{N-1} \exp(j\omega_0 nT) W_N^{-nk} \tag{3.2.5}$$

$$= \exp\left(\frac{-j(N-1)vT}{2}\right) \frac{\sin(NvT/2)}{\sin(vT/2)} \tag{3.2.5}$$

where $v = k\omega_s/N - \omega_0$. For $k \in [0, N-1]$, $\omega_0 = i\omega_s/N$ and $i \in [0, N-1]$, it follows that

$$X(k) = \begin{cases} N, & k = m \\ 0, & \text{otherwise} \end{cases} \tag{3.2.6}$$

The response of an N-point DFT to a harmonic oscillator, tuned to a frequency f_0, which is an integer multiple of the fundamental frequency $f_1 = 1/NT$, is sketched in Fig. 3.2.1. However, if f_0 does not possess this integer relationship, the resulting spectrum will have nonzero harmonic values at locations other than f_0. This phenomenon is called <u>leakage</u> and will be explained further in terms of the following simple example.

EXAMPLE 3.2.1 Leakage appears as a spectral "smear" located about a center frequency f_0. The effect of leakage can be suppressed by increasing the record length N. This conjecture will be developed in the next example. However, at this time a qualitative feeling for leakage will be developed. Two pure harmonic oscillatory signals are chosen as test signals. One completes 10 complete cycles within a 128-sample interval. The other completes 10.5 cycles within that window. The DFTs of both signals are shown in Fig. 3.2.2. It can be noted that the signal that is periodically "registered" in the sample interval exhibits no leakage distortion. However, the signal that is not correctly registered in the sample interval suffers from leakage.

EXAMPLE 3.2.2 Referring to the periodic time series shown in Fig. 3.2.3, several observations can be made. Leakage can be eliminated if

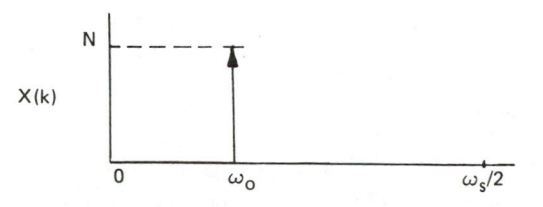

FIG. 3.2.1 Discrete Fourier transform of a pure tone exhibiting frequency domain energies concentrated at a single harmonic.

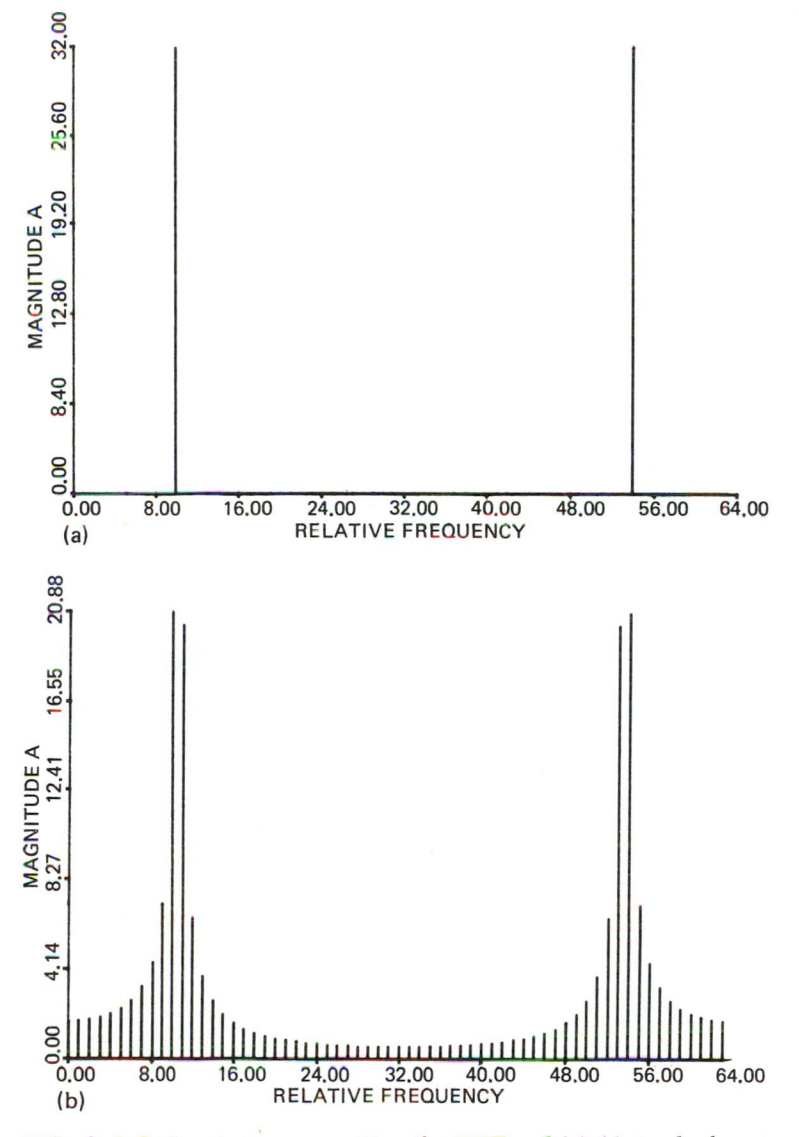

FIG. 3.2.2 Spectra representing the DFTs of (a) 10 (no leakage) and (b) 10.5 leakage cycles of a pure sinusoidal tone.

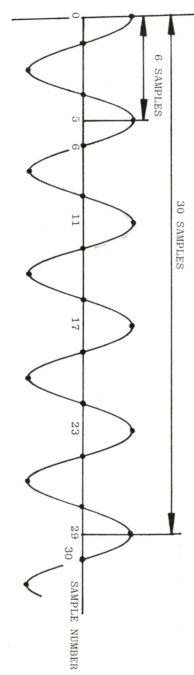

FIG. 3.2.3 Graphical interpretation of the jump discontinuity giving rise to a leakage spectrum.

the periodic function diagrammed does not complete an integer number of cycles in N samples. If N = 5, then 1.25 cycles would be completed. The "jump discontinuity" at sample 5 represents the source of leakage energy (i.e., noise source) and has a normalized value of 1/5. If N = 30, the jump discontinuity has a reduced normalized value of 1/30.

3.3 DFT PROPERTIES

The DFT satisfies a set of important system relationships. They are:

1. Linearity: Here $DFT[x(n) + y(n)] = DFT[x(n)] + DFT[y(n)] = X(k) + Y(k)$.
2. Cyclic time and frequency shifting: Here $DFT[x(n - r)] = X(k) \exp(j2\pi rk/N)$ and $IDFT[X(k - r)] = x(n) \exp(j2\pi nr/N)$.
3. Time reversal: Here $DFT[x(-n)] = X(-k) = X(n - k)$. This provides a mechanism by which negative frequencies can be interpreted.
4. Symmetry for a real input: For $x(n)$ real, $X(k) = X^*(-k)$. Two special cases can be considered. If $\{x(n)\}$ is a real even time series, $X(k)$ is real (i.e., a cosine series). If $\{x(n)\}$ is odd, $X(k)$ is imaginary (i.e., a sine series).
5. Parseval's theorem: Parseval's theorem states that

$$\sum_{n=0}^{N-1} x^2(n) = \frac{1}{2\pi j} \oint X(z)X(z^{-1})z^{-1} \, dz$$

$$= \frac{1}{2\pi j} \oint G(z)z^{-1} \, dz \qquad (3.3.1)$$

where $G(z)$ is sometimes referred to as the power spectrum of $\{x(n)\}$ (see Section 4.6). Through Parseval's theorem, the power of the N-sample process, $\{\Sigma x^2(n)\}$, can be equated to the spectral representation of $\{x(n)\}$. This is an extremely useful result and is essential to the study of many problems where signals are embedded in noise.

6. Circular (cyclic) convolution: If $\{x(n)\}$ and $\{y(n)\}$ are two N-point time series [i.e., $x(i) = x(j)$ mod N and $y(i) = y(j)$ mod N] the circular convolution of these two time series is given by

$$x(n) \circledast y(n) = \sum_{i}^{N-1} x(n)y(n - i) = \sum_{i}^{N-1} x(n - i)y(n) \qquad (3.3.2)$$

The circular convolution theorem states that $DFT[x(n) \circledast y(n)] = X(k)Y(k)$.

The convolution property will allow the DFT to do digital filtering. In practice, a filter's impulse response and input will be transformed, multiplied harmonic by harmonic (complex multiplication), and then inverted. The resulting time series represents the circular convolution of the two original time series. It is important that a distinction be made between circular convolution and the more familiar linear convolution (see Section 4.1) given by

$$x(n) * y(n) = \sum_{i=-\infty}^{\infty} x(n)y(n - i) = \sum_{i=-\infty}^{\infty} x(n - i)y(n) \qquad (3.3.3)$$

It has been noted that circular convolution can be demonstrated by placing the signals to be convolved into a rotating cylinder. Here an N-sample time series, say $\{x_N(n)\}$, is placed inside the rotating cylinder as shown in Fig. 3.3.1. On the stationary sleeve, which the rotating shaft passes through, the second time series is inscribed except in a reverse time direction. The circular convolution sum is formed by summing the N partial products for each discrete cylinder position. It can be seen that this geometry forces the time series to appear to be N-periodic. In Fig. 3.3.1, two general aperiodic signals $\{x(n)\}$ and $\{y(n)\}$ are shown where the first N samples are used to form $x_N(n)$ and $y_N(n)$.

We can easily demonstrate the fact that there exists a difference between circular and linear convolution by example. Referring to Fig. 3.3.2, one notes that the two convolution methods produce different results. When convolution is studied in depth in Chapter 4, this concept will be developed further. At that time, the question of equating circular and linearly convolved time series, using a DFT, will be addressed. However, none of this would be of interest or use to us if the DFT were computationally awkward or inefficient. Fortunately, this is not the case, as will be seen in the next section.

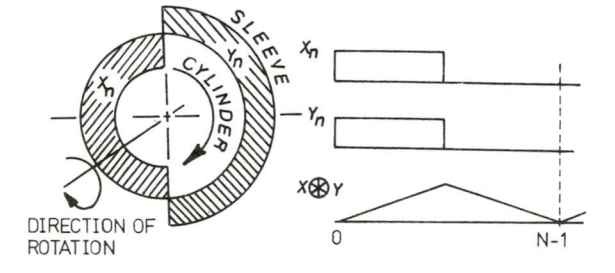

FIG. 3.3.1 Cylinder and sleeve analogy for cyclic convolution.

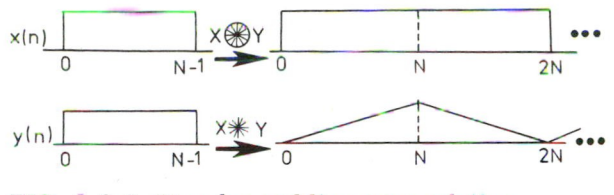

FIG. 3.3.2 Circular and linear convolution.

3.4 THE FAST FOURIER TRANSFORM

The fast Fourier transform (FFT) is a computationally efficient DFT algorithm. The FFT maps an N-point time series into N Fourier harmonics under the operation

$$X(k) = \sum_{n=0}^{N-1} x(n) W_N^{nk} \tag{3.4.1}$$

where $W_N = \exp(j2\pi/N)$. Using direct computational methods, N complex multiplies would be required to define X(k) for each given k. Since, there are N values of k, it follows that the direct method would have a multiplier budget of N^2 multiplies per DFT. Suppose, for the sake of discussion, that we chose to partition the input time series into two groups of N/2 samples composed only of even- or odd-indexed sample values. Using this data structure, Eq. (3.4.1) can be rewritten as

$$X(k) = \sum_{n=0}^{(N/2)-1} x(2n) W_N^{2nk} + \sum_{n=0}^{(N/2)-1} x(2n+1) W_N^{(2n+1)k}$$

$$= \sum_{n=0}^{(N/2)-1} x(2n) W_N^{2nk} + \sum_{n=0}^{(N/2)-1} [x(2n+1) W_N^{2nk}] W_N^k \tag{3.4.2}$$

Equation (3.4.2) illustrates that one N-point FFT can be written as two (N/2)-point FFTs requiring $(N/2)^2$ multiplies per transform. As a result, Eq. (3.4.2) can be realized with a $N^2/2$ multiplier budget. Since this "trick" worked so well, one may consider repeating it again and again. If, in fact, $N = 2^n$, one can partition the input data field into N/2 groups consisting of two sample values each. Using this strategy, the complex multiplier count for a FFT can be shown to be on the order of $(N/2) \log_2(N) = nN/2$. The multiplier savings is significant and is quantified in Table 3.4.1. Since multiplier execution rate is often found to be a limiting factor in establishing the throughput potential of a digital computer, the

TABLE 3.4.1 Multiplication Count of the DFT and FFT

N	DFT = N^2	FFT = (N/2)$\log_2$ N	DFT/FFT
16	256	32	8.0
32	1024	80	12.8
64	4096	192	21.5
128	16384	448	36.6
256	65536	1024	44.0
512	262144	2304	103.8
1024	1048526	5120	204.8

reduced multiplier count will be translated directly into computational speed. It is for this reason that the prefix "fast" is found in the FFT.

A transform is of <u>radix r</u> if $N = r^n$. The structure of a radix-r FFT is architecturally well ordered and appears in either a decimation in time or frequency form. These forms will now be developed in a radix-2 context.

3.5 RADIX-2 FFT (DECIMATION IN TIME)

The radix-2 FFT assumes that r = 2. The input time series is decomposed into N/2 data groups of sample pairs. The sample values are not ordered

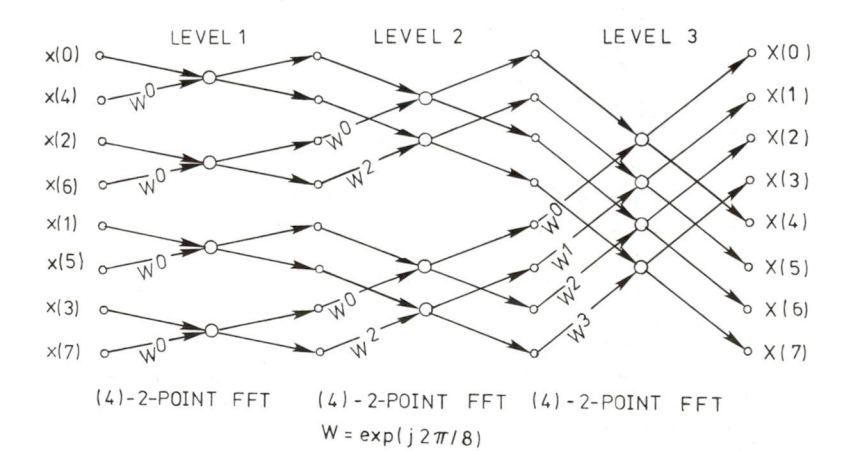

FIG. 3.5.1 Flow diagram for $N = 2^3$ = 8-point radix-2 DIT FFT algorithm.

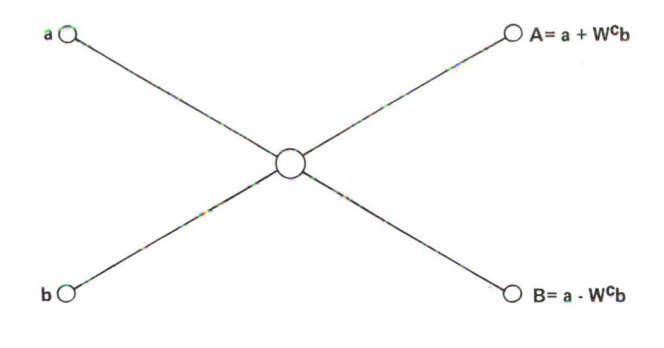

a, b, A, B, W complex

FIG. 3.5.2 Radix-2 DIT FFT butterfly diagram.

sequentially but rather, are reordered in time. This reorder gives rise to the concept of decimating the time axis. For example, for N = 8, the pairings are $\{[x(0),\ x(4)]\}; \{[x(2),\ x(6)]\}; \{[x(1),\ x(5)]\}; \{[x(3),\ x(7)]\}$. The defining DFT equation, based on this partition, is shown in Fig. 3.5.1. Several observations can be abstracted from the data found in this figure. They are:

1. The output Fourier harmonics are sequentially ordered if the input is decimated in time.
2. There are three (in general, n levels where $N = 2^n$) of transform operations.
3. Each two-point FFT found at any level of the regular FFT diagram is referred to as a <u>FFT butterfly</u> (see Fig. 3.5.2).

The complex coefficients W_N^k found in this structure are referred to as <u>twiddle factors</u>. One can gain additional insight into the algebraic structure of a radix-2 decimation in time (DIT) FFT by studying the butterfly found at the top of level 1. Here a = x(0) and b = x(4) are Fourier-transformed into A and B. Specifically, for the first level, $W_N^0 = 1$. Therefore, A = a + b (dc term) and B = a - b (first harmonic term).

The required reordering of the input time series has come to be known as <u>bit reversing</u>. Consider the N = 8 point time series distributed around the periphery of the unit circle as shown in Fig. 3.5.3. Here each location has been given an n-bit representation as well as a bit-reversed image. Notice that the paired sample values [e.g., x(0) and x(4)] are located at diametrically opposed positions on the circle. Furthermore, the order in which the time-series samples appear in the decimated input are simply their decimal bit-reversed value (see Fig. 3.5.3).

All these attributes can be conveniently coded in a high-level language. Shown in Fig. 3.5.4 is a typical FORTRAN source code of a radix-2 DIT FFT.

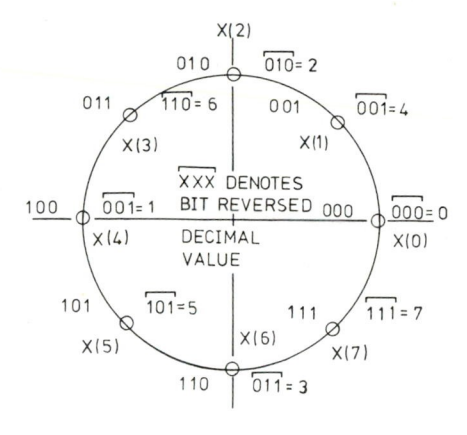

FIG. 3.5.3 Bit-reversing pattern geometry interpreted on the periphery of the unit circle.

```
C
        SUBROUTINE FFTS(A,N,M)
        COMPLEX A(256) ,U,W,T,CMPLX
        N=2**M
C
C       START BIT REVERSAL HERE
C
        NV2=N/2
        NM1=N-1
        J=1
        DO 7 I=1,NM1
        IF(I.GE.J) GO TO 5
        T=A(J)
        A(J)=A(I)
        A(I)=T
5       K=NV2
6       IF(K.GE.J) GO TO 7
        J=J-K
        K=K/2
        GO TO 6
7       J=J+K
C
C       FFT COMPUTATION
C
        PI=3.141592653589793
        DO 20 L=1,M
        LE=2**L
        LE1=LE/2
        U=(1.0,0.0)
        W=CMPLX(COS(PI/FLOAT(LE1)),-SIN(PI/FLOAT(LE1)))
        DO 20 J=1,LE1
        DO 10 I=J,NE,LE
        IP=I+LE1
        T=A(IP)*U
        A(IP)=A(I)-T
10      A(I)=A(I)+T
20      U=U*W
        RETURN
        END


        FFT(A(n))=A(k)

        n,k=0,1,...,N-1;  N=2^M
```

FIG. 3.5.4 FORTRAN IV source code for a DIT FFT radix-2 algorithm.

EXAMPLE 3.5.1 The FFT of a 256-point time series produces a spectrum composed of 128 positive harmonics. The magnitude spectrum of a 0.5 duty cycle pulse is presented in Fig. 3.5.5. In Figs. 3.5.6 and 3.5.7 the magnitude spectra of 34/256 and 18/256 duty cycle pulses are shown.

3.6 RADIX-2 FFT (DECIMATION IN FREQUENCY)

Using the radix-2 DIT FFT routine as a model, the decimation in frequency (DIF) version of the FFT can be easily developed. Consider the data found in Fig. 3.6.1, which differ from the DIT diagram in several key ways. First, one should observe that the input is now sequentially ordered but that the output spectrum is decimated. Also, the butterfly structure shown in Fig. 3.6.2 is different from that for the DIT routine.

3.7 FFT TOPICS

Besides appearing in a DIT or DIF form, the FFT can be realized as in-place or not in-place routines. An in-place program makes efficient use of

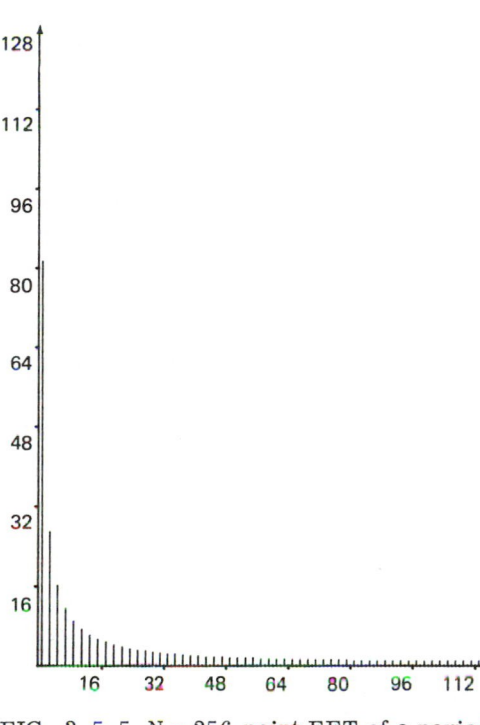

FIG. 3.5.5 N = 256-point FFT of a periodic rectangular pulse process having a duty cycle of 1/2.

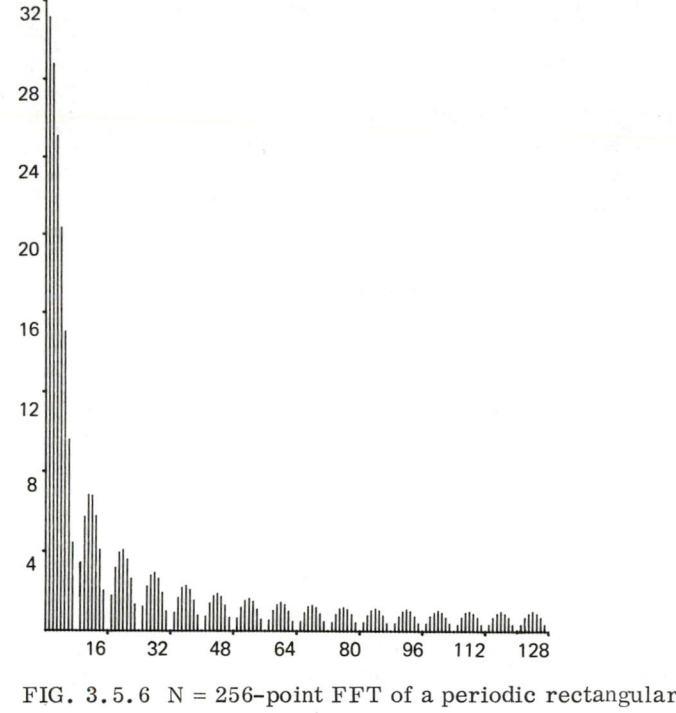

FIG. 3.5.6 N = 256-point FFT of a periodic rectangular pulse process having a duty cycle of 34/256.

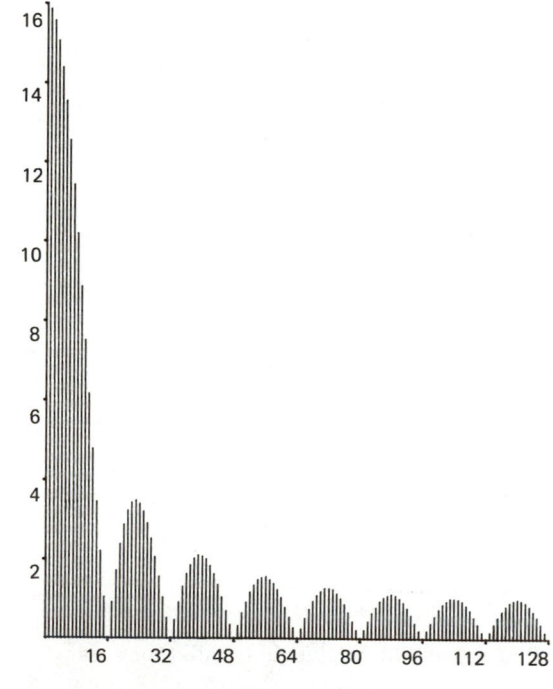

FIG. 3.5.7 N = 256-point FFT of a periodic rectangular pulse process having a duty cycle of 18/256.

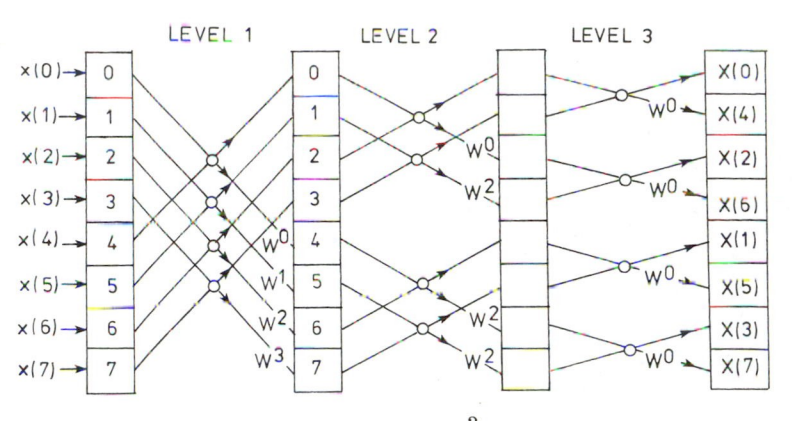

FIG. 3.6.1 Flow diagram for an $N = 2^3 = 8$–point radix-2 DIF FFT.

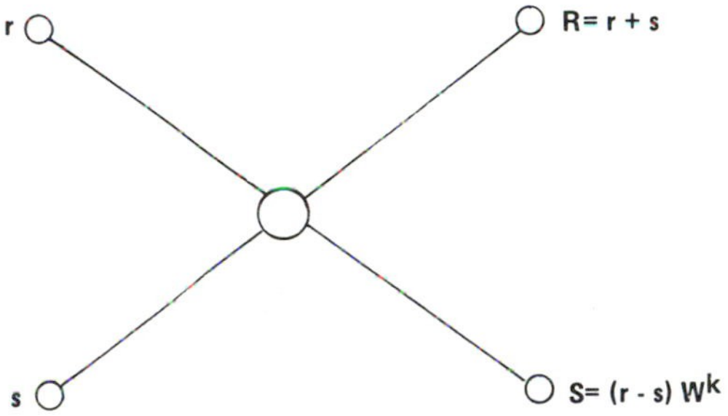

r ◯   ◯ $R = r + s$

◯

s ◯   ◯ $S = (r - s) W^k$

FIG. 3.6.2 Radix-2 DIF FFT butterfly diagram.

a system's storage registers. Consider the data structure shown in Fig.
3.6.1. Registers [0] and [4] are used to define FFT_1, which in turn produces outputs x_i and y_i. These two complex data words can be returned to
registers [0] and [4]. This pattern can be repeated in a pairwise sense.
Managing data in this manner will cause the memory requirements of a N-
point FFT to remain fixed at N complex words. If instead of returning x_i
and y_i to registers [0] and [4], data are stored in temporary locations, then
a not-in-place structure is defined. Here the memory requirements would
be at least 2N complex words.

The fundamental building block for FFT designs is generally the
radix-2 or radix-4 butterfly unit. A radix-4 structure for $N = 4^2 = 16$ is
outlined in Fig. 3.7.1 together with its butterfly structure. A detailed analysis of the multiplier requirements of the radix-2 and radix-4 architecture
would how that a radix-4 configuration can realize an N-point FFT with
fewer multiplications. This would explain why the radix-4 FFT has
achieved a degree of popularity.

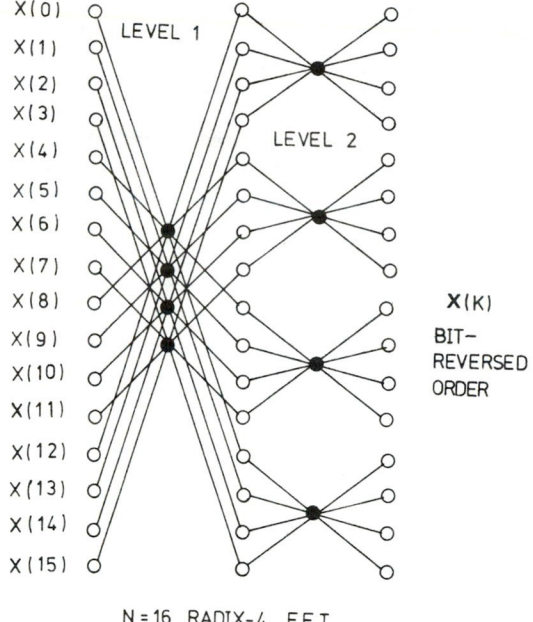

$N = 16$ RADIX-4 FFT

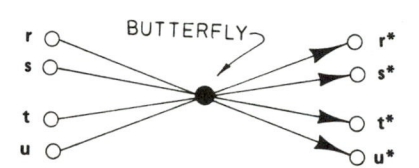

FIG. 3.7.1 Flow diagram for an $N = 4^2 = 16$-point radix-4 FFT and its associated four-point butterfly diagram.

3.8 BLOCK PROCESSING

Whether working with FFT library routines or commercially available FFT systems, the user is often faced with the problem of performing long-record-length transforms. However, there is usually a practical limit to the size of an admissible N-point FFT. This condition is generally imposed on the user by main memory (fast memory) limitations. Therefore, the creation of a $(N = LM)$-point transform can pose an interesting problem. This problem will be approached in a generalized sense. As a result of this analysis, several interesting and useful DFT results will evolve.

Consider partitioning an N-sample time series into blocks of M samples each. Furthermore, let the output spectrum of N harmonics be partitioned into M blocks of L samples. Referring to Fig. 3.8.1, any specific time or frequency domain sample can be represented as

$$x(i) = x(j, m) = x(Mj + m); \quad 0 \le m \le M - 1, \quad 0 \le Mj + m \le N - 1$$

$$X(k) = X(s, r) = X(Lr + s); \quad 0 \le s \le L - 1, \quad 0 \le Lr + s \le N - 1$$

$$(3.8.1)$$

The data displayed in Fig. 3.8.1 consider the data field in sequential and tensor (array) form. Using this data structure, the DFT of $\{x(i)\}$ can be written to read

$$
\begin{aligned}
X(k) = X(s, r) &= \sum_{m=0}^{M-1} \sum_{j=0}^{L-1} x(j, m) W_N^{(Mj+m)(Lr+s)} \\
&= \sum_{m=0}^{M-1} \sum_{j=0}^{L-1} x(j, m) W_N^{Lmr} W_M^{ms} W_N^{Msj} W_N^{MLjr} \\
&= \sum_{m=0}^{M-1} W_N^{Lmr} W_N^{ms} \sum_{j=0}^{L-1} x(j, m) W_N^{Msj}
\end{aligned}
\qquad (3.8.2)
$$

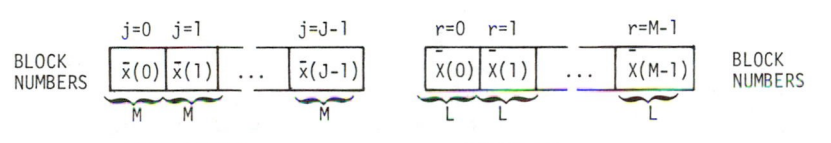

FIG. 3.8.1 Data partitioning of temporal data into an M x L matrix and spectral data into an L x M matrix.

This DFT expression can be algebraically reduced to the following operations:

$$q(s, \ m) = \sum_{j=0}^{L-1} x(j, \ m) W_N^{Msj}; \quad W_N^{Msj} = W_L^{sj} \qquad (3.8.3)$$

$$h(s, \ m) = W_N^{ms} q(s, \ m) \qquad (3.8.4)$$

then

$$X(s, \ r) = \sum_{m=0}^{M-1} h(s, \ m) W_N^{Lmr} = X(k = Lr + s) \qquad (3.8.5)$$

Each term holds a special significance in the transform process. The first [Eq. (3.8.3)] can be recognized to be an L-point FFT of every Mth sample in the time series $\{x(i)\}$ starting with the mth sample. Loosely stated, the temporal parameter j is mapped into the spectral parameter s. In terms of the tensor data structure, $q(s, \ m)$ is computed over the L element columns of [x]. These data would be stored in an N-complex word array, say [Q]. The data offset of m samples between columns of [x] is accounted for by the phase-shifting property of the terms W_N^{ms} [i.e., the twiddle terms of Eq. (3.8.4)]. The phase correction operation can be performed in place by mapping [Q] into [Q]. Finally, Eq. (3.8.5) represents an M-point FFT of the data found in array [Q]. That is, the sth row of [Q] is transformed for all s, $0 \le s \le L - 1$. The result of performing these L-point FFTs would be an N-word data base [the spectrum of $\{x(n)\}$] which would be stored in an N-complex word array, say [H]. The harmonic X(k) would be sequentially accessed from each row of [H] under the rule X(k) = X(Lr + s), where r is the row index and s is the column index. Therefore, an N-point DFT can be performed using nested M-point FFTs. Of course, one would have to support these operations with sufficient memory (e.g., peripheral disk, tape, etc.) to accommodate the arrays [x], [X], [Q], and [H].

Nesting can also be performed by extending this approach to the problem of transforming a time series having indices $m_1, \ m_2, \ \ldots, \ m_L$. Here, instead of two FFT passes over an M × L word array (see Fig. 3.8.1), L passes over a redefined data field would be required. This concept has come to be known as the block process.

A block-processed FFT will interpret an N-sample time series in terms of a multidimensional data field. For $N = r^n$, n represents the number of FFT passes required to form an N-point transform using radix-r building blocks. It will be assumed that multiple small-transform-length FFT units are available for use and they can operate in parallel when

required. The functional requirements of a block processor are listed
below:

1. If $N = r^n$, n radix-r FFT modules will be required.
2. n groups of circulating memory buffers of sizes varying from 1 to
    $(r^n - 1)$ words are required.
3. (n - 1) complex multipliers (used for twiddle factor weighting) are
    required.
4. n buffers of sizes varying from r to r^n words are required.
5. (n - 2) adders are required.

Referring to the partition originally developed in this section, $N = ML$,
one can equate $N = r^n$ to a factorization $L = r^n - 1$ and $M = r$. This factorization policy can be repeated so as to reduce the size of the largest FFT
required in the transformation of an N-sample time series to a value r.
From the factorization we are currently considering, an r-point and
r^{n-1}-point FFT would be required for $N_1 = M_1 L_1$, where $M_1 = r$ and $L_1 =$
r^{n-1}. However, if one were to redefine L_1 to read $L_1 = L_2 M_2$, where $L_2 =$
r^{n-2} and $M_1 = M_2 = r$, the largest FFT required is of length r^{n-2} (rather
than r^{n-1}). This redefinition of $L_i = L_{i+1} M_{i+1}$ can continue until the largest
FFT module is of a convenient size. The block-processing philosophy as
outlined in this section is mechanized in Fig. 3.8.2. Here an N = 16-point
transform is performed using various sizes of radices.

Referring to the data found in Fig. 3.8.2, it can be noted that in one
case a four-pass all-radix-2 FFT can be realized. The data structure of
this example is identical to the radix-2 DIT FFT developed in Section 3.5.
Therefore, the general radix-r FFT algorithm can be developed and defined in terms of the block-processing algorithm.

For the sake of completeness, consider the three-pass transform
in Fig. 3.8.2 based on the factorization, $N = PLM = (4)(2)(2)$ (i.e.,
largest transform of length 4). The resources needed to perform this
transform are:

1. LM radix-P FFT modules at the first level
2. LMP = N twiddles
3. MP radix-L FFT modules at the second level
4. LMP = N twiddles
5. LP radix-M FFT modules at the third level

This pattern, as noted before, can be continued for $N = N_i$. The reason that one would consider such a structure is to accelerate throughput
(i.e., speed). The multiplicity of hardware, in terms of the parallel
processing paths, gives rise to the concept of pipelining. FFT pipelining
can be defined in a vertical, horizontal, or total sense. The pipeline
options are summarized in a subsequent chapter. If only one radix-r FFT

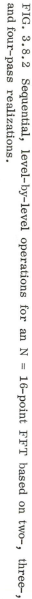
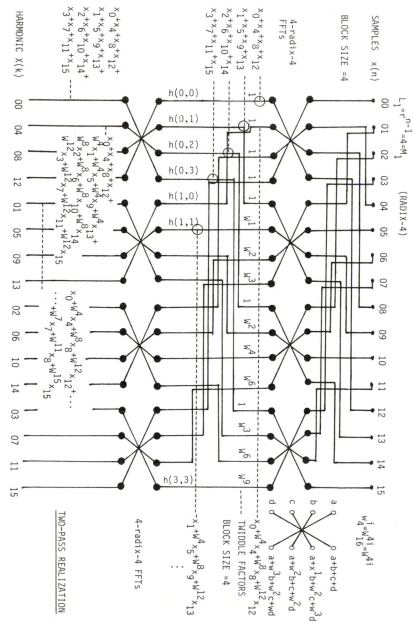

FIG. 3.8.2 Sequential, level-by-level operations for an N = 16-point FFT based on two-, three-, and four-pass realizations.

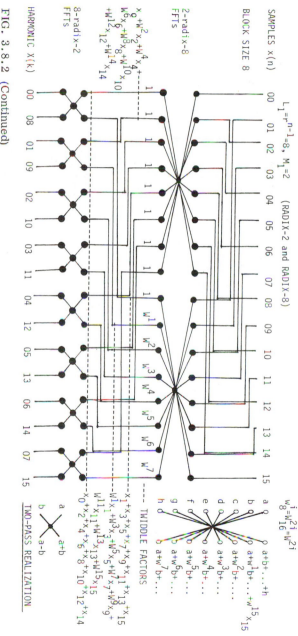

FIG. 3.8.2 (Continued)

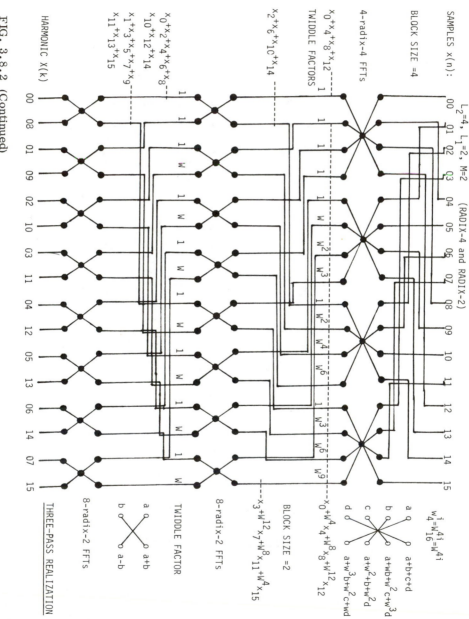

FIG. 3.8.2 (Continued)

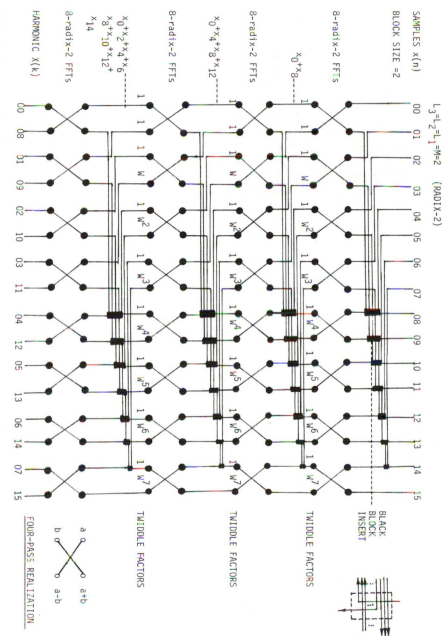

FIG. 3.8.2 (Continued)

module is to be used in a time-multiplexed manner, the FFT algorithm
will run at its lowest speed and have the data flow equivalent to that found
in the FORTRAN FFT in Fig. 3.5.4. However, a FFT operating in this
sequential manner often has sufficient throughput to satisfy many applications. This subject will be reinvestigated when commercial FFT systems
are studied.

3.9 ZOOM FFT

A special interpretation of Eq. (3.8.5) has been suggested by Yip (1976)
and is referred to as the <u>zoom FFT</u>. The zoom FFT should not be confused with the band-selective FFT studied in the next section. Again consider X(s, r) defined in Eq. (3.8.2) except in a juxtaposed order so that

$$
X(s, r) = \sum_{i=0}^{L-1} W_N^{Msj} \sum_{s=0}^{M-1} W_N^{Lmr} W_N^{ms} x(j, m)
$$

$$
\triangleq \sum_{i=0}^{L-1} W_N^{Msj} X_j(r, s) \tag{3.9.1}
$$

Suppose further that the value of ms is small. Then $W_N \simeq 1$ and it follows
that an approximation to $X_j(r, s)$, say $\hat{X}_j(r)$, may be defined as[†]

$$
\hat{X}_j(r) = \sum_{m=0}^{M-1} W_N^{Lmr} x(j, m) \tag{3.9.2}
$$

It can be observed that $\hat{X}_j(r)$ is an M-point Fourier transform of the data
found in the jth data block. This condition requires that s be small,
$0 \leq s \leq L - 1$, which can be guaranteed by requiring L to be small compared
to M. The resulting transform $\hat{X}_j(r)$ possesses a spectral resolution of
1/MT hertz per harmonic. This transform is generally performed using a
suitable data window such as that due to Hanning (data windows are treated
in Chapter 5). The purpose of the zoom FFT is to provide a high-resolution
spectral image about a prespecified frequency location. This thought can
be motivated by assigning the role of the first FFT, namely $\hat{X}_j(r)$, to use
its 1/MT hertz per line resolution to perform a "coarse" FFT of the M-
point block structured data base. That is, if N = ML, then

[†] $W_N^{Lmr} = \exp[-2\pi Lmr/(N = ML)] = \exp(-2\pi mr/M)$.

$$X(s, \ r) = \sum_{j=0}^{L-1} W_N^{Msj} \ {}^{M-1} \ W_N^{Lmr} W_N^{ms} x(j, \ m)$$

$$= \sum_{j=0}^{L-1} W_N^{Msj} \hat{X}_j(n) \tag{3.9.3}$$

where

$$W_N^{Lmr} = \exp\left(\frac{2\pi Lmr}{N}\right) = \exp\left(\frac{2\pi Lmr}{ML}\right) \tag{3.9.4}$$

The spectral series $\{X_j(r)\}$ represents the M Fourier coefficients from data block j, j = 0, . . ., L - 1. Furthermore, the elements of this harmonic series are located at the common frequency $f_r = r/MT$. Finally, the last transformation, found in Eq. (3.9.3), represents another FFT. However, in this case the FFT is performed over a sequence of Fourier coefficients, all representing a common rth coarse harmonic. For example, if r = 0, an L-point FFT of all the dc components obtained from all M data blocks will be performed. Amplitude fluctuations in the dc component, from data block to data block, will be spectrally transformed. The analysis of this spectrum will give insights into the low-frequency behavior of $\{x(k)\}$ about 0 Hz. Of course, this experiment can be conducted for any $r \in [0, \ M/2]$. The improved spectral resolution provided by the second L-point FFT about the rth coarse harmonic is a factor of L over the coarse resolution metric 1/MT. That is, the spectral resolution of X(r, s), for r fixed and $s \in [0, L/2]$, is 1/NT. As a result, one concludes that a zoom factor of L has been applied to the original coarse spectra. The reader should, however, be aware of the approximations used in the development of this strategy as well as the fact that windowing (at topic to be covered) is generally required to ensure satisfactory operation.

EXAMPLE 3.9.1 For the purpose of demonstration, wideband noise was added to a sinusoidal signal centered at f = 541 Hz (see Fig. 3.9.1). An M = 128-point FFT was used to provide a coarse transform of the block-sectioned input time series. The pure tone at 541 Hz can be seen to fall between the coarse harmonic lines located at 505 and 578 Hz. Performing an L = 8-point zoom FFT on a sequence of harmonics located at 505 Hz provides the user with a good indication of spectral activity located at 541 Hz (equivalent to 1024-point resolution).

EXAMPLE 3.9.2 A signal consisting of an additive wideband noise process and eight equally spaced tones was numerically generated. Using a 512-point FFT on 1-Hz centers, the ensemble-averaged magnitude spectrum was computed. The magnitude spectrum using a uniform data window is

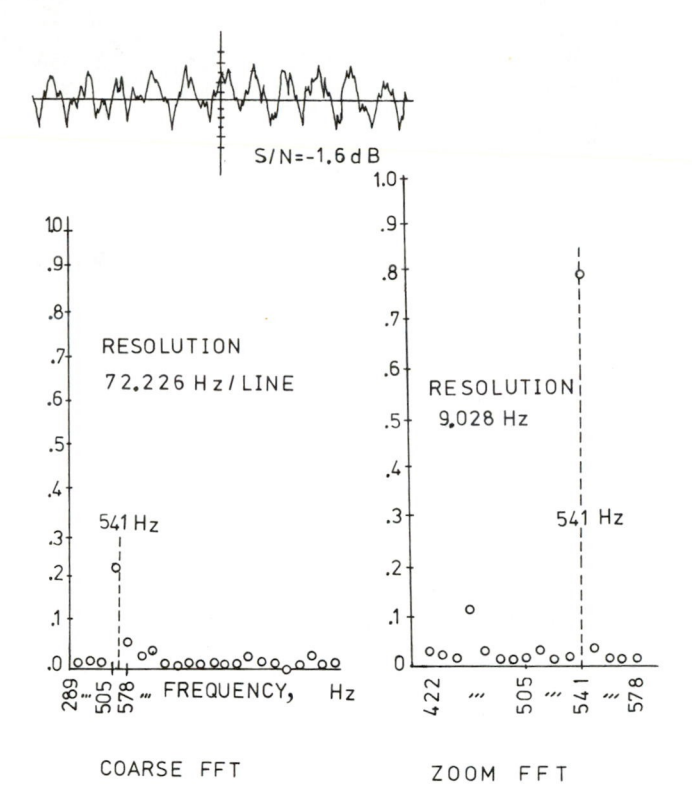

FIG. 3.9.1 Normal and zoom FFT-derived spectra of a pure tone in additive noise with the zoom spectrum centered about 505 Hz.

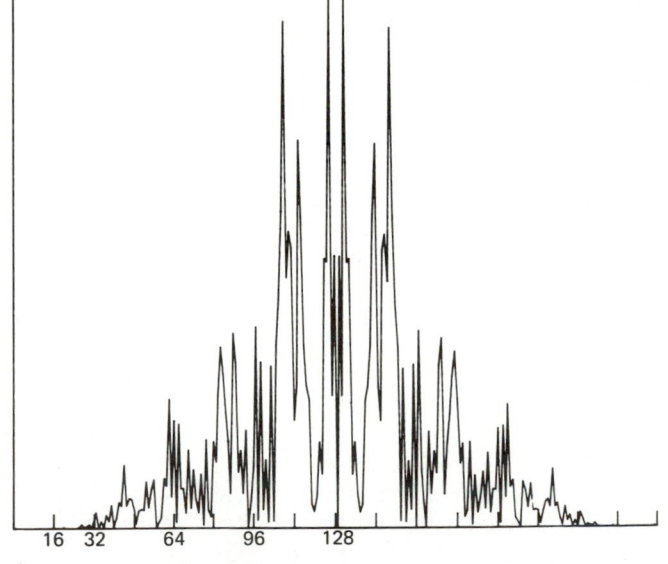

FIG. 3.9.2 Spectrum of a tone complex in additive noise using a FFT and a uniform data window.

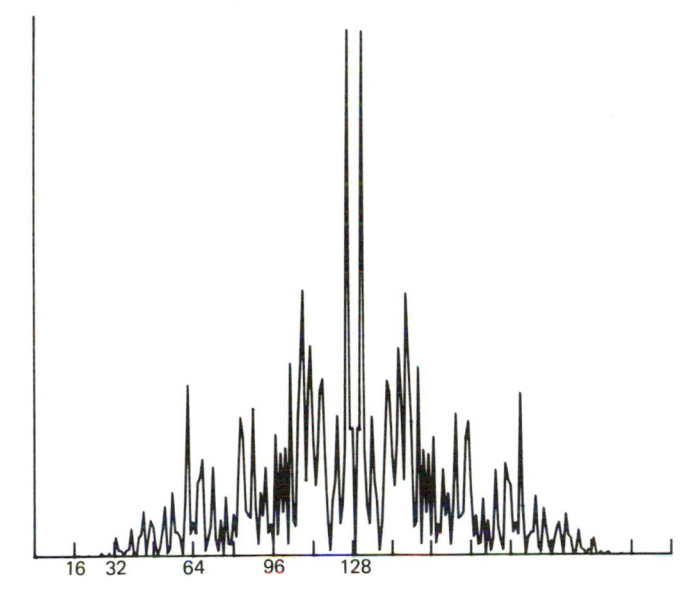

FIG. 3.9.3 Spectrum of a tone complex in additive noise using a FFT and a raised cosine window.

shown in Fig. 3.9.2, and that derived using a raised cosine window is shown in Fig. 3.9.3. A 16-point zoom FFT about the 105- and 114-Hz line was performed with the results as given in Fig. 3.9.4. The pure tones that existed at 105 and 114 Hz are now more sharply defined. The reason the 105-Hz tone is more pronounced is that the numerically synthesized tone at 105 Hz was stronger than the one at 114 Hz.

3.10 BAND-SELECTABLE FOURIER TRANSFORMS

It may be recalled that the FFT is a baseband spectral analysis tool which uniformly resolves the frequency axis into N/2 harmonic values over the range $[0, f_s/2]$. However, there are times when the analysis of the complete baseband range is unnecessary. Recall that the frequency resolution of an N-point FFT is given by $\Delta f = f_s/2N$. Suppose that one is interested only in quantifying the frequency domain behavior of the original time series over a relatively narrow band of frequencies. Unless N is very large, the baseband resolution of the standard FFT may be insufficient to support a detailed analysis of the spectrum over the desired limited range. For example, suppose that the Nyquist frequency of a given signal is 5 kHz. To ensure the absence of aliasing errors, a sample rate above 10 kHz may

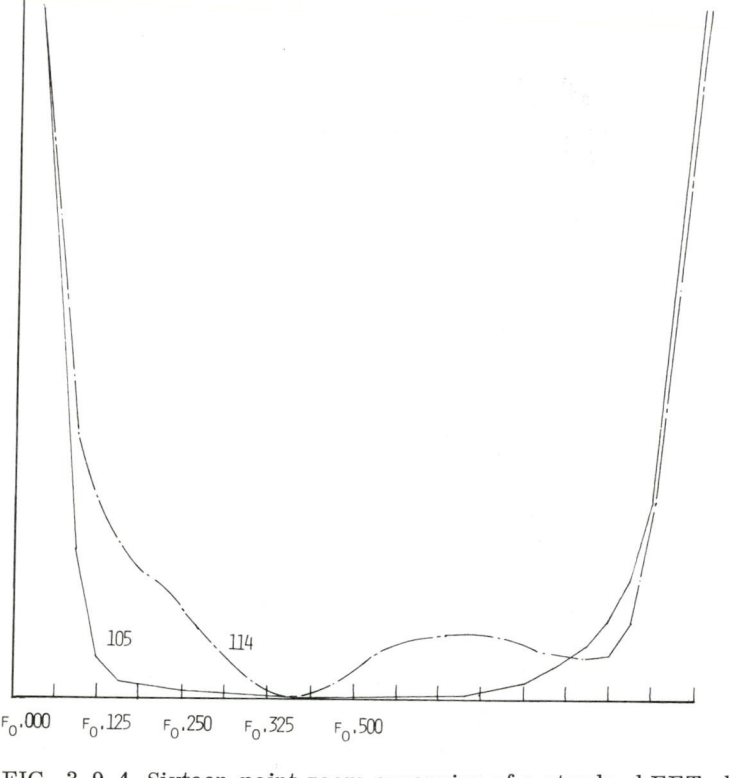

FIG. 3.9.4 Sixteen-point zoom expansion of a standard FFT about 105 and 114 Hz.

be chosen, say 20,480 Hz. Using a 1024-point FFT, a spectral resolution of $f_1 = 20$ Hz per harmonic is established. If one is interested in concentrating on the frequency range found between 1 and 1.1 kHz, only five harmonics would be available for analysis. A "finer" refinement of this frequency range could, of course, be accomplished by increasing N. If N were increased by a factor of 10, then 50 harmonics would be available for analysis. However, the degradation in throughput and increased memory requirements would, in most cases, be prohibitively large. To overcome this problem, the band-selective method has been developed. (Examples of band-selectable transforms are given in Section 5.8.)

The band-selective method is essentially a heterodyne process which provides a localized high-resolution DFT about a prespecified frequency. The structure of such a system is detailed in Fig. 3.10.1.

The value of the resampling frequency, shown to be f_s/K, can be argued in a straightforward manner. Observe that the output of the lowpass

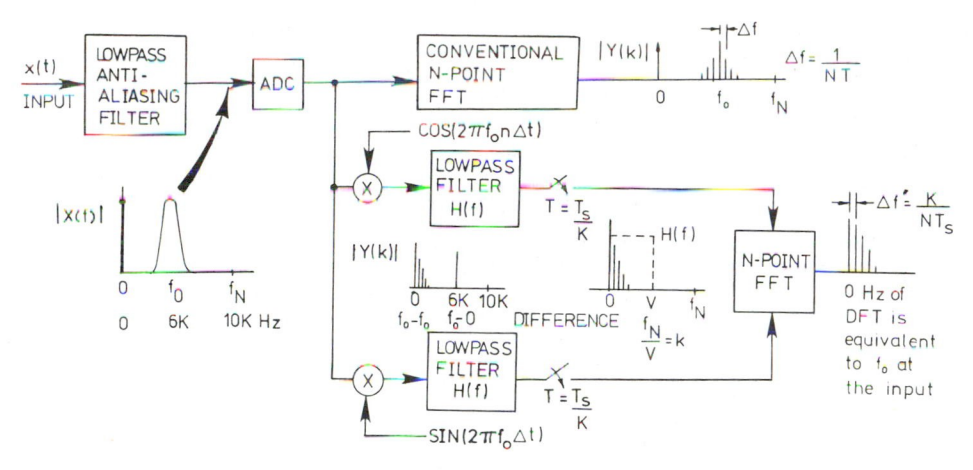

FIG. 3.10.1 Architecture of a band–selectable FFT system using the heterodyne technique.

filter is bandlimited to V hertz. Therefore, a sample rate of only $2V \le f_s/K$ is required to ensure that the synthesized lowpassed signal can be sampled without incurring aliasing errors. The larger the value of K, the "finer" will be the refinement of the spectra about a prespecified frequency. The resolution that one can expect of a band–selective transform about a frequency f_0 is displayed in Fig. 3.10.2. It can be noted the positive base-band FFT spectrum is linearly mapped into the frequency range $[f_0, f_0 + f_s/2K]$, with a frequency resolution of $f_s/2KN$ hertz per spectral line.

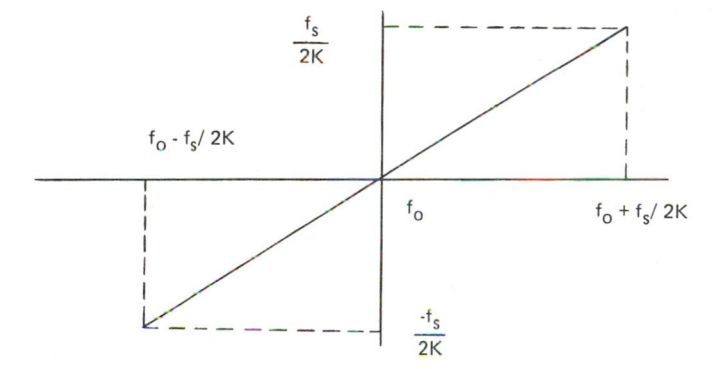

FIG. 3.10.2 Frequency domain resolution for a band–selectable FFT centered about f_0 hertz.

The key elements of the band-selective analyzer are the modulator, the FFT, and the digital lowpass filter. The design of the FFT has been shown to be a straightforward process. Remember, however, that the FFT performs a mapping from the complex number to the complex numbers. Therefore, the sine and cosine nature of the band-selective FFT should pose no problem. The modulator is generally implemented using a digital multiplier. The multiplier will product elements of the incoming time series with samples of sine and cosine series having a center frequency of f_0 (chosen by the user). These trigonometric coefficients are typically stored in a read-only memory unit and randomly accessed during run time. This strategy will allow many center frequencies to be synthesized from one set of sine and cosine tables.

3.11 CHIRP-z TRANSFORM

It has been established that the DFT of a given time series $\{x(n)\}$ is given by

$$X(k) = \sum_{n=0}^{N-1} x(n) W_N^{nk} \tag{3.11.1}$$

The defining DFT equation has an intrinsic inner-product form. However, the DFT can also be written in a convolution form. Using Bluestein's identity, which states that

$$nk = \frac{n^2}{2} + \frac{k^2}{2} - \frac{(n+k)^2}{2} \tag{3.11.2}$$

the basic DFT equation can be rewritten to read

$$Xk) = \sum_{n=0}^{N-1} x(n) W_N^{n^2/2} W_N^{k^2/2} W_N^{-(n+k)^2/2} \tag{3.11.3}$$

or

$$X(k) = W_N^{k^2/2} \sum_{n=0}^{N-1} x(n) W_N^{n^2/2} W_N^{-(n+k)^2/2}$$

Defining a term g(n) to be

$$g(n) = W_N^{n^2/2} x(n) \qquad (3.11.5)$$

allows the DFT form to be expressed as

$$X(k) = W_N^{k^2/2} \sum_{n=0}^{N-1} g(n) W_N^{-(n+k)^2/2}$$

$$= W_N^{k^2/2} [g(n) \circledast W_N^{n^2/2}] \qquad (3.11.6)$$

It can readily be seen that in this form the DFT operation is defined in terms of a convolution sum. The convolution operation accepts time domain samples as inputs and outputs DFT components. The output sample process, which resides in the time domain, can thus be put into one-to-one correspondence with the spectrum of $\{x(n)\}$.[†] The convolution form of the DFT, based on Bluestein's identity, is called the <u>chirp-z transform</u> (CZT). The concept of "chirp" is derived from a radar lexicon where signals of the form $W_N^{n^2}$ are referred to as chirps.

EXAMPLE 3.11.1 For the purpose of clarity, the CZT of a simple sine and cosine are interpreted graphically in Fig. 3.11.1.

In many applications a so-called power spectral representation of some $\{x(n)\}$ is desired. This topic, which is developed in Chapter 4, is formally given by P(k), where $P(k) = X(k)X(k)^*$. It then follows that in a CZT sense,

$$P(k) = X^*(k)X(k)$$

$$= \left[g(n) \circledast W_N^{n^2/2} \right] * W_N^{-n^2/2} W_N^{n^2/2} \left[g(n) \circledast W_N^{n^2/2} \right]$$

$$= \left[g(n) W_N^{n^2/2} * g(n) W_N^{n^2/2} \right]$$

$$= \left| g(n) * W_N^{n^2/2} \right|^2 \qquad (3.11.7)$$

[†] The cochlea in the inner ear serves a similar purpose. This tapered nerve converts acoustic motion into electrical responses.

The CZT and power spectral version of the CZT are shown in Fig. 3.11.2. It can be noted that the postmultiplication operation found in the basic CZT has been replaced by a squaring operation in the power spectral version.

The CZT enjoys limited popularity. The principal reason for this is found in its multiplication count. For use with a general-purpose computer, the convolution operation would normally be performed using the convolution theorem. Here two forward FFTs are taken, produced, and inverted. Therefore, it is obvious that trading three FFTs for a single FFT operation [i.e., FFT $x(n)$] would be counterproductive. However, using a new electronic technology, the charge-coupled device (CCD), efficient nonrecursive linear convolvers (filters) can be realized. Using this technology, CCD-based CZT systems have been designed and, in some cases, are now commercially available (e.g., Reticon R5601). Spectra obtained from a commercial CZT system are shown in Figure 3.11.3.

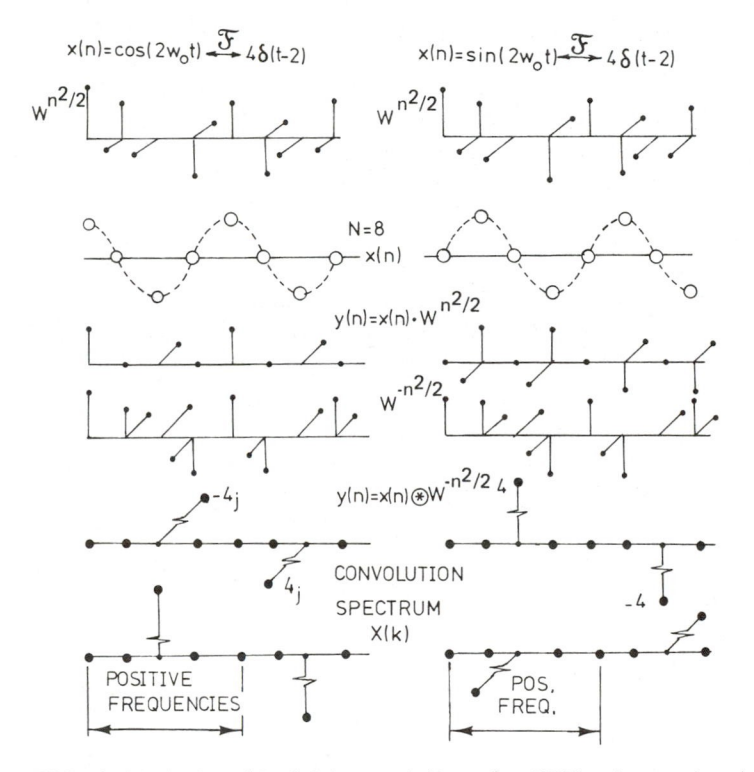

FIG. 3.11.1 Graphical interpretation of a CZT using as inputs a pure cosine and sine wave.

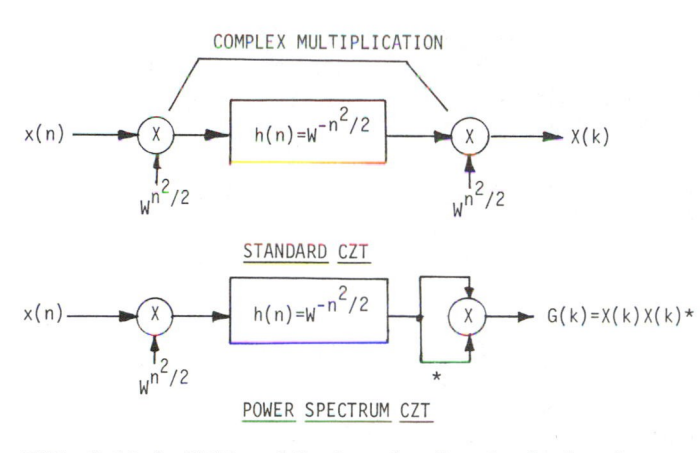

COMPLEX MULTIPLICATION

STANDARD CZT

POWER SPECTRUM CZT

FIG. 3.11.2 CZT architecture for the standard and power transform.

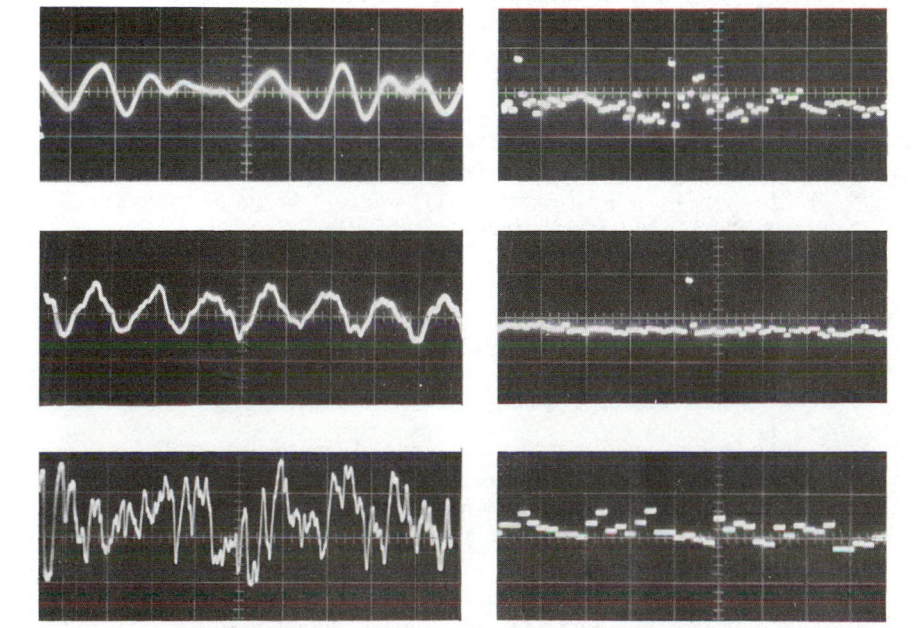

TIME DOMAIN   POWER SPECTRUM

FIG. 3.11.3 Experiments performed using a Reticon 5601 CZT over a class of signal embedded in additive noise.

3.12 ORTHOGONAL TRANSFORMS

It is well known that the Fourier transform belongs to a class of operations known as orthogonal transforms. Besides the FFT, there are other orthogonal transforms that possess useful digital realizations. Some of the more important transforms are:

1. Walsh transform
2. Slant transform
3. Cosine transform

Walsh Transform

It has already been established that a physically realizable signal can be approximated by a linear combination of orthogonal periodic functions (e.g., trigonometric Fourier series). Here one naturally associates the noun "frequency" with the usual concept of periodicity. However, there is a more general definition of frequency and it is known as <u>sequency</u>. Sequency is defined to be one-half the average number of zero crossings per unit time. As previously stated, it is possible to represent a signal as a linear combination of orthogonal functions. Now the functions will be defined in terms of sequency. One of the earliest set of orthogonal functions, based on a sequency metric, was proposed by Haar in 1910. <u>Haar functions</u> have a limited signal processing application and will therefore be sidestepped at this time. A potentially useful class of orthogonal functions developed by Rademacher in 1922 is defined in terms of a function denoted rad(m, t). Here rad(m, t) = rad(m, t + T), where T is the periodicity of the Rademacher functions. They explicitly satisfy

$$\text{rad}(m, \ t) = \text{rad}(1, \ 2^{m-1}t)$$

$$\text{rad}(1, \ t) = \begin{cases} 1, & t \in [0, \ T/2] \\ -1, & t \in (0, \ T) \end{cases} \tag{3.12.1}$$

The attractive feature possessed by this class of functions is their binary-valued amplitude (i.e., +1, -1). The importance of this amplitude restriction is one of arithmetic. Here, only addition (+1) or subtraction (-1) is required to mechanize a Rademacher function. Unfortunately, the Rademacher functions are not complete. That is, an approximating weighted linear combination of Rademacher functions would ideally be used to approximate a function x(t). However, the Rademacher sum may converge to some other value, say y(t), which does not equal x(t). This mathematical shortcoming was overcome by Walsh in 1923. Walsh developed a complete set of binary-valued approximating functions, denoted wal(i, T), referred to as <u>Walsh functions</u>. These functions are also orthogonal and therefore

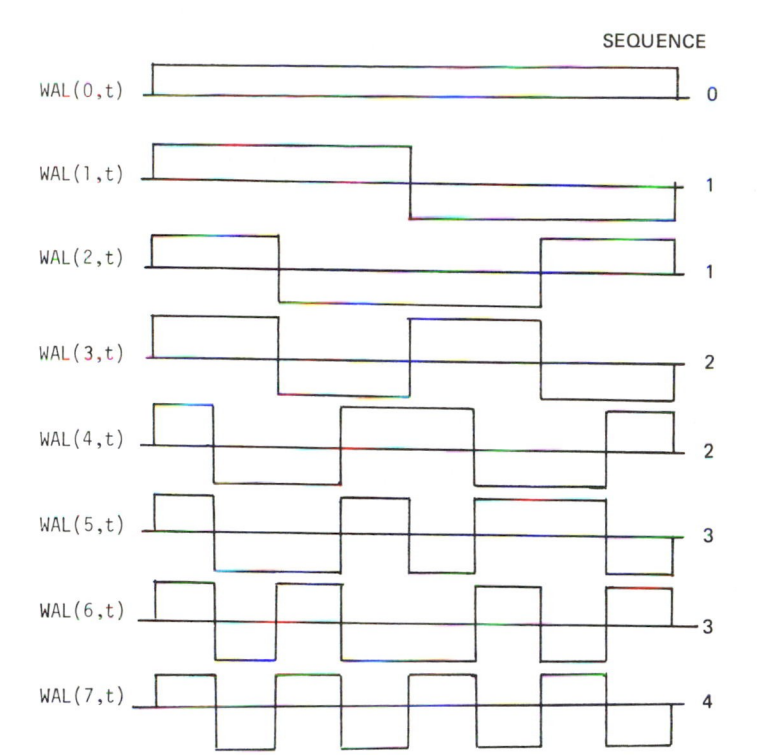

FIG. 3.12.1 Walsh basis functions for an N = 8-point Walsh transform.

useful in interpolation. The first eight Walsh functions, over a period T, are shown in Fig. 3.12.1. It can be noted that wal(5, t) and wal(6, t) are aperiodic in the usual frequency sense. Such aperiodic behavior is not present in the Rademacher approach. However, this aperiodicity was needed to complete the set of binary-valued functions.

The Walsh functions are sometimes symbolically referred to as the "cal" and "sal" functions. Specifically,

$$cal(n, t) = wal(2n, t)$$

$$sal(n, t) = wal(2n - 1, t) \qquad (3.12.2)$$

Other variations (orderings) of the Walsh set have been developed as well. One of the most useful is based on the Hadamard order, which is derived from the structure of the Hadamard matrix. A 2n × 2n Hadamard matrix, say H_{2n}, can be partitioned as follows:

$$H_{2n} = \begin{bmatrix} H_n & H_n \\ \hline H_n & -H_n \end{bmatrix}$$ (3.12.3)

To generate a set of n Walsh functions, one need only define $H_1 = 1$ and recursively generate H_{2n}. Using n = 8 as an example, it can easily be shown that the definition of W_8 is given by

$$W_8 = \begin{bmatrix} 1 & 1 & 1 & 1 & 1 & 1 & 1 & 1 \\ 1 & -1 & 1 & -1 & 1 & -1 & 1 & -1 \\ 1 & 1 & -1 & -1 & 1 & 1 & -1 & -1 \\ 1 & -1 & -1 & 1 & 1 & -1 & -1 & 1 \\ 1 & 1 & 1 & 1 & -1 & -1 & -1 & -1 \\ 1 & -1 & 1 & -1 & -1 & 1 & -1 & 1 \\ 1 & 1 & -1 & -1 & -1 & -1 & 1 & 1 \\ 1 & -1 & -1 & 1 & -1 & 1 & 1 & -1 \end{bmatrix} ;$$

wal(0, t) = cal(0, t)
wal(7, t) = sal(4, t)
wal(3, t) = sal(2, t)
wal(4, t) = cal(2, t)
wal(1, t) = sal(1, t)
wal(6, t) = cal(3, t)
wal(2, t) = cal(1, t)
wal(5, t) = sal(3, t)

As a result, the matrix W_8 may be used to define an isomorphic (invertible) mapping between a time series $\{x(n)\}$ and a coefficient set (Walsh coefficients) $\{w(n)\}$, where

$$\bar{w}_n = \frac{W_N}{2^n} \bar{x}_n; \quad \bar{x}_n = W_n^{-1} \bar{w}_n$$ (3.12.4)

The elements of the coefficient array form what is called the Walsh spectrum.

   Since the orthogonal set of Walsh functions is complete, they can be used to interpolate other functions. An absolutely integrable (i.e., L_1) function x(t) over [0, T] can be expanded in the form

$$x(t) = \sum_{i=0}^{N} a_i \, wal(i, t)$$

$$= a_0 \, wal(0, t) + b_{N/2} \, sal(N/2, t) + \sum_{k=0}^{N/2-1} [a_k \, cal(k, t) + b_k \, sal(k, t)]$$

                                                                                (3.12.5)

where

$$a_i = \int_0^T x(t) \, wal(i, t) \, dt; \quad i = 0, 1, 2, \ldots$$ (3.12.6)

Convergence conditions are similar to those imposed on the Fourier series by Dirichlet. Specifically, the Walsh sum is known to converge uniformly to the value of x(t) wherever x(t) is continuous and converge to the arithmetic mean of the discontinuities of x(t). It is also known that a fast Walsh transform (FWT) algorithm exists which has a structure very similar to the FFT. However, there are some problems with the Walsh transform. Whereas the FFT of a bandlimited signal is bounded in the frequency domain by the Nyquist frequency, the Walsh spectra of the same signal is of infinite bandwidth (in a sequence sense). This is a consequence of the piecewise-constant nature of the Walsh approximation to a continuous signal. The discontinuities found at the sample instances of the Walsh approximation give rise to these high-frequency components. This infinite-bandwidth condition is a drawback to its general use. In practice, a truncated (finite) Walsh spectra is generally used. It is commonly accepted that 2N Walsh coefficients are equivalent to N Fourier coefficients. Since the Fourier coefficients are complex numbers and the Walsh coefficients are real, the 2N-to-N coefficient relationship between the two transforms should be transparent.

The major drawback of the Walsh transform is the absence of a physically interpretable spectrum. Whereas the Fourier spectrum can easily be related to physically interpretable harmonic oscillations, the Walsh spectrum must be interpreted in terms of Walsh functions. Since these functions are generally aperiodic, their material meaning is obscure. Nevertheless, Walsh transforms have been used successfully to model binary-valued communication processes (due to their binary nature) and support some data compression tasks. Here a limited number of Walsh coefficients are used to approximate the entire Walsh spectrum. From these few coefficients an approximation of the original time series is derived.

EXAMPLE 3.12.1 Let $x(n) = \cos(2\pi n/8)$, $n = 0, 1, \ldots, 7$. The DFT of $x(n)$, say $X(k) = \{0 + j0, 1 + j0, 0 + j0, 0 + j0\}$ (see Fig. 3.12.2). The eight-point Walsh transform is given by $W_n = W_8 x_n/8$, or

$$
W_8 = \frac{1}{8}
\begin{bmatrix}
1 & 1 & 1 & 1 & 1 & 1 & 1 & 1 \\
1 & -1 & 1 & -1 & 1 & -1 & 1 & -1 \\
1 & 1 & -1 & -1 & 1 & 1 & -1 & -1 \\
1 & -1 & -1 & 1 & 1 & -1 & -1 & 1 \\
1 & 1 & 1 & 1 & -1 & -1 & -1 & -1 \\
1 & -1 & 1 & -1 & -1 & 1 & -1 & 1 \\
1 & 1 & -1 & -1 & -1 & -1 & 1 & 1 \\
1 & -1 & -1 & 1 & -1 & 1 & 1 & -1
\end{bmatrix}
\begin{bmatrix}
1.000 \\
.707 \\
.000 \\
-.707 \\
-1.000 \\
-.707 \\
.000 \\
.707
\end{bmatrix}
=
\begin{bmatrix}
0 \\
2/8 \\
0 \\
0 \\
2/8 \\
2/8 \\
4.828/8 \\
-.828/8
\end{bmatrix}
$$

with the spectrum interpreted graphically in Fig. 3.12.2.

Slant Transform

Another important signal processing transform in popular use is the slant transform. The slant transform is derived from its sawtooth-like waveform.

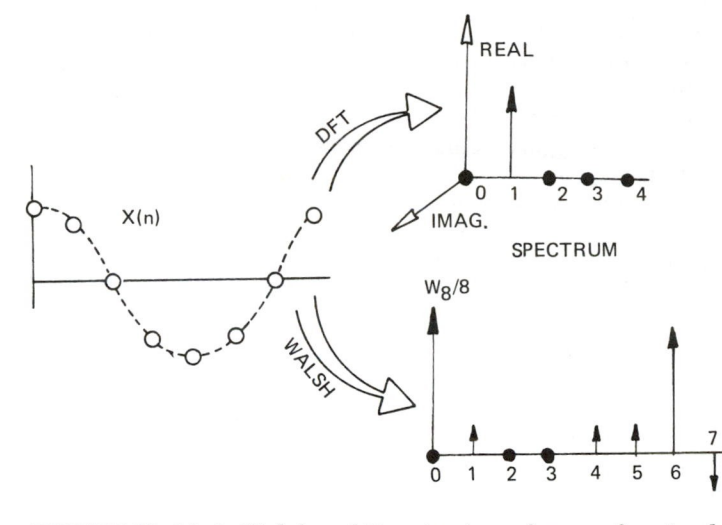

FIGURE 13.12.2 Walsh and Fourier transforms of a simple harmonic process.

This geometric behavior makes this class of transform useful in approximating signals that possess gradual amplitude variations. This is in sharp contrast with the Walsh transform, which, due to its piecewise constant nature, thrives on abrupt amplitude changes. The slant transform can be defined in terms of a slant matrix given by

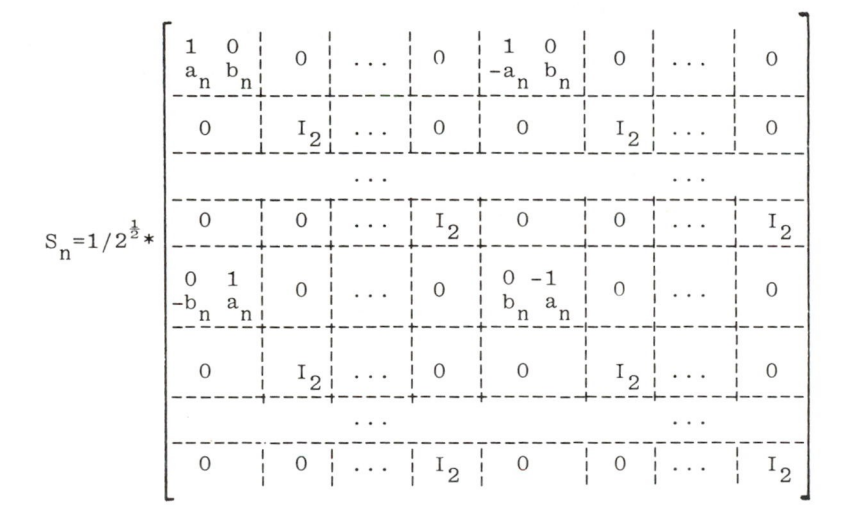

where $a_2 = 1$, $b_n = 1/(1 + 4a_{n/2}^2)^{1/2}$, ..., $a_n = 2b_n a_{n/2}$, and I_2 is the 2×2 identity matrix.

EXAMPLE 3.12.2 For

$$S_1 = \frac{\begin{bmatrix} 1 & 1 \\ 1 & -1 \end{bmatrix}}{2^{\frac{1}{2}}}$$

then

$$S_2 = \frac{1}{2^{\frac{1}{2}}} * \begin{bmatrix} 1 & 0 & 1 & 0 \\ 2/5^{\frac{1}{2}} & 1/5^{\frac{1}{2}} & -2/5^{\frac{1}{2}} & 1/5^{\frac{1}{2}} \\ 0 & 1 & 0 & -1 \\ -1/5^{\frac{1}{2}} & 2/5^{\frac{1}{2}} & 1/5^{\frac{1}{2}} & 2/5^{\frac{1}{2}} \end{bmatrix} \begin{bmatrix} \frac{1}{2^{\frac{1}{2}}} * \begin{bmatrix} 1 & 1 \\ 1 & -1 \end{bmatrix} & \begin{bmatrix} 0 & 0 \\ 0 & 0 \end{bmatrix} \\ \begin{bmatrix} 0 & 0 \\ 0 & 0 \end{bmatrix} & \frac{1}{2^{\frac{1}{2}}} * \begin{bmatrix} 1 & 1 \\ 1 & -1 \end{bmatrix} \end{bmatrix}$$

$$\frac{1}{2} * \begin{bmatrix} 1 & 1 & 1 & 1 \\ 3/5^{\frac{1}{2}} & 1/5^{\frac{1}{2}} & -1/5^{\frac{1}{2}} & -3/5^{\frac{1}{2}} \\ 1 & 1 & 1 & 1 \\ 1/5^{\frac{1}{2}} & -3/5^{\frac{1}{2}} & 3/5^{\frac{1}{2}} & -1/5^{\frac{1}{2}} \end{bmatrix}$$

The graphical form is given in Fig. 3.12.3.

Cosine Transform

The cosine transform is based on the Chebyshev polynomial. The cosine transform can be useful in supporting signal identification and decision tasks.[†] The cosine transformation approximation of a time series $\{x(n)\}$ is given by

---

[†] It has been shown that the cosine transform can perform an "eigenvalue-like" decomposition of a signal.

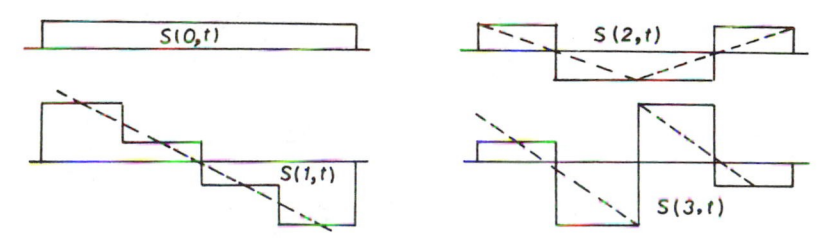

FIG. 3.12.3 First four slant transform basis functions.

$$x(n) = \frac{1}{\sqrt{N}} L_x(0) + \frac{2}{T} \sum_{k=1}^{N-1} L_x(k) \cos\left(\frac{2m+1}{N} kT\right), \quad m = 0, 1, \ldots, N - 1$$

$$(3.12.7)$$

where

$$L_x(0) = \frac{1}{\sqrt{N}} \sum_{m=0}^{N-1} x(n); \quad x(m) = 0 \quad \text{for } m = N, N+1, \ldots, 2N - 1$$

$$L_x(n) = \frac{2}{N} \text{Re}\left[\exp\left(\frac{-jk}{2N}\right) \sum_{m=0}^{2N-1} x(m)W_{2N}^{km}\right]$$

It is apparent from the definition of L_x that it can alternatively be defined in terms of a 2N-point IFFT where $x(m)$ is packed with N zeros. Therefore, the cosine transform possesses an elegant computational media as well.

BIBLIOGRAPHY

Brigham, E. O. (1974), The Fast Fourier Transform, Prentice-Hall, Englewood Cliffs, N.J.

Broderson, R. W., C. R. Hewes, and D. D. Buss (1976), A 500 Stage CCD Transversal Filter for Spectral Analysis, IEEE J. Solid-State Circuits, SC-11, No. 1, February, pp. 75-84.

Oppenheim, A. V., and R. W. Schafer (1975), Digital Signal Processing, Prentice-Hall, Englewood Cliffs, N.J.

Rabiner, L. R., and B. Gold (1975), Theory and Application of Digital Signal Processing, Prentice-Hall, Englewood Cliffs, N.J.

Rabiner, L. R., R. W. Schafer, and C. M. Rader (1969), The Chirp-z Transform Algorithm, IEEE Trans. Audio Electroacoust., AU-17, June, pp. 86-92.

Singleton, R. C. (1969), An Algorithm for Computing the Mixed Radix Fast Fourier Transform, IEEE Trans. Audio Electroacoust., AU-17, June, pp. 93-103.

Taylor, F. J., and S. L. Smith (1965), Digital Signal Processing in FORTRAN, Lexington Books, Lexington, Mass.

Yip, Y. C. (1976), Some Aspects of the Zoom Transform, IEEE Trans. Comput., C-25, March.

4
Systems Concepts

4.1 INTRODUCTION

Linear system theory is at the heart of contemporary engineering. This theory has given us many of the mathematical models of the physical world that we use in design and analysis. Linear systems may be characterized in terms of an impulse response $h(n)$. The system output to an input stimulus, say $x(n)$, is given by the convolution operation $y(n) = x(n) * h(n)$. Many important system concepts can be explained in terms of this convolution. However, evaluating the sum can often be a tedious exercise. Instead, linear systems are often analyzed in the computationally more efficient frequency domain. Here, operations such as the FFT can be used to expedite the computations found troublesome in the time domain. Furthermore, the frequency domain often provides the user with intuitive information about a system and its behavior which would be difficult to obtain in the time domain. Using time and frequency domain methods, many powerful system statements can be derived and put into practical use. However, there are times when the inputs to a linear system are random processes rather than deterministic signals. In such cases, time and frequency domain analysis must be interpreted in a stochastic sense. The stochastic analysis of linear systems requires special interpretations of linear system concepts and theory. Such topics are presented in this chapter.

4.2 CONVOLUTION

The linear convolution of two time series has been previously defined to read

$$y(n) = \sum_{i=-\infty}^{\infty} h(n - i)x(i) = h(n) * x(n) \qquad (4.2.1)$$

75

However, under certain circumstances, it has been shown that the DFT
can be used to replace the convolution sum. In Chapter 3 it was noted that
the circular convolution theorem could produce a result which differed from
that obtained using linear convolution. This problem can be remedied by
using a technique called <u>zero fill</u>.

Suppose that $x(n)$ and $h(n)$ are two N_1- and N_2-point time series.
From the definition of linear convolution it can be seen that $y(n) = 0$ for
$\geq N_1 + N_2 - 1$ since $h(i)$ and $x(n - i) = 0$ for $i \geq N_1 + N_2$. The convolution
operation is graphically reinforced in Figs. 4.2.1 and 4.2.2. Comparing
the data found in these figures, one notes that the circular and linear convolution sums are equal if the component time series are filled with a sufficient number of zeros. This observation will now be formally exploited.

For the purpose of discussion, consider the linear convolution of
the data shown in Fig. 4.2.1. The resulting convolution of the eight- and
four-point signal processes is shown to be of length 11. In order to define
a circular convolution that equals a linear convolution, a periodic time
series of length $(N_1 + N_2 - 1)$ must be created (i.e., period 11 in this case).

Signals of period $N_1 + N_2 - 1$ can be obtained simply by "filling out"
$x(n)$ and $h(n)$ with zeros. More specifically, the N_1-point time series
$\{x(n)\}$, if augmented with $N_2 - 1$ zeros, would form a new time series
$\{\bar{x}(n)\}$. If one considers this time series to be periodically extended for all
time, a periodic process of length $N_1 + N_2 - 1$ is defined. Similarly, if
$\{h(n)\}$ is augmented with $N_1 - 1$ zeros, another periodic time series can be
created. The circular convolution of these new time series, using $(N_1 + N_2 - 1)$-point FFTs, will agree with the linear convolution over the first
$N_1 + N_2 - 1$ sample values. From a practical standpoint, one would gener-
ally choose $N_1 + N_2 - 1$ to have the radix-r value r^n. This will ensure
efficient use of the FFT algorithm.

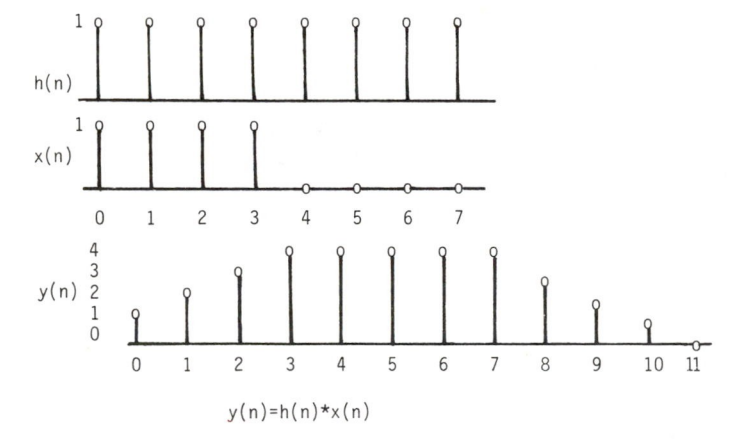

$$y(n) = h(n) * x(n)$$

FIG. 4.2.1 Simple linear convolution.

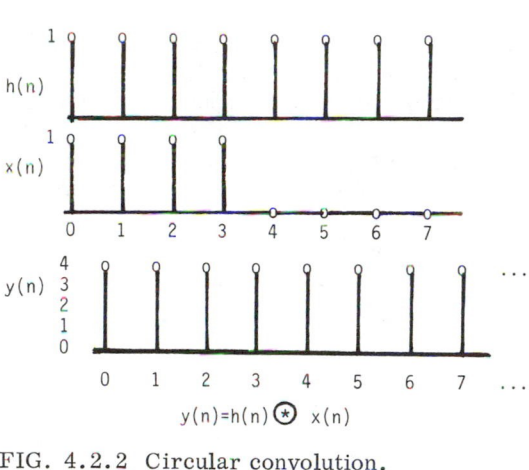

FIG. 4.2.2 Circular convolution.

In a more general setting, suppose that $\overline{x}(n)$ and $\overline{h}(n)$ are N-point and L-point time series defined as follows:

$$\overline{x}(n) = \begin{cases} x(n), & 0 \leq n \leq N - 1 \\ 0, & \text{otherwise} \end{cases} \tag{4.2.2}$$

$$\overline{y}(n) = \begin{cases} y(n), & 0 \leq n \leq N - 1 \\ 0, & \text{otherwise} \end{cases} \tag{4.2.3}$$

Then

$$z(n) = \sum_{i=0}^{N-1} x(n - i)y(i)$$

$$= \sum_{i=0}^{L-1} \overline{x}(n - i)\overline{y}(i) \tag{4.2.4}$$

for $0 \leq n \leq N + L - 2$. Define the $(N + L - 1)$-point FFT of $\overline{x}$ and $\overline{y}$ to be $\overline{X}(k)$ and $\overline{Y}(k)$. It then follows that

$$Z(k) = \overline{X}(k)\overline{Y}(k) \tag{4.2.5}$$

and

$$z(n) = \text{IDFT}[Z(k)] = \overline{x}(n) \circledast \overline{y}(n) \tag{4.2.6}$$

That is, z(n) is the circular convolution of $\bar{x}$ and $\bar{y}$. It is interesting to note
that z(n) also represents the linear convolution of x and h as well. Therefore, Eqs. (4.2.4) and (4.2.5) represent a mechanism by which the power
of the FFT can be used to do linear convolution. All that is required is the
appropriate insertion of zeros (i.e., zero fill) into the appropriate time
series.

Zero-fill methods have been shown to be useful in supporting linear
convolution using circular convolution methods. However, there are practical limitations on the sample size that one FFT can accommodate. Therefore, one may encounter serious computational problems if one attempts to
convolve long record lengths (e.g., greater than 1024). The problem of
convolving long data records can be overcome through the use of specialized
data partitioning methods.

4.3 CONVOLUTION PARTITIONING

A problem often encountered in linear systems is convolving an extremely
long time series, say $\{x(n)\}$, with a finite-length sequence $\{h(n)\}$. For
example, one may define h(n) to be the impulse response of a 1024-sample
filter which is used to scan a long block of raw data. The data field may
be composed of hundreds of thousands of data points obtained from a high-
speed data acquisition system. To convolve these data records, a partitioning of the signal space is required. Two of the most popular partitioning
techniques are the overlap-and-add and the overlap-and-save methods.
These two important methods have been singled out for special attention.

Overlap and Add

For the sake of discussion, assume that $\{x(n)\}$ is a time series consisting
of KN sample values. Therefore, $\{x(n)\}$ can be partitioned into K blocks of
N samples each. Furthermore, assume that the filter function $\{h(n)\}$ is a
L-point process. The circular convolution of h(n) and any two adjacent

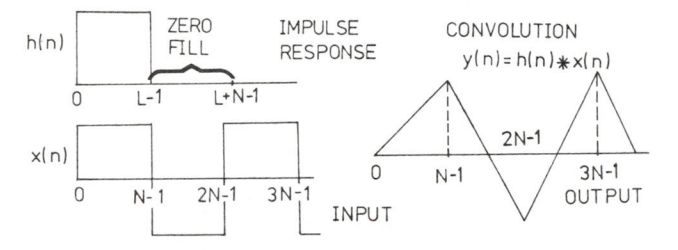

FIG. 4.3.1 Example of convolution using zero fill to overcome the
overlap problem.

data blocks, say $x_i(n)$ and $x_{i+1}(n)$, would overlap or interfere with each other. However, the correct "global" convolution sum can be recovered by properly interpreting the overlapped data. This can best be argued in terms of the following simple example (see Fig. 4.3.1). For convenience, L has been chosen to equal N.

We shall assume that the input time series $\{x(n)\}$ is partitioned into blocks of 2N sample points with the last N sample values consisting of zeros. More specifically, let

$$x_i(n) = \begin{cases} x(n), & iN \le n \le (i+1)N - 1 \\ 0, & \text{otherwise} \end{cases} \qquad (4.3.1)$$

Then it follows that the convolution of $x(n)$ and $h(n)$ is given by

$$y(n) = \sum_{i=0}^{n} h(i) \sum_{j=0}^{\infty} x_j(n - i)$$

$$= \sum_{j=0}^{\infty} h(n) * x_j(n) = \sum_{j=0}^{\infty} y_j(n) \qquad (4.3.2)$$

That is, each convolution over the index set $j \in [0, \infty]$ is of length $N + L - 1$. Over this interval, there are $L - 1$ sample values of the convolution sum with x_j which overlap the convolution sum defined by x_{j+1}. In practice, the convolution operations would be performed using the FFT. Graphically, this process can be explained in terms of the data found in Fig. 4.3.2.

Overlap and Save

The overlap-and-save method overlaps the input rather than output sequences. Here the components of the time series $\{x(n)\}$ are partitioned into $N + L$ sample blocks, as shown in Fig. 4.3.3. Note first that the initial data block is zero-filled. Other values of x_k are formed by $x_k = x[n + k(N - 1)]$, $0 \le n \le N + L - 1$. However, without zero filling there will exist a difference between the linear and circular convolution of the time series being considered. In fact, it can be shown that the first $L - 1$ sample values of these two convolution forms will in general disagree. Therefore, these points are discarded from the circular convolution. The remaining points are accepted as being in agreement with the linear convolution of $x_k(n)$ and $h(n)$, then

$$h(n) = \sum_{k=0}^{\infty} \overline{y}_k (n - k(N - 1)) \qquad (4.3.3)$$

where

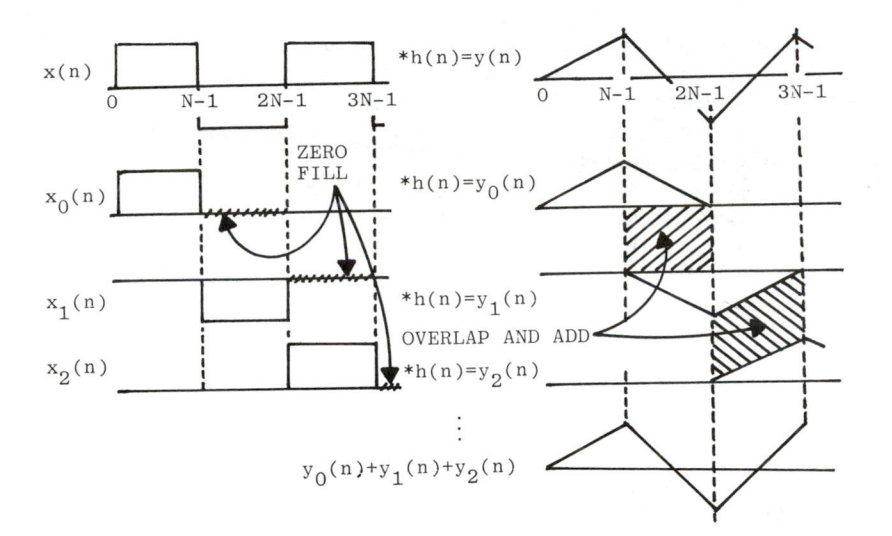

FIG. 4.3.2 Graphic example of the convolution of two signals using the overlap-and-add method.

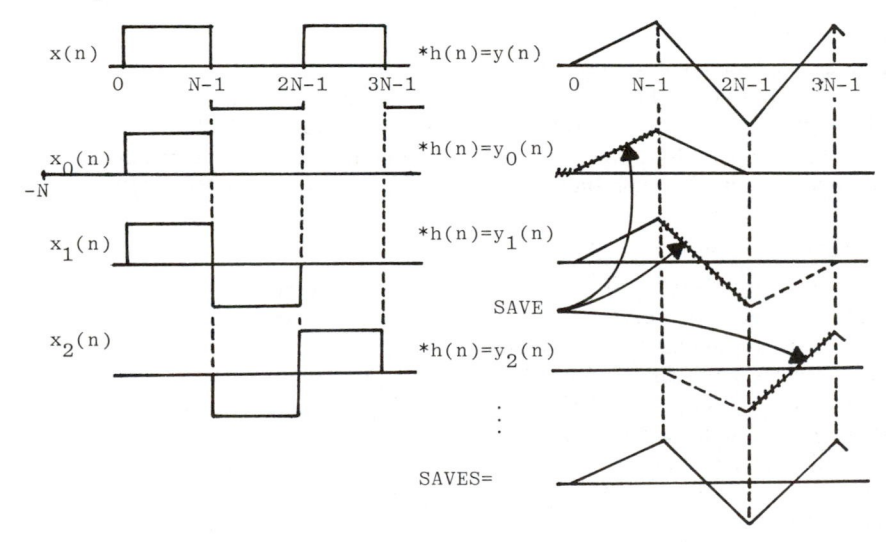

FIG. 4.3.3 Graphic example of the convolution of two signals using the overlap-and-save method.

$$y_k(n) = \begin{cases} \overline{y}_k(n), & L-1 \leq n \leq N-1 \\ 0, & \text{otherwise} \end{cases} \qquad (4.3.4)$$

Both of these methods provide the user with the mechanism to perform long-record-length convolutions. They both require the data to be rigidly managed using either input or output overlap policies.

4.4 CONVOLUTION ALGEBRA[†]

The convolution of two L-point periodic time series is itself periodic, as discussed in the circular convolution section. However, in practice it is rare to find purely periodic time series. Since the convolution of two L-point time series is 2L - 1 points in length, one can conclude that there will be overlap (interference) between the images of two adjacent convolution processes. This problem was resolved previously using zero filling. Other, more technical solutions to this problem also exist. One, based on the algebraic structure of digital machines, will now be developed.

To simplify the development of this method, a simple example will be presented. The vehicle used for this study is a second-order filter and a digital architecture based on a 2-bit-word-length representation of all system variables. The data field and the filter are shown in Fig. 4.4.1. Observe that the 2-bit data words have been separated by 2-bit blocks of zeros.

Consider for the moment the product of filter coefficients a(1) and a(0) with two adjacent input sample values, say x(i) and x(i + 1), to be defined as follows:

$$
\begin{array}{cccccc}
 & a(1) & a(0) & & & \\
\times & x(i + 1) & x(i) & & & \\
\hline
 & a(1)x(i) & a(0)x(i) & & & \\
a(1)x(i + 1) & a(0)x(i + 1) & & & \\
\hline
y(i + 2) & y(i + 1) & y(i) & \quad i \quad i+1 \quad i+2 & & (4.4.1)
\end{array}
$$

where each product of the form a(i)x(i) is a 2-bit × 2-bit full-precision product. For the sake of discussion, consider the following data pattern:

$$\bar{x}_i = \{x(i + 1),\ x(i)\} \longrightarrow \{0,\ 0,\ 2x[i + 1{:}1],\ x[i + 1{:}0],\ 0,\ 0,\ 2x[i{:}1],\ x[i{:}0]\}$$

$$\bar{a} = \{a(1),\ a(0)\} \longrightarrow \{0,\ 0,\ 2a[1{:}1],\ a[1{:}0],\ 0,\ 0,\ a[0{:}1],\ a[0{:}0]\}$$

$$(4.4.2)$$

where x[j:k] is the kth most significant bit (MSB) of x(j). Rather than inserting blocks of zeros between data files (here, two 2-bit words), consider the insertion of an all zero words between each data word in the time series $\{x(n)\}$ (here, a 2-bit word [00]). For example, if x(i + 1) = x(i) = 3, then using the first suggested partition the data string $\{00001111\}$ would be

---

[†]This development makes use of several topics which have not been developed to this point: filter structure and finite digital data fields. However, they are common knowledge to many readers and should therefore pose no problem.

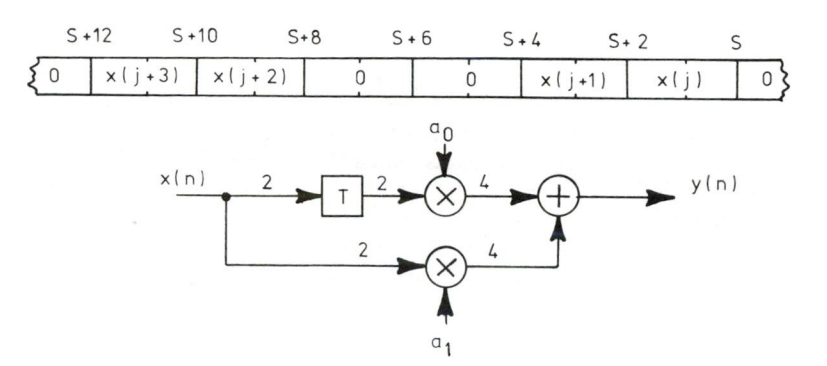

FIG. 4.4.1 Simple shift-register filter exhibiting the data field signature of a convolver (2 and 4 represent word lengths of data path).

synthesized. Using the second method, the data string $\{00110011\}$ would result. With respect to this new ordering, consider again the product $\bar{x}_0$ and $\bar{a}$ [see Eq. (4.4.2)]. Here

$$\bar{y} = \bar{x}_0\bar{a} = (y(2),\ y(1),\ y(0))$$

$$\longrightarrow \{\ldots,\ \underbrace{2y[1:1],\ y[1:0]}_{y(1)},\ \underbrace{8y[0:3],\ 4y[0:2],\ 2y[0:1],\ y[0:0]}_{y(0)}\}$$

That is, the convolution summation can be replaced by the product of two one-dimensional binary data strings of length $(2n)L$, where n is the word length in bits and L is the filter order. For the example under study, $2nL = 8$. For the purpose of demonstration, consider $x(0) = x(1) = a_0 = 3$ and $a_1 = 1$. Then the convolution sum would produce an output process $y(0) = 9$, $y(1) = 12$, and $y(2) = 3$. The product of the two one-dimensional arrays being considered is shown as follows:

```
y=a̅x̅= { 00  01  00  11 }  *  { 00   11   00   11 }   - BINARY VALUE
       { =0  =1  =0  =3 }     { =0   =3   =0   =3 }   - DECIMAL VALUE

                           00  01  00  11      a*x(0:0)
                        0  00  10  01  1       a*x(0:1)
                    00  00  00  00
                 0  00  00  00  0
                00  01  00  11                 a*x(1:0)
             0  00  10  01  1                  a*x(1:1)
            00  00  00  00
         0  00  00  00  0
            _____
       N/A  00  11  11  00  10  01
            y(2)    y(1)    y(0)
            =3      =12     =9
```

Thus a 2nL-bit conventional digital multiplier may be used to do convolution provided that zeros are properly packed into the data stream. This result may be generalized using the following example as a guide.

EXAMPLE 4.4.1 Let $L = 3$, $n = 2$, $a_0 = 7$, $a_1 = 4$, $x_0 = x_2 = 0$, and $x_1 = 0$. Then

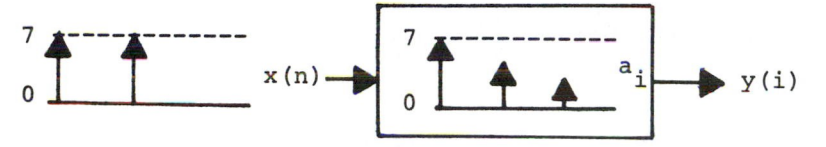

Using the direct method of computing a linear convolution, one obtains

$y(0) = 7 \times 7 = 49$

$y(1) = 7 \times 0 + 4 \times 7 = 28$

$y(2) = 7 \times 7 + 4 \times 0 + 1 \times 7 = 56$

$y(3) = 7 \times 0 + 4 \times 7 + 0 \times 4 = 28$

$y(4) = 7 \times 0 + 0 \times 1 + 1 \times 7 = 7$

Alternatively, using the scalar product form, one obtains

$$\overline{x} = \{ \overbrace{000111}^{x(2)} \; : \; \overbrace{000000}^{x(1)} \; : \; \overbrace{000111}^{x(0)} \}$$

$$\overline{a} = \{ \overbrace{0000}^{a(2)} \; : \; \overbrace{00100}^{a(1)} \; : \; \overbrace{000111}^{a(0)} \}$$

Then

$$\overline{a}\,\overline{x} = \quad 2^5 2^4 2^3 2^2 2^1 2^0 2^5 2^4 2^3 2^2 2^1 2^0 2^5 2^4 2^3 2^2 2^1 2^0$$

$$
\begin{array}{l}
\overline{x}a(0) \; : \left\{ \begin{array}{l}
\;\;\;\;\;\;\;\;\;\;\;\;\;\;\;0\;0\;0\;1\;1\;1\;0\;0\;0\;0\;0\;0\;0\;0\;0\;1\;1\;1 \\
\;\;\;\;\;\;\;\;\;\;\;\;0\;0\;0\;1\;1\;1\;0\;0\;0\;0\;0\;0\;0\;0\;1\;1\;1 \\
\;\;\;\;\;\;\;\;\;\;0\;0\;0\;1\;1\;1\;0\;0\;0\;0\;0\;0\;0\;1\;1\;1
\end{array} \right. \\
\end{array}
$$

$$0\;0\;0\;0\;1\;1\;1\;0\;0\;0\;0\;0\;0\;0\;0\;0\;1\;1\;1 \quad \} \quad : \overline{x}a(1)$$

$$0\;0\;0\;1\;1\;1\;0\;0\;0\;0\;0\;0\;0\;0\;0\;1\;1\;1 \quad \} \quad : \overline{x}a(2)$$

$$\overline{}$$

$$0\;0\;0\;\underbrace{1\;1\;1}_{y(4)=7}\;0\;\underbrace{1\;1\;1\;0\;0}_{y(3)=28}\;\underbrace{1\;1\;1\;0\;0\;0}_{y(2)=56}\;0\;\underbrace{1\;1\;1\;0\;0}_{y(1)=28}\;\underbrace{1\;1\;0\;0\;0\;1}_{y(0)=49}$$

If the foregoing method represents a one-dimensional approach to convolution, a logical question is: What is the dimension of the direct method of computing the linear convolution? The answer is "two" and is justified in the following manner. Suppose that we again consider a and x to be encoded in 2's-complement form (see Section 15.2). Then

$$
y(n) = \sum_{h=0}^{L-1} \left\{ (-a[h:0] + \sum_{i=0}^{n-1} a[h:i]2^{-i}) \right.
$$

$$
\left. \times (-x[n-h:0] + \sum_{j=1}^{n-1} x[n-h:j]2^{-j} \right\}
$$

$$
= \sum_{h=0}^{L-1} a[h:0]x[n-h:0] - \sum_{h=0}^{L-1} (a[h:0] \sum_{j=1}^{n-1} x[n-h:j]2^{-j}
$$

$$
+ x[n-h:0] \sum_{j=1}^{n-1} a[h:j]2^{-j}) + \sum_{h=0}^{L-1}\sum_{i=0}^{n-1}\sum_{j=0}^{n-1} a[h:i]x[n-h:j]2^{-(i+j)}
$$

$$(4.4.4)$$

By defining $m = i + j$ and $P = n$, the two-dimensional representation of $y(n)$ becomes

$$
y(n) = \sum_{h=0}^{L-1} a[h:0]x[n-h:0] - \sum_{i=0}^{n-1} (a[n:0] \sum_{h=0}^{L} x[n-h:i])
$$

$$
- \sum_{n=0}^{L-1} (x[n-h:0] \sum_{i=1}^{n-1} a[h:i]^{-i})
$$

$$
+ \sum_{m=0}^{P-1}\sum_{i=0}^{n-1}\sum_{n=0}^{L-1} a[h:i]x[n-h:m-i]2^{-m}
$$

$$(4.4.5)$$

Therefore, there are at least three models applicable to the problem of linearly convolving two time series. The first is the traditional nesting of delayed multiplies. The second is the multiplication of two one-dimensional arrays. The third is a two-dimensional model of the convolution operation.

4.5 STATISTICAL METRICS

Many of the signals found in modern signal processing are random. Whether they represent noise or intelligence, the statistical nature of these signals must be dealt with. Stochastic methods are used to represent the unquantifiable microscopic behavior of a random process as a set of parameters.

These parameters, which represent macroscopic phenomena, can be manipulated algebraically. As a result of such mathematical operations, added insights into the structure of a random process can be obtained.

There are two common methods used to parameterize a random process. They are called <u>temporal</u> and <u>ensemble</u> averaging. In the temporal case, stochastic parameters are generally measured directly. In the ensemble case, stochastic parameters are computed rather than measured.

Temporal averaging is applied to data defined over a period of time. The treatment of this subject will also be applicable to the problem of averaging spatial data as well as that defined in the frequency domain. The mth temporal moment is defined to be $\overline{x}^m$, where

$$\overline{x}^m = \frac{1}{N} \sum_{i=0}^{N-1} x(i)^m \qquad (4.5.1)$$

as N becomes large. For m = 1 and 2, x and $\overline{x}^2$ are called the <u>mean</u> and <u>mean-squared</u> value of $\tilde{x}$, respectively. The mth <u>central moment</u>, say x^m, is given by

$$\tilde{x}^m = \frac{1}{N} \sum_{i=0}^{N-1} [x(i) - \overline{x}]^m \qquad (4.5.2)$$

For the special case where m = 2, $\tilde{x}^2$ is referred to as mean-squared central moment of x or as the <u>variance of x</u> and is denoted σ_x^2. Computationally, this metric can be computed more efficiently in terms of $\sigma_x^2 = \overline{x}^2 - (\overline{x})^2$. Theoretically, an infinite number of samples are required to define a temporal moment. From a practical viewpoint, this is an unrealistic constraint. Instead, one of several approximate methods is normally used to estimate a temporal moment. The two most popular methods are <u>linear</u> and <u>exponential</u> averaging. Both of these metrics can easily be realized in hardware and can be computed in finite time. The first of these two methods, namely linear averaging, represents the creation of a statistical estimate based on N successive samples. If the statistics are <u>stationary</u> (time invariant), a "good" estimate of the temporal moments can be achieved using the linear method. There are, however, times when the statistics of a signal are slowly changing in time. Under this assumption the exponential approach can prove to be beneficial. The averager, sometimes called a "leaky integrator," processes data through an exponential window similar to that found in RC electric networks. These realizable averaging strategies which offer the user a choice of averaging times (e.g., linear = uniform window versus exponential window).

EXAMPLE 4.5.1 Linear averaging can be performed in software using the algorithm

$$y_n = \frac{(n-1)y_{n-1} + x_n}{n}; \quad 1 \leq n \leq N$$

Exponential averaging can be approximated by

$$y_n = \frac{[(N/2) - 1]y_{n-1} + x_n}{N/2}; \quad y_1 - x_1$$

for N fixed and $n \geq 0$. In Fig. 4.5.1, the result of a numerical experiment can be found. The linear averaging process is terminated at $N = 8$. In general, linear averaging is performed over a block of N samples which may overlap. Alternatively, the approximate exponential routine weights the most current sample most heavily. This effect can be seen in terms of the telescoping series

$$y_{n+1} = \alpha y_n + \beta x_n; \quad \alpha = \frac{N/2 - 1}{n/2} < 1, \quad \beta = \frac{1}{N/2}$$

$$y_{n+2} = \alpha^2 y_n + \alpha x_n + \beta x_{n+1}$$

$$y_{n+3} = \alpha^3 y_n + \alpha^2 \beta x_n + \alpha \beta x_{n+1} + \beta x_{n+2}$$

$$y_{n+j} = \alpha^j y_n + \alpha^{j+1} \beta x_n + \cdots + \alpha^2 \beta x_{n+j-3} + \alpha \beta x_{n+j-2} + \beta x_{n+j-1}$$

where $\alpha^{j-1}\beta < \cdots < \alpha^2 \beta < \alpha \beta < \beta$.

EXAMPLE 4.5.2 The DFT of a time series containing a random component will exhibit a degree of randomness in the frequency domain as well. To improve the interpretability of the resulting spectral display, some form of averaging is often employed. Often, a set of K spectral images of K distinct time series will be averaged. Commercial FFT systems may provide the user with averaging options in addition to the exponential and linear averaging methods discussed in Example 4.5.1. In Fig. 4.5.2 two more are presented. A collection of time series, whose linear average value is found in Fig. 4.5.2a, is interpreted in the frequency domain in terms of a root-mean-square (rms) average and peak value. The rms average (Fig. 4.5.2b) is simply the linear average of the rms value of the spectrum where the rms value of each harmonic is computed in the usual manner. If the peak value of each harmonic over all K spectra is determined, the peak hold metric will result (Figure 4.5.2c). This metric, sometimes referred

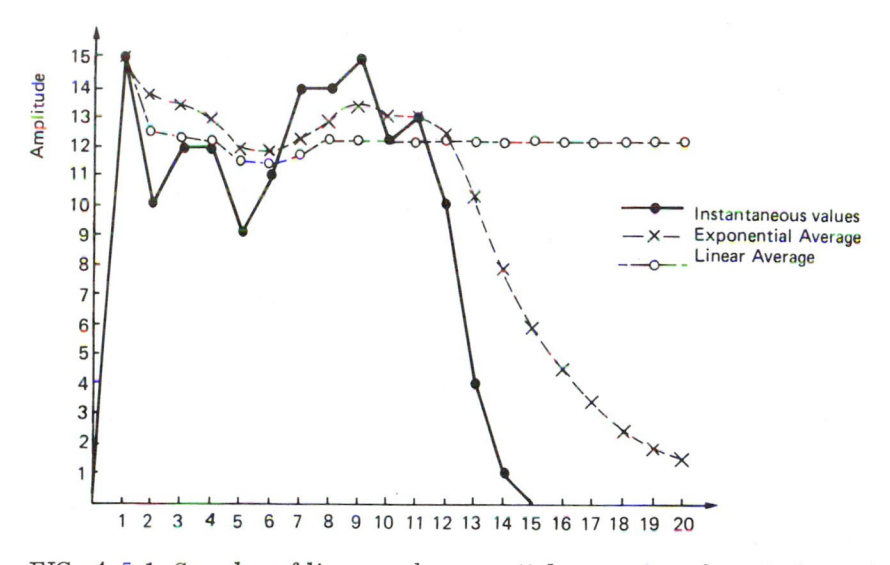

FIG. 4.5.1 Samples of linear and exponential averaging of a set of sample values. (Reprinted with the permission of B and K instruments.)

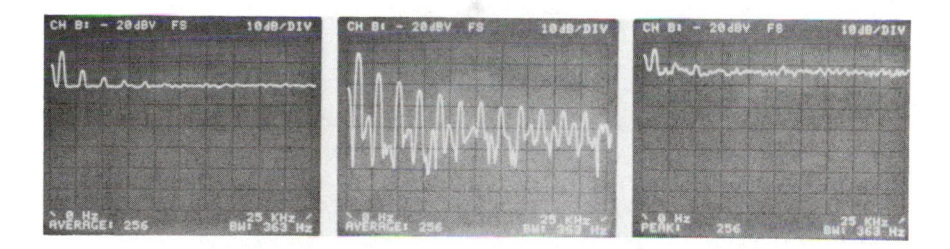

FIG. 4.5.2 Display of the (a) simple, (b) rms, and (c) peak average values of an ensemble of DFTs produced by a commercial FFT system. (Reprinted with the permission of Hewlett-Packard Corporation.)

4.6 AUTOCORRELATION

Correlation is a measure of similarity. The correlation of a signal $\{x(n)\}$ with itself is called <u>autocorrelation</u> and is denoted $R_x(k)$. Here

$$R_x(k) = \frac{1}{N} \sum_{i=0}^{N-1} x(i)x(i+k) \qquad (4.6.1)$$

and it follows that $R_x(0) = \sigma_x^2$, $R_x(k) = R_x(-k)$, and $R_x(k) \le R_x(0)$. The auto-correlation function $R_x(k)$ can be used to uncover several interesting system and signal attributes. First, if $\{x(n)\}$ is periodic, $R_x(k)$ enjoys the same periodicity. Suppose that $x(k)$ has a period N such that $x(n) = x(n + N)$ for all n; then $R_x(k) = R_x(k + N)$. Here the tendency of $x(n)$ to repeat itself every N is mirrored in $R_x(k)$. Therefore, the autocorrelation function can be used to study the periodic behavior of signals and systems. The auto-correlation function also defines how well a given signal maintains its structure over a given interval of time. A signal whose amplitude is slowly varying would possess a strong structure. That is, qualitatively the signal bears a marked similarity to its recent past. In other cases, such as those found in random signal processes, the structure is weak or nonexistent. Signals possessing a strong structure exhibit a slowly changing autocorrelation function about the zero-lag value of $k = 0$. Signals possessing little structure exhibit an autocorrelation function that decays rapidly about $k = 0$.

EXAMPLE 4.6.1 Let $x(i) = A + B \sin (i\omega_0 T_s + \phi)$; then $R_x(k) = A^2/2 + B^2/2 \cos(k\omega_0 T_s) + 1/N * \{\text{other sinusoidally varying terms}\}$. Observe that as $N \longrightarrow \infty$ (equivalently, the averaging interval becomes infinite) the last term in $R_x(k)$ vanishes.

EXAMPLE 4.6.2 Let $x(i)$ equal a pulse train having a duty cycle $d = \tau/T$. The autocorrelation function is easily computed and is summarized in Fig. 4.6.1.

EXAMPLE 4.6.3 Autocorrelation can be used to uncover latent similarities that may exist within a signal structure. This metric can prove to be particularly effective when dealing with signal embedded in noise. Consider the normally distributed noise found in Fig. 4.6.2 added to a pulse train of duty cycle 0.5 (see Fig. 4.6.3). The resulting autocorrelation function, based on the FFT, is displayed in Fig. 4.6.4. Observe that the autocorrelation function exhibits nearly the noise-free triangular shape found in Fig. 4.6.1. This is because the noise is weakly correlated to itself and the original pulse train.

    A sometimes useful variation of the basic autocorrelation concept is the correlation coefficient given by $\rho_x(k)$, where

$$\rho_x(k) = \frac{R_x(k)}{R_x(0)} \qquad\qquad (4.6.2)$$

with $-1 \le \rho_x(k) \le 1$. Here $\rho_x(k) = 0$ indicates no average similarity, while the condition $\rho_x(k) = 1(-1)$ indicates that $x(i)$ and $x(i + k)$ vary in harmony (disharmony).

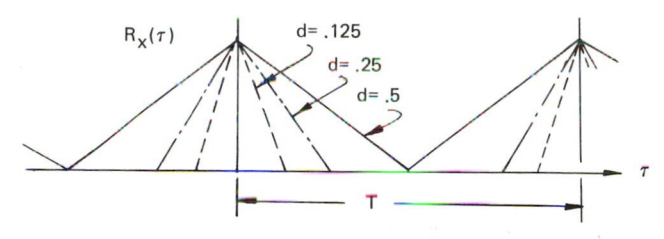

FIG. 4.6.1 Graphical interpretation of the autocorrelation of a rectangular pulse train of duty cycle d.

The autocorrelation function can also be expressed in terms of frequency domain parameters. Suppose that a real $\{x(n)\}$ has a DFT $X(k)$ such that

$$x(i) = \sum_{j=0}^{N-1} x(j)W_N^{ji} \tag{4.6.3}$$

Then

$$R_x(k) = \frac{1}{N} \sum_{i=0}^{N-1} \sum_{j=0}^{N-1} X(j)W_N^{ji} \sum_{p=0}^{N-1} X(p)W_N^{(j+p)}$$

$$= \frac{1}{N} \sum_{p=0}^{N-1} X(p)W_N^{-pk} \sum_{j=0}^{N-1} X(j) \sum_{i=0}^{N-1} W_N^{(j+p)i} \tag{4.6.4}$$

However,

$$\sum_{j=0}^{N-1} W_N^{qi} = \begin{cases} N, & \text{if } q = 0 \\ 0, & \text{otherwise} \end{cases} \tag{4.6.5}$$

since $q = 0$ if $p = -j$, it follows that

$$R_x(k) = \sum_{p=0}^{N-1} X(p)X(-p)W_N^{-pk} \tag{4.6.6}$$

From the symmetry property of the DFT over a real-time series [i.e., $X(k) = X(k)^*$], it follows that

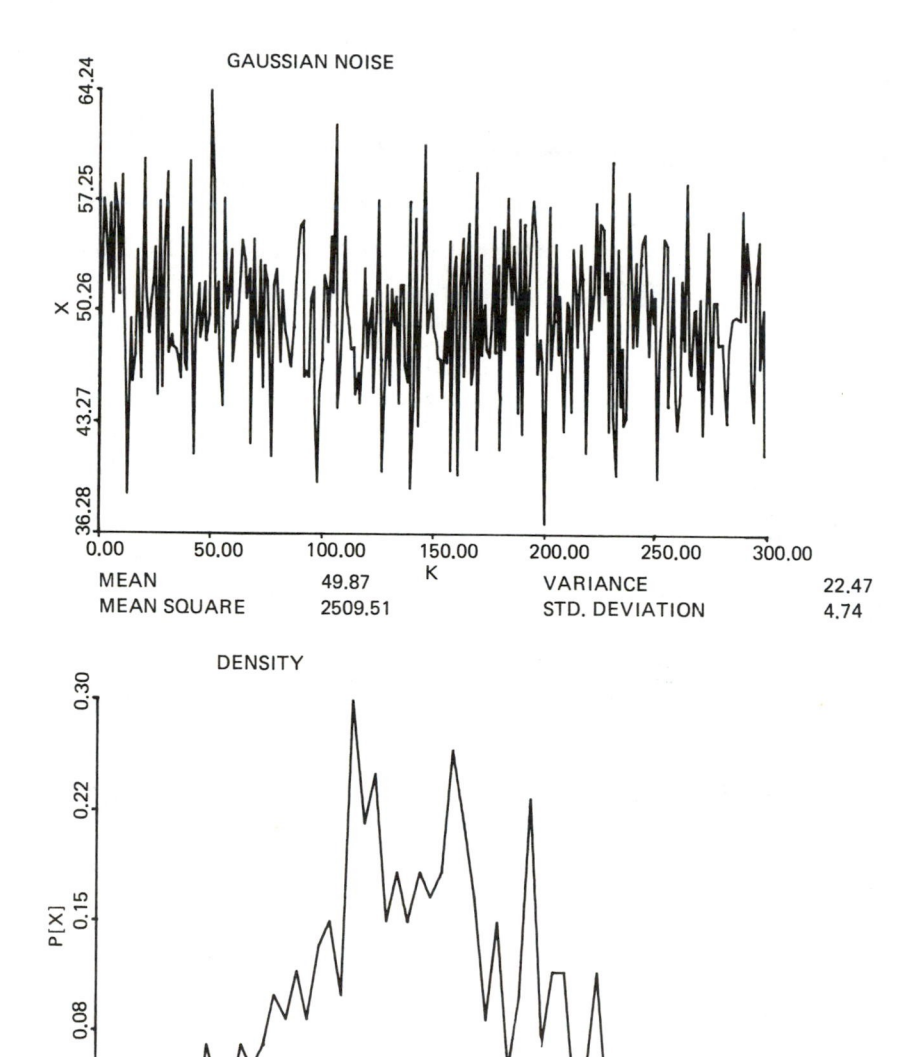

FIG. 4.6.2 Normally distributed time series (top) and amplitude density (bottom).

$$R_x(k) = IDFT[X(k)X^*(k)]$$
$$= IDFT[|X(k)|^2] \qquad (4.6.7)$$

That is, the autocorrelation function can alternatively be computed using the efficiency of the FFT. This methodology will allow autocorrelation data to be computed more rapidly than those obtained using direct sum-of-product methods.

The term $X(k)X(k)^* = |X(k)|^2$ was defined to be $G(k)$ and was referred to as the <u>power spectrum</u> of $\{x(n)\}$. It should be obvious that the power spectrum is real, with all phase information having been destroyed [e.g., the power spectrum of $\cos(\omega t)$ is identical to $\sin(\omega t)$]. The power spectrum communicates the similarity information that is found in the autocorrelation function. For example, if there exists periodic autocorrelation behavior [e.g., $R_x(k) = \cos(k\omega_0 T_s)$], the power spectra will have a strong spectral component located at ω_0 radians per second. Signal processes possessing little structure can be characterized by their short correlation time. That is, $R_x(k)$ is densely defined around $k = 0$ and equal to (or nearly equal to)

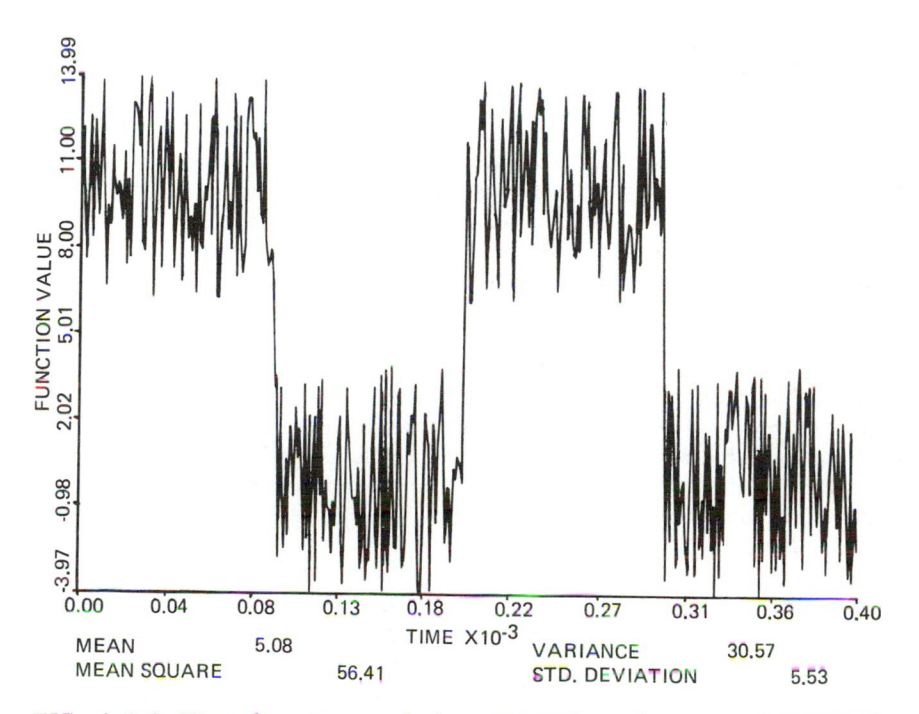

| MEAN | 5.08 | VARIANCE | 30.57 |
| MEAN SQUARE | 56.41 | STD. DEVIATION | 5.53 |

FIG. 4.6.3 Time domain record of a rectangular pulse train of duty cycle $d = 0.5$, with additive pseudo-random normally distributed noise.

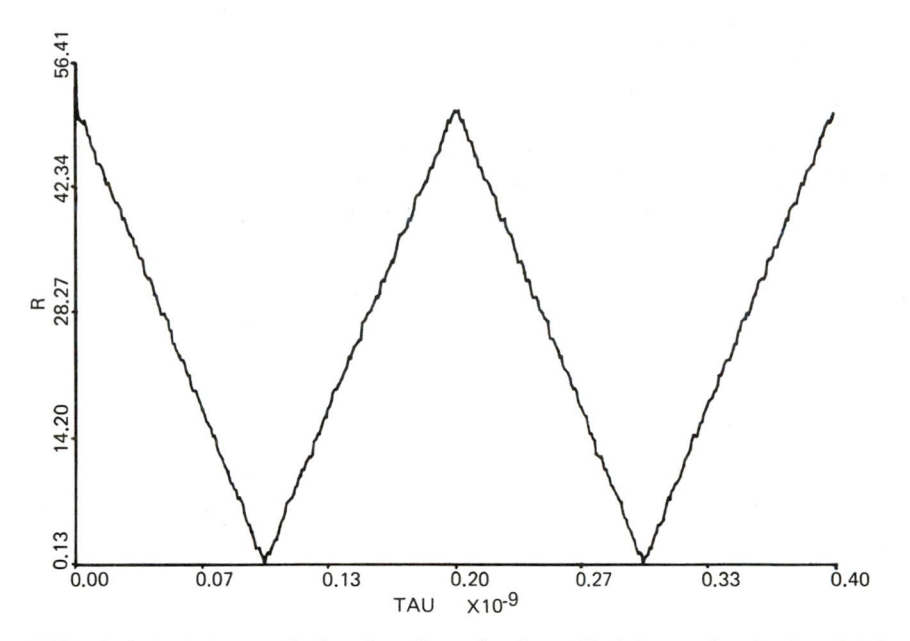

FIG. 4.6.4 Autocorrelation function of noise added to a pulse train of duty cycle d = 0.5 using FFTs.

zero for larger values of k. The power spectrum of this temporal record [i.e., $R_x(k)$] is a broadband process in the frequency domain.

The autocorrelation function and its companion the power spectrum are known to be extremely useful in the study of systems. Later in this chapter, linear filters will be studied in terms of correlation times and power spectra.

The creation of an autocorrelation function using the FFT suffers from the same record-length difficulties uncovered in our study of convolution. This condition was the result of attempting to equate the result of a circular convolution to that of a linear convolution. In a power spectrum application, the circular effects are referred to as a <u>wraparound error</u>. Zero filling can be used to overcome this problem in a manner similar to that used in the problem of convolution. This will be developed later in this chapter.

EXAMPLE 4.6.4 Using the FFT, autocorrelation and the power spectrum of a 100-Hz pure cosine tone, sampled at a 1-kHz rate, are summarized in Figs. 4.6.5 and 4.6.6. The data found in the first figure were obtained using a 512-point FFT and a rectangular window (window functions are discussed in Chapter 5). The data in the second figure were obtained using a 128-point FFT and a Hamming window (see Section 5.3). It can be noted

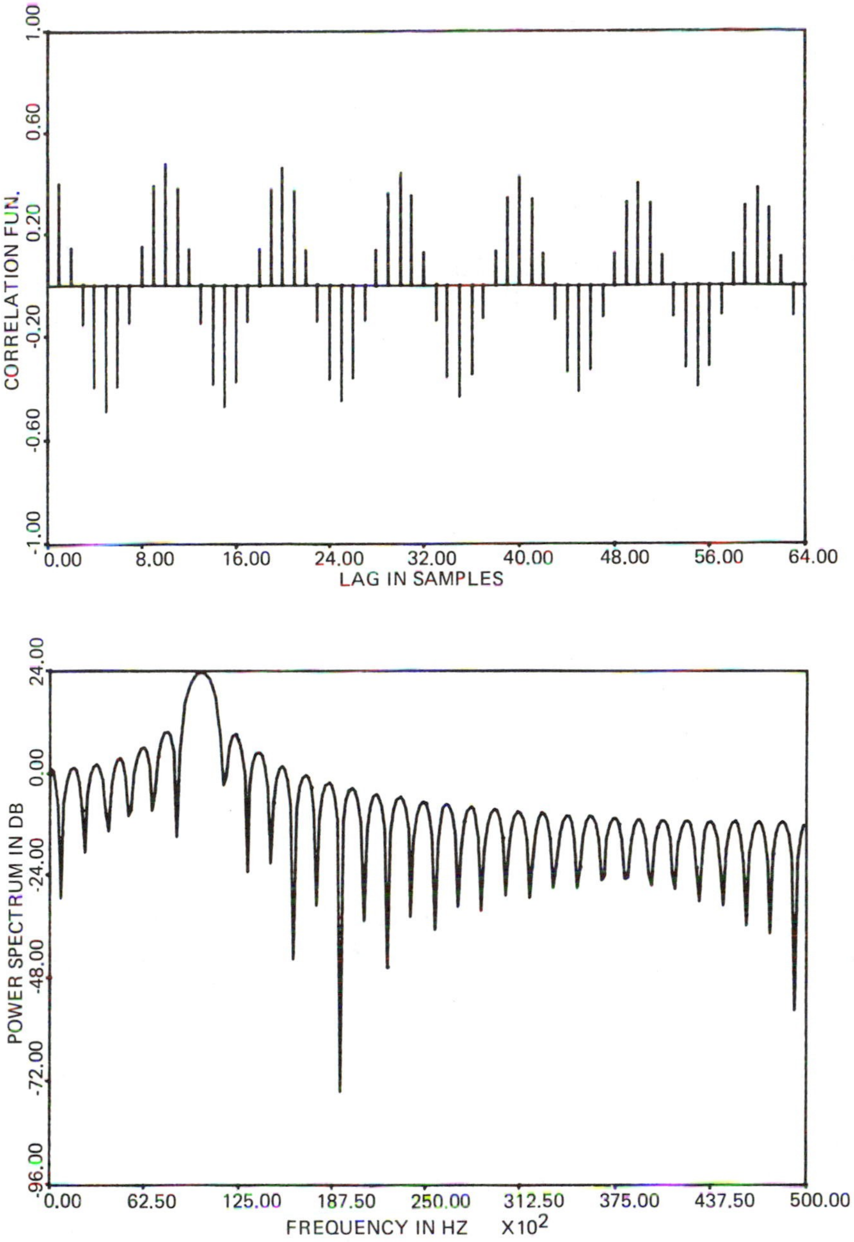

FIG. 4.6.5 Power spectrum and autocorrelation of a pure harmonic oscillation using a 512-point FFT and a rectangular data window.

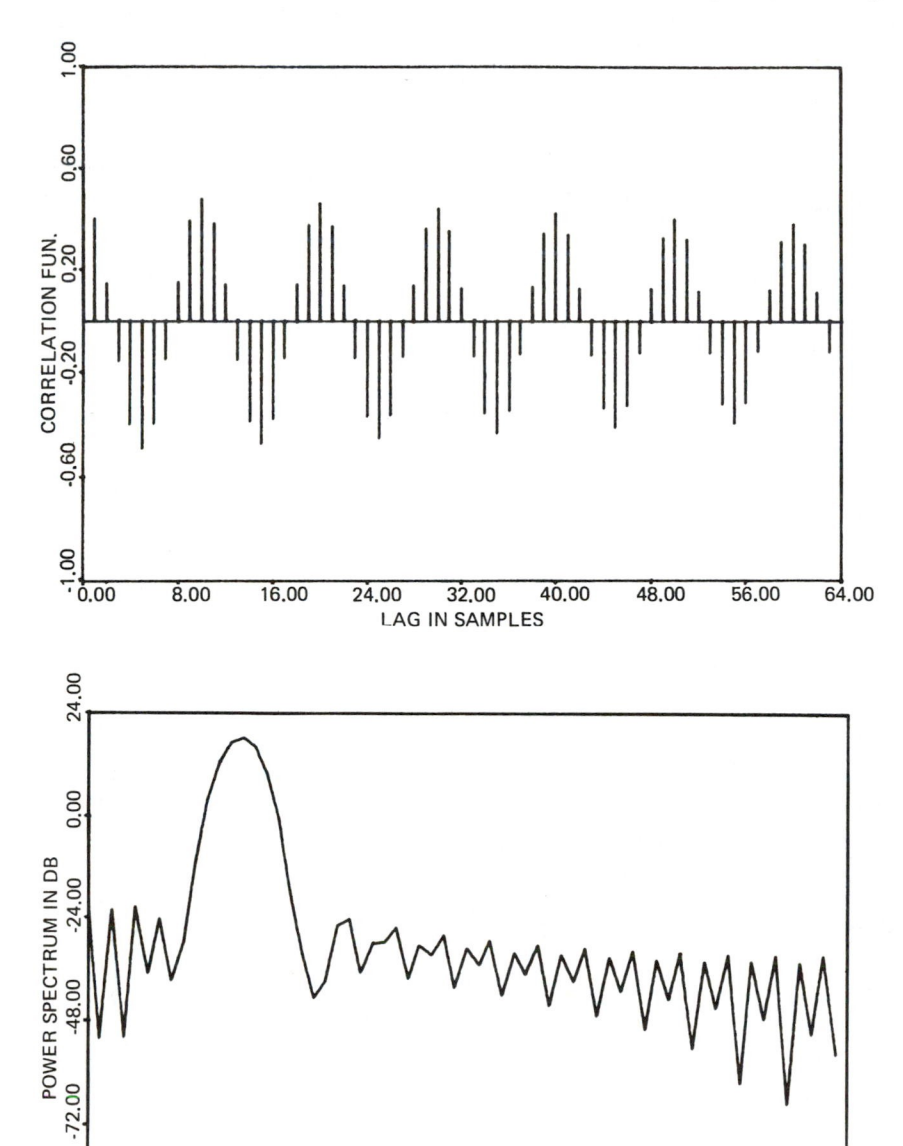

FIG. 4.6.6 Power spectrum and autocorrelation of a pure harmonic oscillation using a 128–point FFT and a Hamming window.

that the longer the transform length (512 versus 128), the more pronounced is the peak in the power spectrum found at 100 Hz. The effect of the Hamming window can be seen in the improved sideband suppression found in Fig. 4.6.6. Data windows, which are discussed in detail in Chapter 5, have the ability to suppress the unwanted sideband activity such as that found in the first figure. As a result, the use of data windows is generally encouraged. The second design parameter, transform length, is influenced by the amount of computer resources available. The use of sectioning techniques is developed later in this chapter.

4.7 CROSS CORRELATION

Cross correlation is used to measure the similarity between two time series, say $\{x(n)\}$ and $\{y(n)\}$, and is denoted $R_{xy}(k)$. Here

$$R_{xy}(k) = \sum_{i=0}^{N-1} \frac{x(i)y(i-k)}{N} \tag{4.7.1}$$

If x and y are real, $R_{xy}(k)$ is also real. A large positive (or negative) value of $R_{xy}(k)$ indicates that $x(i)$ and $y(i-k)$ are varying in harmony (disharmony).

Cross correlation is found in use in many areas of engineering and science. In problem areas, such as radar, sonar, and biological signal processing, cross correlation can be used to determine the propagation delays that may exist within a system. Cross correlation is also known to be useful in extracting features from a complex data field. Here the similarities that exist between signal groups can be uncovered analytically. In other applications, the output of a system can be compared to the input process to determine whether the system decorrelates a signal process.

EXAMPLE 4.7.1 Let $x(i) = a$ for all i and $y(i) = B \sin(i\omega_0 T_s + \phi)$; then $R_{xy}(k) = 0$ for all k if ω_0 is an integer multiple of the sample period NT_s (i.e., $\omega_0 = m/NT_s$).

EXAMPLE 4.7.2 Let $x(i) = \sin(i\omega_0 T_s + \phi_1)$ and $y(i) = \sin(i\omega_0 T_s + \phi_2)$; then $R_{xy}(k) = R_{xy}(-k)$ if $\phi_1 = \phi_2$ and $R_{xy}(k) = 0.5\cos(\omega_0 k + \phi_1 - \phi_2)$ if $\phi_1 \neq \phi_2$. In terms of a spectral concept, cross correlation can be defined in terms of the formula

$$R_{xy}(k) = IDFT(X(k)Y^*(k)) \tag{4.7.2}$$

This DFT relationship is subject to the same zero-fill and partitioning restrictions imposed on the FFT realization of the autocorrelation function.

Also, there exists a dual to the correlation coefficient, knwon as the cross-correlation coefficient, denoted $\rho_{xy}(k)$, where

$$\rho_{xy}(k) = \frac{\text{IDFT}[X(k)Y^*(k)]}{(|X(0)||Y(0)|)^{1/2}} \tag{4.7.3}$$

The concepts developed to this point make use of linear temporal averaging. These metrics, the auto- and cross-correlation functions, are easily computed in hardware or software. In most cases the power of the FFT can be brought to bear on this problem provided that the proper zero-fill management policy is invoked. Since the zero-fill requirements are those studied in the section on convolution, they will not be reinvestigated at this time. However, later in this chapter, block-partitioning methods will be studied in the context of a correlation operation. Before we move away from linear averaging, one additional topic, octave filtering, will be developed.

EXAMPLE 4.7.3 The cross correlation of a cosine and sine 100-Hz tone, sampled at 1 kHz. is summarized in Fig. 4.7.1. It can be noted that the

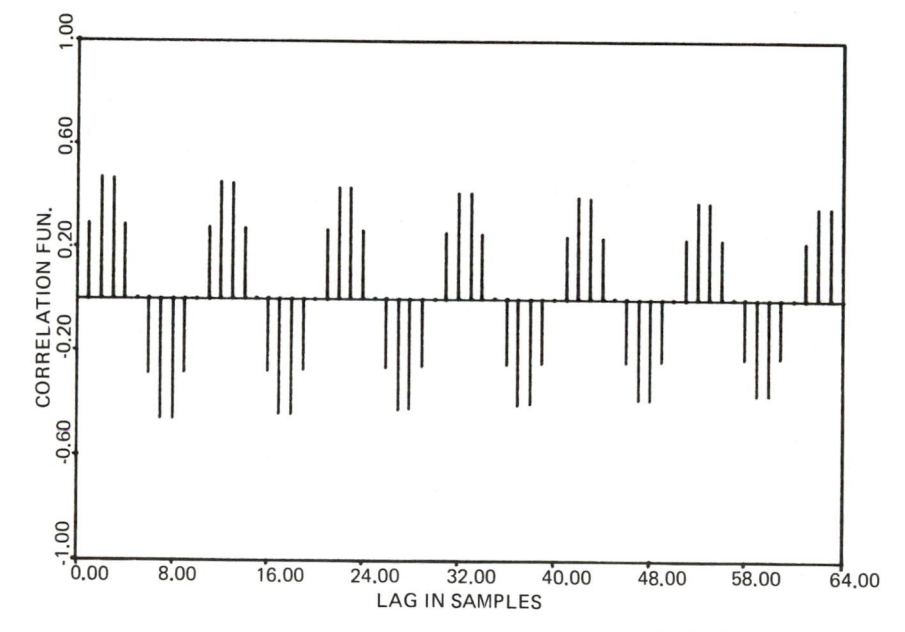

FIG. 4.7.1 Cross correlation of a cosine and sine 100-Hz tone, sampled at 1 kHz.

cross correlation first peaks between samples 3 and 4 (say, 3.5). This corresponds to the 90° phase difference between the sine and cosine waveforms interpreted in terms of sample delays.

4.8 THIRD-OCTAVE FILTERING

Many users of signal processing systems, such as those involved in vibration and modal analysis, traditionally do not uniformly resolve a spectrum into N frequency locations (i.e., DFT). Instead, a technique known as third-octave filtering is often used. Suppose that one models the bandwidth of a given N-point FFT, centered about a given harmonic, by $1/NT_s$. Furthermore, suppose that one wishes to cover the baseband of frequencies $[0, 1/2T_s]$ with K bands of frequencies having a geometrically increasing bandwidth. Then

$$\frac{1}{NT_s}(1 + a + \cdots + a^{k-1}) = \frac{1}{2T_s} \qquad (4.8.1)$$

or $(a^k - 1)/(a - 1) = N/2$. Equivalently, K is given by

$$K = \frac{\log[1 + (a - 1)N/2]}{\log(a)} \qquad (4.8.2)$$

Putting all this together, a third-octave filter has the frequency assignments shown in Fig. 4.8.1.

4.9 ENSEMBLE AVERAGING

Temporal (spatial and frequency) averages are, in general, easily computed in hardware or software. Using these elementary concepts many useful and important linear system theoretic concepts were developed. However, there are areas of this field which require a more rigorous and mathematically complex study. Here the concept known as ensemble averaging is used.

The ensemble statistics of a process $\{x(n)\}$ are defined in terms of a probability distribution function $P_X(x, k)$ and a probability density function $p(x(i))$. Some of the fundamental properties of these statistical parameters are

1. $P_X(x, k) = \text{prob}(x(i) \le x)$

2. $p_X(x(i)) = \dfrac{\partial P_X(x, i)}{\partial x}$

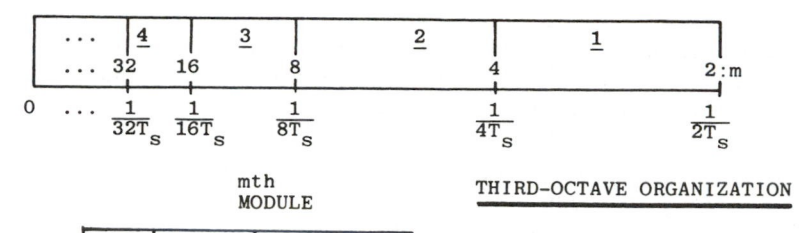

THIRD-OCTAVE ORGANIZATION

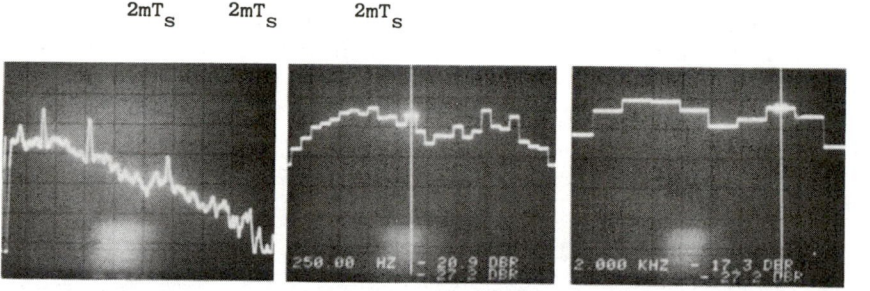

LINEAR   THIRD-OCTAVE   ONE-OCTAVE

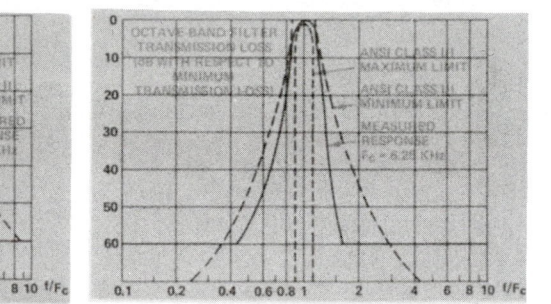

THIRD-OCTAVE FILTER   ONE-OCTAVE FILTER

FIG. 4.8.1 Spectral budget, spectral profile, and experiments demonstrating one- and third-octave filtering. (Reprinted with the permission of Rockland System Corporation.)

3. $P_X(x, k) = \int_{-\infty}^{X} p_X(x(k)) \, dx$

4. $P_X(\infty, k) = 1$

Observe that the ensemble metrics are defined at a <u>sample instant</u>, unlike the temporal metrics, which are defined over an interval of time.

Theoretically, in the temporal case, an infinite number of samples are taken from a single data record $\{x(n)\}$. In the ensemble case, an infinite number of data records $\{x_j(i)\}$ at sample instance i. Therefore, both statistical measures are derived from an infinite-dimensional sample space. However, they represent two different sampling philosophies. This thesis is summarized in Fig. 4.9.1.

The kth ensemble moment is given in terms of the <u>expectation operator</u> (denoted $\mathcal{E}$) as follows:

$$\mathcal{E}[x^k(n)] = \int_{-A}^{A} x^k(n)p[x(n)]\,dx(n) \tag{4.9.1}$$

for $|x(n)| \leq A$. The <u>kth central moment</u> is given by

$$\mathcal{E}\,\overline{(x(n) - x(n))} = \int_{-A}^{A} \overline{(x(n) - x(n))}^k p(x(n))\,dx \tag{4.9.2}$$

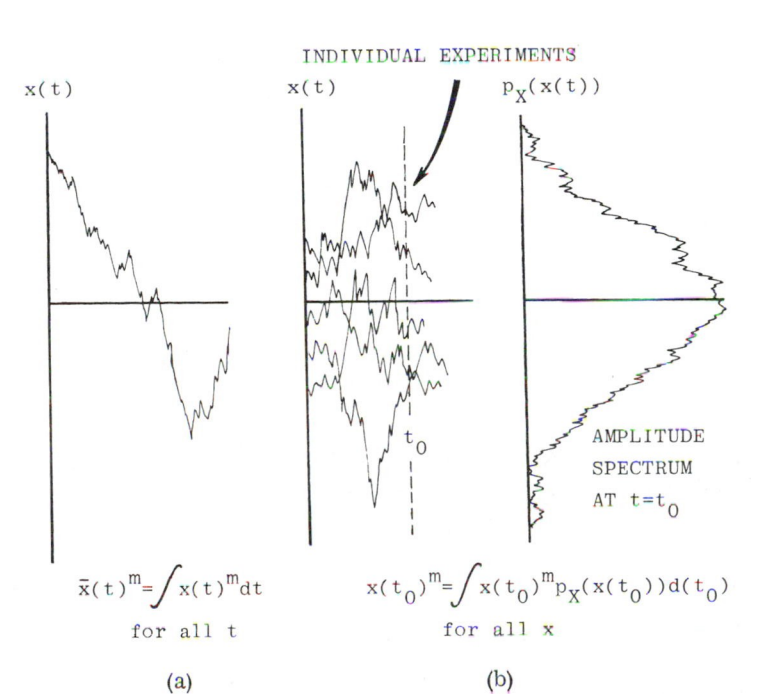

INDIVIDUAL EXPERIMENTS

$x(t)$ $x(t)$ $p_X(x(t))$

t_0

AMPLITUDE SPECTRUM AT $t=t_0$

$$\bar{x}(t)^m = \int x(t)^m dt$$
for all t

$$x(t_0)^m = \int x(t_0)^m p_X(x(t_0))d(t_0)$$
for all x

(a)   (b)

FIG. 4.9.1 Phantom experiment which graphically compares (a) temporal- and (b) ensemble-averaging techniques.

where $\bar{x}(n) = [x(n)]$ [the mean value of $x(n)$]. The second central moment is the <u>variance of $x(n)$</u> and is denoted σ_x^2. The square root of σ_x^2 is called the <u>standard deviation of x.</u>

Often, one is interested in statistically quantifying a random variable at two or more instances of time. Here the <u>joint probability density</u> is used to define this statistical interrelationship. Specifically, if $P_x(\alpha, n, m) = \text{prob}[x(n) \le \alpha(n) \text{ and } x(n) \le \alpha(m)]$, then the joint probability density function satisfies

$$P_x(x(n), \ x(m)) = \frac{\partial^2 P_x(x, \ n, \ m)}{\partial x(n) \ \partial x(m)} \tag{4.9.3}$$

$\{x(n)\}$ and $\{x(m)\}$ are said to be <u>statistically independent</u> if and only if

$$P_x(x(n), \ x(m)) = p_x(x(n)) p_x(x(m)) \tag{4.9.4}$$

A process is said to be <u>stationary</u> if its probability density function is independent of the sample index.

EXAMPLE 4.9.1 One of the more popular probability densities is the uniform density (see Fig. 4.9.2). Here $p_x(x, m)$ is denoted $U[-A/2, A/2]$ and satisfies

$$U\left[-\frac{A}{2}, \ \frac{A}{2}\right] = \begin{cases} \dfrac{1}{A}, & |x| \le \dfrac{A}{2} \\[2mm] 0, & \text{otherwise} \end{cases}$$

It is easily shown that if $x \in U[-A/2, A/2]$, then

$$\mathcal{E}(x(i)) = 0, \qquad \sigma_x^2 = \frac{A^2}{12}$$

EXAMPLE 4.9.2 Another popular density function is the normal or Gaussian density function, denoted $N(\bar{x}, \sigma_x^2)$, where

$$N(\bar{x}, \ \sigma_x^2) = \frac{1}{\sqrt{2\pi\sigma_x^2}} \exp \frac{-(x - \bar{x})^2}{2\sigma_x^2}$$

with $\bar{x}$ and σ_x^2 denoting the mean value and variance of $\{x(n)\}$ (see Fig. 4.9.3).

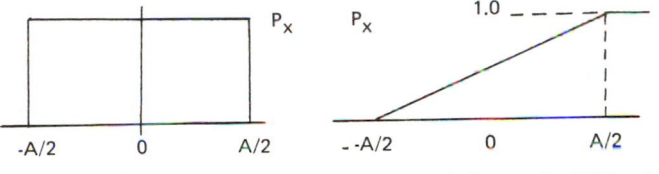

FIG. 4.9.2 Graphical interpretation of the probability density and distribution function of a uniform random process.

EXAMPLE 4.9.3 Using program RANDU, a uniformly distributed pseudo-random process can be synthesized using a 16-bit general-purpose digital computer and FORTRAN. The random data sequence produced by RANDU is uniformly distributed over [0, 1]. In particular, a derived random variable, say YFL, satisfies $0 \leq YFL \leq 1$. To use RANDU, the user must supply an arbitrary noninteger to the computing routine. This integer, say IX, is called the "seed." If a pseudo-random uniformly distributed integer or real sequence is desired, the integer returned by RANDU, say IY, is used to reseed the subroutine (i.e., IY→IX).

Based on the law of large numbers (see the following paragraph), it is known that a large collection of uncorrelated uniformly distributed samples will tend to become normal. Program GAUSS, shown in Fig. 4.9.4, uses 12 uniformly distributed random variables to synthesize an approximately normal random variable V, having mean AM and standard deviation S. The parameter IX is used to seed the uniform number generator.

By identifying the density function in Example 4.9.2 as being normal, one should not conclude that all other density functions are abnormal. The notion of normality can be motivated in the context of the central limit theorem or, as it is sometimes referred to, the law of large numbers. This fundamental theorem can be argued in terms of the sum

$$x = \frac{1}{K} \sum_{i=1}^{E} x_i \qquad (4.9.5)$$

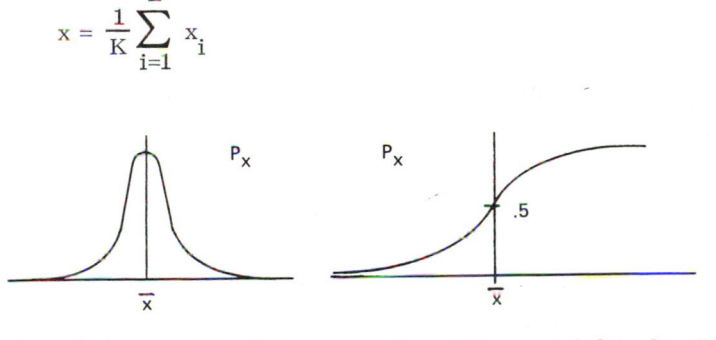

FIG. 4.9.3 Graphical interpretation of the probability density and distribution function of a normal random process.

```
SUBROUTINE GAUSS(IX,S,AM,V)
A=0.0
DO 50 I=1,12
CALL RANDU(IX,IY,Y)
IX=IY
50 A=A+Y
V=(A-6.0)*S+AM
RETURN
END
SUBROUTINE RANDU(IX,IY,YFL)
IY=IX*899
IF(IY)5,6,6
5 IY=IY+32767+1
6 YFL=IY
YFL=YFL/32767.
RETURN
END
```

FIG. 4.9.4 FORTRAN source code for a normally distributed pseudo-random process.

where x_i is the ith member of a set of random variables obtained from an arbitrary population (i.e., density). Furthermore, it shall be assumed that all the samples are statistically independent. Then the central limit theorem states that the distribution of z becomes normal as K tends to infinity. Therefore, it is reasonable to assume that nature is full of normally distributed random processes since, macroscopically, nature deals with large numbers.

Another form of the limit theorem is given in terms of Chebychev's inequality. The inequality states that the probability of the magnitude of a zero-mean random process being greater than s is less than the variance of the random process divided by s^2. That is,

$$P[(|x|) \geq s] \leq \frac{\sigma_x^2}{s^2} \qquad (4.9.6)$$

In terms of Eq. (4.9.6), the weak law of large numbers can be defined to be

$$P(|y - \overline{x}| \geq s) \leq \frac{\sigma_x^2}{Ks^2} \qquad (4.9.7)$$

where $\overline{x}$ is $\mathcal{E}(x_i)$ and y is the arithmetic (or sample) mean given by

$$y = \frac{1}{K} \sum_{i=1}^{K} x_i \qquad (4.9.8)$$

Therefore, the difference between the sample mean y and the ensemble mean x is inversely related to K, the sample size.

4.10 COVARIANCE

The ensemble counterpart of correlation is covariance. The <u>autocovariance</u> of a process $\{x(n)\}$, denoted $\phi_x(i, j)$, is given by

$$\phi_x(i, j) = \mathcal{E}(x(i)x(i + j))$$

$$= \int_{-\infty}^{\infty} \int_{-\infty}^{\infty} x(i)x(i + j)p(x(i), \, x(i + j)) \, dx(i) \, dx(i + j) \qquad (4.10.1)$$

If $j = 0$, then $\phi_x(i, 0) = \sigma^2_{x(i)}$. Similarly, the <u>cross covariance</u> of $\{x(i)\}$ and $\{y(i)\}$ is given by $\phi_{xy}(i, j)$, where

$$\phi_{xy}(i, j) = \mathcal{E}(x(i)y(i + j)) \qquad (4.10.2)$$

Like its counterpart auto- and cross correlation, covariance measures similarity. However, one uses ensemble statistics in this case rather than their temporal versions.

4.11 CONDITIONAL DENSITIES

Many problems found in linear system theory are characterized by more than one random variable (e.g., input-output processes). A <u>conditional probability density function</u>, say $p(x/y)$, represents the probability density of the random variable x(n) given the value of y(n). Formally $p(x/y)$ can be defined in terms of the joint probability density as follows:

$$p(x, \, y) = p\left(\frac{x}{y}\right)p(y) \qquad (4.11.1)$$

This concept can be trivially extended to more than two variables if required. For example, in terms of three variables, one obtains

$$p(x, \, y, \, z) = p\left(x, \, \frac{y}{x}\right)p(z) = p\left(\frac{x}{y}, \, z\right)p(y, \, z)$$

$$= p\left(\frac{x}{y}, \, z\right)p\left(\frac{y}{z}\right)p(z) \qquad (4.11.2)$$

Formally, if x and y are independent, then $p(x, \, y, \, z) = p(x)p(y)p(z)$.

4.12 ERGODICITY

It was previously noted that temporal averaging is technically a simple operation. Using linear or exponential windows, hardware or software routines can be developed which produce these metrics. It was also noted that the mathematically more elegant ensemble averages are difficult to compute in general. Therefore, it would be desirable to find a condition by which the elegance of ensemble averaging and the simplicity of temporal averaging can be unified into a common theory.

A random process is said to be <u>stationary</u> if its ensemble statistics are time invariant. A process is said to be <u>ergodic</u> if <u>all</u> time and ensemble moments are equal. Ergodic processes are always stationary, but not all stationary processes are ergodic. A process is said to be <u>weak sense</u> <u>ergodic</u> if the autocorrelation and autocovariance functions are equal. This is a powerful but restrictive concept. If a process is ergodic, theoretical conjectures can be derived using ensemble-averaging mechanics based on temporal averages.

EXAMPLE 4.12.2   Let $x(i) = \cos(i\omega T_s + \phi_i)$, where $\phi_i \in U[-\pi, \pi]$.   Then

$$\phi_x(i, k) = \mathcal{E}(x(i)x(i+k)) = \frac{\cos(i\omega T_s)}{2} = R_x(k)$$

EXAMPLE 4.12.2   A "<u>white</u>" <u>discrete stochastic process</u> is one whose autocorrelation is given by $\phi_x(k) = \sigma_x^2 \delta_k(k)$, where $\delta_k(k)$ is the <u>Kronecker</u> <u>delta function</u> [i.e., $\delta_k(k) = 1$ if $k = 0$, 0 otherwise]. Loosely stated, a white discrete process is so unpredictable that for any nonzero delay k, $x(i)$ and $x(i+k)$ have nothing at all in common. For $k = 0$, the autocovariance function equals the variance of x. If $x(i)$ is ergodic, $R_x(k)$ also equals $\sigma_x^2 \delta_k(k)$. The spectral representation of $R_x(k)$, in terms of $G(k) = X(k)X(k)^*$, is given by $G(k) = \sigma_x^2$ for all harmonics. It is the appearance of this broadband and "flat" spectrum that gives rise to its name. In physics, a light source that displays uniform energy across the visible spectrum is called white light. It follows that a flat spectrum would suggest that the originating time series should be called a white noise process.

Using a random number generator, a N(0, 1) random data sequence was created. A histogram of the test noise can be found in Fig. 4.12.1. Using a FFT, the autocorrelation coefficient was computed, also shown in Fig. 4.12.1. Note that based on the graphical interpretation of the autocorrelation coefficient, one may conclude that the noise process is essentially white.

Using Parseval's theorem, which states that the power in a time series is given by

$$\sum_{n=0}^{N-1} x^2(n) = \frac{1}{2\pi j} \oint G_N(z) z^{-1} \, dz$$

$$= \frac{\sigma_x^2}{2\pi j} \oint \frac{dz}{z} = \sigma_x^2 \qquad\qquad (4.12.1)$$

one can equate the power in the white spectrum to the signal variance.

White noise is of major importance in the study of linear systems. Since a white process exhibits energy at every frequency location of the discrete spectrum, it represents a potentially good test signal. That is, a white noise source will excite all the natural modes of oscillation of a linear system.

4.13 CORRELATION ESTIMATION

Often, one is interested in quantifying a correlation function over a relatively short delay interval. That is, over a long N-sample data record, it is assumed that only information about $R_x(k)$ for $k \le N/M$ is required. Since many signal and systems possess rather short correlation times, it would be useful to develop a methodology to handle this problem.

One method involves the use of $2M + 1$ DFTs where $M = N/K$. It shall be assumed that $X(k) = DFT[x(n)]$ and $Y(k) = DFT[y(k)]$. It was noted previously that the inverse DFT of $X(k)Y(k)^*$ represents the N-point <u>circular correlation</u> of x and y. Therefore, to ensure that the circular and linear correlation results agree, zero filling is required. Again let us consider that a $K - 1$ lag correlation estimate is required. The data base displayed in Fig. 4.13.1 provides insight into the problem of developing short-lag correlation metrics. In this figure, $N = 9$, $K = 3$, and $M = 3$. That is, there are three blocks consisting of three samples each. Observe that the fundamental sequences $\{x(0), x(1), x(2)\}$, $\{x(3), x(4), x(5)\}$, $\{x(6), x(7), x(8)\}$ are zero filled and staggered (overlapped). Thus BLOCK i is, in fact, $2K - 1$ samples in length. The insertions of these zeros will allow the DFT to be used to form circular correlation functions which can be interpreted as linear correlation functions. Then, with respect to the data found in this diagram,

$$R_x(k) = \sum_{i=0}^{2} R_i(k); \quad k = 0, 1, 2$$

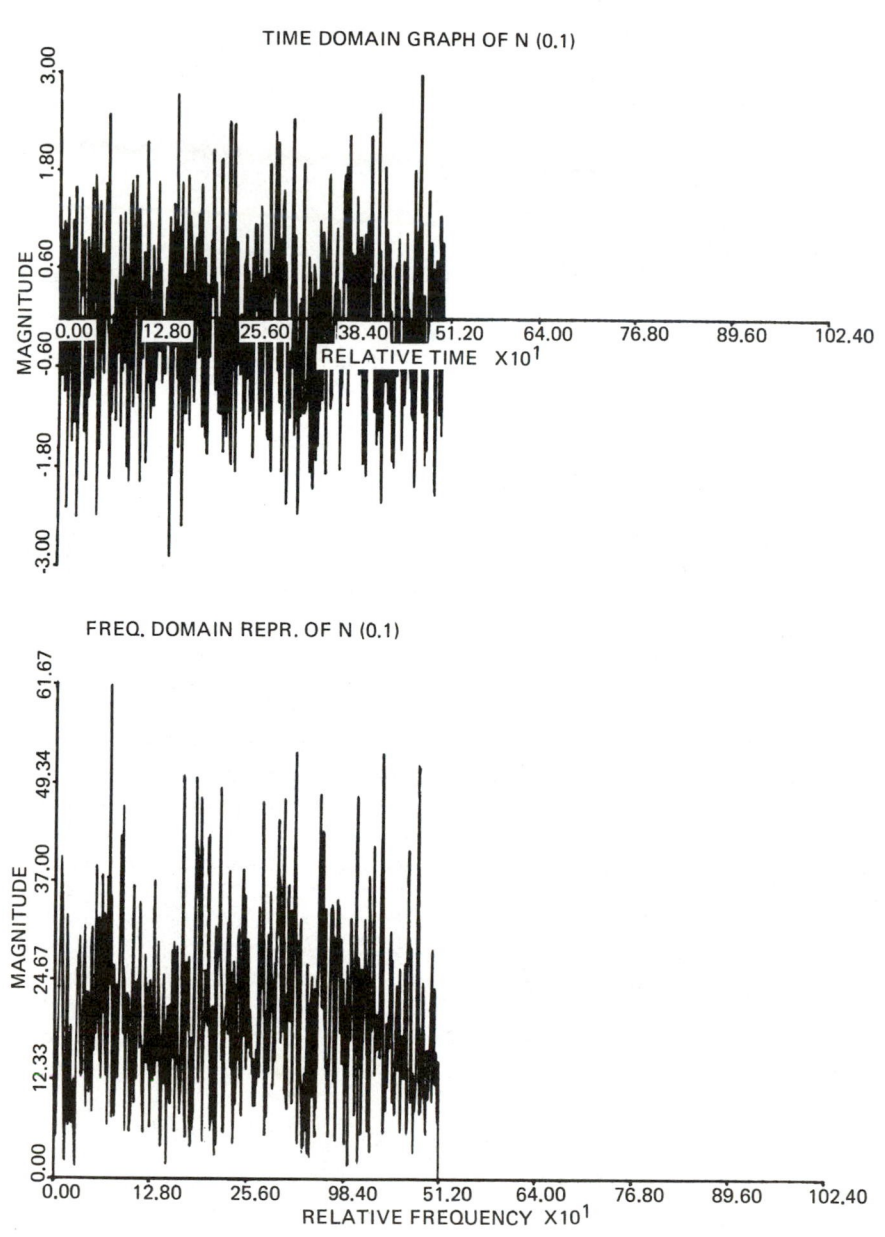

FIG. 4.12.1 Sample of a numerically generated pseudo-random sequence, its spectrum, a histogram of a numerically generated pseudo-random white process, and a "delta distribution-like" autocorrelation function using a FFT.

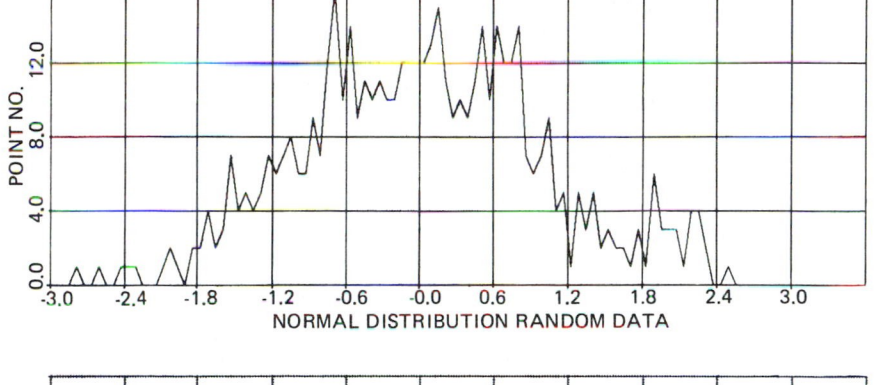

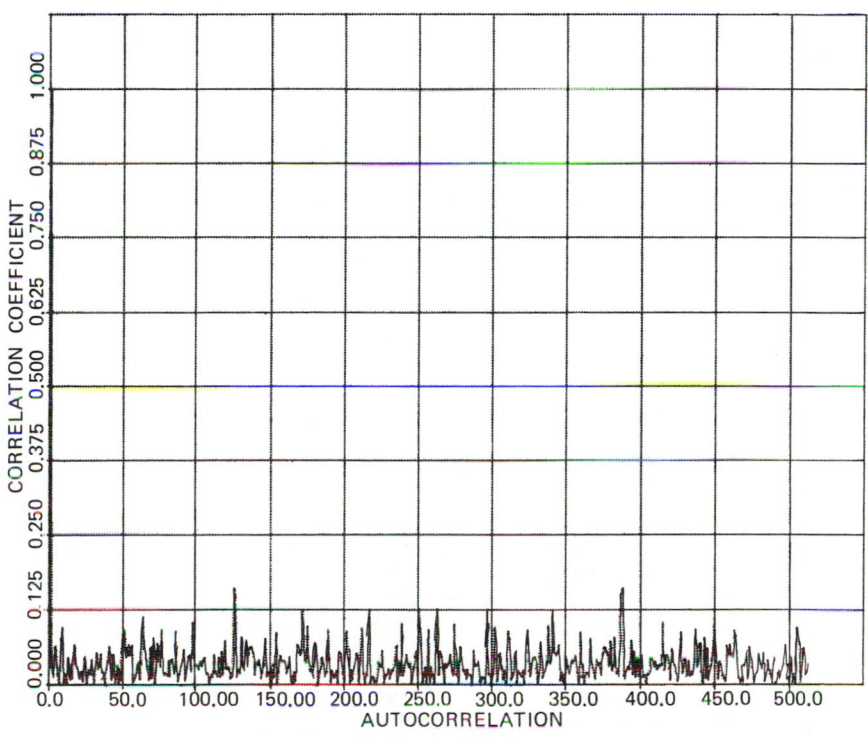

APPEND ZEROS

$\{x(n)\}= x_0 \ x_1 \ x_2 \ x_3 \ x_4 \ x_5 \ x_6 \ x_7 \ x_8 \ \overline{0 \ \ 0} \ \longleftarrow$

BLOCK ZERO X_0:	$x_0 \ x_1 \ x_2 \ 0 \ \ 0$ $0 \ \ x_0 \ x_1 \ x_2 \ 0$ $0 \ \ 0 \ \ x_0 \ x_1 \ x_2$	3FFTs IDFT$(X_0(k)X_0^*(k))$ $R_0(0)=x_0^2+x_1^2+x_2^2$ $R_0(1)=x_0x_1+x_1x_2+x_2x_3$ $R_0(2)=x_0x_2+x_1x_3+x_2x_4$
BLOCK ONE X_1:	$x_3 \ x_4 \ x_5 \ 0 \ \ 0$ $0 \ \ x_3 \ x_4 \ x_5 \ 0$ $0 \ \ 0 \ \ x_3 \ x_4 \ x_5$	$R_1(0)=x_3^2+x_4^2+x_5^2$ $R_1(1)=x_3x_4+x_4x_5+x_5x_6$ $R_1(2)=x_3x_5+x_4x_6+x_5x_7$
BLOCK TWO X_2:	$x_6 \ x_7 \ x_8 \ 0 \ \ 0$ $0 \ \ x_6 \ x_7 \ x_8 \ 0$ $0 \ \ 0 \ \ x_6 \ x_7 \ x_8$	$R_2(0)=x_6^2+x_7^2+x_8^2$ $R_2(1)=x_6x_7+x_7x_8$ $R_2(2)=x_6x_8$

FIG. 4.13.1 Sectioned data field exhibiting the component parts of an auto-correlation computation.

That is,

$$R_x(0) = \underbrace{x_0^2 + x_1^2 + x_2^2}_{\text{Block 0}} + \underbrace{x_3^2 + x_4^2 + x_5^2}_{\text{Block 1}} + \underbrace{x_6^2 + x_7^2 + x_8^2}_{\text{Block 2}}$$

$$R_x(1) = \underbrace{x_0x_1 + x_1x_2 + x_2x_3}_{0} + \underbrace{x_3x_4 + x_4x_5 + x_5x_6}_{1} + \underbrace{x_6x_7 + x_7x_8 + 0}_{2}$$

$$R_x(2) = \underbrace{x_0x_1 + x_1x_3 + x_2x_4}_{0} + \underbrace{x_3x_5 + x_4x_6 + x_5x_7}_{1} + \underbrace{x_6x_8 + 0 + 0}_{2}$$

From a computational standpoint, the foregoing results can be obtained using the following routine.

1. $X(0) = DFT[x(0), \ldots, x(4)]$
2. $Y(0) = DFT[x(0), x(1), x(2), 0, 0]$
3. $X(1) = DFT[x(3), \ldots, x(7)]$
4. $Y(1) = DFT[x(3), x(4), x(5), 0, 0]$
5. $X(2) = DFT[x(6), x(7), x(8), 0 \ 0]$
6. $Y(2) = DFT[x(6), x(7), x(8), 0, 0]$
7. Sum of products of the form $G(k) = \Sigma X(k)Y(k)^*$
8. $R_x(k) = IDFT[G(k)]$

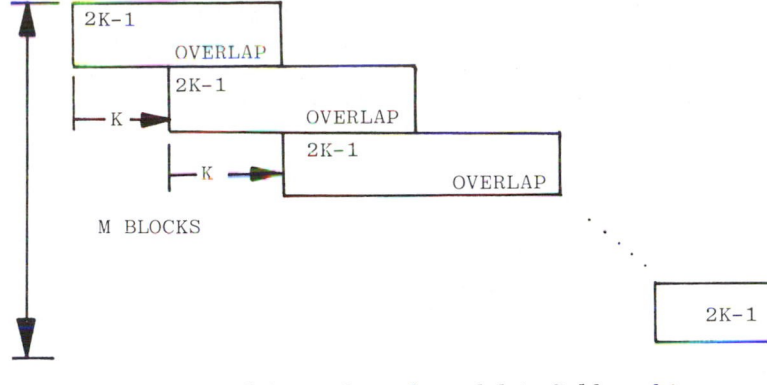

FIG. 4.13.2 General form of overlapped data field used in computing a correlation function.

Only the first K samples will be accepted from $R_x(k)$. The others will be discarded.

In general, from the three-tuple (N, K, M), data would be partitioned into the grouping found in Fig. 4.13.2. Using M + 1 DFTs of length 2K - 1 each, an estimate of the correlation function which is valid from the zeroth to the (K - 1)st delay can be derived in a numerically efficient manner.

Rader has offered an improvement in this basic strategy by requiring that the DFTs be radix 2 (length 2^n) and the data block exhibit a 2:1 overlap. For example, suppose that N = 16 and K = 3; then the Rader sectioning is as shown in Fig. 4.13.3. One can readily see that Y(1), for example, satisfies $Y(1) = DFT[x(4) \cdots x(7)0000]$ and $(-1)^k Y(1) = DFT[0000x(4) \cdots x(7)]$. Then X(0) can be synthesized by computing $X(0) = Y(0) + (-1)^k Y(1)$. Therefore, the number of individual DFT calls can be reduced to K.

| x_0 | x_1 | x_2 | x_3 | x_4 | x_5 | x_6 | x_7 | x_8 | $\cdots$ | x_{11} | x_{12} | x_{13} | x_{14} | x_{15} |

x_0	x_1			$\cdots$			x_7		$x_o(k)$
x_0	x_1	x_2	x_3	0	0	0	0		$y_0(k)$
			x_4	x_5		$\cdots$		x_{11}	$x_1(k)$
			x_4	x_5	x_6	x_7	0	0 $\ldots$ 0	$y_1(k)$

etc.

FIG. 14.3.3 Data-sectioning scheme proposed by Rader.

EXAMPLE 4.13.1 Using a 2:1 overlap, the autocorrelation of the 16-point time series found in Fig. 4.13.4 can be performed. It is readily apparent that the autocorrelation of a periodic time series produces another periodic time series [see $R_X(k)$ of Fig. 4.13.4]. Based on the developed sectioning policy, $R_X(k)$, for $0 \le k \le 4$, can be computed as follows:

$$\{X_0\} = \{X_2\}; \ \{X_1\} = \{X_3\}; \ \{Y_0\} = \{Y_2\}; \ \{Y_1\} = \{Y_3\}$$

$$Y_0(k) = Y_2(k) \rightarrow \{1 + \sqrt{2}, \ -2j, \ -1, \ 0, \ 1 - \sqrt{2}, \ 0, \ -1, \ 2j\}$$

$$Y_1(k) = Y_3(k) \rightarrow \{-(1 + \sqrt{2}), \ 2j, \ 1, \ 0, \ (1 - \sqrt{2}), \ 0, \ 1, \ -2j\}$$

$$X_0(k) = X_2(k) \rightarrow Y_0(k) + (-1)^r Y_1(k) = \{0, \ -4j, \ 0, \ 0, \ 0, \ 0, \ 0, \ 4j\}$$

$$X_1(k) = X_3(k) \rightarrow Y_1(k) + (-1)^r Y_2(k) = \{0, \ 4j, \ 0, \ 0, \ 0, \ 0, \ 0, \ -4j\}$$

$$X_0(k)Y_0^*{} = X_2(k)Y_2^*(k) = \{ \ 0, \ \ -4j, \ 0, \ 0, \ \ \ 0, \ \ \ 0, \ 0, \ \ 4j\}$$
$$\times \{1 + \sqrt{2}, \ 2j, \ -1, \ 0, \ 1 - \sqrt{2}, \ 0, \ -1, \ -2j\}$$
$$\overline{}$$
$$\{ \ \ 0, \ \ \ -8j^2, \ 0, \ 0, \ \ \ 0, \ \ \ 0, \ \ 0, -8j^2\}$$

$$X_1(k)Y_1^*(k) = X_3(k)Y_3^*(k) = \{ \ \ \ 0, \ \ \ \ \ 4j, \ 0, \ 0, \ \ \ \ 0, \ \ \ \ 0, \ 0, -4j\}$$
$$\times \{-(1 + \sqrt{2}), \ -2j, \ 1, \ 0, \ -(1 + \sqrt{2}), \ 0, \ 1, \ \ 2j\}$$
$$\overline{}$$
$$\{ \ \ \ 0, \ \ \ \ -8j^2, 0, \ 0, \ \ \ \ \ 0, \ \ \ \ \ 0, \ 0, -8j^2\}$$

where the power spectrum, after normalizing with respect to the FFT weight of $1/N$ per transform (here $1/64$), satisfies:

$$\text{normalized } \hat{G}_x(k) = \frac{4(8)}{8^2} \{0, \ 1, \ 0, \ 0, \ 0, \ 0, \ 0, \ 1\}$$

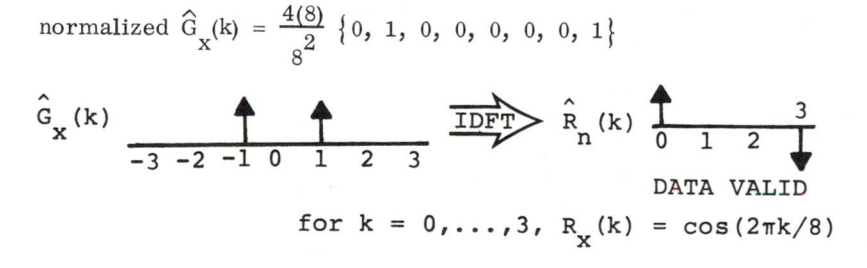

for $k = 0, \ldots, 3$, $R_x(k) = \cos(2\pi k/8)$

A comparison of the computational complexity of these correlation methods can be directly computed. To demonstrate the quantitative nature of this question, a $N = 8$-point time series will be considered. Using a

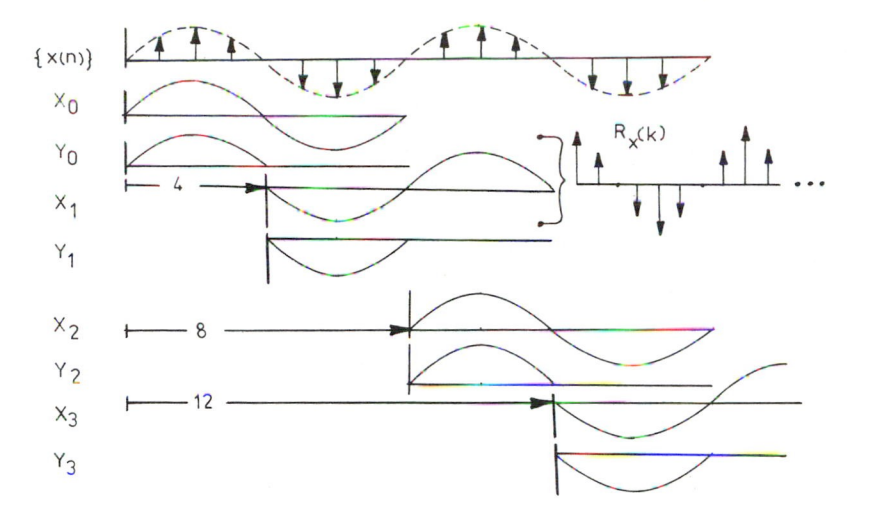

FIG. 4.13.4 Example of a K = 2, two-to-one Rader partitioning.

direct sum-of-products approach, the multiplication count associated with computing $R_X(k)$ is given by[†]

$$\sum_{i=0}^{7} x(n)x(n - m) \xrightarrow[\text{multiplies}]{\text{number of}} 1 + 2 + \cdots + 7 + 8 + 7 + \cdots$$

$$+ 2 + 1 = \frac{8 \times 9}{2} + \frac{7 \times 8}{2}$$

Using 16-point FFTs (eight-sample time series plus an eight-sample zero fill), $X(k) = DFT(x(n))$ would require:

$$\text{number of multiplies} = \frac{16}{2} \log_2 2^4 = 32$$

multiplications. However, for the cross-correlation case, $R_{xy}(k) = IDFT(X(k)Y(k)^{*})$. The multiplication count associated with three DFTs and 16 multiplies is

$$\text{number of multiplies} = 2 \text{ DFTs} + 16 + 1 \text{ IDFT} = 32 * 2 + 16 + 32 + 112$$

[†]Observe that $\sum_{i=1}^{n} i = n(n + 1)/2$.

Using the Rader method, assuming that only the $R_{xy}(k)$ is required for $k = 0$ and 1, data would be sectioned in the manner shown in Fig. 4.13.5.

In general, suppose that one wishes to compute $R_x(k)$ for $k = 0, \ldots,$ $2^m = K$ over a $N = 2^n$-point time series. The direct sum-of-products method would require $N(N + 1)/2 + (N - 1)(N/2) = N^2$ multiplies. Using N-point circular correlation, the multiplication count would be given by $2(2N/2)\log_2 2N + 2N + 2(N/2)\log_2 2N = N(3N + 5) \sim 3nN$ for $n \geq 4$. Using the Rader method, sectioning over $N/2K$ blocks, one observes that there are $2[2(K/2)\log_2 2K] + 2K = 2K(2m + 1)$ multiplications per block. Added to this count would be the IDFT count of $2(K/2)\log_2 2K = 2Km$ multiplies. Therefore, the total multiplication count is given by $6Km + 2K \sim 6Km$. These results are plotted in Fig. 4.13.6 for $m \in [2, 8]$ (i.e., $K \in [4, 256]$) and $N = 1024$.

A word of caution must be voiced, however. These short delay methods can suffer severely from finite aperature effects. For long record lengths, the fidelity of the DFT is improved and the distortion effect reduced. If a power spectrum is to be derived from a short delay correlation function, consisting of only K samples, excessive ripple may result due to a truncated temporal interval. This is developed further in Chapter 5. The quality of the power spectra (PS) can, however, be improved through a zero-augmentation operation. This can be motivated by the example.

EXAMPLE 4.13.2 Suppose that $\{x(n)\}$ is a 1024-point time series. Let it be required that $R_x(k)$ be computed for $0 \leq k \leq 15$. To ensure a limited amount of distortion in a derived power spectrum, let a 64-point transform

$$x_0\ x_1\ x_2\ x_3\ x_4\ x_5\ x_6\ x_7$$

BLOCK 0: $\left\{ \begin{matrix} x_0 & x_1 & 0 & 0 \\ 0 & x_0 & x_1 & 0 \end{matrix} \right\}$ $\left\{ \begin{matrix} \text{4-POINT DFTs; DFT}(x_0 x_1 00)\text{DFT*}(x_0 x_1 00) \\ \text{THEREFORE, } G_0(k) \rightarrow 2((4/2)\log_2 4) + 4 = 12 \text{ MULTIPLIES} \end{matrix} \right\}$

BLOCK 1: $\left\{ \begin{matrix} x_2 & x_3 & 0 & 0 \\ 0 & x_2 & x_3 & 0 \end{matrix} \right\}$ $= 12$ MULTIPLIES

BLOCK 2: $\left\{ \begin{matrix} x_4 & x_5 & 0 & 0 \\ 0 & x_4 & x_5 & 0 \end{matrix} \right\}$ $= 12$ MULTIPLIES

$\text{PLUS IDFT}(\sum_{i=0}^{3} G_i(k))$

(4)*12=48 MULTIPLIES

$= 12$ MULTIPLIES

TOTAL MULTIPLICATION COUNT $= 52$ MULTIPLIES

FIG. 4.13.5 Multiplication count for partitioned correlation computation using DFTs.

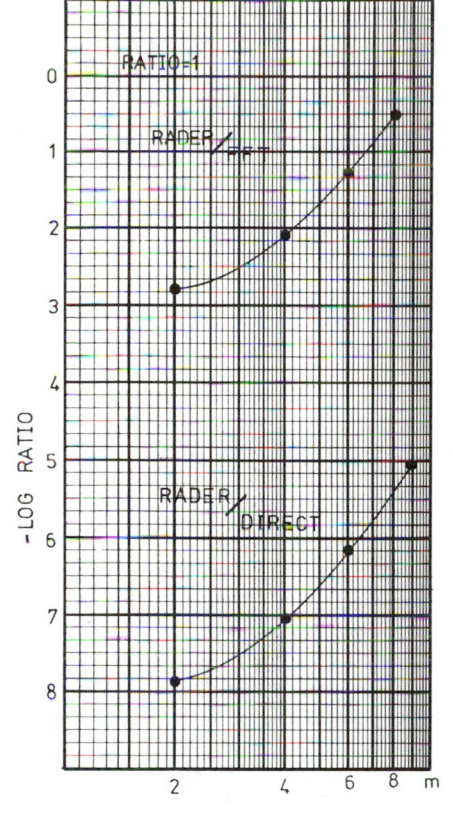

FIG. 4.13.6 Comparison of multiplication complexity of the direct and FFT method computing a correlation function.

be used. The parameters that quantify this example are $N = 2^{10}$, $K = 16$, and a 64-point derived autocorrelation function. Therefore, from the fact that $2K = 32$, we know the original data base will be sectioned into 2^5 zero-filled data blocks (see Fig. 4.13.7). Using the methodology developed previously, a 16-sample image of $R_x(k)$ can be computed using circular correlation. A 64-sample data base can be created by appending 64 - 16 zeros to $R_x(k)$, as suggested in Fig. 4.13.7. The insertion of these zeros is consistent with the knowledge that $R_x(k)$ is assumed to have possible nonzero values for $k \in [0, 15]$ only.

To overcome the effects of short-sample-length DFTs, data windows are usually employed. These windows (e.g., Hamming, Kaiser, etc.) are discussed in Section 5.4. They have a tendency to reduce the undesirable

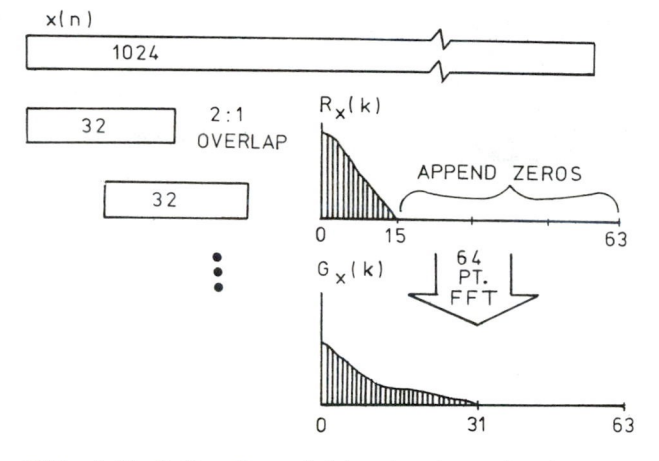

FIG. 4.13.7 Overlapped data structure showing appended zeros.

ripple-like distortion found in short transforms. A typical correlation/
power spectrum estimation model is shown in Fig. 4.13.8. In this diagram
the windows are indicated by w(i).

Using these discussed sectioning schemes, the correlation of one
(auto) or two (cross) variables can be realized efficiently under the hypothesis that only a few delays are required. That is, it is assumed that $R_x(k)$
need only be quantified for k small with respect to the total record length
N. This is particularly useful in the study of white or nearly white stochastic processes where the correlation times are known to be short.

4.14 ONE-BIT CORRELATION

A simple correlator can be realized in digital hardware by using a novel
one-bit or polarity coincident strategy. The hardware efficiency of this

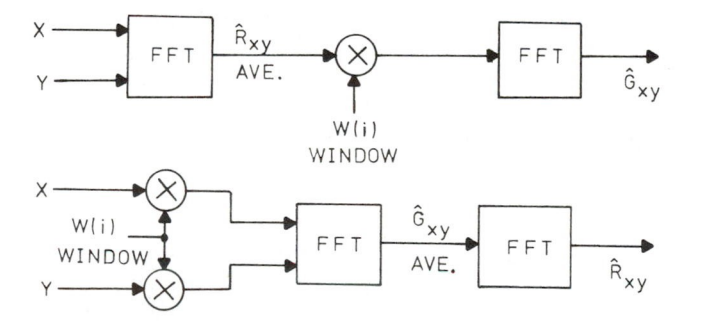

FIG. 4.13.8 Correlation and power spectral system diagrams.

approach is purchased at the expense of accuracy. However, in applications where high-speed correlation estimation is desired without regard to high precision, this method offers some attractive features. The method is predicated on being able to interpret general correlation in terms of one-bit messages.

Suppose that two signals, say $x(t)$ and $y(t)$, are bounded in modulus by one and possess a known cross-correlation function $R_{xy}(\tau)$. It has been shown by Brendt that if (a) uncorrelated (statistically independent) uniformly distributed noise is added to $x(t)$ and $y(t)$ to form $\hat{x}(t) = x(t) + n_1(t)$ and $\hat{y}(t) = y(t) + n_2(t)$, and (b) if $x^c(t)$ and $y^c(t)$ are polarity-detected (severely clipped) versions of $x(t)$ and $y(t)$, then (c) $R_{x^c y^c}(\tau) = R_{xy}(\tau)$.

This has been historically referred to as polarity coincident correlation and can be explained as follows. Uniform noise is used to "dither" both signal processes $x(t)$ and $y(t)$. If the two processes are strongly correlated, they will possess similar zero-crossing patterns. Similarly, two weakly correlated signals will have statistically dissimilar zero-crossing patterns. Since the outputs of the zero-crossing detectors represent a binary-valued process, the actual correlation can be performed with a simple exclusive-OR (ex-OR) gate. The expected value of the ex-OR output, which represents an instantaneous correlation metric, can be computed using an exponential averaging window (i.e., RC filter) or linear window (i.e., up/down counter). A simple 1-bit correlator is shown in Fig. 4.14.1. In this figure, the cross correlation of $x(t)$ and $y(t) = \sin(\omega_0 t)$ or $\cos(\omega_0 t)$ is suggested. In this mode, spectral analysis can be performed. In particular, the kth Fourier coefficients can be written as the following correlation operations:

$$a_k = \frac{1}{T} \int_T x(t) r_{2k}(t + \tau)\, d\tau$$

$$b_k = \frac{1}{T} \int_T x(t) r_{2k+1}(t + \tau)\, d\tau \qquad (4.14.1)$$

for

$$r_{2k}(t) = \cos(\omega_k t)$$

$$r_{2k+1}(t) = \sin(\omega_k t) \qquad (4.14.2)$$

The performance of a 1-bit correlator, using a reference signal $x(t) = \cos(\omega_0 t)$, may be equated to a bandpass filter centered about ω_0. For T sufficiently large, the correlated output can be shown to be

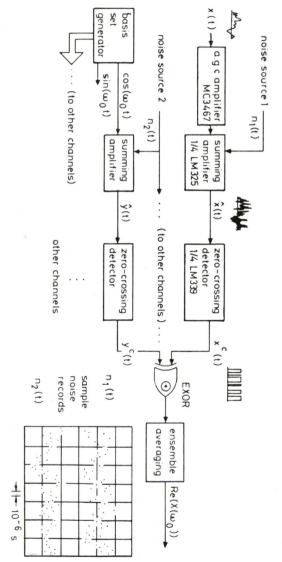

FIG. 4.14.1 System diagram for a 1-bit correlator used in a spectral analysis capacity.

$$R_{xy} = \frac{\sin(\Delta\omega_T)}{2\Delta\omega T}$$

$$\Delta\omega = \omega - \omega_0 \tag{4.14.3}$$

If one adopts the usual 3-dB method of defining bandwidth, denoted BW = $2\Delta f$, then BW = $2\Delta\omega/2\pi$ with Δf given by

$$\Delta f = \left\{ \frac{\Delta\omega}{2} \quad \text{such that} \quad \rho_{xy} = \frac{\sin(\Delta\omega T)}{\Delta\omega T} = \frac{1}{\sqrt{2}} \; ; \; T \text{ given} \right\} \tag{4.14.4}$$

In solving this equation, one concludes that BW satisfies

$$BW = \frac{0.442}{T} \text{ hertz} \tag{4.14.5}$$

A final logical question to ask is how long should the averaging interval be? Using a Chernoff bound, the error in computing the statistics of a binary-valued process, say x_i, $x_i = 0$ or 1, can be quantified. Suppose that the expected value of x_i, over all i, is p. Over N samples the expected value will be assumed to be computed to have a value p'. The probability that p' differs from p by an amount greater than ϵ can be bounded by

$$P(|p - p'| \geq \epsilon) \geq \left[\left(\frac{p}{d} \right)^d \left(\frac{1-p}{1-d} \right)^{1-d} \right]^N \tag{4.14.6}$$

where $d = p + \epsilon$. For a given ϵ, N assumes its largest value for p = 0.5. For this worst-case condition, an N-sample estimate of cross correlation will be within $\pm 5\%$ (i.e., $\epsilon = \pm 0.025$) of its theoretical value $N > 0.5 \times 10^3$.

4.15 PERIODOGRAMS

To this point the question of estimating the auto- and cross-correlation functions over long data records and short delay intervals has been developed. We shall now turn our attention to the problem of estimating the power spectrum of a long time series. The periodogram is used to estimate the power spectrum, namely $G_N(k)$, of a N-sample time series $\{x_N(n)\}$, based on K L-point transforms (KL $\leq$ N). The long-record-length time series is sectioned into a series of K overlapping subsequences composed of L samples each. The overlap of these sequences, denoted $\{x_L^i(n)\}$, will be parameterized by d, where L = dD. The interpretation of these parameters is shown in Fig. 4.15.1. A periodogram is defined in terms of a windowed DFT. The window function, say $\{w(n)\}$, has a value zero for $j < 0$ and $j \leq L$ [e.g., a uniform window is given by $w(n) = 1$ for $0 \leq N \leq L - 1$

and zero elsewhere]. In terms of the window function, the data subsequences can be defined as follows:

$$x_L^i = w(n)x(n + iD); \quad i = 0, 1, \ldots, K - 1 \tag{4.15.1}$$

If the L-point DFT of $\{x_L^i(n)\}$ is $X_L^i(k)$, the <u>periodogram</u> is formally defined to be $P_i(k)$, where

$$P_i(k) = \frac{1}{E} X_L^i(k)^2; \quad k = 0, 1, \ldots, L/2 \tag{4.15.2}$$

$$E = \sum_{i=0}^{L-1} w^2(i); \quad \text{energy of window}$$

The power spectrum estimate of $G_N(k)$, based on averaging K periodograms, is denoted $\hat{G}(k)$ and satisfies

$$\hat{G}(n) = \frac{1}{K} \sum_{i=0}^{K-1} P_i(k) = \frac{1}{KE} \sum_{i=0}^{K-1} |X_L^i(k)|^2$$

It has been shown that if $\{x_N(n)\}$ is Gaussian, and if $G_N(k)$ (the true power spectrum) is "reasonably flat" over the frequency interval where the window response is large, then

$$\text{var } G(k) = \mathcal{E}((\hat{G}(k) - G_N(k))$$

$$= \frac{G_N(k)}{K} \left[1 + 2 \sum_{i=1}^{K-1} \frac{K - i}{K} \rho(i) \right] \tag{4.15.4}$$

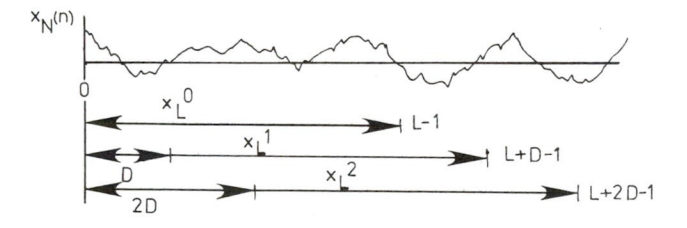

FIG. 4.15.1 Data structure used to compute a periodogram.

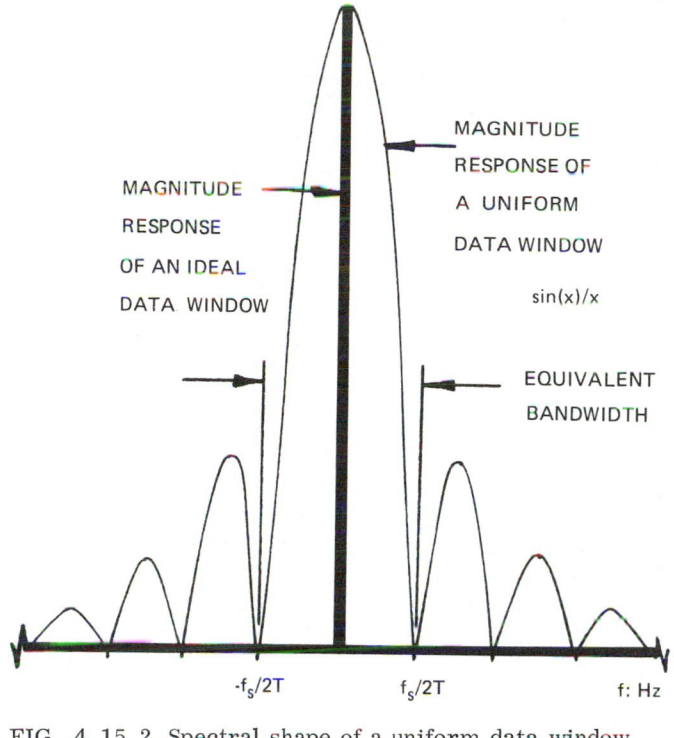

FIG. 4.15.2 Spectral shape of a uniform data window.

for N large and

$$\rho_{(i)} = \frac{\left[\displaystyle\sum_{j=0}^{L-1} w(j)w(j + iD)\right]^2}{\displaystyle\sum_{j=0}^{L-1} w^2(j)} \qquad (4.15.5)$$

Here the variance is defined at those harmonics derived from the L-point DFT rather than the N/2 harmonics found in $G_N(j)$.

The question of a "good" data window will be examined in Section 5.4. At that time it will be argued that a good data window has a DFT, which is approximately a Kronecker delta function (see Example 4.12.2).

For example, the L-sample uniform window has a spectral image (see Fig. 4.15.2). Here the main lobe is found between $\pm f_S/2L$ and defines

the window to have an equivalent bandwidth $B = f_s/L$. It can also be seen
that the side-lobe gains are really too high to qualify as an approximation
to a Kronecker delta function. Later, other windows will be studied which
possess superior side-lobe behavior. Nevertheless, for a "good" window,
it can be argued that the periodogram components located at frequency f_i,
$f_i > B$, are uncorrelated. Thus as L becomes large (equivalently, B decreases), the frequency separation between periodogram components
decreases and the individual periodograms have a tendency to exhibit large
amplitude difference from harmonic to harmonic. These fluctuations will,
of course, be averaged (smoothed) in the creation of the spectral density
estimate G(k).

If, for example, a uniform window is used with $D = L$ (i.e., contiguous disjoint data blocks), then $\rho(j) = 0$ and

$$\text{var } \widehat{G}(k) = \frac{G_N(k)^2}{K} \qquad (4.15.6)$$

That is, the reduction in the estimation error variance is inversely related
to K. However, if the overlap is 50% (i.e., $D = L/2$), and w(n) is a triangular window (which has superior side-lobe behavior), it can be shown that

$$\text{var } \widehat{G}(k) = \frac{G_N^2(k)}{K} \left(\frac{17}{32}\right) \qquad (4.15.7)$$

That is, the variance has been decreased by a factor of $(17/32) \simeq 0.5$, over
the case considered previously. The price paid for the reduced error variance is computational time. First, compared to the 1:1 overlap, a 2:1
overlap would require twice as many periodogram computations. Second,
the use of a nonuniform window will require an additional L multiplications
per subsequence. Obviously, the user must be aware and respond to these
trade-offs. The minimum value of D, however, is established by the maximum throughput rate per periodogram (i.e., minimum of one periodogram
per D samples).

EXAMPLE 4.15.1 The power spectrum of a random noise source can be
estimated using a periodogram. The result of this experiment is reported
in Fig. 4.15.3. A noise record of 1024 points from a N(0, 1) population
was used for the test. The noise time series and histogram are displayed
in this figure. Data were partitioned using a 2:1 overlap for $D = 32$ and
$L = 64$. The power spectrum derived from a single record, denoted $x_0(n)$,
namely $G_0(k)$, is graphically interpreted. It can be seen that a single short
data record produces a power spectrum which possesses a lot of variability.
After averaging $K = 32$ such records, the spectrum GN(k) was obtained.

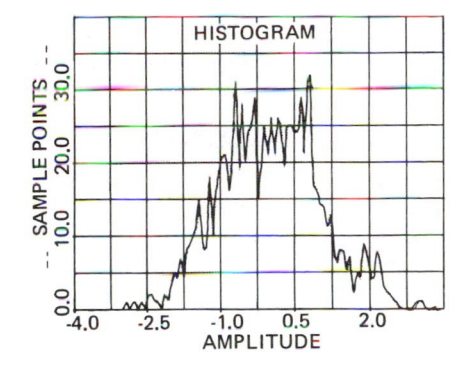
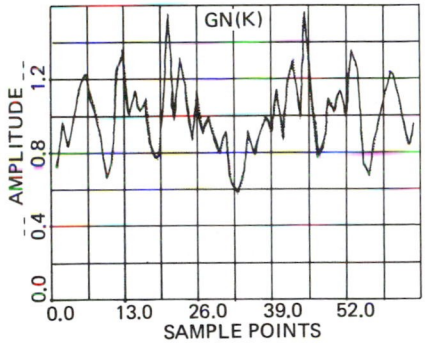
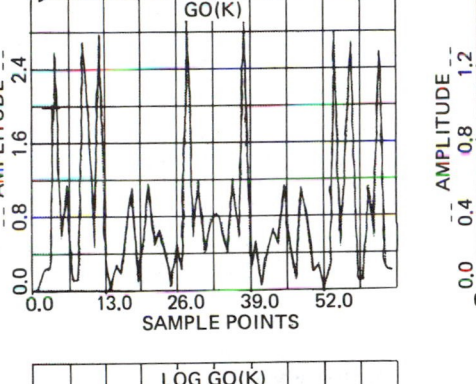
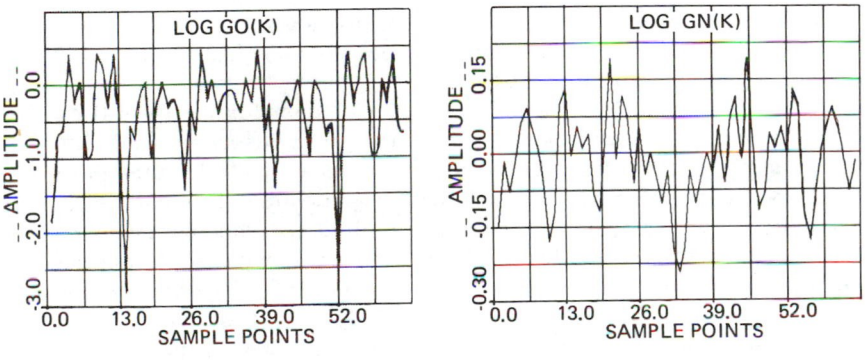

FIG. 4.15.3 Periodogram analysis of individual and ensemble pseudo-random white noise processes.

This spectrum is now tending to flatten and suggests the presence of white noise.

EXAMPLE 4.15.2 Example 4.15.1 is repeated in Fig. 4.15.4. Here the data are displayed as a line graph in the frequency domain.

4.16 POWER TRANSFER

The study of linear discrete systems generally requires the examiner to quantify mathematically the system's transfer function. In general, a transfer function is a ratio of two polynomials having one of the two following forms:

$$H(z) = \frac{\sum_{i=0}^{N} a_i z^{-i}}{\sum_{i=0}^{N} b_i z^{-1}} \tag{4.16.1}$$

$$H(z) = \frac{K' \prod_{i=0}^{N} (z + \gamma_i)}{\prod_{i=0}^{D} (z + \lambda_i)} \tag{4.16.2}$$

It was noted previously that it is generally advantageous to factor H(z) into second-order sections incorporating complex conjugate pole and/or zero pairs. In the frequency domain, H(z) is interpreted as $H[\exp(j\omega_i)]$, where $\omega_i = 2\ i/NT_s$, $i = 0, 1, \ldots, N/2$. The complex values of $H \exp(j\omega_i)$ provide information about the magnitude and phase response of the filter under study. Another important quality of a linear discrete filter is its ability to modify selectively an input power spectrum. This may be a destructive operation (such as the removal of 60-Hz noise power) or constructive (frequency-selective gain). The power spectrum modification capability of a filter can be developed using the DFT. Suppose that the input of a discrete filter is $\{x(n)\}$ having a DFT X(k). It follows that the power spectrum of x(n) is given by $G_x(k) = X(k)X^*(k) = |X(k)|^2$. Suppose further that $\{x(n)\}$ is to be passed through a filter whose impulse response is H(z). Then it follows that the output power spectrum, say $G_x(k)$, satisfies

$$\begin{aligned} G_y(k) &= Y(k)Y^*(k) \\ &= H(k)H^*(k)X(k)X^*(k) \\ &= |H(k)|^2 G_x(k) \end{aligned} \tag{4.16.3}$$

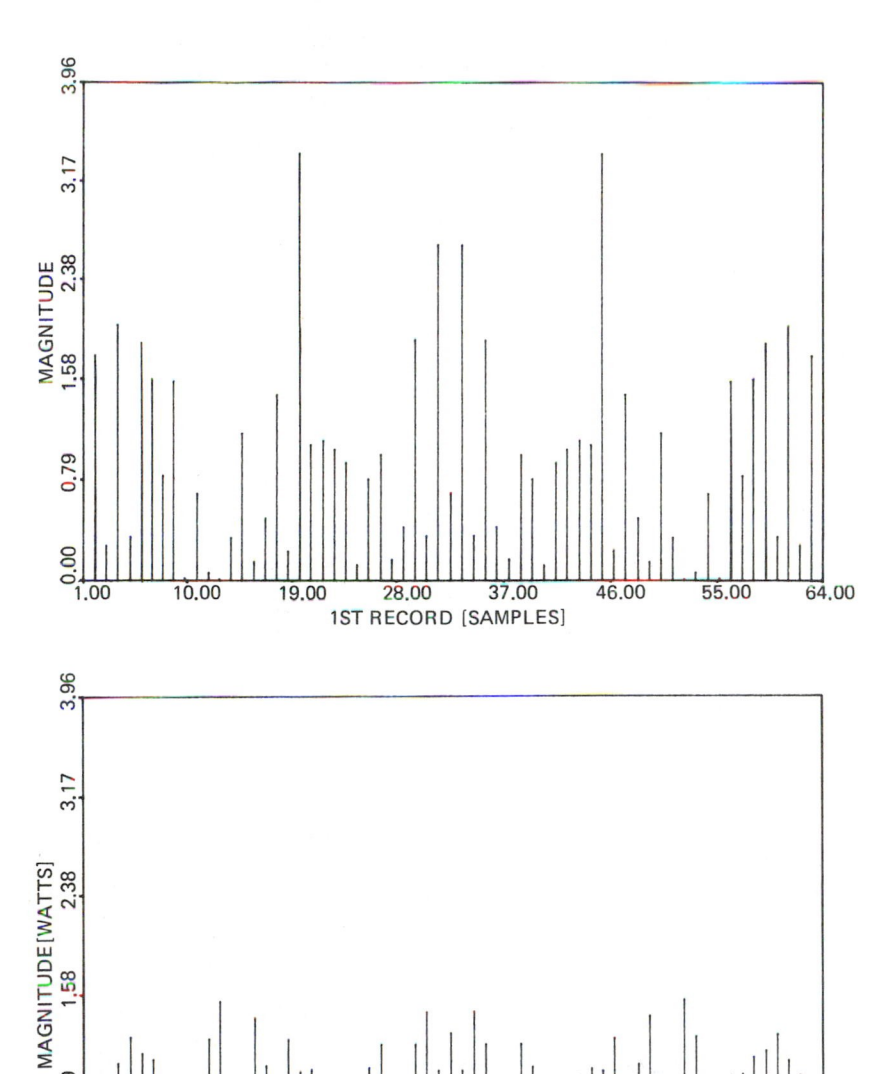

FIG. 4.15.4 Comparison of single- (top) and ensemble- (bottom) averaged periodograms.

The output spectrum can be seen to be a mapping of the input spectrum under $\phi(k) = |H(k)|^2$. Therefore, by properly choosing the filter function H(k), wanted behavior can be reinforced, and unwanted behavior, such as noise corruption, can be removed.

Using Parseval's theorem, the power in the output time series $y(n)$, say P_y, can be computed to be

$$P_y = \sum_{k=0}^{N-1} G_y(k) = \sum_{k=0}^{N-1} |H(k)|^2 G_x(k) \qquad (4.16.4)$$

Using these basic results, various optimal and suboptimal filters can be designed. However, it turns out that the design and analysis of discrete filters is architecture dependent. This will be developed in Chapter 5.

4.17 COHERENCE FUNCTION

It was stated that the transfer function is indispensable to the study of the linear discrete systems. Implicit in the definition of transfer function $H(x) = Y(z)/X(z)$ (where x and y denote input and output processes, respectively) are the following assumptions:

1. The system is linear.
2. The initial state of the system is zero.
3. The system is forced by a single input, $\{x(n)\}$.

However, because of nonlinearities, additive noise, or multiple signal sources, a system may not possess a meaningful transfer function representation. The coherence function will allow the user to test hypotheses 1 to 3 in a computationally efficient manner. As a result, phenomena that may exist within a system which would preclude a linear analysis can be detected.

To measure the linear dependence between an input and output process, cross correlation can be used. It will be assumed that all nonlinear, multiple-input, and additive internal noise distortions can be modeled as an additive output noise component, as suggested in Fig. 4.17.1.

Assume now that the noise process n(i) and system output y(i) are uncorrelated. Then for $z(i) = y(i) + n(i)$, it follows immediately that $Z(k) = X(k) + N(k)$ and that the cross power spectrum satisfies

$$G_{zx}(i) = [Y_i(i) + N(i)]X^*(i)$$
$$= X(i)X^*(i) + N(i)X^*(i)$$
$$= G_{xx}(i) + G_{nx}(i) \qquad (4.17.1)$$

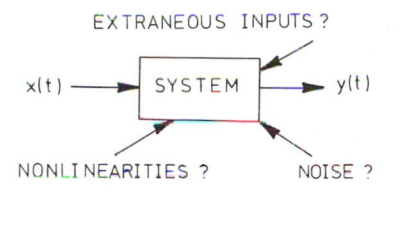

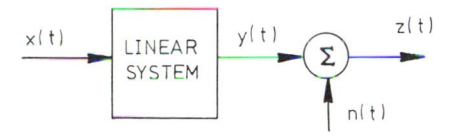

FIG. 4.17.1 System diagram representing sources of coherence information.

Since $G_{zx}(i)$ is the sum of a noise-free and a noisy process [i.e., $G_{yx}(i)$ and $G_{nx}(i)$], it is impossible to measure the relative strength of each looking only at the sum. Instead, frequency domain averaging methods must be used. Letting a bar denote a linear N-sample average value, define $\delta^2(i)$, called the coherence function, as follows:

$$\delta^2(i) = \frac{\overline{G_y(i)}}{\overline{G_y(i)} + \overline{G_n(i)}} \tag{4.17.2}$$

The coherence function represents a real number which is bounded between 0 and 1. If $\delta^2(i)$ is near unity, the noise-free signal dominates the noise process $n(i)$. If $\delta^2(i)$ has a value near zero, the converse is true. Unfortunately, the process $\{y(i)\}$ is generally not available for direct measurement. Therefore, $G_y(i)$ is numerically unknown. To overcome this realization problem, the coherence function needs to be defined in terms of directly measurable parameters. Therefore, consider the alternative form

$$\delta^2(i) = \frac{\overline{G_y(i)}}{\overline{G_y(i)} + \overline{G_n(i)}}$$

$$= \frac{\overline{G_x(i)}\ \overline{G_y(i)}}{\overline{G_x(i)}\ \overline{[G_y(i)} + \overline{G_n(i)]}} \tag{4.17.3}$$

It was assumed that the noise n(i) is uncorrelated with x(i) and y(i). Therefore, if the averaging interval is sufficiently long (i.e., $N \to \infty$), then

$$\overline{G_{nx}(i)} \to 0, \quad \overline{G_{ny}(i)} \to 0, \quad \overline{G_{ny}(i)} \to 0, \quad \overline{G_{nx}(i)} \to 0 \qquad (4.17.4)$$

From z(i) = y(i) + n(i), it follows that

$$\overline{G_z}(i) = [\overline{Y}(i) + \overline{N}(i)][\overline{Y}(i) + \overline{N}(i)]^*$$

$$= \overline{G_y}(i) + \overline{G_n}(i) + \overline{G_{ny}}(i) + \overline{G_{yn}}(i)$$

$$= \overline{G_y}(i) + \overline{G_n}(i) \qquad (4.17.5)$$

and

$$\delta^2(i) = \frac{\overline{G_x(i)} \; \overline{G_y(i)}}{\overline{G_x(i)} \; \overline{G_z(i)}} \qquad (4.17.6)$$

Finally, observe that the numerator of Eq. (4.17.6) can be rewritten to read

$$\overline{G_x(i)G_y(i)} = [\overline{X(i)} \; \overline{X(i)}^*][\overline{Y(i)} \; \overline{Y(i)}^*]$$

$$= \overline{X(i)Y(i)}^* \; \overline{Y(i)X(i)}^*$$

$$= |G_{yx}(i)|^2 \qquad (4.17.7)$$

However, it can be shown that

$$\overline{G_{zx}(i)} = \overline{G_{yx}(i)} + \overline{G_{\!\!\!\nearrow x}^{\;0}(i)} \to \overline{G_{yx}(i)} \qquad (4.17.8)$$

Therefore, one can obtain a definition of the coherence function which is a function of x(i) and z(i) only! The resulting coherence function definition thus becomes

$$\delta^2(i) = \frac{|\overline{G_{zx}(i)}|^2}{\overline{G_z(i)} \; \overline{G_x(i)}} \qquad (4.17.9)$$

Finally, Fig. 4.17.2 shows a flow diagram description of the coherence function.

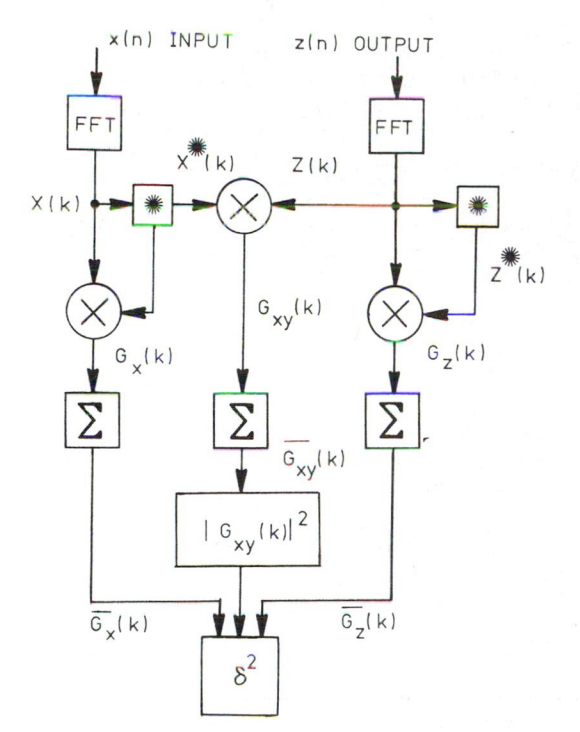

FIG. 4.17.2 Block diagram of a coherence function calculation using the FFT.

EXAMPLE 4.17.1 Suppose that the system satisfies system requirements 1 to 3 [i.e., $n(i) = 0$]; then $G_z(k) = G_x(k)$ and $G_{zx}(i) = G_{yx}(i)$ [see Eq. (4.17.7)]. Then

$$\delta^2(i) = \frac{\left| G_{yx}(i) \right|^2}{G_y(i) G_x(i)} = 1 \quad \text{for all ii}$$

That is, the coherence function would imply that the system being considered can be modeled using a linear constant coefficient transfer function.

EXAMPLE 4.17.2 Suppose that a linear system suffers from internal 60–Hz noise corruption. If the Kth harmonic is located at 60 Hz, it follows that $\delta^2(i) = 1$ for all harmonics not equal to K and $\delta^2(k)$ has a value less than unity for $k = K$.

   More specifically, a linear lowpass system having a transfer function $H(z) = K(z + 1)/(z + 0.5)$ was driven by an impulse. Line–frequency noise

of the form $n(t) = A \cos(2f_0 t + \phi)$, for $f_0 = 60$ Hz, was added to the output
process. The phase angle ϕ was chosen from a uniform population over
$[-\pi, \pi]$. This random phase angle simulates the random triggering of the
data acquisition subsystem of a FFT. Using a 1024-point transform, a
256-Hz Nyquist frequency, and ensemble-averaging 16 experiments, the
coherence function was computed. The result of this experiment is documented in Fig. 4.17.3. Observe that the coherence function is unity for
all frequencies except 60 Hz. Here the depression in coherence value
corresponds to the signal/signal-plus-noise ratio found at that frequency.

EXAMPLE 4.17.3 Consider the simple lowpass filter having an impulse
response $h(n) = a^n$ and the input process shown in Fig. 4.17.4a. To the
output process, broadband noise is added. A typical noise-added output
response is abstracted in Fig. 4.17.4b. Performing a coherence analysis
over a single experiment (i.e., $N = 1$) results in the unity coherence function shown in Fig. 4.17.4c. However, upon averaging 10 experiments
together, one obtains the coherence function shown in Fig. 4.17.4d. At
low frequencies, where the filter gain is high, the coherence function is
near unity. However, at the high end, where the filter gain is low, the

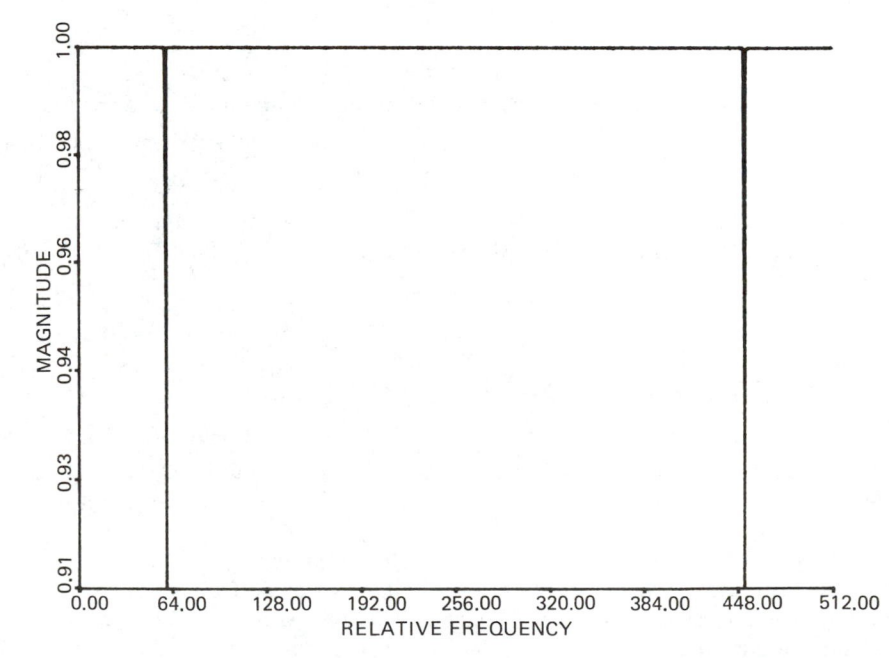

FIG. 4.17.3 Coherence function of a 60-Hz corrupted linear system;
$A = 1.0$.

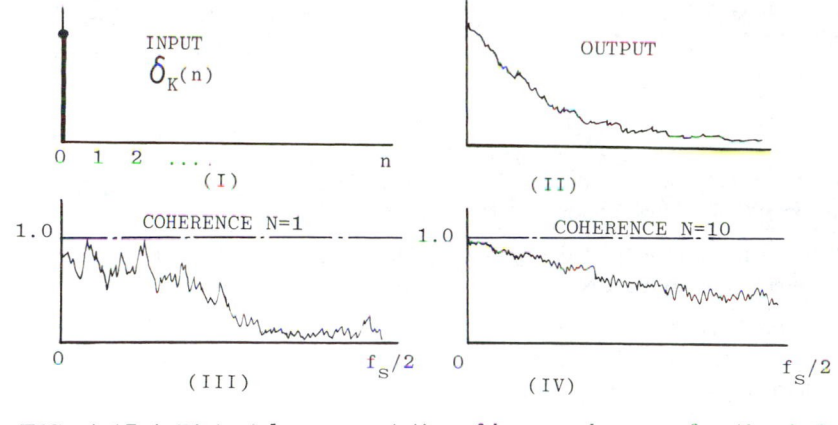

FIG. 4.17.4 Pictorial representation of how a coherence function is improved through ensemble averaging.

coherence function shows the effects of additive noise (equivalently nonlinearities or multiple inputs).

EXAMPLE 4.17.4 Figure 4.17.5a and b show the frequency domain records of L coherence experiments. In the first figure, the input to a noisy system is displayed. Observe that the input signal energy is locally concentrated about 0, 1, 3, 5, 7, and 9 kHz. The output spectrum, shown in Fig. 4.17.5b, shows the energy to be spread over a wide range of frequencies. The coherence function, shown in Fig. 4.17.5c, exhibits how much of the output spectral energy, at a frequency location, is due to energy present at the same frequency at the input. The fact that the coherence function is nearly zero at 2i kHz, i = 1, . . ., 5, does not necessarily mean that the systems exhibit strong nonlinearities or possess a high noise-to-signal

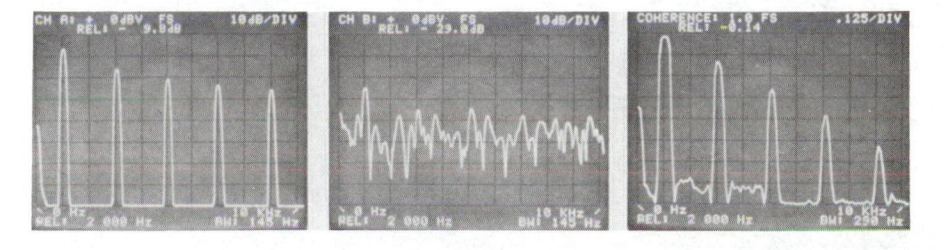

FIG. 4.17.5 Coherence function derived by using a commercial Fourier analyzer. (Reprinted with the permission of Hewlett-Packard Corporation.)

ratio. It may mean, as in this case, that the input was void of energy of certain sections of the spectrum.

BIBLIOGRAPHY

Gold, B., and C. M. Rader (1969), Digital Signal Processing of Signals, McGraw-Hill, New York.

Gold, B., et al. (1969), Theory and Implementation of the Discrete Hilbert Transform, Proc. Symp. Comp. Commun., pp. 235-250.

Hewlett-Packard Staff (1978), The 3582A, Dual-Channel Real-Time Spectrum Analysis and Transfer Function Measurements, Hewlett-Packard Monogr. 5952-8769D, Palo Alto, Calif., April.

Oppenheim, A. V., and R. Schafer (1975), Digital Signal Processing, Prentice-Hall, Englewood Cliffs, N.J.

Rabiner, L. R., and B. Gold (1975), Theory and Application on Digital Signal Processing, Prentice-Hall, Englewood Cliffs, N.J.

Stockham, T. G (1969), High Speed Convolution and Correlation, AFIPS Conf. Proc., 28, pp. 229-233.

Taylor, F. J., and S. L. Smith (1975), Digital Signal Processing in FORTRAN, Lexington Press, Lexington, Mass.

5
Finite Impulse Response Filters

5.1 INTRODUCTION

Digital filters can be partitioned into two distinct classes: the finite and infinite response filters. A finite impulse response filter (FIR) possesses an output response to an input impulse forcing function which is of finite duration. A FIR filter may appear in either a recursive (with feedback) or nonrecursive (without feedback) form. Nonrecursive FIRs are known to have simple architectures, are stable, and have a low sensitivity to round-off errors. The principal advantage of the FIR structure is its ability to exhibit linear phase versus frequency behavior. This property is useful in many applications areas, such as:

1. Phase (delay) equalization for digital communication systems
2. Speech processing
3. Image processing
4. Stochastic filtering

However, there are some negative features of the FIR that should be appreciated as well. The most important of these is its inability to achieve sharp (steep-skirt) magnitude response in the frequency domain. In general, a high-order FIR must be used to achieve sharp frequency response. Therefore, high-Q filters, and those which have nearly a piecewise constant spectral shape, are generally architected as infinite impulse response filters. In such applications, the low-order infinite response filter will generally have a higher throughput than its higher-order finite impulse response counterpart.

Formally, an FIR filter can be characterized in the z domain as H(z), where

$$H(z) = \sum_{n=0}^{N-1} h(n)z^{-n} \qquad (5.1.1)$$

or, in terms of its frequency response,

$$H[\exp(j\omega)] = \sum_{n=0}^{N-1} h(n) \exp(-j\omega)$$

$$= \pm\ H(j\omega)\ \exp[j(\phi)(\omega)] \qquad (5.1.2)$$

The model suggested by Eq. (5.1.1) defines explicitly the filter's impulse response which is given by $h(i)$ for $i \in [0, N-1]$.

5.2 LINEAR PHASE FILTER

An important class of FIR filters are called the linear phase filters. In the linear phase filter, the phase angle is a linear function of frequency and it is given by

$$\phi(\omega) = -a\omega; \quad -\pi \leq \omega \leq \pi \qquad (5.2.1)$$

with

$$H[\exp(j\omega)] = \pm|H(j\omega)|\ \exp(-ja\omega)$$

$$= \pm|H(j\omega)|\ [\cos(a\omega) + j\ \sin(a\omega)] \qquad (5.2.2)$$

Linear phase shifting can be achieved if and only if

$$a = \frac{N-1}{2}; \quad h(N-1-n) = h(n); \quad 0 \leq n \leq N-1 \qquad (5.2.3)$$

These conditions indicate that there exists a high degree of symmetry in the phase space. If N is odd, a is an integer. Here the filter's delay is an integer number of samples. For N odd, the filter's impulse response is centered about sample number $(N-1)/2$. For these reasons, most FIR designs are of odd order.

It is often desired to realize a linear phase filter function having a slope-intercept form given by

$$\phi(\omega) = a\omega + b; \quad a = \frac{N-1}{2}; \quad b = \pm\frac{\pi}{2} \tag{5.2.4}$$

The resulting filter satisfies

$$h(n) = -h(N-1-n); \quad 0 \le n \le N-1 \tag{5.2.5}$$

By virtue of the indicated negative sign, the filter is said to possess anti-symmetry.

The transfer function of a linear phase filter can be expressed as

$$
\begin{aligned}
H(z) &= \sum_{n=0}^{N-1} h(n)z^{-n} \\
&= \sum_{n=0}^{(N-1)/2} h(n)z^{-n} \pm \sum_{n=0}^{(N-1)/2} h(n)z^{-(N-n-1)}
\end{aligned}
\tag{5.2.6}
$$

which can be algebraically manipulated to produce

$$H(z^{-1}) = \pm z^{(N-1)} H(z) \tag{5.2.7}$$

If the zeros of $H(z)$ are located at $z_i = r_i \exp(j\phi_i)$, the zeros of $H(z^{-1})$ must be located at $z_i^{-1} - \exp(-j\phi_i)/r$. For example, if $r_i = 1$ for all i, and if N is odd, the zeros of the filter are located on the periphery of the unit circle in the z domain.

EXAMPLE 5.2.1 In this example the comb filter is introduced. This particular filter will be reinvestigated later in the chapter. An Nth-order comb filter is characterized by the impulse response $\{h(n)\}$, where

$$
h(n) = \begin{cases} 1, & 0 \le n \le N-1 \\ 0, & \text{otherwise} \end{cases}
$$

and is shown in Fig. 5.2.1. The reader can readily see that the comb filter possesses a simple architecture. All that is required to realize a comb filter in hardware is a system of adders and delay registers. In addition, the software realization of this filter is also trivial. Here the delay lines would be replaced with a first-in first-out (FIFO) queue. The transfer function that represents a comb filter is given by $H(z)$, where

$$H(z) = \sum_{i=0}^{N-1} z^{-i} = \sum_{i=0}^{\infty} z^{-i} - \sum_{i=N}^{\infty} z^{-i}$$

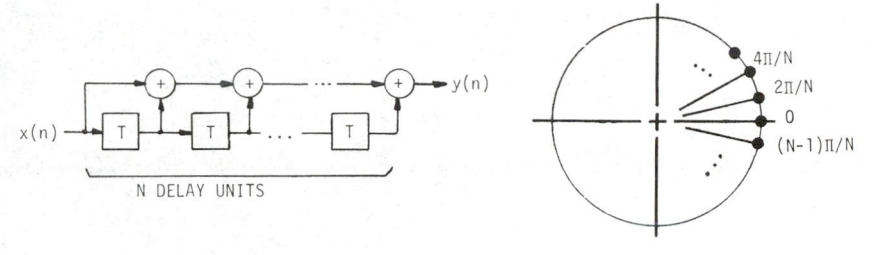

FIG. 5.2.1 Block diagram and pole locations of a comb filter.

$$= \frac{1}{1 - z^{-1}} - \frac{z^{-n}}{1 - z^{-1}} = \frac{1 - z^{-N}}{1 - z^{-1}} \tag{5.2.8}$$

Based on De Moivre's theorem, the zeros of H(z) can be shown to be uniformly distributed along the unit circle on $2\pi/N$ radian centers. It can be
noted that there exists pole-zero cancellation at z = 1. The resulting z-
domain description of an odd-order comb filter is abstracted in Fig. 5.2.1.
Here it can be seen that the extremal frequencies of the filter are located
at $\omega_i^* = 2\pi k/N$. That is, $|H[\exp(\omega_i^* T)]| = N$. Therefore, the dynamic range
of this filter is bounded by N times the maximal value of the input time
series. This parameter will establish the dynamic range limits over which
registers will not overflow. Furthermore, since h(n) = h(N - 1 - n), the
comb filter is also a linear phase filter. The frequency response of a comb

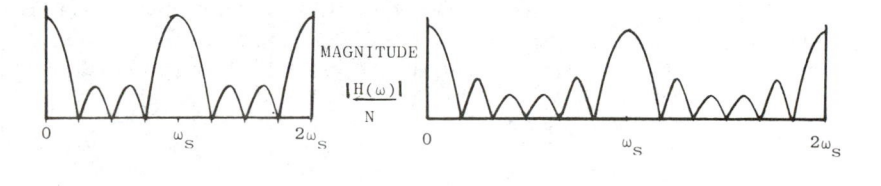

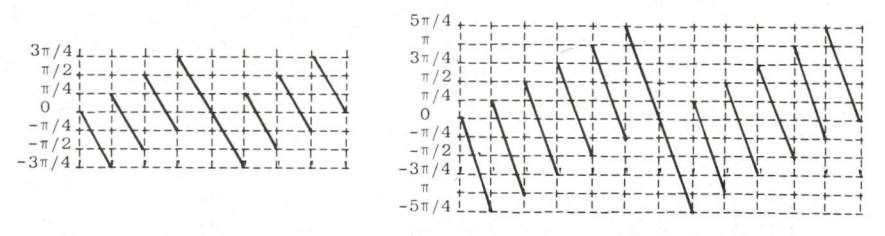

FIG. 5.2.2 Magnitude and phase response of a low-order comb filter.

filter for several values of N is shown in Fig. 5.2.2. It can be noted that
as N increases, the frequency selectivity about the critical frequencies
$\omega_i^* = 2\pi k/T$ becomes sharper and the phase slope increases as well.

EXAMPLE 5.2.2 The magnitude response of a 33rd-order FIR can be
found in Fig. 5.2.3. The impulse response for this typical bandpass filter
can be found in the same figure. It can be noted that the filter under study
possesses symmetry about the center time series (sample location 16.5).
The phase response of the filter can be seen to have the classic linear
phase behavior, suggested by the theory, over the interval $[-\pi/2, \pi/2]$.

EXAMPLE 5.2.3 The telecommunications industry has established design
requirements for modems. For example, a modem may operate at 2400,
4800, or 9600 bits per second (bps) under CCITT recommendations. Least-
squares methods can be used to design a FIR compensator. The compensator's response C(k) will be used to augment the response of the existing
system, which shall be assumed to be given by G(k). The design goal is to
have the cascaded system G(k)C(k) provide an acceptable approximation to
some prespecified idealized response H(k). That is, find a C(k) so that the
magnitude-squared error (i.e., ℓ_2 norm) of $||C(k)G(k) - H(k)||^2$ is mini-
mized. This classic norm minimization problem can be approached in one
of many time-proven ways.

Suppose that the desired spectral shape of a modem transmitter is the
square root of a 90% raised cosine waveform plus some delay compensation.
The delay is required to account for the distributed delay found in the analog sections of the modem. Since the delay is nonlinear, a nonlinear phase
shifting filter is needed. As a consequence, the designed FIR will not
possess the symmetry needed for linear phase shifting. This is summarized
in Fig. 5.2.4.

5.3 WINDOWS

There are several FIR design techniques currently used in general practice. One of the most important of these is the window method. Windows
are an overlay applied to a given time series to improve the spectral quality
of the data base. The windows used most often have been singled out for
specific analysis in this section. They are:

1. Rectangular or uniform window
2. Hamming and Hann window
3. Blackman window
4. Kaiser window.

To assist the reader in understanding the purpose of these data windows,
several introductory comments will be made. Consider first the primitive

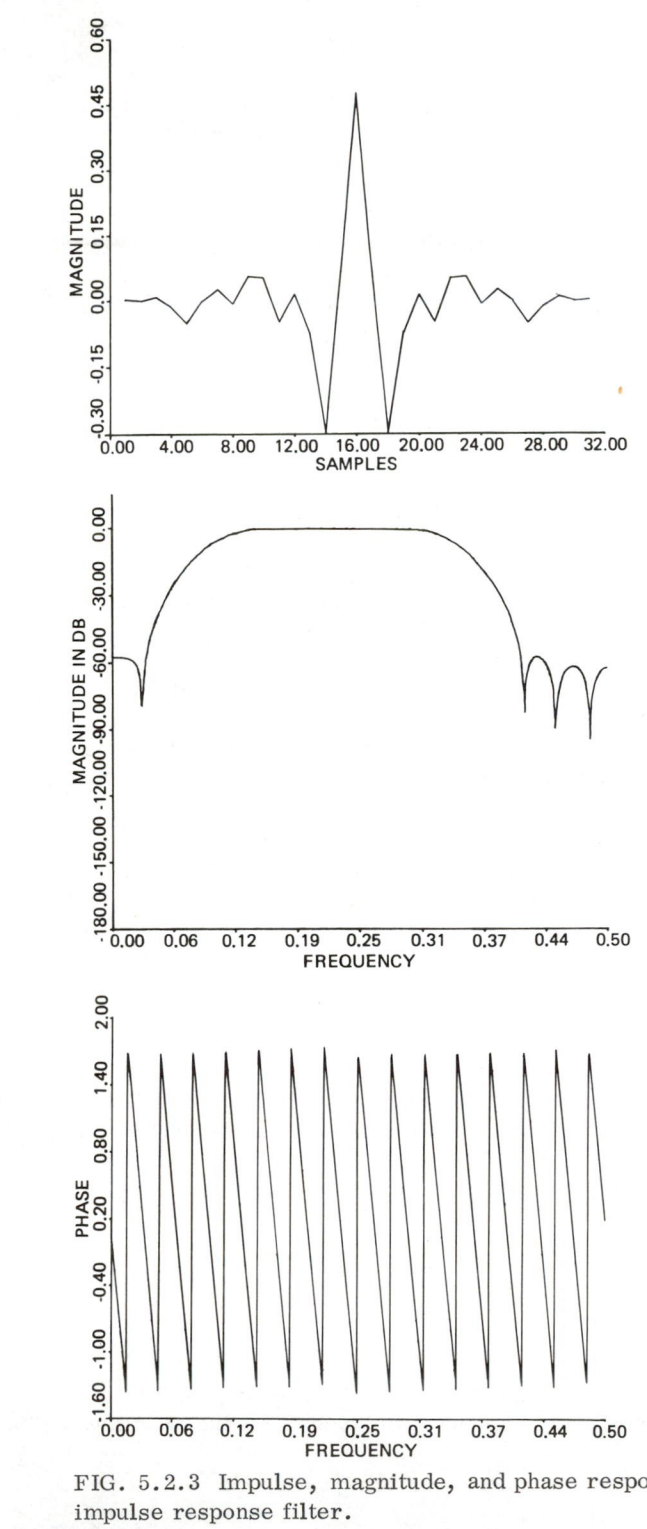

FIG. 5.2.3 Impulse, magnitude, and phase response for a 33rd-order finite impulse response filter.

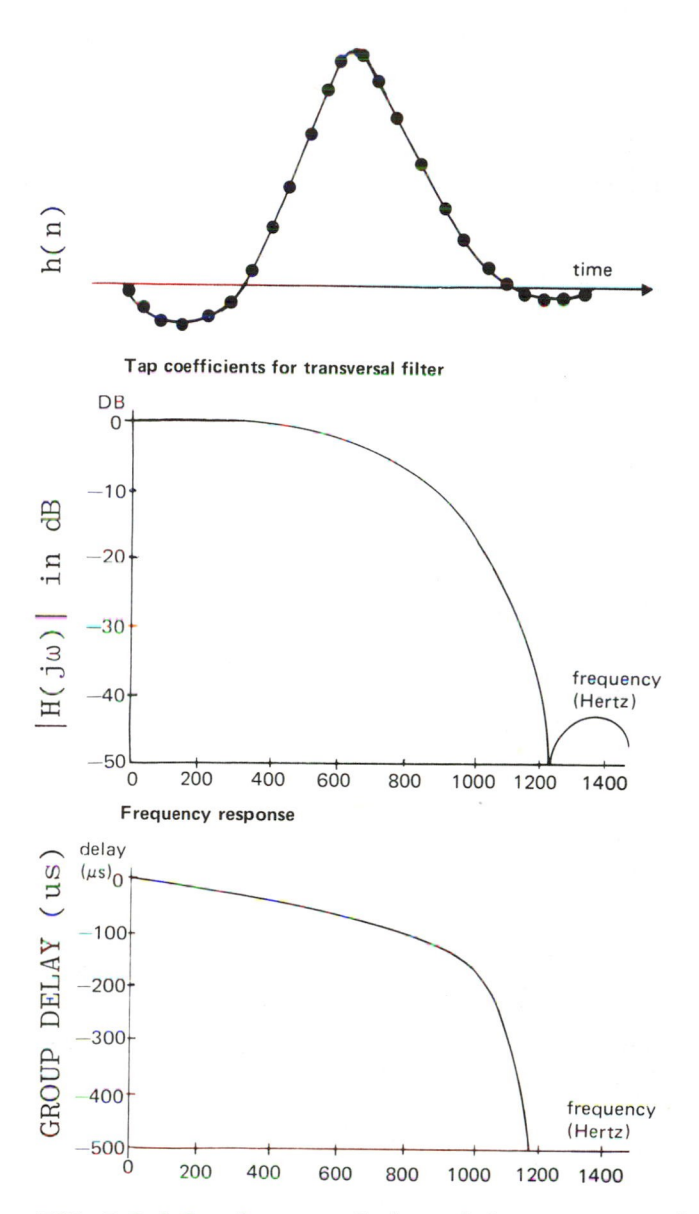

Tap coefficients for transversal filter

Frequency response

FIG. 5.2.4 Impulse, magnitude, and phase response of a commercial communications FIR. (Reprinted with the permission of EDN.)

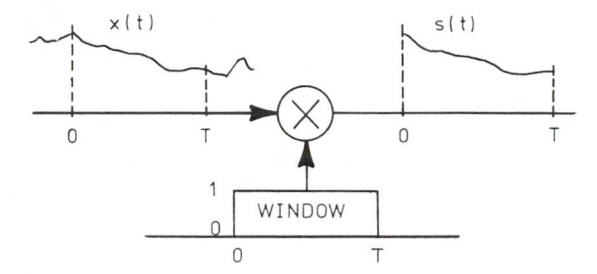

FIG. 5.3.1 System diagram interpreting a temporal window.

switching function shown in Fig. 5.3.1. Here the switching function is a rectangular window and is denoted w(t). The window w(t) is of finite time duration. As a result, a T-second section of the input process, say x(t), is presented to the output. The multiplication operation found in the time domain has a convolution counterpart in the frequency domain (i.e., convolution theorem). That is, the output spectrum, say S(f), is given by W(f) * X(f). It can be noted that if W(f) has a Dirac delta distribution [i.e., W(f) = δ(f)], the output spectrum is identical to the input spectrum. However, for this to be true, the period of the rectangular switching function would have to be infinite [i.e., $\mathcal{F}^{-1}[\delta(f)] \longrightarrow 1$ for all t]. This is obviously unrealistic in a practical setting. However, our design goal should be to choose a window function which has a "delta distribution-like" spectrum.

An N-point discrete window function $\{w(n)\}$ has, in general, the DFT which is abstracted in Fig. 5.3.2. It can be seen that the window function has a main lobe of width $k\omega_s/N$, where k is an integer. Ideally, the main lobe is narrow (like a delta distribution) and the side lobes are of low amplitude (like a delta distribution). The side-lobe attenuation can be quantified in terms of the normalized parameter known as the ripple ratio (RR), satisfying

$$RR = 100\left(\frac{\text{maximum side-lobe gain}}{\text{main-lobe gain}}\right) \qquad (5.3.1)$$

The ripple ratio indicates the depth of the side-lobe activity relative to the high gain of the main lobe.

Rectangular Window

A rectangular window is given by

$$w(n) = \begin{cases} 1, & |n| \leq \dfrac{N-1}{2} \\ 0, & \text{otherwise} \end{cases} \qquad (5.3.2)$$

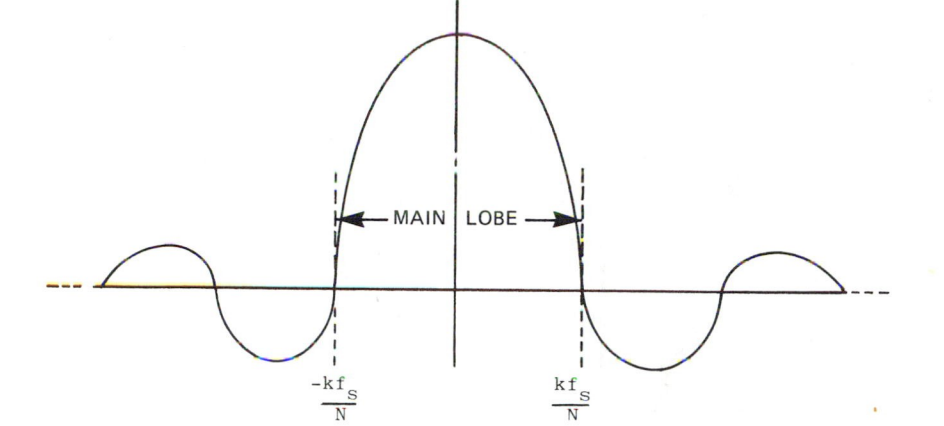

FIG. 5.3.2 Model of a typical magnitude spectrum of a data window.

The spectral shape of the rectangular window is shown in Fig. 5.3.3. As N becomes arbitrarily large, $W(\omega)$ tends to become the idealized delta distribution. However, for N finite, there is distortion due to its finite aperature.

Hamming-Hann Window

A window function may also be defined in terms of

$$w(n) = \begin{cases} a + (1 - a) \cos\left(\dfrac{2\pi n}{N}\right), & |n| \leq \dfrac{N - 1}{2} \\[2ex] 0, & \text{otherwise} \end{cases}$$ (5.3.3)

For $a = 0.54$, a Hamming window is defined. If $a = 0.5$, a Hann window (not Hanning) is defined.

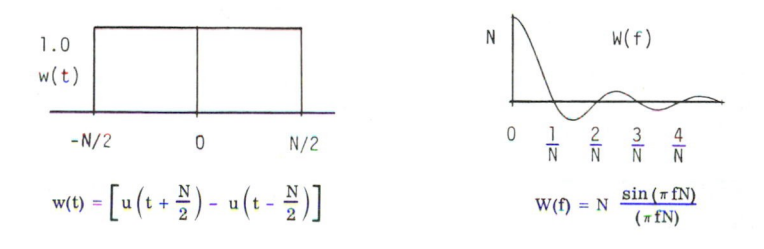

FIG. 5.3.3 Uniform window and its magnitude frequency response.

Blackman Window

The Blackman window is a refinement of the previous windows. It is given by

$$w(n) = \begin{cases} 0.42 + 0.5 \cos\left(\dfrac{2\pi n}{N-1}\right) + 0.08 \cos\left(\dfrac{4\pi n}{N-1}\right), & |n| \leq \dfrac{N-1}{2} \\ 0, & \text{otherwise} \end{cases}$$

(5.3.4)

The four common windows can be compared in terms of their main-lobe width and RRs. This is summarized in Table 5.3.1.

Kaiser Window

The Kaiser window is based on an optimization analysis and results in a window function having the form

$$w(n) = \begin{cases} I_0\left\{ B\left[1 - \left(\dfrac{2n}{N-1}\right)^2 \right]^{1/2} \right\}, & |n| \leq \dfrac{N-1}{2} \\ 0, & \text{otherwise} \end{cases}$$

(5.3.5)

where I_0 represents a zeroth-order Bessel function given by

$$I_0(x) = 1 + \sum_{k=1}^{\infty} \left[\left(\frac{1}{k!} \, \frac{x}{2}\right)^k \right]^2$$

(5.3.6)

The parameter B establishes a trade-off between the main-lobe width and side-lobe attenuation.

EXAMPLE 5.3.1 When applied to an arbitrary time series, windows have a periodicity enhancement ability. Windows can also be applied to a FIR impulse response in order to "smooth" the frequency domain response of the filter. For example, the frequency domain performance of a 31st-order linear phase filter is shown in Fig. 5.3.4. Upon applying a Hamming window to the filter's impulse response, a new filter is derived. The frequency response of the new filter is detailed in Fig. 5.3.5. It can be noted that the passband and stopband behavior of the original filter has been altered. More specifically, passband and stopband ripple have been suppressed. The price paid for this ripple reduction is a reduced slope in the transition band (i.e., selectivity).

TABLE 5.3.1

Type	Main lobe	RR:N = 11	RR:N = 21	RR:N = 31
Rectangular	$4w_s/N$	22.3	22.9	21.8
Hann	$4w_s/N$	2.6	2.7	2.7
Hamming	$4w_s/N$	1.5	.9	.8
Blackman	$6w_s/N$	.09	.12	.12

5.4 COMPARISON OF WINDOWS

By evaluating the DFT of a window function, the main lobe and side lobes
can be analyzed. The DFT of the rectangular, Hamming, Blackman, and
Kaiser windows are summarized in Figs. 5.4.1 through 5.4.4. It can be
noted that the Hamming, Blackman, and Kaiser windows are most "delta-
like." Observe that in Fig. 5.4.4, the effect of parameter a (equivalently,
B in the established notation) effects the passband-stopband trade-offs.
However, it should also be noted that the improved spectral window shape
is purchased at the expense of arithmetic complexity.

EXAMPLE 5.4.1 Window functions are superimposed over a given time
series in order to improve the spectral fidelity of their resulting transform.
To demonstrate this capability, an idealized lowpass filter model will be
used. The ideal filter shown in Fig. 5.4.5 represents a lowpass filter
having a cutoff frequency of 250 Hz with respect to a 1-kHz sample rate.
The inverse Fourier transform of the ideal filter has a known $(\sin x)/x$
form. In a practical FIR sense, this $(\sin x)/x$ curve would be embedded
into the $2N + 1$ coefficients of a 2N-delay finite impulse response filter (see
Fig. 5.2.1). Using $N = 10$, 25, and 50, the DFT of a truncated ideal lowpass filter can be shown to produce the spectral information found in Fig.
5.4.6. The "ripple" found near the edge of the passband is called the Gibbs
phenomenon. Using window functions, this unwanted rippling can be suppressed. For example, rectangular, Hamming, and Kaiser windows were
superimposed over the center of idealized $(\sin x)/x$ impulse response of a
lowpass FIR. The rectangular window function would produce the data
found in Fig. 5.4.6. The Hamming window result is shown in Fig. 5.4.7.
It can be seen that a considerable amount of the unwanted ripple has been
suppressed. Finally, the Kaiser window tests are reported in Fig. 5.4.8.
Here, three values of B (the passband width-side lobe attenuation adjustment
parameter) were used. Observe that for $B = 3$, a steep-skirt filter is derived at the expense of added ripple.

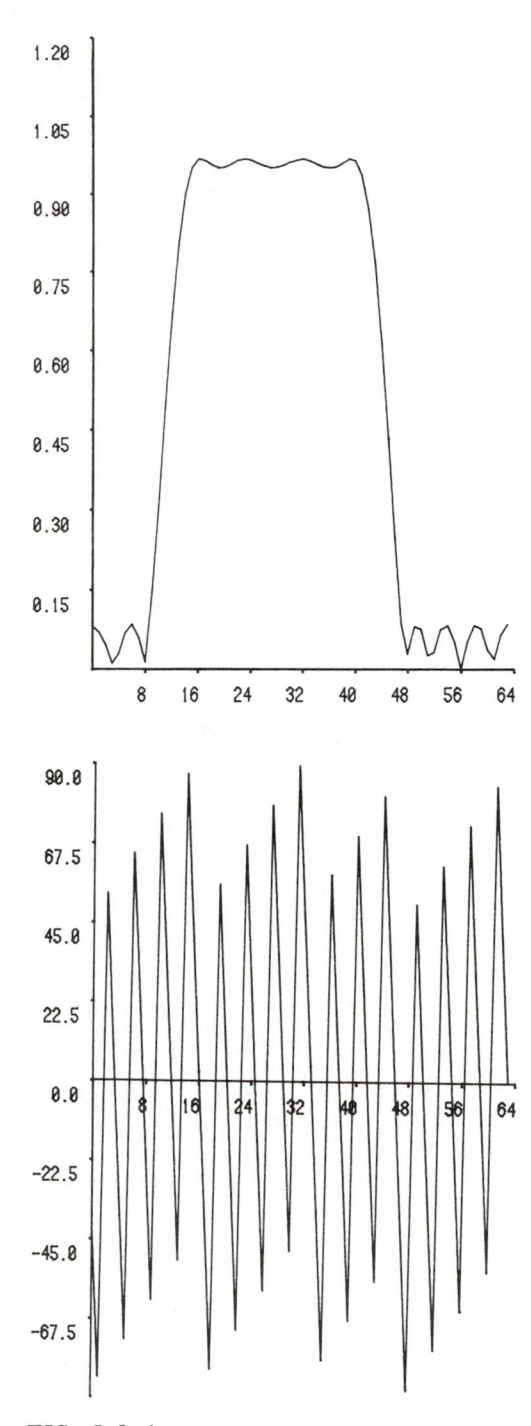

FIG. 5.3.4 Magnitude frequency and phase response of a 31st-order band-pass filter using a rectangular data window.   Sample frequency = 1/128.

142

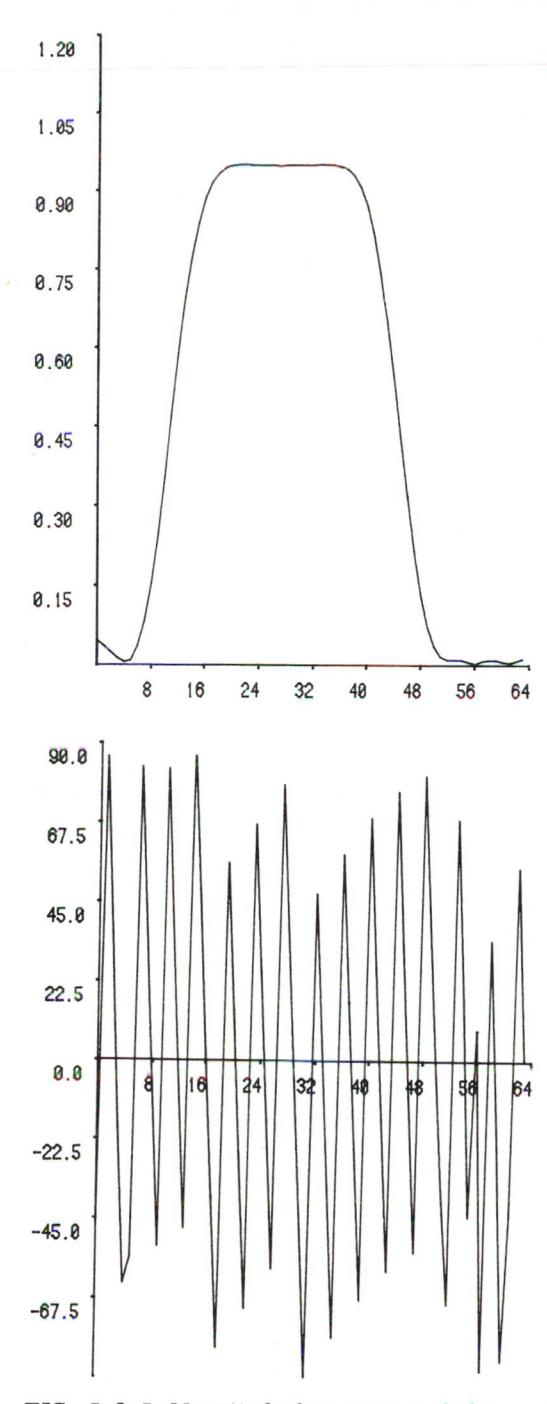

FIG. 5.3.5 Magnitude frequency and phase response of a 31st-order band-pass filter using a Hamming data window. Sample frequency = 1/128.

143

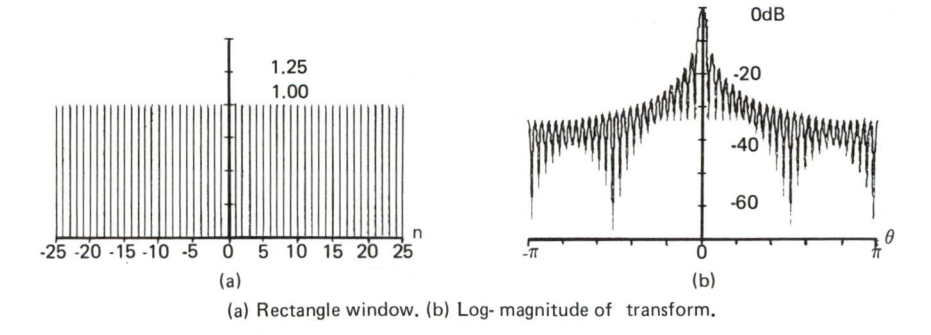

(a) Rectangle window. (b) Log- magnitude of transform.

FIG. 5.4.1 Temporal and spectral profiles of a rectangular data window.
(After Harris, 1978; reprinted with the permission of the IEEE.)

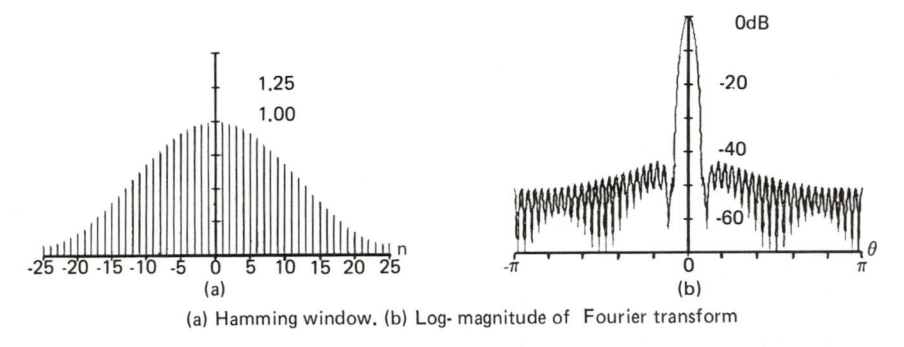

(a) Hamming window. (b) Log- magnitude of Fourier transform

FIG. 5.4.2 Temporal and spectral profiles of a Hamming data window.
(After Harris, 1978; reprinted with the permission of the IEEE.)

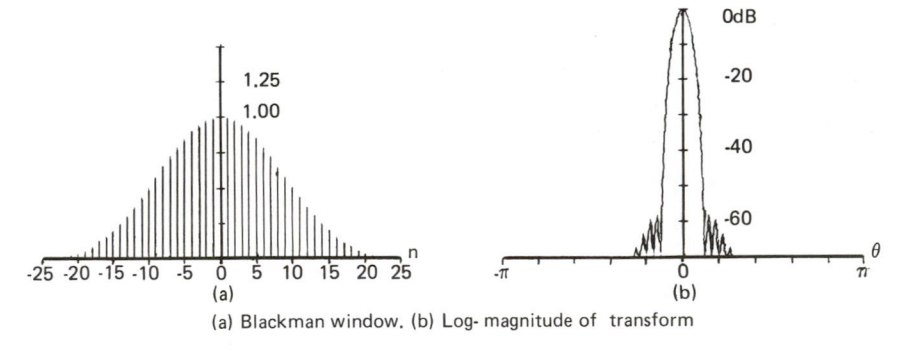

(a) Blackman window. (b) Log- magnitude of transform

FIG. 5.4.3 Temporal and spectral profiles of a Blackman data window.
(After Harris, 1978; reprinted with the permission of the IEEE.)

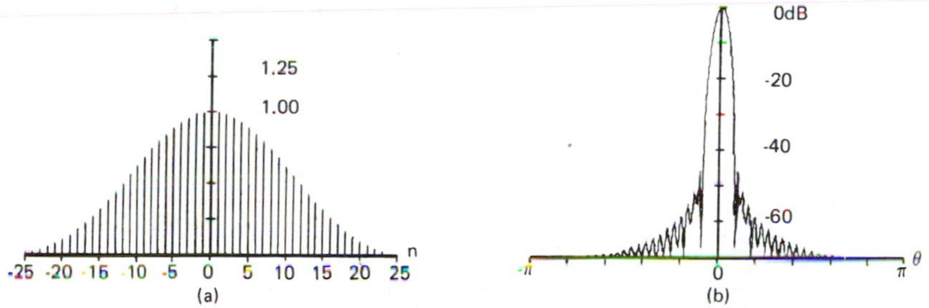

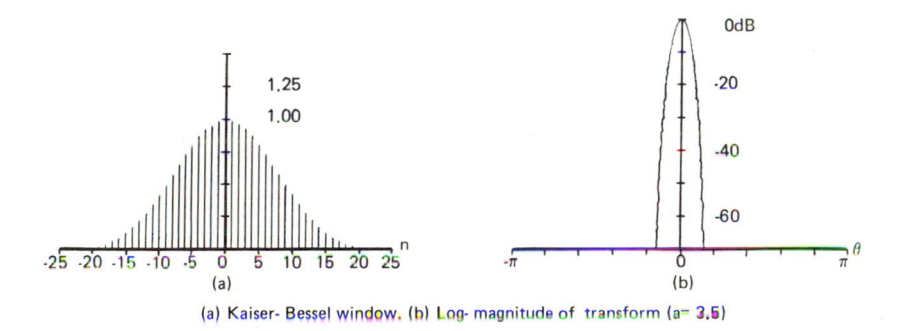

(a) Kaiser- Bessel window. (b) Log- magnitude of transform (a= 2.0)

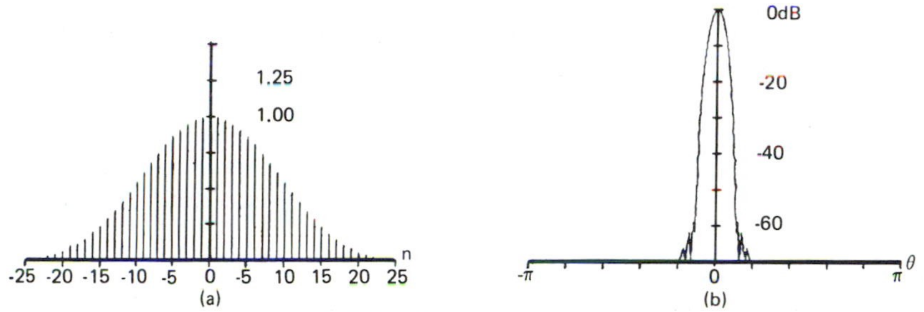

(a) Kaiser- Bessel window. (b) Log- magnitude of transform (a= 2.5)

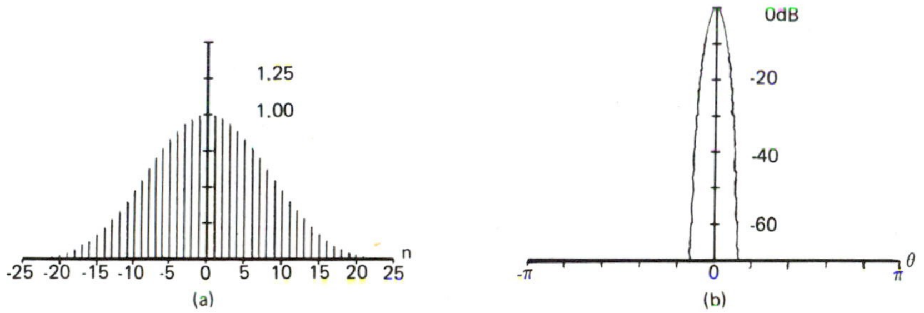

(a) Kaiser- Bessel window. (b) Log- magnitude of transform (a= 3.0)

(a) Kaiser- Bessel window. (b) Log- magnitude of transform (a= 3.5)

FIG. 5.4.4 Temporal and spectral profiles of a Kaiser–Bessel data window. (After Harris, 1978; reprinted with the permission of the IEEE.)

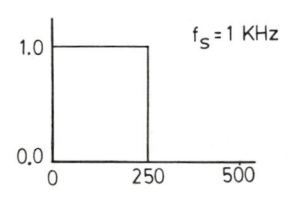

FIG. 5.4.5 Ideal lowpass filter model having a cutoff frequency of 250 Hz.

A FIR may be mechanized in software using a FIFO queue. If realized in a n-bit hardware architecture, multiple n-bit shift registers are required. Execution speed is generally limited by the slowest operation in the signal stream (usually multiplication). It would be a rare (and expensive) situation where N distinct multipliers would be used to mechanize an Nth-order FIR. Instead, one (or a few) multipliers would generally be time-multiplexed

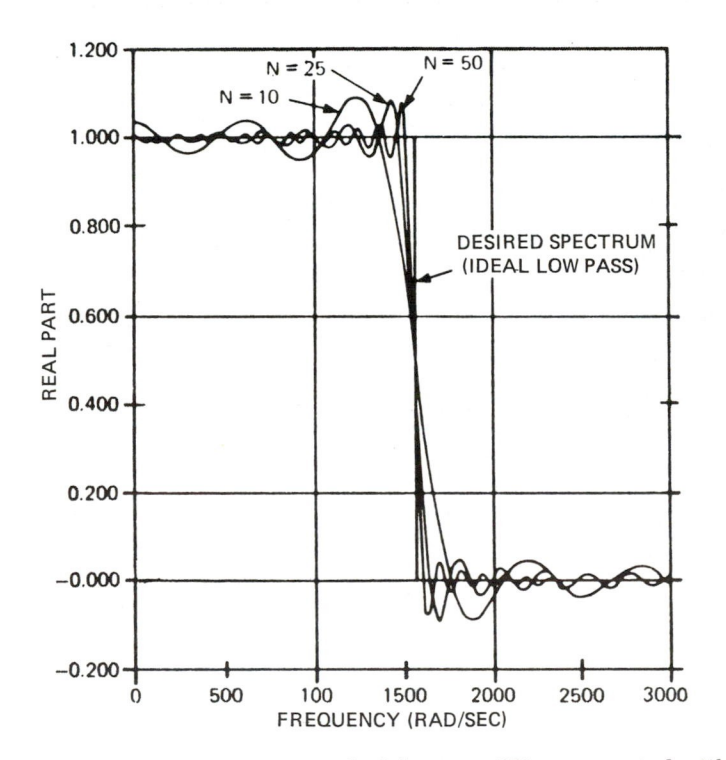

FIG. 5.4.6 Approximate ideal lowpass filter computed with N = 10-, 25-, and 50-point FFTs exhibiting Gibbs phenomena. (After Bass and Leon, 1974; reprinted with the permission of EDN.)

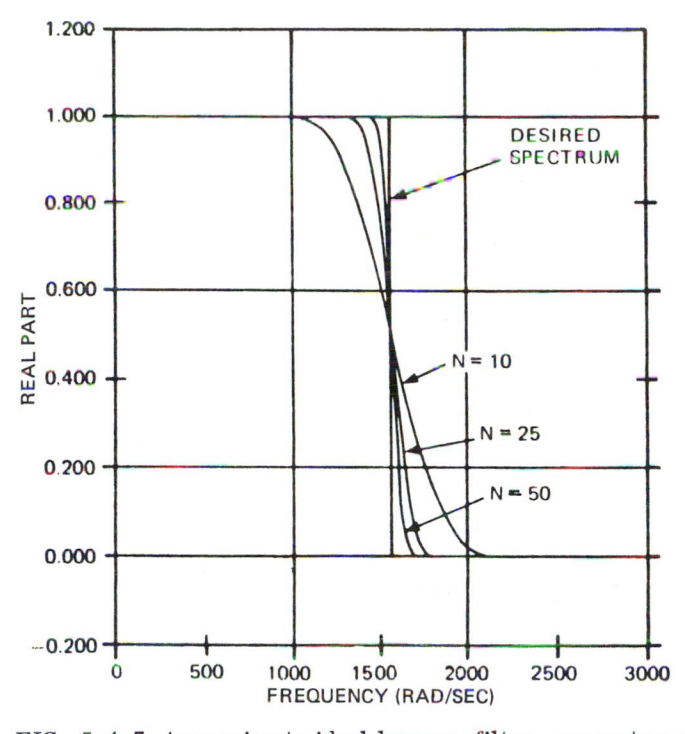

FIG. 5.4.7 Approximate ideal lowpass filter, computer with N = 10, 25-, and 50-point FFTs using a Hamming window. (After Bass and Leon, 1974; reprinted with the permission of EDN.)

into the data stream. The window coefficients would typically be found stored on RAM or ROM. They would be sequentially accessed and presented to the multiplexed multiplier on a delayed sample-by-sample basis.

EXAMPLE 5.4.2 Harris (1978) produced an in-depth study of these basic windows as well as others. Twenty-three windows were tested and summarized in Table 5.4.1. It can be noted that several new high-quality metrics were introduced. One of the more useful of these is the equivalent noise bandwidth (ENBW) parameter, which is given by

$$ \text{ENBW} = \frac{\displaystyle\sum_{n=0}^{N-1} w^2(n)}{\left[\displaystyle\sum_{n-0}^{N-1} w(n)\right]^2} $$

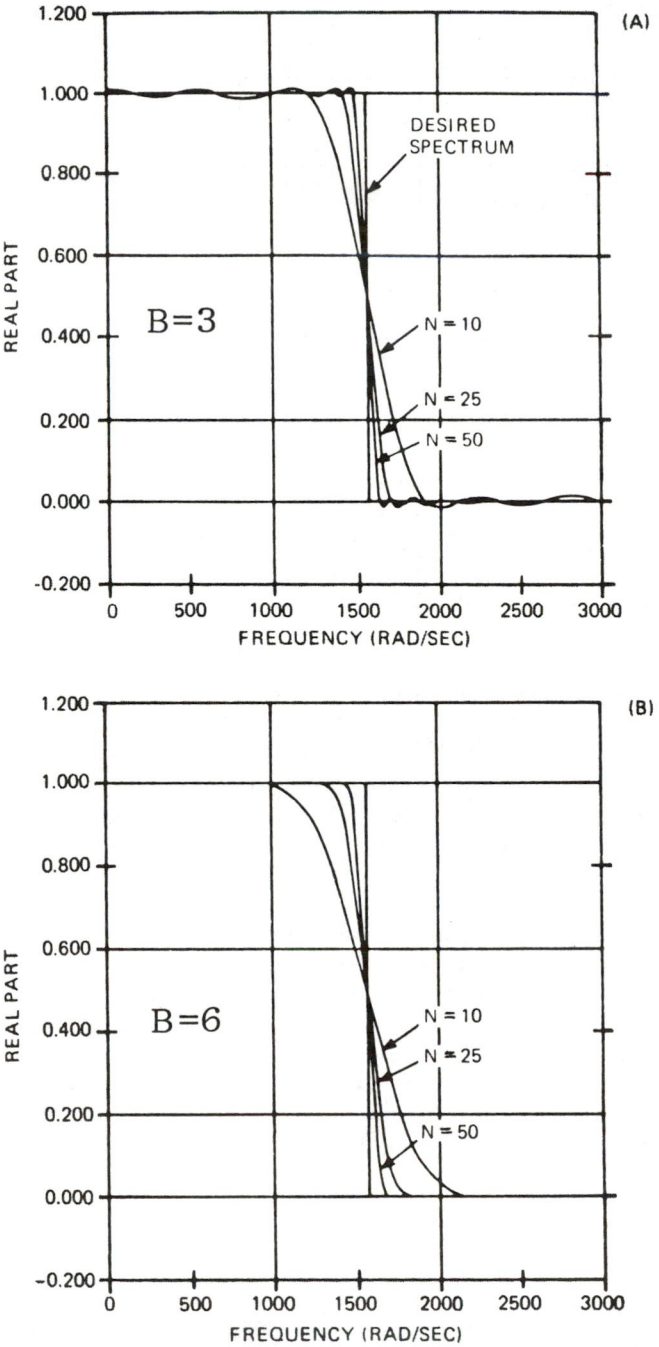

FIG. 5.4.8 Approximate ideal lowpass filter, computed with N = 10-, 25-, and 50-point FFTs using a Bessel window. (After Bass and Leon, 1974; reprinted with the permission of EDN.)

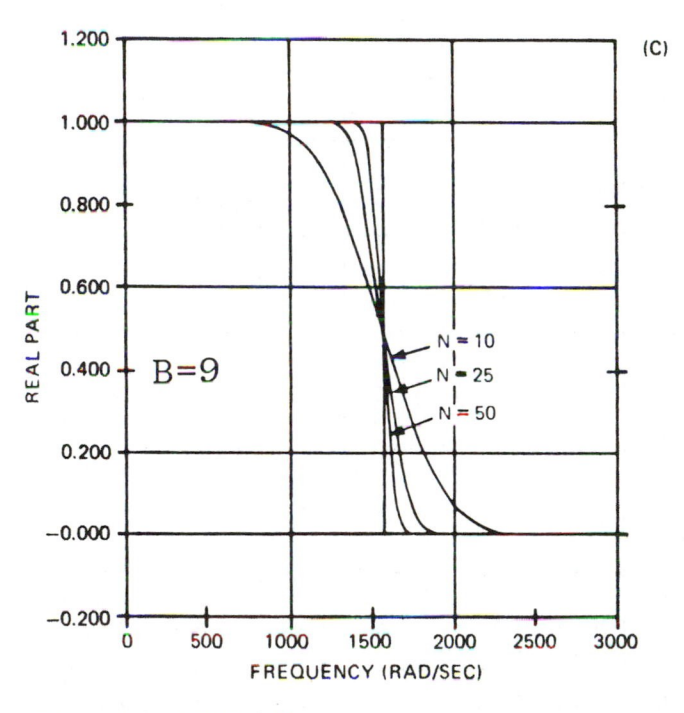

FIG. 5.4.8 (continued)

TABLE 5.4.1 Use of Windows for Harmonic Analysis

WINDOW		HIGHEST SIDE-LOBE LEVEL (dB)	SIDE-LOBE FALL-OFF (dB/OCT)	COHERENT GAIN	EQUIV. NOISE BW (BINS)	3.0-dB BW (BINS)	SCALLOP LOSS (dB)	WORST CASE PROCESS LOSS (dB)	6.0-dB BW (BINS)	OVERLAP CORRELATION (PCNT)	
										75% OL	50% OL
RECTANGLE		-13	-6	1.00	1.00	0.89	3.92	3.92	1.21	75.0	50.0
TRIANGLE		-27	-12	0.50	1.33	1.28	1.82	3.07	1.78	71.9	25.0
COS^a(x) HANNING	α = 1.0	-23	-12	0.64	1.23	1.20	2.10	3.01	1.65	75.5	31.8
	α = 2.0	-32	-18	0.50	1.50	1.44	1.42	3.18	2.00	65.9	16.7
	α = 3.0	-39	-24	0.42	1.73	1.66	1.08	3.47	2.32	56.7	8.5
	α = 4.0	-47	-30	0.38	1.94	1.86	0.86	3.75	2.59	48.6	4.3
HAMMING		-43	-6	0.54	1.36	1.30	1.78	3.10	1.81	70.7	23.5
RIESZ		-21	-12	0.67	1.20	1.16	2.22	3.01	1.59	76.5	34.4
RIEMANN		-26	-12	0.59	1.30	1.26	1.89	3.03	1.74	73.4	27.4
DE LA VALLE-POUSSIN		-53	-24	0.38	1.92	1.82	0.90	3.72	2.55	49.3	5.0
TUKEY	α = 0.25	-14	-18	0.88	1.10	1.01	2.96	3.39	1.38	74.1	44.4
	α = 0.50	-15	-18	0.75	1.22	1.15	2.24	3.11	1.57	72.7	36.4
	α = 0.75	-19	-18	0.63	1.36	1.31	1.73	3.07	1.80	70.5	25.1
BOHMAN		-46	-24	0.41	1.79	1.71	1.02	3.54	2.38	54.5	7.4
POISSON	α = 2.0	-19	-6	0.44	1.30	1.21	2.09	3.23	1.69	69.9	27.8
	α = 3.0	-24	-6	0.32	1.65	1.45	1.46	3.64	2.08	54.8	15.1
	α = 4.0	-31	-6	0.25	2.08	1.75	1.03	4.21	2.58	40.4	7.4
HANNING-POISSON	α = 0.5	-35	-18	0.43	1.61	1.54	1.26	3.33	2.14	61.3	12.6
	α = 1.0	-39	-18	0.38	1.73	1.64	1.11	3.50	2.30	56.0	9.2
	α = 2.0	NONE	-18	0.29	2.02	1.87	0.87	3.94	2.65	44.6	4.7

WINDOW										
CAUCHY	α = 3.0	-31	-6	0.42	1.48	1.34	1.71	3.40	1.90	61.6	20.2
	α = 4.0	-35	-6	0.33	1.76	1.50	1.36	3.83	2.20	48.8	13.2
	α = 5.0	-30	-6	0.28	2.06	1.68	1.13	4.28	2.53	38.3	9.0
GAUSSIAN	α = 2.5	-42	-6	0.51	1.39	1.33	1.69	3.14	1.86	67.7	20.0
	α = 3.0	-55	-6	0.43	1.64	1.55	1.25	3.40	2.18	57.5	10.6
	α = 3.5	-69	-6	0.37	1.90	1.79	0.94	3.73	2.52	47.2	4.9
DOLPH-CHEBYSHEV	α = 2.5	-50	0	0.53	1.39	1.33	1.70	3.12	1.85	69.6	22.3
	α = 3.0	-60	0	0.48	1.51	1.44	1.44	3.23	2.01	64.7	16.3
	α = 3.5	-70	0	0.45	1.62	1.55	1.25	3.35	2.17	60.2	11.9
	α = 4.0	-80	0	0.42	1.73	1.65	1.10	3.48	2.31	55.9	8.7
KAISER-BESSEL	α = 2.0	-46	-6	0.49	1.50	1.43	1.46	3.20	1.99	65.7	16.9
	α = 2.5	-57	-6	0.44	1.65	1.57	1.20	3.38	2.20	59.5	11.2
	α = 3.0	-69	-6	0.40	1.80	1.71	1.02	3.56	2.39	53.9	7.4
	α = 3.5	-82	-6	0.37	1.93	1.83	0.89	3.74	2.57	48.8	4.8
BARCILON-TEMES	α = 3.0	-53	-6	0.47	1.56	1.49	1.34	3.27	2.07	63.0	14.2
	α = 3.5	-58	-6	0.43	1.67	1.59	1.18	3.40	2.23	58.6	10.4
	α = 4.0	-68	-6	0.41	1.77	1.69	1.05	3.52	2.36	54.4	7.6
EXACT BLACKMAN		-51	-6	0.46	1.57	1.52	1.33	3.29	2.13	62.7	14.0
BLACKMAN		-58	-18	0.42	1.73	1.68	1.10	3.47	2.35	56.7	9.0
MINIMUM 3-SAMPLE BLACKMAN-HARRIS		-67	-6	0.42	1.71	1.66	1.13	3.45	1.81	57.2	9.6
*MINIMUM 4-SAMPLE BLACKMAN-HARRIS		-92	-6	0.36	2.00	1.90	0.83	3.85	2.72	46.0	3.8
*61 dB 3-SAMPLE BLACKMAN-HARRIS		-61	-6	0.45	1.61	1.56	1.27	3.34	2.19	61.0	12.6
74 dB 4-SAMPLE BLACKMAN-HARRIS		-74	-6	0.40	1.79	1.74	1.03	3.56	2.44	53.9	7.4
4-SAMPLE KAISER-BESSEL	α = 3.0	-69	-6	0.40	1.80	1.74	1.02	3.56	2.44	53.9	7.4

*REFERENCE POINTS FOR DATA ON FIGURE 12 — NO FIGURES TO MATCH THESE WINDOWS.

After Harris, 1978; reprinted with permission of the IEEE.

EXAMPLE 5.4.3 As a qualitative statement of the subject of windows, the results of a computer experiment will be discussed. Data will be processed and analyzed with respect to rectangular, Hamming, and Kaiser windows. Sample source codes for the later windows can be found in Fig. 5.4.9. The test signal considered is a cosine wave that completes five periods of oscillation in 64 samples (see Fig. 5.4.10). The spectrum of the rectangularly windowed time series has the spectrum shown in Fig. 5.4.11. It can be noted that this spectrum is essentially that of a pure tone. The Hamming and Kaiser windowed data can be found in Figs. 5.4.12 and 5.4.13. It can be noted that these time series are distorted versions of the original. This distortion is echoed in the frequency domain as seen in Figs. 5.4.14 and 5.4.15. As a result, one notes that applying a sophisticated window to a periodic time series is counterproductive. However, since most signals

```
C
C SUBROUTINE TO GENERATE HAMMING WINDOW AND MULTIPLY (REAL)
C
C ARGUMENTS:
C              NPTS-NUMBER OF DATA POINTS IN BLOCK
C              XFORM-DATA BLOCK TO BE WINDOWED
C
       SUBROUTINE HAM(NPTS,XFORM)
       DIMENSION XFORM(NPTS)
       DO 100 I=1,NPTS
100    XFORM(I)=XFORM(I)*(0.54-0.46*COS(2.0*3.1417*(I-1)/NPTS))
       RETURN
       END


C
C KAISER WINDOW ROUTINE
C
C ROUTINE GENERATES KAISER WINDOW AND WINDOWS DATA BLOCK
C
C ARGUMENTS:
C              NPTS-NUMBER OF DATA POINTS IN BLOCK
C              XFORM-DATA BLOCK TO BE WINDOWED (REAL)
C              BETA-BETA, ARGUMENT TO MODIFIED BESSEL FUNCTION
C
       SUBROUTINE KAISER(NPTS,XFORM,BETA)
       DIMENSION XFORM(NPTS)
       TWOPI=3.141592*2.0
       NN=NPTS/2
       NNP=NN+1
       CALL IO(BETA*TWOPI/2.0,XD)
       DO 100 I=1,NN
         X=(TWOPI/2.0)**2.0-(I*TWOPI/NPTS)**2.0
       CALL IO(SQRT(X)*BETA,XN)
       XXF=XN/XD
       XFORM(NNP-I)=XFORM(NNP-I)*XN/XD
       XFORM(NN+I)=XFORM(NN+I)*XN/XD
100    CONTINUE
       RETURN
       END
```

FIG. 5.4.9 FORTRAN source code for generating Hamming and Kaiser data windows.

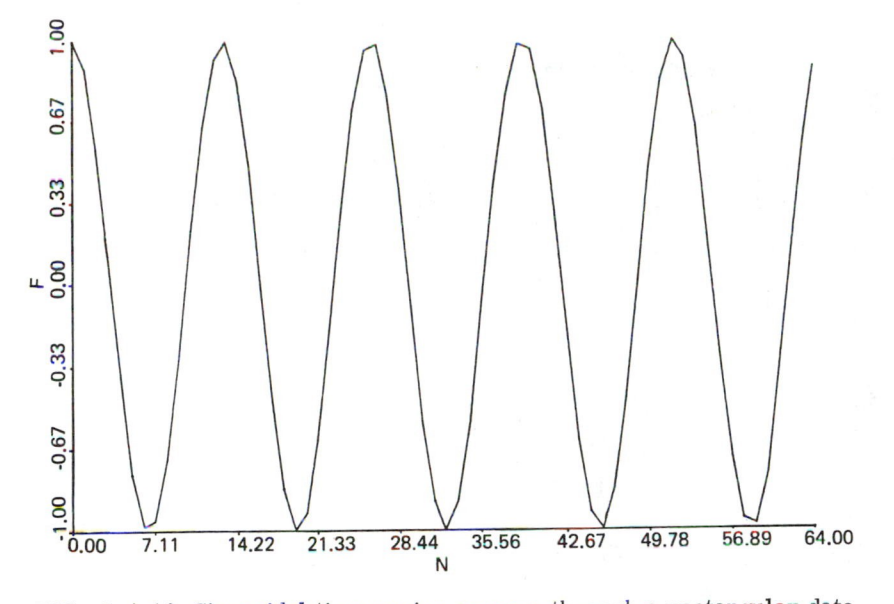

FIG. 5.4.10 Sinusoidal time series as seen through a rectangular data window.

encountered in practice are aperiodic, the use of a window is strongly encouraged.

EXAMPLE 5.4.4 A slightly more detailed experiment was performed using the rectangular, Hamming, triangular, Kaiser, and Chebyshev windows. The triangular and Chebyshev windows where not explicitly studied in this section but are summarized in Table 5.4.1. The unwindowed filter models used in the experiment were ideal lowpass, bandpass, and highpass filters of order N. The interpretation of the results, found in Figs. 5.4.16 through 5.4.22, should be self-evident in light of the analysis provided in Example 5.4.3.

5.5 FREQUENCY SAMPLING FILTERS

Classical time-series analysis places a strong emphasis on interpolation formulas. Interpolation routines, such as those due to Shannon and Fourier, were discussed previously. Shannon's remarkable discovery provides us with a method whereby the temporal profile of a bandlimited signal can be reconstructed from its sample values. Even if the periodically sampled signal is not bandlimited, one can at least find some comfort in the

ONE-SIDED SPECTRUM

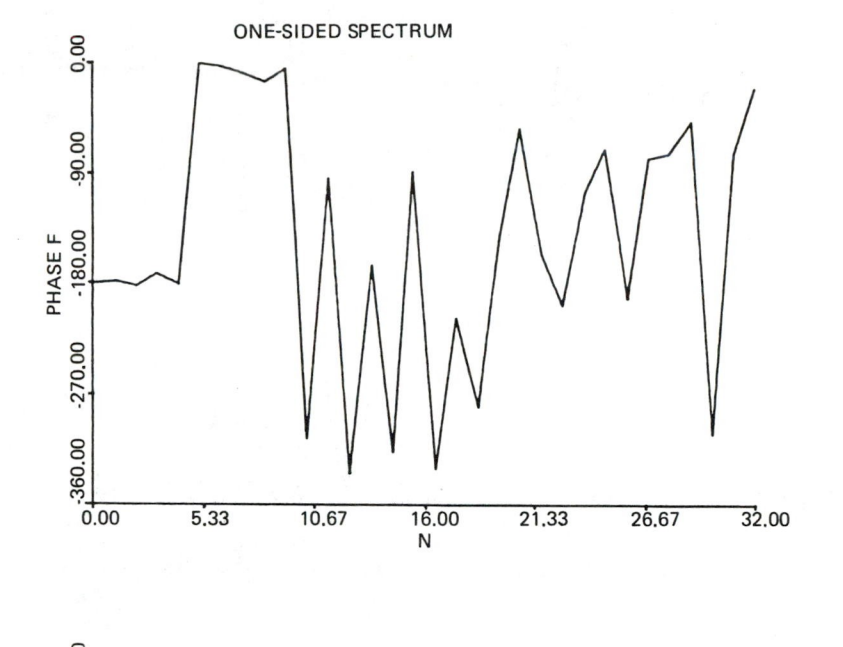

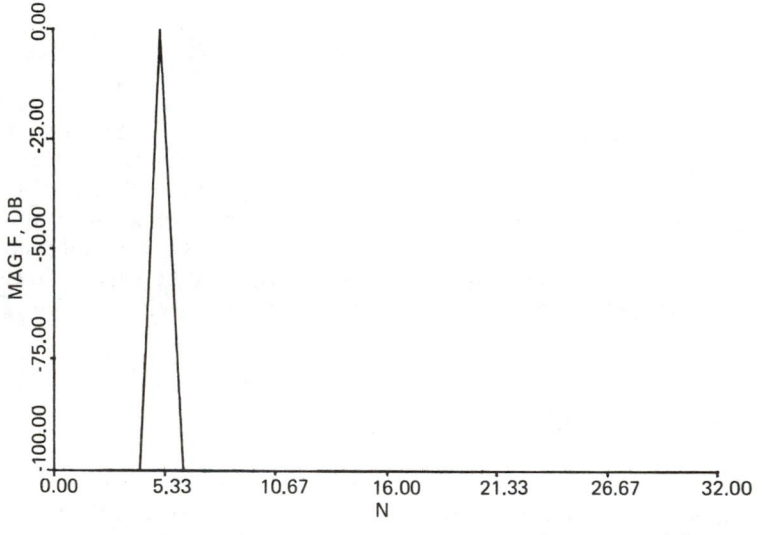

FIG. 5.4.11 Magnitude and phase spectrum of a sinusoidal time series derived using a FFT and a rectangular window.

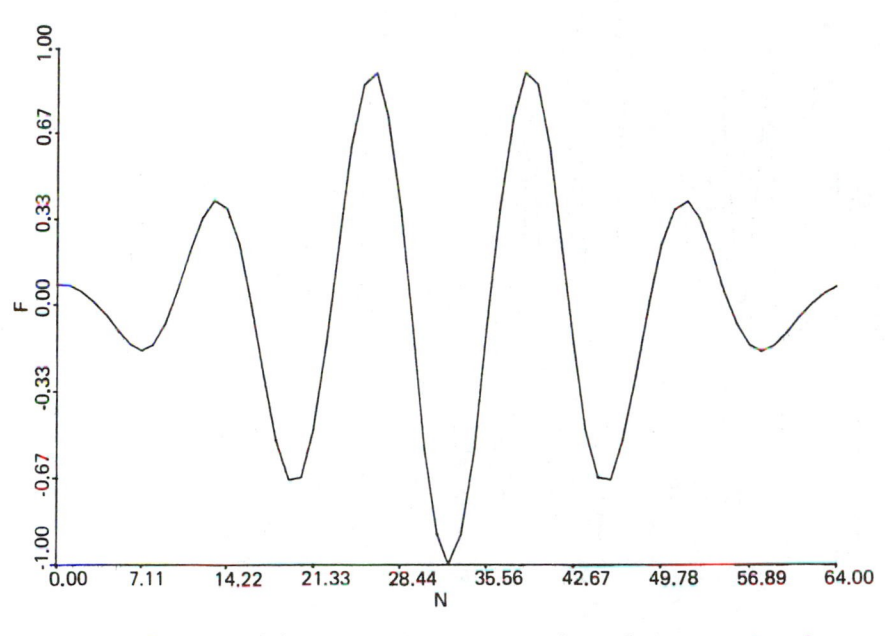

FIG. 5.4.12 Sinusoidal time series as seen through a Hamming data window.

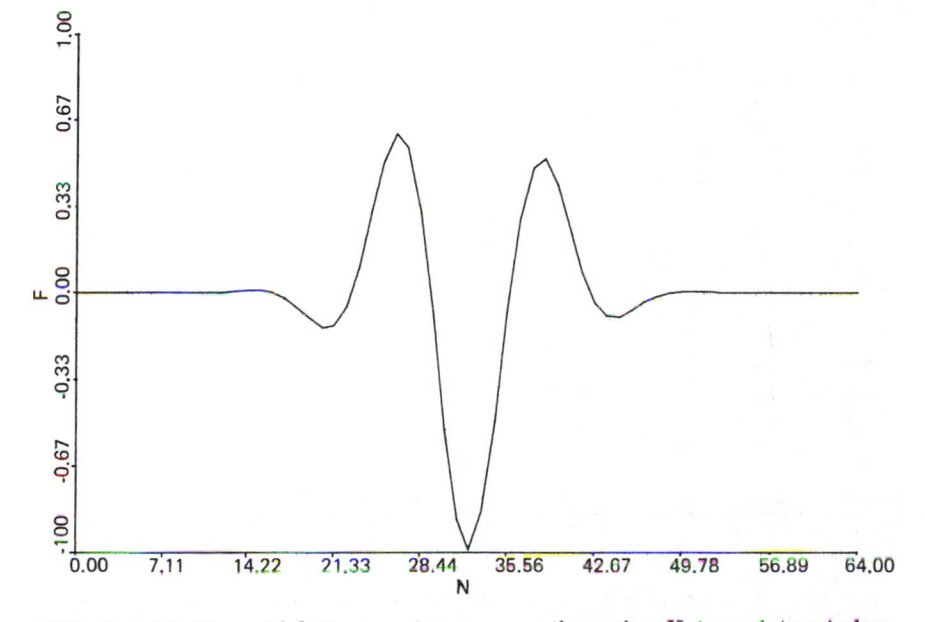

FIG. 5.4.13 Sinusoidal time series as seen through a Kaiser data window. Beta = 0.60.

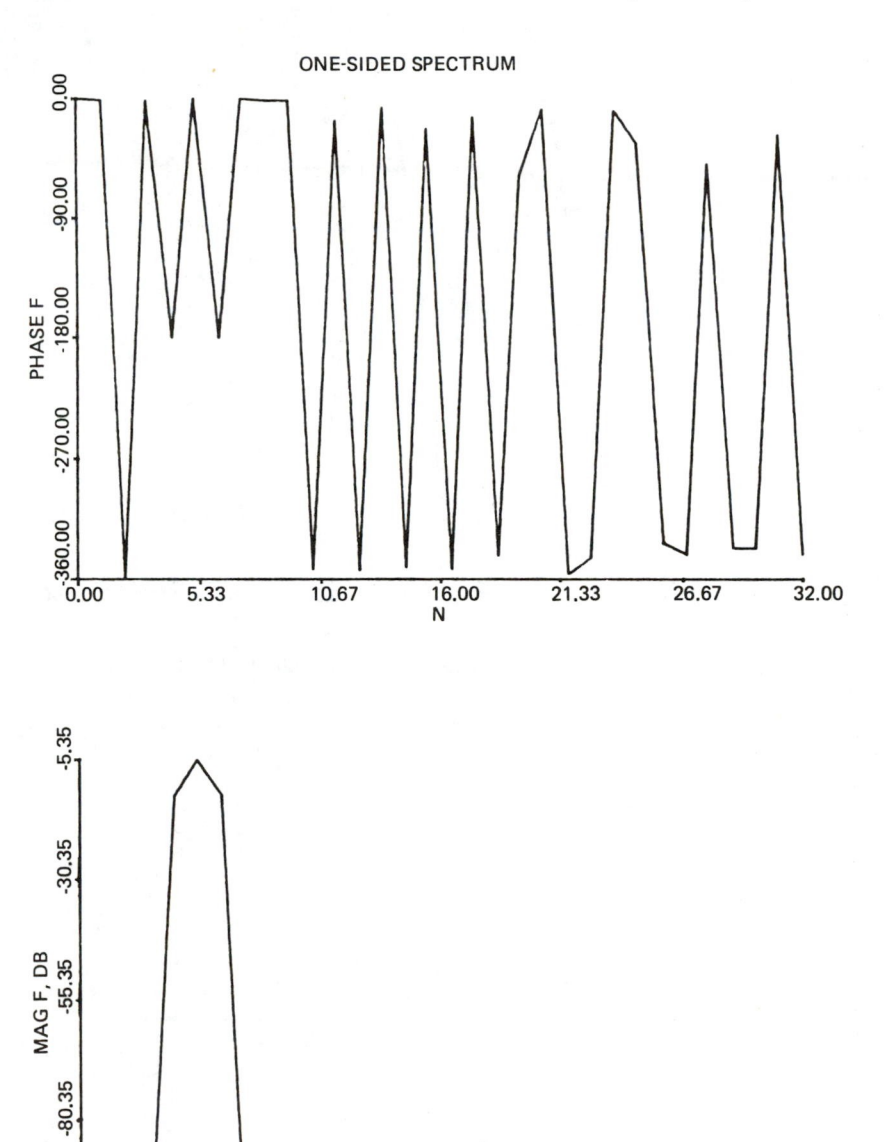

FIG. 5.4.14 Magnitude and phase spectrum of a sinusoidal time series derived using a FFT and a Hamming data window.

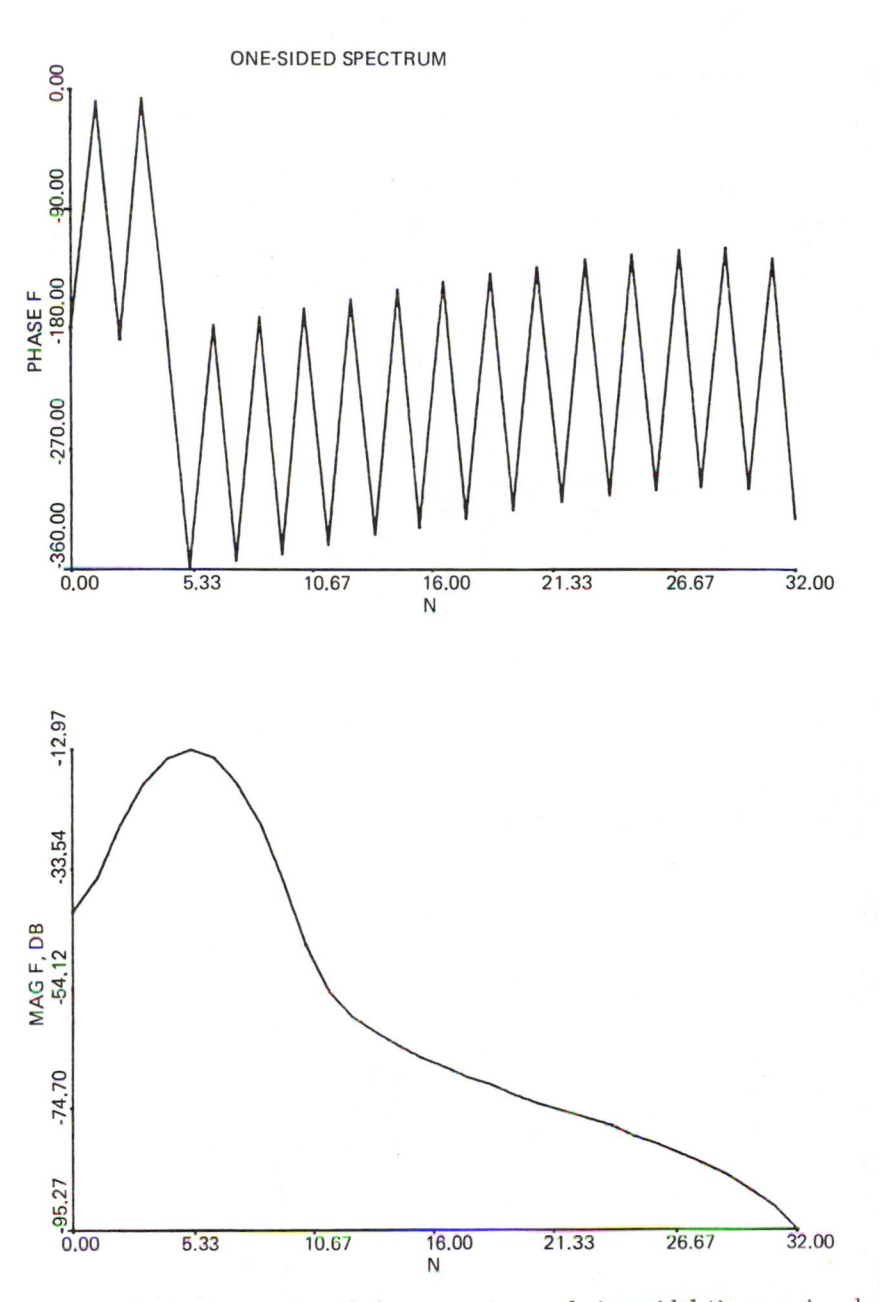

FIG. 5.4.15 Magnitude and phase spectrum of sinusoidal time series de‐
rived using a FFT and a Kaiser data window.

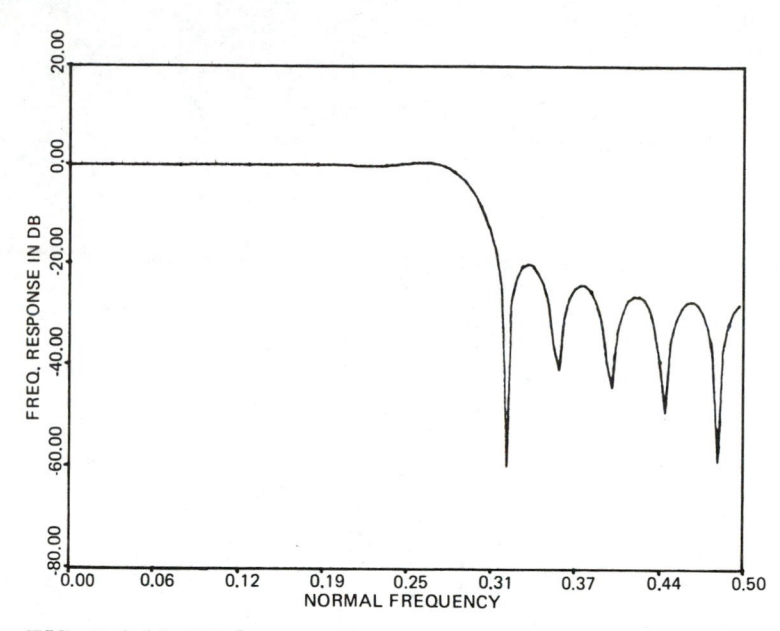

FIG. 5.4.16 FIR lowpass filter magnitude response using a FFT and a rectangular data window. Cutoff frequency = 0.3.

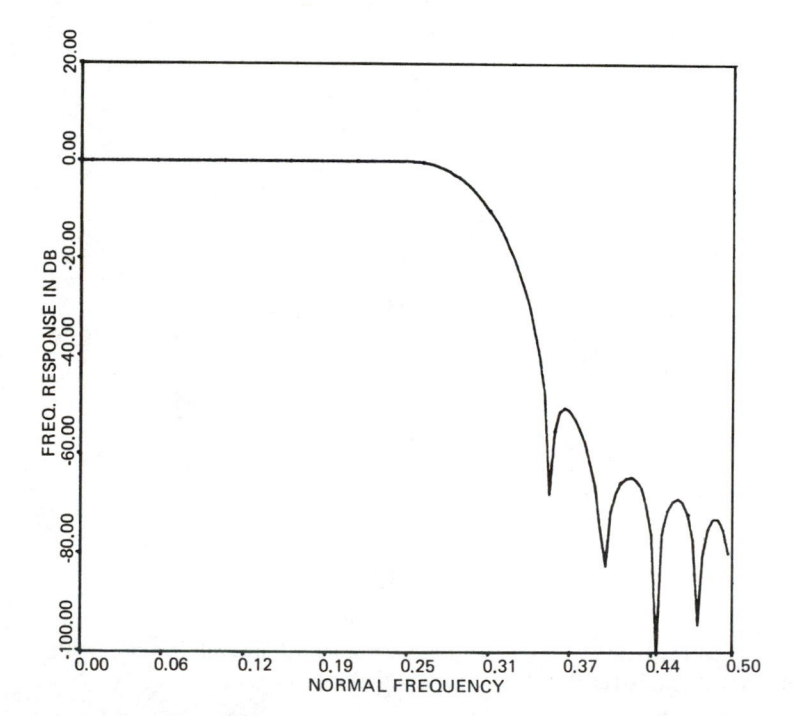

FIG. 5.4.17 FIR lowpass filter magnitude response using a FFT and a Hamming data window. Cutoff frequency = 0.3, N = 30.

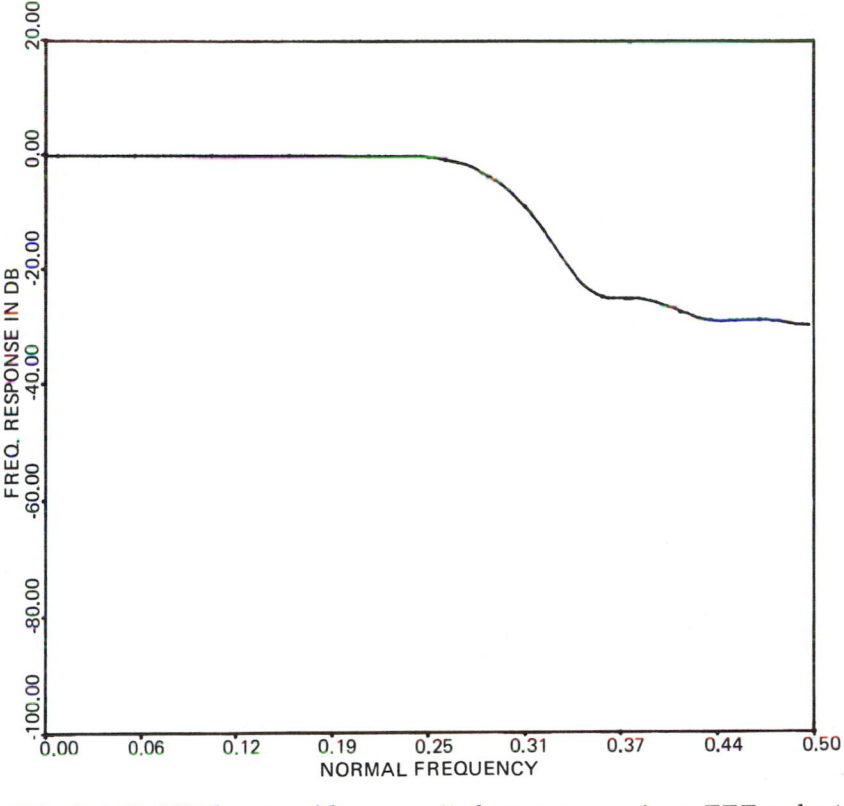

FIG. 5.4.18 FIR lowpass filter magnitude response using a FFT and a triangular data window. Cutoff frequency = 0.3, N = 25.

knowledge that the interpolated signal will agree with the original signal at least at the sample instances. The fundamental work of Shannon can be used to design interpolating filters as well. This will now be investigated.

Consider $\{x(n)\}$ to be an aperiodic time series of length N and $\{x_p(n)\}$ to be the periodic extension of $\{x(n)\}$. That is, $x_p(j) = x(j \mod N)$. The discrete Fourier transform of a periodic time series takes the form

$$X_p(k) = \sum_{n=0}^{N-1} x_p(n) W_N^{kn}$$

$$x_p(n) = \frac{1}{N} \sum_{k=0}^{N-1} X_p(k) W_N^{-kn} \qquad (5.5.1)$$

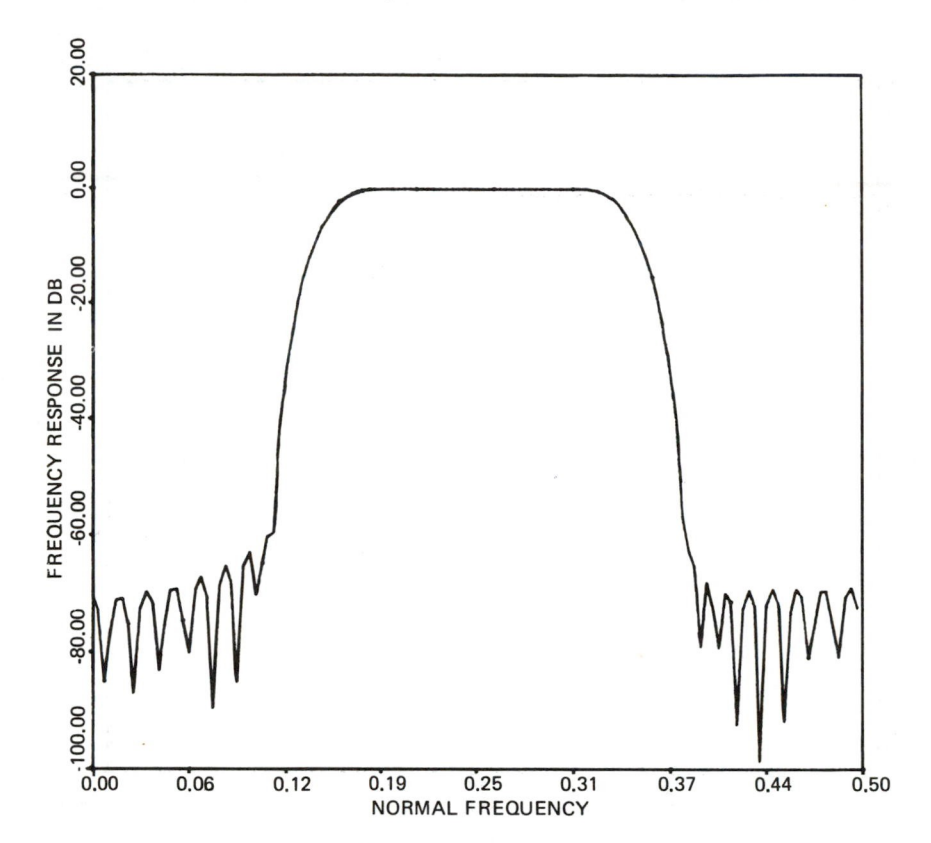

FIG. 5.4.19 FIR bandpass filter magnitude response using a FFT and a Kaiser data window. Cutoff frequencies = 0.15, 0.35; N = 56.

where $W_N = \exp(-j2\pi/N)$. Suppose that the z transform of x(n) is X(z). After some algebraic manipulation, X(z) can be written as

$$X(z) = \sum_{n=0}^{N-1} x(n)z^{-n} = \sum_{n=0}^{N-1} \frac{1}{N}\left[\sum_{n=0}^{N-1} X_p(k)W_N^{-nk}\right]z^{-n}$$

$$= \sum_{k=0}^{N-1} \frac{X(k)}{N} \sum_{n=0}^{N-1}\left(W_N^{-k}z^{-1}\right)^n$$

$$= \frac{1-z^{-N}}{N} \sum_{k=0}^{N-1} \frac{X_p(k)}{1-W_N^{-k}z^{-1}} \qquad (5.5.2)$$

Evaluating formula (5.5.2) along the periphery of the unit circle [i.e., $z = \exp(j\omega)$], one obtains

$$X[\exp(j\omega)] = \sum_{k=0}^{N-1} X_p(k)\phi\left[\omega\left(\frac{N-1}{2}\right)\right] \tag{5.5.3}$$

where the weighting (interpolated) factor $\phi(\omega)$ satisfies

$$\phi(\omega) = \frac{\sin(\omega N/2)}{N \sin(\omega/2)} \exp\left[\frac{-j\omega(N-1)}{2}\right] \tag{5.5.4}$$

The interpolation formula (5.5.4) allows the frequency response over a continuum of frequencies to be synthesized from the N equally spaced

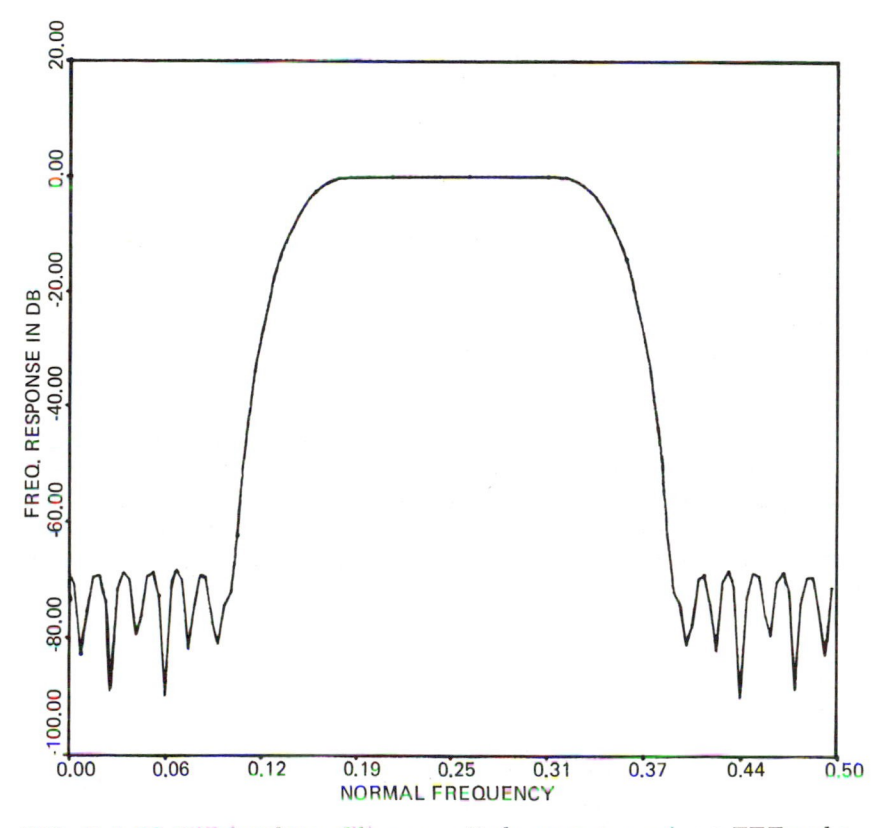

FIG. 5.4.20 FIR bandpass filter magnitude response using a FFT and a Chebyshev data window. Cutoff frequencies = 0.15, 0.35; N = 55.

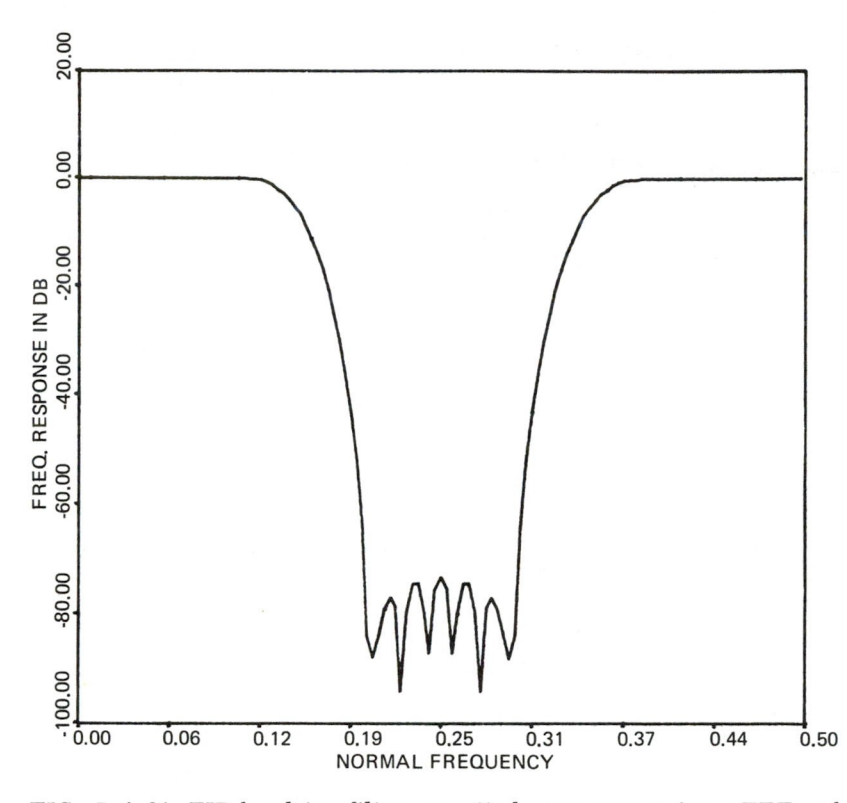

FIG. 5.4.21 FIR bandstop filter magnitude response using a FFT and a Chebyshev data window. Cutoff frequencies = 0.15, 0.35; N = 51.

harmonic samples of $DFT[x_p(n)]$. Observe that $X[\exp(j\omega)]$ agrees with $X_p(\omega)$ whenever $\omega = (2\pi/N)k$. The implication is that a filter with a given frequency response can be synthesized from harmonically related samples.

The <u>frequency sampling approximation</u> method exploits this philosophy. Observe that the kernel of Eq. (5.5.2) can be rewritten in terms of a periodic transfer function, say $H_p(z)$, as follows:

$$\sum_{n=0}^{N-1} x_p(n)z^{-1} = \sum_{k=0}^{N-1} \frac{1}{N}\left[\sum_{n=0}^{N-1} X_p(k)W_N^{-kn}z^{-n}\right]$$

$$= \frac{1}{N}\sum_{k=0}^{N-1} \frac{X_p(k)}{1 - W_N^{-k}z^{-1}} \qquad (5.5.6)$$

Therefore, Eq. (5.5.2) can be expressed as

$$H(z) = (1 - z^{-n})H_p(z); \quad h(n) = [u(nT) - u(nT - NT)]h_p(n) \quad (5.5.7)$$

where $u(\cdot)$ is a unit step function. The effect of all this is one of removing a finite block of N samples from the N-periodic time series $\{x_p(n)\}$. Simply stated, in the frequency sampling method, one specifies a desired frequency response, say $X_p(k)$, which in turn is converted into a periodic time series $\{x_p(n)\}$ using an inverse DFT. Finally, $H_p(z)$ [equivalently, $h_p(n)$] is realized using the architecture shown in Fig. 5.5.1. Using composite filters

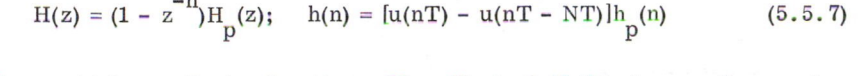

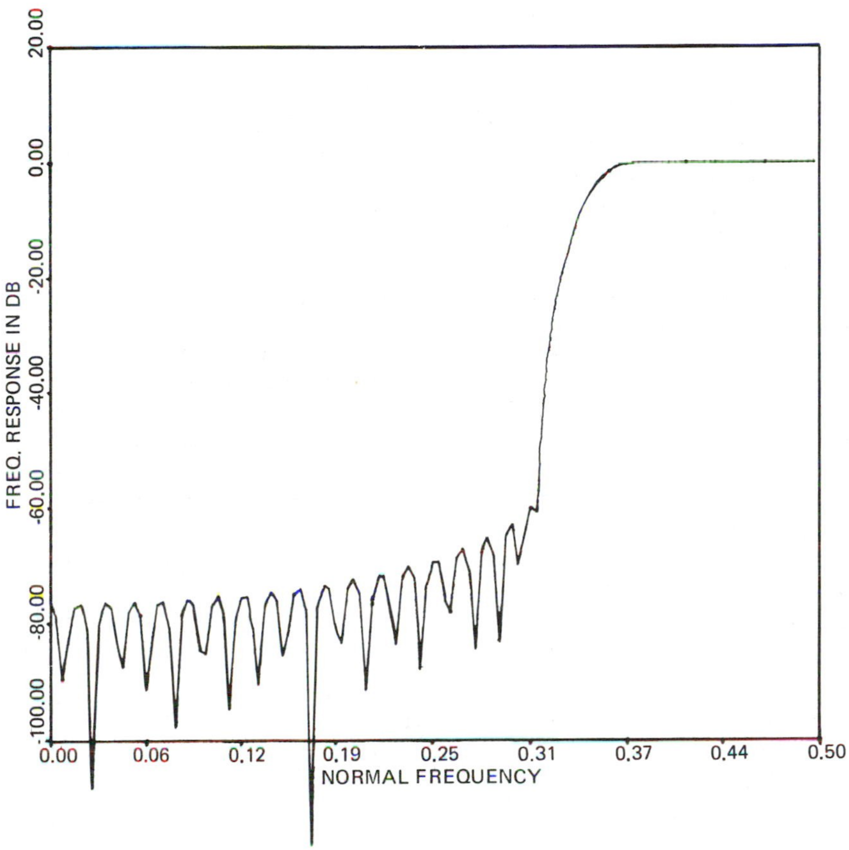

FIG. 5.4.22 FIR highpass filter magnitude response using a FFT and a Kaiser data window. Cutoff frequency = 0.35, N = 55.

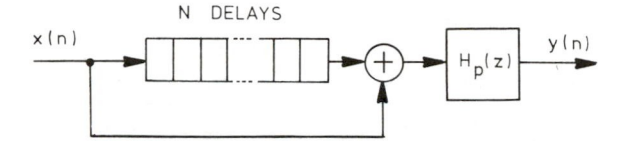

FIG. 5.5.1 Architecture for a frequency-sampling filter.

and the architecture shown in Fig. 5.5.2, the transfer function H(z) can be realized, where

$$H(z) = \frac{1 - z^{-N}}{N} \sum_{n=0}^{N-1} \frac{X(j2\pi n/N)}{1 - W_N^{-n} z^{-1}} \tag{5.5.8}$$

It should be apparent that the N-point interpolated output time-series samples are weighted by the IDFT coefficients $X(k)/N$. It is also apparent that if N is large, the frequency sampling filter becomes complex from a hardware standpoint. However, several modifications have been suggested which permit a more attractive hardware realization of this filter.

It is interesting to note that the topmost signal path found in Fig. 5.5.2 is a simple comb filter. Recall that a comb filter provides the user with a linearly weighted sum of N contiguous sample values. As a result, the frequency domain response, from dc to the Nyquist frequency, is decidedly lowpass in nature. As N increases, the resulting comb filter becomes more intensely lowpass. These thoughts can be extended to the architecture found in Fig. 5.5.2. The jth signal path can now be seen to cancel one of the N zeros (uniformly distributed about the periphery of the unit circle) with a pole at $z_i = W_N^i = \exp(j2\pi i/N)$. As a result, the spectral shape originally associated with the lowpass comb filter is now centered about ω_i.

A variation of this scheme is called the <u>modified comb filter</u> and is characterized by the transfer function

$$H(z) = 1 - \exp(-NaT)z^{-N} \tag{5.5.9}$$

For a = 0, the zeros of $H(z) = 1 - z^{-N}$ can be seen to be distributed about the unit circle. For a less than unity, the zeros are found interior to the unit circle. Nevertheless, Eq. (5.5.9) can be expressed in terms of the difference equation

$$y(n) = x(n) - \exp(-aNT)x(n - N) \tag{5.5.10}$$

In the analysis of the frequency sampling filter, a bank of filters having zero cancellations at $z_i = W_N^i$ were paralleled. A more efficient structure can be derived by grouping conjugate pairs of zeros together. This produces the elemental filter which is given by

$$h_i(n) = \exp(-aT)\ \cos(n\omega_i T)$$

$$H_i(z) = \frac{1 - \exp(-aT)\ \cos(\omega_i T)z^{-1}}{1 - 2\exp(-aT)\ \cos(\omega_i T)z^{-1} + z^{-2}} \qquad (5.5.11)$$

This filter possesses poles at $z = W_N^{\pm i}$. In practice, it may be difficult to achieve exact pole-zero cancellation. However, this is much less a problem in the digital arena than it is in analog designs. Nevertheless, due to roundoff errors and the like, there may exist a mismatch between poles and zeros. To overcome this problem, the stabilizing term $\exp(-aT)$ is chosen to have a value bounded below unity, say $(1 - 2^{-S}) < 1$. This choice will ensure that the poles of the elemental filter are stable. Therefore, any residual component found in the output response due to a pole-zero mismatch will decay to zero. Experimenters have concluded that satisfactory filtering can be achieved for s residing somewhere between 12 and 27.

An interesting property of the composite filter structure is the phase difference that exists between adjacent data paths. The phase difference found between the data path including the pole-zero cancellation at $z = W_N^r$ and $z = W_N^{r+1}$ is π for $\omega_r < \omega < \omega_{r+1}$ and zero elsewhere. Gold and Rader

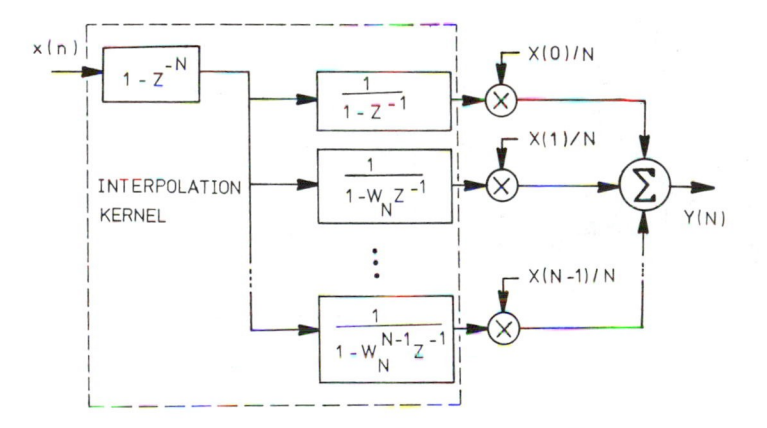

FIG. 5.5.2 Architecture for a frequency-sampling filter.

(1969) have used this principle to design banks of FIR filters by alternating
the signs associated with each adjacent frequency-sampled filter coefficient.
This can best be dramatized in the context of the following example.

EXAMPLE 5.5.1 Suppose that it is desired to design a filter having cutoff
frequencies located at 300 and 700 Hz using the frequency-sampling method.
It shall be further assumed that the gain of the filter at these critical frequencies is to be -3 dB with respect to the midband gain. The design
objective is shown in Fig. 5.5.3. For the purpose of design, it will be
assumed that the elemental filters are to be located on 100-Hz centers. If
the sample period is chosen to be T = 80 μsec, it follows that the Nyquist
frequency is 6.25 kHz. Therefore, with respect to this parameterization,
it can be readily seen that to cover the frequency range [0, 6250] on 100-Hz

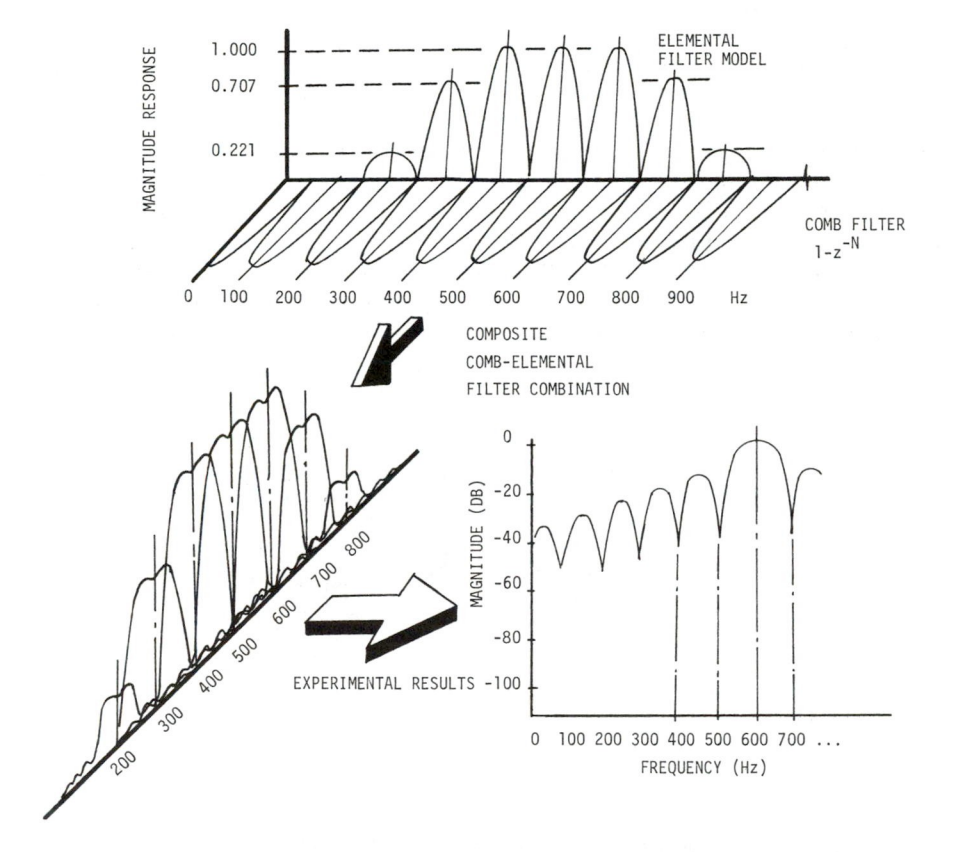

FIG. 5.5.3 Design model for a frequency-sampling bandpass filter and
coefficients.

centers, N must be chosen to be 125 (i.e., NT = 1/100). In Gold and Rader's analysis of this problem, the following system of frequency-sampling
coefficients were chosen (remember the alternating sign requirement):

$a_0 = 0.00$ $a_4 = 1.000$ $a_8 = 0.221$
$a_1 = 0.00$ $a_5 = 1.000$ $a_9 = 0.000$
$a_2 = 0.221$ $a_6 = 1.000$ $a_{10} = 0.000$
$a_3 = -0.707$ $a_7 = -0.707$.   .

The stopband transition gain of 0.221 was experimentally determined to
provide good out-of-band performance. The architected filter is found in
Fig. 5.5.4. It is interesting to note the presence of the stabilizing gain
$\exp(-aT) = 1 - 2^{-26}$, found in the comb section. The interaction between
the comb and elemental filters is motivated in Fig. 5.5.4. Finally, the
frequency response of the frequency-sampling filter is found in Fig. 5.5.5.

The frequency-sampling method can make direct use of the FFT.
Therefore, the design of frequency-sampling filters can be automated.
Another automated design procedure has been developed using Chebyshev
approximating polynomials. This method will be shown to be well suited to
the problem of designing linear-phase-shift FIR filters.

5.6. WEIGHTED CHEBYSHEV METHOD

The weighted Chebyshev method can be used to design optimal linear phase
filters. As in all optimization problems, the criterion of optimality must
be stated. In this method it is assumed that H(exp(jw)) represents the
spectral image of the desired filter independent of any side constraints.
However, there are always side constraints imposed on the design of a
system. For example, the maximal order of a realized filter represents a
practical side constraint. Subject of a system of side constraints, let
$H_N(\exp(j\omega))$ represent the transfer function of the realized filter. Ideally,
one would like to minimize the difference between the ideal and realized
transfer functions in some optimal fashion. An often-used criterion of
optimality is the least-squares (i.e., ℓ_2) metric. However, the historical
problem associated with the ℓ_2 norm is that it will tolerate large localized
error as long as the average squared error is minimal. This philosophy
can result in a filter design that possesses unacceptable errors at certain
critical frequencies.

A sometimes superior criterion of optimality is the <u>minimax</u> norm,
given by

$$| E_N(\exp(j\omega)) | = \min [\max E_N(\exp(j\omega))] \qquad (5.6.1)$$

where $E_N(\exp(j\omega)) = H(\exp(j\omega)) - H_N(\exp(j\omega))$ for a given N. This particular
error metric will minimize the worst-case error. In general, this method

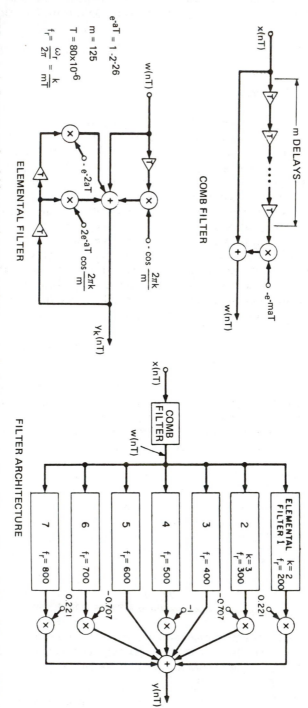

FIG. 5.5.4 System and subfilter architecture of the example frequency sampling filter. (From Gold and Rader, 1969.)

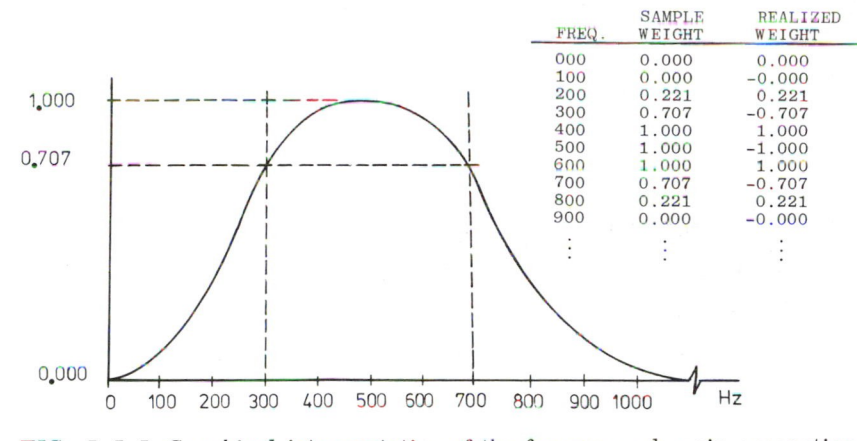

FREQ.	SAMPLE WEIGHT	REALIZED WEIGHT
000	0.000	0.000
100	0.000	-0.000
200	0.221	0.221
300	0.707	-0.707
400	1.000	1.000
500	1.000	-1.000
600	1.000	1.000
700	0.707	-0.707
800	0.221	0.221
900	0.000	-0.000
⋮	⋮	⋮

FIG. 5.5.5 Graphical interpretation of the frequency domain properties of the example frequency-sampling filter.

will offer an overall fit which is inferior to an ℓ_2 synthesized filter but will not suffer from localized large estimation errors.

Fortunately, linear phase filters are mathematically compatible with the minimax philosophy. The frequency response of a general linear phase FIR can be expressed as

$$H(\exp(j\omega)) = \exp -j\omega\left(\frac{N-1}{2}\right)\exp \ j\omega\left(\frac{L}{2}\right) H^*(\exp(j\omega)) \qquad (5.6.2)$$

where L has a value of zero for symmetrical filters and 1 for antisymmetric filters. Using standard trigonometric identities, it can be shown that $H(\exp(j\omega))$ can be expressed as a linear combination of weighted cosines. As a result, a unique Chebyshev approximation of order N [i.e., $H_N(\exp(j\omega))$] of $H(\exp(j\omega))$ can be efficiently computed. The basis for this claim is found in the Chebyshev alternation theorem, which states that in order for a weighted sum of cosines to be uniquely optimal (in a minimax sense), $E_N(\exp(j\omega))$ must have at least N extremal frequencies, say ω_i, such that $E_N(\exp(j\omega_i)) = -E_N(\exp(j\omega_{i+1}))$ and $|E_N(\exp(j\omega_i))| = \max[E(\exp(j\omega))]$. This result is familiar to classical filter theorists who have studied the equiripple concept for decades. Rather than entering a detailed technical discussion of this method, however, it will only be highlighted.

Classical filter design theorists (McGee, 1967; Szentermai, 1963) have used the Remez exchange method (Remez, 1934) to design linear phase filters in an optimal sense. This method represents an iterative algorithm which can be used to determine the location of the required N critical frequencies (local externals) of an equiripple filter. Parks and McClellan (1972a, b) have developed a particularly useful software interpretation of

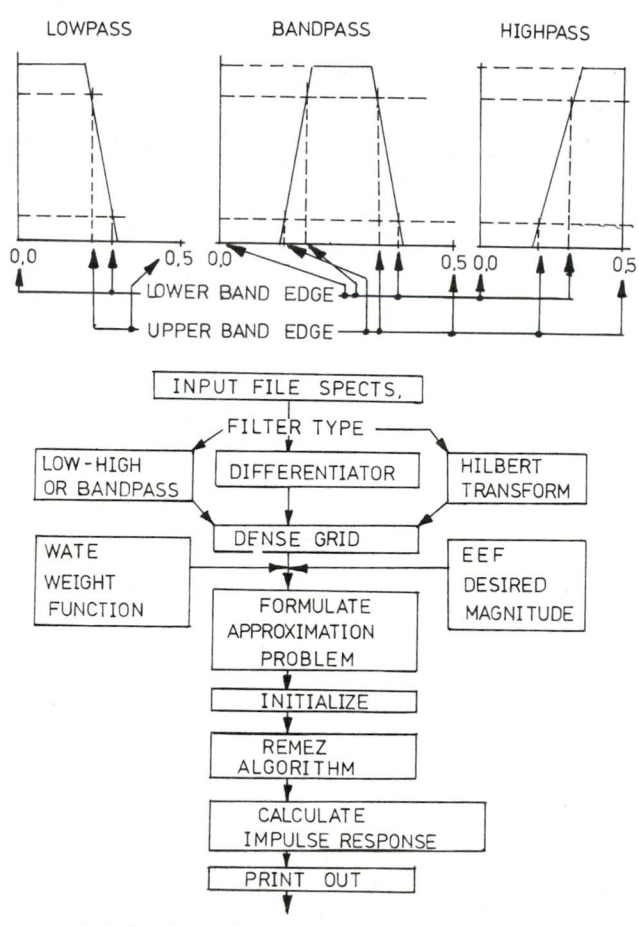

FIG. 5.6.1 Flow diagram for the Remez exchange method of designing optimal FIRs.

the Remez method. This method is shown in flow diagram form in Fig.
5.6.1.

The software system developed by McClellan has the ability to synthesize the following class of filters:

1. Type 1: lowpass, bandpass, highpass, bandstop
2. Type 2: differentiator (i.e., $j\omega$ frequency response)
3. Type 3: Hilbert transform (developed in the next section)

The well-referenced software system can work with filters of order N,
where $3 \leq N \leq 256$. The user can specify the desired magnitude frequency

response in a piecewise-constant fashion over a maximum of 10 contiguous frequency bands. The user can also assign relative weights of importance to each of these bands through the use of user-supplied weighting parameters. The software package, as outlined, can be found reproduced in McClellan et al. (1973). This code was developed to run on a Univac 1106 at MIT. This program, as well as much other useful signal processing software, is available, at a nominal cost, from the IEEE Service Center, 345 East 47th Street, New York, N.Y. The essence of the software, with embedded PLOT-10 (Tektronix) calls has been coded by M. Rengan (University of Cincinnati) for execution on a Digital Equipment PDP-11 series machine. This software can be obtained at a modest cost through the ECE Department, FIR Software, Mail Location 30, University of Cincinnati, Cincinnati, OH 45221. The following results were obtained using this software and a PDP-11/60.

EXAMPLE 5.6.1 A 31st-order lowpass filter was designed using the Remez method. The ideal filter's frequency response, relative to the Nyquist frequency, was chosen to be unity over the frequency range [0, 0.08] hertz and zero over [0.16, 0.5] hertz relative. The transition region therefore resides in the interval [0.08, 0.16]. The passband and stopband were placed on an equal parity by defining their respective band weights to be unity. The designed filter is interpreted in Figs. 5.6.2 through 5.6.6.

EXAMPLE 5.6.2 Here we deal with the design of a bandpass filter having a passband (ideal gain = 1) from 0.125 ro 0.325 Hz with respect to the Nyquest sample frequency. The stopbands (ideal gain zero) over [0.0, 0.075] and [0.375, 0.5] are to be considered of major importance in that they have been assigned a weight of 10. The results of this experiment are summarized in Figs. 5.6.7 through 5.6.11.

EXAMPLE 5.6.3 The effect of the weighting coefficients can be demonstrated through the following example. A simple bandpass filter was designed for two weighting policies. The first, described in Fig. 5.6.12, uses the weighting scheme {10, 1, 10} for first stopband, passband, and second stopband weights. The second, found in Fig. 5.6.13, uses {1, 10, 1}. It can be noted that the first has deeper stopband attenuation due to the higher weight (premium) placed on this attribute. The second sacrifices stopband attenuation for reduced passband ripple.

EXAMPLE 5.6.4 Periodicity inducing windows, such as the Hamming window, can be integrated into the filter design. In Fig. 5.6.14, a 31st-order Chebyshev bandpass FIR filter's magnitude frequency response can be found. A Hamming window was applied to the filter's impulse response. The magnitude frequency response of the windowed filter is also shown in Fig. 5.6.14. It can be seen that much of the ripple found in the original is

```
**************************************************************************

FINITE IMPULSE RESPONSE (FIR)
LINEAR PHASE DIGITAL FILTER DESIGN
REMEZ EXCHANGE ALGORITHM

BANDPASS FILTER

FILTER LENGTH =   31

*****  IMPULSE RESPONSE  *****
H(   1)  =  -0.42180517E-02  =  H(   31)
H(   2)  =  -0.45150425E-02  =  H(   30)
H(   3)  =  -0.24560867E-02  =  H(   29)
H(   4)  =   0.36547068E-02  =  H(   28)
H(   5)  =   0.11415420E-01  =  H(   27)
H(   6)  =   0.15417040E-01  =  H(   26)
H(   7)  =   0.99993944E-02  =  H(   25)
H(   8)  =  -0.63715177E-02  =  H(   24)
H(   9)  =  -0.27752254E-01  =  H(   23)
H(  10)  =  -0.41150242E-01  =  H(   22)
H(  11)  =  -0.31729966E-01  =  H(   21)
H(  12)  =   0.91501521E-02  =  H(   20)
H(  13)  =   0.77349104E-01  =  H(   19)
H(  14)  =   0.15493656E+00  =  H(   18)
H(  15)  =   0.21662518E+00  =  H(   17)
H(  16)  =   0.24008550E+00  =  H(   16)

                         BAND  1           BAND  2           BAND
LOWER BAND EDGE      0.000000000       0.159999996
UPPER BAND EDGE      0.079999998       0.500000000
DESIRED VALUE        1.000000000       0.000000000
WEIGHTING            1.000000000       1.000000000
DEVIATION            0.003963314       0.003963314
DEVIATION IN DB    -48.038837433     -48.038837433

EXTREMAL FREQUENCIES
    0.0234375      0.0507812      0.0722656      0.0800000      0.1600000
    0.1678125      0.1892969      0.2166406      0.2459375      0.2752344
    0.3064844      0.3396875      0.3709375      0.4021875      0.4353906
    0.4685937      0.5000000

**************************************************************************
```

FIG. 5.6.2 Output listing for an example 31st-order lowpass FIR displaying tap weight, or coefficients, and critical frequencies.

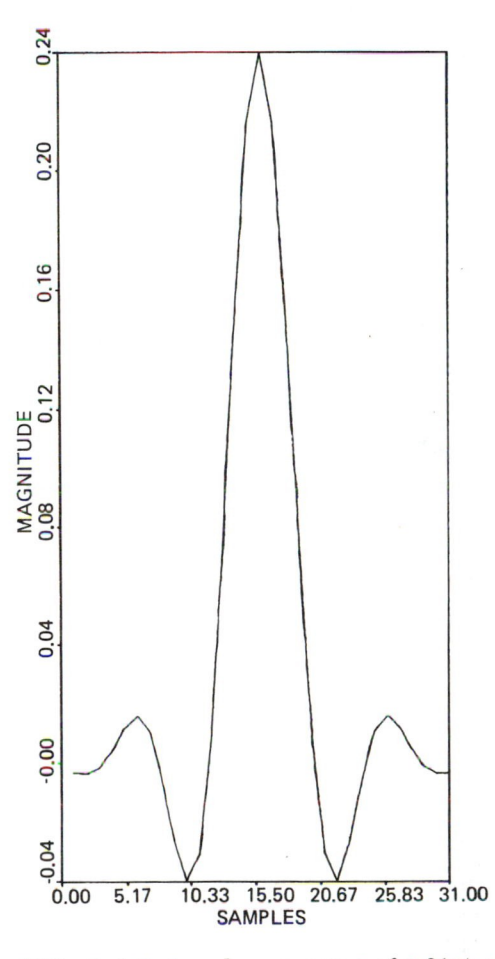

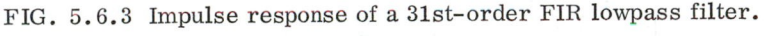

FIG. 5.6.3 Impulse response of a 31st-order FIR lowpass filter.

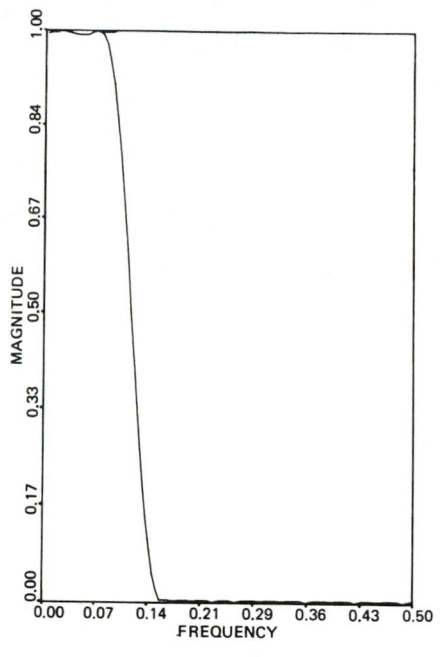

FIG. 5.6.4 Magnitude frequency response of a 31st-order FIR lowpass filter.

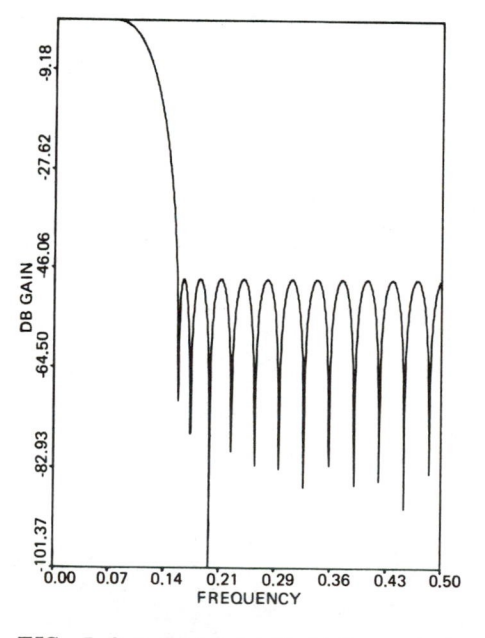

FIG. 5.6.5 Log magnitude frequency response of a 31st-order FIR lowpass filter.

FIG. 5.6.6 Phase response of a 31st-order FIR lowpass filter.

```
****************************************************************************

     FINITE IMPULSE RESPONSE (FIR)
     LINEAR PHASE DIGITAL FILTER DESIGN
     REMEZ EXCHANGE ALGORITHM

     BANDPASS FILTER

     FILTER LENGTH =   31

     ***** IMPULSE RESPONSE *****
          H(  1) =   0.47340328E-02 = H(  31)
          H(  2) =   0.65854663E-03 = H(  30)
          H(  3) =   0.10547641E-01 = H(  29)
          H(  4) = -0.12568651E-01 = H(  28)
          H(  5) = -0.49427792E-01 = H(  27)
          H(  6) =   0.83850534E-03 = H(  26)
          H(  7) =   0.28130272E-01 = H(  25)
          H(  8) = -0.62510259E-02 = H(  24)
          H(  9) =   0.58212180E-01 = H(  23)
          H( 10) =   0.55028792E-01 = H(  22)
          H( 11) = -0.47244899E-01 = H(  21)
          H( 12) =   0.16759349E-01 = H(  20)
          H( 13) = -0.72137393E-01 = H(  19)
          H( 14) = -0.29824528E+00 = H(  18)
          H( 15) =   0.67185961E-01 = H(  17)
          H( 16) =   0.47803843E+00 = H(  16)

                        BAND   1         BAND   2         BAND   3         BAND
     LOWER BAND EDGE    0.000000000     0.125000000     0.375000000
     UPPER BAND EDGE    0.075000003     0.324999988     0.500000000
     DESIRED VALUE      0.000000000     1.000000000     0.000000000
     WEIGHTING         10.000000000     1.000000000    10.000000000
     DEVIATION          0.009521103     0.095211029     0.009521103
     DEVIATION IN DB  -40.426254272   -20.426256180   -40.426254272

     EXTREMAL FREQUENCIES
          0.0000000     0.0371094     0.0644531     0.0750000     0.1250000
          0.1445312     0.1816406     0.2226562     0.2256250     0.3046875
          0.3250000     0.3750000     0.3847656     0.4082031     0.4375000
          0.4687500     0.5000000

****************************************************************************
```

FIG. 5.6.7 Output listing of a FIR bandpass filter.

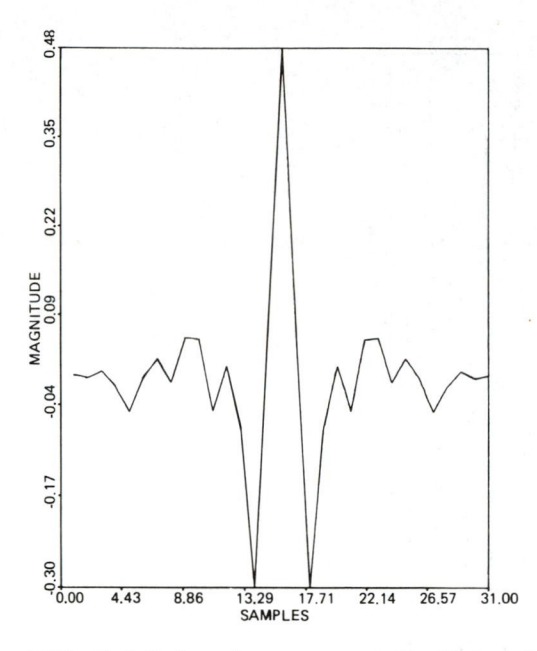

FIG. 5.6.8 Impulse response of a 31st-order bandpass filter.

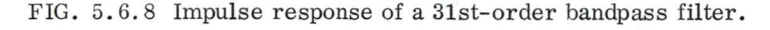

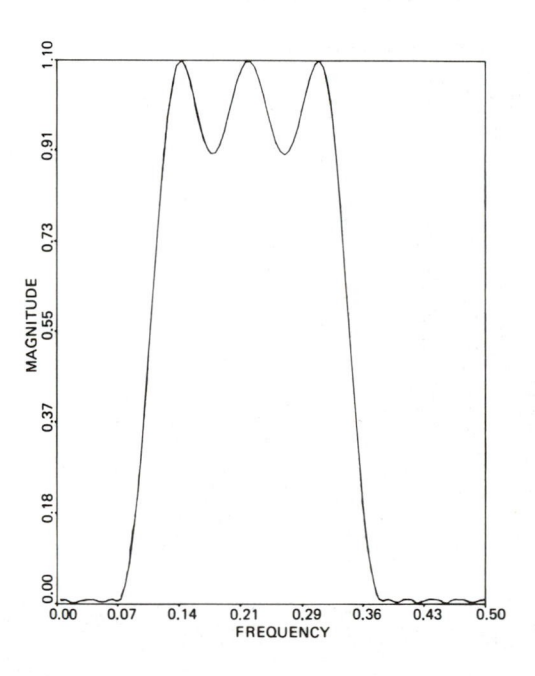

FIG. 5.6.9 Magnitude frequency response of a 31st-order bandpass filter.

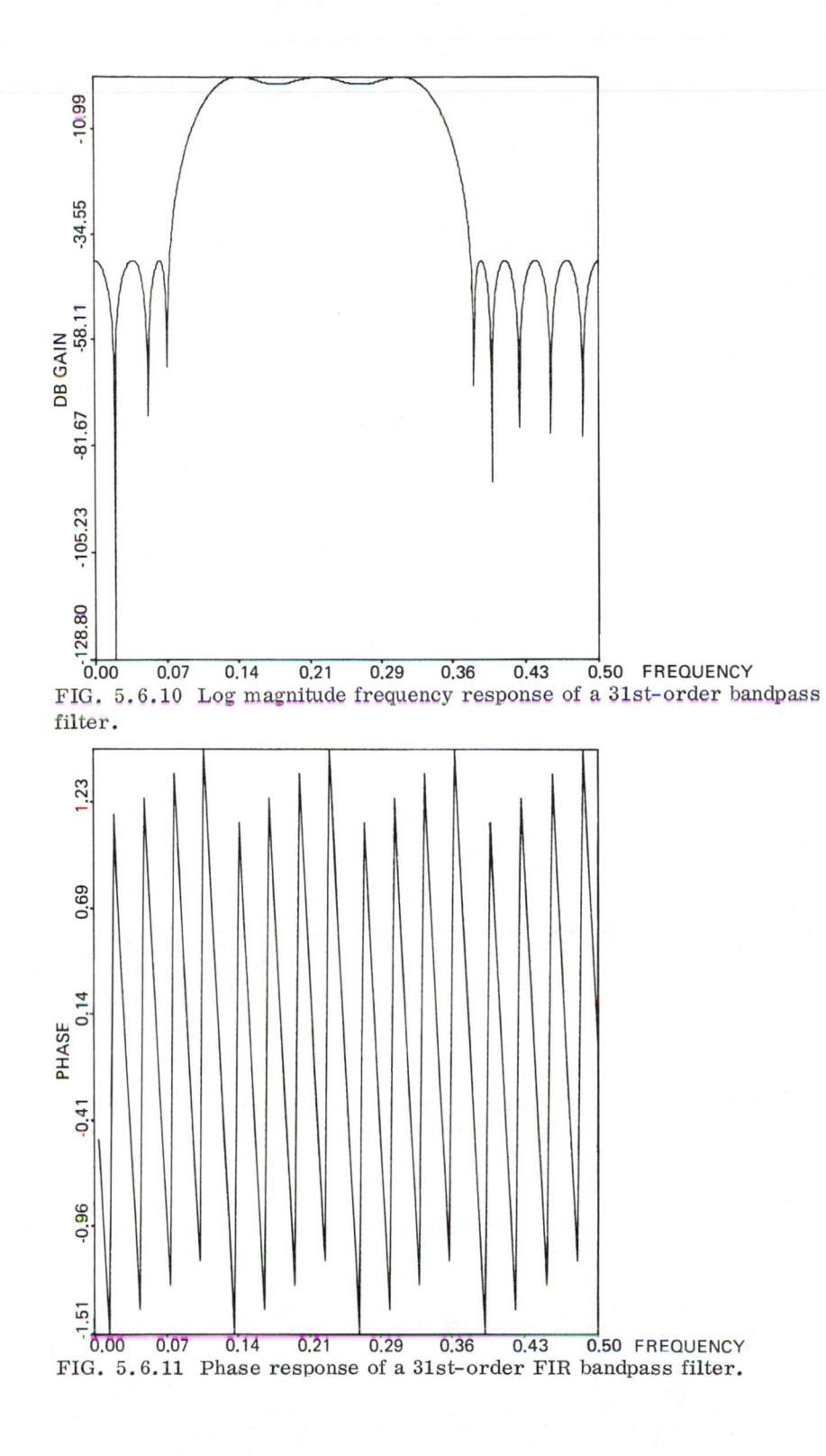

FIG. 5.6.10 Log magnitude frequency response of a 31st-order bandpass filter.

FIG. 5.6.11 Phase response of a 31st-order FIR bandpass filter.

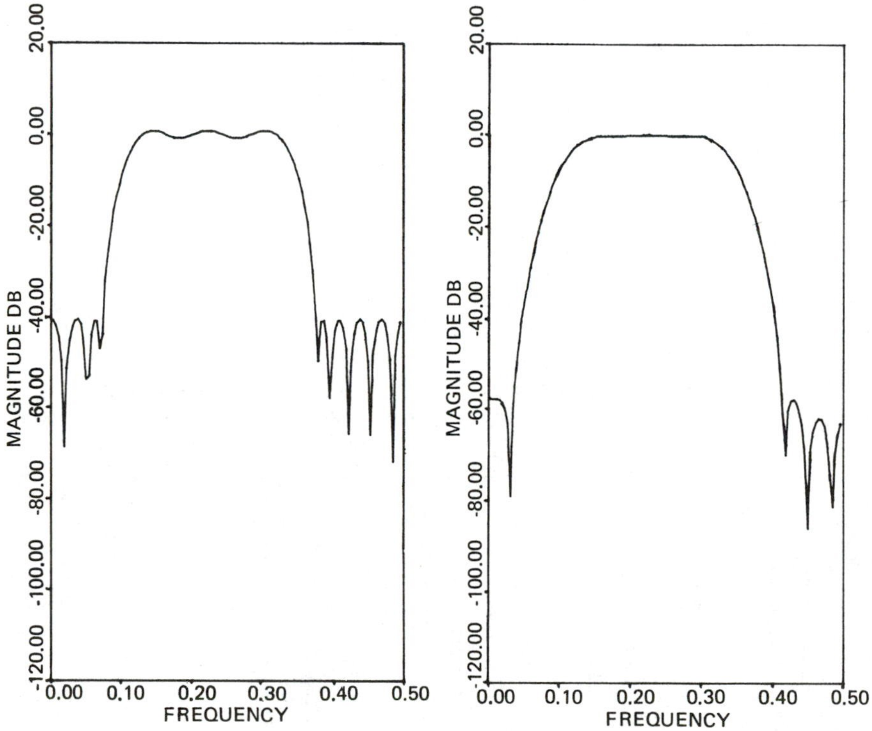

(A)

(B)

FIG. 5.6.12 Bandpass FIR filter with a strong stopband data fit penalty
using (A) rectangular and (B) Hamming windows.

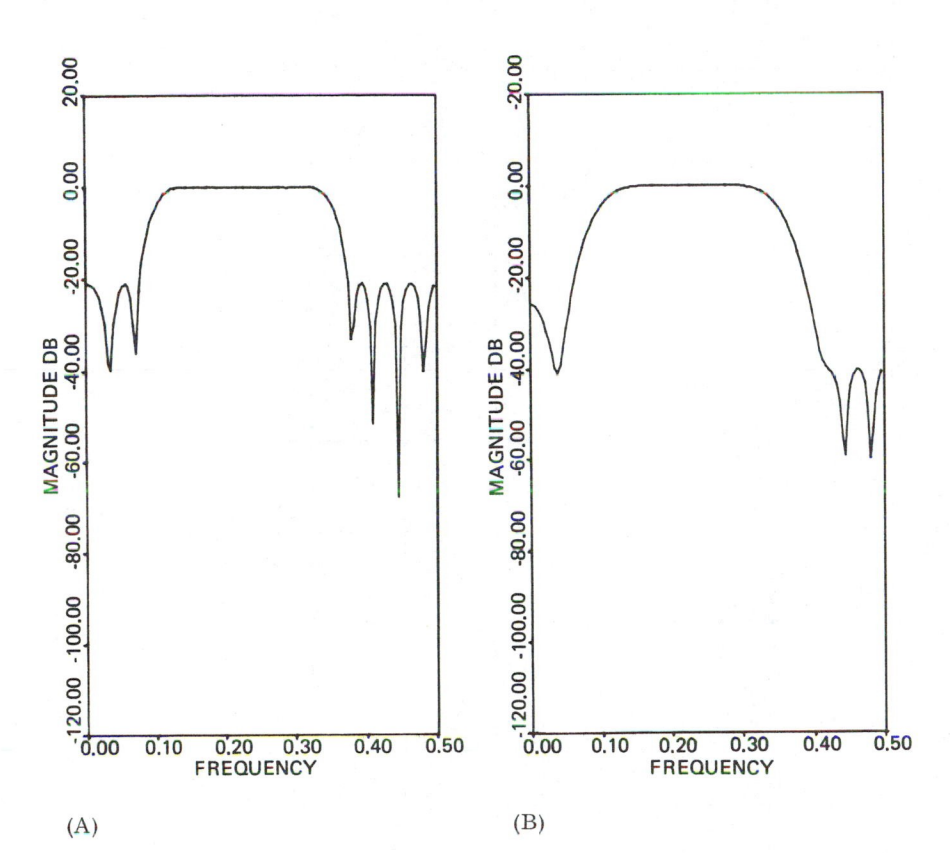

(A)   (B)

FIG. 5.6.13 Bandpass FIR filter with a strong passband data fit penalty using (A) rectangular and (B) Hamming windows.

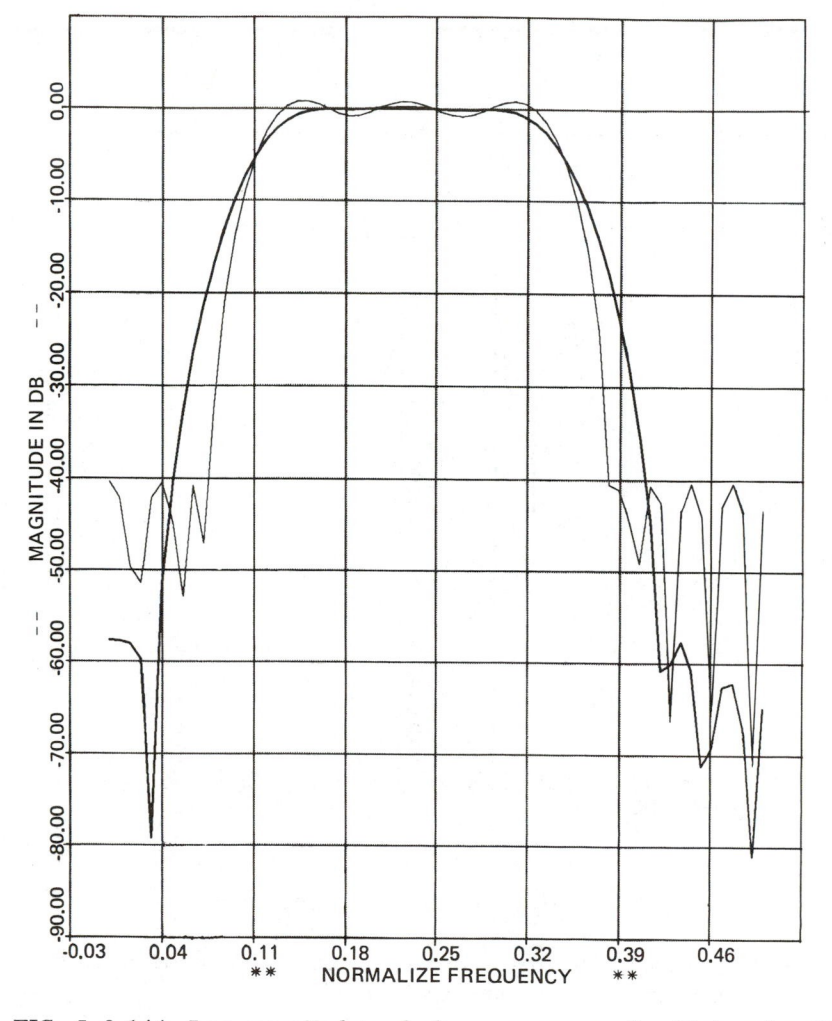

FIG. 5.6.14A Log magnitude and phase response of a 31st-order FIR
Chebyshev filter using a rectangular window.

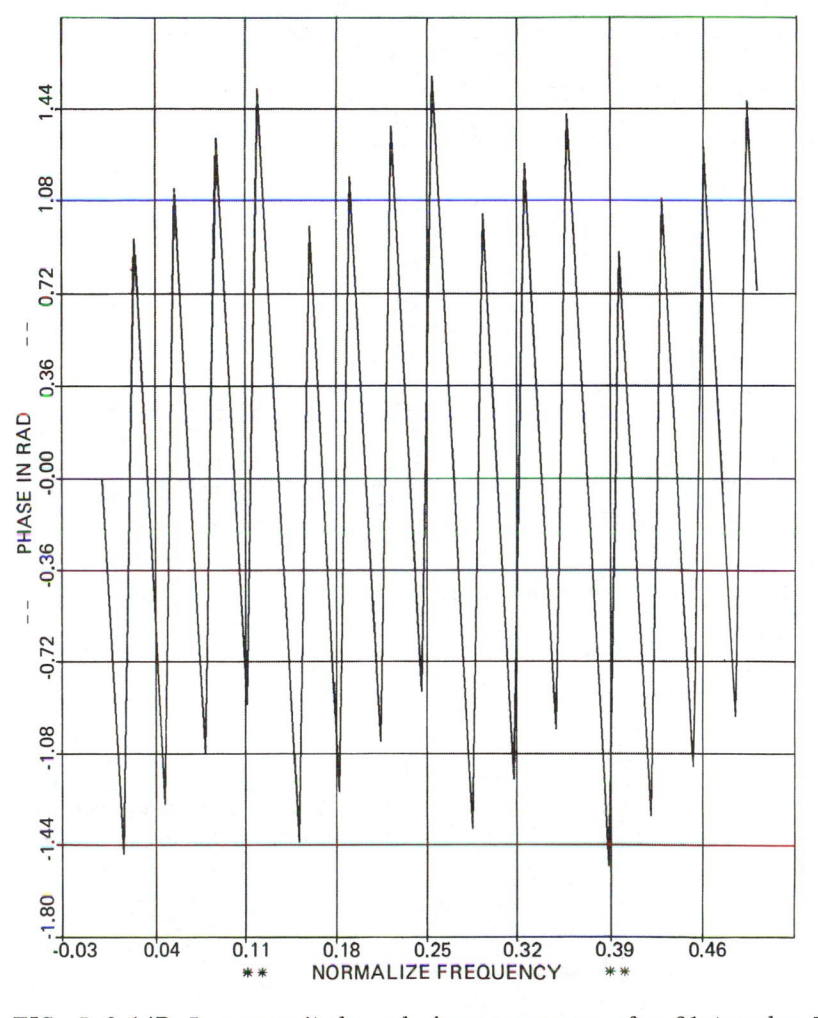

FIG. 5.6.14B Log magnitude and phase response of a 31st-order FIR
Chebyshev filter using a Hamming window.

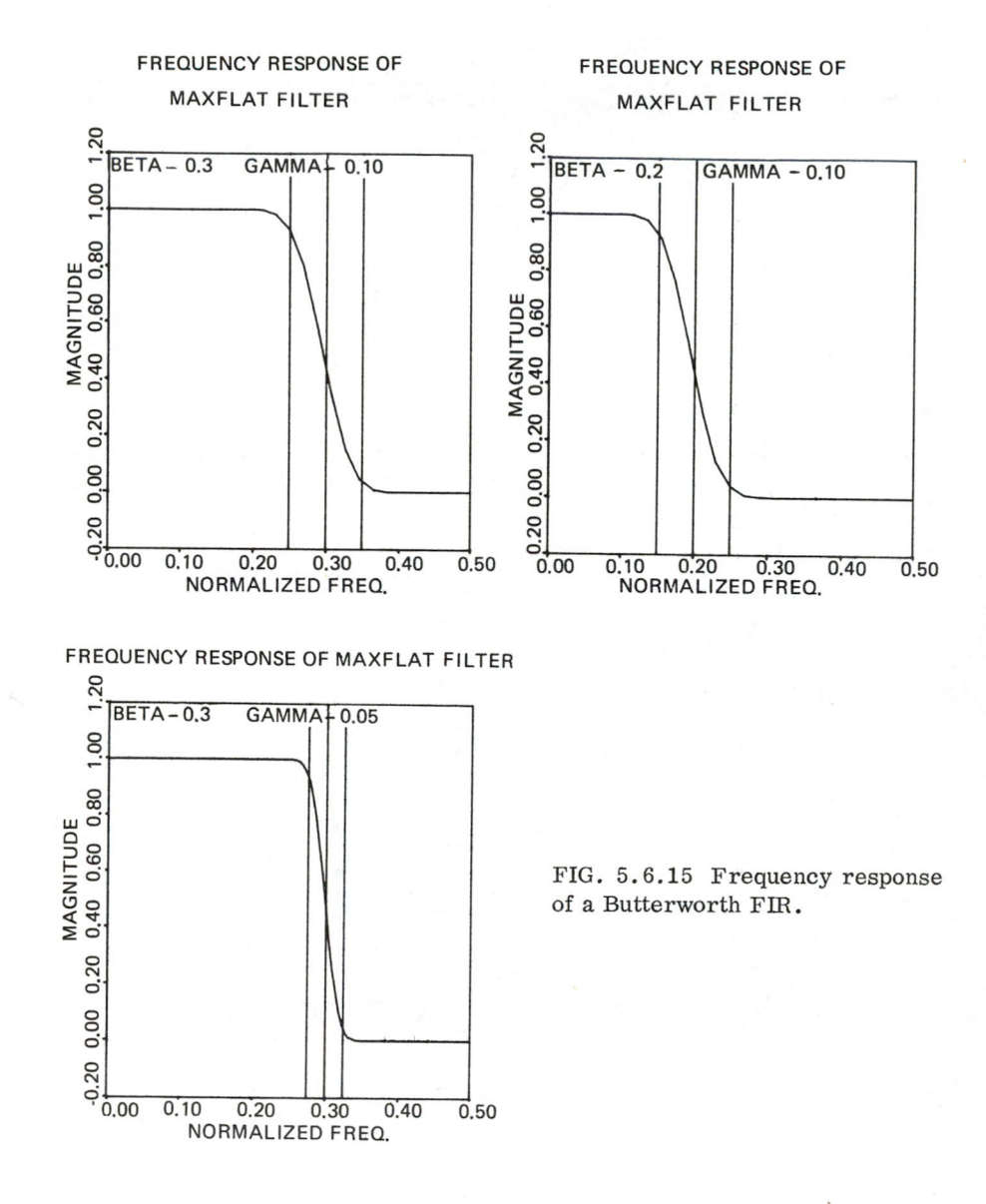

FIG. 5.6.15 Frequency response of a Butterworth FIR.

smoothed. However, there is a slight degradation in the filter's skirt slope.

EXAMPLE 5.6.5 In Chapter 1 the maximally flat polynomial of Butterworth was presented. This polynomial can also be used to design FIR filters. The resulting filter has a maximal flatness in the passband but, compared to the Chebyshev design, suffers in terms of its transition and stopband magnitude characteristics. As a result, the Chebyshev approach has achieved greater popularity.

The magnitude responses of a maximally flat 53rd-order FIR, configured as a lowpass filter, for various critical frequencies and transition bandwidths, are summarized in Fig. 5.6.15.

5.7 HILBERT TRANSFORMS

The ability of the Fourier transform to provide a frequency domain image of a signal is well documented. This standard analysis tool can be used to spectrally decompose a given time series. Another frequency domain transform, known as the Hilbert transform, can also be used to provide this type of information. The Hilbert transform introduces a phase shift of $\pm \pi / 2$ radians to all spectral components. The Hilbert transform of $y(t)$ is formally given by $y(t) * x(t)$, where $x(t) = 1/\pi t$, or

$$y_H(t) = -\int_{-\infty}^{\infty} \frac{y(t)}{t - s} \, ds \qquad (5.7.1)$$

Specifically, if $X(f)$ is the Fourier transform of $x(t)$, its Hilbert transform is represented by $X_H(f)$, where

$$X_H(f) = -j \, \text{sgn}(f) X(f) \qquad (5.7.2)$$

with

$$\text{sgn}(f) = \begin{cases} 1 & \text{if } f > 0 \\ 0 & \text{if } f = 0 \\ -1 & \text{if } f < 0 \end{cases} \qquad (5.7.3)$$

The Hilbert transform provides the user with several useful services. They are its ability to model:

1. Single-sideband modulators
2. Phase distortion in linear systems
3. Bandpass signal and systems

The bandpass analysis capability of the Hilbert transform can prove to be especially useful and is therefore singled out for special study.

Suppose that one chooses to test a system with a modulated signal $s(t) = a(t) \cos(\omega_0 t) - b(t) \sin(\omega_0 t)$, where $a(t)$ and $b(t)$ represent bandlimited intelligence. It will be assumed that these two signals are bandlimited to B hertz, where $\omega_0 > B/2$. Using the frequency-shifting property of Fourier transforms, one can write

$$\mathcal{F}(a(t) \cos(\omega_0 t)) = \frac{A(\omega - \omega_0) + A(\omega - \omega_0)}{2}$$

$$\mathcal{F}(b(t) \sin(\omega_0 t)) = \frac{B(\omega - \omega_0) - B(\omega - \omega_0)}{2}$$

(5.7.4)

A classic theorem, known as the <u>bandpass representation theorem</u>, states that if $s(t)$ is a signal having a Fourier transform $S(f)$, which is nonzero for $\omega_1 \le \omega \le \omega_2$, $0 \le \omega_1 \le \omega_2 < \infty$, then $s(t)$ can be represented as $s(t) = a(t) \cos(\omega_0 t) - b(t) \sin(\omega_0 t)$, with

$$a(t) = s(t) \cos(\omega_0 t) + s_H(t) \sin(\omega_0 t)$$

$$b(t) = s_H(t) \cos(\omega_0 t) - s(t) \sin(\omega_0 t) \tag{5.7.5}$$

$$A(\omega) = \begin{cases} S(\omega + \omega_0) + S^*(\omega - \omega_0), & |\omega| \le \max(|\omega_1 - \omega_0|, |\omega_2 - \omega_0|) \\ 0, & \text{otherwise} \end{cases}$$

$$B(\omega) = \begin{cases} S(\omega + \omega_0) - S^*(\omega - \omega_0), & |\omega| \le \max(|\omega - \omega_0|, |\omega_2 - \omega_0|) \\ 0, & \text{otherwise} \end{cases}$$

Therefore, the message $a(t)$ and $b(t)$ can be reconstructed by modulating $s(t)$ and its Hilbert transform $s_H(t)$ by a locally generated $\cos(\omega_0 t)$ and $\sin(\omega_0 t)$.

The Hilbert transform can be obtained by convolving a signal $s(t)$ with a Hilbert transform filter having a transfer function of $X(\omega) = -j \operatorname{sgn}(\omega_0)$.

$$x(n) = \begin{cases} 0, & n \text{ even} \\ \dfrac{2 \sin^2(\pi n/2)}{\pi n}, & n \text{ odd} \end{cases} \tag{5.7.6}$$

or in graphical terms it would appear as the data shown in Fig. 5.7.1.

Observe that $x(n)$ has odd symmetry about $n = 0$. Therefore, $X(\omega)$ may be alternatively represented as the sine series

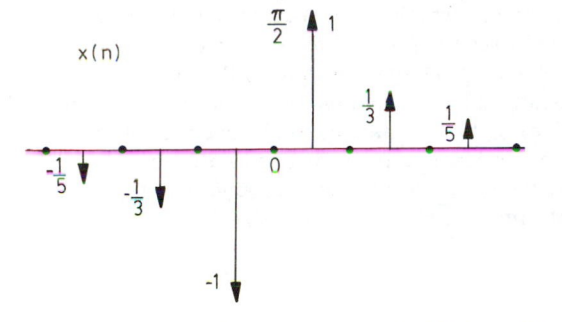

FIG. 5.7.1 Impulse response of a Hilbert filter.

$$X(\omega) = \frac{-j^4}{\pi} \sum_{n=1}^{\infty} \frac{1}{n} \sin(n\omega T) \qquad\qquad (5.7.7)$$

If a FIR is to be used to realize a Hilbert filter, only a finite number of samples of the impulse response suggested in Fig. 5.7.1 would be mechanized. The resulting spectrum of a FIR Hilbert convolving filter would have a shape similar to that suggested in Fig. 5.7.2. The ripple found in the figure, which is due to a truncated impulse response, can of course be smoothed through the use of a window function.

EXAMPLE 5.7.1 The 180° phase inversion about f = 0 Hz is difficult to approximate with a finite-order FIR. If the signal spectrum of the signal to be transformed is devoid (or weak) of low-frequency energy, say over

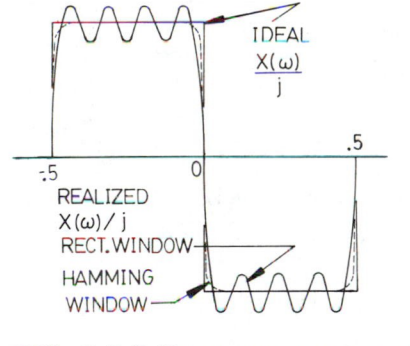

FIG. 5.7.2 Frequency response of an ideal, finite, and smoothed (Hamming) finite Hilbert filter.

[0, 0.05], then the transition band, which originally was 0 Hz in width, can be extended to 0.05 Hz relative to the sampling frequency. As a result, a filter can be more readily designed which will perform a Hilbert transform over the essential frequencies found in the original time series. For example, using the Remez method, such a filter can be designed. The Hilbert transform is basically v alid from $0.05 \times f_s$ to the Nyquist frequency (f_s = sample frequency). The resulting filter is discussed in Figs. 5.7.3 through 5.7.6.

```
*****************************************************************************

FINITE IMPULSE RESPONSE (FIR)
LINEAR PHASE DIGITAL FILTER DESIGN
REMEZ EXCHANGE ALGORITHM

HILBERT TRANSFORMER

FILTER LENGTH =  32

*****  IMPULSE RESPONSE  *****
H(  1) =   0.22249452E-02 =-H(  32)
H(  2) =   0.25339632E-02 =-H(  31)
H(  3) =   0.39164694E-02 =-H(  30)
H(  4) =   0.57545481E-02 =-H(  29)
H(  5) =   0.81385979E-02 =-H(  28)
H(  6) =   0.11190426E-01 =-H(  27)
H(  7) =   0.15060835E-01 =-H(  26)
H(  8) =   0.19955467E-01 =-H(  25)
H(  9) =   0.26181787E-01 =-H(  24)
H( 10) =   0.34217909E-01 =-H(  23)
H( 11) =   0.44886693E-01 =-H(  22)
H( 12) =   0.59750900E-01 =-H(  21)
H( 13) =   0.82171559E-01 =-H(  20)
H( 14) =   0.12093079E+00 =-H(  19)
H( 15) =   0.20831975E+00 =-H(  18)
H( 16) =   0.63531584E+00 =-H(  17)

                           BAND  1          BAND
LOWER BAND EDGE       0.050000001
UPPER BAND EDGE       0.500000000
DESIRED VALUE         1.000000000
WEIGHTING             1.000000000
DEVIATION             0.002502190

EXTREMAL FREQUENCIES
   0.0500000     0.0578125     0.0753906     0.0988281     0.1281250
   0.1554688     0.1867188     0.2179688     0.2472656     0.2785156
   0.3097656     0.3410156     0.3742188     0.4054688     0.4367188
   0.4679688     0.5000000

*****************************************************************************
```

FIG. 5.7.3 Output listing for a Hilbert FIR filter design problem.

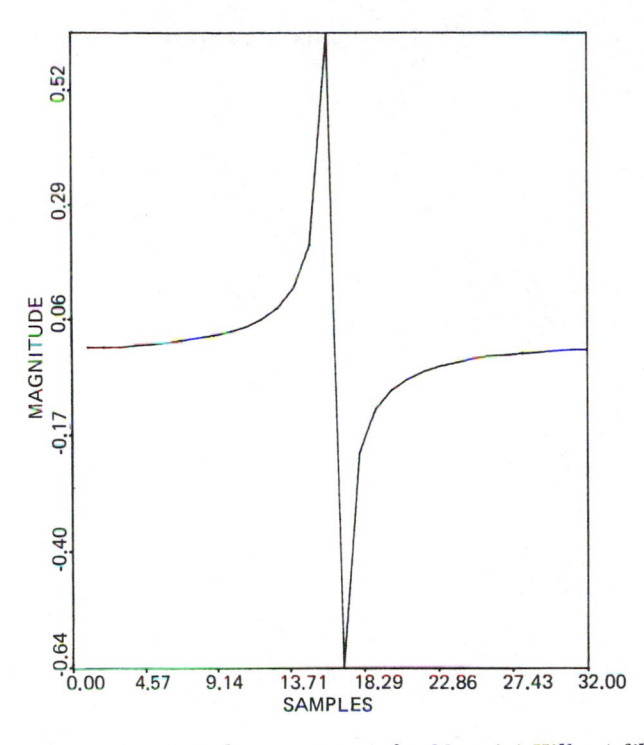

FIG. 5.7.4 Impulse response of a 32-point Hilbert filter approximation.

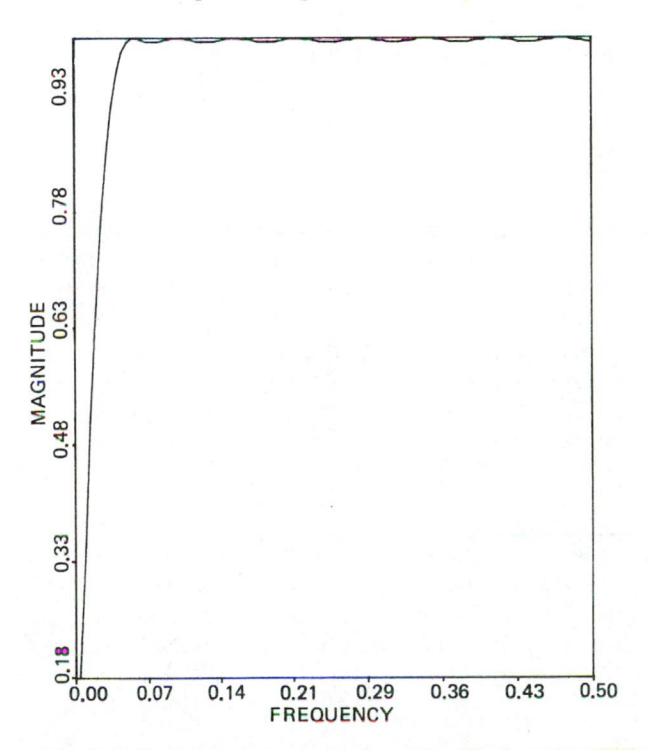

FIG. 5.7.5 Frequency response of a 32nd-order Hilbert filter approximation.

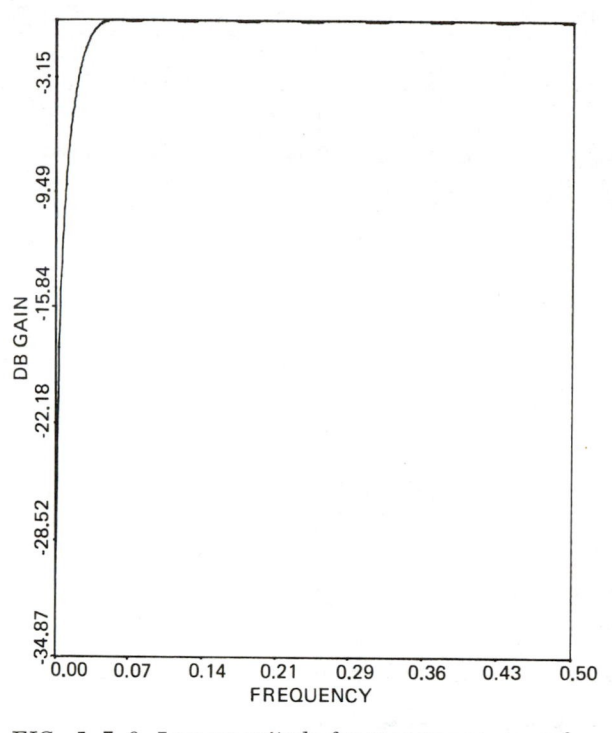

FIG. 5.7.6 Log magnitude frequency response for a 32nd-order Hilbert filter approximation.

5.8 TEMPORAL COMPRESSION

There are many times when the signal processing scientist must (or would like to) alter the system sample rate. One such application is in the mechanization of a band-selectable FFT (see Section 3.10). Here, upon heterodyning a time series with W_N^k and lowpass-filtering the result, a selected portion of an original spectrum can be translated down to baseband. This baseband of frequencies could then be transformed under a FFT. This concept is abstracted in Fig. 5.8.1. It can be noted that the sample rate appearing at the output of the lowpass filters is slower than the input sample rate by a factor of K. In general, the reduction of sample rate is referred to as decimation.

Suppose that the spectrum of a given time series is $X(k)$ and has the shape suggested in Fig. 5.8.2. Decimation can be accomplished by defining a new time series, in a samplewise sense, to be $z(k) = x(kK + s)$, $0 \leq s < K$. That is, the output sample rate is reduced to f_s/K samples per second. This clock rate can easily be achieved by using a clock divider.

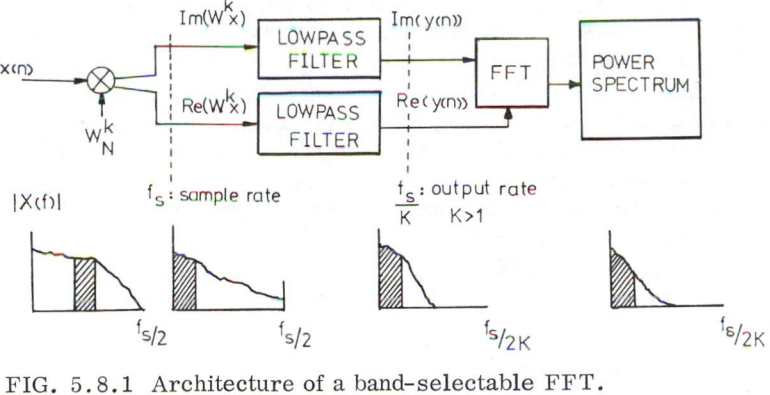

FIG. 5.8.1 Architecture of a band-selectable FFT.

It should be noted that the Nyquist frequency, found at the output of the low-pass filter, is $f_S/2K$. It should therefore be apparent that $f_S/2K$ must be at least as large as M hertz, as shown in Fig. 5.8.2. It is recommended that a FIR be used to realize a decimating filter. Here the filter would have the form

$$z(k) = \sum_{m=0}^{N-1} x(k - m)h(m) \tag{5.8.1}$$

where the filter coefficient $\{h(n)\}$ can be computed using the Remez method (e.g., lowpass filter over [0, M]).

EXAMPLE 5.8.1 Many commercial FFT systems have band selectability. One such system is the Hewlett-Packard 3582A. The digital filter front end of this FFT machine has been realized using four 40-pin LSI chips. Maximal frequency domain expansion, on the order of 1 to 5000, can be achieved. User-supplied data define the extent of the expansion and center

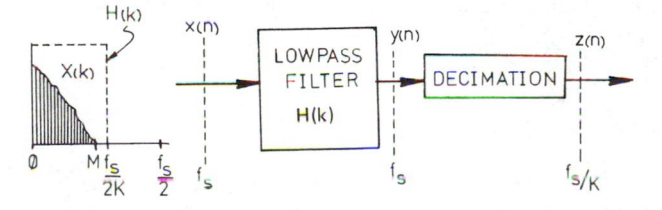

FIG. 5.8.2 Block diagram of a decimation in the time filter system.

frequency. The spectra found in Fig. 5.8.3 demonstrate the mechanics of
a band-selectable analysis. The topmost figure displays a typical baseband
512-point FFT, on nearly 100-Hz centers, and a sample rate of 50 kHz. A
marker, or cursor, can be placed over a user-identified center frequency
for the expanded analysis (i.e., zoom). Choosing a center frequency of
3.60 kHz and processing the data in a band-selectable sense produces the
spectrum shown at the bottom of Fig. 5.8.3. This spectrum provides the
user with a ±2.5 Hz spectral window about the defined center frequency.
Locally, at about 3.60 kHz, the spectral resolution has increased from 100
Hz to 20 mHz between harmonics. The band-selectable analysis (zoom)
methodology can also be applied to the problem of detailed transfer function

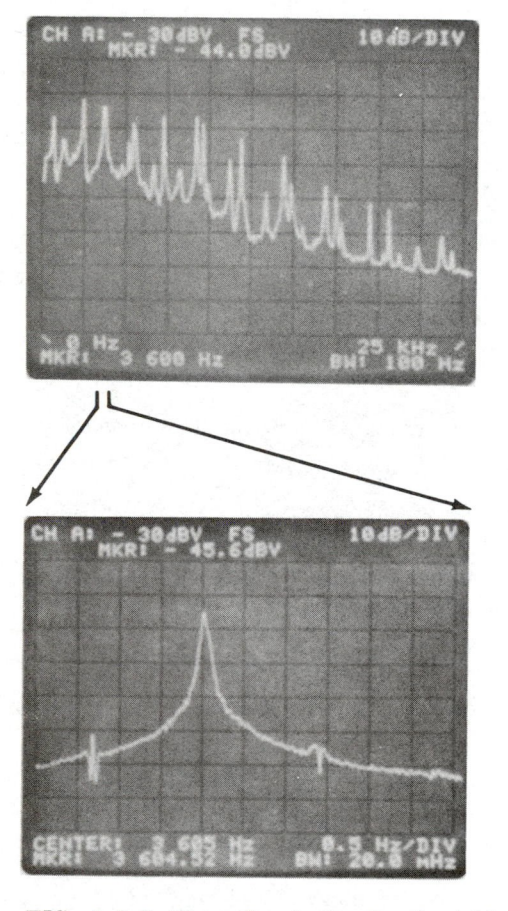

FIG. 5.8.3 Example of a band-selectable expansion of a spectrum using
a commercial FFT system. (Reprinted with the permission of Hewlett-
Packard Corporation.)

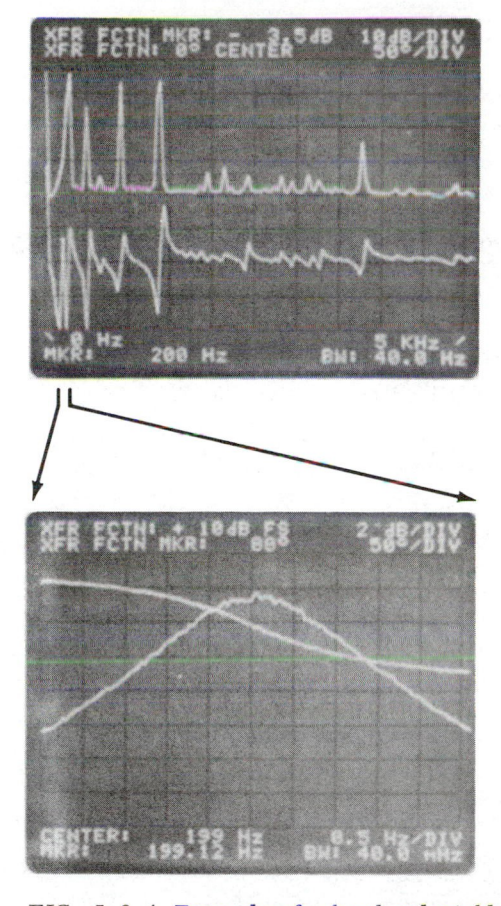

FIG. 5.8.4 Example of a band-selectable expansion of a transfer function. (Reprinted with the permission of Hewlett-Packard Corporation.)

analysis. This is suggested in Fig. 5.8.4. In both cases care should be exercised that noise does not distort the expanded spectral data base. Because of its coarseness, the original baseband spectrum has a natural tendency to smooth the frequency domain display. As spectral resolution increases, this smoothing effect is reduced. Therefore, if the system under study does not possess a reasonable degree of stability, spectral averaging methods should be applied to the output process.

EXAMPLE 5.8.2 Decimation and interpolation are dual operations. They are used whenever a change in the system data rate is advantageous. For example, in speech, data samples or hybrid parameters (see Chapter 20) can be computed at a relatively low real-time data rate. Slow data input rates mean that a plurality of messages can be interleaved (multiplexed) together within a finite bandwidth channel. Analog-quality speech intelligence can be achieved at the receiver through the use of an <u>interpolator</u>. The interpolator will estimate what the intersample values, of a sparsely sampled process, will be.

Oetken et al. (1979) have studied the problem of designing a $2rL + 1$ (odd-)-order discrete FIR interpolating filter. This filter operates under the assumption that a times series $\{x(n)\}$ is bandlimited and decimated by ignoring all but the krth sample, $k = 0, 1, 2, \ldots$. As a result, bandwidth compression, by a factor of r, may be achieved. The FIR interpolator is designed so that the approximation (interpolation) error, given by

$$| x(n)\delta_k(n - kr) * h(n) - x(n)|^2; \quad k = 0, 1, 2, \ldots$$

is minimized. That is every kr sample of $\{x(n)\}$ will be convolved with the FIR linear interpolator's impulse response. The filtered output will, in a least-squares sense, resemble the original densely sampled input process $\{x(n)\}$. Using published optimization methods, based on an autocorrelation study, the coefficients of the interpolating filter can be derived using a general-purpose computer. For example, if the magnitude squared value of the input signal is given by

$$|X(\omega)|^2 = 1; \quad |\omega| < \frac{0.25}{r}$$

(i.e., ideal lowpass filter), with L chosen to be 5, then the results exhibited in Figs. 5.8.5 and 5.8.6 are obtained. In Fig. 5.8.5 is found the interpolated spectrum for $r = 5$ ($|w| < 0.05$ times the Nyquist frequency) for a $2rL + 1 = 51$st-order FIR. For a higher degree of decimation, say $r = 8$ ($|w| < 0.03125$ times the Nyquist frequency), the interpolated spectrum is found in Fig. 5.8.6. It can be seen that as r increases, so will the filter order and complexity. The larger the value of r, the more the bandwidth can be compressed. However, the interpolation error will generally increase with r. Therefore, one is left with a design trade-off decision involving weight speed, complexity, and accuracy.

5.9 ARCHITECTURE

Three FIR design strategies have been presented in this section. Many other design philosophies, based on other approximating polynomials or mathematical programming techniques, are found in use. The three

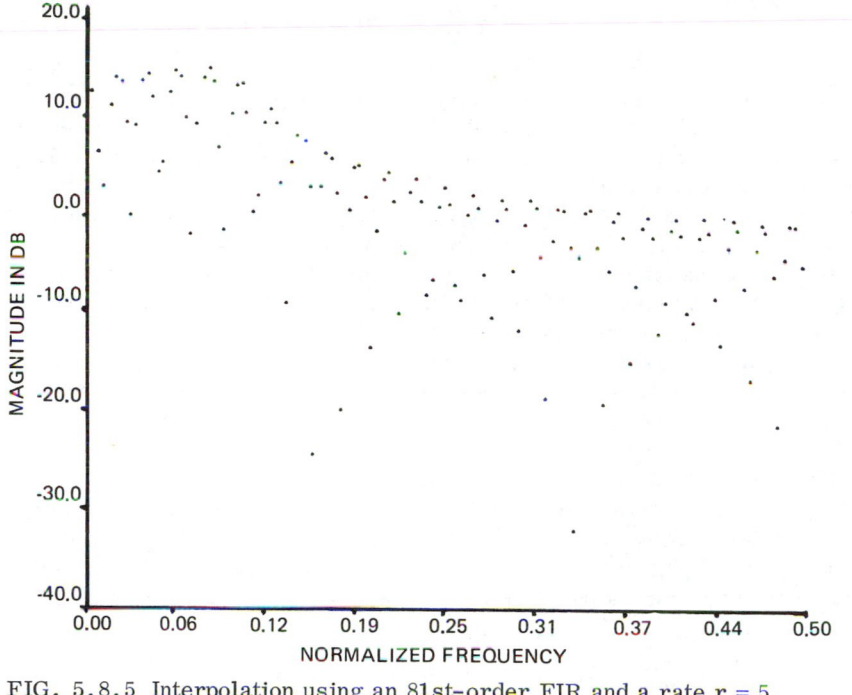

FIG. 5.8.5 Interpolation using an 81st-order FIR and a rate r = 5.

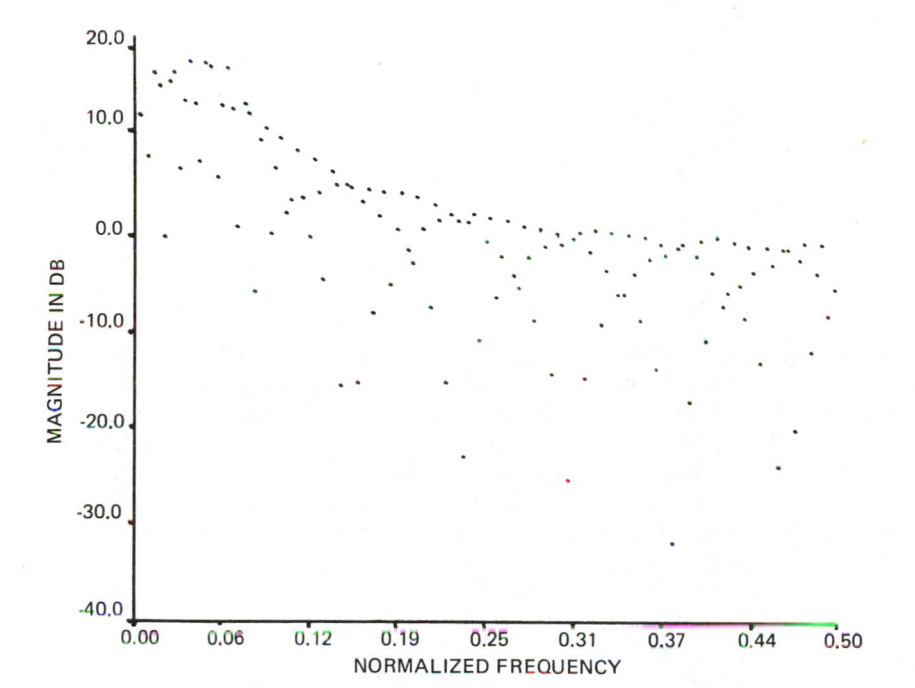

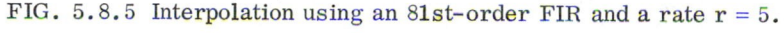

FIG. 5.8.6 Interpolation example using an 81st-order FIR and a rate r = 8.

methods presented here are known to be useful, popular, and well documented. Regardless, once a set of FIR filter coefficients have been determined, the question of filter architecture arises. Referring to Fig. 5.9.1,
one observes that in general L multiplications are needed together with a
memory queue and accumulator. Admittedly some of the filter coefficients
may be zero (e.g., Hilbert transform of odd order), in which case those
multiplies can be ignored. The given realizations can be mechanized in
software, floating point, or fixed point. Efficient software memory management routines are discussed in Section 18. When fixed-point hardware
realizations are used, the memory queues become n-bit shift registers,
sequentially accessed or randomly accessed memory. If one is willing to
purchase speed at the expense of system costs, multiple multipliers may
be considered. If economy is of prime consideration, filter coefficients
would normally be stored on RAM or ROM and time-multiplexed to a single
multiplier. Here a simple sequential indexing scheme can be used to read
the coefficients from memory. Finally, it should be noted that the maximal
value which can be presented to the system accumulator, under the hypothesis that the input x(n) is bounded by unity (e.g., fractional binary representation) is given by

$$V = \max\left(\sum_{i=0}^{L-1} |a_i| \right) \qquad\qquad (5.9.1)$$

Therefore, the accumulator must be able to operate over $[-V, V]$ if it is to
be protected from a register overflow condition. This dynamic range can
be guaranteed by ensuring that the accumulators word length is sufficiently
large or by scaling the input. Such questions are given a more detailed
analysis in Chapter 13.

5.10 THROUGHPUT

The direct I filter found in Fig. 5.9.1 can be realized in the manner found
in Fig. 5.10.1. The throughput of this filter is essentially limited by the
time required to perform L consecutive multiplies. Another structure
uses a parallel architecture and trades off increased throughput with added
cost and hardware complexity. This system, detailed in Fig. 5.10.2, is
called the parallel FIR. The degree of parallelism is a function only of
the desired cost-speed trade-off. Commercially available digital hardware
will readily support most FIR designs of order N, assuming that N is not
outrageously large.

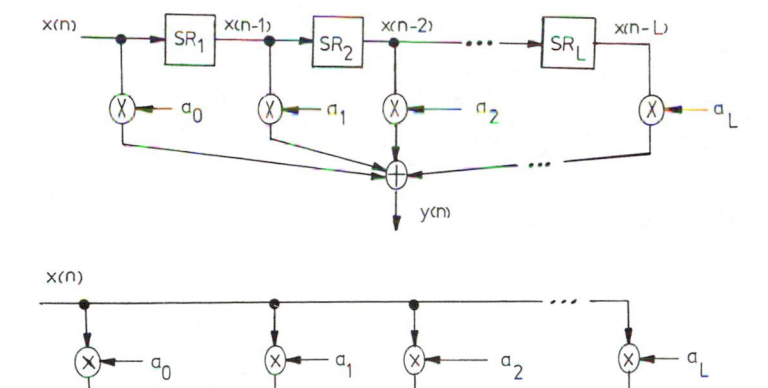

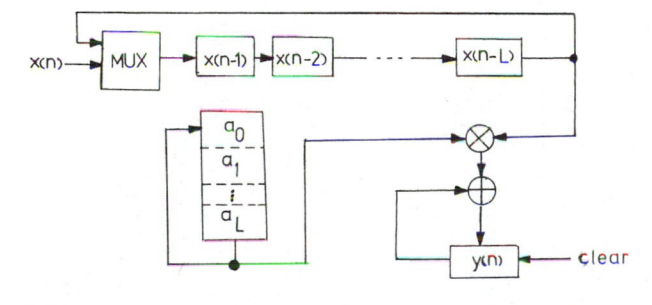

FIG. 5.9.1 Generalized direct FIR filter architectures.

FIG. 5.10.1 Hardware realization of a direct FIR filter.

196

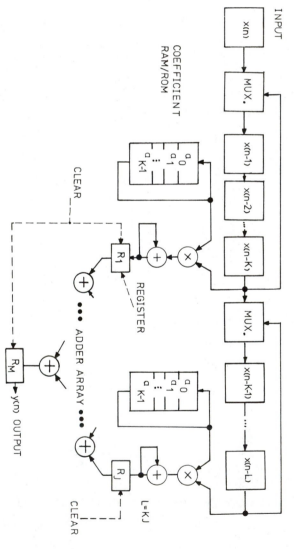

FIG. 5.10.2 Parallel hardware realization of a direct FIR filter.

5.11 COMPARISON

In general, the optimal FIR design procedures will produce an acceptable FIR model of a desired filter response. The historical objection to these methods has been the level of mathematical skill required to understand and code them. Fortunately, many software packages are available which overcome these objections. The use of a window is an open question. However, if the filter is to process aperiodic data, it is recommended that a window be used. Finally, floating-point filters provide high precision at reduced throughput. Fixed-point filters sacrifice precision for speed and a simplified hardware architecture.

BIBLIOGRAPHY

Blackman, R. B., and J. W. Tukey (1958), The Measurement of Power Spectra, Dover, New York.

Gold, B., and C. Rader (1969), Digital Processing of Signals, McGraw-Hill, New York.

Harris, F. J. (1978), On the Use of Windows for Harmonic Analysis with the Discrete Fourier Transform, Proc. IEEE, 66, January, pp. 51-83.

Kaiser, J. F. (1966), Digital Filters, in System Analysis by Digital Computer, Wiley, New York, Chap. 7.

Leon, B. J., and S. C. Bass (1974), Designers' Guide to Digital Filters [6-part series], EDN, January-June.

McClellan, J. H., T. W. Parks, and L. R. Rabiner (1973), A Computer Program for Designing Optimum FIR Linear Phase Digital Filters, IEEE Trans. Audio Electroacoust., AU-21, December, pp. 506-526.

McGee, W. F. (1967), Numerical Approximation Technique for Filter Characteristic Functions, IEEE Trans. Circuit Theory, CT-14, March pp. 92-94.

Oetken, G., T. W. Parks, and H. W. Schussler (1979), A Computer Program for Digital Interpolation Design, in Programs for Digital Signal Processing, IEEE Press, New York, pp. 8.1.1-1.6.

Parks, T. W., and J. H. McClellan (1972a), Chebyshev Approximation for Nonrecursive Digital Filters with Linear Phase, IEEE Trans. Circuit Theory, CT-19, March, pp. 189-194.

Parks, R. W., and J. H. McClellan (1972b), A Program for the Design of Linear Phase Finite Impulse Response Digital Filters, IEEE Trans. Audio Electroacoust., AU-21, August, pp. 195-199.

Remez, E. (1934), Sur le calcul effectif des polynomes de Tchebicheff, Compt. Rend. Acad. Sci. (France), 199, July, pp. 337-340.

Szentermai, G. (1963), Theoretical Basis of a Digital Computer Program for Filter Synthesis, Proc. 1st Allerton Conference on Circuits and Systems, pp. 37-49.

Taylor, F. J., and R. J. Molepske (1973), Optimal Filter Design via Mathematical Programming, IEEE Trans. Syst. Man Cybern., SMC-3, No. 4, July.

6
Infinite Impulse Response Filters

6.1. INTRODUCTION

Unlike the FIR, an infinite impulse response filter (IIR) possesses an impulse response which may persist for all time. The IIR filter does not exhibit the phase linearity of the FIR but compensates for this shortcoming by providing an improved magnitude frequency response. The feedback structure found in this class of filter allows the IIR to reinforce desired frequency domain behavior and attenuate the undesired. Therefore, IIRs are found replacing many bandpass analog filters in communications, controls, biomedicine, geophysics, vibration analysis, radar, sonar, and so on, where high-performance frequency selectivity is required.

The impulse response of an IIR is a data sequence $\{h(n)\}$. In terms of a monic (i.e., leading denominator coefficient equals unity) transfer function $H(z)$, one obtains

$$H(z) = \frac{N(z)}{D(z)} = \sum_{n=0}^{\infty} h(n)z^{-n} = \frac{\displaystyle\sum_{i=0}^{M} b_i z^{-i}}{1 + \displaystyle\sum_{i=1}^{N} a_i z^{-i}} \tag{6.1.1}$$

From a practical viewpoint, a realizable filter must produce a bounded output if stimulated by bounded inputs. Therefore, it is required that the ℓ_1 norm of $\{h(n)\}$, namely

$$\sum_{n=0}^{\infty} |h(n)| < M \tag{6.1.2}$$

This condition can be related to the pole locations of the filter under study. For example, it is well known that a causal discrete system with a rational transfer function H(z) is stable (i.e., bounded inputs produce bounded outputs) if and only if its poles are interior to the unit circle in the z domain. Often referred to as the circle criterion, it can be tested using computer root-finding methods (e.g., POLRT in an IBM or DEC sense). Other algebraic tests—Schur-Cohn, Routh-Hurwitz, and Nyquist—may also be used.

The stability condition is implicit to the FIR as long as all N coefficients are finite. Here the finite sum of magnitudes will be bounded. Also, the zeros and poles of the FIR studied in the preceding section were generally found located along the periphery of the unit circle. However, the polynomial structure of Eq. (6.1.1) [i.e., N(z) and D(z)] allows the poles to be defined to reside anywhere within the unit circle.

The FIR and IIR trade their frequency and phase filtering attributes. They differ in another important area as well. It was noted in the preceding section that the FIR possesses a limited number of architectures. The IIR philosophy, however, is extremely rich in architectural choices. In the following sections, basic IIR design practice and architectural consciousness are developed.

6.2. FLOW DIAGRAMS

Block diagrams may be used to represent symbolically the structure of a system. They are used to display the functional decomposition of a system or filter. However, when analysis is needed, mathematical models must be developed which are based on a set of detailed algebraic operations. Here the signal flow graph of Mason (1953) can assist the user in quantifying these mathematical relationships. The flow graph illustrates the flow of information within a system. A signal flow network is a system of nodes which are interconnected by directed line segments called branches. The elements of a flow diagram are:

1. Input mode (source): a node possessing only outgoing branches
2. Output mode (sink): a node possessing only incoming branches
3. Path: any continuous unidirectional collection of branches
4. Feedback path: path which originates and terminates on a common node in which all nodes are traversed but once
5. Feedforward path: path from input node to output node in which all nodes are traversed but once
6. Path gain: product of branch values along a path
7. Loop gain: product of branch values along a closed path

The signal flow graph provides the user with the following information:

1. Node values are system variables.
2. The directed branch connecting variable x_i to x_j defines the dependence of x_i to x_j.
3. Signals (information) can only flow along directed line segments.
4. Information x_i traversing a branch between x_i and x_j is modified (algebraically) by the branch gain, say p, to form $x_j = px_i$.

One is generally interested in quantifying the input-output transfer function of a given system. This task can be dispatched with relative ease using flow graph methods and Mason's gain formula can be expressed as

$$H(z) = \sum_k \frac{M_k(z)\, \Delta_k(z)}{\Delta(z)} \qquad (6.2.1)$$

where $M_k(z)$ equals the gain of the kth forward path, satisfies

$$\Delta = 1 - \sum_m P_{m1} + \sum_m P_{m2} - \sum_m P_{m3} + \cdots$$

and

$P_{mr}(z)$ = gain of the mth possible combination of r nontouching loops (i.e., no common loops)

$\Delta(z)$ = 1 - [sum of all individual loops (r = 1)] + [sum of gain products of all possible combinations of two nontouching loops (r = 2)] ± etc.

$\Delta_k(k)$ = the value of Δ for the part of the graph not touching the kth forward path.

EXAMPLE 5.2.1 Consider the flow diagram shown in Fig. 6.2.1. The analysis of this network, in a Mason's sense, is summarized in Fig. 6.2.2.

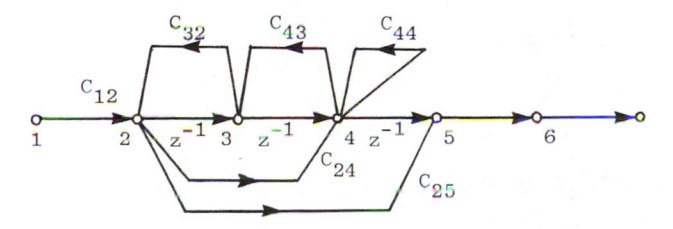

FIG. 6.2.1 Flow diagram for a six-node example problem.

FORWARD PATHS:

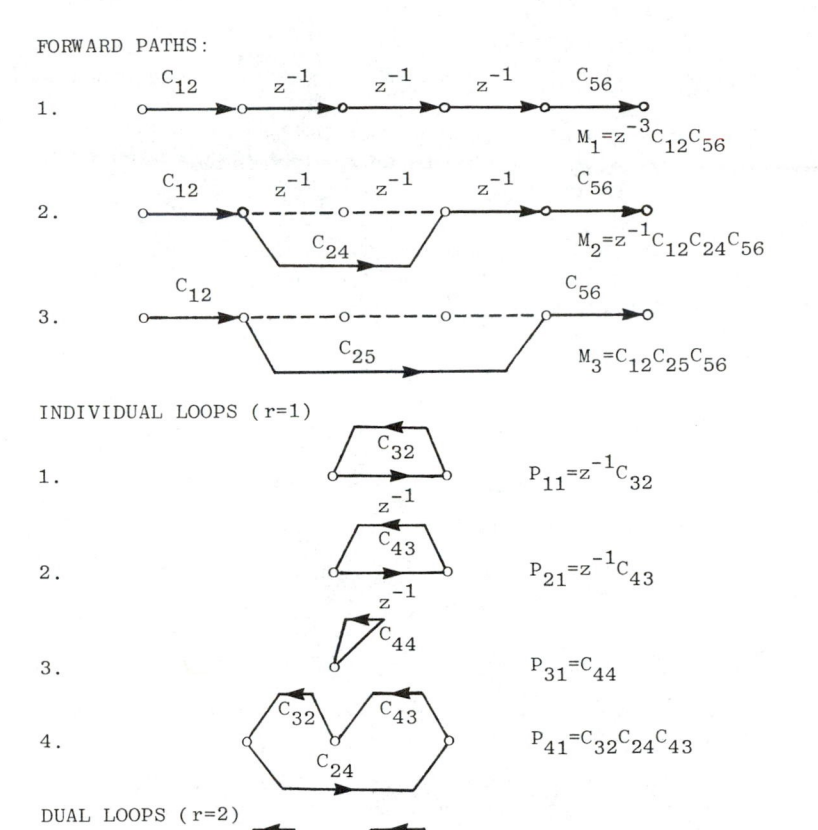

INDIVIDUAL LOOPS (r=1)

DUAL LOOPS (r=2)

MULTIPLE LOOPS (r>2)

FIG. 6.2.2 Description of the forward, individual, and dual loops for the example problem considered in Fig. 6.2.1.

Therefore, the characteristic equation satisfies

$$\Delta = 1 - (z^{-1}C_{32} + z^{-1}C_{43} + C_{44} + C_{32}C_{43}C_{24}) + (z^{-1}C_{32}C_{44})$$

Finally, the gain terms M_i are developed in Fig. 6.2.3, and, as a result, $H(z)$ satisfies

$$H(z) = \frac{\Delta_1(z)M_1(z) + \Delta_2(z)M_2(z) + \Delta_3(z)M_3(z)}{\Delta(z)}$$

$$= \frac{z^{-3}C_{12}C_{56} + z^{-1}C_{12}C_{24}C_{56} + (C_{12}C_{56}C_{25})[1 - (z^{-1}C_{43} + C_{44})]}{1 - (z^{-1}C_{32} + z^{-1}C_{43} + C_{44} + C_{32}C_{43}C_{24}) + (z^{-1}C_{32}C_{44})}$$

Mason's gain formula and flow graphs will be used in this chapter to study the IIR. Since the IIR uses a information feedback principle to achieve a high level of performance, these two analysis methods will be indispensable to their study. Using these principles, information flow can be expressed mathematically in closed form.

There exists another information modeling and management policy that has received wide acceptance in recent times. Called state variables, it is a product of the aerospace era. This method has been successfully used by control scientists to model and analyze complex linear systems. The fundamental concepts of state and state space will be presented in the next section.

FORWARD PATH 1:

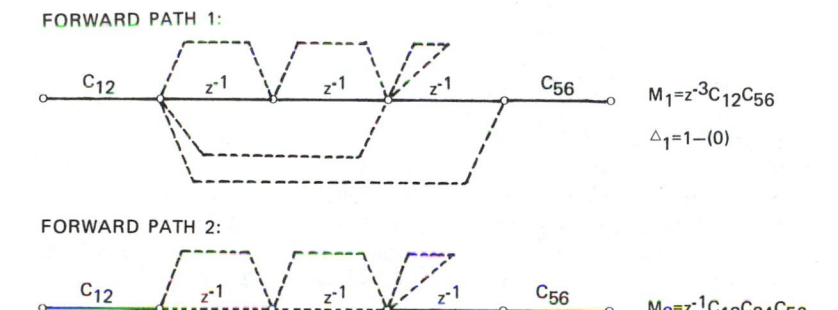

$M_1 = z^{-3}C_{12}C_{56}$

$\Delta_1 = 1 - (0)$

FORWARD PATH 2:

$M_2 = z^{-1}C_{12}C_{24}C_{56}$

$\Delta_2 = 1 - (0)$

FORWARD PATH 3:

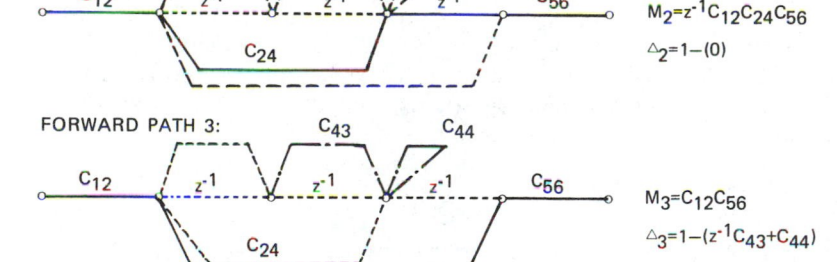

$M_3 = C_{12}C_{56}$

$\Delta_3 = 1 - (z^{-1}C_{43} + C_{44})$

FIG. 6.2.3 Analysis of forward paths 1, 2, and 3 for the example problem considered in Fig. 6.2.1.

6.3. STATE-SPACE METHODS

One of the most convenient methods of providing for a unified FIR modeling philosophy is state space. State-space concepts have been used for several decades to represent linear systems mathematically. They will be used in this section to create a modeling forum by which FIR filters can be analyzed in later sections.

The state of a discrete system, loosely stated, is the minimum amount of knowledge required to determine the future outputs and states of a system given knowledge of the forcing function and the mathematical model of the system. The states of a system, and its outputs, can be expressed in state-variable form as

$$x(k + 1) = f(x(k), u(k)), \quad x(n) \in R^n u(n) \in R^n$$
$$y(k) = g(x(k), u(k)), \quad y(n) \in R^p \tag{6.3.1}$$

where $x(k + 1)$, $v(k)$, and $y(k)$ are n-, m-, and p-dimensional real vectors representing state, forcing function, and output, respectively. The signal processing systems designer is generally concerned with the study of a linear discrete state-determined system of the form

$$x(k + 1) = A(k)x(k) + B(k)u(k)$$
$$y(k) = C(k)x(k) + D(k)u(k) \tag{6.3.2}$$

which is interpreted graphically in Fig. 6.3.1. When the filter's coefficients are constant, a shift-invariant filter results. A linear shift-invariant filter can be uniquely specified in terms of the four-tuple (A, B, C, D). Linear shift-invariant filters are characterized by the property that if the response of the system to a vector-valued times series $\{v(n)\}$ is a vector-valued sequence $\{x(n)\}$, the response to $\{v(n + m)\}$ is $\{x(n + m)\}$.

The response of a linear system to a given stimulus $v(n)$ can be quantified in terms of the state transition matrix. The state transition matrix, as the name implies, will define how a given state vector will evolve in time. For example, the homogeneous (i.e., unforced) solution to the state equation, found in Eq. (6.3.2), is given by

$$x(i) = \Phi(i, j)x(j), \quad \forall i, \forall j \tag{6.3.3}$$

where $\Phi(i, j)$ is called the state transition matrix. By observing that $x(j + 1) = A(j)x(j)$, $x(j + 2) = A(j + 1)x(j + 1) = A(j + 1)A(j)x(j)$, and so forth, it follows that

$$\Phi(i, j) = \prod_{k=j}^{i-1} A(k), \quad \Phi(i, i) = I \tag{6.3.4}$$

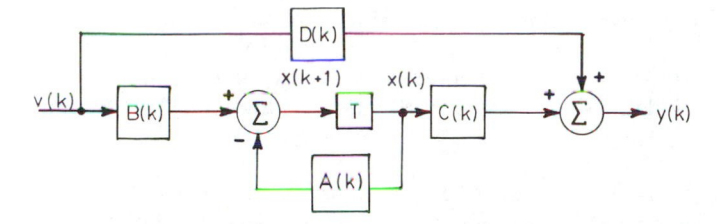

FIG. 6.3.1 Block diagram description of a general linear discrete–state-determined system.

or for the shift-invariant case

$$\Phi(i, j) = \prod_{k=1}^{i-1} A = A^{i-j}, \quad \Phi(i, i) = I \tag{6.3.5}$$

The complete solution to a linear state–determined discrete equation [i.e., Eq. (6.3.2)] is known to be given by

$$x(k) = \Phi(k - k_0)x_0 + \sum_{m=k_0}^{k-1} \Phi(k - m - 1)Bv(m), \quad x(k_0) = x_0 \tag{6.3.6}$$

where the term $\Phi(k - m - 1, 0)$ can be more conveniently represented as $\Phi(k - m - 1)$. For the shift-invariant case, one concludes that for $k_0 = 0$,

$$x(k) = A^k x_0 + \sum_{m=0}^{k-1} A^{k-m-1}Bv(m) \tag{6.3.7}$$

The problem of formalizing a system in state–variable form can therefore be seen as one quantifying the state transition matrix. Many methods have been proposed to compute $\Phi(k)$. The generation of the state transition matrix is essentially a study of classic linear algebraic structures. The standard generation methods are:

1. The Cayley-Hamilton method
2. The transform method
3. The transfer function method

Cayley-Hamilton Method

Cayley-Hamilton's device can be used to compute the shift-invariant transition matrix A^k, where A is a given n × n constant coefficient matrix. It is known that if λ is an eigenvalue of A, and if $f(\lambda) = \Sigma a_i \lambda^i$, i = 0, 1, ..., n − 1, then $f(A) = \Sigma a_i A^i$. Therefore, by choosing $f(\lambda) = \lambda^k$, it follows that $f(A) = A^k$. Since λ is in general a complex number, the algebra of Cayley-Hamilton's device is defined over the complex field. However, $f(\lambda)$ is generally defined to be real.

EXAMPLE 6.3.1 Consider the unstable system defined in terms of the matrix A having eigenvalues of −2 and −3 (i.e., outside the unit circle).

$$A = \begin{bmatrix} 0 & 1 \\ -6 & -5 \end{bmatrix} f(\lambda) = \lambda^k = a_0 + a_1 \lambda$$

$$\lambda_1 = -2, \quad \lambda_2 = -3$$

$$f(\lambda_1) = (-2)^k = a_0 + (-2)a_1 \longrightarrow a_0 \left. \begin{array}{c} 3(-2)^k - 2(-3)^k \\ \\ (-2)^k - (-3)^k \end{array} \right\}$$

$$f(\lambda_2) = (-3)^k = a_0 + (-3)a_1 \longrightarrow a_1$$

Then, from $f(A) = A^k = a_0 I + a_1 A$, one obtains

$$\Phi(k) = \begin{bmatrix} 3(-2)^k - 2(-3)^k & (-2)^k - (-3)^k \\ -6[(-2)^k - (-3)^k] & -2(-2)^k + 3(-3)^k \end{bmatrix}$$

As a numerical check it can immediately be seen that $\Phi(0) = I$ and $\Phi(k)$ [∞] as k tends to infinity (i.e., unstable).

EXAMPLE 6.3.2 Consider now the stable system having eigenvalues located at 1/2 and 1/4 and an A matrix given by

$$A = \begin{bmatrix} 0 & 1 \\ -\dfrac{1}{8} & \dfrac{6}{8} \end{bmatrix}$$

$$\lambda_1 = \frac{1}{2}$$

$$\lambda_2 = \frac{1}{4}$$

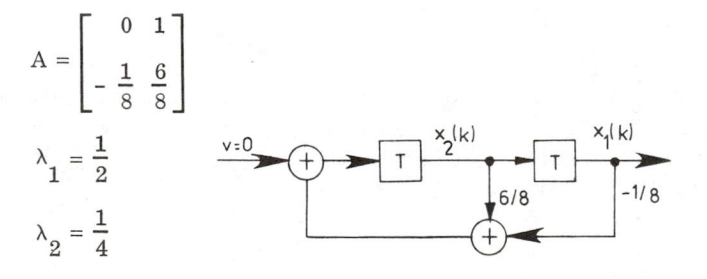

Repeating the analysis found in Example 6.3.1, one concludes that

$$f(\lambda) = \lambda^k = a_0 + a_1 \lambda$$

$$f(\lambda_1) = \left(\frac{1}{2}\right)^k = a_0 + a_1\left(\frac{1}{2}\right) \left.\begin{array}{c} \\ \\ \end{array}\right\} a_0 = 2\left(\frac{1}{4}\right)^4 - \left(\frac{1}{2}\right)^k ,$$

$$f(\lambda_2) = \left(\frac{1}{4}\right)^k = a_0 + a_1\left(\frac{1}{4}\right) \left.\begin{array}{c} \\ \end{array}\right| a_1 = 4\left(\frac{1}{2}\right)^k - \left(\frac{1}{4}\right)^k$$

$$(k) = a_0 I + a_1 A$$

For repeated eigenvalues, the analysis procedure above needs to be modified. If λ_p is an eigenvalue of multiplicity m, λ_p will produce m terms of the form

$$\left.\frac{d^n f(\lambda)}{d\lambda^n}\right|_{\lambda=\lambda_p} = \left.\frac{d^n}{d\lambda^n}\left(\sum_{s=0}^{n-1} a_s \lambda^s\right)\right|_{\lambda=\lambda_p} ; \quad r = 0, 1, \ldots, m-1 \quad (6.3.8)$$

After generating all equations derived from all eigenvalues of all multiplicities, a system of n equations in n unknowns will be defined. These equations are then solved for a coefficient set $\{a_i\}$ and used to synthesize $f(A) = A^k = \Sigma a_i A^i$, $i = 0, 1, \ldots, n-1$.

Transform Method

Since Φ can be defined in terms of the homogeneous state equation $x(k+1) = Ax(k)$, one notes that

$$zX(z) - zx(0) = AX(z) \qquad (6.3.9)$$

or

$$X(z) = (zI - A)^{-1} zX(0) \qquad (6.3.10)$$

Therefore,

$$\Phi(k) = Z^{-1}(zI - A)^{-1} z^k \qquad (6.3.11)$$

which, from the theory of complex variables (i.e., residue theory) is known to equal

$$\Phi(k) = \Sigma \text{ residues of } (zI - A)^{-1} z^k \qquad (6.3.12)$$

(see Section 2.4).

EXAMPLE 6.3.3 Given

$$A = \begin{bmatrix} 0 & 1 \\ -6 & -5 \end{bmatrix}; \quad (zI - A)^{-1} z^k = \frac{\begin{bmatrix} z+5 & 1 \\ -6 & z \end{bmatrix} z^k}{(z+2)(z+3)} = \begin{bmatrix} p_{11}(z) & p_{12}(z) \\ p_{21}(z) & p_{22}(z) \end{bmatrix}$$

one obtains, on using partial fraction expansion,

$$z^k p_{11}(z) = \frac{z+5}{(z+2)(z+3)} = \frac{3(-2)^k}{z+2} - \frac{2(-3)^k}{z+3}$$

$$z^k p_{12}(z) = \frac{1}{(z+2)(z+3)} = \frac{(-2)^k}{z+2} - \frac{(-3)^k}{z+3}$$

$$z^k p_{22}(z) = \frac{z}{(z+2)(z+3)} = \frac{(-2)^{k+1}}{z+2} - \frac{(-3)^{k+1}}{z+3}$$

$$z^k p_{21}(z) = \frac{-6}{(z+2)(z+3)} = \frac{(-2)^{k+1}}{z+2} - \frac{(-3)^{k+1}}{z+3}$$

and

$$\Phi(k) = A^k = \begin{bmatrix} 3(-2)^k - 2(3)^k & (-2)^k - (-3)^k \\ -6(-2)^k + 6(-3)^k & (-2)^{k+1} - (-3)^{k+1} \end{bmatrix}$$

EXAMPLE 6.3.4 Given

$$A = \begin{bmatrix} 0 & 1 \\ -\dfrac{1}{8} & \dfrac{6}{8} \end{bmatrix}; \quad (zI - A)^{-1} z^k = \frac{\begin{bmatrix} z - \dfrac{6}{8} & 1 \\ -\dfrac{1}{8} & z \end{bmatrix} z^k}{(z - 1/2)(z - 1/4)} = \frac{N(z) z^k}{(z - 1/2)(z - 1/4)}$$

and noting that

$$\frac{I}{(z - 1/2)(z - 1/4)} = \frac{-4I}{z - 1/2} + \frac{4I}{z - 1/4}$$

it follows that for all $k \geq 0$,

$$\Phi(k) = \Sigma \text{ residues of } (zI - A)^{-1} z^k$$

$$= 4N\left(z = \frac{1}{2}\right) + 4N\left(z = \frac{1}{4}\right)$$

$$= \left(\frac{1}{2}\right)^k \begin{bmatrix} -1 & 4 \\ -\frac{1}{2} & 2 \end{bmatrix} - \left(\frac{1}{4}\right)^k \begin{bmatrix} -2 & 4 \\ -\frac{1}{2} & 1 \end{bmatrix}$$

Transfer Function Method

We may alternatively write

$$\Phi(k) = Z^{-1}(zI - A)^{-1}z = Z^{-1}(\Phi(z)) \tag{6.3.13}$$

or

$$z\Phi(z) = (zI - A)^{-1} \tag{6.3.14}$$

where $\Phi(z)$ can be inverted using table-lookup techniques. The transform method is generally considered to be superior to the other options since the resolvent matrix $[(zI - A)^{-1}]$ need only be evaluated at simple poles.

6.4 COMPUTER-AIDED METHODS

The inversion of the resolvent matrix can be performed on a general-purpose digital computer. However, normal matrix inversion routines used for inverting constant-coefficient matrices are not applicable here. The presence of the variable z in the resolvent matrix $(zI - A)$ forces one to seek an alternative method. One such alternative can be traced to Leverrier in 1840, Horst in 1935, Souriou in 1948, Frame in 1949, and Faddeev and Sominshii in 1949, with extensions by Taylor (1976). Leverrier's algorithm states that

$$(zI - A)^{-1} = \frac{\text{adj}(zI - A)}{\det(zI - A)} \tag{6.4.1}$$

where adj denotes adjoint and $\det(zI - A) = z^n + a_n z^{n-1} + \cdots + a_{n-1}z + a_n$, and

$$\text{adj}(zI - A) = H_1 z^{n-1} + H_2 z^{n-2} + \cdots + H_n \tag{6.4.2}$$

The matrices H_i satisfy the following system of equations:

$$H_1 = I, \quad d_0 = c_0 = 1.0$$

$$d_1 = -\text{trace } A$$

$$H_k = AH_{k-1} + d_{k-1}I, \quad k = 2, 3, \ldots, n \tag{6.4.3}$$

$$d_k = -\frac{1}{k} \text{ trace}(AH_k)$$

(see Fig. 6.4.1). As a computational check, it is known that $H_{n+1} = [0]$. It should be noted that the H_i's are defined in terms of known coefficients and therefore will support digital computer analysis. For example, consider the simple system given by

$$y(k) = v(k) - \frac{4}{8}v(k-1) + \frac{14}{8}y(k-1) - \frac{7}{8}y(k-2) + \frac{1}{8}y(k-3)$$

and having the flow graph found in Fig. 6.4.2. The system's characteristic equation can be computed using Mason's gain formula dn it is

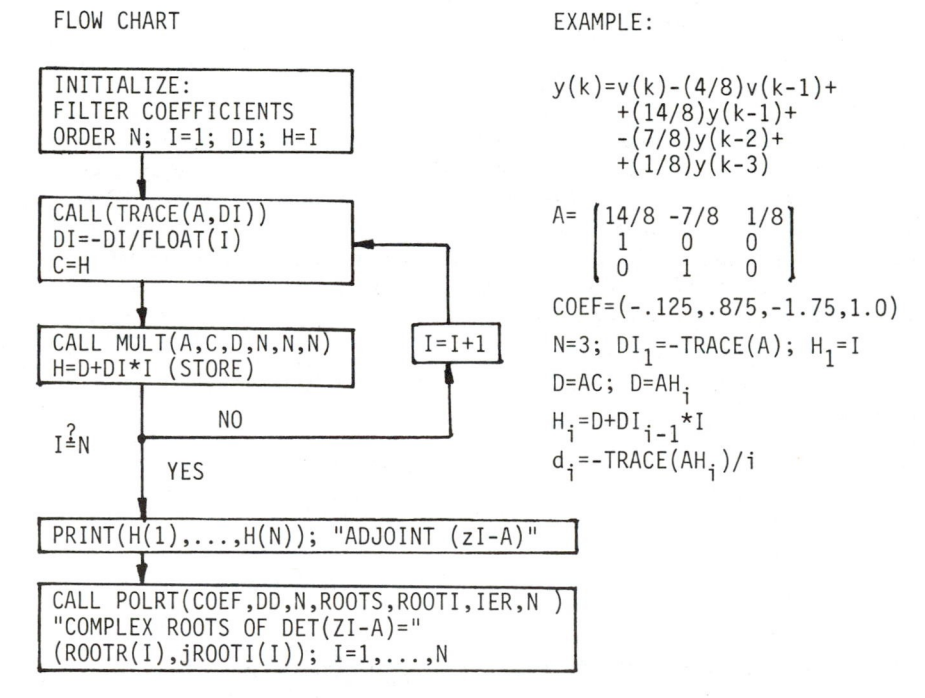

FIG. 6.4.1 Flow diagram for computing a state transition matrix.

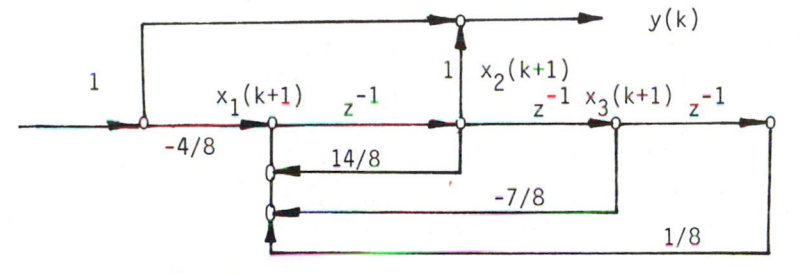

FIG. 6.4.2 Flow diagram for computing a state transition matrix using Leverrier's algorithm.

and having the flow graph found in Fig. 6.4.2. The system's characteristic equation can be computed using Mason's gain formula and it is

$$\Delta = 1 - \frac{14}{8}z^{-1} + \frac{7}{8}z^{-2} - \frac{1}{8}z^{-3}$$

or

$$z^3\Delta = z^3 - \frac{14}{8}z^2 + \frac{7}{8}z - \frac{1}{8}$$

In terms of a state-space description, one obtains

$$x(k + 1) = Ax(k) + bu(k)$$
$$y(k) = cx(i) + d_0u(k) \tag{6.4.4}$$

where

$$A = \begin{bmatrix} \frac{14}{8} & -\frac{7}{8} & \frac{1}{8} \\ 1 & 0 & 0 \\ 0 & 1 & 0 \end{bmatrix}, \quad b = \begin{bmatrix} -\frac{4}{8} \\ 0 \\ 0 \end{bmatrix}, \quad c = (1 \ 0 \ 0), \quad d_0 = 1$$

Direct computation yields

$$d_1 = -\frac{14}{8}, \quad d_2 = \frac{7}{8}, \quad d_3 = \frac{-1}{8}, \quad H_1 = I$$

$$H_2 = \begin{bmatrix} 0 & -\dfrac{7}{8} & \dfrac{1}{8} \\[2ex] 0 & -\dfrac{14}{8} & 0 \\[2ex] 0 & 1 & -\dfrac{14}{8} \end{bmatrix}, \quad H_3 = \begin{bmatrix} 0 & \dfrac{1}{8} & 0 \\[2ex] 0 & 0 & \dfrac{1}{8} \\[2ex] 1 & -\dfrac{14}{8} & \dfrac{7}{8} \end{bmatrix}$$

and it follows that

$$\mathrm{adj}(zI - A) = H_1 z^2 + H_2 z + H_3 = \begin{bmatrix} z^2 & & \dfrac{3}{8} \\[2ex] z & z^2 - \left(-\dfrac{14}{8}\right)z & \dfrac{1}{8} \\[2ex] 1 & z - \dfrac{14}{8} & z^2 - \dfrac{14}{8}z + \dfrac{7}{8} \end{bmatrix}$$

Using standard software support (e.g., Scientific Support Package of IBM, DEC, etc.), the resolvent matrix and pole locations can be easily computed. An output listing for the third-order problem in Fig. 16.4.2 showing the values of H(1), H(2), and H(3) is as follows:

```
FROM D1 TI DN        -1.750   0.875   -0.125

H(1) MATRIX

1.000000       0.000000       0.000000
0.000000       1.000000       0.000000
0.000000       0.000000       1.000000

H(2) MATRIX

0.000000      -0.875000       0.125000
1.000000      -1.750000       0.000000
0.000000       1.000000      -1.750000

H(3) MATRIX

0.000000       0.12500        0.000000
0.000000       0.00000        0.125000
1.000000      -1.75000        0.875000

ADJ(XI-A)=H(1)*Z**(N-1)+H(2)*Z**(N-2)+...+H(N)


THE COEFFICIENTS OF DET(ZI-A) FROM THE LARGEST POWER ORDER OF POLY

-0.125000     0.875000     -1.750000     1.000000

THE COMPLEX ROOTS OF DET(ZI-A) ARE

0.250000      0.000000     0.500000     0.000000     1.000000     0.000000

** INVERSE(ZI-A)=ADJ(ZI-A)/DET(ZI-A)
```

A software package capable of computing state transition matrices for linear shift-invariant systems is presented in Appendix A at the end of this chapter. Sample output, for the example problem considered in this section, is also given in Appendix A.

6.5 PRECEDENCE RELATIONS

Flow diagrams allow linear systems to be efficiently modeled. In general, a system would be decomposed into a network of nodes and branches. A system consists of N nodes and branches. Suppose that $\{y(n)\}$ is a N-dimensional time-series vector of signals found at the nodes of a filter. Let $\{x(n)\}$ be a N-dimensional time-series vector of the same network. Furthermore, let X(z) and Y(z) be their N-dimensional z-transforms. In general, a linear shift-invariant system can be characterized as

$$y(n) = x(n) + F_c y(n) + F_d y(n-1) \tag{6.5.1}$$

or

$$Y(z) = X(z) + F_c Y(z) + F_d Y(z) z^{-1} \tag{6.5.2}$$

where F_c is a $N \times N$ matrix of branch coefficients and F_d is a $N \times N$ matrix of coefficients found in the delay branches. This concept can be developed in the context of the second-order system shown in Fig. 6.5.1. For this specific network, the flow diagram realization can be expressed as

$$
Y(z) =
\begin{bmatrix}
Y_1(z) \\
Y_2(z) \\
Y_3(z) \\
Y_4(z) \\
Y_5(z)
\end{bmatrix}
=
\begin{bmatrix}
X_1(z) \\
0 \\
0 \\
0 \\
0
\end{bmatrix}
+
\begin{bmatrix}
0 & 0 & 0 & 0 & 0 \\
a_{12} & 0 & a_{32} & a_{12} & 0 \\
0 & 0 & 0 & 0 & 0 \\
0 & 0 & 0 & 0 & 0 \\
0 & 0 & a_{35} & a_{45} & 0
\end{bmatrix}
\begin{bmatrix}
Y_1(z) \\
Y_2(z) \\
Y_3(z) \\
Y_4(z) \\
Y_5(z)
\end{bmatrix}
$$

$$
+ z^{-1}
\begin{bmatrix}
0 & 0 & 0 & 0 & 0 \\
0 & 0 & 0 & 0 & 0 \\
0 & 0 & 0 & 0 & 0 \\
0 & 0 & 0 & 0 & 0 \\
0 & 0 & 0 & 0 & 0
\end{bmatrix}
\begin{bmatrix}
Y_1(z) \\
Y_2(z) \\
Y_3(z) \\
Y_4(z) \\
Y_5(z)
\end{bmatrix}
$$

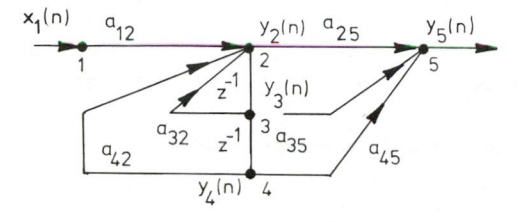

FIG. 6.5.1 Original node ordering for a second-order example problem, showing all five definable nodes.

In terms of a composite system, one obtains

$$Y(z) = T(z)X(z) \tag{6.5.3}$$

where $T(z) = [I - F_c - z^{-1}F_d]^{-1}$. The ijth element of $T(z)$, say $T_{ij}(z)$, is the transfer function that exists between nodes i and j. Those who have designed such systems realize that there exists a temporal hierarchy which underscores their realization. There is an implied <u>precedence</u> relationship found in the realization of Eq. (6.5.1). That is, the order of computation is important. For example, suppose that the example problem considered previously is to be realized as a FORTRAN source code. An excerpt from such a code list would read as follows:

```
        DO 100 I=1,N

        Y1(I)=X1(I)

        Y2(I)=A12*Y1(I)+A32*Y3(I)+A42*Y4(I)

        Y3(I)=Y2(I-1)

        Y4(I)=Y3(I-1)

100     Y5(I)=A25*Y2(I)+A35*Y3(I)+A45*Y4(I)
```

The second line of code contains a potential flaw. It can be noted that Y2(I), which represents $y_2(i)$, is defined in terms of Y1(I) [i.e., $y_1(i)$] and the two variables Y3(I) and Y4(I). The last two variables represent in fact $y_3(i-1)$ and $y_4(i-1)$ (in terms of the code) and not the desired $y_3(i)$ and $y_4(i)$. The condition described in this example is due to a <u>lack of precedence.</u> It has been shown by Crochiere and Oppenheim (1975) that to define a system which is free of such logical errors, the matrix F_c must be lower triangular. If so, YJ(I) will be represented in terms of YK(I), which for $K > J$, represents $y_j(i-1)$.

If a valid precedence relationship for a given system exists, it can be obtained through row-column reordering of Eq. (6.5.1). For example, Eq. (6.5.3) can be seen to have a F_c matrix which is not lower triangular. It may be rewritten to read

$$
\begin{bmatrix} w_1(z) \\ w_2(z) \\ w_3(z) \\ w_4(z) \\ w_4(z) \end{bmatrix} = \begin{bmatrix} x_1(z) \\ 0 \\ 0 \\ 0 \\ 0 \end{bmatrix} + \begin{bmatrix} 0 & 0 & 0 & 0 & 0 \\ 0 & 0 & 0 & 0 & 0 \\ 0 & 0 & 0 & 0 & 0 \\ w_{14} & w_{24} & w_{34} & 0 & 0 \\ 0 & w_{25} & w_{35} & w_{45} & 0 \end{bmatrix} \begin{bmatrix} w_1(z) \\ w_2(z) \\ w_3(z) \\ w_4(z) \\ w_5(z) \end{bmatrix}
$$

$$
+ z^{-1} \begin{bmatrix} 0 & 0 & 0 & 0 & 0 & w_1(z) \\ 0 & 0 & 1 & 0 & 0 & w_2(z) \\ 0 & 0 & 0 & 1 & 0 & w_3(z) \\ 0 & 0 & 0 & 0 & 0 & w_4(z) \\ 0 & 0 & 0 & 0 & 0 & w_5(z) \end{bmatrix}
$$

which does possess an acceptable F_c matrix. The system with reindexed node values is interpreted in Fig. 6.5.2.

The reordering can be performed in a computer-aided sense as well. First observe that the delay branch matrix F_d has no effect on a precedence relationship. One can therefore concentrate on searching network nodes which have no coefficients entering them (here nodes 1, 3, and 4 of Fig. 6.5.1). These nodes can be collected into a common node set, say N_1. Continuing along this line, N_i is formed from those remaining nodes connected to the set N_{i+1}. The node set decomposition of the example network is found in Fig. 6.5.3. The data in this figure are encoded in terms of the node indices of Fig. 6.5.1 and 6.5.2. Observe that the latter node order can be seen to be identical to that suggested in Eq. (6.5.3).

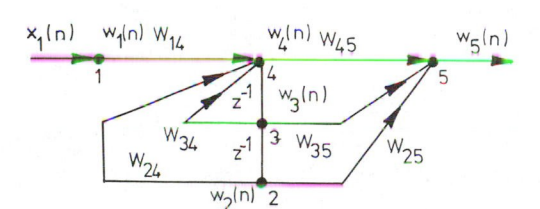

FIG. 6.5.2 Reordering of the nodes found in Fig. 6.5.1 in order to achieve precedence.

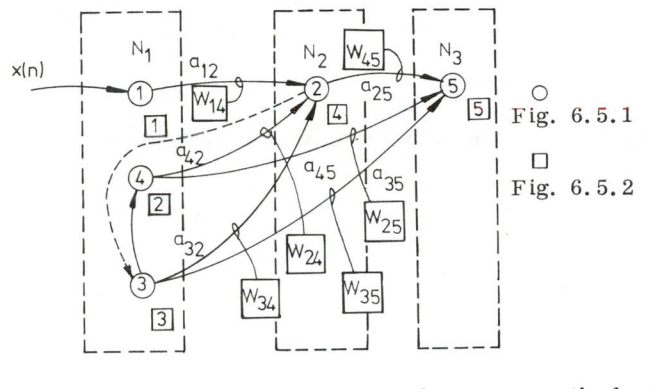

FIG. 6.5.3 Grouping nodes into subsystems on the basis of their precedence relations.

EXAMPLE 6.5.1 The FORTRAN source code found in Appendix B can be used to analyze a discrete system in terms of precedence. Analyzing the systems found in Fig. 6.5.4 in a computer-aided sense resulted in the data given in Fig. 6.5.5. The output listing is interpreted in terms of FC (a filter coefficient) and FD (a delay coefficient) residing between nodes i and j. The nodes are classified in terms of their precedence values. A node value determined to belong to the #j must <u>precede</u> any node belonging to group #j if i is less than j. Ordering within a common group is immaterial. That is, for example network 2, the original nodes 2 and 4 belong to group #1. As a result, both renumbering assignments, (2 = 1, 4 = 2) and (4 = 1, 2 = 2) are valid. The original nodes 1, 3, and 5 belong to groups #2, #3, and #4. As a result, the following reordering schemes may be used:

Original node value	Reordering 2	Reordering 2
1	3	3
2	1	2
3	4	4
4	2	1
5	5	5

6.6 FLOW DIAGRAM-TO-STATE SPACE CONVERSION

The simple system outlined in Fig. 6.6.1 can be interpreted in terms of the following vector-valued system of equations.

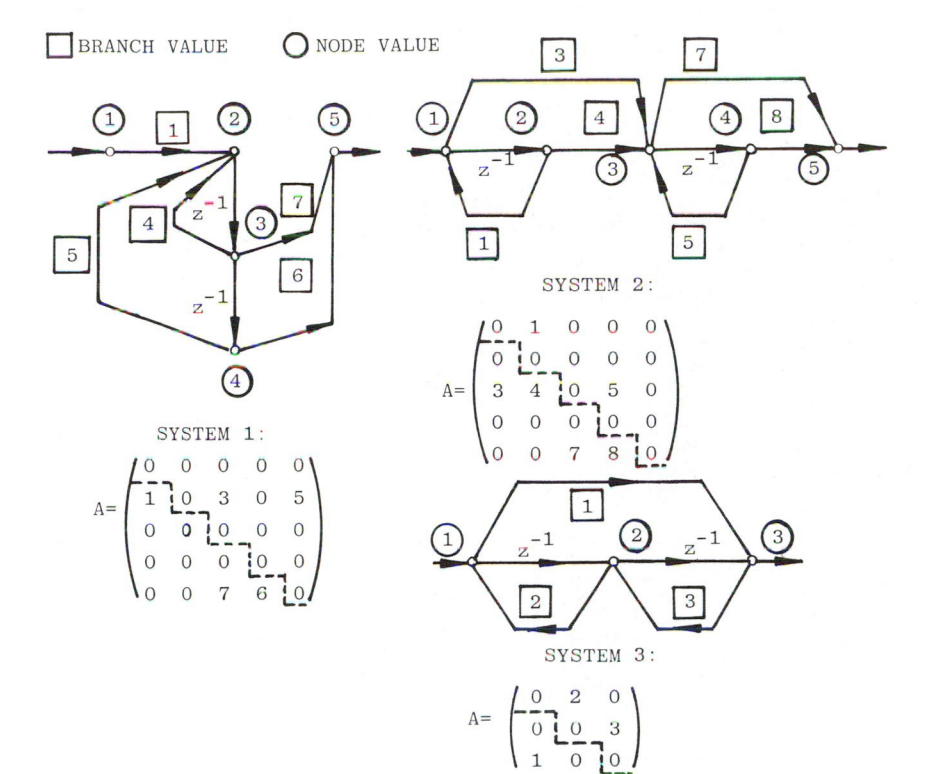

FIG. 6.5.4 Three second-order systems to be studied in terms of precedence relations.

$$W(z) = F(z)W(z) + BV(z)$$

$$F(z) = [F_c + z^{-1}F_d]$$

$$w(n+1) = \begin{bmatrix} 1 & 0 & 0 & a \\ 1 & 0 & 0 & 0 \\ 0 & b_0 & 0 & b_1 \\ 0 & 0 & 0 & 0 \end{bmatrix} w(n) + \begin{bmatrix} 0 & 0 & 0 & 0 \\ 0 & 0 & 0 & 0 \\ 0 & 0 & 0 & 0 \\ 0 & 1 & 0 & 0 \end{bmatrix} w(n-1) + \begin{bmatrix} 1 \\ 0 \\ 0 \\ 0 \end{bmatrix}$$

$$y(n) = (0 \quad 0 \quad 1 \quad 0)w(n) + 0v(n)$$

The output process $y(n)$ is seen to be a function of the two information-bearing values w_2 and w_4. Therefore, one can assign these two parameter state values in the context of a state-determined filter model. For example, let $x_1(n) = w_2(n)$ and $x(n) = w_4(n)$; then

```
                EXAMPLE #1
X  1  N=  1
B  1  N(  1,  2)  FC=  1.000
B  2  N(  2,  3)  FD=  1.000
B  3  N(  3,  4)  FD=  1.000
B  4  N(  3,  2)  FC=  4.000
B  5  N(  4,  2)  FC=  5.000
B  6  N(  4,  5)  FC=  6.000
B  7  N(  3,  5)  FC=  7.000
#  1    CALCULATE NODE    1
#  1    CALCULATE NODE    3
#  1    CALCULATE NODE    4
#  2    CALCULATE NODE    2
#  2    CALCULATE NODE    5
```

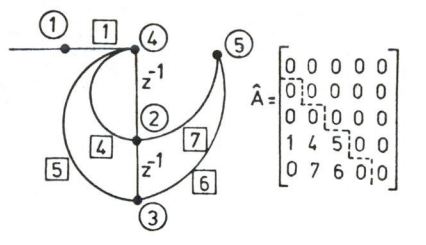

```
                EXAMPLE #2
X  1  N=  1
B  1  N(  2,  1)  FC=  1.000
B  2  N(  1,  2)  FD=  1.000
B  3  N(  1,  3)  FC=  3.000
B  4  N(  2,  3)  FC=  4.000
B  5  N(  4,  3)  FC=  5.000
B  6  N(  3,  4)  FD=  1.000
B  7  N(  3,  5)  FC=  7.000
B  8  N(  4,  5)  FC=  8.000
#  1    CALCULATE NODE    2
#  1    CALCULATE NODE    4
#  2    CALCULATE NODE    1
#  3    CALCULATE NODE    3
#  4    CALCULATE NODE    5
```

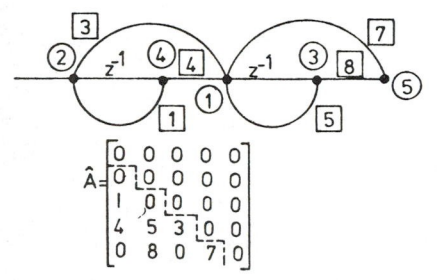

```
            EXAMPLE #3   (THE SYSTEM HAS NO SOLUTION)
X  1  N=  1
B  1  N(  1,  3)  FC=  1.000
B  2  N(  2,  1)  FC=  2.000
B  3  N(  3,  2)  FC=  3.000
B  4  N(  1,  2)  FD=  1.000
B  5  N(  2,  3)  FD=  1.000
 THE SYSTEM DOES NOT HAVE A COMPLETE SOLUTION
```

FIG. 6.5.5 Computer-aided analysis of the systems report in Fig. 6.5.4.

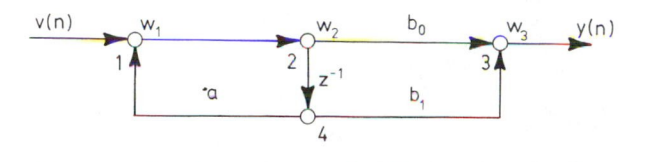

FIG. 6.6.1 Flow diagram for first–order system.

$$[I - F(z)]W(z) = BV(z)$$ (6.6.2)

or

$$(I - F_c - z^{-1}F_d)W(z) = BV(z), \quad F(z)W(z) = BV(z)$$

$$F(z) = \begin{bmatrix} 1 & 0 & 0 & -a \\ -1 & 1 & 0 & 0 \\ 0 & -b_0 & 1 & -b_1 \\ 0 & -z^{-1} & 0 & -b \end{bmatrix}$$ (6.6.3)

Solving for the state variables w_2 and w_4, we obtain (using Cramer's rule)

$$X_1(z) = W_2(z) = \frac{\det \begin{bmatrix} 1 & V(z) & 0 & -a \\ -1 & 0 & 0 & 0 \\ 0 & 0 & 1 & -b_1 \\ 0 & 0 & 0 & 1 \end{bmatrix}}{\det(F)} = \frac{V(z)}{1 - az^{-1}}$$

$$\Rightarrow x_1(n) = ax(n - 1) + v(z)$$

$$X_2(z) = W_4(z) = \frac{\det \begin{bmatrix} 1 & 0 & 0 & V(z) \\ -1 & 1 & 0 & 0 \\ 0 & -b_0 & 1 & 0 \\ 0 & -z^{-1} & 0 & 0 \end{bmatrix}}{\det(F)} = \frac{z^{-1}V(z)}{1 - az^{-1}} = z^{-1}X_1(z)$$

$$\Rightarrow x_2(n) = x_1(n - 1)$$

The output process $y(n)$ is noted to be explicitly given by $w_3(n)$, where

$$Y(z) = W_3(z) = \frac{\det \begin{bmatrix} 1 & 0 & V(z) & -a \\ -1 & 1 & 0 & 0 \\ 0 & -b_0 & 0 & -b_1 \\ 0 & -z^{-1} & 0 & 1 \end{bmatrix}}{} = \frac{z^{-1}b_1 V(z) + b_0 V(z)}{1 - az^{-1}}$$

$$= b_1 X_2(z) + b_0 X_1(z)$$

Therefore, the resulting state equations become

$$\begin{bmatrix} x_1(n+1) \\ x_2(n+1) \end{bmatrix} = \begin{bmatrix} a & 0 \\ 1 & 0 \end{bmatrix} \begin{bmatrix} x_1(n) \\ x_2(n) \end{bmatrix} + \begin{bmatrix} 1 \\ 0 \end{bmatrix} v(n)$$

$$y(n) = (b_0 \quad b_1) \begin{bmatrix} x_1(n) \\ x_2(n) \end{bmatrix}$$

and for the special case where $b_1 = 0$, the equations degenerate to

$$x(n) = ax(n-1) + u(n); \quad y(n) = b_0 u(n)$$

6.7 ARCHITECTURES

During the last decade several digital filter architectures have evolved. In this section the more commonly used architectures will be explored, diagrammed, and a state-variable description provided. A comparative analysis of their performance will be offered in future sections. The filters considered in this section belong to a class of systems referred to as single input-single output filters having a transfer function given by

$$H(z) = \frac{\sum_{i=0}^{n} a_i z^{-i}}{1 + \sum_{i=1}^{n} b_i z^{-i}} \tag{6.7.1}$$

Many filters fall into this broad classification. Some of the more common are:

1. Direct I
2. Direct II
3. Standard
4. Parallel
5. Cascaded

Other filters, such as the ladder and wave systems, will be discussed at another time.

Direct I

Direct I realizations obtain their name from the fact that their architecture is directly deduced from the given transfer function. An admissible direct I architecture is shown in Fig. 6.7.1. These architectures are inefficient from the standpoint of requiring 2N delays. Other structures, which will be studied in this section, require only N delays. However, for the sake of completeness, direct I architectures will be developed in a state-space vein.

First consider the form

$$y(k) = \begin{bmatrix} 0 & 1 & 0 & \cdots & 0 \\ 0 & 0 & 1 & \cdots & 0 \\ \cdot & \cdot & \cdot & & \cdot \\ \cdot & \cdot & \cdot & & \cdot \\ \cdot & \cdot & \cdot & & \cdot \\ b_1 & b_2 & b_3 & \cdots & b_N \end{bmatrix} y(k-1) + \begin{bmatrix} \lambda_1 \\ \lambda_2 \\ \cdot \\ \cdot \\ \cdot \\ \lambda_N \end{bmatrix} u(k)$$

where $y(k)$ is a real n-dimensional vector and

$$y_1(k) = y_2(k-1) + \lambda_1 u(k)$$

$$y_2(k) = y_3(k-1) + \lambda_2 u(k)$$

$$\vdots$$

$$y_N(k) = \sum_{i=1}^{n} b_i y(k-i) + \lambda_N u(k)$$

(6.7.3)

Upon recombining, one obtains

$$y_n(k) = b_1 y_1(k-1) + b_2 y_2(k-1) + \cdots + b_{n-1} y_{n-1}(k-1) + b_n y_n(k-1) + \lambda_n u(n)$$

$$= b_1[y_2(k-2) + \lambda_1 u(k-1)] + b_2[y_3(k-2) + \lambda_2(k-1)u(k-1) + \cdots$$

$$+ b_{n-1}[y_n(k-2) + \lambda_{n-1} u(k)] + b_n y_n(k-1) + \lambda_n u(k)$$

(continued until a function of y_n and u only)

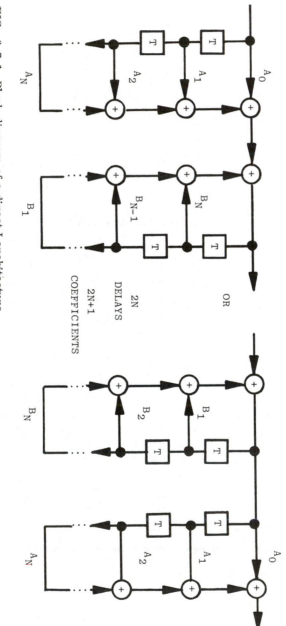

FIG. 6.7.1 Block diagram of a direct I architecture.

$$= \sum_{i=1}^{N} b_i y_N[k + (N + 1) - i] + (b_1 \lambda_N)u(k - N) + \cdots$$

$$+ (b_1 \lambda_{N-1} + b_2 \lambda_N)u(k - N + 1) + \cdots (b_1 \lambda_1 + b_2 \lambda_2 + \cdots$$

$$+ b_N \lambda_N)u(k - 1) + \lambda_N u(k) \tag{6.7.4}$$

which implies that

$$b_1 \lambda_n = a_n; \quad b_1 \lambda_{n-1} + b_2 \lambda_n = a_{n-1}; \quad \cdots; \quad \lambda_n = a_0 \tag{6.7.5}$$

Finally, the observations (output) are given by $z(k) = (0, 0, \ldots, 1)y(k)$. Since this is an inefficient architecture, it will not be pursued further.

Direct II[†]

The direct II form is a popular architecture and is abstracted in Fig. 6.7.2. It requires N delays and in general $2N - 1$ coefficients and therefore $2N - 1$ multipliers. The construction rule used to derive the state-variable filter is given by

$$y(k) = x_n(k) + a_0 v(k)$$

$$x_1(k + 1) = -b_n y(k) + a_n v(k) = -b_n x_n(k) + (a_n - a_0 b_n)v(k)$$

$$x_2(n + 1) = x_1(k) - b_{n-1}x_n(k) + (a_{n-1} - a_0 b_{n-1})v(k) \tag{6.7.6}$$

$$\vdots \qquad \vdots \qquad \vdots \qquad \vdots$$

$$x_n(k + 1) = x_{n-1}(k) - b_1 x_n(k) + (a_1 - a_0 b_1)v(k)$$

or in more compact form,

$$x(k + 1) = \begin{bmatrix} 0 & 0 & \cdots & 0 & 0 & -b_n \\ 1 & 0 & \cdots & 0 & 0 & -b_{n-1} \\ \cdot & \cdot & & \cdot & \cdot & \cdot \\ \cdot & \cdot & & \cdot & \cdot & \cdot \\ 0 & 0 & \cdots & 0 & 1 & -b_2 \\ 0 & 0 & \cdots & 0 & 0 & -b_1 \end{bmatrix} x(k) + \begin{bmatrix} a_n - a_0 b_n \\ a_{n-1} - a_0 b_{n-1} \\ \cdot \\ \cdot \\ a_2 - a_0 b_2 \\ a_1 - a_0 b_1 \end{bmatrix} v(k)$$

where $x(n)$ is a real n-dimensional vector and $y(n) = (0, 0, \ldots, 1)x(n) + a_0 v(n)$.

[†]A digital network is said to be <u>canonic</u> if the number of delays used equals the filter order. The direct II and all following architectures are canonic.

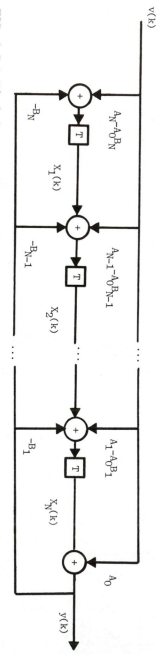

FIG. 6.7.2 Block diagram of a direct II architecture.

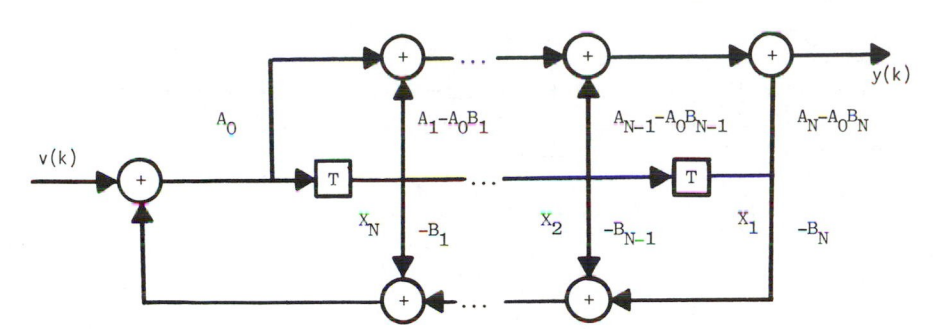

FIG. 6.7.3 Block diagram of an alternative direct II architecture.

An alternative direct II architecture can be found in Fig. 6.7.3. The state construction rule which is applicable to this architecture is outlined below:

$$x_1(k + 1) = x_2(k)$$

$$x_2(k + 1) = x_3(k)$$

$$\vdots \qquad \vdots$$

$$x_{n-1}(k + 1) = x_n(k)$$

$$x_n(k + 1) = -b_n x_1(k) - b_{n-2} x_2(k) - \cdots - b_1 x_2(k) + v(k)$$

$$y(k) = a_n x_1(k) + a_{n-1} x_2(k) + \cdots + a_1 x_n(k)$$

$$+ a_0[v(k) - b_n x_1(k) - \cdots - b_1 x_n(k)]$$

It follows that

$$x(k + 1) = \begin{bmatrix} 0 & \cdots & 0 \\ 0 & \cdots & 0 \\ \vdots & \ddots & \vdots \\ 0 & \cdots & 0 \\ -b_n & -b_n & \cdots & -b_2 & -b_1 \end{bmatrix} x(k) + \begin{bmatrix} 0 \\ 0 \\ \vdots \\ 0 \\ 1 \end{bmatrix} v(k) \qquad (6.7.9)$$

$$y(k) = (a_n - a_0 b_n, \ a_{n-1} - a_0 b_{n-1}, \ \ldots, \ a_1 - a_0 b_1)x(k) + a_0 v(k)$$

Standard Form

The standard form combines the feedback and feedforward structures found in the direct II realizations. Its structure is shown in Fig. 6.7.4, with a construction rule given as follows:

$$x_1(k + 1) = x_2(k) + \alpha_1 v(k)$$
$$\vdots \qquad \vdots \qquad \vdots$$
$$x_j(k + 1) = x_{j+1}(k) + \alpha_j v(k)$$
$$\vdots \qquad \vdots \qquad \vdots \qquad\qquad\qquad (6.7.10)$$
$$x_n(k + 1) = -\beta_n x_1(k) - \beta_{n-1} x_2(k) - \cdots - \beta_1 x_n(k) + \alpha_n v(k)$$

$$y(k + 1) = x_1(k + 1) + \alpha_0 v(k + 1)$$

Upon substituting for $x_1(n + 1)$ in $y(n + 1)$, one obtains $y(n + 1) = x_2(n) + \alpha_1 v(n) + \alpha_0 v(n + 1)$. Continuing, it directly follows that

$$y(k + n) = -\beta_n x_1(k) - \beta_{n-1} x_n(k) - \cdots - \beta_1 x_n(k) + \alpha_n v(k)$$
$$+ \cdots + \alpha_0 v(k + n) \qquad\qquad (6.7.11)$$

Upon collecting these terms algebraically, one obtains

$$\beta_k = b_k; \quad k = 1, \ldots, n$$

$$\alpha_0 = a_0$$

$$\alpha_1 = a_1 - b_1 \alpha_0$$

$$\alpha_2 = a_2 - b_2 \alpha_0 - b_1 \alpha_1$$
$$\vdots \qquad \vdots \qquad \vdots \qquad \vdots$$
$$\alpha_n = a_n - b_n \alpha_0 - b_{n-1} \alpha_1 - \cdots - b_1 \alpha_{n-1}$$

Therefore,

$$x(k + 1) = \begin{bmatrix} 0 & 1 & \cdots & 0 & 0 \\ 0 & 0 & \cdots & 0 & 0 \\ \vdots & \vdots & & \vdots & \vdots \\ 0 & 0 & \cdots & 0 & 1 \\ -b_n & -b_{n-1} & \cdots & -b_2 & -b_1 \end{bmatrix} x(k) + \begin{bmatrix} \alpha_1 \\ \alpha_2 \\ \vdots \\ \alpha_{n-1} \\ \alpha_n \end{bmatrix} v(n) \qquad (6.7.13)$$

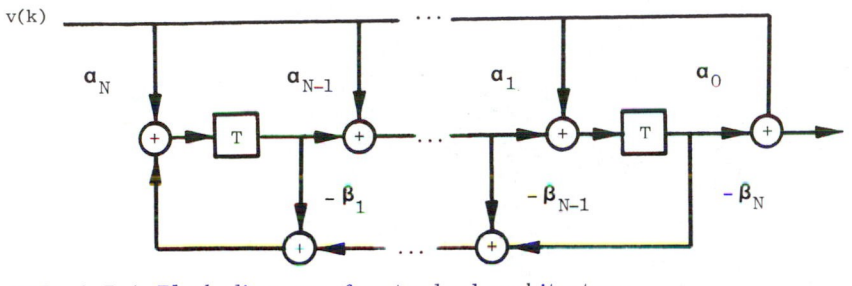

FIG. 6.7.4 Block diagram of a standard architecture.

where

$$
\begin{bmatrix} \alpha_0 \\ \alpha_1 \\ \alpha_2 \\ \vdots \\ \alpha_n \end{bmatrix} = \begin{bmatrix} 1 & 0 & 0 & \cdots & 0 \\ b_1 & 1 & 0 & \cdots & 0 \\ b_2 & b_1 & 1 & \cdots & 0 \\ \vdots & \vdots & \vdots & & \vdots \\ b_n & b_{n-1} & b_{n-2} & \cdots & b_1 \end{bmatrix}^{-1} \begin{bmatrix} \alpha_0 \\ \alpha_1 \\ \alpha_2 \\ \vdots \\ \alpha_n \end{bmatrix}
\qquad (6.7.14)
$$

The hybrid coefficients α_i can be computed using a general–purpose digital computer and a matrix inversion routine.

Parallel Form

A rational transfer function $H(z)$ can be factored to read

$$
H(z) = \frac{N(z)}{D(z)} = \prod_{k=1}^{r} \frac{N(z)}{(z - p_k)^{n(k)}} \text{ such that } \sum_{k=1}^{r} n(k) = n \qquad (6.7.15)
$$

where p_k is a pole of $H(z)$ and $n(k)$ is the multiplicity of that pole. Using the familiar Heaviside method, $H(z)$ can be represented by

$$
H(z) = \sum_{k=1}^{r} \sum_{j=1}^{n(k)} \frac{d_{kj}}{(z - p_k)^{n(k)+1-j}} + a_0 \qquad (6.7.16)
$$

where the Heaviside coefficient d_{kj} is computed in the following manner:

$$a_0 = \lim_{z \to \infty} H(z)$$

$$d_{kj} = \lim_{z \to p_k} \frac{1}{(j-1)!} \frac{d^{j-1}[(z - p_k)^{n(k)} H(z)]}{dz^{j-1}}$$

For the special case where $n(k) = 1$, $d_{k1} = \lim (z - p_k)H(z)$ as $z \to p_k$. In general, $Y(z) = H(z)V(z)$, where

$$Y(z) = a_0 V(z) + \sum_{k=1}^{r} \sum_{j=1}^{n(k)} \frac{d_{kj} V(z)}{(z - p_k)^{n(k)+1-j}}$$

$$= \sum_{k=1}^{r} \sum_{j=1}^{n(k)} d_{kj} Z_{kj}(z) \qquad (6.7.18)$$

A vectored-valued representation of this result can be obtained by defining two n vectors, say X and a, as follows:

$$X(z)^T = [Z_{11}(z), \ldots, Z_{1n(1)}(z); Z_{21}, \ldots, Z_{rn(r)}(z)]^T$$
$$a^T = (d_{11}, \ldots, d_{1n(1)}, d_{21}, \ldots, d_{rn(r)})^T \qquad (6.7.19)$$

Therefore, the output processes are given by

$$Y(z) = a_0 V(z) + a^T X(z) \qquad (6.7.20)$$

For the special case where $n(k) = 1$, $Z_{k1}(z)$ reduces to

$$Z_{k1}(z) = X_j(z) = \frac{V(z)}{z - p_k} \qquad (6.7.21)$$

for some j, or equivalently $zZ_j(z) = p_L X_j(z) + V(z)$. Therefore, $x_j(n)$ may be defined in terms of

$$x_j(n + 1) = p_k x_j(n) + v(n) \qquad (6.7.22)$$

For the case where $n(k) = K$, one obtains for some j,

(1) $\quad X_j(z) = Z_{kK}(z) = \dfrac{V(z)}{z - p_k}$

(6.7.23)

$\qquad$ or $\quad x_j(i + 1) = p_k x_j(i) + v(i)$

(2) $\quad X_{j-1}(z) = Z_{k(K-1)}(z) = \dfrac{Z_{kK}(z)}{z - p_k}$

(6.7.24)

$\vdots \qquad$ or $\quad x_{j-1}(i + 1) = p_k x_{j-1}(i) + x_j(i)$

(K) $\quad X_{j-K+1}(z) = \dfrac{Z_{k2}(z)}{(z - p_k)^K}$

(6.7.25)

$\qquad$ or $\quad x_{j-K+1}(i) = p_k x_{j-K+1}(i) + x_{j-K+2}(i)$

Collecting the foregoing information in matrix-vector form, one obtains the following Jordan representation:[†]

$$J = \begin{bmatrix} A_1 & 0 & 0 & \cdots & 0 \\ 0 & A_2 & 0 & \cdots & 0 \\ 0 & 0 & A_3 & \cdots & 0 \\ \vdots & \vdots & \vdots & \ddots & \vdots \\ 0 & 0 & 0 & \cdots & A_r \end{bmatrix}$$

where the A_i's are called Jordan blocks and have the following attributes:

1. If $n(i) = 1$, then $A_i = p_i$, $b_i = 1$.
2. If $n(i) = K \neq 1$, then

[†] For a rigorous development of this subject, refer to a linear algebra reference under the subject heading of Jordan forms, canonical forms, or minimal polynomials.

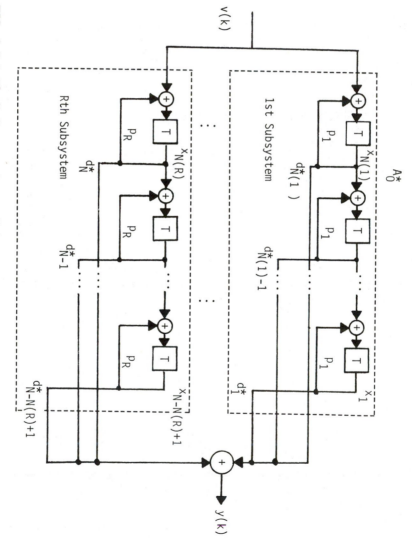

FIG. 6.7.5 Block diagram of a parallel architecture.

$$
A_i = \begin{bmatrix} p_i & 1 & 0 & 0 & \cdots & 0 & 0 \\ 0 & p_i & 1 & 0 & \cdots & 0 & 0 \\ \vdots & \vdots & \vdots & \vdots & & \vdots & \vdots \\ 0 & 0 & 0 & 0 & \cdots & p_i & 1 \\ 0 & 0 & 0 & 0 & \cdots & 0 & p_i \end{bmatrix}
$$

$$
b_i = \left.\begin{bmatrix} 0 \\ 0 \\ \vdots \\ 0 \\ 1 \end{bmatrix}\right\} K
$$

and, in both cases, $Y(z) = a^T X(z) + a_0 V(z)$ or $y(n) = a^T x(n) + a_0 v(n)$. The parallel filter structure, in terms of these derived parameters, is shown in Fig. 6.7.5. In this figure, the d_{ij}'s of Equation (6.7.19) are denoted A_{ij}^*.

Cascaded Form

The antithesis of the parallel form is the cascaded architecture. Whereas the parallel architecture consists of r independent paths, the cascaded form consists of a single path formed by n "chained" subfilters. Again the transfer function will be modeled as

$$
H(z) = N(z) \prod_{k=1}^{n} \frac{1}{z - p_k} \tag{6.7.26}
$$

where $N(z)$ is a polynomial of the form $N(z) = a_0 \Pi(p - q_i)$, $i = 1, \ldots, n$. The construction rule for this filter is given by

$$
x_1(k + 1) = p_1 x_1(k) + a_0 v(k)
$$
$$
x_2(k + 1) = p_2 x_2(k) - q_1 x_1(k) + x_1(k + 1)
$$
$$
\vdots \qquad \vdots \qquad \vdots \qquad \vdots \tag{6.7.27}
$$
$$
x_j(k + 1) = p_j x_j(k) - q_{j-1}(k) x_{j-1}(k) + x_{j-1}(k + 1); \quad 2 \le j \le n
$$

where p_i may or not equal p_j. Finally,

$$
y(k) = x_n(k + 1) - q_n x_n(k) \tag{6.7.28}
$$

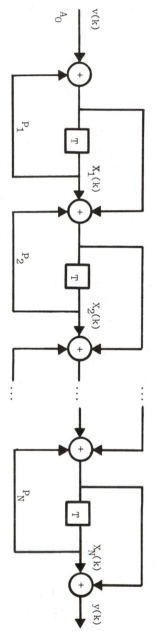

FIG. 6.7.6 Block diagram of a cascaded architecture.

The resulting architecture is found in Fig. 6.7.6. In terms of a vector-matrix format, the equations represent

$$x_j(k + 1) = p_j(k)x_j(k) + \sum_{r=1}^{j} (p_r - q_r)x_r(k) + a_0 v(k)$$

$$(6.7.29)$$

$$y(k) = \sum_{r=1}^{n} (p_r - q_r)x_r(k) + a_0 v(k)$$

where

$$x(k + 1) = \begin{bmatrix} p_1 & & \cdots & 0 & 0 \\ p_1 - q_1 & & \cdots & 0 & 0 \\ \vdots & \vdots & & \vdots & \vdots \\ p_1 - q_1 & p_2 - q_2 & \cdots & p_{n-1} & 0 \\ p_1 - q_1 & p_2 - q_2 & \cdots & p_{n-1} - q_{n-1} & p_n \end{bmatrix} x(k) + \begin{bmatrix} a_0 \\ a_0 \\ \vdots \\ a_0 \\ a_0 \end{bmatrix} v(k)$$

$$y(k) = (p_1 - q_1, \ldots, p_n - q_n)x(k) + a_0 v(k)$$

EXAMPLE 6.7.1 A fourth-order discrete system H(z) will be analyzed in terms of a direct II, cascaded, and parallel architecture, where

$$H(z) = \frac{0.001836(1 + z^{-1})^4}{(1 - 1.499z^{-1} + 0.8482z^{-2})(1 - 1.5548z^{-1} + 0.6493z^{-2})}$$

or

$$H(z) = \frac{0.001836 + 0.00734z^{-1} + 0.011016z^{-2} + 0.007344z^{-3} + 0.001836z^{-4}}{1 - 3.0536z^{-1} + 3.8281z^{-2} - 2.2921z^{-3} + 0.5507z^{-4}}$$

Direct II Filter. The resulting architecture is shown in Fig. 6.7.7. Then

$$a_0 = 0.001836 \qquad b_1 = 3.0538 \qquad a_4 - a_0 b_4 = 0.002847$$

$$a_1 = 0.007344 \qquad b_2 = -3.8281 \qquad a_3 - a_0 b_3 = 0.003136$$

$$a_2 = 0.011016 \qquad b_3 = 2.2921 \qquad a_2 - a_0 b_2 = 0.018044$$

$$a_3 = 0.007344 \qquad b_4 = -0.5507 \qquad a_1 - a_0 b_1 = 0.001737$$

$$a_4 = 0.001836$$

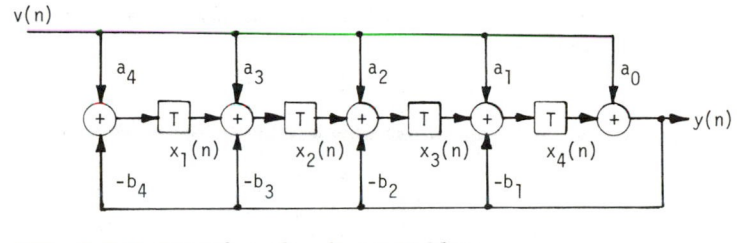

FIG. 6.7.7 Fourth-order direct II filter.

$$x(k + 1) = \begin{bmatrix} 0 & 0 & 0 & 0.5507 \\ 1 & 0 & 0 & -2.2921 \\ 0 & 1 & 0 & 3.8281 \\ 0 & 0 & 1 & -3.0538 \end{bmatrix} x(k) + \begin{bmatrix} 0.002850 \\ 0.003136 \\ 0.018044 \\ 0.001734 \end{bmatrix} v(k)$$

$$y(k) = (0 \quad 0 \quad 0 \quad 1)x(k) + 0.001836v(k)$$

One can use this opportunity to reinforce the study of computer-aided state-determined systems (Section 6.4) by noting that

$$(zI - A)^{-1} = \begin{bmatrix} z^2 + 3.0588z^2 - 3.8281z + 2.291 & 0.5507 \\ z^2 + 3.0538z - 3.8281 & z^3 + 3.0538z^2 - 3.8281z \\ 3.0538z + 3.0538 & z^2 + 3.0538z \\ 1 & z \end{bmatrix}$$

$$* \begin{bmatrix} 0.5507z & 0.5507z^2 \\ -2.291z + 0.5507 & -2.2921z^2 + 0.5507z \\ z^3 + 3.0538z^2 & -3.8281z^2 - 2.2921z + 0.5507 \\ z & z^3 \end{bmatrix}$$

Observe that by using Leverrier's method, the same solution can be derived. That is,

$$H_1 = I; \quad d_1 = 3.0538$$

$$H_2 = AH_1 + d_1 I = \begin{bmatrix} 3.0538 & 0 & 0 & 0.5807 \\ 1 & 3.0538 & 0 & -2.2921 \\ 0 & 1 & 3.0538 & 3.8281 \\ 0 & 0 & 1 & 0 \end{bmatrix}$$

$$d_2 = -\frac{1}{2}\text{trace}[AH_2] = -3.8281$$

$$H_3 = AH_2 + d_2 I = \begin{bmatrix} -3.8281 & 0 & 0.5507 & 0 \\ 3.0538 & -3.8281 & -2.2921 & 0.5507 \\ 1 & 3.0538 & 0 & -2.2921 \\ 0 & 1 & 0 & 0 \end{bmatrix}$$

$$d_3 = -\frac{1}{3}\text{trace}[AH_3] = 2.2921$$

$$\begin{bmatrix} 0.5507 & 0 & 0 & 0 \\ 0 & 0.5507 & 0 & 0 \\ 0 & 0 & 0.5507 & 0 \\ 0 & 0 & 0 & 0.5507 \end{bmatrix}$$

$$d_4 = -\frac{1}{4}\text{trace}[AH_4] = H \ -0.5507$$

Finally, as a check, $H_5 = [0]$. Then

$$(zI - A)^{-1} = \frac{\text{adj}(zI - A)}{\det(zI - A)} = \frac{Iz^3 + H_2 z^2 + H_3 z^1 + H_4}{\det(zI - A)}$$

$$= -Iz^3 + \begin{bmatrix} 3.0538 & 0 & 0 & 0.5507 \\ 1 & 3.0538 & 0 & -2.2921 \\ 0 & 1 & 3.0538 & 3.8281 \\ 0 & 0 & 1 & 0 \end{bmatrix}$$

$$+ \begin{bmatrix} -3.8281 & 0 & 0.5507 & 0 \\ 3.0538 & -3.8281 & -2.2921 & 0.5507 \\ 1 & 3.0538 & 0 & -2.2921 \\ 0 & 1 & 0 & 0 \end{bmatrix}$$

$$+ \begin{bmatrix} 2.2921 & 0.5507 & 0 & 0 \\ -3.8281 & 0 & 0.5507 & 0 \\ 3.0538 & 0 & 0 & 0.5507 \\ 1 & 0 & 0 & 0 \end{bmatrix} \Big/ \det(zI - A)$$

When designing cascade or parallel filters, one is generally restricted to realize filters over a real coefficient field. It was stated previously that this is important for the following reasons:

1. Real coefficients require one word of storage, whereas complex coefficients would require two words of storage.
2. Real multiplies represent a single algebraic operation, whereas complex multiplies are defined in terms of four real multiplies.

   Second-Order Cascade Filters. Since the poles of H(s) are complex, a cascade filter will be realized as a network of second-order filters. Note that H(z) can be factored to read:

$$H(z) = \frac{0.04285(1 + z^{-1})^2}{1 - 1.499z^{-1} + 0.8482z^{-2}} \quad \frac{0.04285(1 + z^{-1})^2}{1 - 1.5548z^{-1} + 0.6493z^{-2}}$$

$$= H_1(z)H_2(z)$$

Figure 6.7.8 shows the block diagram, form, where

$$x_1(k + 1) = A_1 x_1(k) + b_1 v(k); \quad y_1(k) = (0 \quad 0 \quad \cdots \quad 1)x_1(k) + a_0 v(k)$$

$$x_2(k + 1) = A_2 x_2(k) + b_2 y_1(k); \quad y(k) = (0 \quad 0 \quad \cdots \quad 1)x_2(k) + a_0 y_1(k)$$

If one wishes to choose to realize each second-order section as a direct II subfilter, one obtains

$$x_1(k + 1) = \begin{bmatrix} 0 & -b_{21} \\ 0 & -b_{11} \end{bmatrix} x_1(k) + \begin{bmatrix} a_{21} - a_{01}b_{21} \\ a_{11} - a_{01}b_{11} \end{bmatrix} v(n)$$

$$b_{11} = -1.499 \qquad b_{21} = 0.8482$$

$$a_{01} = a_{21} = 0.04855 \qquad a_{11} = 0.0857$$

$$x_2(k + 1) = \begin{bmatrix} 0 & -b_{22} \\ 1 & -b_{12} \end{bmatrix} x_2(k) + \begin{bmatrix} a_{22} - a_{02}b_{22} \\ a_{12} - a_{02}b_{21} \end{bmatrix} y_1(n)$$

$$b_{21} = 0.6493 \qquad b_{12} = -1.5548$$

$$a_{02} = a_{22} = 0.04285 \qquad a_{12} = 0.0857$$

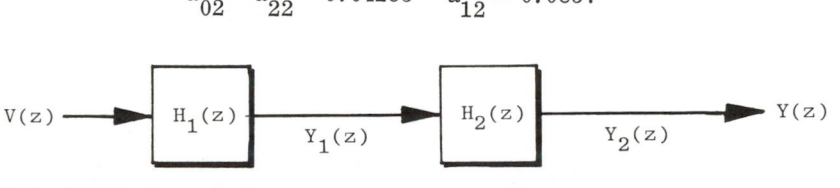

FIG. 6.7.8 Cascade form.

The resolvent matrix and alike can be computed in the manner presented in the direct II example just completed.

Second-Order Parallel Filters. For reasons previously stated, a second-order parallel filter should be configured over a real coefficient field. Therefore, under this side constraint, $H(z)$ will be factored as follows:

$$H(z) = a_1 + \frac{a_2 z^2 + a_3 z}{z^2 - 1.499z + 0.8482} + \frac{a_4 z^2 + a_5 z}{z^2 - 1.5548z + 0.6493}$$

$$= 0.00334 + \frac{0.01498z - 0.00174z^2 + 0.01752z^3 - 0.00149z^4}{(z^2 - 1.499z + 0.8482)(z^2 - 1.5548z + 0.6493)}$$

where the coefficient a_1 is computed, using long divisions, to be

$$a_1 = \frac{0.001836}{(0.8482)(0.6493)} = 0.003334 \quad [\text{i.e., } H(\infty)]$$

The remaining coefficients can be computed by solving a system of linear equations for assumed values of z. For example, consider choosing for values of z, $z = \pm 1$, and $z = \pm 2$. Then compute

$$-0.029265 = 0.0945a_2 + 0.0945a_3 + 0.3492a_4 + 0.3492a_5; \; z = 1$$

$$-0.03573 = 3.2041a_2 - 3.2041a_3 + 3.3492a_4 - 3.3472a_5; \; z = -1$$

$$0.139218 = 6.156a_2 + 3.078a_3 + 7.4008a_4 + 3.7004a_5; \; z = 2$$

$$-0.162333 = +31.0356a_2 - 15.5178a_3 + 25.3881a_4 - 12.6944a_5; \; z = -2$$

Solving for the coefficient set $\{a_i\}$ (i.e., $a = M^{-1}b$) with a general-purpose computer, one obtains

$$a^T = (-.0713, -0.0486, 0.0713, 0.0386)$$

and $H(z)$ follows, where

$$H(z) = 0.003334 + \frac{-0.713(1.0 + 0.6816z^{-1})}{1 - 1.1499z^{-1} + 0.8482z^{-2}}$$

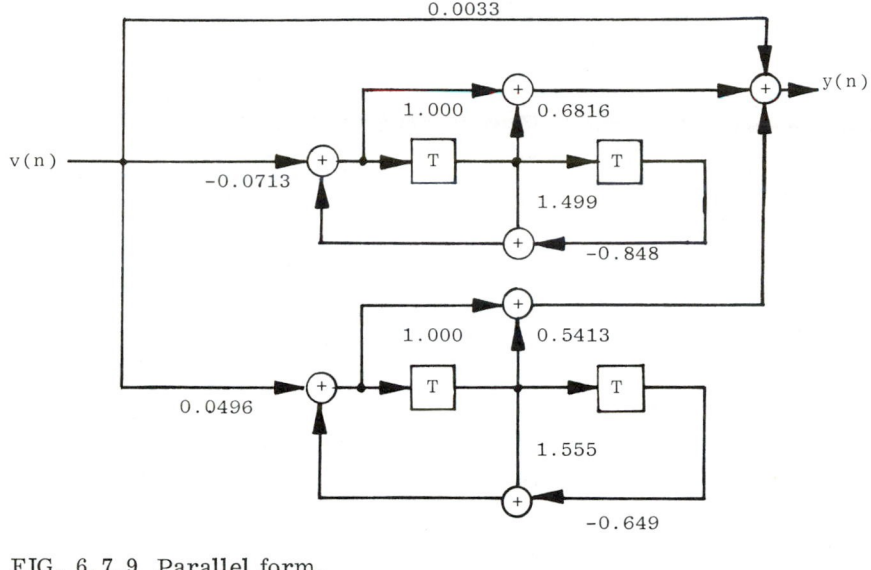

FIG. 6.7.9 Parallel form.

$$+ \frac{0.0486(1.0 + 0.5413z^{-1})}{1 - 1.5548z^{-1} + 0.6493z^{-2}}$$

or in terms of a direct II second-order section realization, one finally obtains the architecture shown in Fig. 6.7.9.

6.8 SUMMARY

In this chapter, the concepts of state, flow graphs, precedence, and rudimentary architectures were developed. This work will serve as the foundation for more advanced FIR topics developed in subsequent chapters. At that time, optimal filters, wave, and ladder structures will be explored. In addition, the FIR and IIR philosophy will be compared and critiqued. However, it should be remembered that these studies are premised on discrete filters rather than pure digital filters. In a later section, the effects of finite word length will be rigorously developed. A comprehensive computer-aided design software package for direct II cascade, parallel, and other architectures is reported in the appendix to Chapter 13. The reader should refer to this appendix when considering the details of filter design.

BIBLIOGRAPHY

Cadzow, J. A. (1974), <u>Discrete Time Systems</u>, Prentice-Hall, Englewood Cliffs, N.J.

Crochiere, R. E., and A. V. Oppenheim (1975), Analysis of Linear Digital Networks, Proc. IEEE, <u>63</u>, April, pp. 518-595.

Taylor, F. J. (1974), On Discrete Transition Matrices, Proc. IEEE, Sept., pp. 242-244.

Gabel, R. A., and R. A. Roberts (1973), <u>Signal and Linear Systems</u>, Wiley, New York.

Mason, S. J. (1953), Feedback Theory—Some Properties of Signal Flow Graphs, Proc. IRE, <u>41</u>, September, pp. 1144-1156.

Tretter, S. A. (1976), <u>Introduction to Discrete Time Signal Processing</u>, Wiley, New York.

APPENDIX A   Infinite Impulse Response Filters

In Section 6.4, a method for developing the discrete transfer function of a linear shift-invariant state-determined system was presented. This algorithm, which is based on Leverrier's algorithm, is matrix intensive. The software used to compute the discrete-state transition matrix (flow-charted in Fig. 6.4.1), was written in FORTRAN IV using matrix support subroutines. The software package is augmented with a line printer plotting routine to allow the user to interpret visually a filter's poles in the z plane.

The program reads user-supplied parameters which quantify the filter order N and the N coefficients of the discrete filter equation

$$y(k) + y(k - 1) * coef(N) * y(k - 1) + coef(N - 1) * y(k - 2) + \cdots$$

$$+ coef(1) * y(k - N) = v(k) + b(1) * v(k - 1) + \cdots$$

EXAMPLE A.1

1. For N = 3, let D(1), ..., D(3) = -0.125, 0.875, -1.75.
2. For N = 9, let D(1), ..., D(9) = 1, -2, 3, -4, 5, -6, 7, -8, 9.

where the d's represent filter coefficients. The parameters are summarized in the card sequence shown below; the results follow.

FORTRAN Source Code for Discrete-State Transition Matrix Computation

BEECHER

```
1.              REAL H(15,15,15),AI(15,15),A(15,15)
2.              REAL B(15,15),C(15,15),D(15,15)
3.              REAL COF(15),DD(15),ROOTR(15),ROOTI(15)
4.              REAL XAX(101),YAX(101),COL(101)
5.              INTEGER II(101)
6.      C          Y(K)+Y(K-1)*COF(N)+Y(K-2)*COF(2)+....
7.      C          +Y(K-N)*COF(1)=V(K)+V(K-1)*B(1)+....
8.              DO 70 KLM=1,2
9.              READ,N
10.             READ,(COF(I),I=1,N)
11.             COF(N+1)=1.
12.             DO 1 I=1,N
13.             DO 1 J=1,N
14.             AI(I,J)=0.
15.             IF(I.EQ.J) AI(I,J)=1.
16.             C(I,J)=AI(I,J)
17.             H(1,I,J)=C(I,J)
18.             A(I,J)=0.
19.           1 CONTINUE
20.             DO 10 I=1,N
21.             A(1,I)=-COF(N-I+1)
22.             A(I+1,I)=1.
23.          10 CONTINUE
24.             CALL TRACE(A,N,TB)
25.             DI=-TR
26.             DD(1)=DI
27.             DO 2 I=2,N
28.             CALL MULT(A,C,D,N,N,N)
29.             DO 3 J=1,N
30.             DO 4 K=1,N
31.             H(I,J,K)=D(J,K)+DI*AI(J,K)
32.           4 C(J,K)=H(I,J,K)
33.           3 CONTINUE
34.             CALL MULT(A,C,D,N,N,N)
35.             CALL TRACE(D,N,TR)
36.             DI=-TR/FLOAT(I)
37.             DD(I)=DI
38.           2 CONTINUE
39.             PRINT5,(DD(I),I=1,N)
40.           5 FORMAT('1',3X,'FROM D1 TO DN',3X,15(F8.3))
41.             DO 6 I=1,N
42.             PRINT9,I
43.           9 FORMAT(///,5X,'H(',I1,') MATRIX')
44.             DO 6 J=1,N
45.             PRINT8,(H(I,J,K),K=1,N)
46.           8 FORMAT(/,15F12.6)
47.           6 CONTINUE
```

FORTRAN Source Code (Continued)

```
48.         PRINT20
49.     20  FORMAT(//,5X,'ADJ(ZI-A) =H(1)*Z**(N-1)+H(2)*Z**(N-2)+.....+H(N)')
50.         PRINT15
51.     15  FORMAT(//)
52.         PRINT11
53.     11  FORMAT(7X,'THE COEFFICIENTS OF DET(ZI-A) FROM SMALLEST TO LARGEST
54.       7 POWER ORDER OF POLYNOMIAL')
55.         N1=N+1
56.         PRINT12,(COF(I),I=1,N1)
57.     12  FORMAT(12F12.6)
58.         CALL POLRT(COF,DD,N,ROOTR,ROOTI,IER,N1)
59.         PRINT14
60.     14  FORMAT(/,5X,'THE COMPLEX ROOTS OF DET(ZI-A) ARE')
61.         PRINT13,(ROOTR(I),ROOTI(I),I=1,N)
62.     13  FORMAT(5(3X,2F11.6))
63.         PRINT22
64.     22  FORMAT(///,5X,'** INVERSE(ZI-A)=ADJ(ZI-A)/DET(ZI-A)')
65.         PRINT16
66.     16  FORMAT('1')
67.         CALL DIAGM(XAX,YAX,ROOTI,ROOTR,N,CCL,DX,II)
68.     70  CONTINUE
69.         STOP
70.         END
71.  C   CALCULATE THE TRACE OF A MATRIX
72.         SUBROUTINE TRACE(B,N,T)
73.         REAL B(15,15)
74.         T=0.
75.         DO 12 I=1,N
76.     12  T=T+B(I,I)
77.         RETURN
78.         END
79.  C   MATRIX ADDITION  B+C=D(N*M)
80.         SUBROUTINE ADD(B,C,D,N,M)
81.         REAL B(15,15),C(15,15),D(15,15)
82.         DO 10 I=1,N
83.         DO 10 J=1,M
84.     10  D(I,J)=B(I,J)+C(I,J)
85.         RETURN
86.         END
87.  C   MATRIX MULTIPLICATION  B(N*M)*C(M*L)=D(N*L)
88.         SUBROUTINE MULT(B,C,D,N,M,L)
89.         REAL B(15,15),C(15,15),D(15,15)
90.         DO 11 I=1,N
91.         DO 11 J=1,L
92.         D(I,J)=0.
93.         DO 11 K=1,M
94.     11  D(I,J)=D(I,J)+B(I,K)*C(K,J)
95.         RETURN
96.         END
97.         SUBROUTINE DIAGM(XAX,YAX,ROOTI,ROOTR,N,CCL,DX,II)
```

FORTRAN Source Code (Continued)

```
98.                INTEGER II(101)
99.                REAL XAX(101),YAX(101),COL(101)
100.               REAL ROOTI(15),ROOTR(15)
101.               DATA BLANK,DOT,STAR/' ','.','*'/
102.               DO 1 I=1,101
103.               II(I)=0
104.          1    YAX(I)=0.
105.               CALL REAR(N,ROOTI,ROOTR)
106.               XI=ROOTI(N)-ABS(ROOTI(N)*0.05)
107.               XF=ROOTI(1)+ABS(ROOTI(1)*0.05)
108.               DX=(XF-XI)/50
109.               K=0
110.               IM=0
111.               DO 2 I=1,N
112.               IF(DX.NE.0.) GO TO 30
113.               M=25
114.               GO TO 31
115.          30   M=IFIX(((ROOTI(J)-XI)/DX)
116.          31   IF(M.EQ.K) GO TO 17
117.               YAX(2*M)=ROOTR(I)
118.               GO TO 18
119.          17   IM=IM+1
120.               II(M)=IM
121.               YAX(2*M+IM)=ROOTR(I)
122.               GO TO 2
123.          18   K=M
124.               IM=0
125.          2    CONTINUE
126.               YMAX=0.
127.               DO 4 I=1,101
128.               IF(ABS(YAX(I))-YMAX)4,4,3
129.          3    YMAX=ABS(YAX(I))
130.          4    CONTINUE
131.               DO 9 K=1,50
132.               DO 10 J=1,101
133.          10   COL(J)=BLANK
134.               COL(51)=DOT
135.               IF(K-25) 8,7,8
136.          7    DO 11 L=1,101
137.               IF((L/2*2-L).NE.0) COL(L)=DOT
138.          11   CONTINUE
139.          8    M=50.*(YAX(2*K)/YMAX+1.)+1.5
140.               IF(YAX(2*K).EQ.0) GO TO 15
141.               COL(M)=STAR
142.               PRINT16,COL
143.               IF(II(K).EQ.0) GO TO 9
144.               JJ=0
145.          19   JJ=JJ+1
146.               M=50.*YAX(2*K+JJ)/YMAX+51.5
147.               COL(M)=STAR
```

FORTRAN Source Code (Continued)

```
148.              PRINT21,COL
149.          21 FORMAT('+',19X,101A1)
150.              YAX(2*K+JJ)=0.
151.              IF(JJ.LT.II(K)) GO TO 19
152.              GO TO 9
153.          15 PRINT16,COL
154.          16 FORMAT(20X,101A1)
155.           9 CONTINUE
156.              PRINT54
157.          54 FORMAT(5X,'THE COORDINATES ARE')
158.              PRINT55,(ROOTR(I),ROOTI(I),I=1,N)
159.          55 FORMAT(5(3X,2F10.6))
160.              RETURN
161.              END
162.      C
163.      C   REARRANGE ROOTI IN ASCENDING ARRAY
164.      C
165.              SUBROUTINE REAR(N,ROOTI,ROOTR)
166.              REAL ROOTI(15),ROOTR(15)
167.              DO 8 I=1,N
168.              CALL SORT(N-I+1,I,ROOTI,ROOTR,J,SMALL,SW)
169.              ROOTI(J)=ROOTI(N-I+1)
170.              ROOTR(J)=ROOTR(N-I+1)
171.              ROOTI(N-I+1)=SMALL
172.              ROOTR(N-I+1)=SW
173.           8 CONTINUE
174.              RETURN
175.              END
176.              SUBROUTINE SORT(N,I,DD,DM,J,SMALL,SW)
177.              REAL DD(15),DM(15)
178.              SMALL=DD(1)
179.              SW=DM(1)
180.              J=1
181.              IF(N.EQ.1) GO TO 7
182.              DO 6 K=2,N
183.              IF(DD(K).GE.SMALL) GO TO 6
184.              SMALL=DD(K)
185.              SW=DM(K)
186.              J=K
187.           6 CONTINUE
188.           7 RETURN
189.              END
```

Third-Order Example Problem

```
FROM D1 TO DN    -1.750    0.875   -0.125

H(1) MATRIX

1.000    0.000    0.000
0.000    1.000    0.000
0.000    0.000    1.000

H(2) MATRIX

0.000   -0.875    0.125
1.000   -1.750    0.000
0.000    1.000   -1.750
```

Third-Order Example Problem (continued)

H(3) MATRIX

```
0.000   0.125   0.000
0.000   0.000   0.125
1.000  -1.750   0.875
```

ADJ(ZI-A)=H(1)*Z**(N-1)+H(2)*Z**(N-2)+...+H(N)

THE COEFFICIENTS OF DET(ZI-A) FROM SMALLEST TO LARGEST POWER ORDER OF POLYNOMIAL
-0.125   0.875   -1.750   10.000
THE COMPLEX ROOTS OF DET(ZI-A) ARE
0.250   0.000   0.500   0.000   1.000   0.000

**   INVERSE(ZI-A)=ADJ(ZI-A)/DET(ZI-A)

Three-Plane Pole Locations

$$y(k)=0.125y(k-1)-0.875y(k-2)+ \\ +1.750y(k-3)$$

```
            θ      ·0·              ·0
           0.25    0.5             1.0

            Z-PLANE
```

THE COEFFICIENTS ARE
0.500000 0.000000 1.000000 0.000000 0.250000 0.000000

Ninth–Order Example Problem (continued)

```
FROM D1 TO DN  9.0 -8.0  7.0 -6.0  5.0 -4.0  3.0 -2.0  1.0
```

H(1) MATRIX

```
1.00   0.00   0.00   0.00   0.00   0.00   0.00   0.00   0.00
0.00   1.00   0.00   0.00   0.00   0.00   0.00   0.00   0.00
0.00   0.00   1.00   0.00   0.00   0.00   0.00   0.00   0.00
0.00   0.00   0.00   1.00   0.00   0.00   0.00   0.00   0.00
0.00   0.00   0.00   0.00   1.00   0.00   0.00   0.00   0.00
0.00   0.00   0.00   0.00   0.00   1.00   0.00   0.00   0.00
0.00   0.00   0.00   0.00   0.00   0.00   1.00   0.00   0.00
0.00   0.00   0.00   0.00   0.00   0.00   0.00   1.00   0.00
0.00   0.00   0.00   0.00   0.00   0.00   0.00   0.00   1.00
```

H(2) MATRIX

```
0.00   8.00  -7.00   6.00  -5.00   4.00  -3.00   2.00  -1.00
1.00   9.00   0.00   0.00   0.00   0.00   0.00   0.00   0.00
0.00   1.00   9.00   0.00   0.00   0.00   0.00   0.00   0.00
0.00   0.00   1.00   9.00   0.00   0.00   0.00   0.00   0.00
0.00   0.00   0.00   1.00   9.00   0.00   0.00   0.00   0.00
0.00   0.00   0.00   0.00   1.00   9.00   0.00   0.00   0.00
0.00   0.00   0.00   0.00   0.00   1.00   9.00   0.00   0.00
0.00   0.00   0.00   0.00   0.00   0.00   1.00   9.00   0.00
0.00   0.00   0.00   0.00   0.00   0.00   0.00   1.00   9.00
```

H(3) MATRIX

```
0.00  -7.00   6.00  -5.00   4.00  -3.00   2.00  -1.00   0.00
0.00   0.00  -7.00   6.00  -5.00   4.00  -3.00   2.00  -1.00
1.00   9.00  -6.00   0.00   0.00   0.00   0.00   0.00   0.00
0.00   1.00   9.00  -8.00   0.00   0.00   0.00   0.00   0.00
0.00   0.00   1.00   9.00  -8.00   0.00   0.00   0.00   0.00
0.00   0.00   0.00   1.00   9.00  -8.00   0.00   0.00   0.00
0.00   0.00   0.00   0.00   1.00   9.00  -8.00   0.00   0.00
0.00   0.00   0.00   0.00   0.00   1.00   9.00  -8.00   0.00
0.00   0.00   0.00   0.00   0.00   0.00   1.00   9.00  -8.00
```

H(4) MATRIX

```
0.00   6.00  -5.00   4.00  -3.00   2.00  -1.00   0.00   0.00
0.00   0.00   6.00  -5.00   4.00  -3.00   2.00  -1.00   0.00
0.00   0.00   0.00   6.00  -5.00   4.00  -3.00   2.00  -1.00
1.00   9.00  -8.00   7.00   0.00   0.00   0.00   0.00   0.00
0.00   1.00   9.00  -8.00   7.00   0.00   0.00   0.00   0.00
0.00   0.00   1.00   9.00  -8.00   7.00   0.00   0.00   0.00
0.00   0.00   0.00   1.00   9.00  -8.00   7.00   0.00   0.00
0.00   0.00   0.00   0.00   1.00   9.00  -8.00   7.00   0.00
0.00   0.00   0.00   0.00   0.00   1.00   9.00  -8.00   7.00
```

H(5) MATRIX

```
0.00  -5.00   4.00  -3.00   2.00  -1.00   0.00   0.00   0.00
0.00   0.00  -5.00   4.00  -3.00   2.00  -1.00   0.00   0.00
0.00   0.00   0.00  -5.00   4.00  -3.00   2.00  -1.00   0.00
0.00   0.00   0.00   0.00  -5.00   4.00  -3.00   2.00  -1.00
1.00   9.00  -8.00   7.00  -6.00   0.00   0.00   0.00   0.00
0.00   1.00   9.00  -8.00   7.00  -6.00   0.00   0.00   0.00
0.00   0.00   1.00   9.00  -8.00   7.00  -6.00   0.00   0.00
0.00   0.00   0.00   1.00   9.00  -8.00   7.00  -6.00   0.00
0.00   0.00   0.00   0.00   1.00   9.00  -8.00   7.00  -6.00
```

Ninth-Order Example Problem (continued)

H(6) MATRIX

```
0.00    4.00   -3.00    2.00   -1.00    0.00    0.00    0.00    0.00
0.00    0.00    4.00   -3.00    2.00   -1.00    0.00    0.00    0.00
0.00    0.00    0.00    4.00   -3.00    2.00   -1.00    0.00    0.00
0.00    0.00    0.00    0.00    4.00   -3.00    2.00   -1.00    0.00
0.00    0.00    0.00    0.00    0.00    4.00   -3.00    2.00   -1.00
1.00    9.00   -8.00    7.00   -6.00    5.00    0.00    0.00    0.00
0.00    1.00    9.00   -8.00    7.00   -6.00    5.00    0.00    0.00
0.00    0.00    1.00    9.00   -8.00    7.00   -6.00    5.00    0.00
0.00    0.00    0.00    1.00    9.00   -8.00    7.00   -6.00    5.00
```

H(7) MATRIX

```
0.00   -3.00    2.00   -1.00    0.00    0.00    0.00    0.00    0.00
0.00    0.00   -3.00    2.00   -1.00    0.00    0.00    0.00    0.00
0.00    0.00    0.00   -3.00    2.00   -1.00    0.00    0.00    0.00
0.00    0.00    0.00    0.00   -3.00    2.00   -1.00    0.00    0.00
0.00    0.00    0.00    0.00    0.00   -3.00    2.00   -1.00    0.00
0.00    0.00    0.00    0.00    0.00    0.00   -3.00    2.00   -1.00
1.00    9.00   -8.00    7.00   -6.00    5.00   -4.00    0.00    0.00
0.00    1.00    9.00   -8.00    7.00   -6.00    5.00   -4.00    0.00
0.00    0.00    1.00    9.00   -8.00    7.00   -6.00    5.00   -4.00
```

H(8) MATRIX

```
0.00    2.00   -1.00    0.00    0.00    0.00    0.00    0.00    0.00
0.00    0.00    2.00   -1.00    0.00    0.00    0.00    0.00    0.00
0.00    0.00    0.00    2.00   -1.00    0.00    0.00    0.00    0.00
0.00    0.00    0.00    0.00    2.00   -1.00    0.00    0.00    0.00
0.00    0.00    0.00    0.00    0.00    2.00   -1.00    0.00    0.00
0.00    0.00    0.00    0.00    0.00    0.00    2.00   -1.00    0.00
0.00    0.00    0.00    0.00    0.00    0.00    0.00    2.00   -1.00
1.00    9.00   -8.00    7.00   -6.00    5.00   -4.00    3.00    0.00
0.00    1.00    9.00   -8.00    7.00   -6.00    5.00   -4.00    3.00
```

H(9) MATRIX

```
0.00   -1.00    0.00    0.00    0.00    0.00    0.00    0.00    0.00
0.00    0.00   -1.00    0.00    0.00    0.00    0.00    0.00    0.00
0.00    0.00    0.00   -1.00    0.00    0.00    0.00    0.00    0.00
0.00    0.00    0.00    0.00   -1.00    0.00    0.00    0.00    0.00
0.00    0.00    0.00    0.00    0.00   -1.00    0.00    0.00    0.00
0.00    0.00    0.00    0.00    0.00    0.00   -1.00    0.00    0.00
0.00    0.00    0.00    0.00    0.00    0.00    0.00   -1.00    0.00
0.00    0.00    0.00    0.00    0.00    0.00    0.00    0.00   -1.00
1.00    9.00   -8.00    7.00   -6.00    5.00   -4.00    3.00   -2.00
```

$ADJ(ZI-A)=H(1)*Z**(N-1)+H(2)*Z**(N-2)+...+H(N)$

THE COEFFICIENTS OF DET(ZI-A) FROM SMALLEST TO LARGEST POWER OF ORDER OF POLYNOMIAL
```
1.00   -2.00    3.00   -4.00    5.00   -6.00    7.00   -8.00    9.00    1.00
```

THE COMPLEX ROOTS OF DET(ZI-A) ARE

```
0.68   -0.23    0.68    0.23   -0.56   -0.57   -0.56    0.57    0.39   -0.61
0.39    0.61   -0.07   -0.75   -0.07    0.75   -9.88    0.00
```

**   INVERSE(ZI-A)=ADJ(ZI-A)/DET(ZI-A)

Ninth–Order Example Problem (continued)

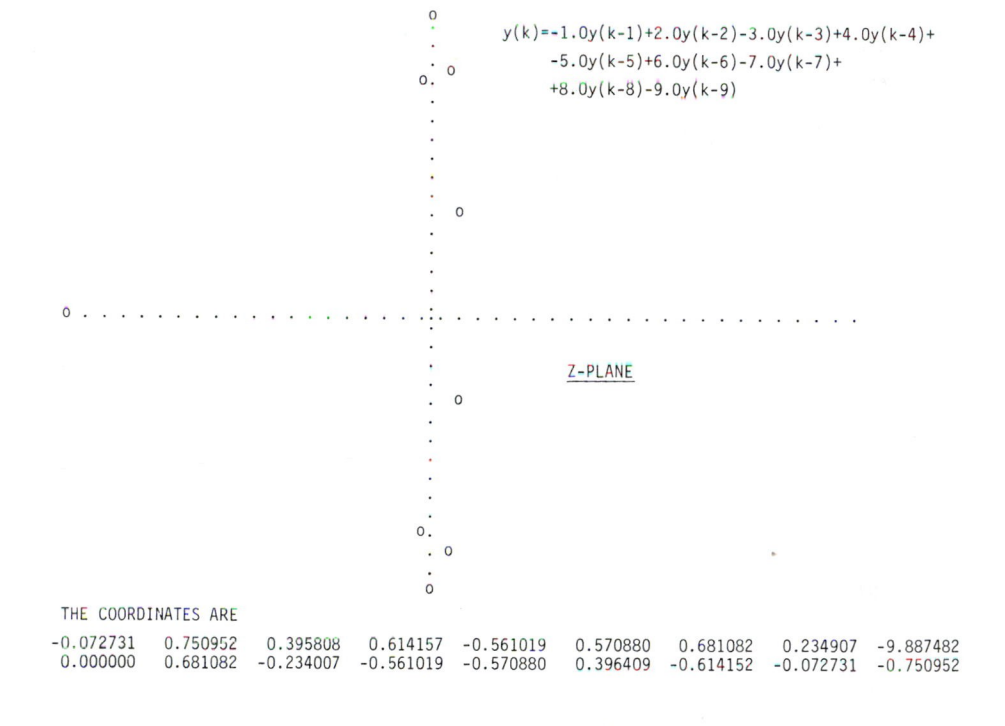

$y(k)=-1.0y(k-1)+2.0y(k-2)-3.0y(k-3)+4.0y(k-4)+$
$-5.0y(k-5)+6.0y(k-6)-7.0y(k-7)+$
$+8.0y(k-8)-9.0y(k-9)$

Z–PLANE

THE COORDINATES ARE

-0.072731	0.750952	0.395808	0.614157	-0.561019	0.570880	0.681082	0.234907	-9.887482
0.000000	0.681082	-0.234007	-0.561019	-0.570880	0.396409	-0.614152	-0.072731	-0.750952

APPENDIX B   FORTRAN Source Code for Flow Diagram Analysis

```
  C*
C*****************************************************
C*    " Z TRANSFORM FLOWGRAPH ANALYSIS "          *
C*****************************************************
C*
C*
C*    THIS PROGRAM IS DESIGNED TO ANALYZE FLOWGRAPHS INVOLVING
C* 'Z' COEFICIENTS.  THE SOLUTION SHOWS THE ORDER WHICH THE NODES
C* OF THE FLOWGRAPH ARE TO BE CALCULATED.
C*
C*   INPUT FORMAT:
C*   XII N=II
C*   BII N(II,II) FC=RRRR.RR
C*   BII N(II,II) FD=RRRR.RR
C*   E
C*
C*   VARIABLE NAMES:
C* X........SOURCE CARD.
C* B........BRANCH CARD.
C* E........THE END OF INPUT DATA
C* II.......INTEGER. SHOWS THE NUMBER OF BRANCH OR SOURCE.
C* N=II.....INTEGER. REPRESENTS THE NODE THAT THE SOURCE IS GOING.
C* N(II,II).INTEGERS. THE FIRST IS THE INITIAL NODE AND THE SECOND
C*          THE FINAL NODE OF THE BRANCH.
C* FC.......COEFICIENT OF BRANCH.
C* FD.......'Z' COEFICIENT OF BRANCH.
C* RRRR.RR..REAL NUMBERS OF G10.4 FORMAT.
C* NOTES:
C*   1. SOURCE CARDS AND BRANCH CARDS MUST BE IN ARITHEMETIC SEQUENCE.
C*   2. THE MAXIMUM NUMBER OF CARDS MUST BE 30.
C*   3. THE FIRST CHARACTER STARTS AT COL. 3 AND THE FORMAT MUST
C*      BE FOLLOWED EXACTLY THE WAY IS SHOWN.
C*
C*
C*   OUTPUT FORMAT:
C*   FIRST IT PRINTS THE INPUT DATA. THEN IT PRINTS:
C*   #II CALCULATE NODE PP
C* II....THE ORDER OF WHICH NODE 'PF' TO BE CALCULATED.
C*
C*
C*   ERROR DETECTED:
C* 1. 'B' AND 'X' CARDS ARE IN ARITHEMETIC SEQUENCE.
C* 2. OTHER LETTER THAN B, X, E IN THE THIRD COLUMN.
C* 3. IF INITIAL NODE IS THE SAME AS FINAL.
C* 4. OTHER LETTERS THAN 'FC' OR 'FD'.
C* 5. IF SOLUTION DOES NOT EXIST.
C* NOTE: IF AN INPUT ERROR IS DETECTED THE PROGRAM WILL
C*       NOT RUN, BUT THE  OUTPUT WILL SHOW THE TYPE OF
C*       ERROR  AND  THE CARD WHERE OCCURED.
C*
C*
C*.  VARIABLES:
C* D(30,30)........REAL; ARRAY(N,N); REPRESENTS THE 'Z' COEFFICIENTS.
C* SOURCE(30)......INTEGER; ARRAY(M); REPRESENTS THE NODES THAT ARE
C*                 CONECTED WITH SOURCES.
C* COEF(30,30).....REAL; ARRAY(N,N); REPRESENTS THE COEFICIENTS
C* IB(30)..........INTEGER; ARRAY(M); USED TO IDENTIFY THE  BRANCH AND
C*                 SOURCE NUMBERS.
C* NI(30)..........INTEGER; ARRAY(M); TO IDENTIFY THE INITIAL NODE OR
C*                 THE SOURCE NODE.
C* NF(30)..........INTEGER; ARRAY(M); TO IDENTIFY THE FINAL NODE.
C* VAL(30).........REAL; ARRAY(M); TO IDENTIFY THE VALUES OF THE
C*                 COEFICIENTS
C* ERROR...........INTEGER; COUNTER; TO COUNT ALL ERRORS MADE IN
```

```
C*                    THE INPUT DATA .
C* NMAX............INTEGER;  REPRESENTS THE MAXIMUM NUMBER OF NODES.
C* KCOUNT...:......INTEGER; COUNTER; TO CHECK IF ALL NODES HAVE BEEN
C*                    SOLVED.
C* L..............INTEGER; COUNTER; COUNTS THE ORDER OF WHICH THE NODES
C*                    ARE SOLVED.
C* NC(30,30)......INTEGER; ARRAY(L,N); TO STORE THE PROCEDURES OF
C*                    THE SOLUTION.
C* NG(30).........INTEGER; ARRAY(N); TO STORE THE SOLVED NODES.
C* NK(30).........INTEGER; ARRAY(N); TO STORE WRITTEN NODES.
C*INI............INTEGER; COUNTER; TO DETECT ANY NON ZERO VALUE
C*                    IN THE  'COEF(N,N)'.
C* CB(30).........2*CHARACTER; ARRAY(M); TO IDENTIFY THE INPUT DATA
C* WORD(30).......2*CHARACTER; ARRAY(M); TO IDENTIFY THE TYPE OF THE

C*                    COEFICIENTS.
C*
C*    NOTE: IN THE ARRAYS ABOVE;
C*       N;  NODES
C*       M;  INPUT DATA CARDS
C*       L;  ORDDER OF CALCULATING NODES
C*
C*************************************************
C*DESIGNED BY: GEORGE M. PAPADOURAKIS
C*************************************************
C*
C*
        DIMENSION D(30,30), SOURCE(30), COEF(30,30),
     #  IB(30), NK(30), NI(30), NF(30), VAL(30)
        INTEGER ERROR,NMAX,KCOUNT,L,NC(30,30),NG(30)
        INTEGER*2 CB(30),WORD(30)
        INI=0
        KCOUNT=0
C*
C*    CALL SUBROUTINE 'DATA'
C*
        CALL DATA(ERROR,D,NMAX,SOURCE,COEF,CB,IB,NI,NF,WORD,VAL)
C*
C*    CHECK IF THERE ARE ANY ERRORS. IF THERE ARE THE PROGRAM WILL
C*    STOP.
C*
        IF(ERROR.GT.0) GO  TO 231
        GO TO 233
231     WRITE(6,232) ERROR
232     FORMAT(2X,'ERRORS:',I2)
        GO TO 230
C*
C* INITIALIZE THE ORDER OF CALCULATING NODES.
C*
233     L=1
C*
C*    THESE TWO DO LOOPS FIND THE NODES THAT CAN BE
C*    CALCULATED BY EXAMINING THE COEF(N,N)
C*
340     DO 240 I=1,NMAX
        DO 250 N=1,NMAX
        IF(COEF(N,I).EQ.0) GO TO 250
        INI =INI+1
250     CONTINUE
        IF(INI.EQ.0) GO TO 260
        INI=0
        GO TO 240
260     NC(L,I) = I
        NG(I)=I
240     CONTINUE
C*
```

```
C*   THE ORDER OF CALCULATION IS INCREMENTED.
C*   CHECK IF THE SYSTEM HAS A SOLUTION.
C*
        L=L+1
        IF(L.LE.NMAX) GO TO 900
        WRITE(6,931)
931     FORMAT(2X,' THE SYSTEM DOES NOT HAVE A COMPLETE SOLUTION ')
        GO TO 230
C*
C*   THESE TWO DO LOOPS RECONSTRUCT THE MATRIX COEF(N,N) TAKING
C*   IN CONSIDERATION THE NODES THAT ARE ALREADY SOLVED.
C*
900     DO 270 K=1, NMAX
        IF(NG(K).EQ.K) GO TO 270
        KCOUNT=KCOUNT+1
        DO 280 N=1, NMAX
        IF(COEF(N,K).EQ.0) GO TO 280
        IF(NG(N).NE.N) GO TO 280
        COEF(N,K)=0
280     CONTINUE
270     CONTINUE
C*
C*   TO CHECK IF ALL THE NODES HAVE BEEN SOLVED.
C*
        IF(KCOUNT.EQ.0) GO TO 300
        KCOUNT=0
        GO TO 340
C*
C*   THESE TWO DO LOOPS PRINT THE SOLUTION.
C*
300     DO 310 I=1, L
        DO 320 K=1,NMAX
        IF (NK(K).EQ.K) GO TO 320
        IF(NC(I,K).EQ.0)GO TO 320
        WRITE(6,330) I, NC(I,K)
330     FORMAT(2X,'#',I2,2X,'CALCULATE NODE ',I2)
        NK(K)=K
320     CONTINUE
310     CONTINUE
        GO TO 230
230     STOP
        END
C*
C*********************************************************
C*
C*   THIS IS SUBROUTINE 'DATA'. IT ANALYZES THE INPUT DATA AND THEN
C*   TRANSFERS THE PROPER ARRAYS TO THE MAIN PROGRAM.
C*
C*   INTERNAL VARIABLES OF SUBROUTINE:
C*   NM(30).....INTEGER; ARRAY(N); IS USED FOR DETERMINIG NMAX.
C*   MCOUNT.....INTEGER; COUNTER; IS USED FOR COUNTING THE SOURCES.
C*   BCOUNT.....INTEGER; COUNTER; IS USED FOR COUNTING THE BRANCHES.
C*   NJ.........INTEGER; REPRESENTS INITIAL NODE.
C*   NQ.........INTEGER; REPRESENTS FINAL NODE.
C*
C*
        SUBROUTINE DATA(ERROR,D,NMAX,SOURCE,COEF,CB,IB,NI,NF,WORD,VAL)
        DIMENSION  IB(30), NI(30), NF(30),
     *    VAL(30), SOURCE(30), NM(30), D(30,30), COEF(30,30)
        INTEGER MCOUNT, NJ, NQ, ERROR, BCOUNT
        INTEGER*2 CB(30), WORD(30)
        MCOUNT =0
        ERROR =0
        BCOUNT =0
C*
C*   THIS BIG DO LOOP ANALYZES EACH DATA AT A TIME
C*
        DO 10 I=1,30
        TYPE 12
12      FORMAT(2X,'ENTER ONE DATA AT A TIME')
        ACCEPT 20, CB(I), IB(I), NI(I), NF(I), WORD(I), VAL(I)
20      FORMAT(2X,A1,I2,3X,I2,1X,I2,2X,A2,1X,G10.4)
C*
C*   TO EXAMINE WHAT KIND OF A LETTER IS IN COLUMN 3. IF NOT
C*   A KNOWN LETTER, ERROR WILL BE PRINTED.
C*
```

```
           IF(CB(I).EQ.'X') GO TO 40
           IF(CB(I).EQ.'B') GO TO 60
           IF(CB(I).EQ.'E') GO TO 70
           WRITE(6,80) CB(I)
80         FORMAT(2X,A1,2X,'ERROR.UNKNOWN VARIABLE')
           ERROR=ERROR+1
           GO TO 10
C*
C*   THE DATA CARD HAS BEEN IDENTIFIED AS A SOURCE CARD. THE CARD
C*   IS PRINTED. THE SEQUENCE IS CHECKED AND THE NODE WHERE THE
C*   SOURCE IS GOING IS STORED IN THE ARRAY "SOURCE(M)".
C*
40         WRITE(6,30) CB(I), IB(I), NI(I)
30         FORMAT(2X,A1,I2,1X,'N=',I2)
           MCOUNT=MCOUNT+1
           IF (MCOUNT.EQ.IB(I)) GO TO 50
           WRITE(6,110) IB(I)
110        FORMAT(2X,I2, 2X,'ERROR. SOURCES NOT IN SEQUENCE')
           ERROR=ERROR+1
           GO TO 10
50         SOURCE(I)=NI(I)
           GO TO 10
C*
C*   THE DATA CARD HAS BEEN IDENTIFIED AS BRANCH CARD.
C*   THE CARD IS PRINTED. THE SEQUENCE IS CHECKED.
C*
60         WRITE(6,90) CB(I), IB(I),NI(I),NF(I),WORD(I),VAL(I)
90         FORMAT(2X,A1,I2,1X,'N(',I2,',',I2,')',1X,A2,'=',G10.4)
           BCOUNT=BCOUNT+1
           IF(BCOUNT.EQ.IB(I)) GO TO  100
           WRITE(6,120) IB(I)
120        FORMAT(2X,I2,2X,'ERROR. BRANCHES NOT IN SEQUENCE')
           ERROR=ERROR+1
           GO TO 10
C*
C*   CHECK IF THE INITIAL NODE IS THE SAME AS THE FINAL.
C*
100        IF(NI(I).EQ.NF(I)) GO TO 130
C*
C*   TO DETERMINE THE MAXIMUM NUMBER OF NODES.
C*
           NJ=NI(I)
           NQ=NF(I)
           IF (NI(I).LT.NF(I)) GO TO 140
           NM(I)=NI(I)
           GO TO 190
140        NM(I)=NF(I)
           GO TO 190
190        IF(NMAX.LT.NM(I)) GO TO 150
           GO TO 200
150        NMAX=NM(I)
           GO TO 200
C*
C*   TO FIND WHAT TYPE OF COEFICIENT IS THE BRANCH CARD
C*   AND STORED IN THE APROPRIATE ARRAYS.
C*
200        IF(WORD(I).EQ.'FC') GO TO 160
           IF(WORD(I).EQ.'FD') GO TO 170
           WRITE(6,180) WORD(I)
180        FORMAT(2X,A2,2X,'ERROR. UNKNOWN VARIABLE')
           ERROR=ERROR+1
           GO TO 10
170        D(NJ,NQ)=VAL(I)
           GO TO 10
160        COEF(NJ,NQ)=VAL(I)
           GO TO 10
130        WRITE(6,210) NI(I), NF(I)
210        FORMAT(2X,I2,'?',I2,2X,'ERROR. THE TWO NODES ARE EQUAL')
           ERROR=ERROR+1
10         CONTINUE
           GO TO 70
70         RETURN
           END
```

7
Optimal IIR Filters

7.1 INTRODUCTION

In Chapter 1 it was noted that there exists a rich class of classic optimal
active and passive analog filters (e.g., Butterworth, Chebyshev, etc.).
Classic analog filters are generally defined in terms of a normalized
prototype transfer function H(s). The magnitude response of such a filter,
in decibel units, is given by

$$G(\Omega) = 20 \log_{10}|H(j\Omega)| = 10 \log_{10}|H(j\Omega)|^2 \qquad (7.1.1)$$

The frequency response of the discrete form of an analog prototype filter is
specified in terms of a cutoff frequency ω_p, a stopband frequency ω_s, and
stop and passband delimiters denoted ϵ and A. These parameters are
graphically interpreted in Fig. 7.1.1. The critical frequencies ω_p and ω_a
represent the end of the passband and start of the stopband, respectively.
In decibels, the gains at these critical frequencies are given by $-A_p = -10$
$\log_{10}(1 + \epsilon^2)$ (passband ripple constraint) and $-A_a = -10 \log_{10}(A^2)$ (stopband
attenuation). For the case where $\epsilon = 1$, the common 3-dB filter is realized.
Other classic design parameters found in general use are the gain transition
ratio $\eta = \epsilon/(A^2 - 1)^{1/2}$ and the discrete transition frequency ratio k = ω_p/ω_a. The parameters η and k define the steepness of the filter's skirt. One
often refers to the set of hybrid parameters defined below:

$$\epsilon^2 = 10^{0.1A_p} - 1 \quad \text{[from } 0.1A_p = \log_{10}(1 + \epsilon^2)]$$

$$\delta^2 = 10^{0.1A_a} - 1 \quad \text{[from } 0.1A_a = \log_{10}(1 + \delta^2)] \qquad (7.1.2)$$

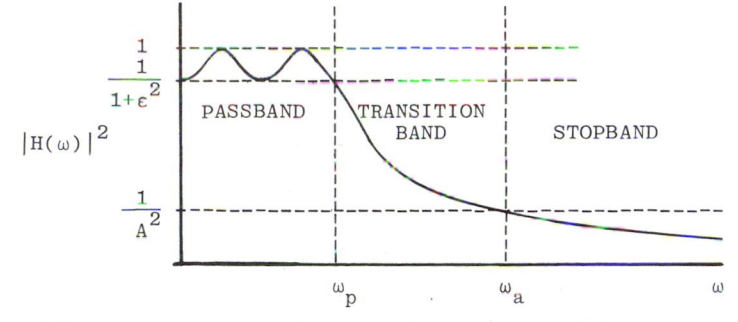

FIG. 7.1.1 Prototype lowpass filter design model.

where $\delta^2 = A^2 - 1$. The last hybrid parameters to be considered are d and D, where

$$d = \frac{\delta}{\epsilon}; \quad D = d^2 \tag{7.1.3}$$

The metrics d and D are a measure of the depth of the transition region. In terms of these parameters, classically defined filters can be designed. These filters form a very important class of discrete and digital filters since they possess outstanding magnitude frequency attributes. Furthermore, since they are defined in terms of well-known approximating polynomials and analog design practices, they can usually be computed in a computer-aided design sense.

7.2 DESIGN PRELIMINARIES

In terms of the introduced parameters, the classic Butterworth, Chebyshev, and elliptic filters are summarized in Table 7.2.1. Data found in this table will be referred to throughout this chapter. The formulas found in this table will translate a set of frequency domain design objectives into a filter. The filter order (i.e., n), in terms of the design parameters A, ϵ, k(analog), analog frequencies Ω, and so forth, are summarized as follows:

Butterworth:

$$n = \frac{\log_{10}[(A^2 - 1)/\epsilon^2]}{2 \log_{10}(\Omega_r)} = \frac{\log_{10}(D)}{\log_{10}(1/k)} = \frac{\log_{10}(\eta)}{\log_{10}(k)}$$

TABLE 7.2.1 Design Summary for Butterworth, Chebyshev, and Elliptic Filters

Filter type	Transfer function squared $|H(j\Omega)|^2$	Filter order estimate n	Example: A = 100(-40dB), $\epsilon = 1.0$, $\Omega_r = 3.0$ (transition ratio = 1/3)
Butterworth	$|H(\Omega)|^2 \quad \dfrac{1}{1+\epsilon^2}=(1-\delta)^2$	$n = \dfrac{\log_{10}(A^2-1)/\epsilon^2}{2*\log_{10}(\Omega_r)}$ $= \dfrac{\log_{10}(d)}{\log_{10}(1/k)}$ $= \dfrac{\log_{10}(\eta)}{\log_{10}(k)}$	$n = \dfrac{\log_{10}(9999)}{2*\log_{10}(3)}$ $= 4.2$ ~ 5
Chebyshev	$\epsilon^2 = \dfrac{1}{(1-\delta)^2} - 1$ $\delta = 1 - (1+\epsilon^2)^{1/2}$	$n = \dfrac{\log_{10}(1+(1-\eta^2)^{1/2})}{\log_{10}\;\Omega_r+(\Omega^2-1)^{1/2}}$ $= \dfrac{\cosh^{-1}(1/\,)}{\ln(1+(1-k^2)^{1/2})/k}$	$1/\eta = (9999/1)^{1/2}$ $n = \dfrac{\log_{10}(999^{1/2}+998^{1/2})}{\log_{10}(3+8^{1/2})}$ ~ 3
Inverse Chebyshev		Note: $\dfrac{\ln(a)}{\ln(b)} = \dfrac{\log_2(a)}{\log_2(b)}$ $= \dfrac{\cosh^{-1}D^{1/2}}{\cosh^{-1}(1/k)}$	

Elliptic

Chebyshev rational function

$R_n(\Omega, \Omega_0)$

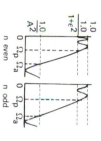

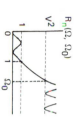

$$|H(j\Omega)|^2 = \dfrac{1.0}{1 + \epsilon^2 R_n^2(\Omega, \Omega_0)}$$

$$n = \dfrac{K(k)K(1 - k_1^2)^{1/2}}{K(k_1)K(1 - k^2)^{1/2}}$$

where K is a complete elliptic integral of the form

$$\int_0^\theta \dfrac{ds}{1 - k^2 \sin^2 s}$$

To avoid tedious computation of n, based on evaluating this definite integral, use approximation methods. They are (1) algebraic, (2) nomographic. The order of an elliptic filter is approximately:

$$n \geq \log_{10} 16D / \log_{10} 1/q$$

where

$$k = \Omega_p / \Omega_a \; ; \; k' = (1 - k^2)^{1/2}$$

$$q_0 = \dfrac{0.5*(1 - (k')^{1/2})}{(1 + (k')^{1/2})}$$

$$q = q_0 + 2q_0^5 + 15q_0^9 + 150q_0^{13}$$

$$D = d^2$$

$k = 1/3$

$k' = (1 - (1/9))^{1/2}$
$= .9428$

$q_0 = \dfrac{0.5(1 - 0.9428)}{(1 - 0.9428)}$
$= .0147$

$q = 0.0417$
$+ 2(6.86 \times 10^{-10})$
$+ 15(3.20 \times 10^{-17})$
$+ 150(1.49 \times 10^{-24})$
~ 0.0147

$D = 0.9999$

$n \geq \dfrac{\log_{10} 159{,}984}{\log_{10} 67.9}$

$= \dfrac{5.204}{2.83}$

~ 3

Chebyshev:

$$n = \frac{\log_{10}[(1 + \sqrt{1 - \eta^2})/\eta]}{\log_{10}[\Omega_r + \sqrt{\Omega_r^2 - 1}]} = \frac{\cosh^{-1}(1/\eta)}{\ln[(1 + \sqrt{1 - k^2})/k]} = \frac{\cosh^{-1}(\sqrt{D})}{\cosh^{-1}(1/k)}$$

(7.2.1)

Elliptic:

$$n = \frac{K(k)K(\sqrt{1 - k_1^2})}{K(k_1)K(\sqrt{1 - k^2})} \sim \frac{\log_{10}(16D)}{\log_{10}(1/q)} ;$$

where K is a rational Chebyshev function

These deterministic formulas can easily be programmed on a general-purpose computer. Rabiner et al. (1974) have converted these data into a set of nomographs found in Figs. 7.2.1 through 7.2.5. These data will support the following operations:

1. Conversion of η and A into δ units.
2. Conversion of η and transition ratio into a filter-order estimate for Butterworth, Chebyshev, or elliptic filters.
3. Assuming that a bilinear z transform is used, the prewarping map is defined as a function of the transition ratio (see Section 2.8).

The use of these nomographs can best be explained in terms of an example.

EXAMPLE 7.2.1 Let the passband ripple (where applicable) of a discrete lowpass filter be bounded by -1 dB, stopband attenuation to be -80 dB, and the critical discrete frequencies by $f_p = 480$ Hz, $f_a = 540$ Hz, $f_s = 8000$ Hz (see Fig. 7.2.6). The design of Butterworth, Chebyshev, or elliptic filters would progress as follows:

Step 1: Normalize the critical frequencies.

$$f_p = \frac{480}{8000} = 0.06; \quad f_a = \frac{540}{8000} = 0.0675$$

Compute passband bounds.

(1) -1 dB = $-10 \log_{10}(1 + \epsilon^2)$; $(1 + \epsilon^2) = 1.2589$ and $\epsilon = 0.50884$

or

(2) -1 dB = $-20 \log_{10}(1 - \delta)$; $(1 - \delta) = 0.89125$ and $\delta = 0.10875$

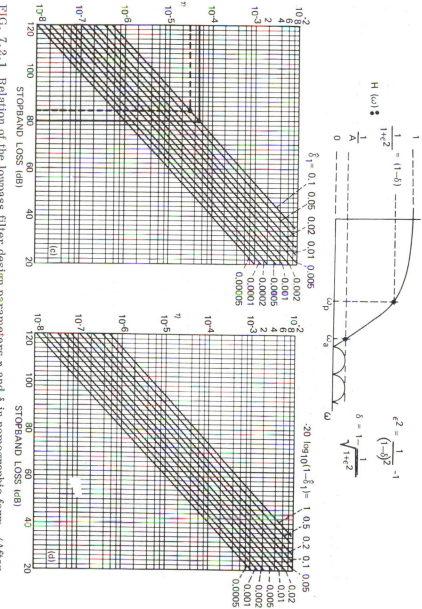

FIG. 7.2.1 Relation of the lowpass filter design parameters η and δ in nomographic form. (After Rabiner et al., 1974. Copyright 1974 American Telephone and Telegraph Company. Reprinted by permission from the Bell System Technical Journal.)

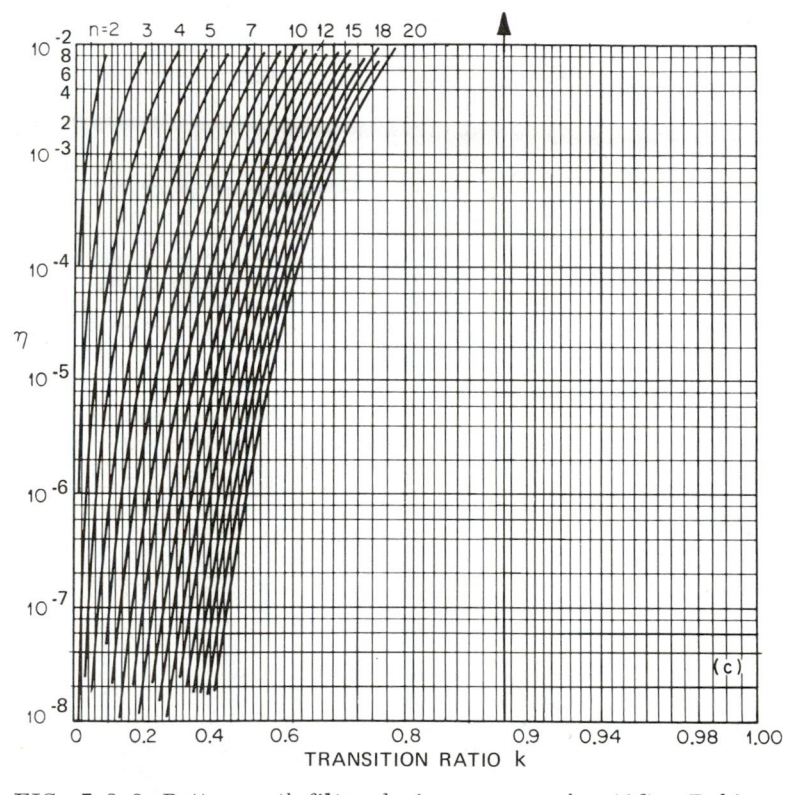

FIG. 7.2.2 Butterworth filter design nomograph. (After Rabiner et al., 1974. Copyright 1974 American Telephone and Telegraph Company. Reprinted by permission from the Bell System Technical Journal.)

Step 2: From $\eta = 5 \times 10^{-5}$ (Fig. 7.2.1), compute $v = f_a - f_p = 0.0675 - 0.06 = 0.0075$. From knowledge that the normalized passband width is 0.06, both k(analog) and k(discrete) are estimated to be k = 0.885 (Fig. 7.2.5) (direct computation produces k(discrete) = 0.06/0.0675 = 0.8888).

Step 3: Determine filter order.

Butterworth	Chebyshev	Elliptic
Figs. 7.2.2 and 7.2.6 off-scale	Figs. 7.2.3 and 7.2.6	Figs. 7.2.4 and 7.2.6

$$n = \frac{\log_{10} \eta}{\log_{10} k} = \frac{-4.31}{-5.119 \times 10^{-3}}$$

$$= 84.3 \rightarrow \boxed{85}$$

$n > 20$

$n = 9.5 \rightarrow 10$

EXAMPLE 7.2.2 Reconsider Example 7.2.1 in terms of a -0.5 dB passband, -40 dB stopband, f_p = 350 Hz, f_a = 450 Hz, and f_s = 1000 Hz.
  Step 1: Normalize the critical frequencies

$$f_p = \frac{350}{1000} = 0.35; \quad f_a = \frac{450}{1000} = 0.45$$

Compute passband bounds.

$$-0.5 \text{ dB} = -10 \log_{10}(1 + \epsilon^2); \quad 1 + \epsilon^2 = 1.225 \text{ and } \epsilon = 0.35$$

  Step 2: From   = 0.35 (Fig. 7.2.7), compute   = 0.45 - 0.35 = 1.
From knowledge that the normalized discrete passband width is 0.35, k(analog) = 0.32 and k(discrete) = 0.78 (see Fig. 7.2.5).

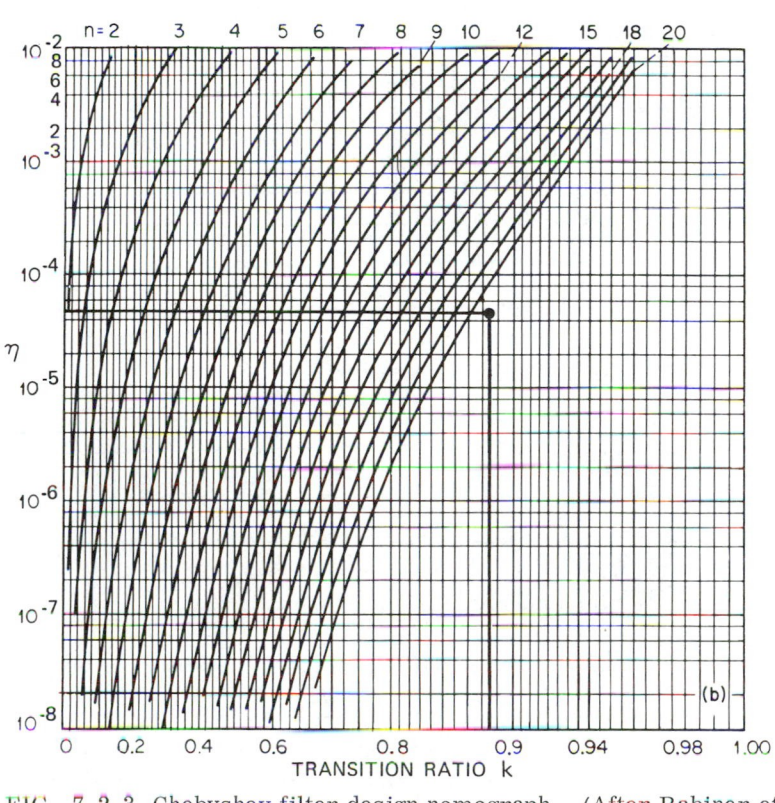

FIG. 7.2.3 Chebyshev filter design nomograph. (After Rabiner et al.,
1974. Copyright 1974 American Telephone and Telegraph Company. Reprinted by permission from the Bell System Technical Journal.)

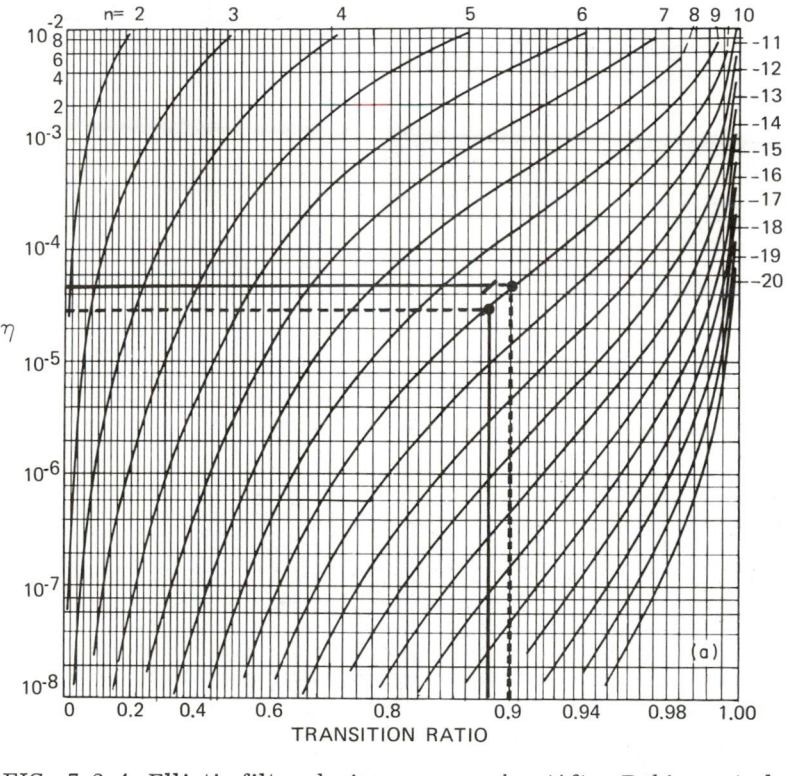

FIG. 7.2.4 Elliptic filter design nomograph. (After Rabiner et al., 1974. Copyright 1974 American Telephone and Telegraph Company. Reprinted by permission from the Bell System Technical Journal.)

Step 3: Determine filter order (Fig. 7.2.7).

Butterworth	Chebyshev	Elliptic
~6	~4	~3
(interpolation)	(interpolation)	(interpolation)

Observe that the elliptic filter satisfies the given design constraints with a lower-order filter. This is a common occurrence. Several decades ago, the superiority of the elliptic filter was established for most general applications. That is, for a common set of constraint parameters, the elliptic filter would generally produce the lowest-order realization of the three filters considered. This is important from a design and economy viewpoint. However, deriving the pole-zero locations for this class of

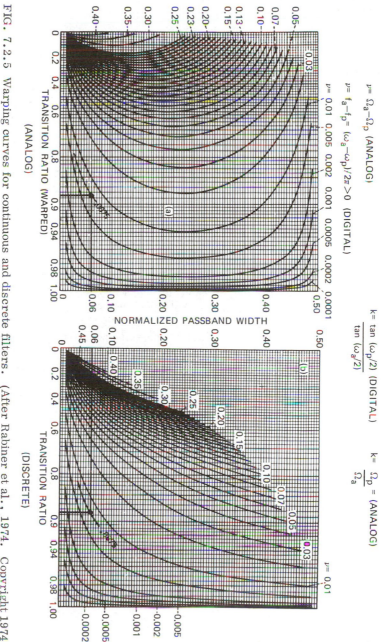

FIG. 7.2.5 Warping curves for continuous and discrete filters. (After Rabiner et al., 1974. Copyright 1974 American Telephone and Telegraph Company. Reprinted by permission from the Bell System Technical Journal.

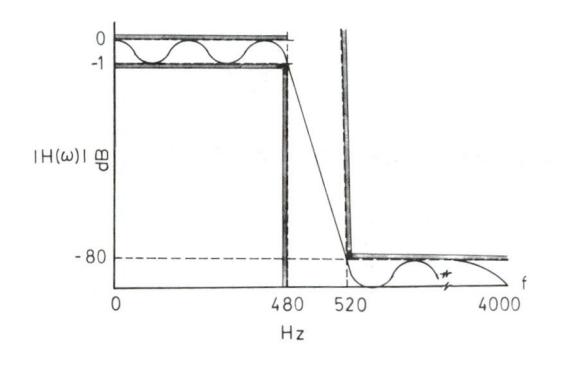

FIG. 7.2.6 Lowpass filter design example model.

filter, using traditional analytic tools, was a most formidable task. In more recent times, the general-purpose digital computer was introduced to this problem area. As a result, the design of elliptic filters has for the most part been automated. Therefore, based on the elliptic filter's superior performance and reduced filter complexity, it is becoming the most popular of the three candidates.

The nomograph approach was used to determine that a 9.5th-order order elliptic filter is required. On rounding this to 10, one observes that the nomograph data indicate that the realized filter produces the following design bonus (i.e., beyond the original design objectives):

for k = 0.885   for $\eta = 5 \times 10^{-5}$

$\eta = 3 \times 10^{-5}$ (vs. 5×10^{-5})   k = 0.9 vs. 0.8888

$\delta = 0.1 \rightarrow$ stopband = -84 dB $\rightarrow$ transition band is [480, 533] Hz
    vs. -80 dB

EXAMPLE 7.2.3 A lowpass filter, having a maximum -1-dB passband gain over [0, 0.35] rad/sec and a minimum -15-dB stopband gain over [0.45, 0.5] rad/sec, using a 1-rad/sec sample rate, can be realized as a third-order Butterworth filter or a second-order Chebyshev or elliptic filter. Furthermore, the minimum stopband gains for these filters compute to be -15, -20, and -26 dB, respectively. The magnitude response of the tested filters is shown in Fig. 7.2.8.

Suppose that an analog Butterworth, Chebyshev, or elliptic prototype filter is to be designed. Suppose further that the bilinear z transform is to be used to map the analog filter into a discrete realization. Then according to the material found in Sections 2.7 and 2.8, there exists a warping between the analog and discrete frequency domain. In terms of the example problem, the desired discrete critical frequencies are 480 and 540 Hz. These must be related to the critical frequencies of a prewarped analog

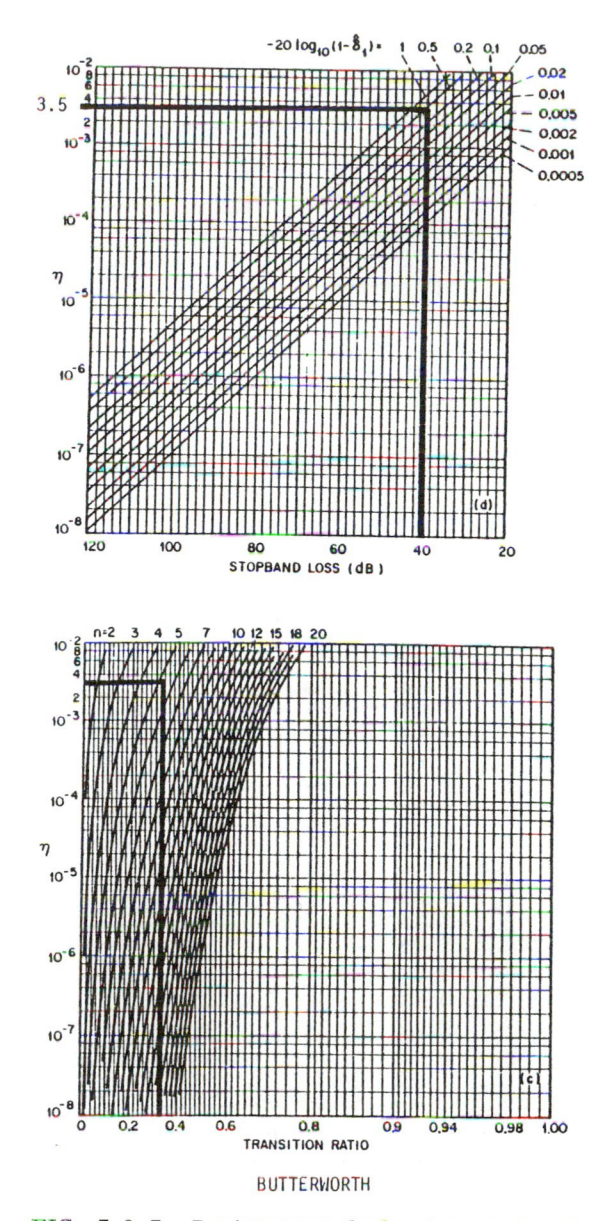

BUTTERWORTH

FIG. 7.2.7a Design example for determining the order of a Butterworth
filter. (After Rabiner et al., 1974. Copyright 1974 American Telephone
and Telegraph Company. Reprinted by permission from the Bell System
Technical Journal.)

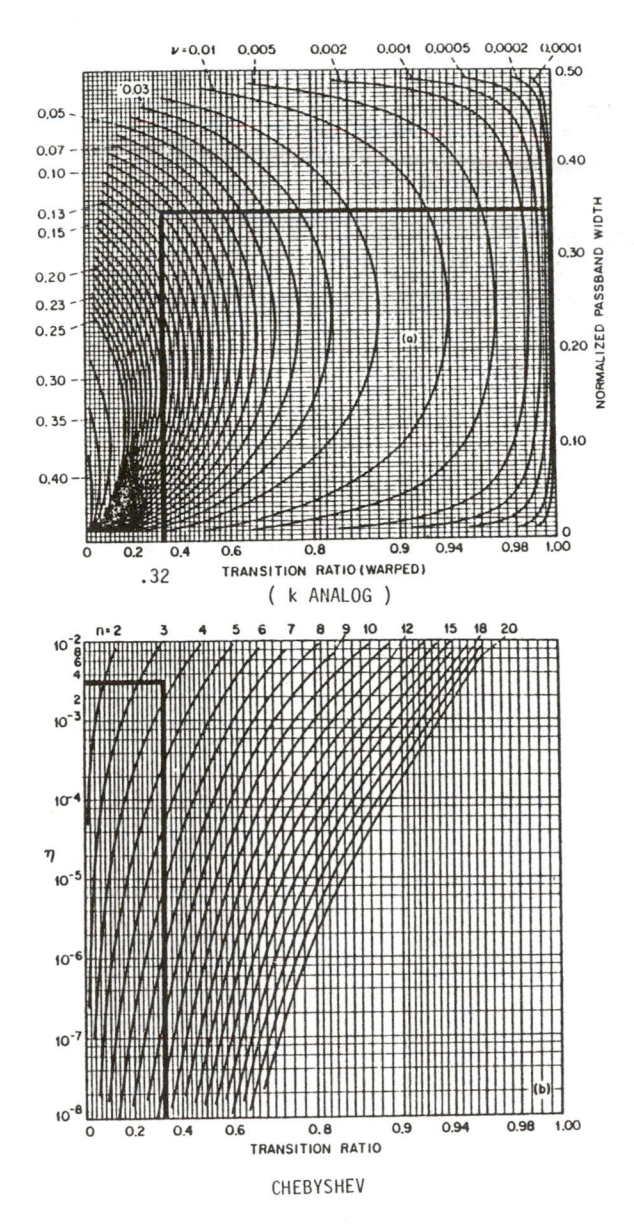

(k ANALOG)

CHEBYSHEV

FIG. 7.2.7b Design example for determining the order of a Chebyshev filter. (After Rabiner et al., 1974. Copyright 1974 American Telephone and Telegraph Company. Reprinted by permission from the Bell System Technical Journal.)

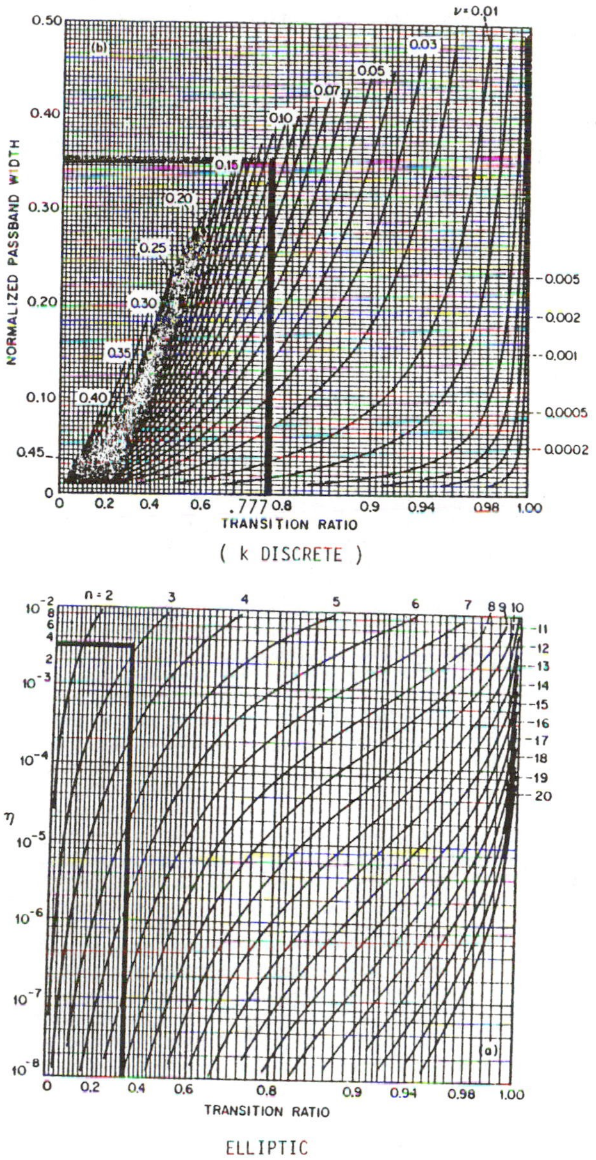

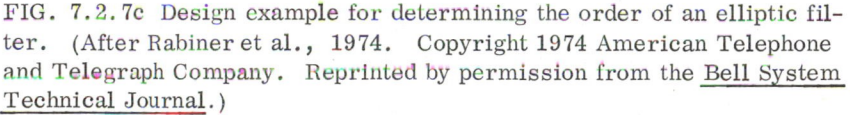

FIG. 7.2.7c Design example for determining the order of an elliptic filter. (After Rabiner et al., 1974. Copyright 1974 American Telephone and Telegraph Company. Reprinted by permission from the Bell System Technical Journal.)

COMPARISON	BUTTERWORTH	CHEBYSHEV	ELLIPTIC
PASSBAND GAIN (dB)	-1	-1	-1
STOPBAND GAIN (dB)	-15	-20	-26
TRANSITION RATIO (ANALOG)	.31	.31	.31
TRANSITION RATIO (DISCRETE)	.78	.78	.78
PASSBAND FREQUENCY (Hz)	.35	.35	.35
STOPBAND FREQUENCY (Hz)	.45	.45	.45
SAMPLE FREQUENCY (Hz)	1.00	1.00	1.00
FILTER ORDER	3	2	2

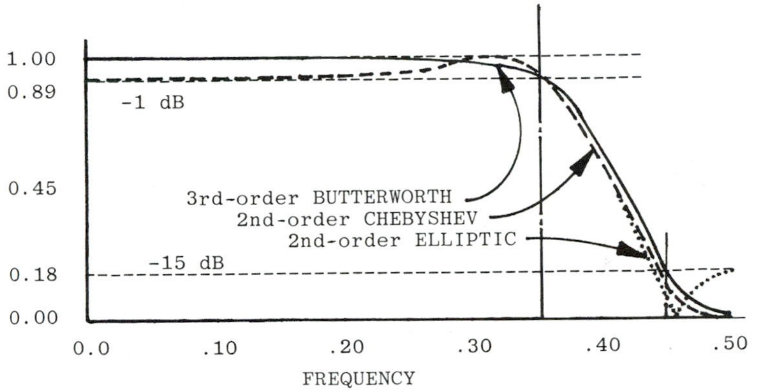

FIG. 7.2.8 Comparison of Butterworth, Chebyshev, and elliptic realizations of a lowpass filter.

filter. Here, for a normalized passband of 0.06, and $v = 0.0075$, k for the analog filter may be graphically interpreted to be 0.885 (see Fig. 7.2.5). Through direct computation, one obtains

$$\Omega_p = \frac{2}{T}\tan\left(\frac{\omega_p}{2}\right) = 480.14$$

$$\Omega_a = \frac{2}{T}\tan\left(\frac{\omega_a}{2}\right) = 540.20$$

with k given by $\Omega_p/\Omega_a = 0.888$. This value is in good agreement with the graphically determined value of k. The critical analog frequencies for a Butterworth prototype filter would be $\Omega_p = 1$ and $\Omega_a = 1.125$ in order to achieve the required value for k. This is summarized in Fig. 7.2.9.

EXAMPLE 7.2.4 Consider again the problem studied in Example 7.2.2. The warped passband and stopband critical frequencies are given by

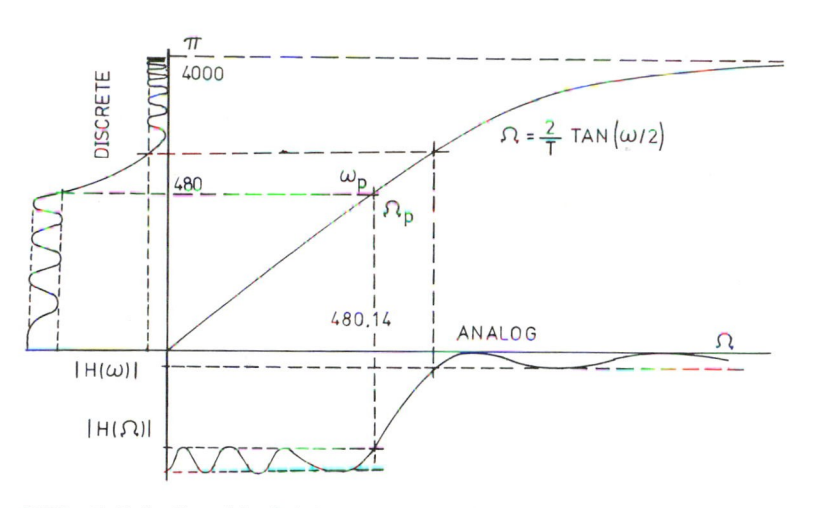

FIG. 7.2.9 Graphical interpretation of a warping curve.

$$\Omega_p = \frac{2}{T} \tan\left[\frac{2\pi(0.35)}{2}\right] = \frac{2}{T}(1.9626)$$

$$\Omega_a = \frac{2}{T} \tan\left[\frac{2\pi(0.45)}{2}\right] = \frac{2}{T}(6.3137)$$

The transition coefficient k for the analog filter computes to be k = Ω_p/Ω_a = 0.3108. The transition coefficient for the discrete filter is given by k = 0.35/0.45 = 0.777. Thus the step-skirt discrete filter can be seen to be derived from a shallow-skirt analog prototype filter.

7.3 DESIGN PROTOCOL

The classic filters, such as the Butterworth, Chebyshev, and elliptic, are well known in terms of their analog models. These filters are known to possess excellent magnitude frequency responses. In Chapter 2 it was argued that if such frequency domain behavior is to be mapped into the z domain, the bilinear z transform should be used. This transform was able to preserve the frequency domain attributes of an analog filter that were usually distorted under an impulse invariant transform (e.g., standard z transform). As a result, the mappings found in this chapter will be assumed to be the bilinear z transform. It may be recalled that the bilinear z transform established a "warped" map between the analog and discrete frequency axes. Such distortion must be accounted for in any accepted design protocol. The two most popular design methodologies are abstracted in Figs. 7.4.1 and 7.5.1. Here it is assumed that a discrete filter has a magnitude

TABLE 7.3.1 Summary of the Discrete Frequency-to-Frequency Transform

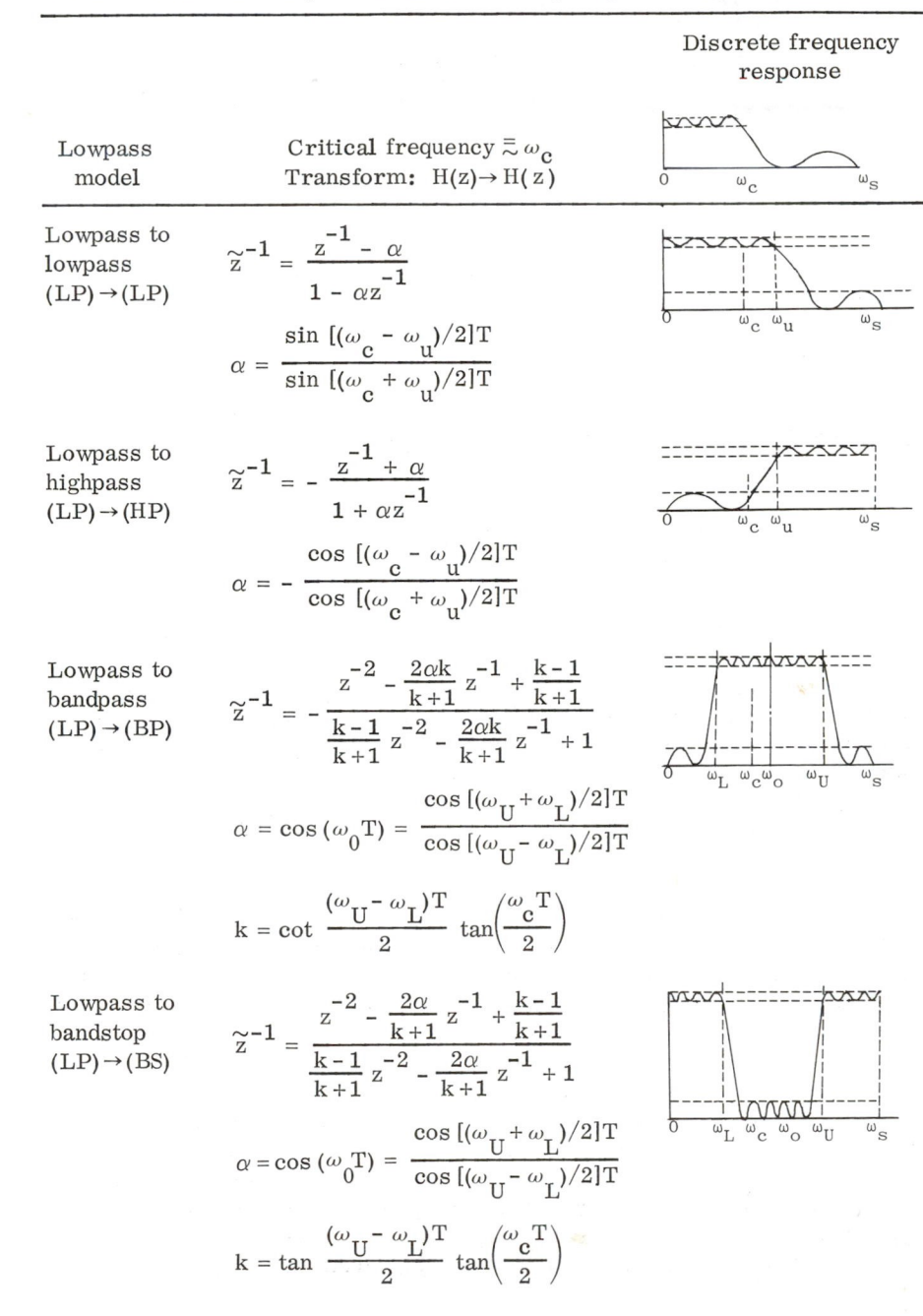

Lowpass model	Critical frequency $\cong \omega_c$ Transform: $H(z) \to H(\tilde{z})$	Discrete frequency response
Lowpass to lowpass (LP)→(LP)	$\tilde{z}^{-1} = \dfrac{z^{-1} - \alpha}{1 - \alpha z^{-1}}$ $\alpha = \dfrac{\sin\,[(\omega_c - \omega_u)/2]T}{\sin\,[(\omega_c + \omega_u)/2]T}$	
Lowpass to highpass (LP)→(HP)	$\tilde{z}^{-1} = -\dfrac{z^{-1} + \alpha}{1 + \alpha z^{-1}}$ $\alpha = -\dfrac{\cos\,[(\omega_c - \omega_u)/2]T}{\cos\,[(\omega_c + \omega_u)/2]T}$	
Lowpass to bandpass (LP)→(BP)	$\tilde{z}^{-1} = -\dfrac{z^{-2} - \dfrac{2\alpha k}{k+1}z^{-1} + \dfrac{k-1}{k+1}}{\dfrac{k-1}{k+1}z^{-2} - \dfrac{2\alpha k}{k+1}z^{-1} + 1}$ $\alpha = \cos\,(\omega_0 T) = \dfrac{\cos\,[(\omega_U + \omega_L)/2]T}{\cos\,[(\omega_U - \omega_L)/2]T}$ $k = \cot\,\dfrac{(\omega_U - \omega_L)T}{2}\,\tan\!\left(\dfrac{\omega_c T}{2}\right)$	
Lowpass to bandstop (LP)→(BS)	$\tilde{z}^{-1} = \dfrac{z^{-2} - \dfrac{2\alpha}{k+1}z^{-1} + \dfrac{k-1}{k+1}}{\dfrac{k-1}{k+1}z^{-2} - \dfrac{2\alpha}{k+1}z^{-1} + 1}$ $\alpha = \cos\,(\omega_0 T) = \dfrac{\cos\,[(\omega_U + \omega_L)/2]T}{\cos\,[(\omega_U - \omega_L)/2]T}$ $k = \tan\,\dfrac{(\omega_U - \omega_L)T}{2}\,\tan\!\left(\dfrac{\omega_c T}{2}\right)$	

frequency response which most resembles some desired $H(\exp(j\omega))$. The synthesis problem is then one of designing, in some optimal sense, an approximately ideal filter of a given order. In both design methodologies that have been outlined, it is required that an analog prototype filter be specified. Once specified, classical analytic and computer-aided methods can be used to synthesize a realizable prototype filter. The methods differ on how they approach the problem of converting an arbitrary discrete filter set of specifications into an analog prototype model. Method I considers the conversion of $H(z)$ into a prewarped filter $H(s)$. This $H(s)$ can then be reduced to an analog prototype filter through the use of analog frequency-to-frequency conversion formulas (see Table 7.2.1). The second method uses the discrete frequency-to-frequency formulas given in Table 7.3.1 to map $H(z)$ into a discrete prototype filter. This filter is, in turn converted into a prewarped analog prototype filter.

Both methods are equivalent. The choice of which method to use is one of personal convenience and experience. In the next section, a number of design examples are found. These examples should provide the reader with a mechanical appreciation of the problem of designing optimal discrete filters.

7.4 METHOD I FILTER DESIGN

Design method I is, for reader convenience, interpreted graphically in Fig. 7.4.1. Here Ω_p is the prewarped frequency given by $\Omega_p = 2\tan(\omega_p T/2)/T$. The lowpass-to-lowpass analog transform is known to be given by $\bar{s} = s/\Omega_p$. Suppose for the sake of argument that the normalized analog prototype filter is given by $H_N(s)$, where

$$H_N(s) = \frac{\displaystyle\prod^{M}(s + z_i)}{\displaystyle\prod^{N}(s + p_i)} \qquad (7.4.1)$$

with z_i and p_i representing the filter zeros and poles. The lowpass-to-lowpass version of $H_N(s)$, say $H(s)$, satisfies

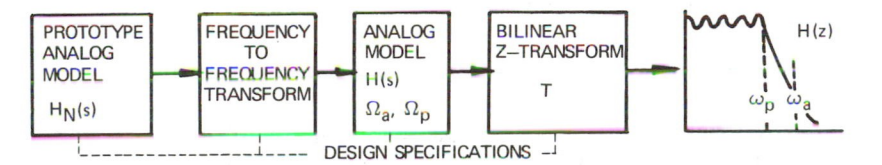

FIG. 7.4.1 System block diagram of the IIR method I design procedure.

$$H(s) = H_N(s)\Big|_{\substack{s=s/\Omega_p \\ \Omega_p=(2/T)\tan(\omega_p T/2)}} = \frac{\prod\limits^{M}(s + \Omega_p z_i)\Omega_p^{N-M}}{\prod\limits^{N}(s + \Omega_p p_i)} \qquad (7.4.2)$$

In the z domain, under $s = 2(z - 1)/T(z + 1)$, $H(z)$ satisfies

$$H(z) = H(s)\Big|_{s=(2/T)[(z-1)/(z+1)]} = \frac{\prod\limits^{M}\left(\frac{2}{T}\frac{z-1}{z+1} + \Omega_p z_i\right)\Omega_p^{N-M}}{\prod\limits^{N}\left(\frac{2}{T}\frac{z-1}{z+1} + \Omega_p p_i\right)}$$

$$= \tan\left(\frac{\omega_p T}{2}\right)^{N-M} \frac{\prod\limits^{M}\frac{z-1}{z+1} + \tan\left(\frac{\omega_p T}{2}\right)z_i}{\prod\limits^{N}\frac{z-1}{z+1} + \tan\left(\frac{\omega_p T}{2}\right)p_i}$$

$$= \alpha^{N-M} \frac{\prod\limits^{M}\frac{z-1}{z+1} + \alpha z_i}{\prod\limits^{N}\frac{z-1}{z+1} + \alpha p_i} \qquad (7.4.3)$$

where

$$\Omega_p = \frac{2}{T}\tan\left(\frac{\omega_p T}{2}\right)$$

$$\alpha = \tan\left(\frac{\omega_p T}{2}\right)$$

It can be seen from this development that the 2/T scale factor found in the prewarping bilinear z-transform formula is superfluous. Therefore, computational simplification can be achieved by ignoring this scale factor. As a result, a modified prewarping equation can be derived which is given by

$$\Omega = \tan\left(\frac{\omega T}{2}\right) \qquad (7.4.4)$$

Furthermore, the 2/T scale factor associated with the bilinear z transform can now be dropped since the modified prewarping formula corresponds to a bilinear z transform satisfying

$$s = \frac{z-1}{z+1} \qquad\qquad (7.4.5)$$

EXAMPLE 7.4.1 The problem is to design a lowpass Butterworth filter having the discrete filter response found in Fig. 7.4.2. Referring to Table 7.2.1, one notes that $n = \log_{10}[(A^2 - 1)/\epsilon^2]/2 \log_{10}(\Omega_r)$. Suppose that $\epsilon = 1$, $A^2 = 10$, and $\Omega_r = 1/k(\text{analog}) = 2$. Then it follows that $n = \log_{10}(9)/2 \log_{10}(2) = 1.58$ (say $n = 2$). Alternatively, using the nomograph, the filter's order could be verified to have a value of 2. The normalized prototype filter $H_N(s)$ can be designed using standard tables. The resulting filter satisfies

$$H_N(s) = \frac{1}{(s + 0.707 + j0.707)(s + 0.707 - j0.707)} = \frac{1}{s^2 + 1.414s + 1}$$

The prewarped bandpass cutoff frequency, for $f_p = 10^3$, $f_s = 10^4$, is given by

$$\Omega_p = \tan\left(\frac{\omega_p T}{2}\right) = \tan\left(\frac{2\pi \times 10^3}{10^4 \times 2}\right) = \tan(0.1\pi) = 0.325$$

where, in terms of a lowpass-to-lowpass transform, the prewarped analog model to be

$$H(s) = H_N(s)\Big|_{s=s/0.325} = \frac{(0.325)^2}{s^2 + 0.46s + 0.106}$$

Finally, the desired $H(z)$ is directly computed to be

$$H(z) = H(s)\Big|_{s=(z-1)/(z+1)} = \frac{0.0676(z^2 + 2z + 1)}{z^2 - 1.142z + 0.412}$$

$$= \frac{0.0676(z + 1)^2}{(z - 0.57 + j0.295)(z - 0.57 - j0.295)}$$

This result from each of these steps is graphically interpreted in Fig. 7.4.2 as well.

A computationally more efficient approach to the problem of prewarping is to integrate the warping factor into the bilinear z transform. Consider the problem abstracted in Fig. 7.4.3. Here the analog frequency is related to the discrete axis through the relationship $\Omega = m \tan(\omega T/2)$, where

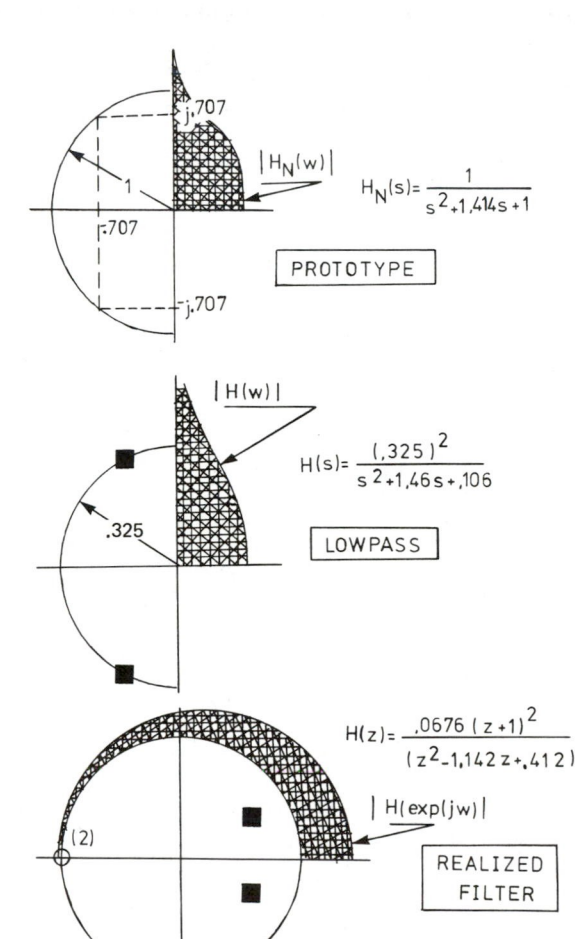

FIG. 7.4.2 Graphical interpretation of the steps found in a method I design.

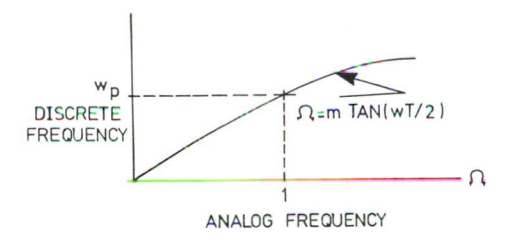

FIG. 7.4.3 Diagram showing an integrated warping curve.

m is to be determined. If one evaluates this equation at $\Omega = 1$ and $\omega = \omega_p$, then m can be seen to equal $1/\tan(\omega_p T/2)$. As a result, the integrated bilinear z transform can be interpreted to read $s = m(z - 1)/(z + 1)$. Continuing along this line of reason, one observes that if $H(s) = 1/(s^2 + 1.414s + 1.0)$, and $m = 1/\tan(0.1\pi) = 3.08$, then $H(z)$ is given by

$$H(z) = \frac{1}{(0.308)^2[(z - 1)/z + 1)]^2 + 1.414(0.308)(z - 1)/(z + 1) + 1}$$

$$= \frac{0.0676(z + 1)^2}{z^2 - 1.142z + 0.412}$$

Since the technical narrative is now considering the question of transform efficiency, the special topic of the Smith chart used by electromagnetic field theorists will now be examined.

EXAMPLE 7.4.2 The conformal map, given by $w = (z - 1)/(z + 1)$, has been well studied by electromagnetic field theorists. It can be seen that this important conformal map is structurally related to the inverse of the bilinear z transform relationship $s = c(z + 1)/(z - 1)$ [see Eq. (2.7.5)]. The conformal map is so widely used in field theory that it has been formalized in terms of the Smith chart. To use the Smith chart in the design of bilinear z-transformed filters, it must be rotated by 180° to account for the reciprocal relationship existing between the z and w domains. For example, suppose that the poles of $H(z)$ are located along the $j\omega$ axis. On the rotated Smith chart, the z-domain poles would be found along the arc having a resistance value of 0 and a capacitive reactance ranging from 0 to ∞ (see Fig. 7.4.4a). If the s-domain poles are located along the negative real axis, the loci of pole locations is found on the 0° line over the range $[0, -\infty]$ (see Fig. 7.4.4b). A specific point, say $s = 0.23 + -j0.23$, would be found at the intersection of a resistive value of 0.23 and capacitive reactance of 0.23 (see Fig. 7.4.4c). The resulting pole locations in the z domain would be read in terms of the two-dimensional coordinates of a uniform grid that is placed over the unit circle. Therefore, referring to Fig. 7.4.4, the z-domain s-domain pole pairings are:

s domain	z domain
$s = 0 + j1$	$z = 0 + j1$
$s = 0 + j0.8$	$z = 0.2 + j0.95$
$s = 0 + j0$	$z = 1.0$
$s = -1 + j0$	$z = 0.0$
$s = 0.23 \pm j0.23$	$z = 0.57 \pm j0.29$

EXAMPLE 7.4.3 Consider again the prewarped transfer function H(s) given in Example 7.4.1. Note that there are two zeros at $j\omega \to \infty$ (i.e., implicit zeros) and poles at $0.23 \pm j0.23$. The Smith chart interpretation of H(s) is diagrammed in Fig. 7.4.5. The resulting H(z) can be seen to have two zeros at $z = -1$ and poles at $z = 0.57 \pm j0.29$. That is, H(z) satisfies

$$H(z) = \frac{K(z + 1)^2}{(z - 0.57 + j0.29)(z - 0.57 - j0.29)}$$

which is in agreement with the previous calculation.

EXAMPLE 7.4.4 The dc sign of a Chebyshev lowpass filter having a -1-dB passband over [0, ω_p] rad/sec and a -20-dB stopband over [ω_a, $\omega_s/2$] rad/sec will be detailed step by step in this example. Using normalized frequencies, consider $\omega'_p = \omega_p/\omega_s = 0.2$ and $\omega'_a = \omega_a/\omega_s = 0.3$. The 1-dB passband establishes ϵ to be $\epsilon = 0.50885$ (i.e., $10^{-0.05}$). The order of the Chebyshev filter can be determined by using approximating polynomials,

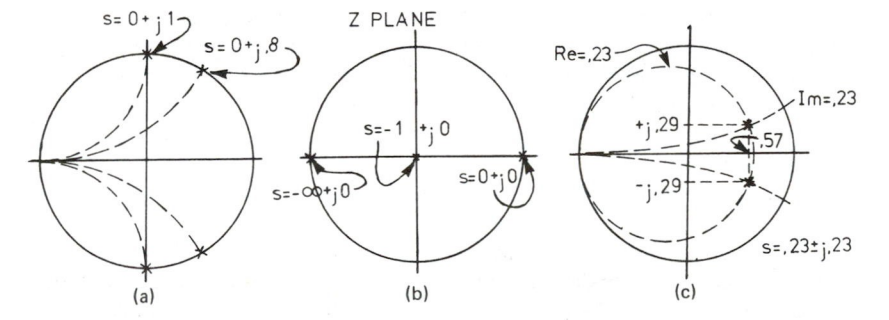

FIG. 7.4.4 Interpretation of the z plane using a Smith chart.

$$H(s) = \frac{(.325)^2}{(s+.325(.707+j.707))(s+.325(.707-j.707))}$$

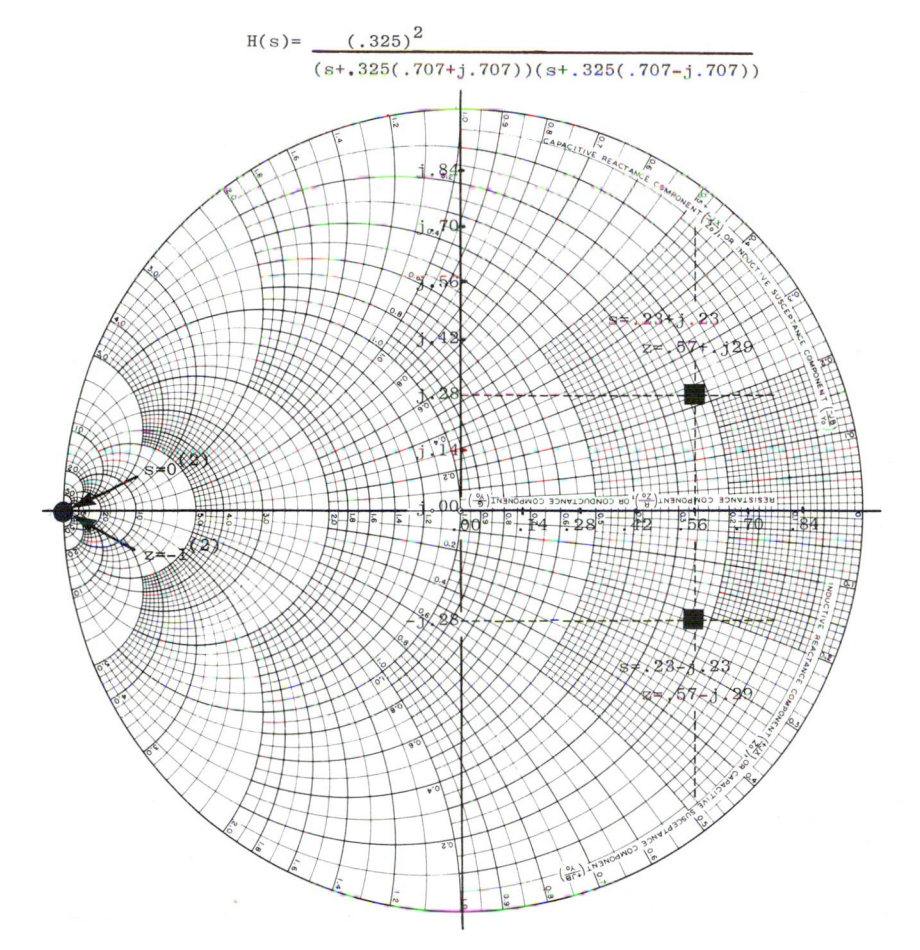

FIG. 7.4.5 Smith chart interpretation of a simple second-order filter having poles located at $0.23 \pm j0.23$.

nomographs, or direct computation. For expository reasons, the latter will be pursued. Referring to Eqs. (1.2.2) and (1.2.3), it follows that

$$|H(jx)|^2 = \frac{1}{1 + {}^2C_n^2(x)} \qquad \text{(analog model)}$$

$$n = 0, \quad C_0(x) = 1$$

$$n = 1, \quad C_1(x) = x$$

$$n = 2, \quad C_2(x) = 2x^2 - 1$$
$$n = 3, \quad C_3(x) = 4x^3 - 3x$$
$$n = 4, \quad C_4(x) = 8x^4 - 8x^2 + 1$$

where $x = \Omega_p/\Omega_a$ [reciprocal of prewarped analog transition coefficients, i.e., $1/k$(analog]

$$x = \frac{2 \tan(0.3\pi/2)}{2 \tan(0.2\pi/2)} = \frac{1.0191}{0.650} = 1.567$$

For $n = 3$, $C_3^2(x)$, $= (4x^3 - 3x)^3 = 114.93$, it follows that at $\omega_a = 0.3\pi$,

$$20 \log_{10}\left[\left|H\left(j2 \tan\left(\frac{0.3\pi}{2}\right)\right)\right|\right] = 10 \log_{10}\left[\left|H\left(j2 \tan\left(\frac{0.3\pi}{2}\right)\right)\right|^2\right]$$

$$= 10 \log_{10}\left[\frac{1}{1 + (0.50885)^2(114.93)}\right]$$

$$= -14.978 > -20 \text{ dB}$$

For $n = 4$, $C_4(x) = (8x^4 - 8x^2 + 1)^2 = 882.4$. At $\omega_a = 0.3\pi$,

$$20 \log 10\left[\left|H\left(j2 \tan\left(\frac{0.3\pi}{2}\right)\right)\right|\right] = -23.6075 < -20 \text{ dB}$$

Therefore, a fourth-order Chebyshev filter is required.

The poles of the fourth-order Chebyshev filter can be found in standard tables (e.g., Budak, 1974), or derived. The poles of a nth-order Chebyshev filter are found on an ellipse that satisfies the equation $x^2/a^2 + y^2/b^2 = 1$ (see Fig. 7.4.6).

$$a = \frac{1}{2}(\alpha^{1/N} - \alpha^{-1/N})$$

$$b = \frac{1}{2}(\alpha^{1/N} + \alpha^{-1/N})$$

$$\alpha = \overset{-1}{} + \sqrt{1 + \epsilon^{-2}} \qquad \text{defined in Eq. (7.2.1)}$$

$$\theta_i = \frac{180 + 360i}{2N}$$

θ_i = angle of ith pole location with respect to the x axis

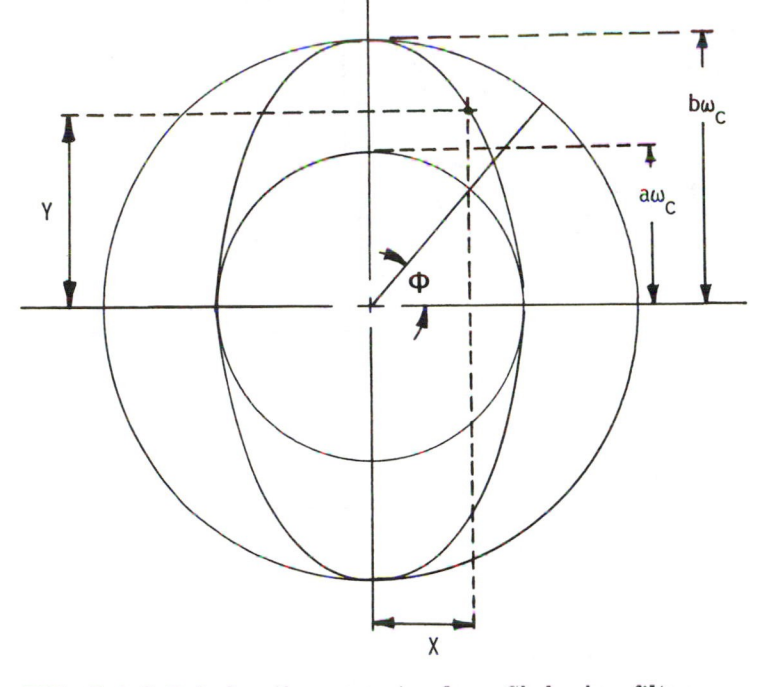

FIG. 7.4.6 Pole location geometry for a Chebyshev filter.

For N = 4 and ϵ = 0.50885, it follows that

$$a = \frac{1}{0.50885} + \sqrt{1 + \frac{1}{(0.50885)^2}} = 4.170226494$$

$$a = \frac{1}{2}(4.17^{1/4} - 4.17^{-1/4}) = 0.364623769$$

$$b = \frac{1}{2}(4.17^{1/4} + 4.17^{-1/4}) = 1.064401481$$

The poles of $|H(s)|^2$ are the roots of the eighth-order polynomial $H(s) * H(-s)$ and they are located at θ_i degrees from the real axis. In particular:

n	θ_i (deg)
0	$\theta_0 = 22.5$
1	$\theta_1 = 67.5$
2	$\theta_2 = 112.5$
3	$\theta_3 = 157.5$
4	$\theta_4 = 202.5$
5	$\theta_5 = 247.5$
6	$\theta_6 = 292.5$
7	$\theta_7 = 337.5$

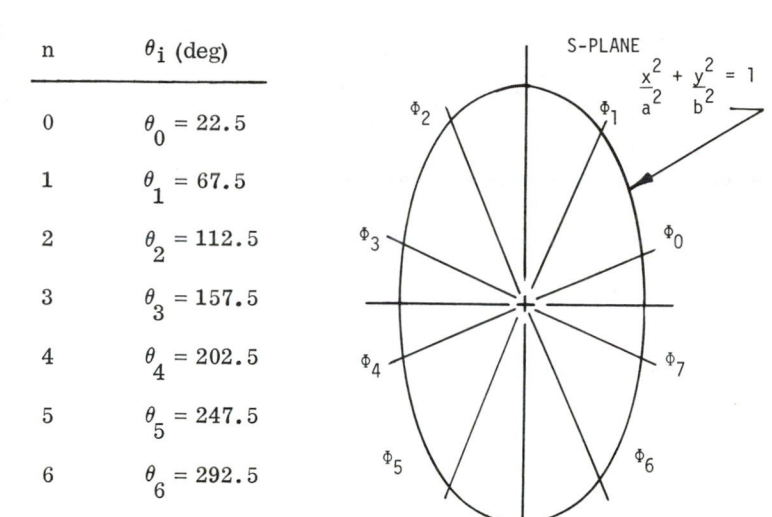

From a stability standpoint, only the poles in the left-hand plane can be used to synthesize a practical filter [i.e., left-hand plane poles belong to H(s) and right-hand plane poles belong to H(-s)]. The poles appear as complex conjugate pairs which combine to give $(s - a \cos \theta_i + jb \sin \theta_i)(s - \cos \theta_i - jb \sin \theta_i) = s^2 - (2a\Omega_p \cos \theta_i)s + (a + \Omega_p)^2 \cos^2 \theta_i + (b\Omega_p)^2 \sin^2 \theta_i$. For a = 0.36462796, b = 1.06440148, $\Omega_p = 2 * \tan(0.2\pi/2) = .6498$, one obtains

$$\theta_i = 112.5^0 \Rightarrow X_i(s) = s^2 + 0.18145s + 0.4166$$
$$\theta_i = 157.5^0 \Rightarrow X_i(s) = s^2 + 0.437s + 0.1180$$

The lowpass Chebyshev analog model can now be deduced to have the form

$$H(s) = \frac{k}{(s^2 + 0.1814s + 0.4166)(s^2 + 0.4378s + 0.1180)}$$

The gain parameter k is chosen so that in the passband, $1 \geq |H(j\Omega)|^2 \geq 1/\Omega$ $(1 + \epsilon^2)$, $0 \leq \Omega \leq \Omega_p$. Since $|H(j\Omega_p = j0.6498)| = k/0.04916$, it follows that k = 0.04381.

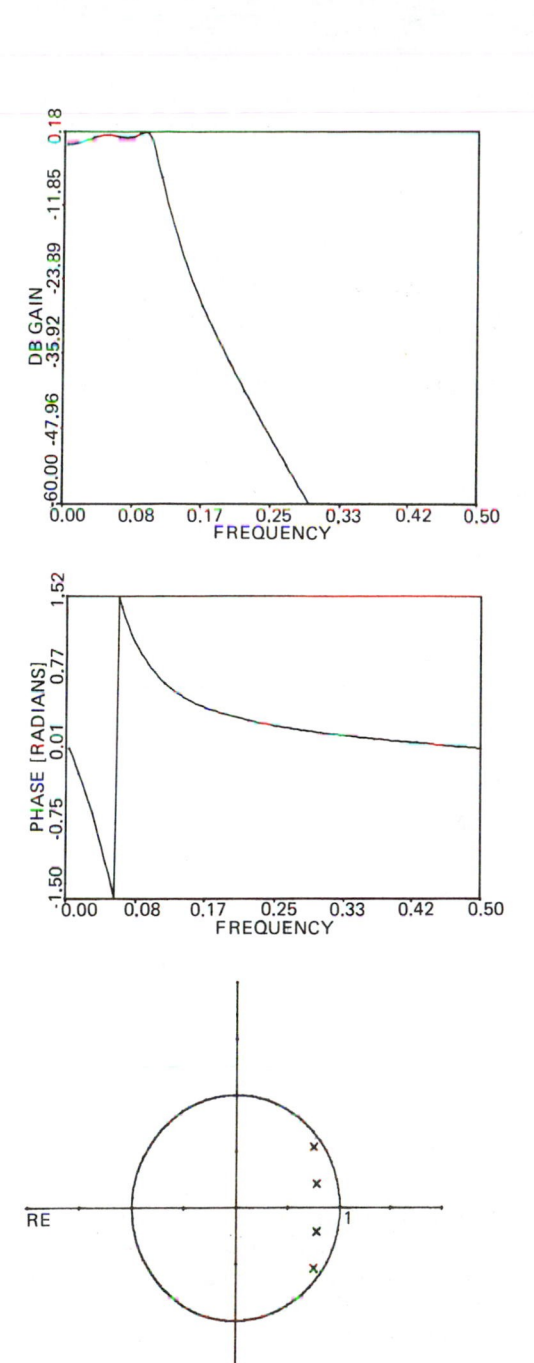

FIG. 7.4.7 Frequency, phase, and pole–zero locations of a fourth–order Chebyshev filter.

In terms of the bilinear transform, $s = (2/T)(1 - z^{-1})/(1 + z^{-1})$. It follows immediately that

$$H(z) = \frac{0.001836(1 + z^{-1})^4}{(1 - 1.499z^{-1} + 0.8482z^{-2})(1 - 1.5548z^{-1} + 0.6493z^{-2})}$$

The frequency response of the filter derived is presented in Fig. 7.4.7.

7.5 METHOD II FILTER DESIGN

Method II is abstracted in Fig. 7.5.1. The mechanics of a method II solution can be demonstrated in terms of the preceding example problem.
  Recall that

$$H_N(s) = \frac{1}{(s + 0.707 + j0.707)(s + 0.707 - j0.707)} = \frac{1}{s^2 + 1.414s + 1}$$

Then

$$H(s) = H_N(s)\Big|_{s=(z-1)/(z+1)}$$

and, given that the digital lowpass-to-lowpass transform (Table 7.3.1) satisfies

$$z = \frac{z - \alpha}{1 - \alpha z}$$

where

$$\alpha = \frac{\sin[(\omega_a - \omega_p)/2T]}{\sin[(\omega_a + \omega_p)/2T]}$$

a desired lowpass filter can be derived from $H_N(z)$. In the equation above, ω_a and ω_p represent the critical frequencies (normalized with respect to the sampling frequency) of the desired discrete lowpass filter. The critical frequency ω_i is equivalent to the warped frequency given by $\omega_i = 2 \tan^{-1}(\Omega_i/2)$, mapped into a discrete filter under $s = 2(z - 1)/(z + 1)$ under the hypothesis that $T = 1$.[†] Given that the analog prototype filter has a critical

---

[†]For reasons previously developed, the scale factor $2/T$ can be omitted.

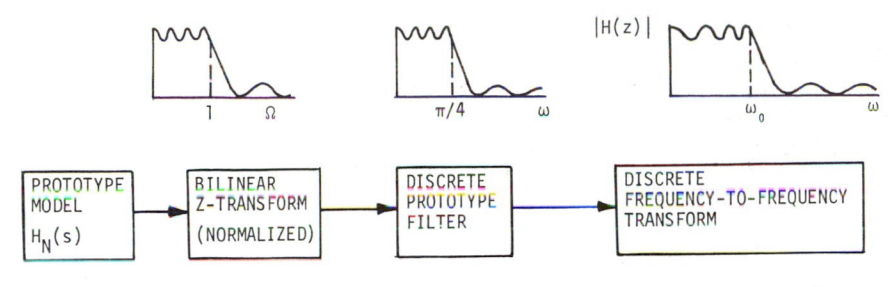

FIG. 7.5.1 System block diagram of the IIR method II design procedure.

frequency located at $\Omega_p = 1$, the discrete prototype filter's critical frequency can be found at $\omega_p = \tan^{-1}(1) = \pi/4 = 0.7854$ rad (under the assumption that the term $2/T$ is to be ignored. Therefore, the discrete prototype filter, say $H_N(z)$, satisfies (see Fig. 7.5.2)

$$H_N(z) = H_N(s)\bigg|_{s=(z-1)/(z+1)} = \frac{(z+1)^2}{(3.414)^2(z^2 + 0.1716)}$$

Before considering another example, the reader should be made aware of a growing trend in the field of filter design. General-purpose digital computers are becoming an increasingly important design tool. Warping, prototype filter derivation, bilinear z transform, and so forth, can be programmed. However, the study and analysis of these filters can be accomplished using one's wits, a hand-held calculator, and a sharp pencil (with a large eraser). Through the manual manipulation of filter design data, one can gain added insights into the filter synthesis problem. This will pay large dividends in the future when the reader must create new software or modify existing software.

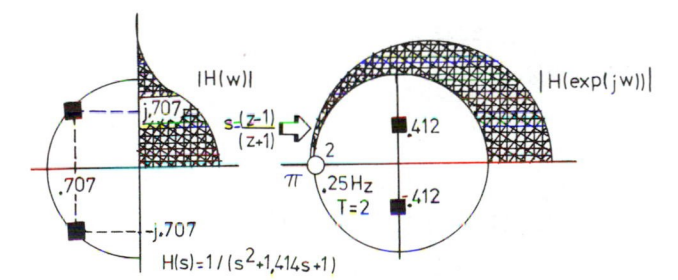

FIG. 7.5.2 Example of the pole mapping associated with the method II design method.

EXAMPLE 7.5.1 A first-order prototype Butterworth filter is given by
$H(s) = 1/(s + 1)$. It is desired to design a lowpass filter having a cutoff frequency of $\omega_p = 2\pi(100)$ radians per second and a sample rate of 1000 samples per second.

Method I:

$$\omega_p = 2\pi \times 100 \quad \text{(discrete critical frequency)}$$

$$\Omega_p = \tan\left(\frac{2\pi \times 100}{2 \times 1000}\right) = \tan(0.1\pi) = 0.325 \quad \text{(warped critical frequency)}$$

$$H(s) = H_N(s)\Big|_{s=s/0.325} = \frac{0.325}{s + 0.325}$$

$$H(z) = H(s)\Big|_{s=(z-1)/(z+1)} = \frac{0.325(z + 1)}{1.325z - 0.675}$$

$$= \frac{0.245(z + 1)}{z - 0.51}$$

Method II:

$$H_N(z) = H_N(s)\Big|_{s=(z-1)/(z+1)} = \frac{z + 1}{2z}$$

Lowpass-to-lowpass:

$$z^{-1} = \frac{z^{-1} - \alpha}{1 - \alpha a^{-1}}; \quad \omega_p = \frac{2\pi \times 100}{1000} = \frac{\pi}{5} \quad \text{(normalized)}$$

$$\alpha = \frac{\sin[(\pi/4) - (\pi/10)]}{\sin[(\pi/4) + (\pi/10)]} = \frac{\sin(3\pi/10)}{\sin(7\pi/10)} = 0.51$$

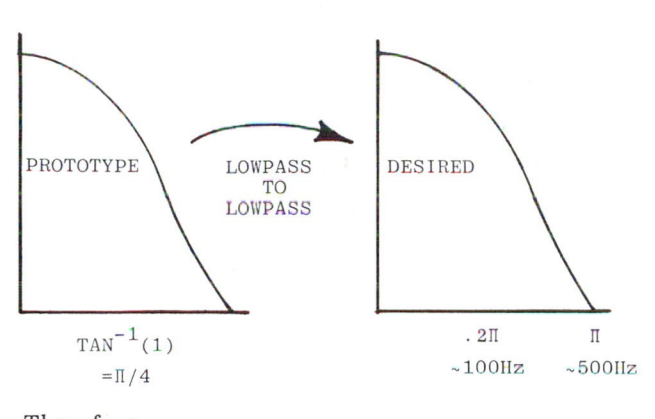

$$\text{TAN}^{-1}(1)$$
$$= \Pi/4$$

$.2\Pi$ Π

$\sim 100\text{Hz}$ $\sim 500\text{Hz}$

Therefore,

$$z = \frac{z - 0.51}{1 - 0.51z}$$

and

$$H(z) = H_N(z)\Big|_{z=(z-0.51)/(1-0.51z)} = \frac{0.245(z + 1)}{z - 0.51}$$

EXAMPLE 7.5.2 The design of a Chebyshev filter possessing the following attributes will now be considered (see Fig. 7.5.3).

1. Passband ripple: -5 dB
2. Stopband gain: -19 dB
3. Passband critical frequency: $f_p = 100$ Hz
4. Stopband critical frequency: $f_a = 200$ Hz
5. Sampling frequency: 1000 Hz

From the fact that $1 + \epsilon^2 = 10^{0.05}$, one concludes that $\epsilon = 0.35$. Also $k(\text{analog}) = \tan(0.1\pi)/\tan(0.2\pi)$ and $\Omega_r = 1/k(\text{analog}) = 2.235$ and $A^2 = 79$. The filter order may be estimated to be:

$$n = \frac{\log_{10}[(1 + \sqrt{1 - \eta^2})/\eta]}{\log_{10}[\Omega_r = \sqrt{\Omega_r^2 - 1})} = \frac{1.7}{0.626} = 2.72 \Rightarrow n = 3$$

$$\eta = \frac{\epsilon}{\sqrt{A^2 - 1}} = \frac{0.35}{\sqrt{78}} = 0.03963$$

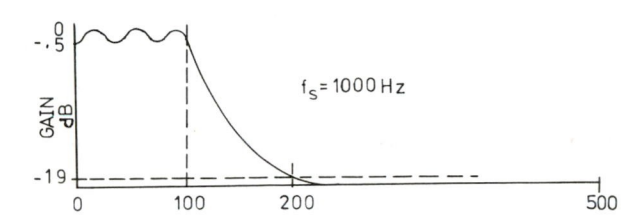

FIG. 7.5.3 Frequency magnitude response model for the example problem.

The computed prewarped passband frequency is given by

$$\Omega_p = \tan\left(\frac{\omega_p T}{2}\right) = \tan\frac{2\pi \times 100}{2 \times 1000} = 0.325$$

Therefore, the normalized prototype filter (obtained from standard analog filter design tables) satisfies

$$H_N(s) = \frac{1}{s^2 + 1.253s^2 + 1.535s + 0.716}$$

Upon performing a lowpass-to-lowpass conversion (i.e., $s = s/\Omega_p = s/0.325$), one obtains

$$H(s) = H_N(s)\Big|_{s=s/0.325} = \frac{0.0256}{s^3 + 0.413s^2 + 0.167s + 0.026}$$

which can be converted to H(z) under $s = (z - 1)/(z + 1)$. That is,

$$H(z) = H(s)\Big|_{s=(z-1)/(z+1)} = \frac{0.0156(z + 1)^3}{z^3 - 1.975z^2 + 1.52z + 0.454}$$

EXAMPLE 7.5.3 The problem is to design the elliptic lowpass filter diagrammed in Fig. 7.5.4a. The key parameters in this example are $\epsilon = 0.5$, $\eta = \epsilon/(A^2 - 1)^{1/2} = 2 \times 10^{-3}$, and k(analog) = 0.3. Using the nomograph method, one can determine that the filter order should be n = 3 (see Fig. 7.5.4b). Using modified method I, the parameter m is computed to be

$$m = \frac{1}{\tan[(2\pi \times 250)/(2 \times 1000)]} = \frac{1}{\tan(0.7854)} = 1.0$$

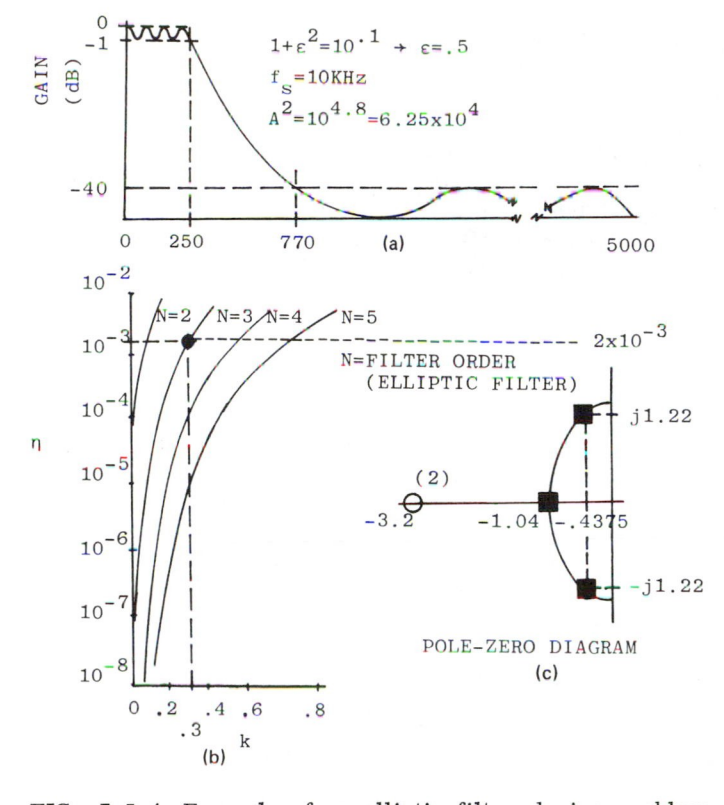

FIG. 7.5.4 Example of an elliptic filter design problem.   (Portions of these data are redrawn after Rabiner et al., 1974.)

Thus the analog filter becomes (using standard analog filter design tables)

$$H_N(s) = \frac{s^2 + 10.21}{(s + 1.04)(s^2 + 0.87s + 1.67)}$$

which has the pole-zero configuration shown in Fig. 7.5.4c.

7.6 FREQUENCY-TO-FREQUENCY TRANSFORMS

Under the transformations found in Tables 7.2.1 and 7.3.1, the following filter attributes are preserved (invariant):

1. Passband ripple (i.e., ϵ)
2. Transition coefficient (i.e., k)
3. Stopband attenuation (i.e., A)

Obviously, all design parameters, which are explicit functions of ϵ, k, and A, are also invariant under these transforms. Finally, the transition co-efficient k, and stopband attenuation A are equal to the leading and trailing edges of the bandpass and bandstop filters. This is reinforced in Fig. 7.6.1. These transforms will now be studied through the vehicle examples using a method I approach.

7.7 HIGHPASS FILTERS

A highpass filter may be derived in terms of a normalized lowpass analog filter. A highpass filter's magnitude response is abstracted in Fig. 7.7.1. The design methodology is essentially that used to design a lowpass analog filter from its prototype model. Specifically, in terms of ϵ, A, and k, a normalized filter of order n can be derived. This filter, in turn, can be mapped into a prewarped highpass filter (in a method I sense) or digital prototype filter (in a method II sense). Finally, the prewarped analog filter or digital prototype filter will be mapped into its final digital form under the appropriate transform. The design of a highpass filter, using a method I format, can be motivated by Example 7.7.1.

EXAMPLE 7.7.1 In Example 7.4.1, a lowpass Butterworth filter satis-fying $\epsilon = 1$, $A^2 = 10$, and $\Omega_r = 1/k(\text{analog}) = 2$ was designed. Based on this parameterization, the order of the filter was estimated using several tech-niques. Using a formula approach, one obtains

$$n = \frac{\log_{10}[(A^2 - 1)/\epsilon]}{2 \log_{10}(\Omega_r)} = 1.58 \Rightarrow n = 2$$

Then

$$H_N(s) = \frac{1}{(s + 0.707 + j0.707)(s + 0.707 - j0.707)} = \frac{1}{s^2 + 1.414s + 1}$$

One needs to compute the prewarped critical cutoff frequency which, for $f_p = 1000$, $f_s = 10,000$, equates to

$$\Omega_p = \tan\left(\frac{\omega_p T}{2}\right) = \tan(0.1\pi) = 0.325$$

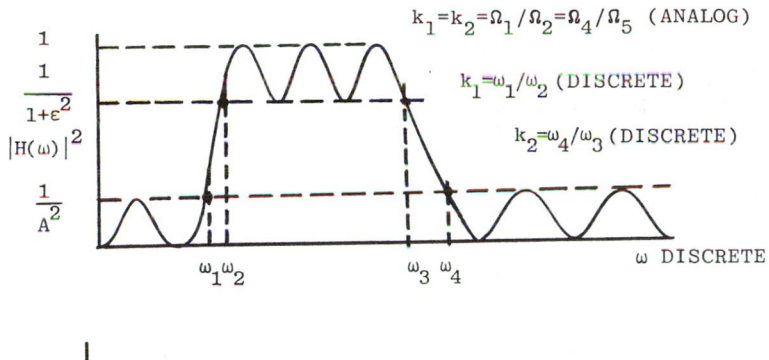

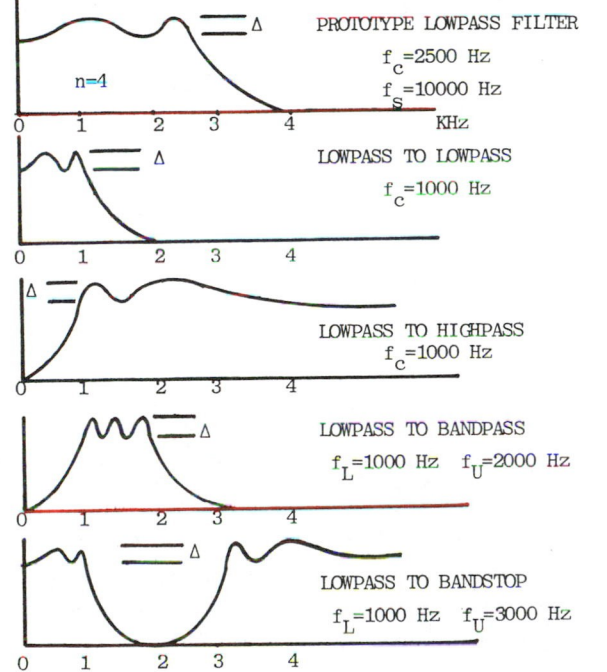

FIG. 7.6.1 Parameterization of a general filter's magnitude frequency response and examples of the conversion of a prototype lowpass filter into a specific lowpass, highpass, bandpass, and bandstop filters.

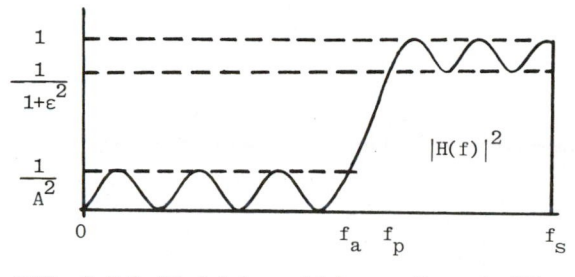

FIG. 7.7.1 Model for a highpass discrete filter.

Using method I, the highpass filter (found in Fig. 7.7.2) can be derived.
First one performs a lowpass-to-lowpass conversion under the rule $s = \Omega_p/$
$s = 0.325/s$. Here

$$H(s) = H_N(s)\Big|_{s=0.325/s} = \frac{1}{(0.325/s)^2 + [1.414(0.325)s] + 1}$$

$$= \frac{s^2}{s^2 + 0.46s + 0.106}$$

Finally, one defines H(z) in terms of the bilinear transform of H(s) under
$s = (z - 1)/(z + 1)$. That is,

$$H(z) = H(s)\Big|_{s=(z-1)/(z+1)} = \frac{(z - 1)^2}{(z^2 - 1.142z + 0.412)}$$

$$= \frac{(z - 1)^2}{(z - 0.57 + j0.295)(z - 0.57 - j0.295)}$$

The pole-zero configuration for this example is shown in Fig. 7.7.3.

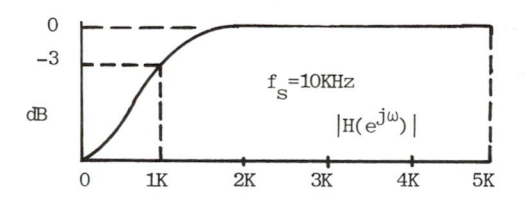

FIG. 7.7.2 Example of a highpass filter design problem.

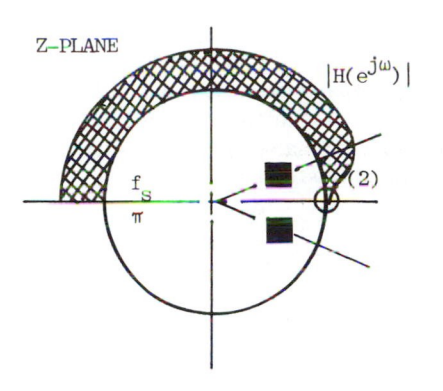

FIG. 7.7.3 Magnitude frequency response of a highpass filter interpreted in the z plane.

EXAMPLE 7.7.2 The modified method I approach can also be used to design a highpass filter. Consider again the problem outline in Example 7.7.1 The lowpass-to-lowpass transform was integrated into the statement $s = m(z + 1)/(z - 1)$, where $m = \tan(\omega T/2)$. That is, the transform is the inverse of the modified lowpass-to-lowpass conversion. Graphically, these operations are summarized in Fig. 7.7.4.

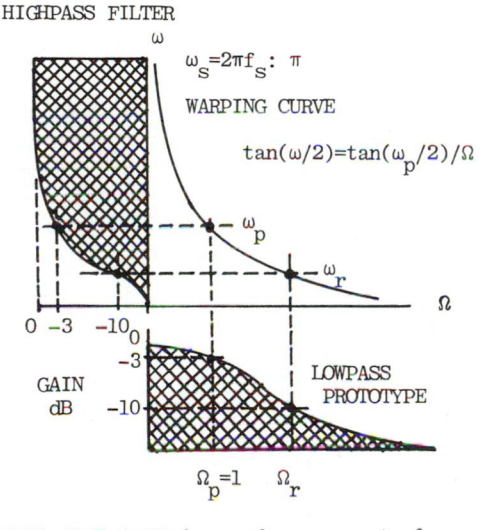

FIG. 7.7.4 Highpass frequency-to-frequency conversion warping curve.

7.8 BANDPASS (BANDSTOP) FILTERING

It will be seen that the nth-order prototype lowpass filter is mapped into a second-order bandpass filter under a lowpass-to-bandpass transform. The conversion of a lowpass prototype filter is conceptually straightforward but can be algebraically complex. The mapping philosophy is abstracted in Fig. 7.8.1.

EXAMPLE 7.8.1   The problem is one of designing a Butterworth bandpass filter having the magnitude response found in Fig. 7.8.2. Here, $\epsilon = 1$, $A^2 = 10$, $\Omega_r = 1/k(\text{analog}) = 2$. The order of the prototype filter was previously computed to be 2. Therefore,

$$H_N(s) = \frac{1}{s^2 + 1.414s + 1}$$

Using the lowpass-to-bandpass transform, one notes that

$$s = \frac{s^2 + \Omega_L \Omega_U}{s(\Omega_u - \Omega_L)}$$

where Ω_L and Ω_U are the prewarped frequencies given by

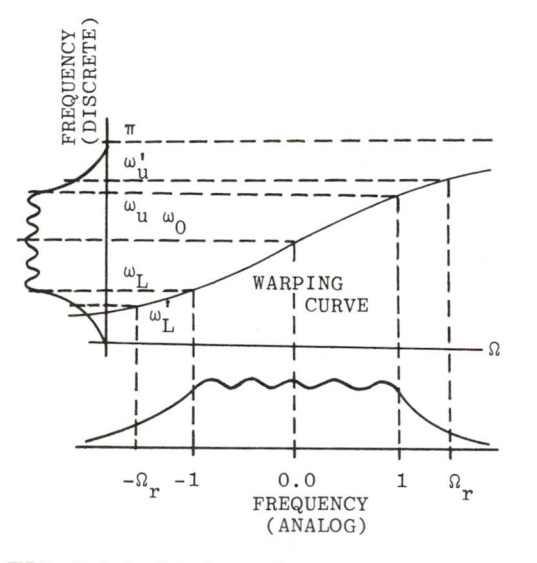

FIG. 7.8.1 Bandpass design and warping philosophy.

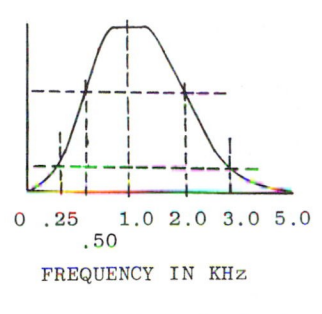

0 .25   1.0 2.0 3.0 5.0
    .50

FREQUENCY IN KHz

FIG. 7.8.2 Model for the example bandpass filter design problem.

$$\Omega_U = \tan(0.2\pi) = 0.7265$$

$$\Omega_L = \tan(0.05\pi) = 0.1584$$

Then

$$s = \frac{s^2 + 0.1151}{0.5285s}$$

and

$$H(s) = H_N(s)\Big|_{s=(s^2+0.1151)/0.5681s} = \left[\left(\frac{s^2 + 1151}{0.5681s}\right)^2 + 1.414\left(\frac{s^2 + 0.1151}{0.5681s}\right) + 1\right]^{-1}$$

When dealing with high-order filters, it is virtually impossible to compute the pole-zero locations manually. Instead, a general-purpose digital computer is often used. Shown in Fig. 7.8.3 is a source code that uses a common all-purpose root-finding routine called POLRT (e.g., IBM and DEC). This software will be used to support the analysis of data found in this example.

The derived fourth-order bandpass filter can be analyzed in a computer-aided sense. The result of this study is shown in Fig. 7.8.4. This result can be substantiated through the use of the Smith chart. Referring to Fig. 7.8.5, it can be seen that the poles and zeros of the fourth-order H(s) map into

$$\underline{\text{Poles:}} \quad \left\{ \begin{array}{c} 0.8 \pm j0.26 \\ 0.3 \pm j0.55 \end{array} \right\}; \quad \underline{\text{zeros:}} \quad \left\{ \begin{array}{c} +1, \ +1 \\ -1, \ -1 \end{array} \right\}$$

```
        DIMENSION A(37),W(37),ROOTR(37),ROOTI(37)
C       POLYNOMIAL ROOTS FINDING MAIN CALLING ROUTINE
C       USES SCIENTIFIC SUPPORT SUBROUTINE POLRT
C       ORDER: L .GE. IORD .LE. 36
C       IIN=LOGICAL INPUT DEVICE; IOUT=LOGICAL OUTPUT DEVICE
        WRITE(1,10)
10      FORMAT(1H,'INPUT LOGICAL INPUT-OUTPUT UNIT NUMBERS')
        READ(1,*) IIN,IOUT
        WRITE(IOUT,15)
15      FORMAT(1H,'POLYNOMIAL ORDER')
20      READ(IIN,*)IORD
        WRITE(IOUT,40)
40      FORMAT(1H,'READ COEFFICIENTS A(0),A(1),...,A(N)')
        DO 100 I=1,1+IORD
        WRITE(IOUT,110),I-1
110     FORMAT(1H,'A(',I3,')=')
        READ(IIN,*)A(I)
100     CONTINUE
        CALL POLRT(A,W,IORD,ROOTR,ROOTI,IER)
        IF(IER-1)190,160,170
160     WRITE(IOUT,165)
165     FORMAT(//1H,'ORDER .LT. 1')
        GO TO 20
170     IF(IER-3)175,180,178
175     WRITE(IOUT,177)
177     FORMAT(//1H,'ORDER .GT. 36')
        GO TO 20
178     WRITE(IOUT,179)
179     FORMAT(//1H,'A(N)=0')
        GO TO 20
180     WRITE(IOUT,185)
185     FORMAT(//1H,'UNABLE TO DETERMINE ALL ROOTS')
190     WRITE(IOUT,195)
195     FORMAT(//1H,5X,'REAL ROOT        COMPLEX ROOT',//)
        DO 200 I=1,IORD
200     WRITE(IOUT,210)ROOTR(I),ROOTI(I)
210     FORMAT(1H,2E16.7)
        STOP
        END
```

FIG. 7.8.3 Source code of calling routines used to compute the complex root of a polynomial.

Therefore, H(z) satisfies

$$H(z) = \frac{K(z + 1)^2(z - 1)^2}{(z - 0.8 + j0.26)(z - 0.8 - j0.26)(z - 0.3 + j0.55)(z - 0.3 - j0.55)}$$

To this point, the analysis mechanism used in this chapter has been the bilinear z transform. It was noted that this transform had the ability to preserve the magnitude response of an analog model. To emphasize the interconnection between the choice of transform and the performance of a filter, the following example is offered.

EXAMPLE 7.8.2 A second-order example frequency-selective filter (i.e.,
high Q) can be modeled as

$$H(s) = \frac{s + a}{(s + a)^2 + \Omega_0^2}$$

For the purpose of demonstration, consider $a = 100$, $\omega_0 = 200\pi$, and
$f_s = 1000$ Hz. The standard z transform of H(s), denoted by $H_1(z)$, is
given by

$$H_1(z) = \frac{z(z - z_1)}{(z - p_1)(z - p_1^*)}; \qquad \begin{aligned} z_1 &= \exp(-aT)\cos(\omega_0 T) \\ p_1 &= \exp(-aT)[\cos(\omega_0 T) + j\sin(\omega_0 T)] \end{aligned}$$

Therefore,

$$H_1(z) = \frac{z(z - 0.732)}{(z - 0.732 - j0.531)(z - 0.732 + j0.531)}$$

Several methods have been presented by which H(s) may be bilinearly
mapped into, say $H_2(z)$. For the purpose of demonstration, a Smith chart

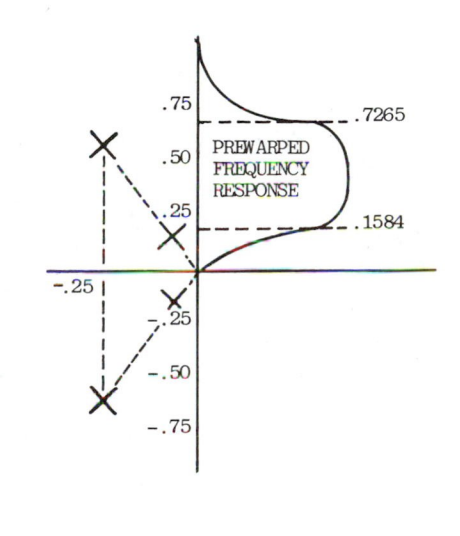

```
INPUT LOGICAL INPUT-OUTPUT UNIT NUMBER
1,1
POLYNOMIAL ORDER
4
READ COEFFICIENTS A(0),A(1),...,A(N)
A(   0)=
.01325
A(   1)=
.0925
A(   2)=
.5529
A(   3)=
.8033
A(   4)
1
```

REAL ROOT	COMPLEX ROOT
−0.8806811E−01	−0.1566879E+00
−0.8806811E−01	0.1566879E+00
−0.3135819E+00	−0.5583844E+00
−0.3135819E+00	0.5583844E+00

FIG. 7.8.4 Output listing displaying the complex pole locations for the
fourth-order bandpass design example.

approach will be used. The bilinear z transform is given by $s = 2(z - 1)/(z + 1)T$, whereas the rotated Smith chart is defined in terms of $(z - 1)/(z + 1)$. Therefore, $H(s)$ must be redefined in terms of $H_2 = H(2s/T)$ for a Smith chart interpretation. That is,

$$H_2(s) = H(s)\Big|_{s=2s/T} = \frac{1}{2 \times 10^3} \frac{s + 0.05}{s^2 + 0.1s + 0.1012}$$

$$= \frac{1}{2 \times 10^3} \frac{s + 0.05}{(s + 0.05 + j0.3142)(s + 0.05 - j0.3142)}$$

The poles and zeros of this filter, together with those obtained using the standard z transform, are interpreted in Fig. 7.8.6. It can be noted that the poles are located approximately in the same location. Their close proximity to the periphery of the unit circle guarantees a strong frequency selectivity about the normalized frequency of 0.2π. However, there is a

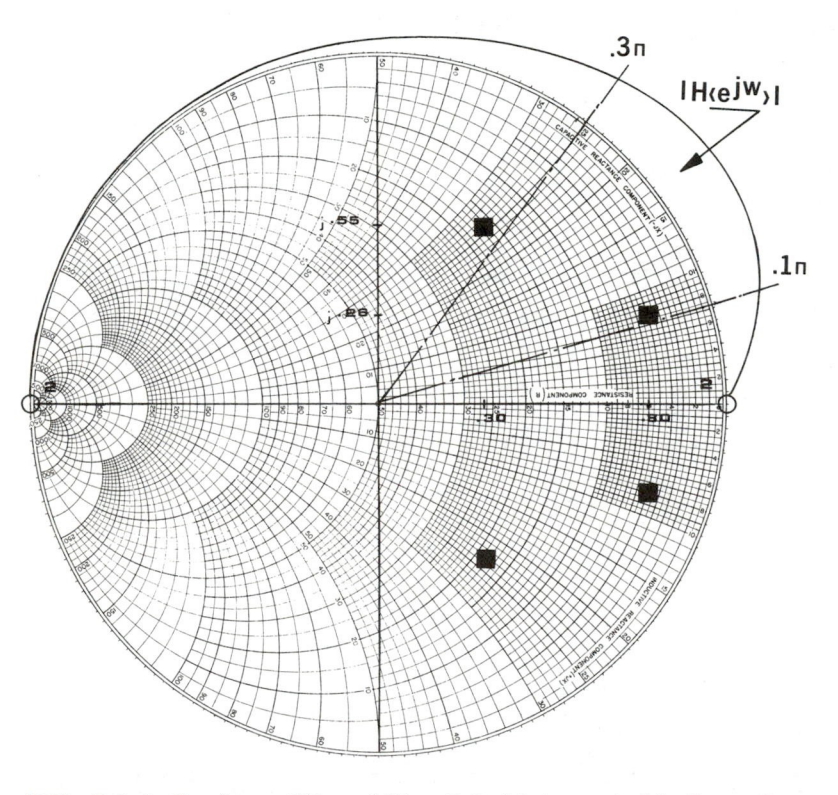

FIG. 7.8.5 Bandpass filter of Fig. 7.8.4 interpreted in the z plane.

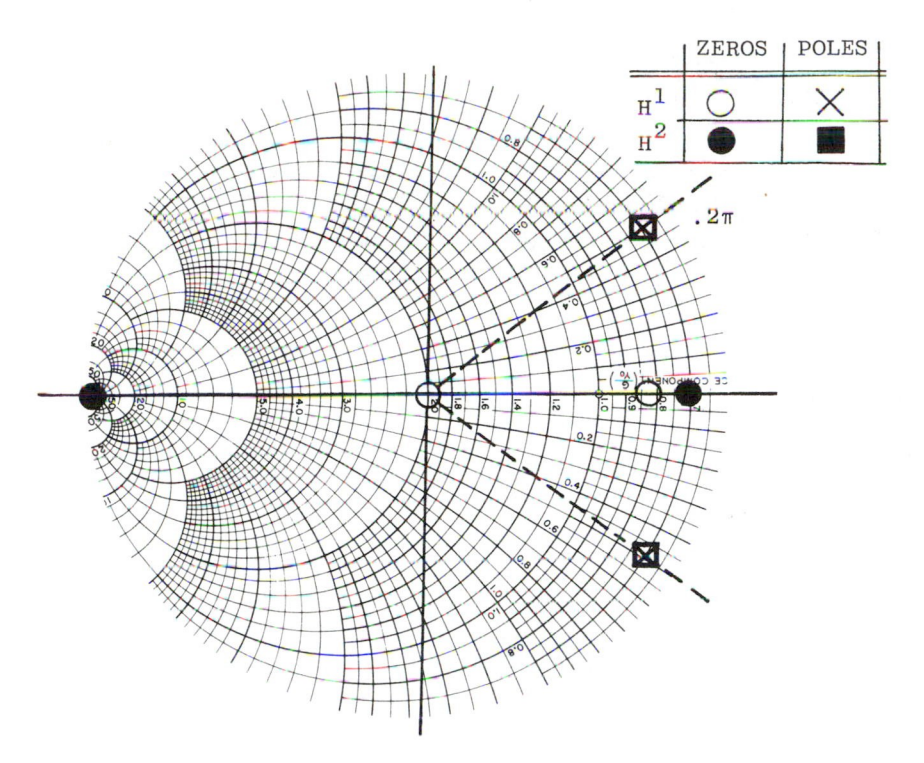

FIG. 7.8.6 Comparison of the pole locations of a bandpass filter using standard and bilinear transforms.

significant difference in their frequency response due to dissimilar zero locations. While the standard z-transform filter exhibits a relatively large residual gain, $\omega = 0$ and π, the bilinear z-transform filter's gain is zero.

An integrated approach may be taken to the problem of designing bandpass and bandstop filters. Referring again to Fig. 7.8.1, it can be seen that

$$\Omega = \frac{a[b - \cos(\omega)]}{\sin(\omega)} \qquad (7.8.1)$$

where

$$a = \cot\left(\frac{\omega_U - \omega_L}{2}\right) \qquad (7.8.2)$$

$$b = \frac{\sin(\omega_U - \omega_L)}{\sin(\omega_U) + \sin(\omega_L)}$$

This specific mapping produces the required prewarped information. The resulting <u>modified bilinear z transform</u> associated with the curve discussed is a mapping given by

$$s = \frac{a(z^2 - 2bz + 1)}{z^2 - 1} \tag{7.8.3}$$

This transform is used in the next example.

EXAMPLE 7.8.3 Reconsider Example 7.8.1 in a modified bilinear sense. Here

$$a = \cot\left(\frac{0.3\pi}{2}\right) = 1.96$$

$$b = \frac{\sin(0.5\pi)}{\sin(0.4\pi) + \sin(0.1\pi)} = \frac{1}{1.26} = 0.794$$

with

$$s = \frac{1.96(z^2 - 1.23z + 1)}{z^2 - 1}$$

and

$$H(z) = \left[(1.96)^2 \left(\frac{z^2 - 1.588z + 1}{z^2 - 1}\right)^2 + (1.414)(1.96)\left(\frac{z^2 - 1.588z + 1}{z^2 - 1}\right) + 1 \right]^{-1}$$

$$= \frac{(z - 1)^2(z + 1)^2}{(1.96)^2(1.98z^4 - 4.319z^3 + 4.002z^2 - 2.033z + 0.54)}$$

Using the pole-zero determining software discussed previously, the pole-zero locations were determined and presented in Fig. 7.8.7. It can be seen that the pole locations agree with those previously computed for this fourth-order example.

7.9 COMPUTER-AIDED DESIGN

Computer-aided filter design routines generally fall into one of two groups. The first of these concentrates on the design of optimal classic filters (i.e.,

```
INPUT LOGICAL INPUT-OUTPUT UNIT NUMBERS
1,1
POLYNOMIAL ORDER
4
READ COEFFICIENTS A(0),A(1),...,A(N)
A(    0)=
.54
A(    1)=
-2.033
A(    2)=
4.002
A(    3)=
-4.319
A(    4)=
1.98
```

REAL ROOT	COMPLEX ROOT
0.8012689E+00	-0.2589415E+00
0.8012689E+00	0.2589415E+00
0.2893879E+00	-0.5485206E+00
0.2893879E+00	0.5485206E+00

FIG. 7.8.7 Output listing displaying the pole locations of the filter described in Example 7.8.2.

Butterworth, Chebyshev, or elliptic). The second group considers designing an arbitrary filter using optimal curve-fitting techniques. Specifically,
the rational transfer function $H(s) = N(s)/D(s)$ is modeled to be

$$H(s) = K \prod_{i=1}^{M_1} \frac{a_i s + b_i}{s + \lambda_i} \prod_{i=1}^{M_2} \frac{c_i s^2 + d_i s + e_i}{s^2 + f_i s + g_i} \qquad (7.9.1)$$

where all coefficients are real and $N = M_1 + 2M_2$. The poles of $H(s)$ are
$s = \lambda_i$ (λ_i real) and $s = s_i^0 = -\sigma_i - j\omega_i$, where $(s - s_i^0)(s - s_i^{0*}) = s^2 + f_i s +
g_i$. In a similar manner, a filter $H(z)$ could be modeled by simply replac-
ing s with z. If the design procedure originates with a normalized lowpass
filter, then $H(s) = 1/D(s)$. The design of optimal classic filters is embedded in the characterization of $H(s) = 1/D(s)$ and a suitable frequency-to-
frequency conversion. In this area numerous numerical techniques have
been developed which allow the polynomial $D(s)$ to be derived using a digital
computer. They generally involve root-finding or polynomial approximation routines. Standard FORTRAN support packages are available to the
user to simplify this analysis. Software systems such as those published
by Gray and Markel (1973) and Antoniou (1979), can be used to design discrete filters with a minimum amount of programming knowledge.

7.10 COMPUTER DESIGN OF CLASSIC FILTERS

An optimal computer-aided design software package is discussed in Appendix A. The FORTRAN source code presented in this appendix can be merged with other routines developed later in this text. As a result, a comprehensive filter design and analysis computer-based software system can be configured. The user is required to supply to the optimal filter design software the data found in Table 7.3.1. The required data set includes the sampling frequency f_s, critical frequencies f_c (given in Fig. 7.6.1), transition ratio k, passband minimum gain (AP in decibels), and the stopband attenuation (AA in decibels).

EXAMPLE 7.10.1 A set of filter design parameters are reported in Table 7.10.1. The result of processing these filter data sets through the software found in Appendix A results in the information presented in Figs. 7.10.1 through 7.10.12.

7.11 COMPUTER DESIGN OF ARBITRARY FILTERS

The optimal estimate of a rational transfer function $H(j\omega)$ will be denoted $G(j\omega)$. The optimization criterion encountered most often is the familiar "least squares" of the L_2 functional.[†] In particular, let $\phi(\omega)$ be the ℓ_2 functional satisfying

$$\phi(\omega) = \sum_{n=1}^{N} [|H(j\omega)| - |G(j\omega)|]^2 \tag{7.11.1}$$

Since $H(j\omega) = N(j\omega)/D(j\omega)$, the resulting optimization problem is nonlinear. There are several relatively slowly converging algorithms (e.g., Newton's method) which can be used to synthesize an optimal filter parameter set. Steiglitz (1970) used the Fletcher-Powell method and Deczky extended this minimization procedure to design filters having a minimum group delay. Taylor and Molepske (1973) used the sequential unconstrained minimization technique (SUMP) to optimize a filter subject to both magnitude and phase constraints. These constraints take the form of those diagrammed in Fig. 7.11.1. They are algebraically stated in terms of a set of side constraints $g_i(x)$, where

$$g_i(x) = |H(j\omega_i)|^2 \le \epsilon_i \quad \pm \tan^{-1} \frac{\text{Im}[H(j\omega_i)]}{\text{Re}[H(j\omega_i)]} \le \delta_i \text{ or } \tau_i \tag{7.11.2}$$

[†] A functional is a mapping from a vector space onto the real line.

TABLE 7.10.1 Summary of FIR Butterworth, Chebyshev, and Elliptic Designs Using the IIR Design Software Reported in Appendix A

Type	Class	$w_s(f_s)$	$w_{c_1}(f_{c_1})$	$w_{c_2}(f_{c_2})$	k(analog prototype)	AP(dB)	AA(dB)	N	Figure 7.10
Butterworth	Lowpass	1.0 r/s	0.30	0.40	0.477214	-1.0	-30.0	6	0.1
Butterworth	Highpass	1.0 r/s	0.30	0.40	0.447214	-1.0	-30.0	6	0.2
Butterworth	Bandpass $f_0 = 0.2128$	1.0 r/s	0.10 / 0.35	0.30 / 0.50	0.60612	-1.0	-30.0	18	0.3
Butterworth	Bandstop $f_0 = 0.3183$	1.0 r/s	0.075	0.150	0.370192	-1.0	-30.0	10	0.4
Chebyshev	Lowpass	1.0 r/s	0.30	0.325	0.843448	-1.0	-45.0	11	0.5
Chebyshev	Highpass	1.0 r/s	0.30	0.40	0.447214	-2.0	-45.0	5	0.6
Chebyshev	Bandpass $f_0 = 0.2128$	1.0 r/s	0.10 / 0.35	0.30 / 0.50	0.60612	-1.0	-45.0	14	0.7
Chebyshev	Bandstop $f_0 = 0.1774$	1.0 r/s	0.10	0.30	0.637691	-1.0	-45.0	14	0.8
Elliptic	Lowpass	1.0 r/s	0.30	0.40	0.447214	-1.0	-45.0	4	0.9
Elliptic	Highpass	3.5 kHz	500.0	700.0	0.700	-2.0	-35.0	4	0.10
Elliptic	Bandpass	3.0 kHz	350.0 / 900.0	400.0 / 994.8	0.815438	-1.0	-45.0	14	0.11
Elliptic	Bandstop	3.0 kHz	300.0 / 900.0	350.0 / 994.8	0.82934	-1.0	-45.0	12	0.12

INPUT DATA

BUTTERWORTH APPROXIMATION
AP, DB: 1.000000
AA, DB: 30.000000
K: 0.447214
OUTPUT DATA

ORDER: 6
ACTUAL AA   30.00000
H(S)=H*H1(S)*H2(S)...
H1(S)=(A0+A11S+A21S**2)/(B0+B11S+B21S**2)

SECTION #: 1

A01   0.1000000E+01   B01   0.1000000E+01
A11   0.0000000E+00   B11   0.1931852E+01
A21   0.0000000E+00   B21   0.1000000E+01

SECTION #: 2

A02   0.1000000E+01   B02   0.1000000E+01
A12   0.0000000E+00   B12   0.1414214E+01
A22   0.0000000E+00   B22   0.1000000E+01

SECTION #: 3

A03   0.1000000E+01   B03   0.1000000E+01
A13   0.0000000E+00   B13   0.5176380E+00
A23   0.0000000E+00   B23   0.1000000E+01

 H: 0.100000E+01
POLE- ZERO OUTPUT

COMPLEX ZERO PAIRS: 0
#REAL ZEROS: 0
#COMPLEX POLE PAIRS: 3
REAL POLES : 0
H: 0.100000E+01
POLES:
(-0.965926   0.258819)
(-0.707107   0.707107)
(-0.258819   0.965926)
COMPLEX ZERO PAIRS: 0
#REAL ZEROS: 0
#COMPLEX POLE PAIRS: 3
REAL POLES : 0
H: 0.100000E+01
POLES:
(-0.965926E+00   0.258819E+00)
(-0.707107E+00   0.707107E+00)
(-0.258819E+00   0.965926E+00)

DIGITAL FILTER DEVELOPMENT

AP, DB: 0.100000E+01
AA, DB: 0.300000E+02
K   : 0.447214E+00
ACTUAL AA (DB): 0.300000E+02
SAMPLING FREQUENCY (R/S): 0.100000E+01
ORDER : 6

H(Z)=H*(A01+A11Z+A21Z^2/B01+B11Z+B2Z^2)...
H: 0.138978E-01

300

BUTTERWORTH LP FILTER

FREQUENCY EDGES AT :   0.400000E+00   0.300000E+00
LAMDA :   2.039430
SECTION #: 1

A01   0.1554564E+01   B01   0.6253770E-01
A11   0.3109129E+01   B11   0.4324965E+00
A21   0.1554564E+01   B21   0.1000000E+01

SECTION #: 2

A02   0.1777856E+01   B02   0.2151567E+00
A12   0.3555713E+01   B12   0.4946187E+00
A22   0.1777856E+01   B22   0.1000000E+01

SECTION #: 3

A03   0.2366642E+01   B03   0.6175892E+00
A13   0.4733285E+01   B13   0.6584253E+00
A23   0.2366642E+01   B23   0.1000000E+01

COMPLEX ZERO PAIRS: 0
#REAL ZEROS: 0
#COMPLEX POLE PAIRS: 3
REAL POLES :   0
H: 0.138978E-01
POLES:
(-0.216248   0.125596)
(-0.247309   0.392422)
(-0.329213   0.713588)

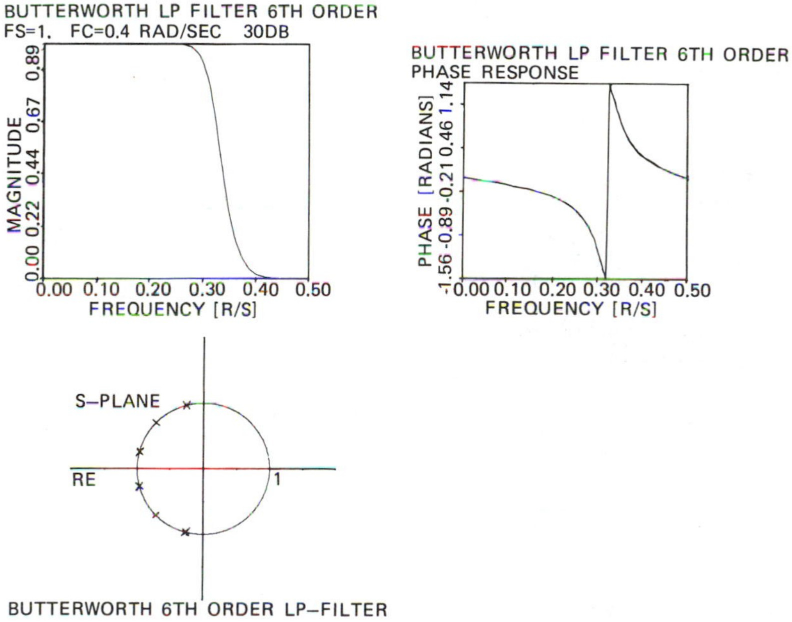

BUTTERWORTH LP FILTER 6TH ORDER
FS=1. FC=0.4 RAD/SEC   30DB

BUTTERWORTH LP FILTER 6TH ORDER
PHASE RESPONSE

BUTTERWORTH 6TH ORDER LP-FILTER

FIG. 7.10.1 Butterworth IIR lowpass filter design summary.

```
INPUT DATA
----------

BUTTERWORTH APPROXIMATION
AP, DB:        1.000000
AA, DB:       30.000000
K:        0.447214
OUTPUT DATA
-----------

ORDER:  6
ACTUAL AA       30.00000
H(S)=H*H1(S)*H2(S)...
H1(S)=(A0+A11S+A21S**2)/(B0+B11S+B21S**2)

SECTION #: 1
-----------

A01   0.1000000E+01      B01   0.1000000E+01
A11   0.0000000E+00      B11   0.1931852E+01
A21   0.0000000E+00      B21   0.1000000E+01

SECTION #: 2
-----------

A02   0.1000000E+01      B02   0.1000000E+01
A12   0.0000000E+00      B12   0.1414214E+01
A22   0.0000000E+00      B22   0.1000000E+01

SECTION #: 3
-----------

A03   0.1000000E+01      B03   0.1000000E+01
A13   0.0000000E+00      B13   0.5176380E+00
A23   0.0000000E+00      B23   0.1000000E+01

 H:     0.100000E+01

POLE- ZERO OUTPUT
--------------------
# COMPLEX ZERO PAIRS:  0
#REAL ZEROS:   0
#COMPLEX POLE PAIRS:  3
# REAL POLES :  0
H:    0.100000E+01
POLES:
(-0.965926E+00    0.258819E+00)
(-0.707107E+00    0.707107E+00)
(-0.258819E+00    0.965926E+00)
```

BUTTERWORTH HP FILTER 6TH ORDER
FS=1. FC=0.3 RAD/SEC 30DB

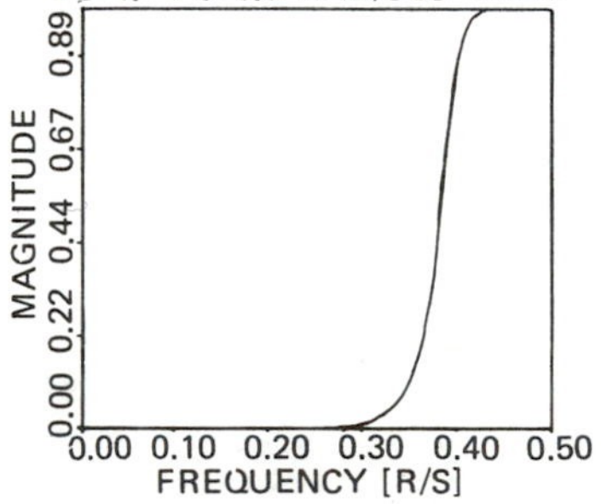

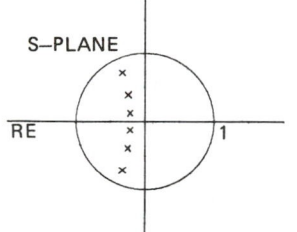

BUTTERWORTH 6TH ORDER LP—FILTER

```
DIGITAL FILTER DEVELOPMENT
---------------------------
AP, DB:  0.100000E+01
AA, DB:  0.300000E+02
K   :  0.447214E+00
ACTUAL AA (DB):  0.300000E+02
SAMPLING FREQUENCY (R/S):  0.100000E+01
ORDER : 6

H(Z)=H*(A01+A11Z+A21Z^2/B01+B11Z+B2Z^2)...
H:     0.100000E+01
BUTTERWORTH HP FILTER
---------------------
```

```
FREQUENCY EDGES AT :  0.300000E+00   0.400000E+00
LAMDA :        0.875330
SECTION #: 1
-----------
```

```
A01    0.7207423E-01     B01    0.2342170E+00
A11   -0.1441485E+00     B11    0.9459201E+00
A21    0.7207423E-01     B21    0.1000000E+01
```

```
SECTION #: 2
-----------
```

```
A02    0.8031409E-01     B02    0.3753185E+00
A12   -0.1606282E+00     B12    0.1054062E+01
A22    0.8031409E-01     B22    0.1000000E+01
```

```
SECTION #: 3
-----------
```

```
A03    0.1001443E+00     B03    0.7148954E+00
A13   -0.2002885E+00     B13    0.1314318E+01
A23    0.1001443E+00     B23    0.1000000E+01
```

```
# COMPLEX ZERO PAIRS:  0
#REAL ZEROS:  0
#COMPLEX POLE PAIRS:  3
# REAL POLES :  0
H:     0.100000E+01
POLES:
(     -0.472960         0.102595)
(     -0.527031         0.312341)
(     -0.657159         0.532012)
```

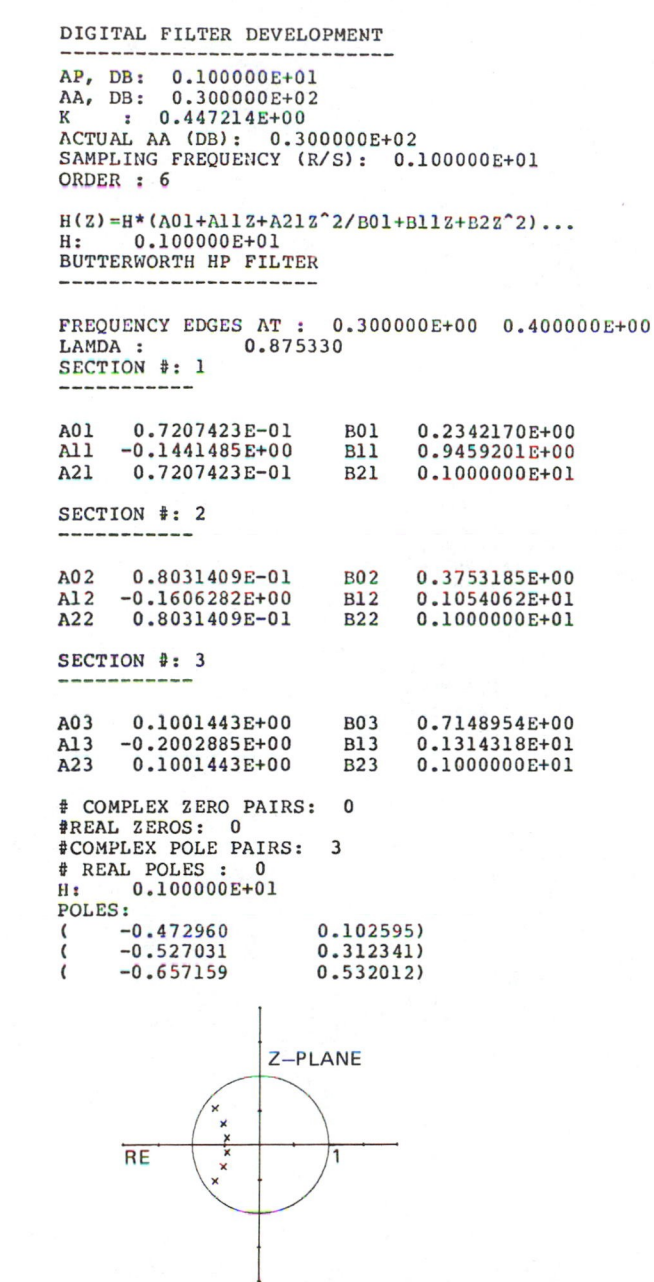

BUTTERWORTH HP 6TH ORDER FILTER

FIG. 7.10.2 Butterworth IIR highpass filter design summary.

DIGITAL FILTER DEVELOPMENT

AP, DB: 0.100000E+01
AA, DB: 0.300000E+02
K   :   0.606120E+00
ACTUAL AA (DB):   0.300000E+02
SAMPLING FREQUENCY (R/S):   0.100000E+01
ORDER :18

$H(Z)=H*(A01+A11Z+A21Z^2/B01+B11Z+B2Z^2)...$
H: 0.103564E-03
BUTTERWORTH BP FILTER

PASSBAND EDGES AT: 0.100000E+00   0.300000E+00   R/S
STOPBAND EDGES AT: 0.500000E-01   0.350000E+00   R/S
CENTRE FREQ (R/S):   0.212867
BANDWIDTH (R/S):   0.360782
SECTION #: 1

A01 -0.1217367E+01   B01   0.1215908E+00
A11   0.0000000E+00   B11 -0.4284095E+00
A21   0.1217367E+01   B21   0.1000000E+01

SECTION #: 2

A02 -0.8961076E+00   B02   0.1543274E+00
A12   0.0000000E+00   B12   0.1336790E-01
A22   0.8961076E+00   B22   0.1000000E+01

SECTION #: 3

A03 -0.1704436E+01   B03   0.2971244E+00
A13   0.0000000E+00   B13 -0.8730310E+00
A23   0.1704436E+01   B23   0.1000000E+01

SECTION #: 4

A04 -0.8142742E+00   B04   0.3047724E+00
A14   0.0000000E+00   B14   0.2680064E+00
A24   0.8142742E+00   B24   0.1000000E+01

SECTION #: 5

A05 -0.2057081E+01   B05   0.4822314E+00
A15   0.0000000E+00   B15 -0.1136925E+01
A25   0.2057081E+01   B25   0.1000000E+01

SECTION #: 6

A06 -0.8295186E+00   B06   0.5168502E+00
A16   0.0000000E+00   B16   0.4606744E+00
A26   0.8295186E+00   B26   0.1000000E+01

SECTION #: 7

A07 -0.2370397E+01   B07   0.6702340E+00
A17   0.0000000E+00   B17 -0.1347849E+01
A27   0.2370397E+01   B27   0.1000000E+01

SECTION #: 8

```
A08   -0.9287346E+00      B08    0.8087881E+00
A18    0.0000000E+00      B18    0.6262865E+00
A28    0.9287346E+00      B28    0.1000000E+01
```

SECTION #: 9

```
A09   -0.2691203E+01      B09    0.8796701E+00
A19    0.0000000E+00      B19   -0.1546876E+01
A29    0.2691203E+01      B29    0.1000000E+01
```

```
# COMPLEX ZERO PAIRS:   0
#REAL ZEROS:   0
#COMPLEX POLE PAIRS:   9
# REAL POLES :   0
H:    0.103564E-03
POLES:
(      0.214205        0.275149)
(     -0.006684        0.392788)
(      0.436515        0.326464)
(     -0.134003        0.535552)
(      0.568463        0.398851)
(     -0.230337        0.681025)
(      0.673924        0.464823)
(     -0.313143        0.843048)
(      0.773438        0.530532)
```

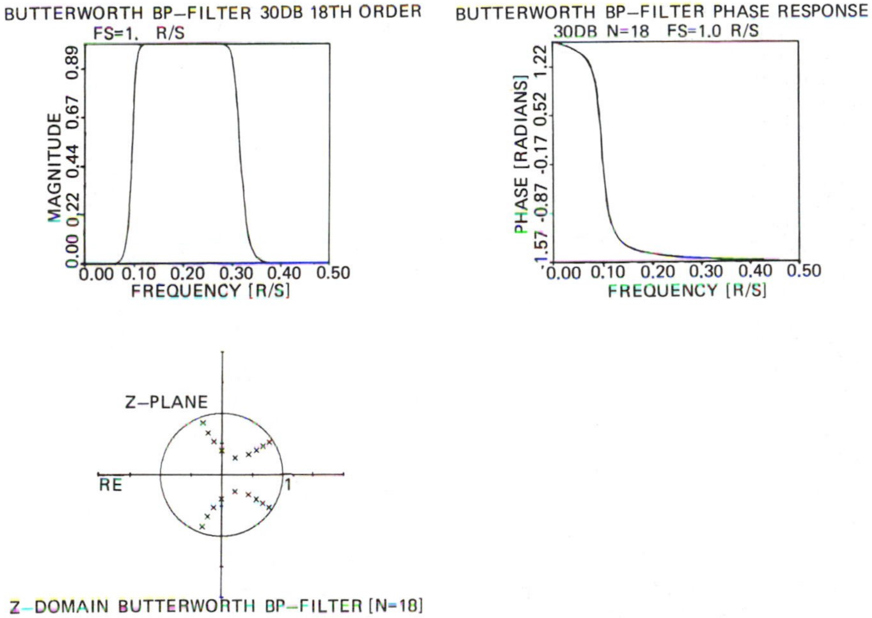

BUTTERWORTH BP–FILTER 30DB 18TH ORDER
FS=1. R/S

BUTTERWORTH BP–FILTER PHASE RESPONSE
30DB N=18 FS=1.0 R/S

Z–DOMAIN BUTTERWORTH BP–FILTER [N=18]

FIG. 7.10.3 Butterworth IIR bandpass filter design summary.

```
INPUT DATA
----------

BUTTERWORTH APPROXIMATION
AP, DB:        1.000000
AA, DB:       30.000000
K:        0.370192
OUTPUT DATA
----------

ORDER:  5
ACTUAL AA       30.00000
H(S)=H*Hl(S)*H2(S)...
Hl(S)=(A0+AllS+A21S**2)/(B0+BllS+B21S**2)

SECTION #: 1
-----------

A01   0.1000000E+01    B01   0.1000000E+01
All   0.0000000E+00    Bll   0.1000000E+01
A21   0.0000000E+00    B21   0.0000000E+00

SECTION #: 2
-----------

A02   0.1000000E+01    B02   0.1000000E+01
A12   0.0000000E+00    B12   0.1618034E+01
A22   0.0000000E+00    B22   0.1000000E+01

SECTION #: 3
-----------

A03   0.1000000E+01    B03   0.1000000E+01
A13   0.0000000E+00    B13   0.6180339E+00
A23   0.0000000E+00    B23   0.1000000E+01

  H:     0.100000E+01
POLE- ZERO OUTPUT
-----------------
# COMPLEX ZERO PAIRS:  0
#REAL ZEROS:  0
#COMPLEX POLE PAIRS:  2
# REAL POLES :  1
H:     0.100000E+01
POLES:
(-0.100000E+01    0.000000E+00)
(-0.809017E+00    0.587785E+00)
(-0.309017E+00    0.951057E+00)
```

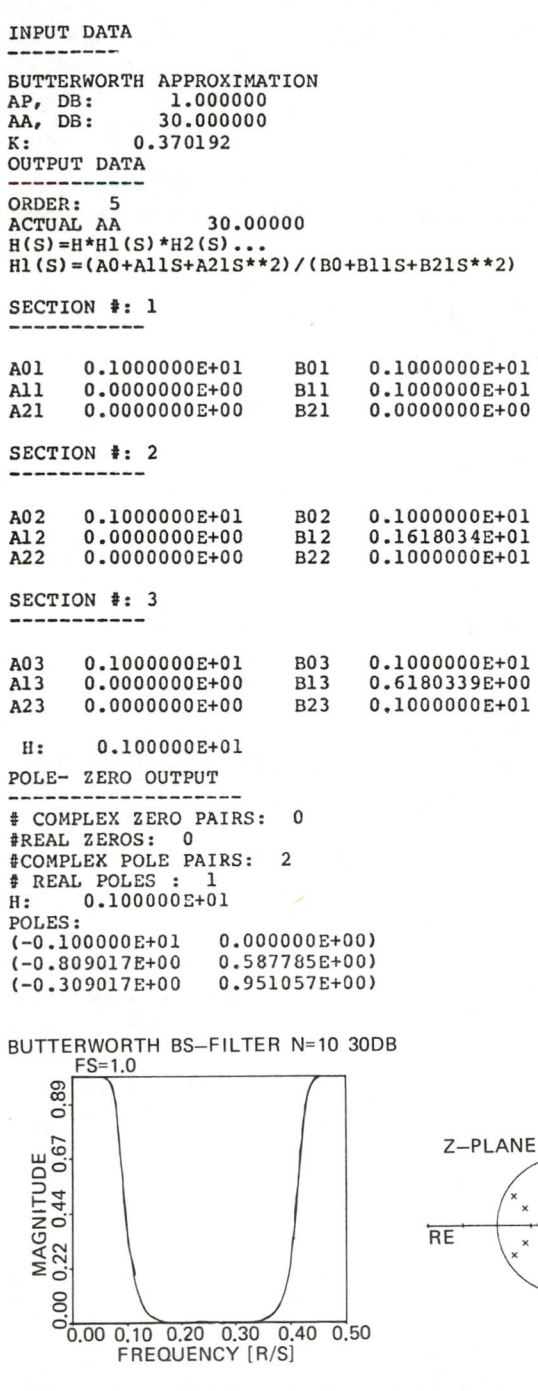

FIG. 7.10.4 Butterworth IIR bandstop filter design summary.

DIGITAL FILTER DEVELOPMENT

AP, DB: 0.100000E+01
AA, DB: 0.300000E+02
K   : 0.370192E+00
ACTUAL AA (DB): 0.300000E+02
SAMPLING FREQUENCY (R/S): 0.100000E+01
ORDER :10

H(Z)=H*(A01+A11Z+A21Z^2/B01+B11Z+B2Z^2)...
H: 0.100000E+01
BUTTERWORTH BS FILTER

PASSBAND EDGES AT: 0.750000E-01   0.425000E+00   R/S
STOPBAND EDGES AT: 0.150000E+00   0.350000E+00   R/S
CENTRE FREQ (R/S):   0.318310
BANDWIDTH (R/S):   1.091520
SECTION #: 1

A01   0.3683844E+00   B01   -0.2632312E+00
A11   0.7093803E-07   B11   0.7093803E-07
A21   0.3683844E+00   B21   0.1000000E+01

SECTION #: 2

A02   0.1143727E+00   B02   0.4166457E+00
A12   0.2202420E-07   B12   0.1187900E+01
A22   0.1143727E+00   B22   0.1000000E+01

SECTION #: 3

A03   0.1302273E+01   B03   0.4166459E+00
A13   0.2507726E-06   B13   -0.1187901E+01
A23   0.1302273E+01   B23   0.1000000E+01

SECTION #: 4

A04   0.1222057E+00   B04   0.7590019E+00
A14   0.2353256E-07   B14   0.1514591E+01
A24   0.1222057E+00   B24   0.1000000E+01

SECTION #: 5

A05   0.1636797E+01   B05   0.7590023E+00
A15   0.3151902E-06   B15   -0.1514591E+01
A25   0.1636797E+01   B25   0.1000000E+01

COMPLEX ZERO PAIRS: 5
#REAL ZEROS: 0
#COMPLEX POLE PAIRS: 4
REAL POLES : 0
H: 0.100000E+01
ZEROS:
(0.000000   1.000000)
(0.000000   1.000000)
(0.000000   1.000000)
(0.000000   1.000000)
(0.000000   1.000000)
POLES:
(-0.593950   0.252723)
(0.593950   0.252723)
(-0.757295   0.430704)
(0.757296   0.430704)

INPUT DATA

CHEBBYCHEV   APPROXIMATION
AP, DB: 1.000000
AA, DB: 45.000000
K: 0.843448
OUTPUT DATA

ORDER: 11
ACTUAL AA   45.45983
H(S)=H*H1(S)*H2(S)...
H1(S)=(A0+A11S+A21S**2)/(B0+B11S+B21S**2)

SECTION #: 1

A01	0.1000000E+01	B01	0.1301809E+00
A11	0.0000000E+00	B11	0.1000000E+01
A21	0.0000000E+00	B21	0.0000000E+00

SECTION #: 2

A02	0.1000000E+01	B02	0.9632027E-01
A12	0.0000000E+00	B12	0.2498152E+00
A22	0.0000000E+00	B22	0.1000000E+01

SECTION #: 3

A03	0.1000000E+01	B03	0.3092395E+00
A13	0.0000000E+00	B13	0.2190302E+00
A23	0.0000000E+00	B23	0.1000000E+01

SECTION #: 4

A04	0.1000000E+01	B04	0.5881045E+00
A14	0.0000000E+00	B14	0.1705007E+00
A24	0.0000000E+00	B24	0.1000000E+01

SECTION #: 5

A05	0.1000000E+01	B05	0.8443774E+00
A15	0.0000000E+00	B15	0.1081582E+00
A25	0.0000000E+00	B25	0.1000000E+01

SECTION #: 6

A06	0.1000000E+01	B06	0.9966935E+00
A16	0.0000000E+00	B16	0.3705334E-01
A26	0.0000000E+00	B26	0.1000000E+01

H: 0.191917E-02

POLE- ZERO OUTPUT

COMPLEX ZERO PAIRS: 0
#REAL ZEROS: 0
#COMPLEX POLE PAIRS: 5
REAL POLES :   1
H: 0.191917E-02
POLES:
(-0.130181E+00   0.000000E+00)
(-0.124908E+00   0.284110E+00)
(-0.109515E+00   0.545203E+00)
(-0.852503E-01   0.762127E+00)
(-0.540791E-01   0.917307E+00)
(-0.185267E-01   0.998173E+00)

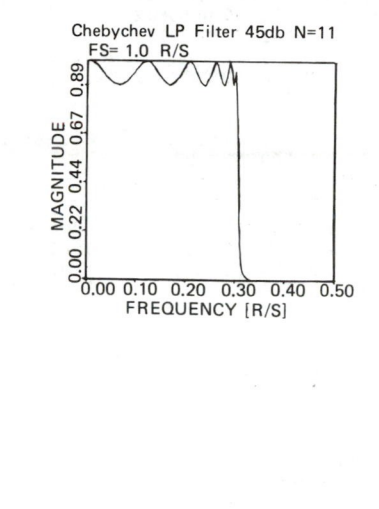

Chebychev LP Filter 45db N=11
FS= 1.0 R/S

```
DIGITAL FILTER DEVELOPMENT
--------------------------

AP, DB:  0.100000E+01
AA, DB:  0.450000E+02
K   :  0.843448E+00
ACTUAL AA (DB):  0.454598E+02
SAMPLING FREQUENCY (R/S):  0.100000E+01
ORDER :11

H(Z)=H*(A01+A11Z+A21Z^2/B01+B11Z+B2Z^2)...
H:   0.219075E-06
CHEBBYCHEV  LP FILTER
---------------------
```

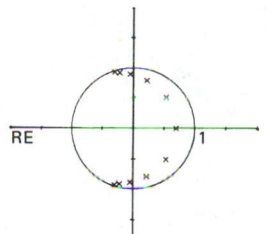

Z—Domain Chebychev LP—Filter, N=11

```
FREQUENCY EDGES AT :  0.325000E+00  0.300000E+00
LAMDA :       2.282501
SECTION #: 1
-----------

A01   0.2664221E+01    B01  -0.6960959E+00
A11   0.2664221E+01    B11   0.1000000E+01
A21   0.0000000E+00    B21   0.0000000E+00


SECTION #: 2
-----------

A02   0.6466305E+01    B02   0.5494486E+00
A12   0.1293261E+02    B12  -0.1071246E+01
A22   0.6466305E+01    B22   0.1000000E+01


SECTION #: 3
-----------

A03   0.5229481E+01    B03   0.6805287E+00
A13   0.1045896E+02    B13  -0.4388999E+00
A23   0.5229481E+01    B23   0.1000000E+01


SECTION #: 4
-----------

A04   0.4201986E+01    B04   0.8001750E+00
A14   0.8403972E+01    B14   0.9717424E-01
A24   0.4201986E+01    B24   0.1000000E+01


SECTION #: 5
-----------

A05   0.3590934E+01    B05   0.8916732E+00
A15   0.7181868E+01    B15   0.4363226E+00
A25   0.3590934E+01    B25   0.1000000E+01


SECTION #: 6
-----------

A06   0.3357964E+01    B06   0.9652966E+00
A16   0.6715929E+01    B16   0.6043649E+00
A26   0.3357964E+01    B26   0.1000000E+01

# COMPLEX ZERO PAIRS:  0
#REAL ZEROS:  0
#COMPLEX POLE PAIRS:  5
# REAL POLES :  1
H:    0.219075E-06
POLES:
(     0.696096        0.000000)
(     0.535623        0.512403)
(     0.219450        0.795217)
(    -0.048587        0.893204)
(    -0.218161        0.918738)
(    -0.302182        0.934870)
```

FIG. 7.10.5 Chebyshev IIR lowpass filter design summary.

INPUT DATA

CHEBBYCHEV   APPROXIMATION
AP, DB: 2.000000
AA, DB: 45.000000
K: 0.447214
OUTPUT DATA

ORDER: 5
ACTUAL AA   54.34648
H(S)=H*H1(S)*H2(S)...
H1(S)=(A0+A11S+A21S**2)/(B0+B11S+B21S**2)

SECTION #: 1

A01	0.1000000E+01	B01	0.2183083E+00
A11	0.0000000E+00	B11	0.1000000E+01
A21	0.0000000E+00	B21	0.0000000E+00

SECTION #: 2

A02	0.1000000E+01	B02	0.3931500E+00
A12	0.0000000E+00	B12	0.3532303E+00
A22	0.0000000E+00	B22	0.1000000E+01

SECTION #: 3

A03	0.1000000E+01	B03	0.9521670E+00
A13	0.0000000E+00	B13	0.1349220E+00
A23	0.0000000E+00	B23	0.1000000E+01

  H: 0.817225E-01
POLE- ZERO OUTPUT

COMPLEX ZERO PAIRS: 0
#REAL ZEROS: 0
#COMPLEX POLE PAIRS: 2
REAL POLES : 1
H: 0.817225E-01
POLES:
(-0.218308   0.000000)
(-0.176615   0.601629)
(-0.067461   0.973456)
COMPLEX ZERO PAIRS: 0
#REAL ZEROS: 0
#COMPLEX POLE PAIRS: 2
REAL POLES : 1
H: 0.817225E-01
POLES:
(-0.218308E+00   0.000000E+00)
(-0.176615E+00   0.601629E+00)
(-0.674610E-01   0.973456E+00)

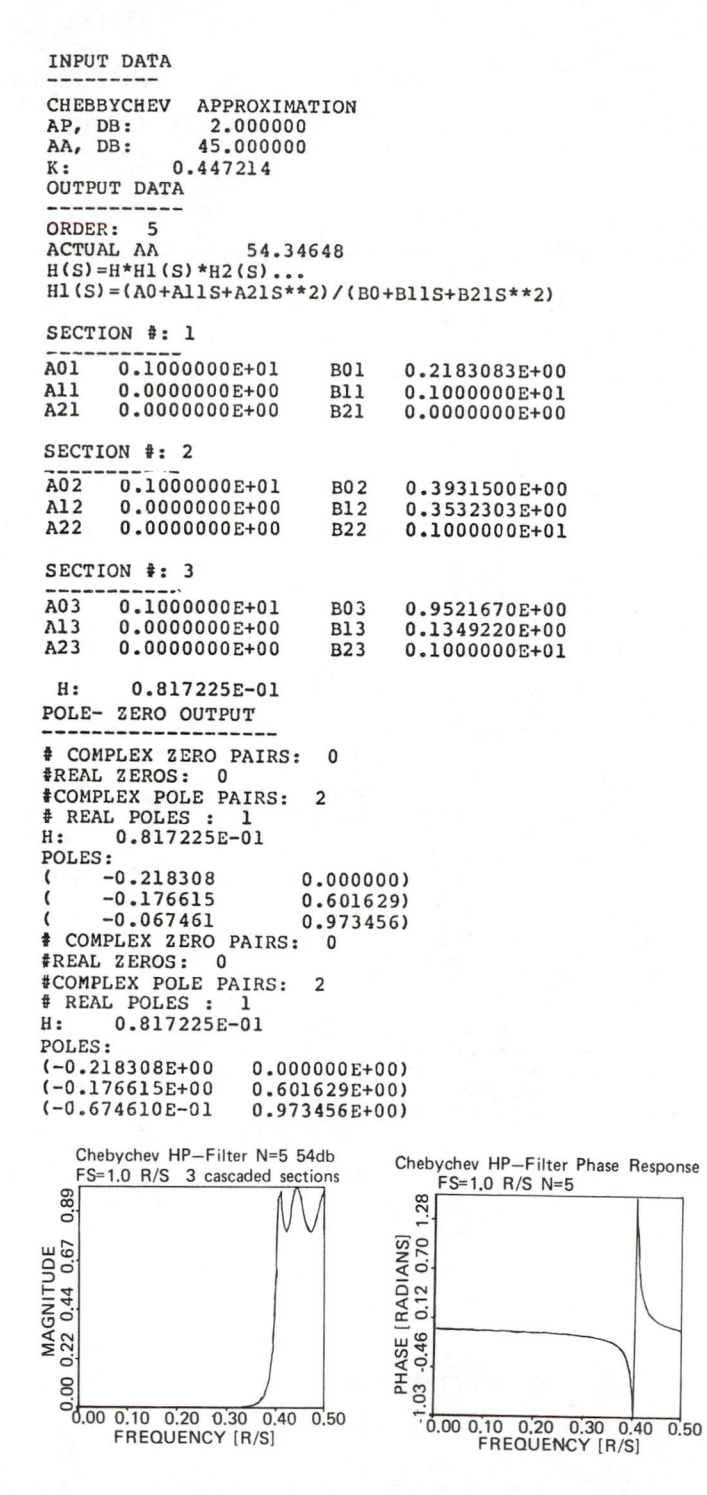

Chebychev HP–Filter N=5 54db
FS=1.0 R/S 3 cascaded sections

Chebychev HP–Filter Phase Response
FS=1.0 R/S N=5

310

```
DIGITAL FILTER DEVELOPMENT
--------------------------
AP, DB:   0.200000E+01
AA, DB:   0.450000E+02
K    :   0.447214E+00
ACTUAL AA (DB):   0.543465E+02
SAMPLING FREQUENCY (R/S):   0.100000E+01
ORDER : 5

H(Z)=H*(A01+A11Z+A21Z^2/B01+B11Z+B2Z^2)...
H:   0.100000E+01
CHEBBYCHEV  HP FILTER
---------------------
```

FREQUENCY EDGES AT :   0.300000E+00   0.400000E+00
LAMDA :   0.979657
SECTION #: 1

A01	−0.6623448E−01	B01	0.8675310E+00
A11	0.6623448E−01	B11	0.1000000E+01
A21	0.0000000E+00	B21	0.0000000E+00

SECTION #: 2

A02	0.3589618E−01	B02	0.8014811E+00
A12	−0.7179236E−01	B12	0.1657896E+01
A22	0.3589618E−01	B22	0.1000000E+01

SECTION #: 3

A03	0.8784191E−01	B03	0.9233830E+00
A13	−0.1756838E+00	B13	0.1572015E+01
A23	0.8784191E−01	B23	0.1000000E+01

```
# COMPLEX ZERO PAIRS:   0
#REAL ZEROS:  0
#COMPLEX POLE PAIRS:   2
# REAL POLES :   1
H:   0.100000E+01
POLES:
(     −0.867531          0.000000)
(     −0.828948          0.338121)
(     −0.786008          0.552788)
```

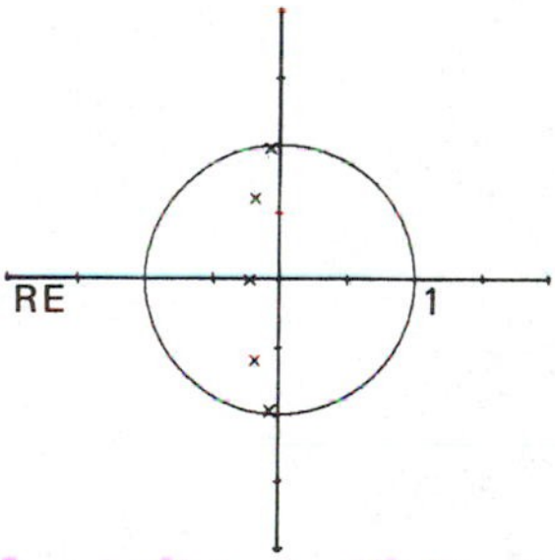

S—Domain Chebychev HP filter N=5 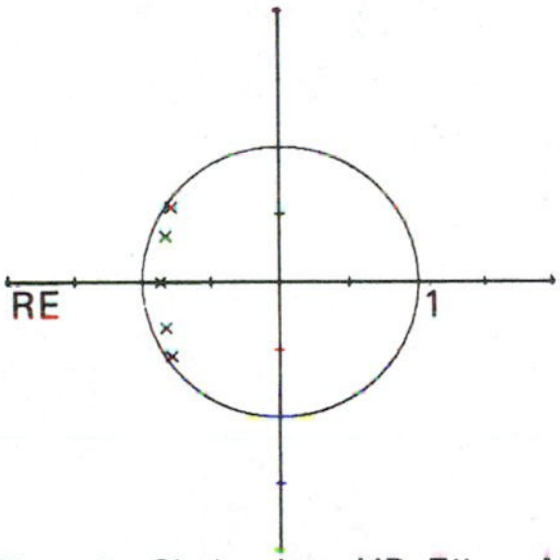   Z—Domain Chebychev HP Filter N=5

FIG. 7.10.6 Chebyshev highpass filter design summary.

```
INPUT DATA
---------

CHEBBYCHEV   APPROXIMATION
AP, DB:        1.000000
AA, DB:       45.000000
K:        0.606120
OUTPUT DATA
-----------

ORDER:  7
ACTUAL AA      54.13468
H(S)=H*H1(S)*H2(S)...
H1(S)=(A0+A11S+A21S**2)/(B0+B11S+B21S**2)


SECTION #: 1
-----------

A01   0.1000000E+01      B01   0.2054143E+00
A11   0.0000000E+00      B11   0.1000000E+01
A21   0.0000000E+00      B21   0.0000000E+00


SECTION #: 2
-----------

A02   0.1000000E+01      B02   0.2304501E+00
A12   0.0000000E+00      B12   0.3701438E+00
A22   0.0000000E+00      B22   0.1000000E+01


SECTION #: 3
-----------

A03   0.1000000E+01      B03   0.6534555E+00
A13   0.0000000E+00      B13   0.2561474E+00
A23   0.0000000E+00      B23   0.1000000E+01


SECTION #: 4
-----------

A04   0.1000000E+01      B04   0.9926795E+00
A14   0.0000000E+00      B14   0.9141796E-01
A24   0.0000000E+00      B24   0.1000000E+01

 H:     0.307067E-01
POLE- ZERO OUTPUT
--------------------

# COMPLEX ZERO PAIRS:  0
#REAL ZEROS:  0
#COMPLEX POLE PAIRS:  3
# REAL POLES :  1
H:     0.307067E-01
POLES:
(-0.205414E+00    0.000000E+00)
(-0.185072E+00    0.442943E+00)
(-0.128074E+00    0.798156E+00)
(-0.457090E-01    0.995284E+00)
```

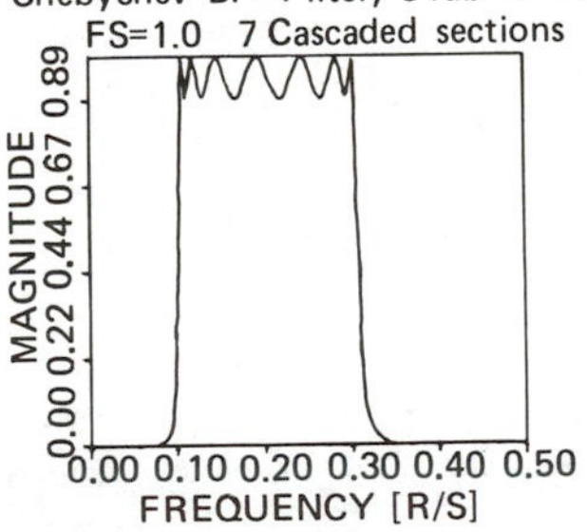

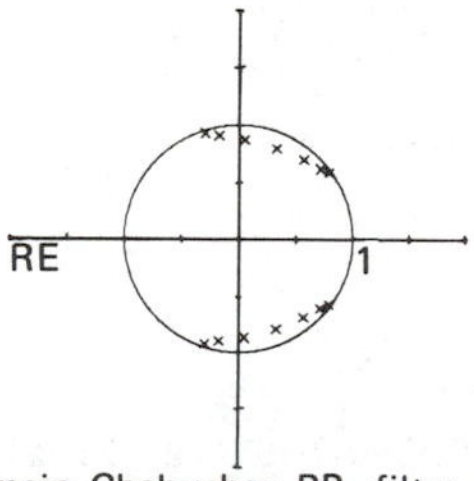

Chebychev BP—Filter, 54db N=14
FS=1.0   7 Cascaded sections

Z—domain Chebychev BP—filter N=14

FIG. 7.10.7 Chebyshev IIR bandpass filter design summary.

```
DIGITAL FILTER DEVELOPMENT
------------------------------
AP, DB:  0.100000E+01
AA, DB:  0.450000E+02
K    :  0.606120E+00
ACTUAL AA (DB):  0.541347E+02
SAMPLING FREQUENCY (R/S):  0.100000E+01
ORDER :14

H(Z)=H*(A01+A11Z+A21Z^2/B01+B11Z+B2Z^2)...
H:     0.144457E-04
CHEBBYCHEV  BP FILTER
--------------------------

PASSBAND EDGES AT:  0.100000E+00    0.300000E+00   R/S
STOPBAND EDGES AT:  0.500000E-01    0.350000E+00   R/S
CENTRE FREQ (R/S):          0.212867
BANDWIDTH (R/S):            0.334691
SECTION #: 1
------------

A01  -0.1888886E+01    B01   0.7402772E+00
A11   0.0000000E+00    B11  -0.6647267E+00
A21   0.1888886E+01    B21   0.1000000E+01

SECTION #: 2
------------

A02  -0.1461081E+01    B02   0.7589656E+00
A12   0.0000000E+00    B12  -0.1013404E+00
A22   0.1461081E+01    B22   0.1000000E+01

SECTION #: 3
------------

A03  -0.2319353E+01    B03   0.8079633E+00
A13   0.0000000E+00    B13  -0.1145129E+01
A23   0.2319353E+01    B23   0.1000000E+01

SECTION #: 4
------------

A04  -0.1174851E+01    B04   0.8456067E+00
A14   0.0000000E+00    B14   0.3497403E+00
A24   0.1174851E+01    B24   0.1000000E+01

SECTION #: 5
------------

A05  -0.2619787E+01    B05   0.8950903E+00
A15   0.0000000E+00    B15  -0.1440526E+01
A25   0.2619787E+01    B25   0.1000000E+01

SECTION #: 6
------------

A06  -0.1060624E+01    B06   0.9475457E+00
A16   0.0000000E+00    B16   0.5971167E+00
A26   0.1060624E+01    B26   0.1000000E+01

SECTION #: 7
------------

A07  -0.2793581E+01    B07   0.9672105E+00
A17   0.0000000E+00    B17  -0.1589688E+01
A27   0.2793581E+01    B27   0.1000000E+01

# COMPLEX ZERO PAIRS:  0
#REAL ZEROS:   0
#COMPLEX POLE PAIRS:  7
# REAL POLES :   0
H:     0.144457E-04
POLES:
(      0.332363          0.793607)
(      0.050670          0.869712)
(      0.572564          0.692916)
(     -0.174870          0.902789)
(      0.720263          0.613443)
(     -0.298558          0.926503)
(      0.794844          0.579166)
```

```
INPUT DATA
----------
CHEBBYCHEV  APPROXIMATION
AP, DB:        1.000000
AA, DB:       45.000000
K:        0.637691
OUTPUT DATA
-----------
ORDER:  7
ACTUAL AA        50.19208
H(S)=H*H1(S)*H2(S)...
H1(S)=(A0+A11S+A21S**2)/(B0+B11S+B21S**2)
SECTION #: 1
-----------
A01   0.1000000E+01      B01   0.2054143E+00
A11   0.0000000E+00      B11   0.1000000E+01
A21   0.0000000E+00      B21   0.0000000E+00
SECTION #: 2
-----------
A02   0.1000000E+01      B02   0.2304501E+00
A12   0.0000000E+00      B12   0.3701438E+00
A22   0.0000000E+00      B22   0.1000000E+01
SECTION #: 3
-----------
A03   0.1000000E+01      B03   0.6534555E+00
A13   0.0000000E+00      B13   0.2561474E+00
A23   0.0000000E+00      B23   0.1000000E+01
SECTION #: 4
-----------
A04   0.1000000E+01      B04   0.9926795E+00
A14   0.0000000E+00      B14   0.9141796E-01
A24   0.0000000E+00      B24   0.1000000E+01
 H:    0.307067E-01
POLE- ZERO OUTPUT
-------------------
# COMPLEX ZERO PAIRS:  0
#REAL ZEROS:   0
#COMPLEX POLE PAIRS:  3
# REAL POLES :  1
H:    0.307067E-01
POLES:
(-0.205414E+00    0.000000E+00)
(-0.185072E+00    0.442943E+00)
(-0.128074E+00    0.798156E+00)
(-0.457090E-01    0.995284E+00)
```

Chebychev BS–Filter 50db N=14

Chebychev BS–filter Phase Response N=14

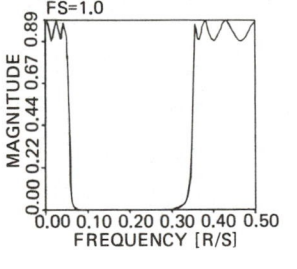

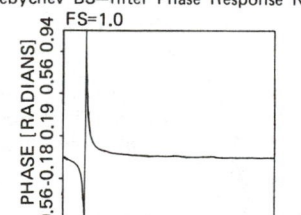

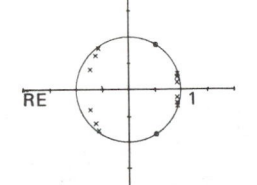

Z–Domain Chebychev BS Filter N=14

```
DIGITAL FILTER DEVELOPMENT
----------------------------
AP, DB:   0.100000E+01
AA, DB:   0.450000E+02
K    :   0.637691E+00
ACTUAL AA (DB):   0.501921E+02
SAMPLING FREQUENCY (R/S):   0.100000E+01
ORDER :14
H(Z)=H*(A01+A11Z+A21Z^2/B01+B11Z+B2Z^2)...
H:    0.100000E+01
CHEBBYCHEV  BS FILTER
----------------------------
PASSBAND EDGES AT:   0.500000E-01   0.350000E+00   R/S
STOPBAND EDGES AT:   0.100000E+00   0.300000E+00   R/S
CENTRE FREQ (R/S):        0.177469
BANDWIDTH (R/S):          0.574303
SECTION #: 1
-----------
A01   0.1298614E+00      B01  -0.7402772E+00
A11  -0.1365444E+00      B11  -0.1365444E+00
A21   0.1298614E+00      B21   0.1000000E+01
SECTION #: 2
-----------
A02   0.7121770E-01      B02   0.6916942E+00
A12  -0.7488271E-01      B12   0.1474376E+01
A22   0.7121770E-01      B22   0.1000000E+01
SECTION #: 3
-----------
A03   0.1228338E+01      B03   0.8865429E+00
A13  -0.1291551E+01      B13  -0.1861683E+01
A23   0.1228338E+01      B23   0.1000000E+01
SECTION #: 4
-----------
A04   0.1812915E+00      B04   0.8147330E+00
A14  -0.1906212E+00      B14   0.1261529E+01
A24   0.1812915E+00      B24   0.1000000E+01
SECTION #: 5
-----------
A05   0.1242724E+01      B05   0.9290093E+00
A15  -0.1306677E+01      B15  -0.1863117E+01
A25   0.1242724E+01      B25   0.1000000E+01
SECTION #: 6
-----------
A06   0.2607143E+00      B06   0.9388161E+00
A16  -0.2741312E+00      B16   0.1143256E+01
A26   0.2607143E+00      B26   0.1000000E+01
SECTION #: 7
-----------
A07   0.1263731E+01      B07   0.9762041E+00
A17  -0.1328765E+01      B17  -0.1880024E+01
A27   0.1263731E+01      B27   0.1000000E+01
# COMPLEX ZERO PAIRS:   7
#REAL ZEROS:   0
#COMPLEX POLE PAIRS:   6
# REAL POLES :   0
H:    0.100000E+01
ZEROS:
(      0.525731         0.850651)
(      0.525731         0.850651)
(      0.525731         0.850651)
(      0.525731         0.850651)
(      0.525731         0.850651)
(      0.525731         0.850651)
(      0.525731         0.850651)
POLES:
(     -0.737188         0.385030)
(      0.930842         0.141692)
(     -0.630764         0.645654)
(      0.931558         0.247404)
(     -0.571628         0.782341)
(      0.940012         0.304272)
```

FIG. 7.10.8 Chebyshev IIR bandstop filter design summary.

```
   INPUT DATA
   ----------

ELLIPTIC      APPROXIMATION
AP, DB:        1.000000
AA, DB:       45.000000
K:        0.447214
OUTPUT DATA
-----------

ORDER:  4
ACTUAL AA       56.31649
H(S)=H*H1(S)*H2(S)...
H1(S)=(A0+A11S+A21S**2)/(B0+B11S+B21S**2)

SECTION #: 1
-----------

A01   0.1390575E+02     B01   0.1386236E+00
A11   0.0000000E+00     B11   0.4656277E+00
A21   0.1000000E+01     B21   0.1000000E+01

SECTION #: 2
-----------

A02   0.2578837E+01     B02   0.4435645E+00
A12   0.0000000E+00     B12   0.1677004E+00
A22   0.1000000E+01     B22   0.1000000E+01

 H:    0.152818E-02

POLE- ZERO OUTPUT
-----------------

# COMPLEX ZERO PAIRS:  2
#REAL ZEROS:  0
#COMPLEX POLE PAIRS:  2
# REAL POLES :  0
H:    0.152818E-02
ZEROS:
( 0.000000E+00    0.372904E+01)
( 0.000000E+00    0.160588E+01)
POLES:
(-0.232814E+00    0.290554E+00)
(-0.838502E-01    0.660707E+00)
```

ELLIPTIC LP—FILTER 4TH ORDER 56DB

```
DIGITAL FILTER DEVELOPMENT
---------------------------


AP, DB:   0.100000E+01
AA, DB:   0.450000E+02
K    :   0.447214E+00
ACTUAL AA (DB):  0.563165E+02
SAMPLING FREQUENCY (R/S):  0.100000E+01
ORDER : 4

H(Z)=H*(A01+A11Z+A21Z^2/B01+B11Z+B2Z^2)...
H:    0.152818E-02
ELLIPTIC    LP FILTER
--------------------


FREQUENCY EDGES AT :  0.400000E+00  0.300000E+00
LAMDA :         1.526400
SECTION #: 1
-----------


A01   0.2353341E+02     B01   0.2470488E+00
A11   0.4549545E+02     B11  -0.3243142E+00
A21   0.2353341E+02     B21   0.1000000E+01


SECTION #: 2
-----------


A02   0.3698407E+01     B02   0.7858917E+00
A12   0.6156168E+01     B12   0.5452453E+00
A22   0.3698407E+01     B22   0.1000000E+01

# COMPLEX ZERO PAIRS:   2
#REAL ZEROS:   0
#COMPLEX POLE PAIRS:   2
# REAL POLES :   0
H:    0.152818E-02
ZEROS:
(     -0.966614          0.256236)
(     -0.832273          0.554366)
POLES:
(      0.162157          0.469845)
(     -0.272623          0.843545)
```

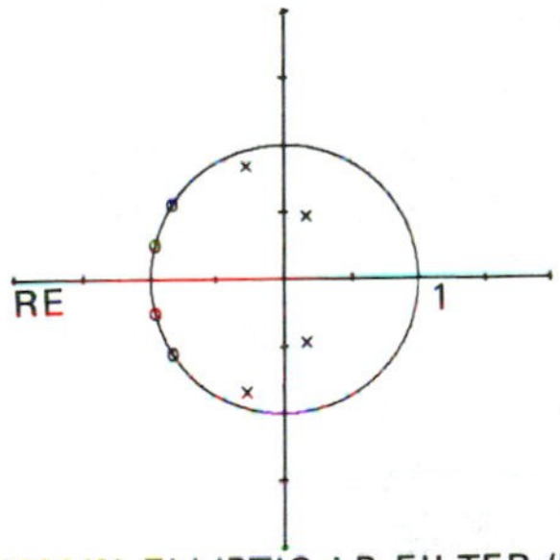

Z–DOMAIN ELLIPTIC LP FILTER (N=4)

FIG. 7.10.9 Elliptic IIR lowpass filter design summary.

DIGITAL FILTER DEVELOPMENT

AP, DB: 0.200000E+01
AA, DB: 0.350000E+02
K   :   0.700000E+00
ACTUAL AA (DB):   0.407061E+02
SAMPLING FREQUENCY (R/S):   0.350000E+04

H(Z)=H*H1(Z)*H2(Z)*...
H: 0.794328
ELLIPTIC HP FILTER

FREQUENCY EDGE AT :   0.700000E+03
SECTION #: 1

A01   0.2533439E+00   B01   0.3443515E+00
A11   -0.4587061E+00   B11   0.3789576E+00
A21   0.2533439E+00   B21   0.1000000E+01

SECTION #: 2

A02   0.7366657E+00   B02   0.8596443E+00
A12   -0.9210235E+00   B12   -0.5347105E+00
A22   0.7366657E+00   B22   0.1000000E+01

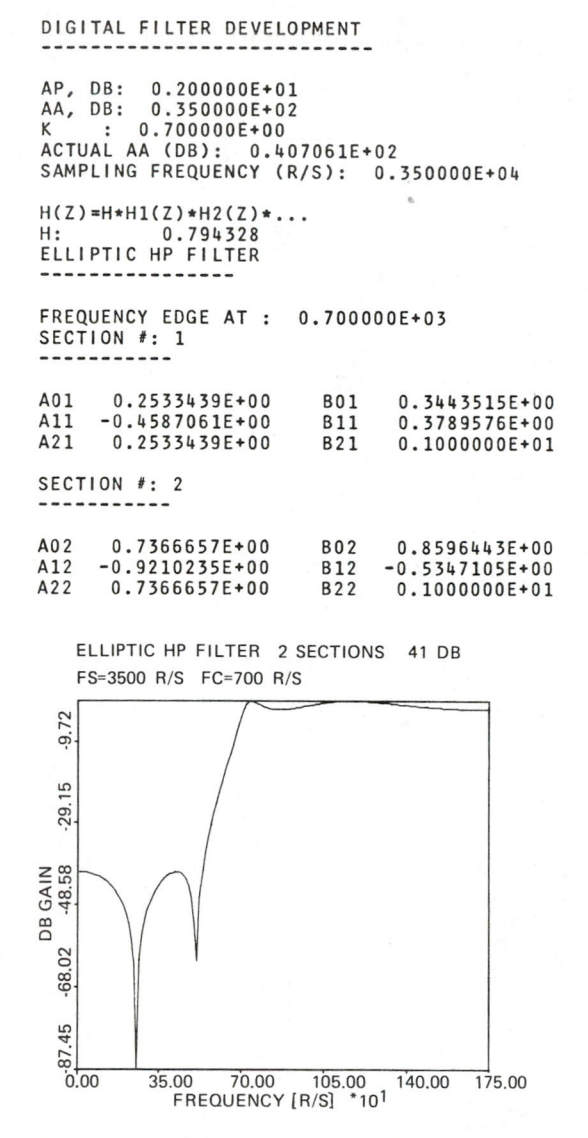

FIG. 7.10.10 Elliptic IIR highpass filter design summary.

```
DIGITAL FILTER DEVELOPMENT
--------------------------

AP, DB:   0.100000E+00
AA, DB:   0.450000E+02
K    :   0.815438E+00
ACTUAL AA (DB):   0.533909E+02
SAMPLING FREQUENCY (R/S):   0.300000E+04

H(Z)=H*H1(Z)*H2(Z)*...
H:        12.661623
ELLIPTIC BP FILTER
------------------


FREQUENCY EDGES AT:
   0.350000E+03    0.400000E+03    0.900000E+03    0.994838E+00 R/S
SECTION #: 1
-----------

A01   -0.5021752E-03    B01    0.5468152E+00
A11    0.0000000E+00    B11   -0.3713528E+00
A21    0.5021752E-03    B21    0.1000000E+01

SECTION #: 2
-----------

A02    0.2482850E+01    B02    0.6601638E+00
A12    0.3504221E+01    B12    0.1986849E+00
A22    0.2482850E+01    B22    0.1000000E+01

SECTION #: 3
-----------

A03    0.7008387E+00    B03    0.7050429E+00
A13   -0.1231081E+01    B13   -0.9277158E+00
A23    0.7008386E+00    B23    0.1000000E+01

SECTION #: 4
-----------

A04    0.1297509E+01    B04    0.8456067E+00
A14    0.1255656E+01    B14    0.5062455E+00
A24    0.1297509E+01    B24    0.1000000E+01

SECTION #: 5
-----------

A05    0.8735239E+00    B05    0.8756751E+00
A15   -0.1343380E+01    B15   -0.1214753E+01
A25    0.8735238E+00    B25    0.1000000E+01

SECTION #: 6
-----------

A06    0.1134734E+01    B06    0.9589587E+00
A16    0.9391376E+00    B16    0.6286277E+00
A26    0.1134734E+01    B26    0.1000000E+01

SECTION #: 7
-----------

A07    0.9531493E+00    B07    0.9679358E+00
A17   -0.1392653E+01    B17   -0.1331016E+01
A27    0.9531492E+00    B27    0.1000000E+01
```

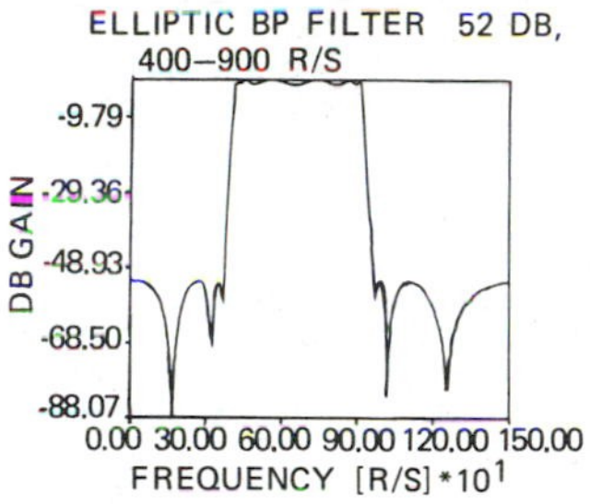

ELLIPTIC BP FILTER 52 DB,
400–900 R/S

DB GAIN: -9.79, -29.36, -48.93, -68.50, -88.07

FREQUENCY [R/S]$*10^1$: 0.00 30.00 60.00 90.00 120.00 150.00

FIG. 7.10.11 Elliptic IIR bandpass filter design summary.

```
DIGITAL FILTER DEVELOPMENT
--------------------------
AP DB:   0.100000E+01
AA DB:   0.450000E+02
K    :   0.829341E+00
ACTUAL AA (DB):   0.505592E+02
SAMPLING FREQUENCY (R/S):   0.300000E+04

H(Z)=H*H1(Z)*H2(Z)*...
H:         0.891251
ELLIPTIC BS FILTER
------------------

FREQUENCY EDGES AT:
   0.300000E+03    0.350000E+03    0.900000E+03    0.994838E+00 R/S
SECTION #: 1
-----------
A01    0.1640273E+00      B01    0.4544659E+00
A11   -0.3081503E-01      B11    0.1095596E+01
A21    0.1640273E+00      B21    0.1000000E+01

SECTION #: 2
-----------
A02    0.1056653E+01      B02    0.6426947E+00
A12   -0.1047899E+01      B12   -0.1518509E+01
A22    0.1056653E+01      B22    0.1000000E+01

SECTION #: 3
-----------
A03    0.5770359E+00      B03    0.8195701E+00
A13    0.2318664E+00      B13    0.8973646E+00
A23    0.5770360E+00      B23    0.1000000E+01

SECTION #: 4
-----------
A04    0.1017873E+01      B04    0.8795223E+00
A14   -0.1397954E+01      B14   -0.1554178E+01
A24    0.1017873E+01      B24    0.1000000E+01

SECTION #: 5
-----------
A05    0.8298657E+00      B05    0.9615858E+00
A15    0.4968736E+00      B15    0.7987278E+00
A25    0.8298658E+00      B25    0.1000000E+01

SECTION #: 6
-----------
A06    0.1016086E+01      B06    0.9740259E+00
A16   -0.1500449E+01      B16   -0.1558596E+01
A26    0.1016086E+01      B26    0.1000000E+01
```

FIG. 7.10.12 Elliptic IIR bandstop filter design summary.

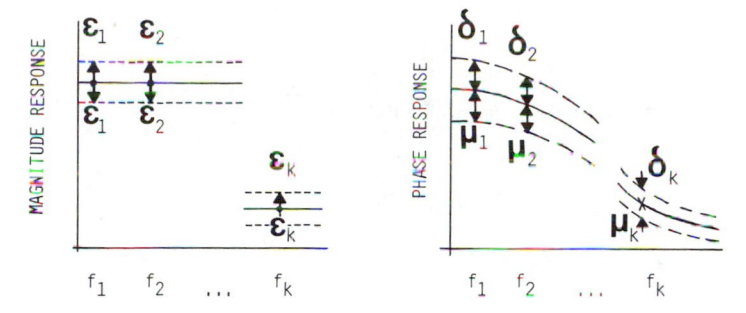

FIG. 7.11.1 Summary of filter design parameters.

where the frequencies ω_i are chosen by the user. Using the SUMP method, an optimal L_2 filter which satisfies all imposed side constraints can be computed (assuming that a solution exists).

EXAMPLE 7.11.1 A filter can be specified in terms of the following data set and the desired response found in Fig. 7.11.2:

Index i	Frequency (rad/s)	Magnitude ε	Constraints	
			Upper Phase λ	Lower Phase δ
1	0.1	0.001	−25	−10
2	0.5	0.001	−70	−50
3	1	0.001	−100	−80
4	1.6	0.001	−110	−90
5	2.5	0.001	−125	−105
6	5.0	0.001	−140	−125
7	9.0	0.001	−155	−140

A filter consisting of three zeros and five poles was chosen as a model with its coefficient set x determined using the SUMT process. The objective function f(x) was given to be

$$f(x) = \sum_{i=1}^{7} \left| G(j\omega_i) - H(x, j\omega_i) \right|^2 V(i)$$

$$= \sum_{i=1}^{7} \left| e_i(x) \right|^2 V(i) \qquad\qquad (7.11.3)$$

with respect to a set of weights V(i) which reflect the designer's subjective feeling regarding the relative importance of e(l). Here V(i) was chosen to be

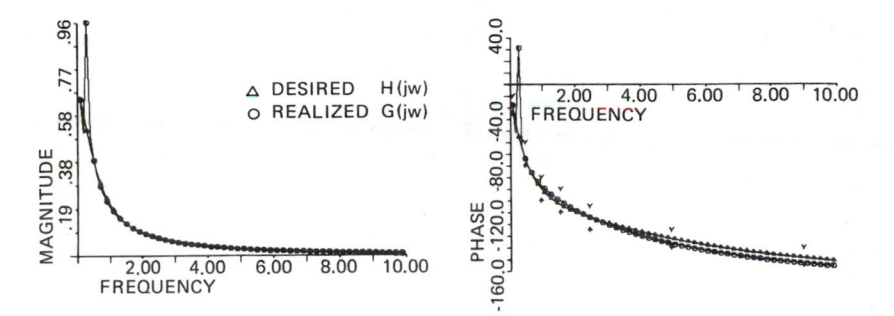

FIG. 7.11.2 Magnitude and phase response desired and optimized filter.

$$V(i) = (50,000, 10,000, 20,000, 10,000, 10,000, 10,000, 50,000)$$

The realized filter was computed to be

$$H(s) = \frac{1849.74s^3 + 0.0013s^2 + 146.44s + 0.2019}{s^5 + 2104s^4 + 9399.09s^3 + 2815.73s^2 + 907.57s + 218.228}$$

which is interpreted graphically in Fig. 7.11.2. The resulting analog model would then be converted into a discrete filter through the use of a bilinear z transform.

7.12 COMPARISON OF FIR AND IIR FILTERS

The comparison of the two basic filter forms, the IIR and FIR filters, is at best difficult. There is a natural inclination to make the comparisons on the basis of case studies. Based on a comprehensive set of case studies, some generalizations emerge. If one is interested in achieving high performance and low, high, or bandpass magnitude frequency response, the classic IIR is probably the best design choice. In particular, the elliptic filter is usually best. If phase linearity is the design objective, the FIR is the obvious choice. Between these two concepts exists a gray area in which the design choice must be based on a case-by-case study.

The design of optimal classic filters relies on decades of study and analysis. Filters of reasonably low order can be found prototyped in filter design references. In this chapter several methods have been proposed by which an analog model could be converted into a desired discrete filter. In all cases, the analysis was straightforward. The filter design problem can also be supported with computer software. This applies to both the IIR and FIR design problems. Here high-order filters can be synthesized with relative ease. Experience has shown, however, that high-order IIR filters are often sensitive to roundoff errors, may become unstable, and sometimes limit-cycle. Specialized analysis procedures have been developed to quantify these undesirable filter characteristics. These errors are either non-existent or minor in FIRs. Therefore, the added analysis time should be factored into a IIR design budget of design time.

The FIR stresses linear phase behavior. It possesses a simple architecture and readily accepts windowing. The coefficients of a FIR are determined in terms of a given or derived impulse response. When synthesized, the impulse response is generally defined in terms of a truncated Fourier series or recursively defined polynomials. When the application requires linear phase behavior (e.g., digital line equalization, voice communication, FFT antialiasing, etc.), the FIR must be given preference.

Rabiner et al. (1974) have published comparative data that demonstrate design trade-offs. The metric of comparison is a common prespecified magnitude filter response. The empirically derived formula for estimating the order of the FIR found in Fig. 7.12.1 is shown in Fig. 7.12.2. The estimated order of the required FIR is given by:

$$n \simeq \frac{-10 \log_{10}(\delta_1 \delta_2) - 15}{14 \Delta f} + 1 \qquad (7.12.1)$$

For example, using the nomograph method, an elliptic filter having a -80-dB ($\delta_2 = 0.0001$) stopband, a passband ripple margin of -0.086 dB ($\delta_1 = 0.01$), and $f_p = 480$, $f_a = 520$, $f_s = 8000$ Hz, the order of the filter can be determined to be 12. For an equivalent FIR, noting that $f_p' = 480/8000 = 0.06$ and $f_a' = 520/8000 = 0.065$, one concludes that

$$n \simeq \frac{-10 \log_{10}(\delta_1 \delta_2) - 15}{14 \Delta f} + 1$$

$$= \frac{-10 \log_{10}(10^{-6})}{14(0.005)} + 1 = 643$$

That is, a 643rd-order filter would be required to do the job of a 12th-order elliptic filter. One may continue this comparison on the basis of required multiplications. Since multiplication time is one of the principal considerations in determining throughput, it is a valid comparison parameter. The number of multiplication required to realize an nth-order FIR, using a direct architecture, is approximated by $[(n + 1)/2]$, where [v] denotes the integer part of v. An nth-order cascaded elliptic filter's multiplier count can be estimated to be on the order of $[(3n + 3)/2]$. Therefore, for an Nth-order FIR to run at the rate of an nth-order IIR, assuming that multiplication time is the key determinate, it follows that

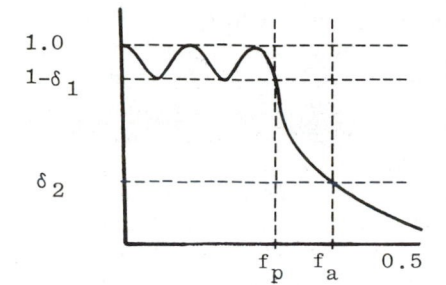

FIG. 7.12.1 Magnitude frequency response of a lowpass design model.

$$\left\lfloor \frac{3n + 3}{2} \right\rfloor = \left\lfloor \frac{N + 1}{2} \right\rfloor \qquad\qquad (7.12.2)$$

or $N/n = 3 + 1/n$. However, this rarely occurs in practice. For example, in the preceding example, one notes that

$$\text{FIR} = \frac{644}{2} = 322 \text{ multiplies}; \quad \text{IIR} = \frac{39}{2} = 20 \text{ multiplies}$$

The ratio of N to n has been experimentally derived for a linear phase filter having a direct-form architecture and an elliptic IIR filter using a cascaded architecture. The experiment was parameterized in terms of the four-tuple $(\delta_1, \delta_2, f_p, f_a)$. Using the filter order estimation formula and multiplication count relationships, architecture complexity can be compared. The result of this experimentation is summarized in Fig. 7.12.3. Here it can be observed that for the explicit architecture considered:

1. For $f \geq 0.3$, $N/n > 3 + (1/n)$ for all δ's and n.
2. For $n \geq 7$, $N/n > 3 + (1/n)$ for all δ's and f_p.
3. The smaller the f_p, the larger the range of δ_1, δ_2, and n for which $N/n < 3 + (1/n)$.

    Based on this analysis, the virtue of the elliptic filter can be dramatized. However, as stated previously, this analysis was biased in favor of the IIR since the filters were compared on the basis of realizing a classical magnitude response form. A case may be made for the FIR by comparing filters on the basis of their phase performance. In particular, one may chose to use the group delay metric as a comparative tool. Here the optimal FIR filters are at a disadvantage since they are known to possess phase nonlinearities in the transition region. This is dramatized in Fig. 7.12.3. Here the group delay properties of a sixth-order elliptic filter are displayed. If a "reasonably flat" system group delay is required, a <u>group delay equalizer</u>

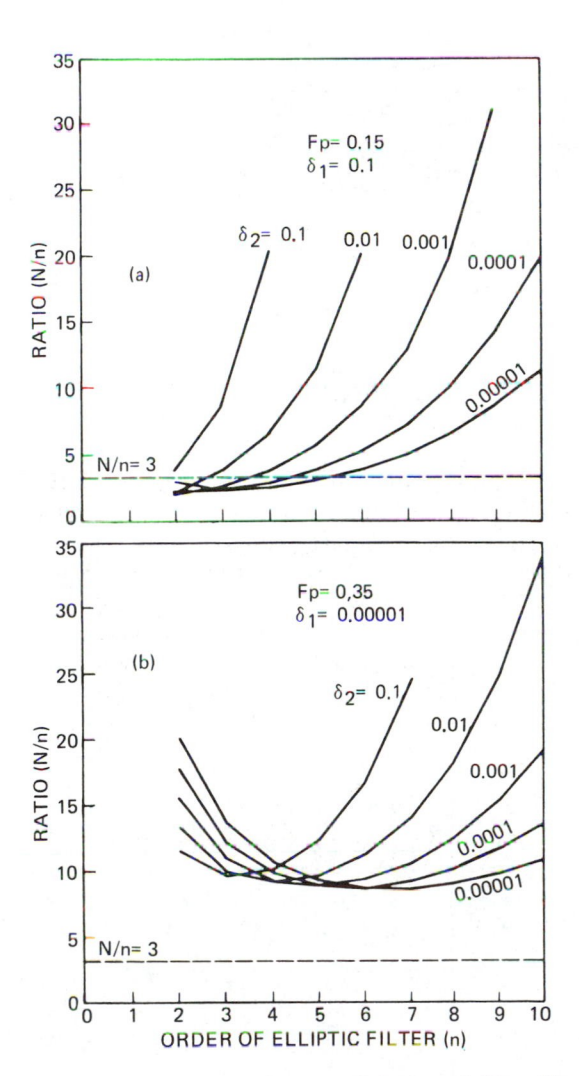

FIG. 7.12.2 Comparison of FIR and IIR elliptic filters in terms of filter order. (After Rabiner et al., 1974. Copyright 1974 American Telephone and Telegraph Company. Reprinted by permission from the Bell System Technical Journal.)

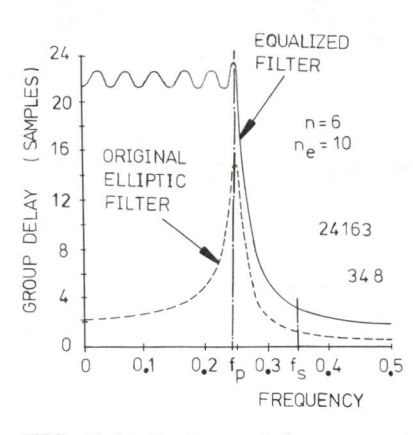

FIG. 7.12.3 Group delay properties of equalized and unequalized elliptic filters. (Data suggested by Rabiner et al., 1974.)

must be used. An equalizer is an allpass filter that preserves the magnitude response of the system it is cascaded with, but alters the phase. The phase profile of the allpass equalizer would be designed so that its phase will add to the uncompensated filter in such a manner that the overall phase will be approximately "flat." A 10th-order equalizer was required to achieve a group delay having approximately a 0.1 sample delay deviation from some average value (see Fig. 7.12.3). The total number of multipliers required by just the equalizer was 20. It is known that the resulting

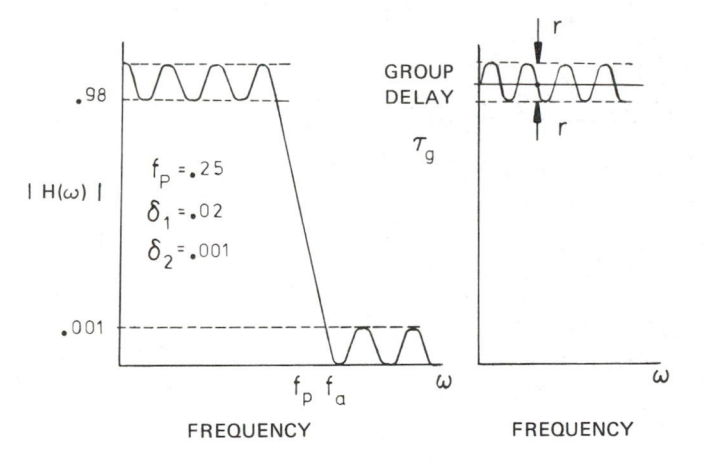

FIG. 7.12.4 Group delay properties of an elliptic filter. (Data suggested by Rabiner et al., 1974.)

delay of an equalized system must be greater than or larger than the largest delay group delay value of the unequalized filter. Referring again to Fig. 7.12.3, one notes that the largest group delay value of the elliptic filter is approximately 15 samples, while the equalized system has a phase delay with an average value on the order of 22 samples. It is interesting to note that an 11-multiplier FIR would have achieved the same design objective. Therefore, if group delay is of major concern, a strong case can be made for the FIR. Finally, as a design guide, it has been argued that the minimum order for an equalizing filter can be bounded by

$$n_0 = 2\tau_{max} f_{max} \qquad\qquad\qquad (7.12.3)$$

In the simple problem under study, $n_0 = [15(2)(0.25)] = 7.5 \Rightarrow 8$. Generally, the realized equalizer is of slightly higher order that the bound n_0 would suggest. Of course, the higher the order, the "flatter" would be the resulting system group delay.

In considering the foregoing example problem, an all-FIR realization was found to provide the desired group delay behavior with a smaller multiplier budget. Rabiner et al. (1974) studied the design of systems having a bound group delay deviation in the context of the following example (see Fig. 7.12.4 and Table 7.12.1). It was noted that with a ripple distortion of 5.2% or greater, an equalized elliptic filter would require fewer multipliers

TABLE 7.12.1 Summary of IIR and FIR Equalizer Designs

f_a	n	N	n_e	τ_g	r	N_{FIR}	N_{eq}
0.30639	6	45	4	13.8	34.7	23	14
			6	16.0	25.0		16
			8	18.7	16.9		18
			10	22.0	11.7		20
			12	25.5	7.9		22
			14	29.4	5.2		24
			16	32.8	3.2		26
			18	36.3	1.8		28

Note: n = elliptic IIR order, N = FIR order, n_e = equalizer order, τ_g = group delay, r = group delay percent ripple, N_{FIR} = number of multipliers for FIR, N_{eq} = number of multipliers for equalized elliptic filter.
Source: Data suggested by Rabiner et al., 1974.

than an equivalent FIR. For higher fidelity (i.e., $r \rightarrow 0$), equalized systems having a higher multiplier budget must be used. Also note that the higher the order of the equalized filter, the higher the realized group delay. It would therefore seem important that the filter designer study these trade-offs on a case-by-case basis when the choice of an IIR or FIR is in question. Using computer-aided methods, this can generally be accomplished with a minimum of effort.

BIBLIOGRAPHY

Antonious, A. (1979), Digital Filters: Analysis and Design, McGraw-Hill, New York.

Budak, A. (1974), Passive and Active Network Analysis and Synthesis, Houghton Mifflin, Boston.

Burris, C. S., and T. W. Parks (1970), Time Domain Design of Recursive Digital Filters, IEEE Trans. Audio Electroacoust., AU-18, June, pp. 137-141.

Constantinides, A. G. (1970), Spectral Transforms of Digital Filters, Proc. IEE, 117, No. 8, pp. 1585-1590.

Daniels, R. W. (1974), Approximation Methods for the Design of Passive, Active, and Digital Filters, McGraw-Hill, New York.

Deczky, A. G. (1972), Synthesis of Recursive Digital Filters Using the Minimum p-Error Criterion, IEEE Trans. Audio Electroacoust., AU-20, October, pp. 257-263.

Fetweis, A. (1972), A Simple Design of Maximally Flat Delay Digital Filters, IEEE Trans. Audio Electroacoust., AU-20, June, pp. 112-114.

Gray, A. H., and Markel, J. D. (1973), Digital Lattice and Ladder Filter Synthesis, IEEE Trans. Audio Electroacoust., AU-21, December, pp. 491-500.

Kaiser, J. F. (1963), Design Methods for Sampled Data Filters, Proc. First Allerton Conf. Circuits Syst., November, pp. 221-235.

Oppenheim, A. V., and R. S. Schafer (1975), Digital Signal Processing, Prentice-Hall, Englewood Cliffs, N.J.

Rabiner, L. R., and B. Gold (1975), Theory and Application of Digital Signal Processing, Prentice-Hall, Englewood Cliffs, N.J.

Rabiner, L. A., J. F. Kaiser, O. Herrmann, and M. T. Dolan (1974), Some Comparisons Between FIR and IIR Digital Filters, Bell Syst. Tech. J., 53, No. 2, February, pp. 305-331.

Taylor, F. J., and R. J. Molepske (1973), Optimal Filter Design via Mathematical Programming, IEEE Trans. Syst. Man Cybern., SMC-3, No. 4, July, pp. 382-388.

APPENDIX A   IIR Filter Design

The software presented in this appendix was written in FORTRAN-IV PLUS on a Digital Equipment PDP-11/60 using double-precision floating-point arithmetic. The admissible filter approximation options are:

1. Butterworth
2. Chebyshev
3. Elliptic

The filter types supported by the software packagd are:

1. Lowpass (LP)
2. Highpass (HP)
3. Bandpass (BP)
4. Bandstop (BS)

The filter design process begins with the user providing option and type data plus key magnitude and frequency domain parameters of the required digital filter. The filter process begins with the design of a prewarped analog lowpass prototype filter which agrees with the target discrete filters passband ripple (AP) and stopband attenuation (AA). The quantities are to be specified in decibels. The order of the prototype filter is computed within the software system. The critical frequencies are specified in radians per second and they are defined as follows:

1. Lowpass and highpass: passband edge = WP1; stopband edge = WA1
2. Bandpass and bandstop: passband edges = WP1 and WP2; stopband edges = WA1 and WA2

The frequency sets are prewarped and used to define the frequency domain attributes of the analog filter. The formal design procedure is shown in Fig. A.1.

The software package will return to the user the following information:

1. Echo-print user-supplied filter attributes
2. Transition ratio k
3. Filter order
4. Computed stopband attenuation (AA)

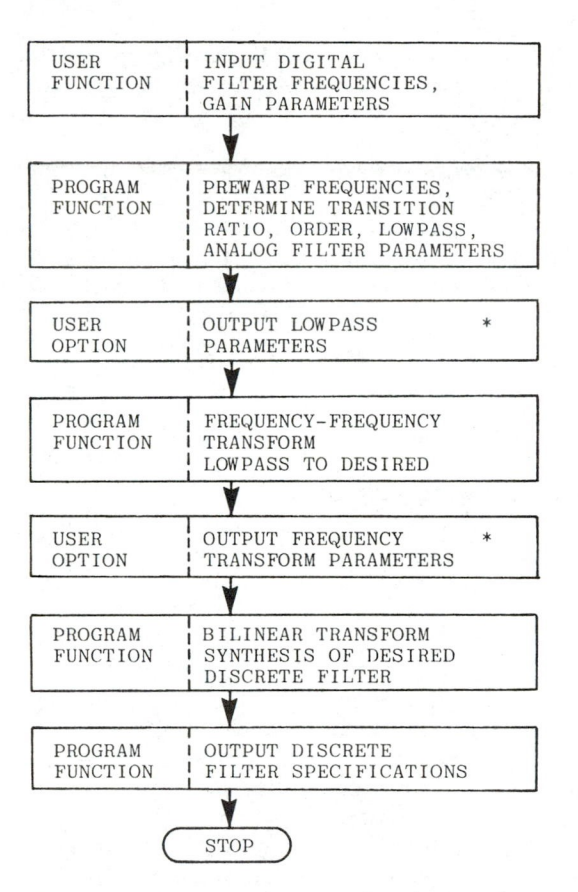

FIG. A.1 IIR design protocol and procedures. The asterisk indicates that pole-zero values may optionally be listed.

5. Coefficients of the analog prototype lowpass filter, based on a second-order decomposition having the form

$$H(s) = H * H_1(s) * H_2(s) * \cdots * H_L(s)$$

$$H_i(s) = \frac{a_{0i} + a_{1i}s + a_{2i}s^2}{b_{0i} + b_{1i}s + b_{2i}s^2}$$

6. Pole and zero locations in the s plane
7. Analog frequency-to-frequency transformation attributes (including new critical frequencies and filter coefficients section by section)

8. Discrete filter design attributes [including the filter gain H from H(z) = H * $H_1(z)$ * $H_2(z)$, . . .], second-order section filter coefficients [from $H_i(z) = a_{0i} + a_{1i}z + a_{2i}z^2/b_{0i} + b_{1i}z + b_{2i}z^2$], critical frequdncies, and pole-zero locations in the z plane

9. Interpret H(z), list, and (optional) plot

A comprehensive example is presented at the end of this appendix to demonstrate how to use and interpret the software. The structure of the software was influenced by the design programs found in Antoniou's software, written in HPL to run on a Hewlett-Packard 9825, has been augmented and reprogrammed by Marco Rengan to run in ANSI standard FORTRAN on a general-purpose 16-bit mini/microcomputer. The software can be obtained, at a modest cost, by contacting IIR Design, EE Department, University of Florida, Gainesville, FL 32611.

Output Listing for an Elliptic Lowpass Filter with a 2-dB Passband and a 15-dB Stopband Normalized over the Frequency Range [0, 0.25] to [0.4, 0.5] Hz.

```
>RUN MtEST.T
INPUT LUN OF PRINTER 3
INPUT APPROXIMATION (# ,TYPE)
1,BUTTERWORTH 2, CHEBYSHEV
3, ELLIPTIC
3,Elliptic
INPUT FUNCTION (BP,BS,LP,HP):LP
INPUT HEADING (DATE ETC 20 CHARS):Example Run
INPUT AP,AA 2.,15.
INPUT SAMPLING FREQUENCY  (R/S) :1.
INPUT PASS BAND EDGE WP1 (R/S):
STOP BAND EDGE WA1 (R/S):
0.25,0.40
** OUT PUT LISTING (Y,N) *** Y
INPUT DATA
---------

Elliptic    APPROXIMATION
AP, DB:      2.000000
AA, DB:     15.000000
K:          0.324920
OUTPUT DATA
-----------
ORDER:  2
ACTUAL AA       28.76722
H(S)=H*H1(S)*H2(S)...
H1(S)=(A0+A11S+A21S**2)/(B0+B11S+B21S**2)

SECTION #: 1
-----------

A01   0.5988377E+01      B01   0.2747946E+00
A11   0.0000000E+00      B11   0.4473636E+00
A21   0.1000000E+01      B21   0.1000000E+01

 H:    0.364501E-01
```

```
POLE- ZERO OUTPUT
-------------------
OUTPUT LISTING?N
# COMPLEX ZERO PAIRS:  1
#REAL ZEROS:  0
#COMPLEX POLE PAIRS:  1
# REAL POLES :  0
H:    0.364501E-01
ZEROS:
( 0.000000E+00    0.244712E+01)
POLES:
(-0.223682E+00    0.474090E+00)

** OUT PUT LISTING (Y,N) *** Y

FILTER TRANSFORMATIONS
------------------------------
LOW PASS FILTER
LAMDA:  0.179076E+01
SECTION #:  1
-----------

A01    0.1867383E+01       B01    0.8569046E-01
A11    0.0000000E+00       B11    0.2498174E+00
A21    0.1000000E+01       B21    0.1000000E+01


H:    0.364501E-01

DIGITAL FILTER DEVELOPMENT
---------------------------
Example Run

AP, DB:  0.200000E+01
AA, DB:  0.150000E+02
K    :  0.324920E+00
ACTUAL AA (DB):  0.287672E+02
SAMPLING FREQUENCY (R/S):  0.100000E+01
ORDER :  2

H(Z)=H*(A01+A11Z+A21Z^2/B01+B11Z+B2Z^2)...
H:    0.364501E-01
Elliptic    LP FILTER
---------------------

FREQUENCY EDGES AT :  0.400000E+00   0.250000E+00
LAMDA :        1.790762
SECTION #:  1
-----------

A01    0.7386399E+01       B01    0.4033012E+00
A11    0.1325221E+02       B11   -0.1172901E+00
A21    0.7386399E+01       B21    0.1000000E+01

OUTPUT LISTING?Y
# COMPLEX ZERO PAIRS:  1
#REAL ZEROS:  0
#COMPLEX POLE PAIRS:  1
# REAL POLES :  0
H:    0.364501E-01
ZEROS:
(    -0.897068       0.441892)
POLES:
(     0.058645       0.632346)
```

Output Listing for a Chebyshev Bandpass Design with a
1-dB Passband and a 15-dB Stopband Normalized over
the Frequency Range [0, 0.1] to [0.4, 0.5]

```
>RUN MTEST.T
INPUT LUN OF PRINTER 3
INPUT APPROXIMATION (# ,TYPE)
1,BUTTERWORTH 2, CHEBYSHEV
3, ELLIPTIC
2,CHEBYCHEV
INPUT FUNCTION (BP,BS,LP,HP):BP
INPUT HEADING (DATE ETC 20 CHARS):TEST RUN #2
INPUT AP,AA 1.0,15.0
INPUT SAMPLING FREQUENCY  (R/S) :1.0
INPUT PASSBAND EDGES WP1,WP2 (R/S):0.2,0.30
INPUT STOPBAND EDGES WA1,WA2 (R/S):0.10,0.40
** OUT PUT LISTING (Y,N) *** Y
INPUT DATA
---------

CHEBYCHEV   APPROXIMATION
AP, DB:     1.000000
AA, DB:     15.000000
K:       0.236068
OUTPUT DATA
-----------
ORDER:  2
ACTUAL AA        24.99916
H(S)=H*H1(S)*H2(S)...
H1(S)=(A0+A11S+A21S**2)/(B0+B11S+B21S**2)

SECTION #: 1
-----------

A01   0.1000000E+01     B01   0.1102510E+01
A11   0.0000000E+00     B11   0.1097734E+01
A21   0.0000000E+00     B21   0.1000000E+01

  H:    0.982613E+00

POLE- ZERO OUTPUT
--------------------
OUTPUT LISTING?
# COMPLEX ZERO PAIRS:  0
#REAL ZEROS:  0
#COMPLEX POLE PAIRS:  1
# REAL POLES :  0
H:    0.982613E+00
POLES:
(-0.548867E+00   0.895129E+00)

** OUT PUT LISTING (Y,N) *** Y

FILTER TRANSFORMATIONS
-----------------------------

BAND PASS FILTER
CENTER FREQUENCY (HZ)  0.318310E+00
BANDWIDTH (HZ)  0.206850E+00
SECTION #: 1
-----------

A01   0.0000000E+00     B01   0.1813503E+00
A11   0.1000000E+01     B11   0.1456766E+00
A21   0.0000000E+00     B21   0.1000000E+01
```

```
SECTION #: 2
-----------

A02   0.0000000E+00      B02   0.5660857E-01
A12   0.1000000E+01      B12   0.8139006E-01
A22   0.0000000E+00      B22   0.1000000E+01

H:    0.420431E-01

DIGITAL FILTER DEVELOPMENT
--------------------------
TEST RUN #2

AP, DB:  0.100000E+01
AA, DB:  0.150000E+02
K    :   0.236068E+00
ACTUAL AA (DB):  0.249992E+02
SAMPLING FREQUENCY (R/S):  0.100000E+01
ORDER : 4

H(Z)=H*(A01+A11Z+A21Z^2/B01+B11Z+B2Z^2)...
H:    0.420431E-01
CHEBYCHEV   BP FILTER
---------------------

PASSBAND EDGES AT:  0.200000E+00    0.300000E+00   R/S
STOPBAND EDGES AT:  0.100000E+00    0.400000E+00   R/S
CENTRE FREQ (R/S):          0.318310
BANDWIDTH (R/S):         0.206850
SECTION #: 1
-----------

A01  -0.9673842E+00     B01    0.7181494E+00
A11   0.0000000E+00     B11    0.4864376E+00
A21   0.9673842E+00     B21    0.1000000E+01

SECTION #: 2
-----------

A02  -0.1731479E+01     B02    0.7181497E+00
A12   0.0000000E+00     B12   -0.4864376E+00
A22   0.1731479E+01     B22    0.1000000E+01

OUTPUT LISTING?Y
# COMPLEX ZERO PAIRS:  0
#REAL ZEROS:  0
#COMPLEX POLE PAIRS:  2
# REAL POLES :  0
H:    0.420431E-01
POLES:
(    -0.243219        0.811784)
(     0.243219        0.811785)
```

APPENDIX B   IIR Design Nomographs

As a convenience to the reader, a set of IIR design nomographs is included in this appendix (Figs. B.1 through B.9). The nomographs were provided by courtesy of Bell Laboratories, with special thanks going to Jim Kaiser.

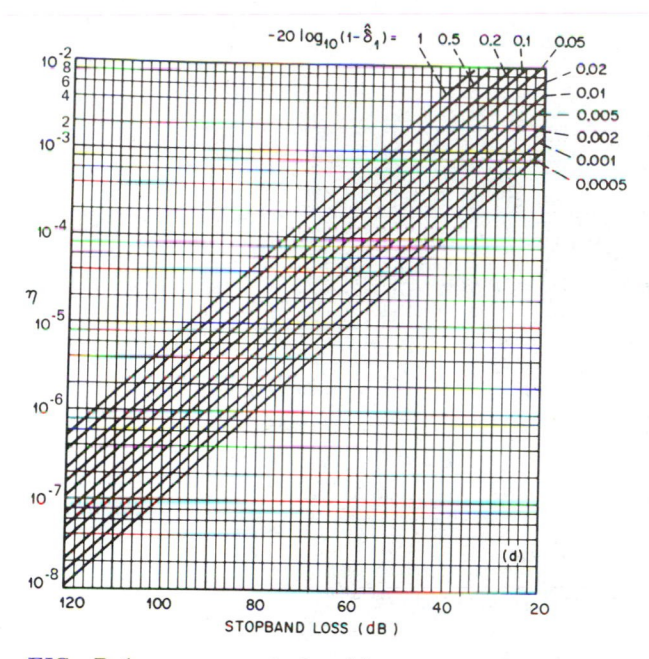

FIG. B.1 η versus stopband loss versus δ_1.

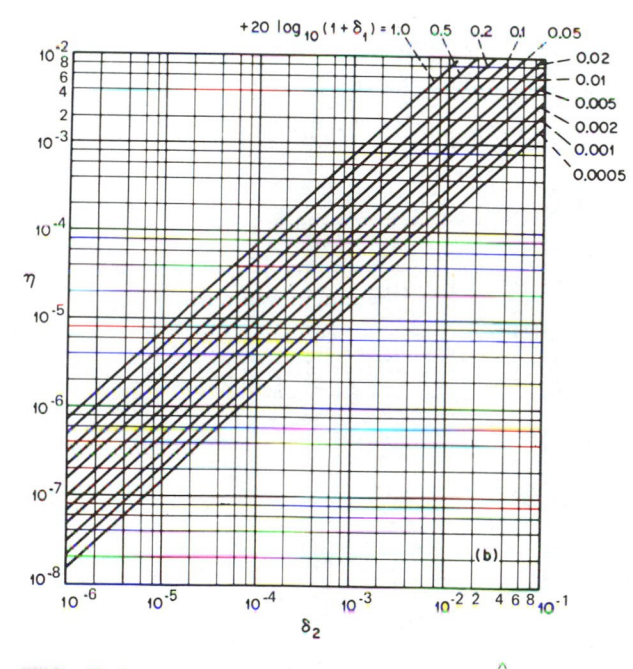

FIG. B.2 η versus stopband loss versus $\hat{\delta}_1$.

335

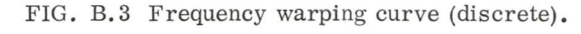

FIG. B.3 Frequency warping curve (discrete).

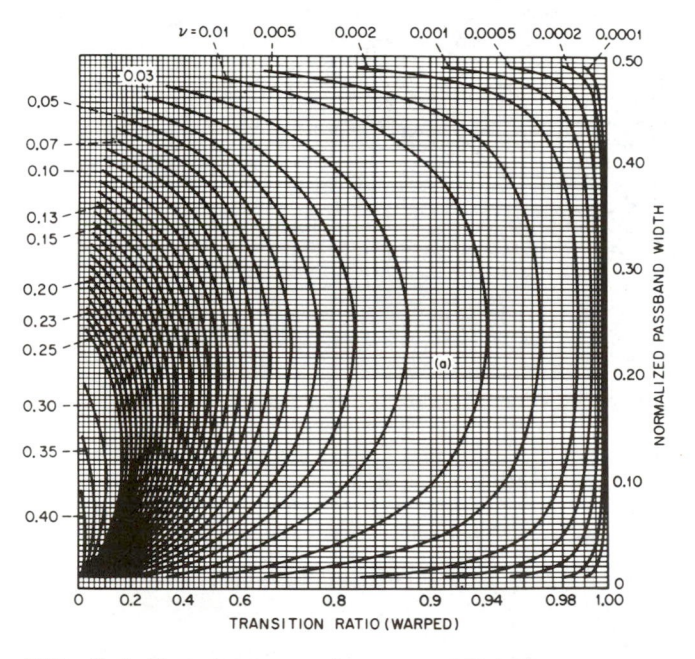

FIG. B.4 Frequency warping curve (analog).

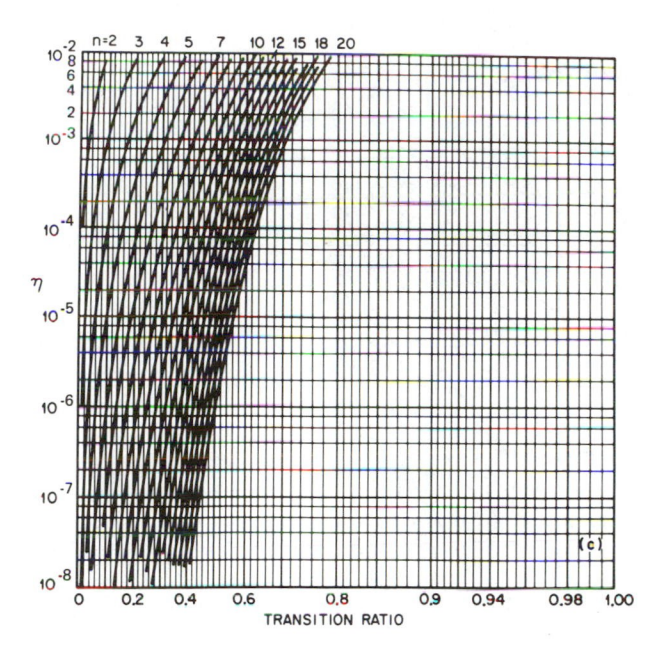

FIG. B.5 Butterworth filter design.

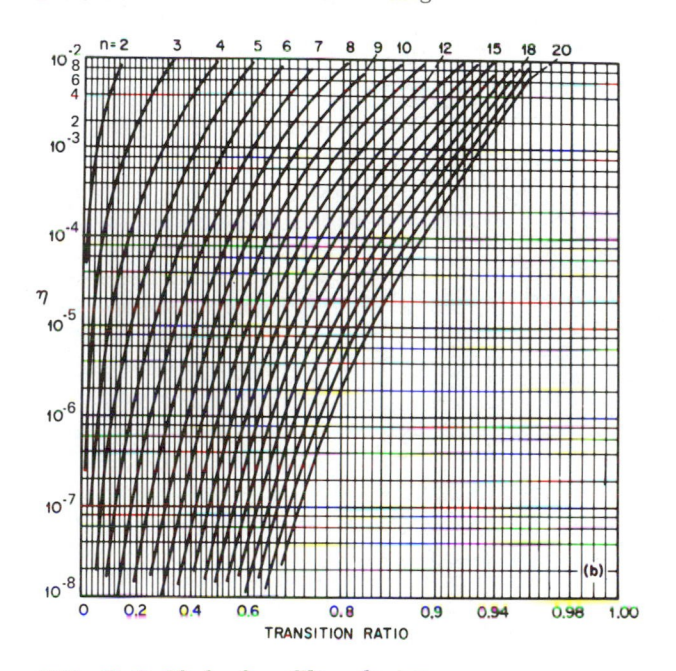

FIG. B.6 Chebyshev filter design

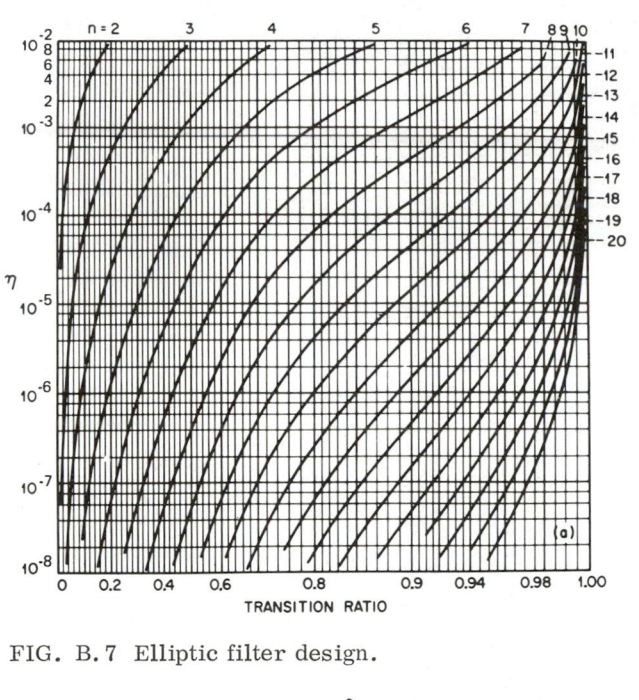

FIG. B.7 Elliptic filter design.

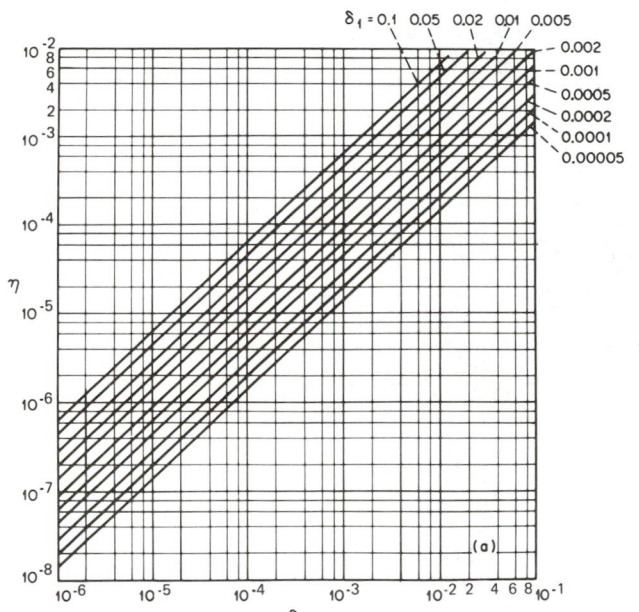

FIG. B.8 η versus δ_2 versus δ_1.

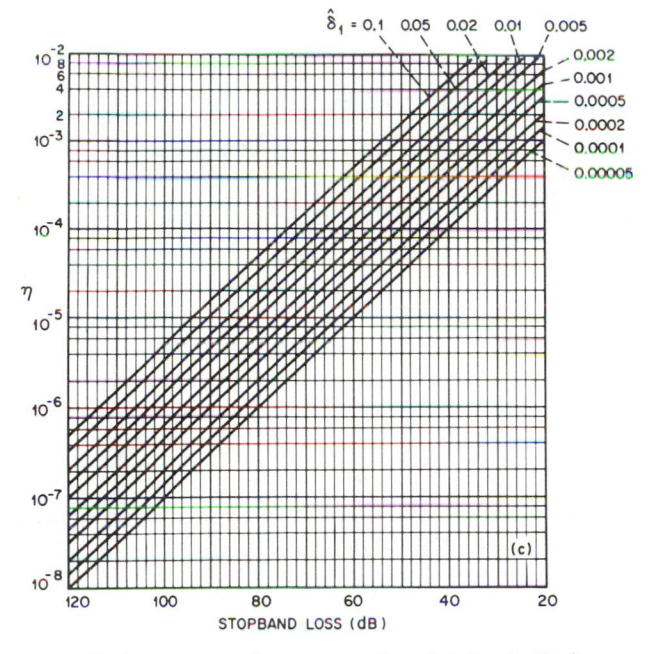

FIG. B.9 η versus δ_2 versus $(1 - \delta_1)$ in decibels.

8
Ladder Filters

8.1 INTRODUCTION

Analog ladder filters have enjoyed a long history of popularity because of their low sensitivity to parameter variations. One could, therefore, hope to translate this attribute into low-sensitivity discrete filters. Ladder filters possess a very regular architecture. Using long division, one can decompose a transfer function of the form

$$H(z) = \frac{N_s(z)}{D_s(z)} = \frac{\displaystyle\sum_{i=0}^{S} n_{s,i} z^{-i}}{\displaystyle\sum_{i=0}^{S} d_{s,i} z^{-i}} = \frac{Y(z)}{X(z)} \qquad (8.1.1)$$

into relationships having structures similar to

$$H(z) = A_0 + \cfrac{1}{B_1 + \cfrac{1}{A_1 + \cfrac{1}{B_2 z + \cdots}}} \qquad (8.1.2)$$

or

$$H(z) = A_0 + \cfrac{1}{B_1 z^{-1} + \cfrac{1}{A_1 + \cfrac{1}{B_2 z^{-2} + \cdots}}}$$

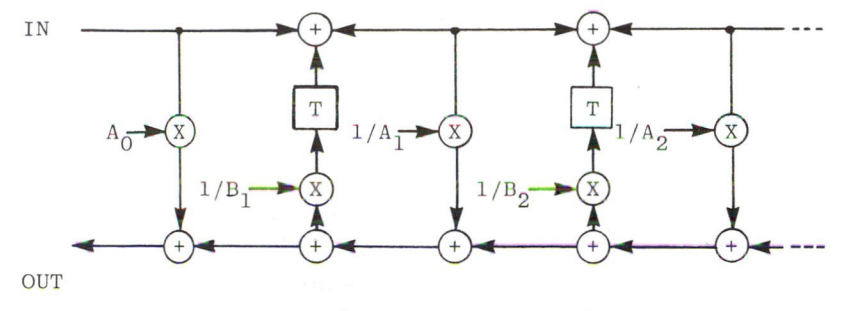

FIG. 8.1.1 General form of a discrete ladder filter.

depending on the order of $N_s(z)$ and $D_s(z)$. Some commonly used architectures are presented in Fig. 8.1.1. The realizability conditions associated with this class of filter have been published by Mitra and Sherwood (1973).

EXAMPLE 8.1.1 Given

$$H(z) = \frac{1 + 2z}{1 + z} = 1 + \frac{1}{z^{-1} + 1/1}$$

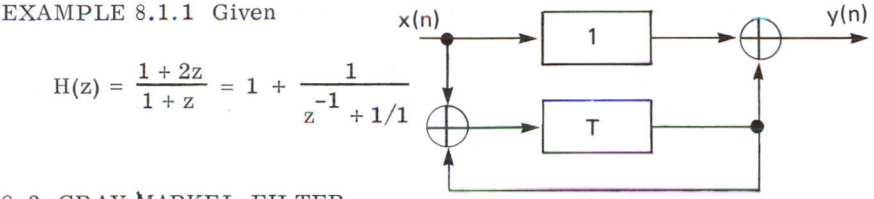

8.2 GRAY-MARKEL FILTER

A popular and special ladder structure has been developed by Gray and Markel. It has the ability to decompose a signal into a set of orthogonal components, which makes this filter useful in some decision-making roles. In addition, this filter has the desired attribute of being a low-sensitivity filter. That is, it can tolerate roundoff errors better than many other classes of digital filters. A stable Gray-Markel filter will also guarantee that the resulting filter coefficients are bounded by unity. This is desirable from a filter realization standpoint since it simplifies the coefficient encoding problem.

The Gray-Markel filter structure can be found in Fig. 8.2.1. In this figure the subsystem $G_i(z)$ can be realized in either the two- or one-multiplier methods suggested in Fig. 8.2.2. Although it may appear to be irrelevant which architecture is used, it will be shown that upon properly mixing the two- and one-multiplier forms, a reduced roundoff error may result.

The two-multiplier form admits a state-variable representation given by

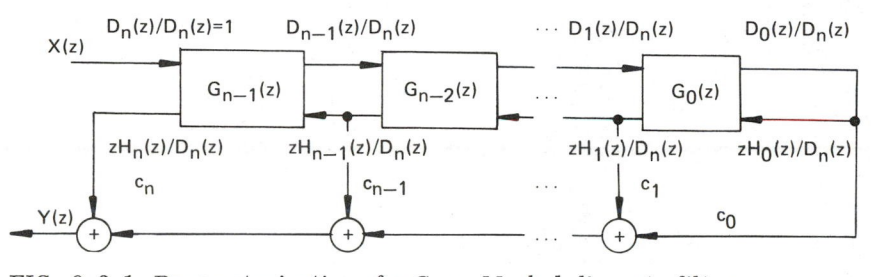

FIG. 8.2.1 Parameterization of a Gray-Markel discrete filter.

$$\begin{bmatrix} D_{i+1}(z) \\ H_{i+1}(z) \end{bmatrix} = \begin{bmatrix} 1 & a_i \\ a_i z^{-1} & z^{-1} \end{bmatrix} \begin{bmatrix} D_i(z) \\ H_i(z) \end{bmatrix} \qquad (8.2.1)$$

or

$$\begin{bmatrix} D_i(z) \\ H_i(z) \end{bmatrix} = \begin{bmatrix} 1 & -a_i \\ -a_i z & z \end{bmatrix} \begin{bmatrix} D_{i+1}(z) \\ H_{i+1}(z) \end{bmatrix} \Bigg/ (1 - a_i^2)$$

The distributed transfer functions, denoted $D_i(z)$ and $H_i(z)$, can be computed recursively as follows:

$$zH_i(z) = D_i(z^{-1})z^{-1} \qquad (8.2.2a)$$

$$a_{i-1} = d_{i,i} \qquad \text{(filter coefficient)} \qquad (8.2.2b)$$

$$D_{i-1}(z) = \frac{D_i(z) - a_{i-1} zH_i(z)}{1 - a_{i-1}^2} = \sum_{j=0}^{i-1} d_{i-1,j} z^{-j} \qquad (8.2.2c)$$

$$c_i = n_{i,i} \qquad \text{(output coefficient)} \qquad (8.2.2d)$$

$$N_{i-1}(z) = N_i(z) - zH_i(z)c_i = \sum_{j=0}^{i-1} n_{i-1,j} z^{-j} \qquad (8.2.2e)$$

From the theory of ladder filters it is known that the numerator of the desired transfer function is a linear combination of each subfilter output transfer function. That is,

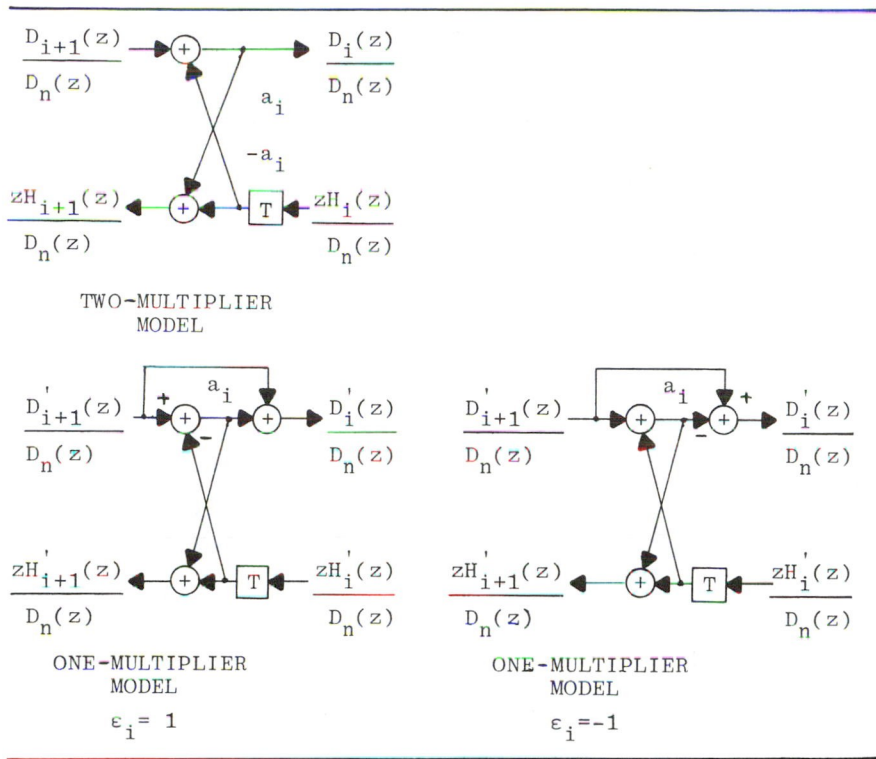

FIG. 8.2.2 Two- and one-multiplier Gray-Markel subfilters.

$$N_n(z) = \sum_{i=0}^{N} c_i zH_i(z) \qquad\qquad (8.2.3)$$

relates to the original transfer function specification as follows:

$$G(z) = \frac{N_n(z)}{D_n(z)} = \frac{\displaystyle\sum_{i=0}^{N} c_i zH_i(z)}{D_n(z)} \qquad\qquad (8.2.4)$$

EXAMPLE 8.2.1 The following is a third-order Chebyshev filter design motivated by Mitra and Sherwood. Let

$$G(z) = \frac{N_3(z)}{D_3(z)} = \frac{0.0154 + 0.0462z^{-1} + 0.0462z^{-2} + 0.0154z^{-3}}{1 - 1.990z^{-1} + 1.572z^{-2} - 0.4583z^{-3}}$$

That is,

$$n_{30} = n_{33} = 0.0154, \ n_{31} = n_{32} = 0.0462$$

$$d_{30} = 1, \ d_{31} = -1.990, \ d_{32} = 1.572, \ d_{33} = -0.4583$$

For i = 3, it follows that

$$a_2 = d_{33} = -0.4583, \ c_3 = n_{33} = 0.0155$$

$$zH_3(z) = D_3(z^{-1})z^{-3} = -0.4583 + 1.572z^{-1} - 1.990z^{-2} + z^{-3}$$

$$N_2(z) = N_3(z) - zH_3(z)c_3 = 0.0224578 + 0.0219912z^{-1} + 0.076886z^{-2}$$

$$D_2(z) = \frac{D_3(z) - a_2 zH_3(z)}{1 - a_2^2} = 1 - 1.607107z^{-1} + 0.835463z^{-2}$$

That is,

$$n_{20} = 0.022458, \ n_{21} = 0.0219912, \ n_{22} = 0.0768446$$

$$d_{20} = 1.0, \ d_{21} = -1.6071075, \ d_{22} = 0.8354626$$

For i = 2, it follows that

$$a_1 = d_{22} = 0.8354626, \ c_2 = n_{22} = 0.076846$$

$$zH_2(z) = D_2(z^{-1})z^{-2} = 0.83546264 - 1.6071075z^{-1} + z^{-2}$$

$$N_1(z) = N_2(z) - zH_2(z)c_2 = -0.04174415 + 0.14549098z^{-1}$$

$$D_1(z) = \frac{D_2(z) - a_1 zH_2(z)}{1 - a_1^2} = 1 - 0.87558719z^{-1}$$

That is,

$$n_{10} = -0.04174415, \ n_{11} = 0.14549098$$

$$d_{10} = 1.0, \ d_{11} = -0.87558713$$

For i = 1, it follows that

$$z_0 = d_{11} = -0.87558713, \quad c_1 = n_{11} = 0.14549098$$

$$zH_1(z) = D_1(z^{-1})z^{-1} = 0.87558713 + z^{-1}$$

$$H_0(z) = N_1(z) - zH_1(z)c_1 = 0.08564588$$

$$D_0(z) = \frac{D_1(z) - a_0 zH_1(z)}{1 - a_0^2} = 1.0$$

For i = 0, it follows that

$$c_0 = 0.085645883$$

$$zH_0(z) = 1.0$$

and as a check, one notes that

$$G(z) = \frac{\sum\limits_{i=0}^{3} c_i zH_i(z)}{D(z)}$$

A two–multiplier subfilter may be modeled in the manner suggested in Fig. 8.2.3. In precedence form, the subfilter is given by

$$z(n) = Ax(n) + Bx(n - 1) + Cu(n) \qquad (8.2.5)$$

where

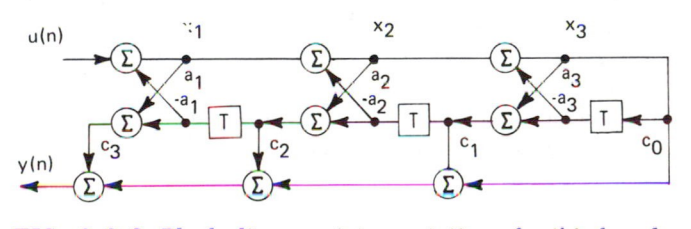

FIG. 8.2.3 Block diagram interpretation of a third–order two–multiplier filter realization.

$$
\begin{bmatrix} x_1(n) \\ x_2(n) \\ x_3(n) \\ x_4(n) \\ x_5(n) \\ x_6(n) \end{bmatrix} =
\begin{bmatrix}
0 & 0 & 0 & 0 & 0 & 0 \\
1 & 0 & 0 & 0 & 0 & 0 \\
0 & 1 & 0 & 0 & 0 & 0 \\
a_2 & 0 & 0 & 0 & 0 & 0 \\
0 & a_1 & 0 & 0 & 0 & 0 \\
0 & 0 & a_0 & 0 & 0 & 0
\end{bmatrix}
\begin{bmatrix} x_1(n) \\ x_2(n) \\ x_3(n) \\ x_4(n) \\ x_5(n) \\ x_6(n) \end{bmatrix} +
\begin{bmatrix}
0 & 0 & 0 & 0 & -a_2 & 0 \\
0 & 0 & 0 & 0 & 0 & -a_1 \\
0 & 0 & -a_0 & 0 & 0 & 0 \\
0 & 0 & 0 & 0 & 1 & 0 \\
0 & 0 & 0 & 0 & 0 & 1 \\
0 & 0 & 0 & 0 & 0 & 0
\end{bmatrix}
\begin{bmatrix} x_1(n-1) \\ x_2(n-1) \\ x_3(n-1) \\ x_4(n-1) \\ x_5(n-1) \\ x_6(n-1) \end{bmatrix}
$$

$$
+ \begin{bmatrix} 1 \\ 0 \\ 0 \\ 0 \\ 0 \\ 0 \end{bmatrix} u(n); \quad \alpha_3 = a_0, \quad \alpha_2 = a_1, \quad \alpha_1 = a_2
$$

with

$$
y(n) = c_0 x_3(n) + c_1 x_6(n) + c_2 x_5(n) + c_3 x_4(n)
$$

The filter coefficients can be computed using a general-purpose digital computer. The results of such an analysis are reported in Fig. 8.2.4. In this figure the filter coefficients are listed and the magnitude frequency and phase response (using floating-point arithmetic) interpreted graphically.

EXAMPLE 8.2.2 The Gray-Markel filter is reported to have a certain insensitivity to coefficient rounding. To motivate this property, Example 8.2.1 will be reinvestigated in the context of fixed-point finite-precision arithmetic. The simulated unit impulse for a floating-point realization of the filter under consideration is summarized in Fig. 8.2.5. The fixed-point realizations are found in Fig. 8.2.6. The fixed-point word format was assumed to have the form $\pm xxx \cdots x$. It can be noted that the filter's performance is acceptable over a wide range of word lengths. The data found in Fig. 8.2.6 indicate that for word lengths of 4 bits or larger, the error variance due to fixed-point roundoff errors is negligible. The problem of roundoff errors is studied in depth in Chapter 12.

8.3 COMPUTER-AIDED DESIGN

The coefficient determination algorithm can be programmed on a general-purpose digital computer provided that division overflow does not occur in

Eq. (8.2.2c). It is known that if the ladder filter is stable, all its coefficients are bounded by unity (i.e., $|a_i| \leq 1$). Therefore, the derived filter coefficients will be guaranteed to be compatible with a fractional binary word representation. Also, it can be noted that the characteristic polynomial, namely $D_i(z)$, need not be factored as a precondition for computing filter coefficients. This is important from a computational, efficiency standpoint. A ladder filter design option has been added to a comprehensive filter design software package reported in the appendix to Chapter 13. The reader should refer to this appendix, or the appendix to this chapter, for computer-aided ladder filter design support.

8.4 FILTER PROPERTIES

Before the question of filter sensitivity is addressed, the structure of the Gray-Markel filter will be explored. First, the transform of a two-multiplier subfilter into a one-multiplier subfilter will be derived. Consider the defining equation:

$$
c_i' = \frac{c_i}{v} \qquad\qquad v_i = \begin{cases} 1 & \text{if } i = 1 \\[2mm] \displaystyle\prod_{i=1}^{n-1} (1 + \epsilon_j a_j) & \text{otherwise} \end{cases}
$$
$$
D_i'(z) = v_i D_i(z),
$$
$$
H_i'(z) = v_i H_i(z)
$$
$$\tag{8.4.1}$$

which induces, as a final form, the transfer function given by

$$
G_i(z) = \frac{\displaystyle\sum_{i=0}^{n} c_i' z H_i'(z)}{D_n(z)}
$$
$$\tag{8.4.2}$$

In terms of the example problem, one obtains (under the hypothesis that $\epsilon_i = +1$ for all i)

$$
v_i = \begin{cases} v_3 = 1 \\[2mm] v_2 = (1 + a_2) = 0.5417 \\[2mm] v_1 = (1 + a_2)(1 + a_1) = 0.994 \\[2mm] v_0 = (1 + a_2)(1 + a_1)(1 + a_0) = 0.1237 \end{cases}
$$

*** COEFFICIENTS OF GRAY—MARKEL FILTER ***

$N_3(Z)$= 0.0153999999 + 0.0462000072 Z^{-1} + 0.0462000072 Z^{-2} + 0.0153999999 Z^{-3}

$D_3(Z)$= 1.0000000000 −1.9899997711 Z^{-1} + 1.5720000267 Z^{-2} −0.4582999945 Z^{-3}

A_2= -0.4582999945 C_3= 0.0153999999

$ZH_3(Z)$= -0.4582999945 + 1.5720000267 Z^{-1}− 1.9899997711 Z^{-2} + 1.0000000000 Z^{-3}

$N_2(Z)$= 0.0224578157 + 0.0219912082 Z^{-1} + 0.0768460035 Z^{-2}

$D_2(Z)$= 1.0000000000 − 1.6071071625 Z^{-1} + 0.8354628086 Z^{-2}

A_1= 0.8354628086 C_2= 0.0768460035

$ZH_2(Z)$= 0.8354628086 − 1.6071071625 Z^{-1} + 1.0000000000 Z^{-2}

$N_1(Z)$= -0.0417441428 + 0.1454909444 Z^{-1}

$D_1(Z)$= 1.0000000000 − 0.8755855560 Z^{-1}

A_0= -0.8755855560 C_1= 0.1454909444

$ZH_1(Z)$= -0.8755855560 + 1.0000000000 Z^{-1}

N_0= 0.0856456459 D_0= 1.0000000000

C_0= 0.0856456459 ZH_0= 1.0000000000

$N_n(Z)$= 0.0153999850 + 0.0461999625 Z^{-1} + 0.0462000072 Z^{-2} + 0.0153999999 Z^{-3}

$D_n(Z)$= 1.0000000000 − 1.9899997711 Z^{-1} + 1.5720000267 Z^{-2}− 0.4582999945 Z^{-3}

FIG. 8.2.4 Magnitude and phase response of a simple third-order filter.

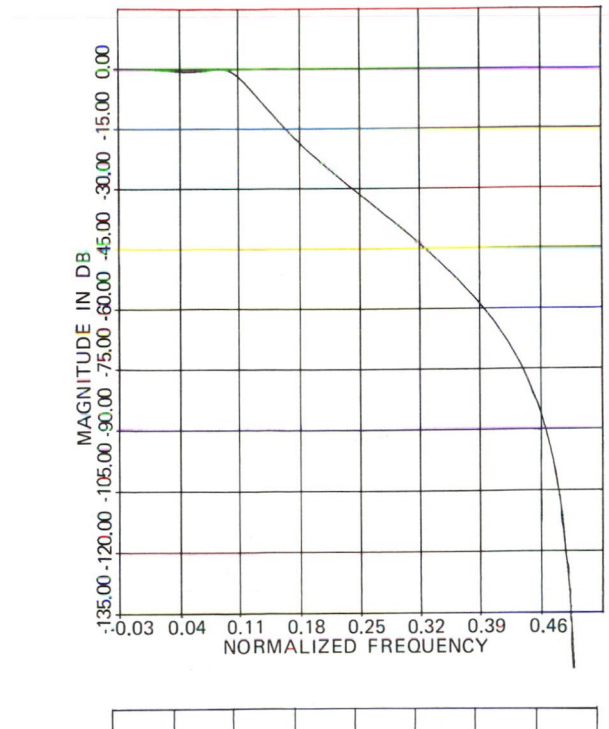

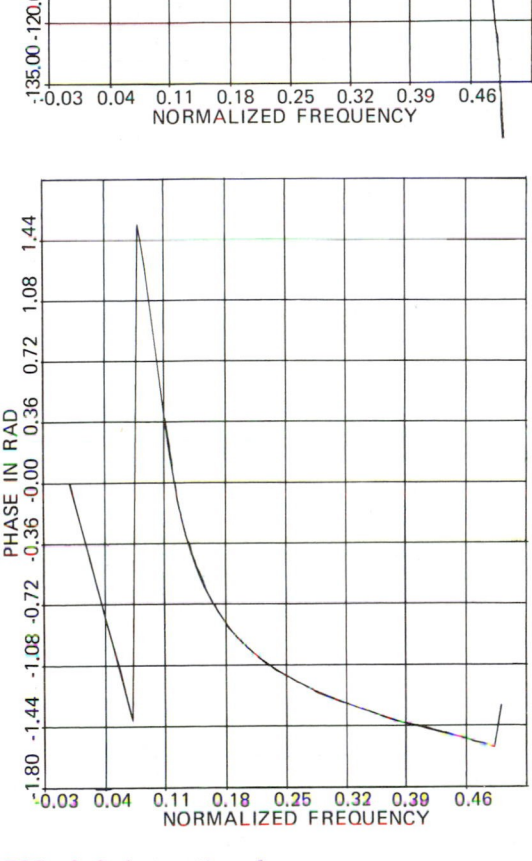

FIG. 8.2.4 (continued)

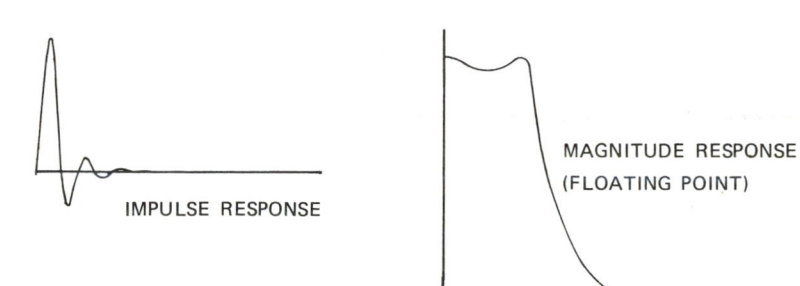

FIG. 8.2.5 Impulse and magnitude frequency response of an example filter.

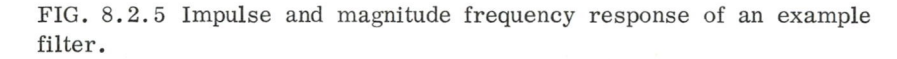

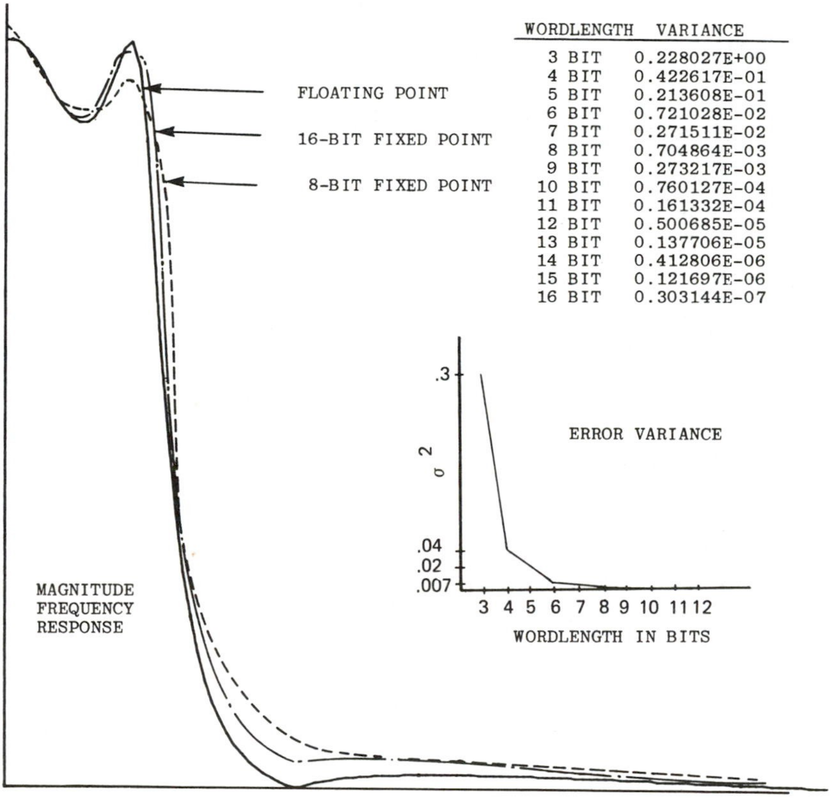

WORDLENGTH	VARIANCE
3 BIT	0.228027E+00
4 BIT	0.422617E-01
5 BIT	0.213608E-01
6 BIT	0.721028E-02
7 BIT	0.271511E-02
8 BIT	0.704864E-03
9 BIT	0.273217E-03
10 BIT	0.760127E-04
11 BIT	0.161332E-04
12 BIT	0.500685E-05
13 BIT	0.137706E-05
14 BIT	0.412806E-06
15 BIT	0.121697E-06
16 BIT	0.303144E-07

FLOATING POINT

16-BIT FIXED POINT

8-BIT FIXED POINT

ERROR VARIANCE

σ^2

.3

.04
.02
.007

3 4 5 6 7 8 9 10 11 12

WORDLENGTH IN BITS

MAGNITUDE
FREQUENCY
RESPONSE

FREQUENCY

FIG. 8.2.6 Fixed-point performance of the example filter.

$$c_i' = \begin{cases} c_3' = c_3 = 0.0154 \\[2ex] c_2' = \dfrac{v_2}{c_2} = 0.14185 \\[2ex] c_1' = \dfrac{v_0}{c_2} = 0.14636 \\[2ex] c_0' = \dfrac{v_0}{c_0} = 0.69236 \end{cases}$$

As a result, the filter found in Fig. 8.4.1 is obtained.

One of the more interesting properties of this class of filters is its orthogonal structure. It has been shown that for the usual definition of inner product over a complex field, it follows that:[†]

1. $D_i(z)$ is orthogonal to z^{-1}, $j = 1, \ldots, i$; that is,

$$(D_i(z), z^{-j}) = (z^j, D_i(z^{-1})) = 0 \qquad (8.4.3)$$

[†]The inner product of $F(z)$ and $K(z)$ is given by

$$<F(z), K(z)> = \int_{-\pi}^{\pi} \frac{R[\exp(j\phi)]F^*(\exp(j\phi)\, d\phi}{2\pi}$$

where

$$R(z) = [D_m(z)D_m(z^{-1})]^{-1}, \qquad R(z) \geq 0$$

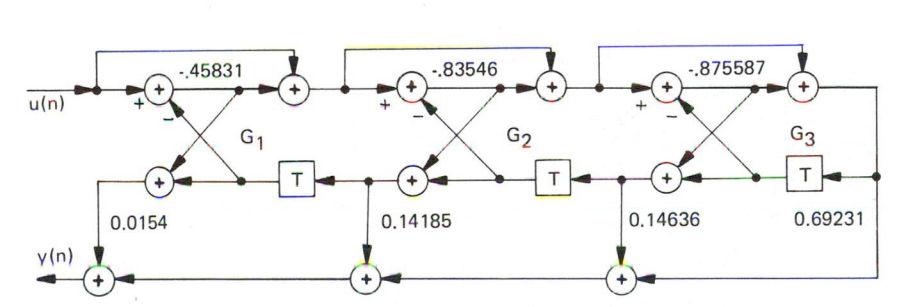

FIG. 8.4.1 One-multiplier architecture for the example filter.

2. $H_i(z)$ are orthogonal to z^j, $j = 1, \ldots, i$; that is,

$$(H_i(z), z^{-j}) = (z^j, H_i(z^{-1})) = 0 \qquad (8.4.4)$$

3. $zH_i(z)$ from an orthogonal set; that is,

$$(zH_i(z), zH_j(z)) = \sigma_i^2 \delta_k(i - j) \qquad (8.4.5)$$

EXAMPLE 8.4.1 Consider $H(z) = N_2(z)/D_2(z) = z^2/(z - 0.5)(z + 0.5) = 1/(1 - 0.25z^{-2})$. Following the analysis method developed for the Example 8.2.2, one notes that

$$D_2(z) = 1 - 0.25z^{-2}$$

$$zH_2(z) = 0.25 + z^{-2}$$

$$a_2 = -0.25$$

$$D_1(z) = \frac{D_2(z) - a_2 zH_2(z)}{1 - a_2^2} = 1$$

$$zH_1(z) = D_1(z^{-1})z^{-1} = z^{-1}$$

$$a_1 = 0$$

$$D_0(z) = D_1(z) - \frac{0zH_1(z)}{1} = 1$$

$$D_2(z) = 1 - 0.25z^{-2} = (z - 0.5)(z + 0.5)$$

$$= \frac{-0.25 + z^{-2}}{D_2(z)} \qquad \frac{z^{-1}}{D_2(z)} \qquad \frac{1}{D_2(z)}$$

The inner products are interpreted in Fig. 8.4.2.

Using elementary linear system theory it can be shown that if $x(n)$ is a white, ergodic random process having a variance σ_x^2, the output error variance is given by

$$\sigma_y^2 = \sigma_x^2 \sum_{n=0}^{N} c_n^2 \sigma_n^2; \qquad \sigma_x^2 = \frac{Q^2}{12} \qquad (8.4.6)$$

INNER PRODUCT TERM	POLE-ZERO PATTERN	INNER PRODUCT VALUE
1. $F_2^*(z)F_2(z)=F_2(z^{-1})F_2(z)=1$		1.0
2. $F_2^*(z)F_1(z)=z/(1-.5z)(1+.5z)$		0.0
3. $F_2^*(z)F_0(z)=z^2/(1-.5z)(1+.5z)$		0.0
4. $F_1^*(z)F_1(z)=1/(1-.5z)(1+.5z)x$ $(z-.5)(z+.5)$		$2/(.75)(1.25)$ $\neq 0.0$
5. $F_1^*(z)F_0(z)=z/(1-.5z)(1+.5z)x$ $(z-.5)(z+.5)$		0.0
6. $F_0^*(z)F_0(z)=1/(1-.5z)(1+.5z)x$ $(z-.5)(z+.5)$		$2/(.75)(1.25)$ $\neq 0.0$

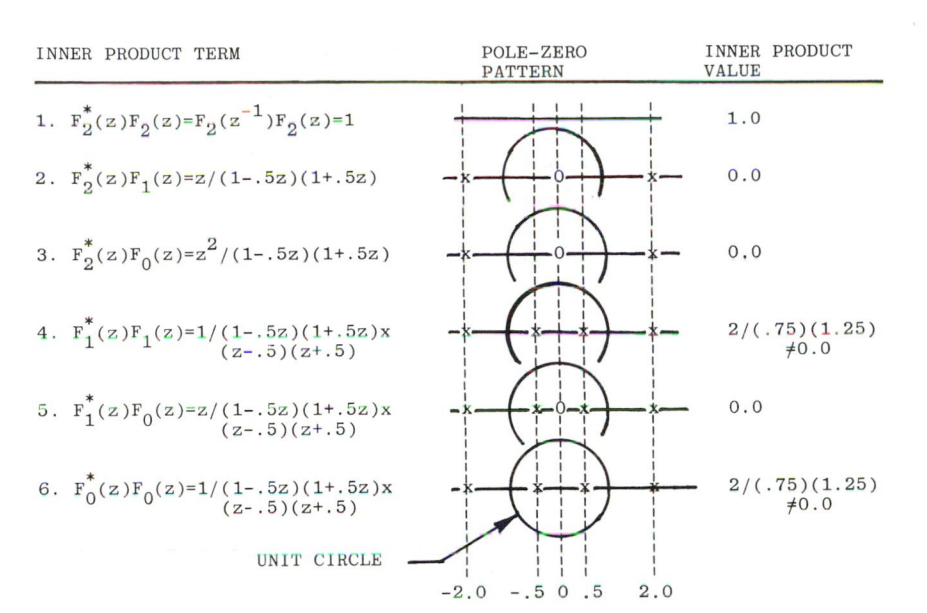

UNIT CIRCLE

-2.0   -.5 0 .5   2.0

FIG. 8.4.2 Analytic and graphical interpretation of the orthogonal properties of a Gray-Markel filter.

where c_i and σ_i^2 have been defined previously [Eq. (8.4.5)]. For a one-multiplier model, $H_i(z)$ and c_i are replaced by $H_i(z) = v_i H_i(z)$ and $c_i' = c_i/v_i$. On substituting into Eq. (8.4.6), the following results:

$$\sigma_y^2 = \left(\sum_{i=1}^N c_1'^2 v_i'^2\right)\sigma_x^2 = \left[\sum_{i=0}^N \left(\frac{c_i}{v_i}\right)^2 (\sigma_i v_i)^2\right]\sigma_x^2$$

$$= \left(\sum_{i=1}^N c_i^2 v_i^2\right)\sigma_x^2 \qquad (8.4.7)$$

Finally, the established orthogonal properties can be used to show that

$$\sigma_{i+1}^2 = \sigma_i^2(1 - a_i^2); \qquad \sigma_N^2 = 1 \qquad (8.4.8)$$

The magnitude of the output error variance for the one-multiplier model can be seen to be sensitive to the choice of σ_i (the sign control parameter). A useful choice algorithm, offered by Gray and Markel (1973), conjectures that a reduced output error variance results from matching the largest statistical uncertainty (i.e., σ_i^2) with the smallest output gain c_i, and vice versa. Defining $|a_L|$ to be the largest $|a_i|$, $0 \leq L \leq N - 1$, and

$$Q_j = \frac{v_j \sigma_j}{V_L \sigma_L}; \quad q_j = \left(\frac{1 + |a_j|}{1 - |a_j|}\right)^{1/2} \tag{8.4.9}$$

the selection process would be given by

Range	Choice	Condition
$i = L - 1, L - 2, \ldots, 0$	$\epsilon_i = \text{sgn}(a_i)$	$Q_{i+1} < 1/q_i$
$i = L, L + 1, \ldots, N - 1$	$\epsilon_i = -\text{sgn}(a_i)$	$Q_{i+1} \geq 1/q_i$
	$\epsilon_i = -\text{sgn}(a_i)$	$Q_i < 1/q_i$
	$\epsilon_i = \text{sgn}(a_i)$	$Q_i \geq 1/q_i$

EXAMPLE 8.4.2 Consider again the third-order example found in Example 8.2.1. Here $a = (a_2, a_1, a_0) = (-0.4583, 0.8355, -0.8756)$ and $c = (c_3, c_2, c_1, c_0) = (0.0154, 0.0768, 0.1455, 0.0836)$. Noting that $|a| \geq |a_i|$, $i = 0, 1, 2$, it follows that $Q_0 = 1$ and $q_0 = \text{SQRT}((1 + |a_0|)/(1 - |a_0|)) = 3.883$. Therefore,

$L = 0$	Choice rule	Result	Q rule	Q_{i+1}
$i = 0$	$i = L; Q_i \geq \dfrac{1}{q_i}$	$\epsilon_0 = \text{sgn}(a_0) = -1$	$\dfrac{Q_i}{Q_{i+1}} = q_i$	$Q_1 = 0.2576$
$i = 1$	$i = L + 1; Q_i < \dfrac{1}{q_i}$	$\epsilon_1 = -\text{sgn}(a_1) = -1$	$\dfrac{Q_i}{Q_{i+1}} = q_i^{-1}$	$Q_2 = 0.8602$
$i = 2$	$i = L + 2; Q_i \geq \dfrac{1}{q_i}$	$\epsilon_2 = \text{sgn}(a_2) = 1$	N.A.	N.A.

Using Eq. (8.4.10), one obtains $\epsilon_0 = -1$, $\epsilon_1 = -1$, $\epsilon_2 = -1$. With this set of ϵ_i's, one can compute the set v to be $v = (1, 1.4583, 0.2399, 0.4500)$ with $c' = (0.0154, 0.05269, 0.60635, 0.1903)$. Structurally, this filter is similar to that found in Fig. 8.2.1, which was based on $\epsilon_i = 1$ for all i. In the new structure, new coefficients c_i are induced by $\epsilon_i = -1$. The resulting filter will provide the user with a low-sensitivity filter having a lower value of σ_y^2 when compared to other admissible realizations. In addition, the orthogonal properties of this filter make it attractive for use in certain speech processing applications (Gray and Markel, 1973).

EXAMPLE 8.4.3 The problem considered in Example 8.4.2 represents the trial data used in the general purpose software reported in Fig. 8.4.3 and in the appendix to this chapter. The software accepts the coefficients of H(z) as input data and returns the optimal one–multiplier coefficients to the user. The error variance is also computed and displayed.

8.5 STATE–SPACE REPRESENTATIONS

It has been previously argued that the second-order section is a fundamental building block for digital filters. The second-order ladder filter will now be studied with respect to its state-variable format. A general flow diagram for a general ladder filter is given in Fig. 8.5.1. It is identified by the four–tuple (A, b, c, d). The resolvent matrix $(zI - A)^{-1}b$ satisfies

$$A = \begin{bmatrix} -a_1 a_2 & | & -a_3 \\ \hline a_1 a_3 a_4 & | & -a_2 a_4 \end{bmatrix}$$

$$b = \begin{bmatrix} ba_2 \\ ba_3 a_4 \end{bmatrix}$$

$$c = (c_1,\ c_2) \tag{8.5.1}$$

$$(zI - A)^{-1} = \frac{\begin{bmatrix} z + a_2 a_4 & a_3 \\ -a_1 a_3 a_4 & z + a_1 a_2 \end{bmatrix}}{z^2 + (a_1 a_2 + a_2 a_4)z + a_1 a_3 a_4 + a_1 a_2 a_4} \tag{8.5.2}$$

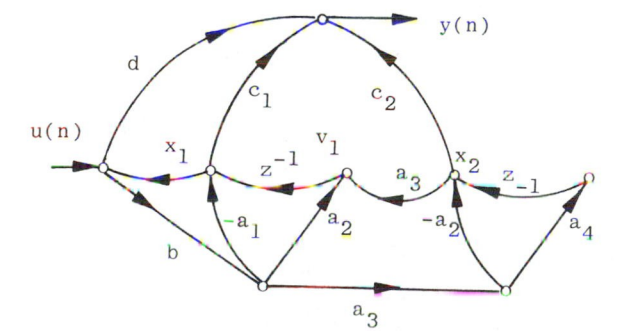

FIG. 8.5.1 Data flow diagram of a simple second–order filter.

and

$$(zI - A)^{-1}b = \frac{(zba_2 + a_2^2 a_4 b + a_3^2 a_4 b, \; za_3 a_4 b)^T}{\Delta(z)}$$

$$= [H_1(z), \; H_2(z)]^T \tag{8.5.3}$$

with $\Delta(z)$ denoting the determinant of $(zI - A)$. Using Mason's gain formula as a check, one obtains from the input to state locations x_1 and x_2 the following:

	To x_1	To x_2
Feedforward	$M_1 = z^{-1}a_2 b$ $M_2 = a_3^2 a_4 bz^{-2}$	$M_2 = a_3 a_4 bz^{-1}$
Individual loops	$\Delta_{11} = -a_1 a_2 z^{-1}$ $\Delta_{12} = -a_1 a_3^2 a_4 z^{-1}$ $\Delta_{13} = -a_2 a_4 z^{-1}$	Same
Dual nontouching loops	$\Delta_{21} = (-a_1 a_2 z^{-1})(-a_2 a_4 z^{-1})$ $\quad = a_1 a_2^2 a_4 z^{-2}$	Same
Characteristic equations	$\Delta(z) = 1 - (\Delta_{11} + \Delta_{12} + \Delta_{13}) + \Delta_{21}$ (as a check) $\quad = z^2 + (a_1 a_2 + a_2 a_4)z + a_1 a_3^2 a_4 + a_1 a_2^2 a_4$	

Induced loops

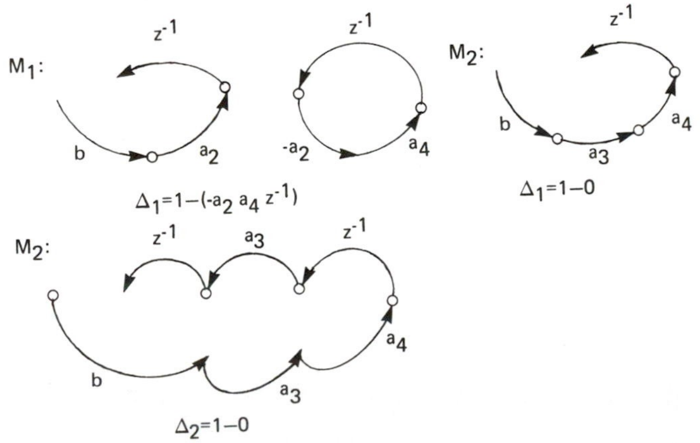

M_1:

$\Delta_1 = 1 - (-a_2 a_4 z^{-1})$

$\Delta_1 = 1 - 0$

M_2:

$\Delta_2 = 1 - 0$

Therefore, $H_1(z)$ and $H_2(z)$ satisfy

$$
H_1(z) = \frac{M_1 \Delta_1 + M_2 \Delta_2}{\Delta(z)}
$$

$$
= \frac{a_2^2 a_4 bz^{-2} + a_3^2 a_4 bz^{-2} + a_2 bz^{-1}}{\Delta(z)} \tag{8.5.4}
$$

$$
H_2(z) = \frac{a_3 a_4 bz^{-1}}{\Delta(z)}
$$

which is in complete agreement with the transfer functions derived using the resolvent matrix approach. Finally, the input-output transfer function is given by $H(z)$, where

$$
H(z) = d + c\underline{H}(z); \quad \underline{H}(z)^T = [H_1(z),\ H_2(z)]
$$

$$
= 1 + \frac{c_1(ba_2 + a_3 a_4 b)z + c_2(a_2^2 a_4 b + a_3^2 a_4 b)}{\Delta(z)}
$$

In order to achieve an overall transfer function of the form $H(z) = z^2/(z^2 - 1.272z + 0.81)$, for example, one notes that

$$
H(z) = \frac{z^2}{z^2} - 1.272z + 0.81 = 1 + \frac{2.272z - 0.81}{z^2 - 1.272z + 0.81} \tag{8.5.6}
$$

A nonunique set of parameters (i.e., a's, b's, and c's) will satisfy this design constraint. One set is given by $b = 1$, $a_1 = 0.9133$, $a_2 = a_3 = 0.9$, $a_4 = 0.5$, $c_1 = 1.683$, $c_2 = -1$.

A ladder filter having a fixed structure was developed in Section 8.2 and referred to as the Gray-Markel filter. The general form of this class of filter is abstracted in Fig. 8.2.1. A second-order multiplier design is presented in Fig. 8.5.2. Its analysis provides the data given in Fig. 8.5.3. Using Mason's gain formula, one obtains

To x_2	To x_1	To b_0
$M_1 = \alpha_1$	$M_1 = \alpha_1$	$M_1 = 1$
$\Delta_1 = 1 + \alpha_0 z^{-1}$	$M_2 = z^{-1}$	$\Delta_1 = 1$

(continued)

To x_2	To x_1	To b_0
$M_2 = \alpha_0 z^{-1}$	$\Delta_1 = 1$	
$\Delta_2 = 1$	$\Delta_2 = 1$	
$M_3 = z^{-2}$		
$\Delta_3 = 1$		

$$H_2(z) = \frac{\alpha_1 + (\alpha_1\alpha_0 + \alpha_0)z^{-1} + z^{-2}}{\Delta(z)} \qquad H_1(z) = \frac{\alpha_0 + z^{-1}}{\Delta(z)} \qquad H_0(z) = \frac{1}{\Delta(z)}$$

The state-variable form of this filter can be derived in the context of the material found in Fig. 8.5.3. The states can be defined in the following manner:

$$x_2(n + 1) = -\alpha_0 x_2(n) - \alpha_1 x_1(n) + u(n)$$

$$x_1(n + 1) = x_2(n) - \alpha_0\alpha_1 x_1(n) + \alpha_0 u(n) \qquad\qquad (8.5.9)$$

or

$$x(n + 1) = \begin{bmatrix} -\alpha_0\alpha_1 & 1 \\ -\alpha_1 & -\alpha_0 \end{bmatrix} x(n) + \begin{bmatrix} \alpha_0 \\ 1 \end{bmatrix} u(n)$$

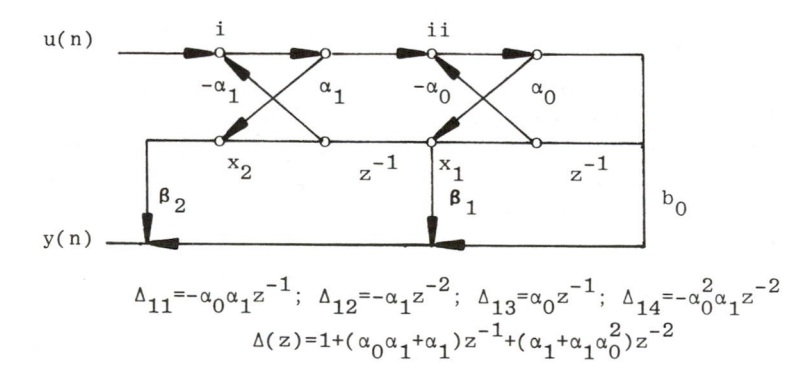

$$\Delta_{11} = -\alpha_0\alpha_1 z^{-1}; \quad \Delta_{12} = -\alpha_1 z^{-2}; \quad \Delta_{13} = \alpha_0 z^{-1}; \quad \Delta_{14} = -\alpha_0^2\alpha_1 z^{-2}$$

$$\Delta(z) = 1 + (\alpha_0\alpha_1 + \alpha_1)z^{-1} + (\alpha_1 + \alpha_1\alpha_0^2)z^{-2}$$

FIG. 8.5.2 Second-order filter realized using the two-multiplier architecture.

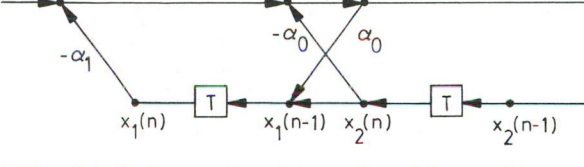

FIG. 8.5.3 Recursive data paths of the architecture displayed in Fig. 8.5.2.

Comparing the A matrix to that belonging to the general ladder filter, it follows that $b = 1$, $a_1 = \alpha_0$, $a_2 = \alpha_1$, and $a_3 = a_4 = 1$. Under this hypothesis, the resolvent form becomes

$$(zI - A)^{-1}b = \begin{bmatrix} z + \alpha_0\alpha_1 & -1 \\ \alpha_1 & z + \alpha_0 \end{bmatrix}^{-1} \begin{bmatrix} \alpha_0 \\ 1 \end{bmatrix} z = \begin{bmatrix} H_1(z) \\ H_2(z) \end{bmatrix}$$

$$= \frac{\begin{bmatrix} z\alpha_0(z + \alpha_0) + z \\ -\alpha_1 z + z^2 + \alpha_0 z \end{bmatrix}}{z^2 + (\alpha_0\alpha_1 + \alpha_0)z + \alpha_0^2\alpha_1 + \alpha_1^2} \qquad (8.5.10)$$

and the definition of $H_1(z)$ and $H_2(z)$ follows.

The computation of the coefficients α_1 and α_2, for this special architecture is accomplished using the formulas found in Section 8.2. Here it may be recalled that if $H(z) = N(z)/D(z)$, such that

$$N(z) = \sum_{i=0}^{n} c_i z^{-i}; \quad D(z) = \sum_{i=0}^{n} a_i z^{-1} \qquad (8.5.11)$$

then for $N_j(z) = \sum_{i=0}^{j} c_{ji}z^{-i}$ and $D_j(z) = \sum_{i=0}^{j} a_{ji}z^{-i}$, it follows that

$$\alpha_{j-1} = a_{ji}, \quad \beta_j = c_{jj} \text{ with } \beta_0 = c_{00}$$

$$D_{j-1}(z) = \frac{D_j(z) - a_{j-1}B_j(z)z}{1 - \alpha_{j-1}^2}; \quad N_{j-1}(z) = N_j(z) - zD_j(z)\beta_j$$

where $zD_j(z) = D_j(1/z)z^{-j}$.

EXAMPLE 8.5.1 The ladder filter realization for the transfer function
$H(z) = z^2/(z^2 - 1.2726z + 0.81) = 1/(1 - 1.2726z^{-1} + 0.81z^{-2})$. Then:

1. $\alpha_1 = 0.81$; $\beta_2 = 0$; $j = 2$

2. $zB_2(z) = D_2(1/z)z^{-2} = z^{-2} - 1.2726z^{-1} + 0.81$

3. $N_1(z) = N_2(z) - zB_2(z)\beta_2 = N_2(z)$

4. $D_1(z) = [D_2(z) - \alpha_1 zB_2(z)]/(1 - \alpha_1^2) = 1 - 0.703z^{-1}$

5. $\alpha_0 = -0.703$; $\beta_1 = 0$; $j = 1$

6. $\beta_0 = 1$

Therefore, upon direct substitution, $\Delta(z)$ and $H_i(z)$ becomes

$$\Delta(z) = 1 + [0.81(-0.703) - 0.703]z^{-1} + [0.81 + 0.81(-0.703)^2 z^{-2}]$$

$$= 1 - 1.2726z^{-1} + 0.81z^{-2}$$

$$H_2(z) = \frac{0.81 - 1.2726z^{-1} + z^{-2}}{\Delta(z)} \quad \beta_2 = 0$$

$$H_1(z) = \frac{0.703 + z^{-1}}{\Delta(z)} \quad \beta_1 = 0$$

$$H_0(z) = \frac{1}{\Delta(z)} \quad \beta_0 = \frac{z^2}{z^2 - 1.2726z + 0.81}$$

which is known to be the desired result.

BIBLIOGRAPHY

Antoniou, A. (1979), Digital Filters: Analysis and Design, McGraw-Hill,
    New York

Gray, A. H., Jr., and J. D. Markel (1973), Digital Lattice and Ladder
    Synthesis, IEEE Trans. Audio Electroacoust., AU-21, December, pp.
    491-500.

Mitra, S. K., and R. J. Sherwood (1973), Digital Ladder Networks, IEER
    Trans. Audio Electroacoust., AU-21, February, pp. 30-36.

APPENDIX Computer–Aided Gray–Markel
Filter Design

A FORTRAN–based ladder filter design and analysis computer–aided design
program has been added to the comprehensive filter design package discussed in the appendix to Chapter 13. The source code to the Gray–Markel
design and analysis software is presented below, followed by the data used
in the analysis of Example 8.5.1. Here

$$A(1) = \alpha_1, \qquad C(2) = \beta_2$$

$$A(0) = \alpha_2, \qquad C(1) = \beta_1$$

$$C(0) = \beta_0$$

The matrix of coefficients following the listing of N_i, α_i, β_i is interpreted
as follows (see Section 8.2):

row 1 = coefficients of $zH_i(z)$

row 2 = coefficients of $N_{i-1}(z)$

row 3 = coefficients of $D_{i-1}(z)$

The values of q and Q [Eq. (8.4.9)] are listed next. The optimal sign
choice parameters are shown to be $\epsilon_1 = \epsilon_2 = 1.0$. The values of v_i and c_i
[Eq. (8.4.1)] are then listed. Finally, the error analysis discussed in the
appendix to Chapter 13 is performed and outputted in the form of a set of
parameters labeled SIGMA. Last in this appendix we give an analysis of the
filter presented in Example 8.2.1.

Source Listing of Ladder Design Software

```
C           PROGRAM TO COMPUTE THE FILTER COEFFICIENTS, A(I),
C           AND THE LINEAR COMBINATION COEFFICIENTS, C(I),
C           FOR A TWO-MULTIPLIER REALIZATION OF A GRAY-
C           MARKEL LADDER FILTER. ALSO COMPUTES A SET OF
C           MODIFIED COEFFICIENTS FOR A ONE-MULTIPLIER
C           REALIZATION AND AN OUTPUT ERROR VARIANCE
C           FACTOR, SIGMA(Y) SQUARED.
C
C           JEFFREY W. MARSHALL
C
            DIMENSION A(16),C(16),PNUM(16),REV(16),SAVER(16),DENOM(16)
            DIMENSION BIGQ(16),SMALLQ(16),E(16),V(16),CPRIME(16)
            DIMENSION SIGMA(16)
            INTEGER*2 NAME(10)
            DO 5 I=1,16
            E(I)=1
```

```
5          CONTINUE
C
C          ENTRY OF NUMERICAL DATA: FILTER ORDER, FILTER COEFFICIENTS
C
           TYPE 10
10         FORMAT (/, ' ENTER FILTER ORDER, N')
           READ (1,*) N
           ISAVE=N
           JUMP=N-1
           TYPE 20
20         FORMAT (/, ' ENTER NUMERATOR COEFF. IN DECREASING
      C ORDER OF Z(-N)')
           READ (1,*) (PNUM(I), I=1,N+1)
           TYPE 30
30         FORMAT (/, ' ENTER DENOMINATOR COEFF. IN DECREASING
      C ORDER OF Z(-N)')
           READ (1,*) (DENOM(I), I=1,N+1)
           TYPE 31
31         FORMAT (/, ' PLEASE ENTER YOUR NAME')
           READ (1,39) NAME
39         FORMAT (10A2)
           WRITE (2,42) NAME
42         FORMAT (' ',10A2)
           WRITE (2,32) N
32         FORMAT (/, ' FILTER ORDER:',I3,/)
           WRITE (2,33)
33         FORMAT (' THE INPUT NUMERATOR COEFFICIENTS ARE:',/)
           DO 35 I=1,N+1
           WRITE (2,34) PNUM(I)
34         FORMAT (F10.4)
35         CONTINUE
           WRITE (2,36)
36         FORMAT (/, ' THE INPUT DENOMINATOR COEFFICIENTS ARE:',/)
           DO 38 I=1,N+1
           WRITE (2,37) DENOM(I)
37         FORMAT (F10.4)
38         CONTINUE
C
C          CALCULATE MODIFIED FILTER COEFFICIENTS AND LINEAR
C          COMBINATION COEFFICIENTS (A(N), C(N))
C
40         A(N)=DENOM(N+1)
           C(N+1)=PNUM(N+1)
           WRITE (2,50) N, JUMP, A(N), N, C(N+1)
50         FORMAT (/, ' FOR N=',I2,' A(',I1,')=',F10.7,5X,
      C ' C(',I1,')=',F10.7,/)
           M=N+1
           DO 60 I=1,N+1
           REV(M)=DENOM(I)
           SAVER(M)=REV(M)
           M=M-1
60         CONTINUE
           WRITE (2,*) (REV(I), I=1,N+1)
           DO 70 I=1,N+1
           REV(I)=REV(I)*C(N+1)
           SAVER(I)=SAVER(I)*A(N)
70         CONTINUE
           DO 80 I=1,N+1
           PNUM(I)=PNUM(I)-REV(I)
           DENOM(I)=(DENOM(I)-SAVER(I))/(1-(A(N)**2))
80         CONTINUE
           WRITE (2,*) (PNUM(I), I=1,N+1)
           WRITE (2,*) (DENOM(I), I=1,N+1)
```

```
            N=N-1
            JUMP=JUMP-1
            IF(N)  90,90,40
90          C(N+1)=PNUM(N+1)
            WRITE (2,100) N, N, DENOM(N+1), N, C(N+1)
100         FORMAT (/, ' FOR N=',I2,' ZH(',I1,')=',F10.7,5X,
     C     ' C(',I1,')=',F10.7,/)
C
C           CALCULATE OPTIMAL SIGN CHOICE AND MODIFIED TAP
C           PARAMETERS, C'(N), FOR THE ONE MULTIPLIER STRUCTURE
C
            N=ISAVE
            INDEX=1
            DO 101 I=1,N
            SMALLQ(I)=SQRT((1+ABS(A(I)))/(1-ABS(A(I))))
101         CONTINUE
            ALARGE=0.0
            DO 105 I=1,N
            ALARGE=AMAX1(ALARGE,ABS(A(I)))
105         CONTINUE
            DO 110 I=1,N
            IF (ABS(ALARGE).EQ.ABS(A(I))) GO TO 120
            INDEX=INDEX+1
110         CONTINUE
120         L=INDEX
            BIGQ(L)=1.0
            IF ((L.EQ.1).OR.(L.EQ.N)) GO TO 121
            CALL CHOIC3(SMALLQ,BIGQ,E,A,L,N)
            CALL CHOIC2(SMALLQ,BIGQ,E,A,L,N)
            GO TO 151
121         IF (L.EQ.1) GO TO 122
            CALL CHOIC1(SMALLQ,BIGQ,E,A,L,N)
            CALL CHOIC3(SMALLQ,BIGQ,E,A,L,N)
            GO TO 151
122         CALL CHOIC2(SMALLQ,BIGQ,E,A,L,N)
            GO TO 151
151         WRITE (2,152)
152         FORMAT (' THE VALUES OF LITTLE Q ARE:',/)
            DO 154 I=1,N
            WRITE (2,153) SMALLQ(I)
153         FORMAT (F10.7)
154         CONTINUE
            WRITE (2,155)
155         FORMAT (/, ' THE VALUES OF CAPITAL Q ARE:',/)
            DO 165 I=1,N
            WRITE (2,160) BIGQ(I)
160         FORMAT (F10.7)
165         CONTINUE
            WRITE (2,170)
170         FORMAT (/, ' THE OPTIMAL SIGN CHOICE PARAMETERS ARE:',/)
            DO 180 I=1,N
            WRITE (2,175) E(I)
175         FORMAT (F6.1)
180         CONTINUE
            GO TO 250
C
C           CALCULATE THE V(N)'S
C
250         V(N+1)=1
            V(N)=(1+A(N)*E(N))
            DO 260 I=N-1,1,-1
            V(I)=V(I+1)*(1+A(I)*E(I))
260         CONTINUE
            WRITE (2,262)
```

```
262        FORMAT (/, ' THE VALUES OF THE MULTIPLIERS, V(N), ARE:',/)
           DO 266 I=1,N+1
           WRITE (2,265) V(I)
265        FORMAT (F10.7)
266        CONTINUE
C
C          CALCULATE THE CPRIME(N)'S
C
           CPRIME(N+1)=C(N+1)
           DO 270 I=N,1,-1
           CPRIME(I)=C(I)/V(I)
270        CONTINUE
           WRITE (2,275)
275        FORMAT (/,' THE MODIFIED ONE MULT. PARAMETERS ARE:',/)
           DO 285 I=1,N+1
           WRITE (2,280) CPRIME(I)
280        FORMAT (F10.7)
285        CONTINUE
C
C          CALCULATE THE VALUES OF SIGMA(I)
C
           SIGMA(N+1)=1.0
           DO 290 I=N,1,-1
           SIGMA(I)=SIGMA(I+1)/(1-(A(I)*A(I)))
290        CONTINUE
           WRITE (2,292)
292        FORMAT (/, ' THE VALUES OF SIGMA(I) ARE:',/)
           DO 300 I=N+1,1,-1
           WRITE (2,295) I-1, SIGMA(I)
295        FORMAT (' SIGMA(',I1,')=',F11.7)
300        CONTINUE
C
C          CALCULATE THE OUTPUT ERROR VARIANCE
C
           SIGMAY=0.0
           DO 310 I=N+1,1,-1
           SIGMAY=SIGMAY+SIGMA(I)*(C(I)**2)
310        CONTINUE
           WRITE (2,315)
315        FORMAT (/, ' THE OUTPUT ERROR VARIANCE IS:',/)
           WRITE (2,320) SIGMAY
320        FORMAT (' SIGMA(Y)=',F10.7,' *SIGMA(X)')
           STOP
           END
C
C
C
           SUBROUTINE CHOIC1(SMALLL,BIGL,EL,AL,M,K)
           DIMENSION SMALLL(K+1),BIGL(K+1),EL(K+1),AL(K+1)
           IF (BIGL(M).GT.1/SMALLL(M)) GO TO 10
           EL(M)=-SIGN(EL(M),AL(M))
           GO TO 20
10         EL(M)=SIGN(EL(M),AL(M))
20         RETURN
           END
C
C
C
           SUBROUTINE CHOIC2(SMALLL,BIGL,EL,AL,M,K)
           DIMENSION SMALLL(K+1),BIGL(K+1),EL(K+1),AL(K+1)
           DO 30 I=M,K
           IF (BIGL(I).LT.1/SMALLL(I)) GO TO 15
```

```
          EL(I)=SIGN(EL(I),AL(I))
          BIGL(I+1)=BIGL(I)/SMALLL(I)
          GO TO 30
15        EL(I)=-SIGN(EL(I),AL(I))
          BIGL(I+1)=BIGL(I)*SMALLL(I)
30        CONTINUE
          RETURN
          END
C
C
C
          SUBROUTINE CHOIC3(SMALLL,BIGL,EL,AL,M,K)
          DIMENSION SMALLL(K+1),BIGL(K+1),EL(K+1),AL(K+1)
          DO 30 I=M-1,1,-1
          IF (BIGL(I+1).LT.1/SMALLL(I)) GO TO 15
          EL(I)=-SIGN(EL(I),AL(I))
          BIGL(I)=BIGL(I+1)/SMALLL(I)
          GO TO 30
15        EL(I)=SIGN(EL(I),AL(I))
          BIGL(I)=BIGL(I+1)*SMALLL(I)
30        CONTINUE
          RETURN
          END
```

Output Listing for the Second-Order Detailed in
Example 8.5.1

FILTER ORDER: 2

THE INPUT NUMERATOR COEFFICEIENTS ARE:

     1.0000
     0.0000
     0.0000

THE INPUT DENOMINATOR COEFFICIENTS ARE:

     1.0000
    -1.2726
     0.8100

FOR N= 2 A(1)= 0.8100000   C(2)= 0.0000000

     0.8100000   -1.272600   1.000000
     1.000000   0.0000000   0.0000000
     1.000000   -0.7030938   0.0000000

FOR N= 1 A(0)=-0.7030938   C(1)= 0.0000000

    -0.7030938   1.000000
     1.000000   0.0000000
     1.000000   0.0000000

FOR N= 0 ZH(0)= 1.0000000   C(0)= 1.0000000

THE VALUES OF LITTLE Q ARE:

2.3950229
3.0864732

THE VALUES OF CAPITAL Q ARE:

0.3239944
1.0000000

THE OPTIMAL SIGN CHOICE PARAMETERS ARE:

   1.0
   1.0

THE VALUES OF THE MULTIPLIERS, V(N), ARE:

0.5374002
1.8100001
1.0000000

THE MODIFIED ONE MULT. PARAMETERS ARE:

1.8608106
0.0000000
0.0000000

THE VALUES OF SIGMA(I) ARE:

SIGMA(2)= 1.0000000
SIGMA(1)= 2.9078219
SIGMA(0)= 5.7505579

THE OUTPUT ERROR VARIANCE IS:

SIGMA(Y)= 5.7505579 *SIGMA(X)

Output Listing for the Third–Order Filter Detailed in
Example 8.2.1

FILTER ORDER: 3
THE INPUT NUMERATOR COEFFICIENTS ARE:
   0.0154
   0.0462
   0.0462
   0.0154

THE INPUT DENOMINATOR COEFFICIENTS ARE:
   1.0000
  -1.9900
   1.5720
  -0,4583

FOR N= 3 A(2)=-0.4583000   C(3)= 0.0154000
  -0.4583000   1.572000   -1.990000   1.000000
   2.2457819E-02   2.1991199E-02   7.6846004E-02   0.0000000
   1.000000   -1.607108   0.8354627   0.0000000

FOR N= 2 A(1)= 0.8354627   C(2)= 0.0768460
   0.8354627   -1.607108   1.000000
  -4.1744146E-02   0.1454910   0.0000000
   1.000000   -0.8755870   0.0000000

```
FOR N= 1 A(0)=-0.8755870        C(1)= 0.1454910
  -0.8755870         1.000000
   8.5645869E-02   0.0000000
   1.000000        0.0000000

FOR N= 0 ZH(0)= 1.0000000       C(0)= 0.0856459
THE VALUES OF LITTLE Q ARE:
3.8827167
3.3399546
1.6407561

THE VALUES OF CAPITAL Q ARE:
1.0000000
0.2575516
0.8602108

THE OPTIMAL SIGN CHOICE PARAMETERS ARE:
 -1.0
 -1.0
 -1.0

THE VALUES OF THE MULTIPLIERS V(N) ARE:
0.4500373
0.2399448
1.4583000
1.0000000

THE MODIFIED ONE MULT. PARAMETERS ARE:
0.1903084
0.6063520
0.0526956
0.0154000

THE VALUES OF SIGMA(I) ARE:
SIGMA(3)=  1.0000000
SIGMA(2)=  1.2658851
SIGMA(1)=  4.1916437
SIGMA(0)= 17.9631023

THE OUTPUT ERROR VARIANCE IS:
SIGMA(Y)= 0.2282030 *SIGMA(X)
```

9
Wave Filters

9.1 INTRODUCTION

Digital signal processing scientists have found wave filters to be an extreme-ly interesting class of systems. They represent the discrete counterpart of a resistively terminated analog wave filter, which is, in turn, a model of an LC transmission line. The class title "wave filter" is derived from the transmission-line analogy, where the wave propagates along the line. This class of filter is known to possess a low sensitivity to parameter variations in the classical sense. However, to interpret properly the concept of low sensitivity, specific filter attributes must be developed.

9.2 PROPERTIES

The wave filter belongs to a class of systems known as <u>maximal available power (MAP) filters</u>. In terms of the model shown in Fig. 9.2.1, it can be shown that the maximum power delivered to the load is given by

$$P_0 = \frac{|V_s|^2}{4R_1} \tag{9.2.1}$$

and the power dissipated in the load resistor is

$$P_L = \frac{|V_2|^2}{R_L} \tag{9.2.2}$$

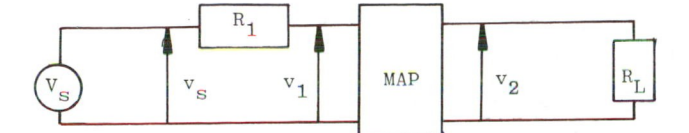

FIG. 9.2.1 Maximum available power filter and peripheral circuitry.

The MAP transfer function, denoted $t(j\omega)$, is sometimes referred to as the power insertion ratio, and

$$|t(j\omega)|^2 = \frac{P_L}{P_0} = \frac{4R_1}{R_L} \frac{V_2(\omega)}{V_s(\omega)} \tag{9.2.3}$$

Defining $T(s)$ to be the usual voltage transfer function [i.e., $T(s) = V_1(s)/V_s(s)$], it follows that $t(s)$ is given by

$$t(s) = 2\sqrt{\frac{R_1}{R_L}}\,T(s) \tag{9.2.4}$$

9.3 ARCHITECTURE

The architecture of the MAP wave filter is somewhat unique and will therefore require some additional explanation. The simplest filter, called a two-port adapter, is shown in Fig. 9.3.1. The port inputs are composed of two distinct waveforms referred to as incident and reflected waves. The quantities denoted R_i represent port impedances. The defining port voltage-current relationship is given by

$$a_i = v_i + R_i i_i; \quad b_i = v_i - R_i i_i \tag{9.3.1}$$

Since the wave filter, as presented, is composed of an incident and a reflected wave, a normalized relationship between the filter's reflected power and insertion loss can be shown to satisfy

$$|t(j\omega)|^2 + |r(j\omega)|^2 = 1 \tag{9.3.2}$$

This MAP relationship provides the reader with an interesting interpretation. Returning to the maximum available power argument for $R_1 = R_2 = 1$, it can be noted that maximum power will occur when the reflected power $|r(j\omega)|^2$ is zero [alternatively, $|t(j\omega)|^2 = 1$ for all ω]. When $|t(j\omega)|^2 = 1$, it follows that $t(j\omega) = \pm 1$ for some ω. For example, the local maximum

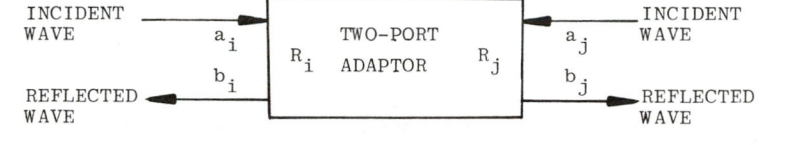

FIG. 9.3.1 Two-port adaptor showing location of input-output incident and reflected waves.

[i.e., $t(j\omega)$] for Butterworth and Chebyshev lowpass filters occurs at the frequencies shown in Fig. 9.3.2.

If p_i is a filter parameter, it has been shown that locally about the MAP frequencies,

$$\left.\frac{\partial |t(j\omega)|}{\partial P_i}\right|_{j\omega^*} = 0; \quad \omega^* = \text{MAP frequency} \qquad (9.3.3)$$

That is, locally about the frequencies where maximum power transfer occurs, small variations in parameter values can be tolerated without altering the MAP policy. Again, this insensitivity is desired since it would suggest that an equivalent digital filter might be insensitive to roundoff errors.

9.4 DESIGN MODELS

Analog wave filters are designed with specific frequency domain objectives. The analog wave filters are designed to make extensive use of inductors

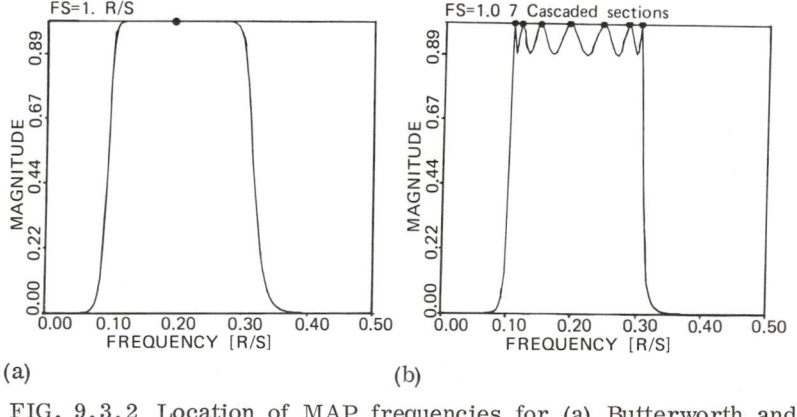

(a)   (b)

FIG. 9.3.2 Location of MAP frequencies for (a) Butterworth and (b) Chebyshev filters.

and capacitors. If we assume that an impedance, say R_ϕ, is capacitive, then the voltage–current relationship at steady state is given by $V = IZ(s)$, where $Z(s)$ denotes the complex impedance $Z(s) = R/s$, $R = 1/C$. Similarly, if the input impedance is inductive, then $V = IZ(s)$, where $Z(s)$ now is given by $Z(s) = sR$ (where $R = L$). Finally, if $Z(s)$ is resistive, then $Z(s) = R$. These three impedances can be compactly summarized as follows:

$$Z(s) = s^\phi R_\phi = \begin{cases} \phi = 1 & \text{if R is inductive} \\ \phi = 0 & \text{if R is resistive} \\ \phi = -1 & \text{if R is capacitive} \end{cases} \tag{9.4.1}$$

Once a continuous filter is realized, it must be translated into a discrete form. Since frequency domain fidelity is desired, a bilinear transform would normally be used. In particular, $s = 2(z - 1)/T(z + 1)$ or $z = [(2/T + s]/[(2/T) - s]$. Substituting these transform definitions into the wave equations, one obtains

$$\left. \begin{array}{l} A = V + IR \\ B = V - IR \\ V = IZ(s) \end{array} \right\} \quad \frac{B}{A} = \frac{IZ(s) - IR}{IZ(s) + IR} = \frac{Z(s) - R}{Z(s) + R} \tag{9.4.2}$$

Replacing the Laplace operators with its bilinear equivalent and the defining R by $R = (2/T)^\phi R_\phi$, it follows that

$$\frac{B}{A} = \frac{s^\phi - (2/T)^\phi}{s^\phi + (2/T)^\phi} = \begin{cases} -z^{-1} & \text{if } \phi = 1 \\ 0 & \text{if } \phi = 0 \\ z^{-1} & \text{if } \phi = -1 \end{cases} \tag{9.4.3}$$

These fundamental impedance forms are summarized in Table 9.4.1.

TABLE 9.4.1 Impedance Representations and Values

Value	Network	Element	Discrete Value	Representation	Equation
1	L	L	2L/T	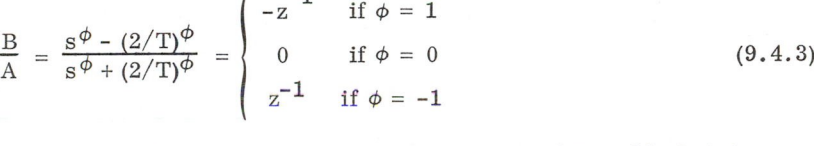	(9.4.4)
0	R	R	R		(9.4.5)
-1	C	C	T/2C	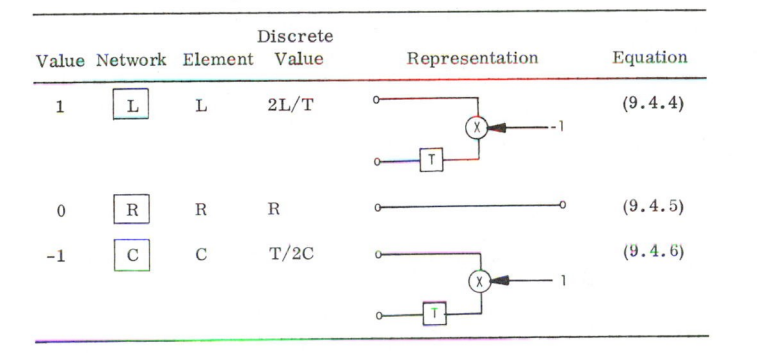	(9.4.6)

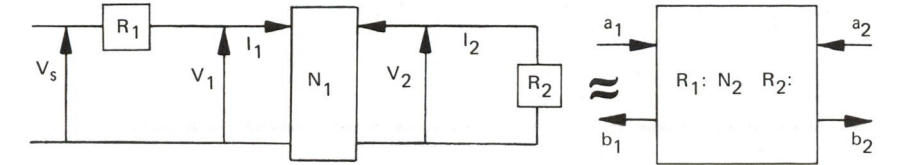

FIG. 9.4.1 Diagram showing the equivalence between a MAP system and its external parameters a and b.

One last network property will be developed at this time in terms of the simple networks shown in Fig. 9.4.1. Direct analysis produces the data found in this figure, where

$$V_1 = V_s - I_1 R_1 = a_1 + b_1; \quad V_2 = -I_2 R_2 = a_2 + b_2$$

$$I_1 = \frac{a_1 - b_1}{R_1}; \quad\quad\quad I_2 = \frac{a_2 - b_2}{R_2} = \frac{-V_2}{R_2} \quad\quad (9.4.7)$$

Therefore,

$$a_1 = \frac{V_1 + I_1 R_1}{2}; \quad a_2 = V_2 + I_2 R_2 = 0$$

$$b_1 = \frac{V_1 - I_1 R_1}{2}; \quad b_2 = -I_2 R_2 = V_2 \quad\quad (9.4.8)$$

or

$$\frac{b_2}{a_1} = \frac{2V_2}{V_1 + I_1 R_1} = \frac{2V_2}{V_s} = 2T(s) \quad\quad (9.4.9)$$

where T(s) is as defined in Eq. (9.4.2). Here it appears in the form $H(j\omega)$ [i.e., $|t(j\omega)|^2$ = transfer power]. The reflected power, previously denoted $|r(j\omega)|^2$, is defined in terms of the reflection coefficient r, where $r = b_1/a_2$. At the load, $a_a + b_2 = v_2$ and $i_2 = a_2/R_0 - b_2/R_0 = v_2/R_2$, with R_0 denoting the characteristic impedance of the network. Alternatively, one notes that

$$r = \frac{b_1}{a_1} = \frac{(V_1 - I_1 R_1)/2}{(V_1 + I_1 R_1)/2} = \frac{Z_{in}(s) + 1}{Z_{in}(s) - 1} \quad\quad (9.4.10)$$

or

$$Z_{in}(s) = \frac{1 + r(s)}{1 - r(s)} \qquad (9.4.11)$$

That is, the input impedance of the network can be expressed in terms of the reflection ratio r(s).

9.5 WAVE FILTER STRUCTURE

Fettweis (1971) formalized these concepts in terms of an n-port filter. In particular, the n-port filter (or adaptor) was shown to be realizable in parallel or series form. The basic building blocks are summarized in Fig. 9.5.1. In this figure the block and flow diagram structure of parallel and series units are presented together with the algebraic relationships that defined the external port behavior of the subfilters. Care must be taken when interconnecting two arbitrary adaptors (series or parallel). The interconnection of an n-port wave filter can be made only if their port re-sistances are the same (see Fig. 9.5.1). Matching of R_i and R_j will ensure a proper impedance match and, therefore, maintain the MAP policy. This can be guaranteed in the following manner. First note the definition of α_i and β_i [see Eq. (9.5.1)]. Now assume that $\alpha_N = 1$ (parallel) and $\beta_N = 1$ (series). It follows that

$$G_N = \sum_{i=0}^{N-1} G_i \quad \text{or} \quad R_N = \sum_{i=0}^{N-1} R_i \qquad (9.5.1)$$

which in turn suggests that Eqs. (c) and (d) (from Fig. 9.5.1) can be combined to read:

Parallel:
$$a_0 = \sum_{i=1}^{N-1} \alpha_i a_i + a_N \qquad (9.5.2)$$

$$b_n = a_0 - a_N = \sum_{i=1}^{N-1} \alpha_i a_i$$

Series:
$$a_0 = \sum_{i=1}^{N} a_i$$

$$b_n = a_N - \beta_N a_0 = -\sum_{i=1}^{N-1} a_i$$

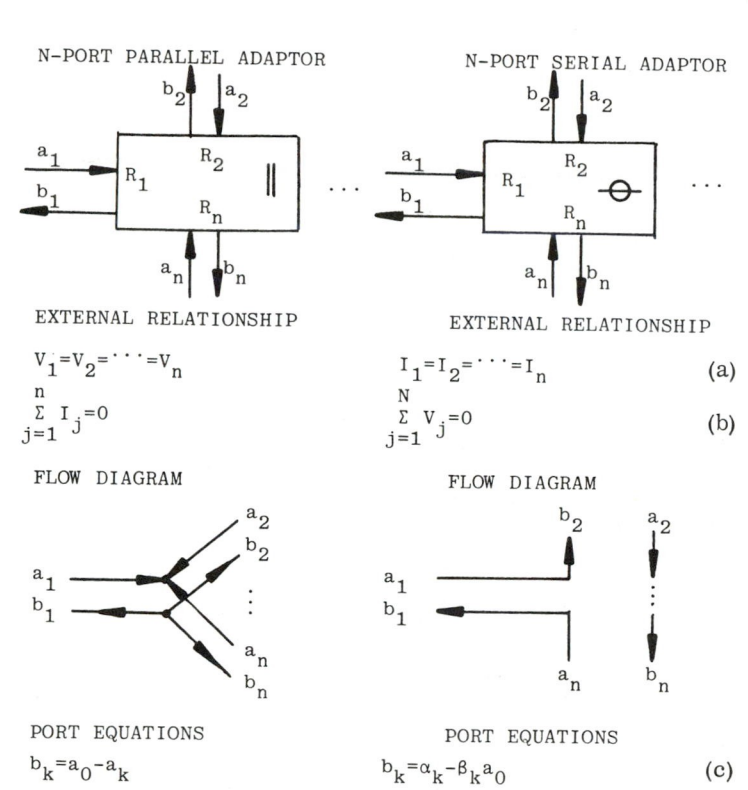

FIG. 9.5.1 Fundamental subsystems, flow diagrams, and port equations belonging to a MAP system.

Thus b_n can be defined independent of a_n. That is, a_n and b_n are physically detached. Therefore, port n can be connected to another network without risk of introducing a closed feedback path. The remaining parameters, namely α_i and β_i, remain interconnected and satisfy

$$\sum_{i=1}^{N-1} \alpha_i = 1 \quad \text{or} \quad \sum_{i=1}^{N-1} \beta_i = 1 \tag{9.5.3}$$

A further simplification can be achieved by choosing one of the remaining ports (numbered 1 through N – 1), say the ith port, to be a dependent port. Then the parameter sets $\{\alpha_i\}$ and $\{\beta_i\}$ can be defined to be

$$\alpha_k = 1 - \sum_{\substack{i=1 \\ i \neq k}}^{N-1} \alpha_i \quad \text{or} \quad \beta_k = 1 - \sum_{\substack{i=1 \\ i \neq k}}^{N-1} \beta_i \qquad (9.5.4)$$

This simplification will result in wave filters having N – 2 multipliers (i.e., α_N and $\beta_N = 1$) instead of the previous budget of N – 1 multipliers.

EXAMPLE 9.5.1   Suppose that N = 3; then

Parallel	Serial
$\alpha_N = 1$	$\beta_N = 1$
$b_i = a_0 - a_i$	$b_i = a_i - \beta_i a_0$
$b_N = \sum_{i=1}^{N-1} \alpha_i a_i$	$b_N = \sum_{i=1}^{N-1} a_i$
$a_0 = \sum_{i=1}^{N-1} \alpha_i a_i + a_N$	$a_0 = \sum_{i=1}^{N} a_i$
For N = 3, $\alpha_3 = 1$:	$\sum_{i=1}^{N-1} \beta_i = 1$
$\alpha_1 + \alpha_2 = 1$, $a_0 = \alpha_1 a_i$	For N = 3, $\beta_3 = 1$:
$\quad + \alpha_2 a_2 + a_3$	$\beta_1 + \beta_2 = 1$

$$\left\{ \text{realization without dependent port} \right\}$$

$$a_0 = a_1 + a_2 + a_3$$

$b_1 = a_0 - a_1$	$b_1 = a_1 - \beta_1 a_0$
$b_2 = a_0 - a_2$	$b_2 = a_2 - \beta_2 a_0$
$b_3 = \alpha_1 a_1 + \alpha_2 a_2$	$b_3 = -a_1 - a_2$

These results are summarized in Fig. 9.5.2.

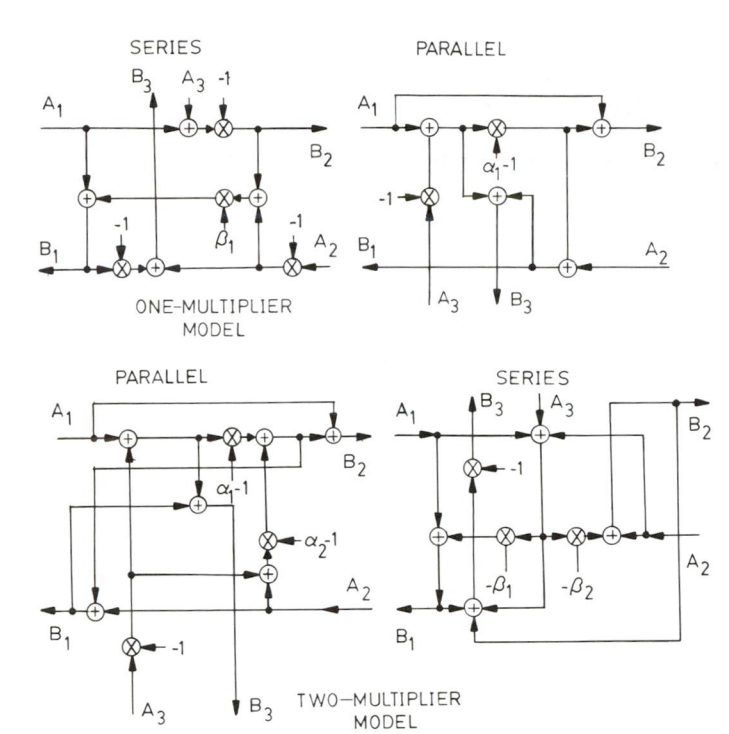

FIG. 9.5.2 Parallel and serial one- and two-multiplier architectures considered in Example 9.5.1.

9.6 WAVE FILTER SYNTHESIS

The wave filter realization of an analog prototype filter requires the nesting of series and parallel adapters. The hierarchy of a wave filter synthesis procedure is detailed below.

1. Design the desired MAP LC filter using any acceptable passive or active analog synthesis procedure. Identify the series and parallel interconnections. Decompose the LC network in a system of three-port adapters. For notational uniformity, let port 1 be the input, port 2 be an output port, and port 3 be the port to which the reactive elements are attached.

2. Assign port impedance or input voltage sources to their proper ports. When tying two wave filters together, the output port impedance of one wave filter must be matched to the input port impedance of the other. Furthermore, in order to interconnect two wave filters, the matched impedance must also be isolated. The one-multiplier adapter was

shown to have this isolation property. Therefore, those wave filter
elements having output ports tied to another wave filter must be one-
multiplier units. However, a system's output stage is not so constrained.
For the purpose of standardization, the output port impedance will be
defined in terms of its own port parameters. The port impedances of
two connected wave filters will be equated.

3. Replace interconnections with an appropriate one- or two-multiplier,
series or parallel, adapter.

4. Calculate the adapter gains.

A wave filter design option has been added to the computer-aided filter design package reported in the appendix to Chapter 13. The reader should
refer to this appendix for assistance in designing wave filters.

EXAMPLE 9.6.1 Consider the design of a lowpass elliptic filter having a
unit source and load resistance with the magnitude response found in Fig.
9.6.1. Using classic design procedures, it can be shown that a fourth-
order elliptic filter, having an input impedance $Z_{in}(s)$, will accomplish this.
The realized filter is diagrammed in Fig. 9.6.2 and prototyped in Fig.
9.6.3. The resulting wave filter is shown in Fig. 9.6.4 with port impedances computed as follows:

Network 1
(parallel)
(one-multiplier)

$$G_{11} = \frac{1}{R_1} = 1 \qquad G_{13} = \frac{2C_1}{T} \qquad G_{12} = G_{11} + G_{13}$$

(input port)   (device port)   (output port)

Network 2
(series)
(one-multiplier)

$$R_{21} = \frac{1}{G_{12}} \qquad R_{23} = R_{31} \qquad R_{22} = R_{21} + R_{23}$$

(input port)   (device port)   (output port)

Network 3
(parallel)
(one-multiplier)

$$G_{31} = \frac{T}{2L} \qquad G_{32} = G_{31} + G_{33} \qquad G_{33} = \frac{2C_2}{T}$$

(input port)   (device port)   (output port)

Network 4
(parallel)
(one-multiplier)

$$G_{41} = \frac{1}{R_{22}} \qquad G_{43} = \frac{2C_3}{T} \qquad G_{42} = \frac{1}{R} = 1$$

(input port)   (device port)   (output port)

To provide some numerical insight into the design problem, suppose
that the sample period is chosen to be 0.628 sec (i.e., $2\pi/10$ or approximately 1.6 Hz); then

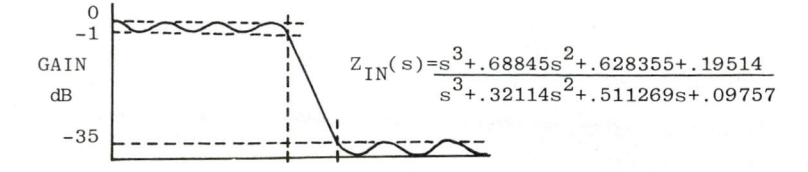

FIG. 9.6.1 Magnitude frequency response for the filter considered in Example 9.6.1.

$$G_{11} = 1, \quad G_{13} = 8.3397, \quad G_{12} = 9.3397$$

$$R_{21} = 0.1071, \quad R_{23} = 0.7823, \quad R_{22} = 0.8894$$

$$G_{31} = 0.25968, \quad G_{32} = 1.27827, \quad G_{33} = 1.0186$$

$$G_{41} = 1.1244, \quad G_{42} = 1, \quad G_{43} = 8.3397$$

It follows that

Network 1 (one-multiplier): $G_{12} = G_{11} + G13$

$$\alpha_{11} = \frac{2G_{11}}{G_{11} + G_{12} + G_{13}} = \frac{G_{11}}{G_{12}} = 0.1071$$

Check: $\alpha_{21} = \dfrac{2G_{12}}{G_{11} + G_{12} + G_{13}} = 1$

Network 2 (one-multiplier): $R_{22} = R_{21} + R_{23}$

$$\beta_{21} = \frac{2R_{21}}{R_{21} + R_{22} + R_{23}} = \frac{R_{21}}{R_{22}} = 0.1204$$

Check: $\beta_{22} = \dfrac{2R_{22}}{R_{11} + R_{22} + R_{23}} = 1$

Network 3 (one-multiplier)

$$\alpha_{31} = \frac{G_{31}}{G_{32}} = 0.2031$$

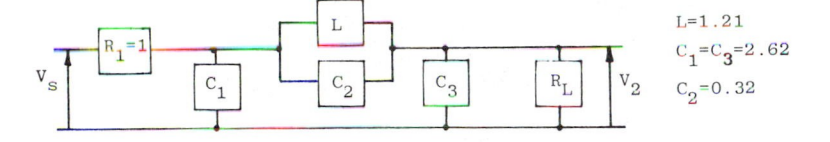

FIG. 9.6.2 RLC model of the example lowpass elliptic filter.

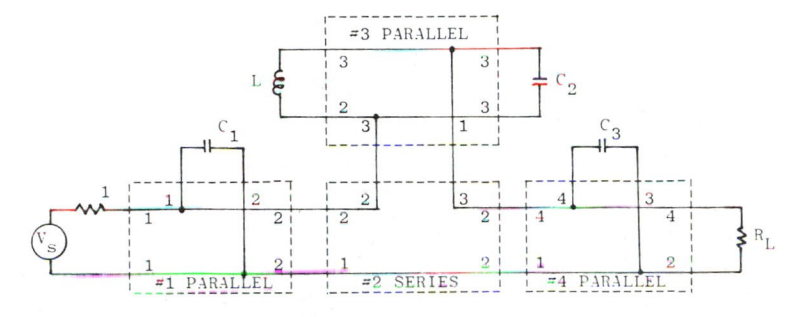

FIG. 9.6.3 Parallel and series network decomposition of the elliptic lowpass filter example.

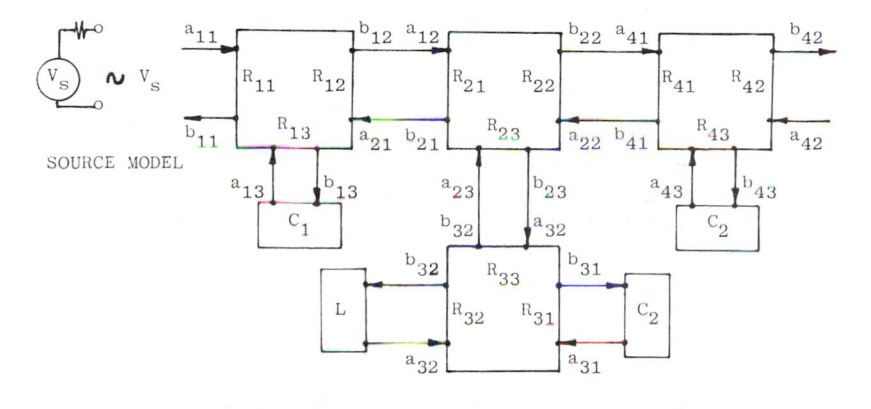

FIG. 9.6.4 Wave filter model of the elliptic lowpass filter example problem.

Network 4 (two-multiplier)

$$\alpha_{41} = \frac{2G_{41}}{G_{41} + G_{42} + G_{43}} = 0.2149$$

$$\alpha_{42} = \frac{2G_{42}}{G_{41} + G_{42} + G_{43}} = 0.1911$$

In terms of a network diagram, the resulting wave filter is abstracted in Fig. 9.6.5. The filter being analyzed can be seen to require four multiplications (equivalently, scaling) if one disregards multiplication by -1. Twenty adders are needed. This large adder budget is the price that must be paid for a low-sensitivity filter. In addition, if the system's multiplies and adds are performed in one time-shared arithmetic unit, throughput may suffer when compared to an architecture possessing fewer operations.

EXAMPLE 9.6.2 Consider the wave filter found in Fig. 9.6.5 with the 16 nodes identified indicated by their node values. The node-value relationships are defined in DO LOOP 211 of Fig. 9.6.6. Using simulation methods, a 16- and 8-bit experiment was performed on the third-order Chebyshev wave filter derived. The discrete magnitude frequency response and the filter's impulse response are displayed in Fig. 9.6.7. It can be noted that the 16-bit, and in fact the 8-bit design, hold up well to a reduction in word length. Notice that there exists an above-average dc component in the 8-bit configuration and, to a lesser degree, in the 16-bit version.

As a comprehensive study of the effects of finite word length on frequency fidelity, a 1- to 16-bit filter was tested, with the results shown in Fig. 9.6.7. It should be readily apparent that filters having word lengths of 13 bits or more provide acceptable spectral resolution.

Finally, the ladder and wave filter may be compared with respect to word length sensitivity. Again the third-order Chebyshev lowpass filter will be used as a model. It can be seen from the error variance data presented in Fig. 9.6.8 that for this example the wave filter is superior to the ladder. However, other filter models may produce different results.

EXAMPLE 9.6.3 Design a third-order 3-dB discrete Butterworth filter having a 100-Hz bandwidth and a sample frequency of 10 kHz (i.e., $T = 10^{-4}$). In this example, explicit use of the transferred and reflected power relationships will be used. In particular, the 3-dB third-order filter satisfies

$$|t(j\omega)|^2 = \frac{1}{1 + (\omega^2)^3} = \frac{1}{1 + \omega^6}$$

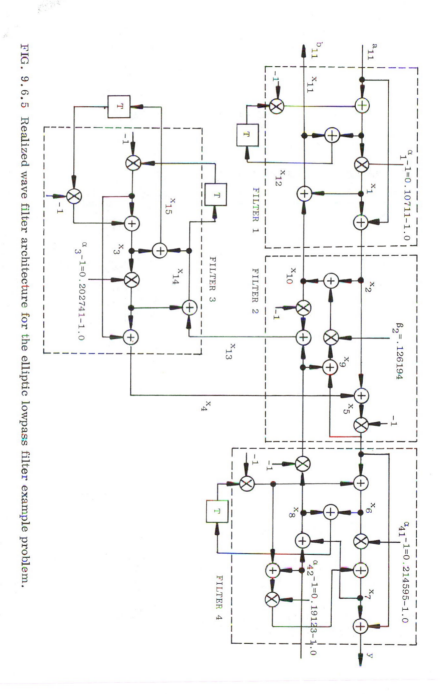

FIG. 9.6.5 Realized wave filter architecture for the elliptic lowpass filter example problem.

```
0001        DIMENSION X1(257),X2(257),X3(257),X4(257),X5(257),X6(257)
0002        DIMENSION X7(257),X9(257),X10(257),X11(257),X12(257)
0003        DIMENSION X13(257),X14(257),X15(257),X16(257),X(257),Y(257)
0004        DIMENSION ST(257),SIGA(16)
0005        COMPLEX YY(256),T,U,W
0006        X(2)=1.
0007        DO 211 I=2,257
0008        X1(I)=X(I)-X12(I-1)
0009        X2(I)=X(I)+(0.10711-1.)*X1(I)
0010        X3(I)=-X15(I-1)-X14(I-1)
0011        X4(I)=(0.202741-1.)*X3(I)-X14(I-1)
0012        X5(I)=X2(I)+X4(I)
0013        X6(I)=-X5(I)-X16(I-1)
0014        X7(I)=(0.214595-1.)*X6(I)-(0.191234-1.)*X16(I-1)
0015        X9(I)=-X7(I)-X5(I)
0016        X10(I)=X2(I)+0.120194*X9(I)
0017        X11(I)=(0.10711-1.)*X1(I)+X10(I)
0018        X12(I)=X11(I)+X1(I)
0019        X13(I)=-X10(I)-X7(I)
0020        X14(I)=X13(I)+(0.202741-1.)*X3(I)
0021        X15(I)=X14(I)+X3(I)
0022        X16(I)=X7(I)+X6(I)
0023 211    ST(I)=X7(I)-X5(I)
0024        DO 213 I=1,256
0025        ST(I)=ST(I+1)
0026 213    YY(I)=CMPLX(ST(I),0.)
0027        LL=31
0028        GO TO 222
0029 2      LL=0
0030        DO 397 NN=1,16
0031        X(2)=1.
0032        DO 111 I=2,257
0033        X1(I)=X(I)-X12(I-1)
0034        A=0.10711-1.
0035        CALL QUANT(A,A,NN)
0036        A=A*X1(I)
0037        CALL QUANT(A,A,NN)
0038        X2(I)=X(I)+A
0039        X3(I)=-X15(I-1)-X14(I-1)
0040        A=0.202741-1.
0041        CALL QUANT(A,A,NN)
0042        A=A*X3(I)
0043        CALL QUANT(A,A,NN)
0044        X4(I)=A-X14(I-1)
0045        X5(I)=X2(I)+X4(I)
0046        X6(I)=-X5(I)-X16(I-1)
0047        A=0.214595-1.
0048        CALL QUANT(A,A,NN)
0049        B=0.191234-1.
0050        CALL QUANT(B,B,NN)
0051        A=A*X6(I)
0052        CALL QUANT(A,A,NN)
0053        B=B*X16(I-1)
0054        CALL QUANT(B,B,NN)
0055        X7(I)=A-B
0056        X9(I)=-X7(I)-X5(I)
```

FIG. 9.6.6 Source program used to simulate the response of a finite-word-length wave filter.

which is factorable (using standard tables) into

$$t(s) = \frac{1}{s^3 + 2s^2 + 2s + 1}$$

It was argued previously that the reflected power satisfies

$$|r(j\omega)|^2 = 1 - |t(j\omega)|^2 = \frac{\omega^6}{1 + \omega^6}$$

or r(s) is given by $s^3/(s^3 + 2s^2 + 2s + 1)$. It has also been argued that the network's input impedance is related to r(s) through

$$Z_{in}(s) = \frac{1 + r(s)}{1 - r(s)} = \frac{2s^3 + 2s^2 + 2s + 1}{2s^2 + 2s + 1}$$

Using standard ladder synthesis methods, one obtains an analog prototype. This filter possesses a lowpass critical frequency which must be equated to a prewarped frequency of $\omega_p = 2\pi \times 100$. However, the prewarping of a 100-Hz discrete critical frequency for T = 10 kHz is approximately an analog frequency 100 Hz. Therefore, using a lowpass-to-lowpass conversion, one obtains the filter diagrammed in Fig. 9.6.9. This network can again be partitioned into a collection of series and parallel three-port networks (see Fig. 9.6.10). The wave filter parameters (see Fig. 9.6.11) are defined as follows:

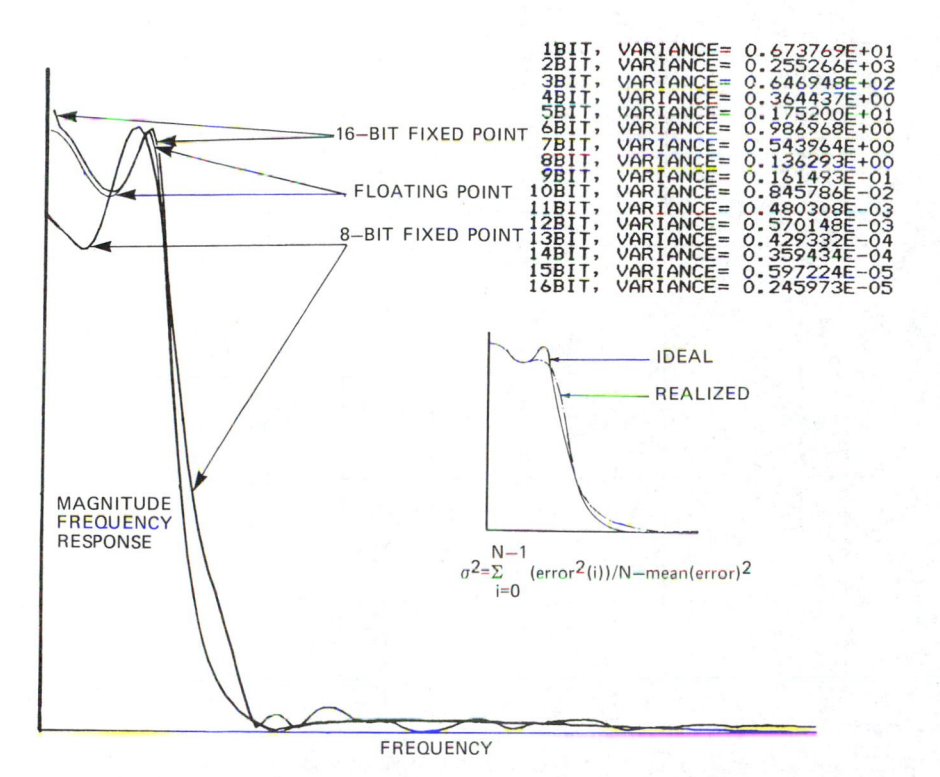

FIG. 9.6.7 Comparison of floating-point and fixed-word-length realization of a third-order lowpass Chebyshev filter.

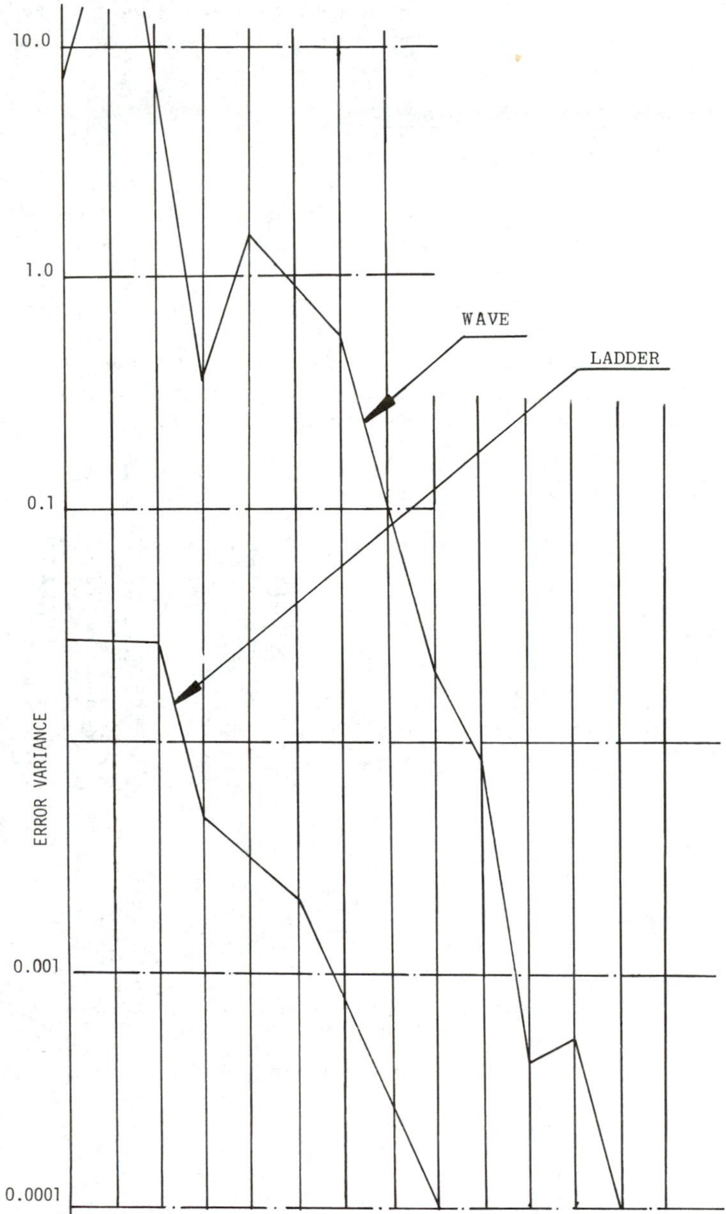

FIG. 9.6.8 Comparison of the error variance of wave and ladder filters as a function of word length.

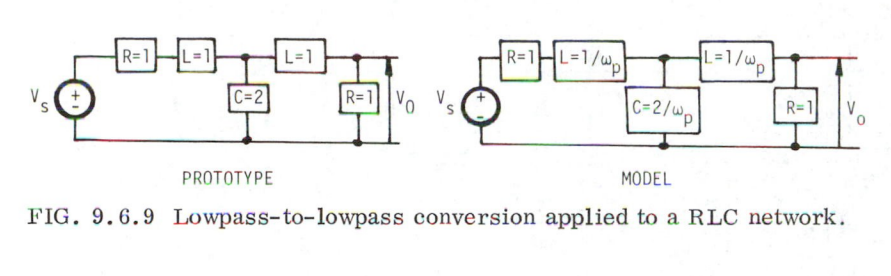

PROTOTYPE   MODEL

FIG. 9.6.9 Lowpass-to-lowpass conversion applied to a RLC network.

$$R_{11} = \frac{1}{R}, \quad R_{13} = \frac{2L_1}{T}, \quad R_{12} = R_{11} + R_{13} \Rightarrow R_{11} = 1, \quad R_{13} = 3.183, \quad R_{12} = 4.183$$

(input)   (device)   (output)

$$G_{21} = \frac{1}{R_{12}}, \quad G_{23} = \frac{2C}{T}, \quad G_{22} = G_{21} + G_{23} \Rightarrow G_{21} = 0.239, \quad G_{23} = 6.366, \quad G_{22} = 6.605$$

(input)   (device)   (output)

$$R_{31} = \frac{1}{G_{22}}, \quad R_{33} = \frac{2L_2}{T}, \quad R_{32} = \frac{1}{R} \Rightarrow R_{31} = 0.1514, \quad R_{33} = 3.183, \quad G_{32} = 1$$

(input)   (device)   (output)

or in terms of the parameter set $\{\alpha\}$, one obtains

Network 1: $\alpha_{11} = \dfrac{R_{21}}{R_{22}} = 0.23906$ (one multiplier)

Network 2: $\alpha_{21} = \dfrac{G_{21}}{G_{22}} = 0.03754$ (one multiplier)

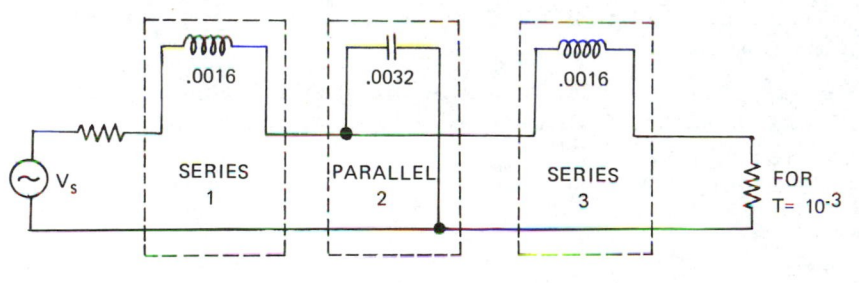

FIG. 9.6.10 Three-port RLC representation of the filter found in Example 9.6.3.

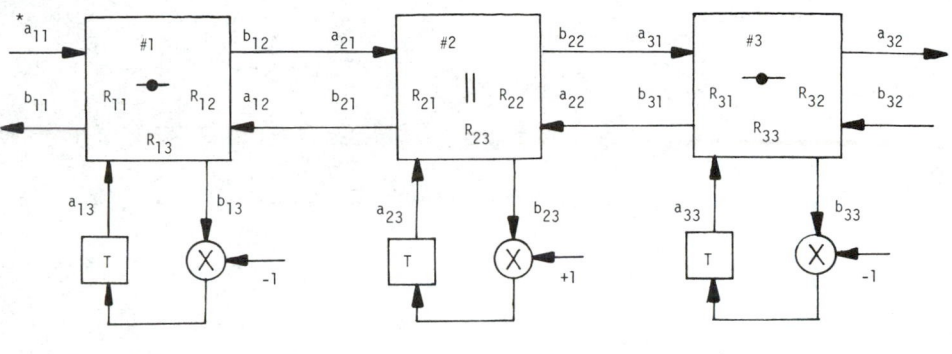

FIG. 9.6.11 Wave filter representation of the RLC network shown in Fig. 9.6.10.

$$\text{Network 3:} \quad \alpha_{31} = \frac{2R_{31}}{R_{31} + R_{32} + R_{33}} = 0.06986 \left. \begin{matrix} \\ \\ \\ \\ \\ \end{matrix} \right\} \quad \text{(two multipliers)}$$

$$\alpha_{32} = \frac{R_{32}}{R_{31} + R_{32} + R_{33}} = 0.461415$$

BIBLIOGRAPHY

Antonious, A. (1979), Digital Filters: Analysis and Design, McGraw-Hill, New York.

Fettweis, A. (1971), Digital Filter Structures Related to Classical Filter Networks, Arch. Elek. Ubertragung, 25, February, pp. 79-89.

Fettweis, A., and K. Meerkotter (1975), Suppression of Parasitic Oscillations in Wave Digital Filters, IEEE Trans. Circuits Syst., CAS-22, March, pp. 239-246.

Sedlmeyer, A., and A. Fettweis (1973), Digital Filters with True Ladder Configuration, Int. J. Circuit Theory Appl., 1, March, pp. 5-10.

10
Finite-Word-Length Effects

10.1 INTRODUCTION

Digital signal processing is a science that is often confused (unfortunately) with the study of discrete or sampled data systems. In discrete system theory, filters and transforms are defined algebraically over a real coefficient field. This means that the familiar rules of addition, multiplication, differentiation, integration, and so forth, hold true. Therefore, operations such as convolution and spectral transformation pose no particular mathematical challenge. However, because of the quantization effects of a finite-word-length digital computer, digital filters may behave differently from their discrete counterparts. This difference is related to the coefficient field constraints imposed by a finite-word-length (technically, the coefficients are defined in terms of a finite ring instead of a real field). The disagreement between a digital filter and its discrete model is said to be due to finite-word-length effects.

Finite-word-length effects are found in many forms. They can be broadly categorized into the following classes:

1. Quantization Errors
   a. Input quantization effects
   b. Coefficient quantization effects
2. Arithmetic Errors

The complete and rigorous analysis of these error sources requires a detailed technical development. The level of mathematical skill required may be beyond the current state of the reader. For this reason this section has been divided into two major parts. The first provides an overview of the topic. It attempts to familiarize the reader with the problem and suggests remedies. After this somewhat tutorial discussion, a more rigorous analysis of these effects is pursued. It is left to the reader to meter his or her interest and competence on this subject and choose, from the next several chapters, that information which will serve his or her needs.

10.2 QUANTIZATION ERRORS

Quantization errors such as those due to analog-to-digital (A/D) data conversion are common to most digital signal processing systems. Errors of this type are influenced by the numbering system used to encode data (e.g., 2's complement, floating point, etc.). Numbering systems are developed in Chapter 15, and A/D conversion is treated in Chapter 17. At that time it will be assumed that a fixed-point word or floating-point mantissa is representable by a signed $(n + 1)$-bit binary-valued data string, say x_i. If the dynamic range required by the encoding scheme is $[-V, V]$, the width of the quantization policy is given by

$$Q = \frac{V}{2^n} = V2^{-n} \quad \text{volts/bit} \tag{10.2.1}$$

Suppose that only the $(m + 1)$st most significant bits (MSBs) of x_n are used to form x_m. Then the truncation error, denoted $Q[x_n]$, is given by

$$E = Q[x_n] - x_n = x_m - x_n \tag{10.2.2}$$

If x_n is a positive number, $Q(2^{-m} - 2^{-n}) \le E \le 0$. The bound on E, for x_n negative, is a function of the numbering system used. For a sign-magnitude and 1's-complement code, it can be shown that $0 \le E \le Q(2^{-m} - 2^{-n})$. If a 2's-complement system is used, $-Q(2^{-m} - 2^{-n}) \le E \le 0$. The truncation processes is interpreted graphically in Fig. 10.2.1.

Another quantization policy that is often used is called rounding. A $(m + 1)$-bit rounded version of x_n, say $\hat{x}_m$, corresponds to assigning the value of $\tilde{x}_m$ to be that of the nearest quantization level, as shown in Fig. 10.2.1. As a result, the roundoff error is bounded by $-Q(2^{-m} - 2^{-n})/2 \le E \le Q(2^{-m} - 2^{-n})/2$. This is one-half of the maximal error bound found in truncation policies.

In an A/D operation, a discrete real variable $x(n)$ is mapped into a quantized digital word $Q[x(n)]$. The resulting conversion error can be modeled to read

$$Q[x(n)] = x(n) + e(n) \tag{10.2.3}$$

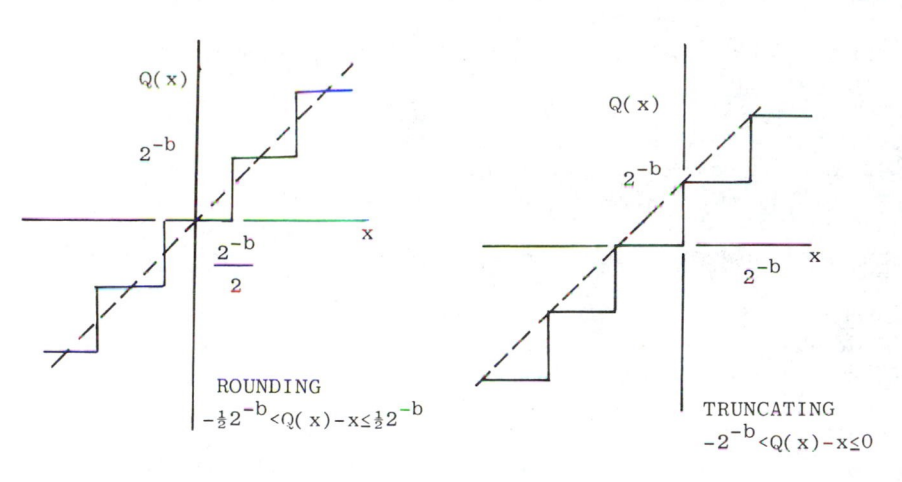

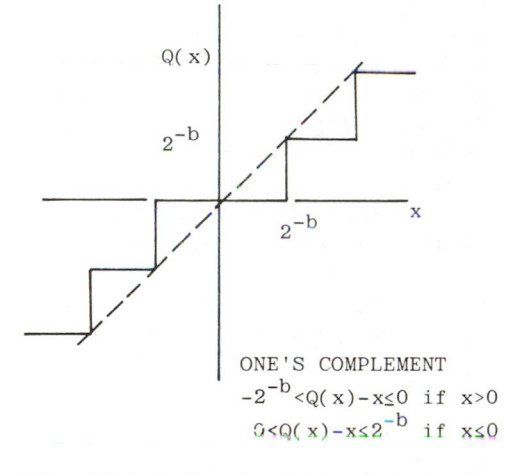

FIG. 10.2.1 Graphical interpretation truncation and rounding rules.

where

$$\frac{-Q}{2} \leq e(n) \leq \frac{Q}{2}$$ (10.2.4)

if rounding is used. Here e(n) denotes the quantization error. The error model consists of adding two real numbers to produce a quantized value (i.e., defined over a finite ring)

$$Q[x(n)] = \pm L\left(\frac{Q}{2^n}\right); \quad L = 0, 1., \ldots, 2^n - 1$$ (10.2.5)

In most applications, the value of x(t) is not known a priori. Shannon has argued that if:

1. the error process $\{x(n)\}$ is statistically stationary, and
2. $\{e(n)\}$ is uncorrelated with $\{x(n)\}$, and
3. $\{e(n)\}$ is a white stationary random process,

then the probability density function p(e(n)) is <u>uniformly distributed</u> over [-Q/2, Q/2]. The quantization error model method is graphically interpreted in Fig. 10.2.2. Macroscopically, these assumptions are generally valid if x(t) is a highly dynamic and complex signal process. However, there are exceptions. For example, if x(t) is a constant, the quantized error process is also constant. A constant error process cannot be uniformly distributed.

By accepting this quantization error model, we have admitted our inability to model the error process rigorously using finite algebraic concepts. Our lack of understanding of this complex process has been embedded into a stochastic model (Fig. 10.2.3). The stochastic model will allow us to draw statistical conclusions about the quantization phenomenon. For example, if we accept the uniform density hypothesis, then the mean round-off error is zero and the variance is $Q^2/12 = V^2 2^{-2n}/12$ (see Example 4.9.1). In terms of a decibel metric, it can be seen that $10 \log_D (V^2 2^{-2n}/12) = K - 6.02n$, where K is a constant. That is, the error variance can be reduced on the order of 6 dB for each additional bit added to the data words.

Quantization errors are also found in the encoding of filter coefficients. It will be assumed that a digital filter coefficient, say a_i, is represented by a finite binary-valued word. As a result, the real coefficients derived in the sections of discrete filter design will be realized with a quantization

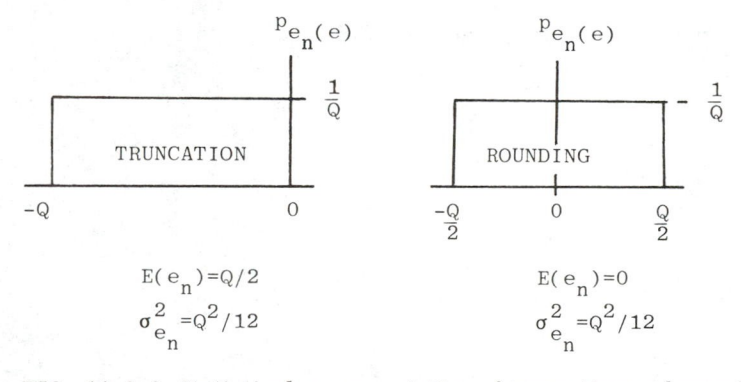

FIG. 10.2.2 Statistical representation of truncation and rounding probability density function.

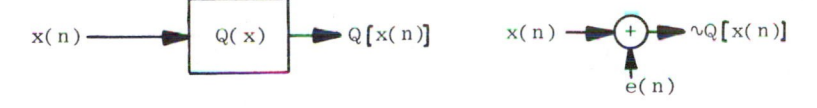

FIG. 10.2.3 Stochastic error model for a finite-word-length quantizer.

error. These approximation errors can affect the performance of a digital filter and in some cases its stability. For example, suppose that one is charged with the problem of designing a digital filter to emulate the response of the discrete system given by

$$H(z) = \underbrace{\frac{z}{z - 3/4} - \frac{z}{z - 1/4}}_{\substack{\text{parallel} \\ \text{filter}}} = \underbrace{\frac{z/4}{z^2 - z + 3/16}}_{\text{canonical filter}} = \underbrace{\frac{z/4}{z - 3/4} \frac{z/4}{z - 1/4}}_{\substack{\text{cascade} \\ \text{filter}}}$$

Suppose that one chooses, for the sake of discussion, to encode all system data as a signed 3-bit word having the form $\pm XX$. The coefficients of the realized digital filter would then take the form

$$\left(\frac{-3}{4}\right)R \rightarrow \frac{3}{4}$$

$$\left(\frac{-1}{4}\right)R \rightarrow \frac{1}{4}$$

$$\left(\frac{1}{2}\right)R \rightarrow \frac{1}{2}$$

$$\left(\frac{1}{4}\right)R \rightarrow \frac{1}{4}$$

$$(-1)R \rightarrow -1$$

$$\left(\frac{3}{16}\right)R \rightarrow \frac{1}{4}$$

where $(z)_R$ denotes the rounding of z. Therefore, the parallel and cascade architectures will be unaffected by coefficient rounding, but the canonical filter will be realized as

$$H'(z) = \frac{z/2}{z^2 - z + 1/4}$$

which has poles located at 1/2 and 1/2 rather than the original locations of 1/4 and 3/4. This dislocation of poles will severely alter the filter's frequency response.

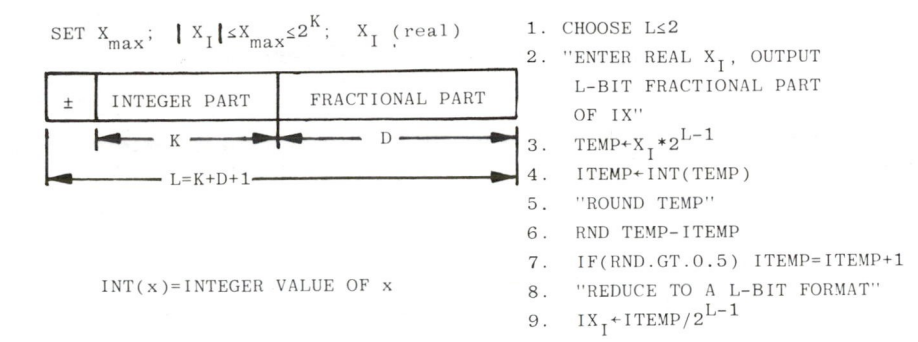

SET X_{max}; $|X_I| \le X_{max} \le 2^K$; X_I (real)

| ± | INTEGER PART | FRACTIONAL PART |

$\longleftarrow$ K $\longrightarrow$ $\longleftarrow$ D $\longrightarrow$

$\longleftarrow$ L=K+D+1 $\longrightarrow$

INT(x)=INTEGER VALUE OF x

1. CHOOSE L≤2
2. "ENTER REAL X_I, OUTPUT L-BIT FRACTIONAL PART OF IX"
3. TEMP←X_I*2^{L-1}
4. ITEMP←INT(TEMP)
5. "ROUND TEMP"
6. RND TEMP-ITEMP
7. IF(RND.GT.0.5) ITEMP=ITEMP+1
8. "REDUCE TO A L-BIT FORMAT"
9. IX_I←ITEMP/2^{L-1}

FIG. 10.2.4 Roundoff error algorithm which maps a real number into one of finite precision.

One can always test the effects of quantization through simulation. Here a filter would be modeled as a difference equation, coded in software, and simulated with an impulse [i.e., $x(0) = 1$, $x(n) = 0$ otherwise]. The simulated impulse response would be analyzed using an FFT. A double-precision floating-point study may be assumed to be equivalent to a discrete filter. This response would be compared to that obtained using fixed-point (integer) computation. The routine, found in Fig. 10.2.4, can be used to convert a decimal digit into a finite-precision fractional number which would be used in the simulation.

EXAMPLE 10.2.1 Using an $L = 8$-bit architecture, simulate the response of

$$H(z) = \frac{z^2 + 0.51z + 0.765}{z^2 - 1.372z + 0.9338}$$

Assume that the input $u(n)$ is bounded by unity (i.e., $|u(n)| \le 1$); then the maximal system variable can be shown to be bounded by $x_{MAX} < 2^2$ or $K = 2$ (maximal bounds are covered in Chapter 11). It follows that if we wish to assign 5 fractional bits of accuracy to the 8-bit data word, -1.372 is converted as follows:

$$\text{TEMP} = -1.372 \times 2^7 = -175.616$$

$$\text{ITEMP} = -176$$

$$\text{RND} = 0.384 \quad \text{and} \quad \text{ITEMP} = -176$$

$$IX_i = \frac{176}{2^7} = -1.375 \quad \text{error} = 0.003 < \frac{\text{LSB}}{2} = 0.015625$$

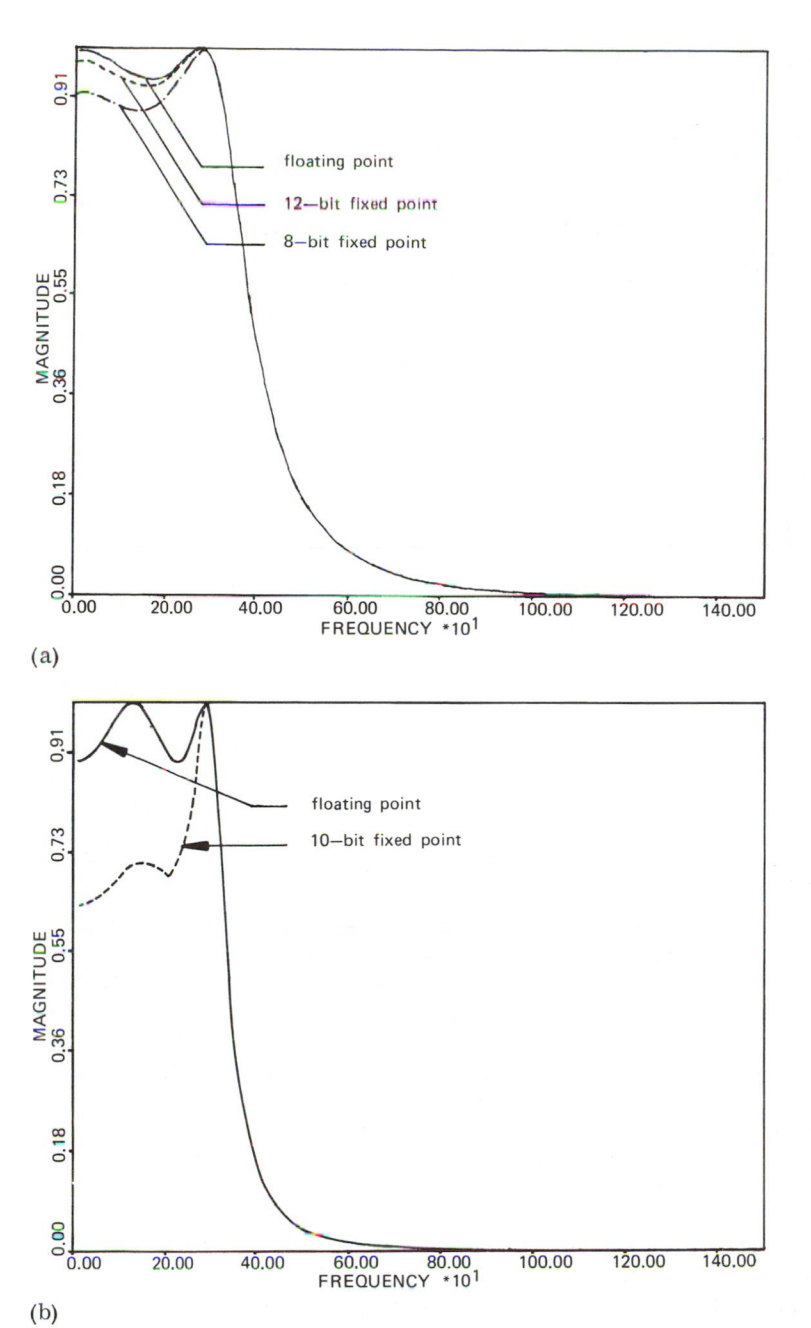

(a)

(b)

FIG. 10.2.5 Comparison of the magnitude frequency response of (a) a third-order and (b) a fourth-order Chebyshev lowpass filter showing the effect of finite-word-length arithmetic.

For the coefficient 0.9338, the conversion would produce

$$TEMP = 0.9338 \times 2^7 = 119.5264$$

$$ITEMP = 119$$

$$RND = 0.5264 \quad \text{and} \quad ITEMP = 120$$

$$IX_i = \frac{120}{2^7} = 0.9375 \quad error = 0.0037 < \frac{LSB}{2}$$

EXAMPLE 10.2.2 Using simulation, the coefficient quantization effects on third- and fourth-order Chebyshev filters are shown in Fig. 10.2.5.

10.3 ARITHMETIC ERRORS

Another error source found in digital systems is due to the reduction of data precision derived from rounding full-precision multiplications. These errors, referred to as roundoff errors, are the result of rounding the 2n-bit product of two n-bit numbers to their n most significant bits (MSBs). The errors are modeled as a zero-mean, uniformly distributed random process over $[-Q/2, Q/2]$, where Q, as before, represents the quantization step size (i.e., $Q = V2^{-n}$). Each error source is considered to be statistically independent from all others. For example, the second-order filter previously considered would possess the roundoff error model described in Fig. 10.3.1. In this figure, e_i denotes the roundoff error source, which is assumed to be white uniformly distributed noise. This additive noise corrupts the ideal real multiplication, which is modeled to have infinite precision.

The effect of these distributed noise sources can be quantified using linear system theoretic methods or thorough simulation. This thesis is developed in more detail in subsequent chapters. Generally, the effect of multiplier roundoff error is one of reduced precision. The cumulative effect of this error can be controlled by adjusting the filter's word length and/or choosing a filter architecture which is minimally sensitive to round-off errors. The results of an experiment conducted by Leon and Bass using a third-order Chebyshev lowpass filter model are shown in Fig. 10.3.2. From this simple experiment it can be seen that several decibels of roundoff error distortion can be saved by using the cascaded architecture. However, in many cases the roundoff error noise floor is so low (e.g., -80 dB) over a wide selection of admissible architectures that the question of roundoff error minimization becomes moot. In such cases, other design attributes, such as cost and complexity, become dominant.

Generally, the question of architecture is very technical and demands a detailed analysis. Certain filters are known to possess a degree of

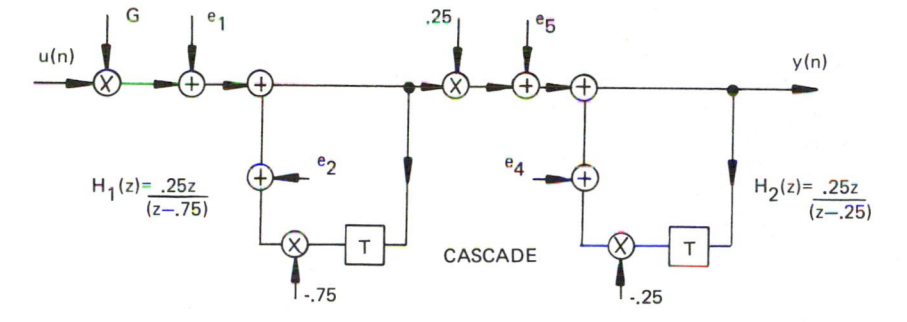

FIG. 10.3.1 Simple second–order discrete filter showing the random error models in place.

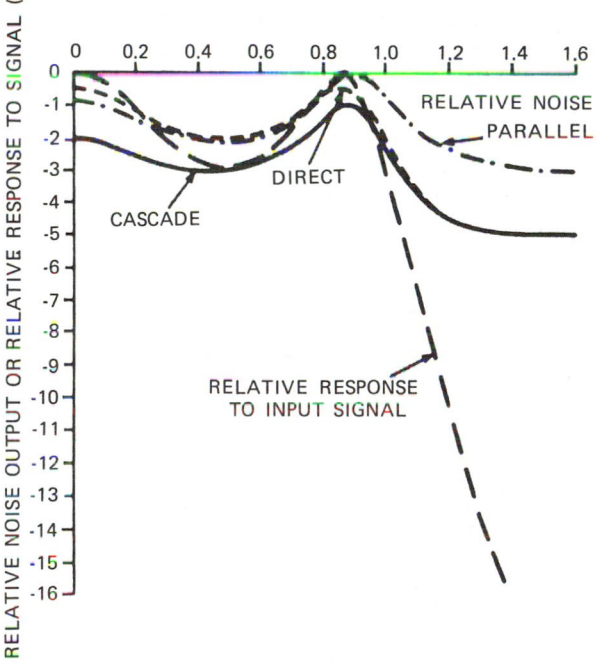

FIG. 10.3.2 Example of the effect of architecture on the performance of a finite–word–length filter. (After Leon and Bass, 1974; reprinted with the permission of EDN.)

roundoff immunity, and these filters are summarized in Section 13.5.
There are, however, some empirical observations which may be made at
this time. It is generally thought to be better to realize an nth-order filter
as a collection of first- and/or second-order subfilters having real coefficients. The roundoff error resulting from this system decomposition is
generally smaller than that obtained using higher-order subfilters.

10.4 ZERO–INPUT LIMIT CYCLING

Arithmetic errors may take the form known as <u>zero input limit cycling</u>,
which is characterized by oscillations appearing at the output of the filter
even though the input is zero. The energy for such oscillatory behavior is
derived from the filter's initial conditions. Theoretically, if the system
is asymptotically stable, this energy should decay to zero. However, in
some cases, due to quantization effects, the output will not converge to
zero but instead limit-cycle. Sometimes these oscillations will tend to de-
stabilize a filter. Even when the oscillations are of low amplitude, they
can prove to be very distracting in voice and audio applications. This limit-
cycling phenomenon can best be motivated in terms of a simple first-order
example.

EXAMPLE 10.4.1 Consider the filter given by $y(n + 1) = -0.75y(n) + u(n)$.
The response of this filter to the stimuli $u(n) = (-1)^n$ is given by $\{1, -1.75,$
$2.3125, \ldots, 4.0\}$. The unforced (homogeneous) response for $y(0) = 1$ and
$u(n) = 0$ for all n is given by $\{-0.75, 0.5625, -0.421875, \ldots, 0\}$. The
discrete filter stability is therefore apparent. However, if the filter is to
be realized digitally using a 5-bit word having the form $\pm X.XXX$, then the
zero-input (homogeneous) response, with respect to rounding, is given by

Event i	$\bar{y}(i + 1) = \dfrac{-3}{4}\,\bar{y}(i)_R; \; y(i) = y(i)_R$
1	$\left(\dfrac{-3}{4} * 1\right)_R \to -3/4$
2	$\left(\dfrac{-3}{4} * \dfrac{-3}{4}\right)_R = \left(\dfrac{9}{16}\right)_R \to \dfrac{1}{2}$
3	$\left(\dfrac{-3}{4} * \dfrac{-1}{4}\right)_R = \left(\dfrac{9}{16}\right)_R \to \dfrac{1}{4}$
4	$\left(\dfrac{-3}{4} * \dfrac{-1}{4}\right)_R = \left(\dfrac{3}{16}\right)_R \to \dfrac{1}{4}$
5	$\left(\dfrac{-3}{4} * \dfrac{1}{4}\right)_R = \left(\dfrac{-3}{16}\right)_R \to \dfrac{1}{4}$

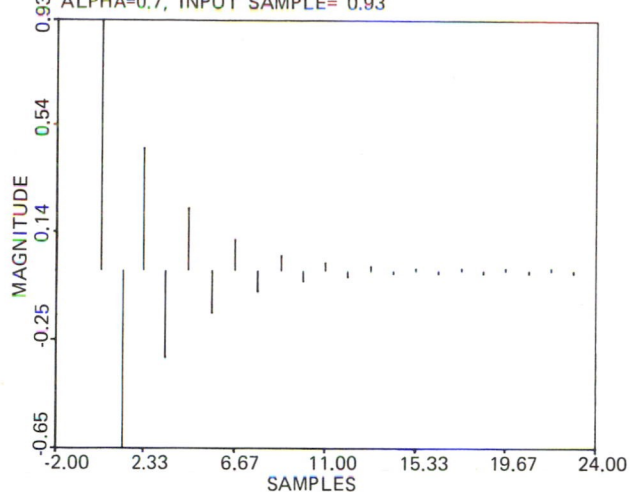

FIG. 10.4.1 Impulse response of a simple first-order filter exhibiting limit cycling.

The finite-word-length effects cause the LSB to oscillate with a magnitude of ±1/4 and a period of 2T. The amplitude and period of these oscillations will, of course, vary from design to design.

EXAMPLE 10.4.2 The problem in Example 10.4.1 can be modified slightly by letting the filter coefficient be -0.7 (rather than -0.75) and establishing a filter word length of 8 bits (rather than 5). The impulse response of the reconfigured filter is presented in Fig. 10.4.1. Observe that for large-sample indices (>12), the output process is dominated by low-amplitude zero-input limit cycling.

The effects and conditions under which zero-input limit cycling will occur can be predicted algebraically. This analytic study is presented in Chapter 11.

10.5 OVERFLOW LIMIT CYCLES

Large-scale overflow limit cycles are another source of error. This source can be a major problem if it is not successfully treated. Whereas zero-input limit cycling may cause a cosmetic problem, large-scale over-flow limit cycles can be disasterous. They are caused by, as the name would suggest, a system state (or variable) exceeding a prespecified dynamic range. The dynamic range limitations are established by the adopted

word length and numbering system. For example, consider the previous
filter realized using a 2's-complement numbering system. If a 4-bit word
is used, having a format $\pm X.XX$, then the output of one of the filter's 2's-
complement adders is diagrammed in Fig. 10.5.1. It can be noted that the
adder is actually a modular adder in that the transition from $011 \rightarrow 1.75$ to
its 2's-complement successor is not the decimal 2. Instead, 100 has a
decimal value of -2. With respect to this constraint, the output of our simple filter to an input given by $u(n) = (-1)^n$, for $y(0) = 0$, becomes

Event	$y(i) = \left(-\dfrac{3}{4} \cdot y(i-1)\right)_R + (-1)^i$	$y(i)$ Ideal	Error
0	$\left(-\dfrac{3}{4} \cdot 0\right)_R \boxed{+} 1$	1	0
1	$\left(-\dfrac{3}{4} \cdot 1\right)_R \boxed{+} -1 = -\dfrac{3}{4} + -1 = -\dfrac{7}{4}$	-1.75	0
2	$\left(-\dfrac{3}{4} \cdot -\dfrac{7}{4}\right)_R \boxed{+} 1 = \dfrac{5}{4} + 1 = -\dfrac{1}{4}$	2.3125	2.5625
3	$\left(-\dfrac{3}{4} \cdot -\dfrac{1}{4}\right)_R \boxed{+} -1 = \dfrac{1}{4} + -1 = -\dfrac{3}{4}$	-2.734375	1.984375
4	$\left(-\dfrac{3}{4} \cdot -\dfrac{3}{4}\right)_R \boxed{+} 1 = \dfrac{1}{2} + 1 = 1.5$	3.05078	1.55078

$\boxed{+}$ Addition in the manner of Fig. 10.5.1.

These large errors are generally intolerable. To overcome their effects,
several remedies are available. One fix is to extend the register length of
the system adders, accumulators, and so on. This will increase their
dynamic range and thereby avoid the overflow problem. This idea has
been used successfully in filter designs. For example, the VLSI $(n \times n)$-
bit multipliers marketed by TRW have a 2n-bit full-precision product. The
accumulator, which is attached to the multiplier, has a $(2n + 3)$-bit dynamic
range. These additional 3 bits allow up to $2^3 = 8$ worst-case multiplies to
be added (accumulated) without fear of overflow.

   Another approach to the problem of overflow prevention is scaling.
This subject is studied in detail in Chapter 11. Here the input process to
a filter is scaled so that no internal variable (state) will exceed its allocated dynamic range. Such an approach has long been used in analog computer programming, for the purpose of precluding amplifier saturation. In
digital filters, it serves the same purpose. For example, suppose that the
first-order filter encountered earlier in this chapter is reinvestigated.
The worst-case output will occur when the input is $u(n) = (-1)^n$, which produces a $y(\text{steady state}) = 4$. If one adopts a data format of $\pm XXX.X$, overflow will possibly occur for an arbitrary forcing function. However, if the

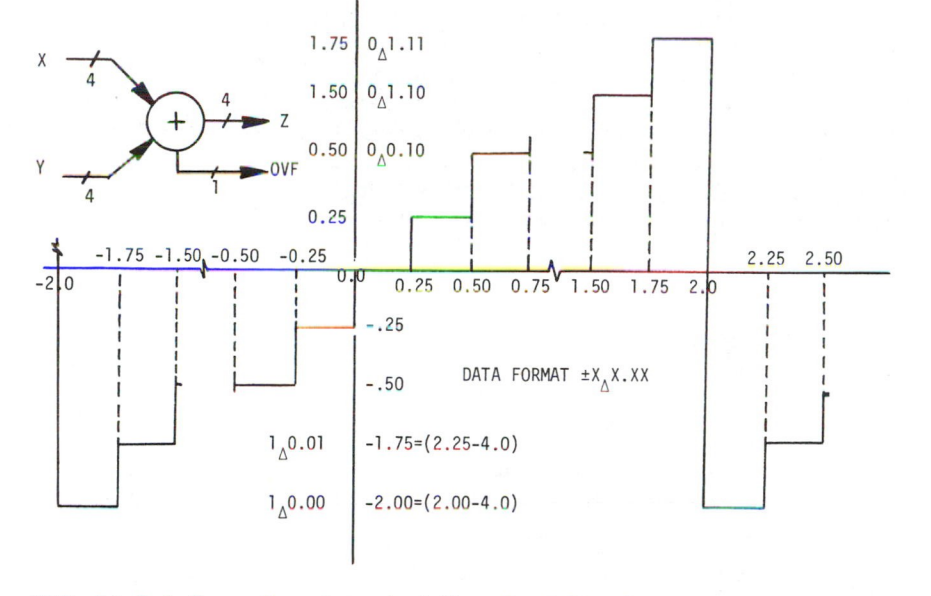

FIG. 10.5.1 Operating characteristics of a 4-bit 2's-complement adder.

input is scaled by 1/4 [i.e., $u'(n) = (-1)^n/4$], overflow cannot occur even under worst-case conditions.

These scaling bounds are historically difficult to determine. However, in Chapter 11 some powerful computer-aided techniques are presented which will simplify this process immensely.

Saturating arithmetic can also be used to reduce the effects of overflow limit cycling. A saturating arithmetic unit will set its output equal to the

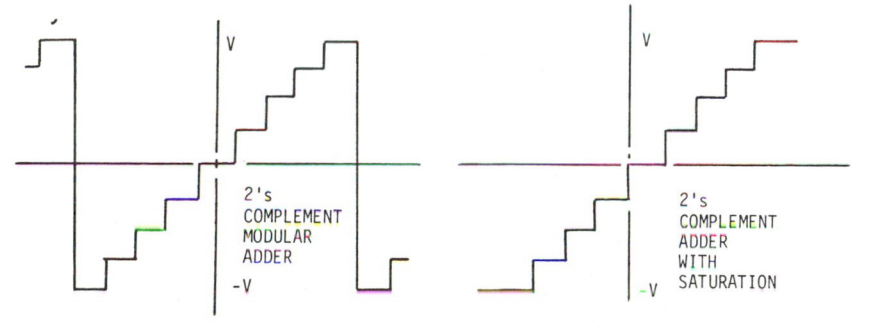

FIG. 10.5.2 Saturating and modular behavior of a finite-word-length digital adder.

most positive (negative) value permitted by the numbering system whenever an overflow is detected. The performance of a saturating 2's-complement adder is compared to its modular version (normal) in Fig. 10.5.2. Saturating arithmetic will often eliminate or suppress the intolerably large oscillations that may occur during overflow periods in a nonsaturating system. Instead of large errors, smaller errors in the form of a nonlinearly distorted output process (i.e., "clipped") will occur. This is often a price worth paying when one considers the alternative.

BIBLIOGRAPHY

Fettweis, A. (1973), Roundoff Noise and Attenuation Sensitivity in Digital Filters with Fixed-Point Arithmetic, IEEE Trans. Circuit Theory, CT-20, March, pp. 174-175.

Jackson, L. B. (1976), Roundoff Noise Bounds Derived from Coefficient Sensitivities for Digital Filters, IEEE Trans. Circuits Syst., CAS-23, August, pp. 481-485.

Leon, B. J., and S. C. Bass (1974), Designers' Guide to Digital Filters [six-part series], EDN, January-June.

Oppenheim, A. V., and R. W. Schafer (1975), Digital Signal Processing, Prentice-Hall, Englewood Cliffs, N. J.

Rabiner, L. R., and B. Gold (1975), Theory and Application of Digital Signal Processing, Prentice-Hall, Englewood Cliffs, N. J.

11
Scaling

11.1 INTRODUCTION

The ability of a filter to provide controlled gain and phase behavior is of course the essence of digital filtering. However, the gains found within a digital filter may be such that they exceed the dynamic range limitations imposed on a data word by its finite word length. For example, it was argued previously that as the poles of a simple shift-invariant digital filter approach the periphery of the unit circle, its gain increases. This gain obviously cannot increase indefinitely. At some point, register overflow will occur. To inhibit or reduce the probability of register overflow, input scaling is often employed.

The problem of establishing an input scaling policy is one of computing, or in some cases estimating, the maximal gain of a specific filter. This study requires that magnitude bounds be computed in a prespecified normed vector space. To explore this problem, a series of motivational examples are presented. The purpose of these examples is to provide the reader with a set of conceptual and rudimentary algebraic tools by which scaling policies can be established. Later in this section, these ideas are developed in a deeper and more rigorous sense. Other chapters, which discuss the problem of roundoff errors, limit cycling, and so forth, build on the results presented in this chapter.

11.2 FIRST-ORDER SYSTEMS

A first-order filter satisfying $x(n) = ax(n - 1) + bu(n)$ and $y(n) = x(n)$ is shown in Fig. 11.2.1. It will be assumed that all system variables are bounded by unity. Using elementary algebra it can easily be shown that

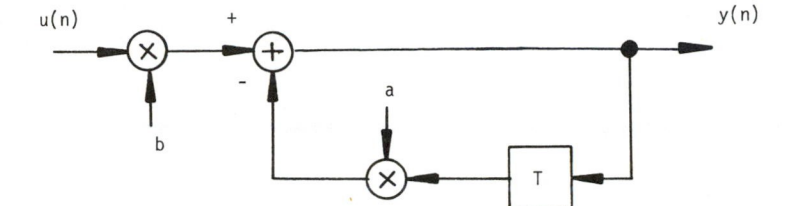

FIG. 11.2.1 Block diagram of a simple first-order filter having an input scale factor b.

$$h(n) = \sum_{i=0}^{n} a^i \tag{11.2.1}$$

which, at steady state, simplifies to

$$\sum_{i=0}^{\infty} a^i = \frac{1}{1 - |a|} \tag{11.2.2}$$

provided that $|a| < 1$ (i.e., stable filter). The resulting bound on the filter state and output variable, namely $x(n) = y(n)$, is therefore given by

$$|y(n)| = |h(n) * bu(n)| \leq |h(n)||bu(n)| \leq |h(n)||b| \leq 1 \tag{11.2.3}$$

If register overflow is to be inhibited at the filter's adder (see Fig. 11.2.1), then b, the input scale factor, must be bounded as follows:

$$\frac{|b|}{|h(n)|} \leq 1 \quad \text{or} \quad |b| \leq 1 - |a| \tag{11.2.4}$$

where it is assumed that $|u(n)| < 1$. Observe that as the filter pole moves toward the periphery of the unit circle (i.e., $a \to \pm 1$), $|b|$ must tend to zero. Since the filter takes on the form of an ideal integrator as $a \to +1$, the input scaling policy should be intuitively obvious.

11.3 SECOND-ORDER FILTERS

Second-order filter sections are one of the mainstays of filter design. It should be self-evident that a general shift-invariant filter, of the form $H(z) = N(z)/D(z)$, can be realized as a system of first- and second-order subfilters. It has been previously noted that the first-order subfilters will be used to synthesize second-order sections. Second-order subfilters may be modeled in terms of a linear combination of the following two impulse responses:

$$h_1(t) = \exp(-aT) \sin(\omega't) \overset{\mathcal{L}}{\longleftrightarrow} \frac{s + a}{(s + a)^2 + \omega}$$

$$\overset{Z}{\longleftrightarrow} \frac{1 - \exp(-aT) \cos(\omega'T)z^{-1}}{\Delta(z)} \tag{11.3.1}$$

$$h_2(t) = \exp(-aT) \cos(\omega'T) \overset{\mathcal{L}}{\longleftrightarrow} \frac{\omega'}{(s + a)^2 + \omega'^2}$$

$$\overset{Z}{\longleftrightarrow} \frac{\exp(-aT) \sin(\omega'T)z^{-1}}{\Delta(z)} \tag{11.3.2}$$

where

$$\Delta(z^{-1}) = 1 - 2 \exp(-aT) \cos(\omega'T)z^{-1} + \exp(-2aT)z^{-2}$$

An arbitrary second-order system can be constructed through the proper choice of a and ω'. The notation found in Eqs. (11.3.1) and (11.3.2) is often considered awkward. Instead, the impulse responses can be more conveniently represented in terms of $r = \exp(-aT)$ and $\cos(\omega'T) = \cos(\phi)$. Then

$$h_1(t) \overset{Z}{\longleftrightarrow} \frac{1 - r \cos(\phi)z^{-1}}{\Delta(z^{-1})}$$

$$= \frac{z^2 - r \cos(\phi)z}{\Delta(z)} \tag{11.3.3}$$

$$h_2(t) \overset{Z}{\longleftrightarrow} \frac{r \sin(\phi)z^{-1}}{\Delta(z^{-1})}$$

$$= \frac{r \sin(\phi)z}{\Delta(z)} \tag{11.3.4}$$

where

$$\Delta(z) = z^2 - 2r \cos(\phi)z + r^2$$

which has the canonical realization shown in Fig. 11.3.1. To realize $h_1(n)$ and $h_2(n)$, the coefficients (a_0, a_1, a_2) would have to be defined in terms of $(1, -r \cos(\phi), 0)$ and $(0, r \cos(\phi), 0)$, respectively. In general,

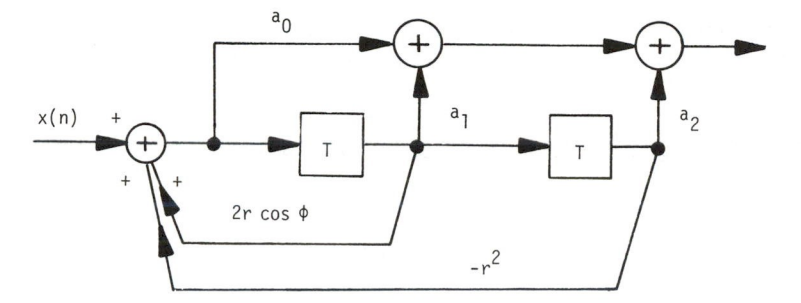

FIG. 11.3.1 Second-order canonic discrete filter architecture.

a number of the form $N(z) = a_0 + a_1 z^{-1} + a_2 z^{-2}$ can be synthesized through various linear combinations of $h_1(z)$ and $h_2(z)$.

The intrinsic advantage of this parameterization can be found in terms of the z-domain representation of second-order filters. The characteristic equation of the second-order filter is given by $D(z) = z^2 - 2r \cos(\phi)z + r^2$, which is factorable into two poles located at an angle of $\pm \phi$ and radius r (see Fig. 11.3.2). The natural resonant frequency is given by $\omega^* = \phi/T$ radians per second. Here T is the filter's sample rate. Furthermore, the real and imaginary projections of the poles onto the real and imaginary axis are given by $r \cos(\phi)$ and $\pm r \sin(\phi)$. Using this obvious geometry the z-domain description of $H_1(z)$ and $H_2(z)$ immediately follows and is displayed in Fig. 11.3.3. The responses of these two filters are evaluated at three key frequencies and are summarized as follows:

$H(j\omega T)$	$z = \exp(j0) = 1$ (dc)	$z = \exp(j\phi)$	$z = \exp(j\pi) = -1$
$H_1(e^{(j\omega T)})$	$\left(\dfrac{1 - r \cos \phi}{1 - 2r \cos \phi + r^2}\right)^2$	$\dfrac{1 - r^2 \cos^2 \phi}{(1 - r^2)[1 + r^2 - 2r \cos(2\phi)]}$	$\left(\dfrac{1 + r \cos \phi}{1 + 2r \cos \phi + r^2}\right)^2$
$H_2(e^{(j\omega T)})$	$\left(\dfrac{r \sin \phi}{1 + 2r \cos \phi + r^2}\right)^2$	$\dfrac{r^2 \sin^2 \phi}{(1 - r^2)[1 + r^2 - 2r \cos(2\phi)]}$	$\left(\dfrac{r \sin \phi}{1 + 2r \cos \phi + r^2}\right)^2$

It may be recalled that in the study of a first-order system, a maximal bound on the impulse response h(n) was computed. It was concluded that $|h(n)| = 1/(1 - |a|)$. This metric, derived from time domain considerations, provided the user with a scale factor which ensured that any input, bounded by unity, would not overflow the dynamic range of the system's registers. Consider again the first-order filter under the assumption that a is positive. Under this hypothesis, the realized filter is decidedly lowpass. Therefore, a worst-case input for this filter is obviously dc. The steady-state response of this filter, to a dc stimulus, can also be derived using frequency domain

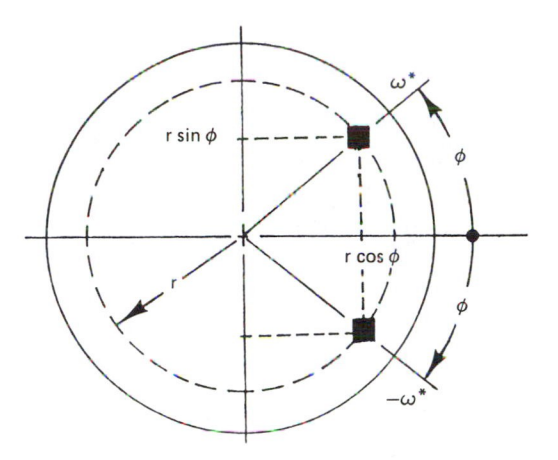

FIG. 11.3.2 Angular orientation of the pole pairs of a second-order system.

methods. In particular, it can be immediately seen that the dc gain of the filter is $|H(\exp(j0))| = |H(1)| = 1/(1 - a)$. This bound is in agreement with that derived from a time domain viewpoint. Therefore, it may be possible to develop a maximal bound estimate based on a profile of the frequency response for a specific filter. From this profile the filter's maximal gain can be determined and a suitable scale factor extracted.

Returning to the second-order example, consider the input time series x(n) to be bounded by unity. Furthermore, let the magnitude response of the filter be globally maximal (not necessarily unique) at $\omega = \omega^0$. That is, $|H(\exp(j\omega^0))| \geq |H(\exp(j\omega))|$ for all $\omega \in [0, \omega_s/2]$. If x(n) is a single tone

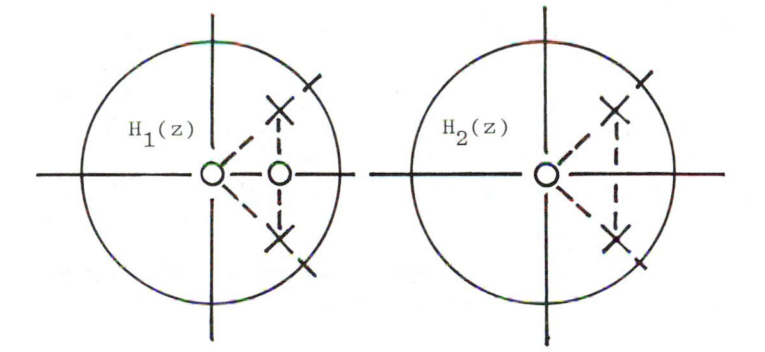

FIG. 11.3.3 z-Plane representation of filters H_1 and H_2.

of the form $x(n) = \cos(n\omega T_s)$, it follows that at steady state, $y(n) = A \cos(n\omega T_s + \phi)$, where A is given by $|H(\exp(j\omega))|$. Then

$$|y(n)| = |H(\exp(j\omega))| = A \leq |H(\exp(j\omega^0))| \qquad (11.3.5)$$

In terms of specific second-order parameters, for $H(z) = z^i/(z^2 - cz + d)$, it follows immediately from $|H(z)|^2 = |H(z)H(z^*)|$ that

$$|H(\exp(j\omega T))|^2 = \left. \frac{z^i z^{i*}}{(z^2 - cz + d)(z^{2*} - cz^* + d)} \right|_{z=\exp(j\omega T)}$$

$$= \frac{1}{(1 + c^2 + d^2 + 2d\cos(2\omega T) - 2c(d+1)\cos(\omega T))}$$

The local maxima of Eq. (11.3.6) can be obtained by using unconstrained differentiation. Upon differentiating, one obtains

$$\omega^0 = \begin{cases} \dfrac{\cos^{-1}(p)}{T} & \text{if } |p| \leq 1 \\[2mm] 0 & \text{if } p > 1 \\[2mm] \dfrac{\pi}{T} & \text{if } p < -1 \end{cases} \qquad (11.3.7)$$

where ω^0 is the maximizing frequency and $p = c(d+1)/4d$. The multiplicity of conditions are summarized (in the first quadrant) in Fig. 11.3.4. In terms of the parameterization $c = 2r\cos(\phi)$ and $d = r^2$, Eq. (11.3.7) can be alternatively written to read

$$p = \frac{2r\cos(\phi)(r^2 + 1)}{4r^2} = \frac{\cos(\phi)(r^2 + 1)}{2r} \qquad (11.3.8)$$

The data found in this figure point out a rather important fact—that the natural resonant frequency, denoted ω^*, does not in general define the maximizing frequency. Only when $z = \pm j$ (i.e., $c = 0$) will these two frequencies agree. However, when the poles of a linear filter are near the periphery of the unit circle, the more easily computed natural resonant frequency [i.e., $z = r\cos(\phi) \pm jr\sin(\phi)$] is often used to estimate the maximal gain. Under this assumption, the maximal filter gain can be estimated to be given by

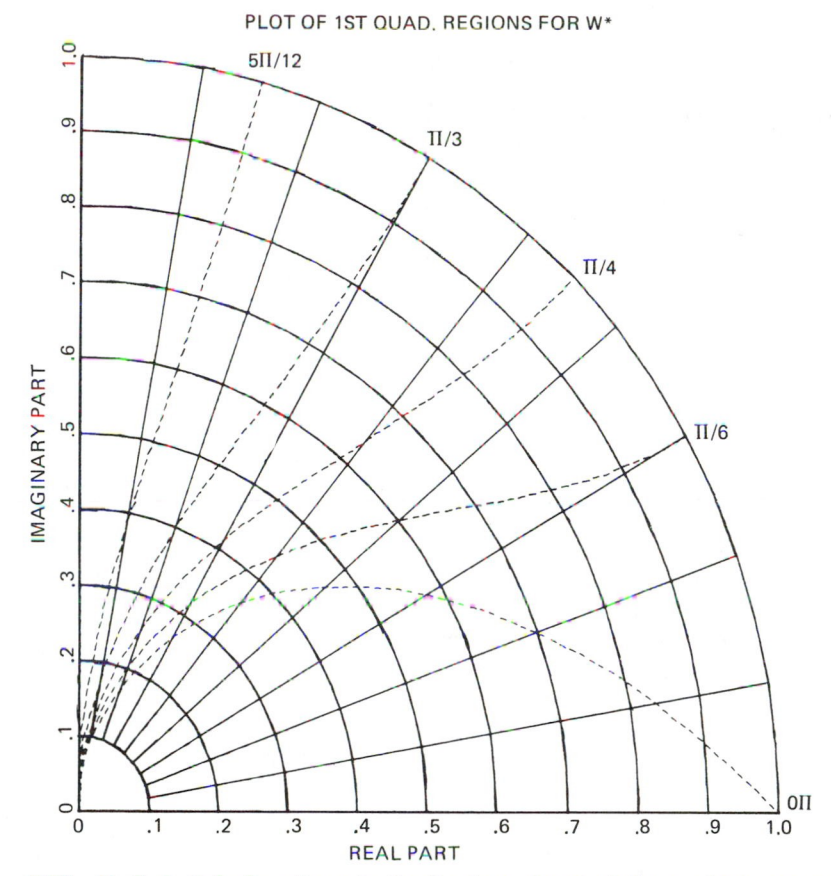

FIG. 11.3.4 Pole locations in the first quadrant of the z-plane, as a function of natural resonate frequency and damping factor. Dashed line indicates loci of resonate frequency for specified value of W^*/T.

$$|H(\exp(j\phi)|^2 = \frac{1}{z - r\cos(\phi) + jr\sin(\phi)} \; \frac{1}{z - r\cos(\phi) - jr\sin(\phi)}$$

$$= \frac{1}{(1 - r^2)[1 + r^2 - 2r\cos(2\phi)]} \qquad (11.3.9)$$

with the pole located at the polar coordinates $(r, \pm\phi)$ and $\omega^* = \phi/T$. From the previous analysis, for $\omega^0 \neq 0$, one observes that at the theoretical resonant frequency

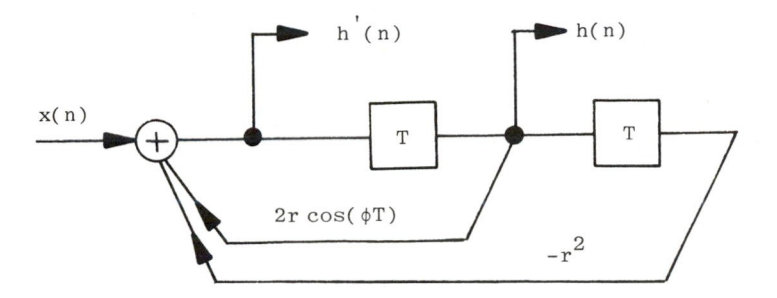

FIG. 11.3.5 Second-order model exhibiting two possible output ports.

$$|H(\exp(j\omega^0 T))|^2 = \frac{1}{[1 - \cos^2(\phi)](1 - r^2)} = \frac{1}{(1 - r^2) \sin^2(\phi)}$$

$$> |H(\exp(j\omega T))|^2 \quad \text{for all } \omega \qquad (11.3.10)$$

Returning to the basic second-order model (see Fig. 11.3.5), one notes that h(n), as diagrammed, is given by $h(n) = r^n \cos(n\phi T)$. The other indicated time series, denoted h'(n), is given by $h'(n) = h(n + 1)/r \sin(\phi T)$. If one defines the output process to be $z(n) = h'(n)$, then

$$|z(n)|^2 \leq x_{max} \sum_{n=0}^{\infty} |h(n)|^2 = \sum_{n=0}^{\infty} \frac{|r^{n+1} \cos[(n + 1)\varphi T]|}{|r \sin(\phi T)|^2}$$

$$\leq \frac{1}{\sin^2(\phi T)} \sum_{n=0}^{\infty} r^{2n} = \frac{1}{(1 - r^2) \sin^2(\phi T)} \qquad (11.3.11)$$

It should be interesting to note that the maximal magnitude bound established through this time-series analysis is that obtained based on a theoretical resonate frequency approach. Therefore, to ensure that no register overflow will occur, x(n) should be scaled by $\sin^2(\phi T)/(1 - r^2)$.

EXAMPLE 11.3.1 For the values a = 0.99, 0.9, 0.8, 0.5, 0.25 and $\omega' = \pi/6$, the frequency response $H_2(\exp(j\omega))$ [for $h_2(n)$ given in Eq. (11.3.2)] is interpreted graphically in Fig. 11.3.6. Observe that for small values of the parameter a, the filter becomes essentially lowpass. As the parameter a approaches unity, the realized resonant frequency approaches the natural resonant frequency $\pi/6$.

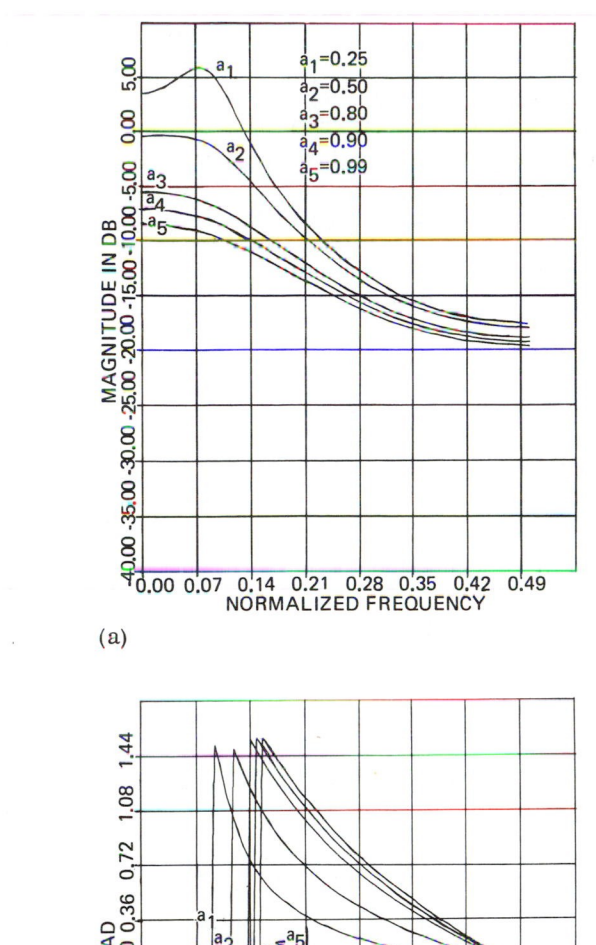

(a)

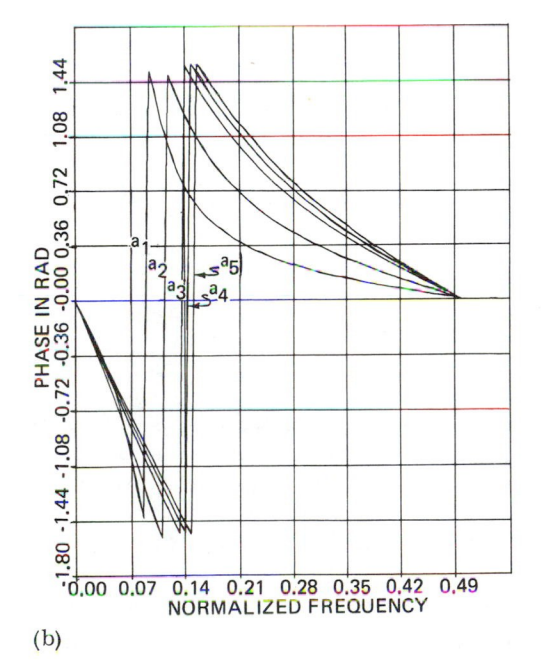

(b)

FIG. 11.3.6 (a) Magnitude frequency and (b) phase response of a second-order filter as a function of the design parameter a and a fixed resonate frequency.

EXAMPLE 11.3.2 The magnitude response of filter h_2 is summarized in Fig. 11.3.7 for the values r = 0:99, 0.9, 0.8, 0.5, 0.25 and $\phi = \pi/6$. It can be seen that only when the poles of $H_2(z)$ are near the periphery of the unit circle (i.e., $r \longrightarrow 1$) will the realized maximal gain near the natural resonate frequency $\pi/6$. The computed maximal (resonate) frequencies are:

$$r = 0.99: \quad p = \frac{\cos(\pi/6)(0.99^2 + 1)}{2(0.99)} = 0.8660692;$$

$$\omega^0 = \cos^{-1}(p) = 0.99983 \, \frac{\pi}{6}$$

$$r = 0.90: \quad p = \frac{\cos(\pi/6)(0.9^2 + 1)}{2(0.9)} = 0.8708367;$$

$$\omega^0 = \cos^{-1}(p) = 0.98146 \, \frac{\pi}{6}$$

$$r = 0.80: \quad p = \frac{\cos(\pi/6)(0.8^2 + 1)}{2(0.8)} = 0.887676;$$

$$\omega^0 = \cos^{-1}(p) = 0.91391 \, \frac{\pi}{6}$$

$$r = 0.50: \quad p = \frac{\cos(\pi/6)(0.5^2 + 1)}{2(0.5)} = 1.0825318 > 1; \quad \omega^0 = 0 \, \frac{\pi}{6}$$

$$r = 0.25: \quad p = \frac{\cos(\pi/6)(0.25^2 + 1)}{2(0.25)} = 1.840304 > 1; \quad \omega^0 = 0 \, \frac{\pi}{6}$$

which is in agreement with the data plotted.

11.4 ARCHITECTURE DEPENDENCE

It should now be apparent that it is desirable to reduce the worst-case magnitude bound associated with a set of filter variables. It turns out that these bounds are architecture dependent. As a result, one can often influence the magnitude of an input scaling policy by choosing alternative architectures. Since scaling and precision are inversely related, the accuracy of a given digital realization of some given H(z) can often be extended by simply experimenting with different filter configurations.[†] For example, one may

---

[†] As V, equivalently Q, found in Fig. 10.2.2, increases, the error variance due to roundoff or truncation also increases. This increased error reduces the precision of the data.

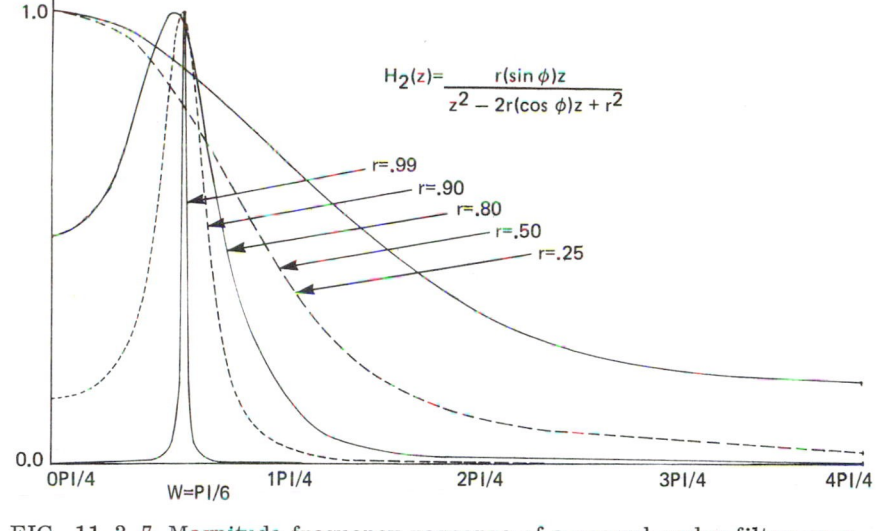

$$H_2(z) = \frac{r(\sin\phi)z}{z^2 - 2r(\cos\phi)z + r^2}$$

r=.99
r=.90
r=.80
r=.50
r=.25

FIG. 11.3.7 Magnitude frequency response of a second-order filter as a function of the design parameter r and a fixed resonate frequency.

seek to realize the second-order filters diagrammed in Fig. 11.4.1. These filters differ in terms of their feedforward paths and operations count. The more complex filter requires two additional adders. Therefore, it would be desirable to know a priori whether or not the additional cost and reduced throughput of the more complex filter will provide improved resolution. The two filters are analyzed side by side using Mason's gain formula (see Table 11.4.1), where $H_i(\exp(j\omega T))$ is the transfer function from input to y_i.

EXAMPLE 11.4.1 Let $\Delta(z) = z^2 - 2r\cos(\theta)z + r^2$ such that $p_2 = r^2 = 0.9698$, $-p_1 = 2r\cos(\theta) = 1.94$, $\theta = 9.94$ (natural resonant frequency). Then

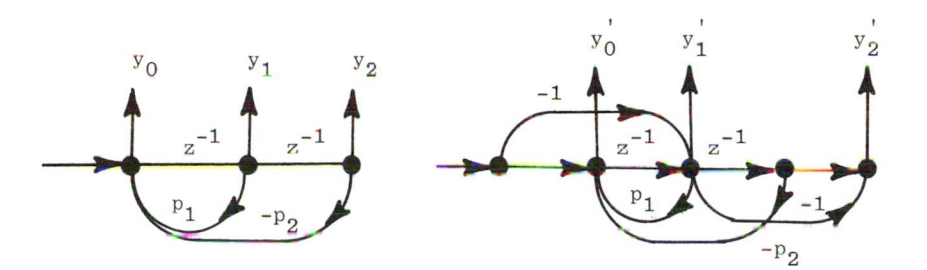

FIG. 11.4.1 Flow diagram of two realizations of a second-order discrete FIR filter.

TABLE 11.4.1 Comparative Analysis of the Architecture Shown in Fig. 11.4.1

To:	Y_0	Y_1	Y_2	Y'_0	Y'_1	Y'_2
Feedforward paths	$M_1 = 1$	$M_1 = z^{-1}$	$M_1 = z^{-2}$	$M'_1 = 1$ $M'_2 = -p_1$ $M'_3 = p_2 z^{-1}$	$M'_1 = z^{-1}$ $M'_2 = -1$	$M'_1 = z^{-2}$ $M'_2 = -z^{-1}$ $M'_3 = -z^{-1}$ $M'_4 = 1$
Induced feedback paths	M_1: $\Delta_1 = 1$	M_1: $\Delta_1 = 1$	M_1: $\Delta_1 = 1$	M'_1: $\Delta'_1 = 1$ M'_2: $\Delta'_2 = 1$ M'_3: $\Delta'_3 = 1$	M'_1: $\Delta'_1 = 1$ M'_2: $\Delta'_2 = 1$	M'_1: $\Delta'_1 = 1$ M'_2: $\Delta'_2 = 1$ M'_3: $\Delta'_3 = 1$ M'_4: $\Delta'_4 = 1$
Nontouching feedback paths	None	None	None	None	None	None
Transfer function	$H_0(z) = \dfrac{z^2}{\Delta(z)}$	$H_1(z) = \dfrac{z}{\Delta(z)}$	$H_2(z) = \dfrac{1}{\Delta(z)}$	$H'_0(z) = (1 - p_1)*$ $\dfrac{z[z + p_2/(1 - p_1)]}{\Delta(z)}$	$H'_1(z) = \dfrac{1 - z}{\Delta(z)}$	$H'_2(z) = \dfrac{(1 - z)^2}{\Delta(z)}$
$\Delta(z)$	$\Delta(z^{-1}) = 1 - p_1 z^{-1} + p_2 z^{-2}$ or $\Delta(z) = z^2 - p_1 z + p_2$					
$\lvert H(e^{j\omega T})\rvert^2$	$\lvert H_0\rvert^2 = \lvert H_1\rvert^2 = \lvert H_2\rvert^2 = \dfrac{1}{\lVert\Delta(\exp(j\omega T))\rVert^2}$			$\lvert H'_0\rvert^2 = \dfrac{(1 + p_1)^2 + 1 + 2(1 + p_1)\cos(\omega T)}{\lVert\Delta(\exp(j\omega T))\rVert^2}$	$\lvert H'_1\rvert^2 = \dfrac{2(1 + \cos(\omega T))}{\lVert\Delta(\exp(j\omega T))\rVert^2}$	$\lvert H'_2\rvert^2 = \dfrac{2(1 + \cos(\omega T))}{\lVert\Delta(\exp(j\omega T))\rVert^2}$

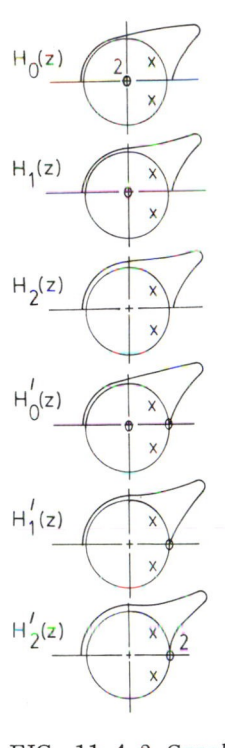

FIG. 11.4.2 Graphical interpretation of the pole-zero and magnitude frequency response of the sample second-order transfer functions.

the pole-zero diagrams for the two filter architectures under study are as suggested in Fig. 11.4.2. The maximal gains are found to be

$$|H_0|^2_{max} = |H_1|^2_{max} \Rightarrow |H_2|^2_{max} \simeq 194$$

$$|H'_0|^2_{max} \simeq 35 > |H'_1|^2_{max} = |H'_2|^2_{max}$$

Therefore, by choosing the second architecture, the maximal internal filter gain is reduced by a factor greater than 2 [i.e., $(194/35)^{0.5} = 2.35$].

In general, a second-order filter section can realize a transfer function of the form $H(z) = (n_0 z^2 + n_1 z + n_2)/(z^2 + p_1 z + p_2)$. Using the first architecture found in Fig. 11.4.1, if $H(z) = k_0 H_0(z) + k_1 H_1(z) + k_2 H_2(z)$, then $k_i = n_i$ for $i = 1$, 2, and 3. If the second architecture is used, and if $H(z) = m_0 H'_0(z) + m_1 H'_1(z) + m_2 H'_2(z)$, it follows that

$$\frac{\begin{bmatrix} 1.0 & 1.0 & 1.0 \\ p_2 & 1.0 - p_1 & -p_2 - 2p_1 + 2.0 \\ 0.0 & 0.0 & p_2 + p_1 - 1.0 \end{bmatrix} \begin{bmatrix} n_0 \\ n_1 \\ n_2 \end{bmatrix}}{p_2 - p_1 + 1.0} = \begin{bmatrix} m_0 \\ m_1 \\ m_2 \end{bmatrix} \qquad (11.4.1)$$

provided that $p_2 - p_1 + 1.0 \neq 0$. In terms of the numerical values found in this example, the set $\{n_i\}$ may be solved for in terms of the set $\{m_i\}$ by inverting the following matrix:

$$\frac{\begin{bmatrix} 1.0 & 1.0 & 1.0 \\ 0.9698 & 2.94 & 4.9102 \\ 0 & 0 & -1.97 \end{bmatrix} \begin{bmatrix} n_0 \\ n_1 \\ n_2 \end{bmatrix}}{3.9098} = \begin{bmatrix} m_0 \\ m_1 \\ m_2 \end{bmatrix} \qquad (11.4.2)$$

If the absolute value of n_i exceeds unity, these output coefficients could be treated as an extended-precision word.

There remains one flaw in this development. It can be observed that some of the adders found in Fig. 11.4.1 have more than two inputs. Therefore, there may be the potential for a partial sum to exceed the computed maximal bound. For example, in those cases where there are three inputs to an adder, there are $(3:2) = 3$ ways to combine these variables two at a time [in general, $(n:j)$, where $(n:j)$ denotes the binomial coefficient]. If a 2's-complement system is used, it is a moot point. It will be demonstrated that as long as the final sum does not overflow, it makes no difference what the partial-sum values are. That is, a 2's-complement system is tolerant of partial-sum overflows. However, in a sign-magnitude system, partial sum overflows are disastrous.

EXAMPLE 11.4.2 Using a 4-bit adder, add $-7/8$, $-3/8$, and $5/8$.

Entry	2's complement	1's complement	Sign-magnitude
$-7/8$	1.001	1.000	1.111
$-3/8$	1.101	1.100	1.011
$-10/8$	1 0.110 = 6/8	1 0.100 = 4/8	1 1.010 = $-2/8$
		└─ end-around carry	
		+ 0.001	
		0.101 = 5/8	
$+5/8$	0.101	0.101	0.101
$-5/8$	1.011 = $-5/8$	1.010 = $-5/8$	1.111 = $-7/8 \neq -5/8$

This simple example demonstrates the fact that if a sign-magnitude encoding scheme is used, the derived scaling policy must also ensure that partial sums do not overflow.

EXAMPLE 11.4.3 Example 11.4.1 will now be repeated in terms of various orderings of partial sums. The results of this analysis are summarized in Fig. 11.4.3. Since, in all cases, the magnitude of $A = |A_2(j\omega)|^2$ exceeds that of $|A_1(j\omega)|^2$, A will define the maximal bound and therefore the scaling policy.

In higher-order systems, it is extremely difficult, if not impossible, to derive analytically the location of the maximizing frequency. The maximal gain of a filter can often be approximated by using the FFT. Here the magnitude response would be scanned and the maximal gain recorded.

Important Observation

It has been observed in this section that the magnitude bounds on the states of a filter are influenced by architecture and pole-zero pairing. High-order filters are generally realized in terms of a system of second-order sub-filters. That is, an Nth-order filter, say H(z), would be realized as

$$H(z) = \prod_{i=1}^{N/2} H_i(z) = \prod_{i=1}^{N/2} \frac{(z - z_i)(z - z_i^*)}{(z - p_i)(z - p_i^*)} \qquad (11.4.3)$$

where the filter's poles and zeros are given by p_i and z_i, respectively. It has become accepted design practice to group the poles and zeros of a filter on the basis of their proximity. That is, second-order sections would be formed by choosing p_i to be the closest zero to z_i. This is suggested in Fig. 11.4.4. Using this empirical design rule, the internal magnitude bound on the states of the filter will assume a minimal maximal value (i.e., minimax).

11.5 SECOND-ORDER SCALING IN STATE SPACE

For the purpose of gaining additional insight into the scaling problem, a state-variable approach will be used. The state-determined model will be given by

$$x(k + 1) = Ax(k) + bu(k); \quad y(k) = cx(k) + du(k - 1) \qquad (11.5.1)$$

while the transfer function satisfies

$$H(z) = (zI - A)^{-1}bz \qquad (11.5.2)$$

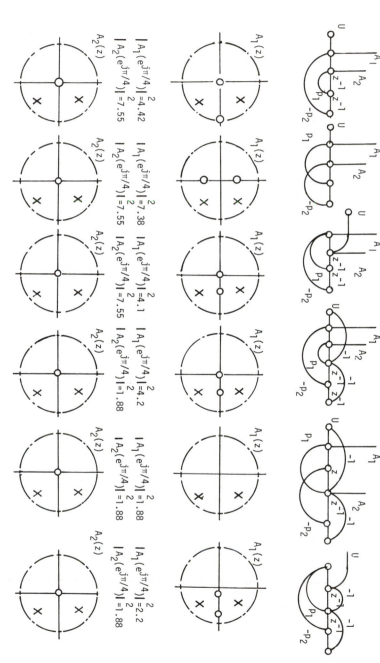

FIG. 11.4.3 Detailed analysis of the magnitude sensitivity of a second-order filter to the order in which arithmetic is performed.

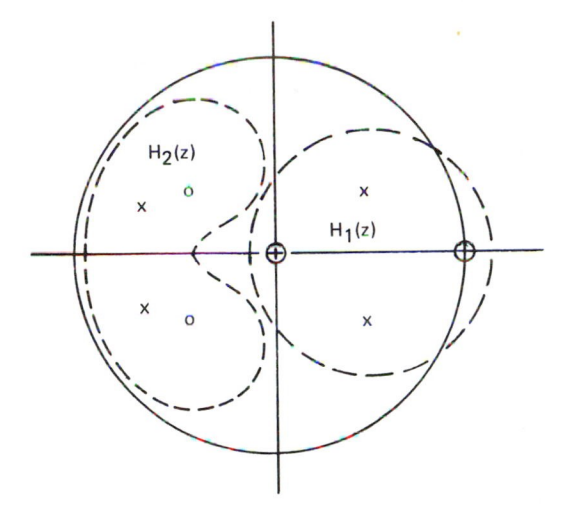

FIG. 11.4.4 Empirical optimal organization of poles and zeros of a discrete filter.

The input-output transfer function is defined by $G(z) = Y(z)/U(z)$, where $H(z)$ satisfies

$$G(z) = d + z^{-1}cH(z) \tag{11.5.3}$$

This is summarized in Fig. 11.5.1. In terms of the example problem discussed in the preceding section, it can be shown that

$$A = \begin{bmatrix} 1.2726 & -0.81 \\ 1.0 & 0.0 \end{bmatrix}; \quad b = \begin{bmatrix} 1 \\ 0 \end{bmatrix}; \quad c = (2.272, -0.81); \quad d = 1.0$$

Using Mason's gain formula, it was shown that

$$H_1(z) = \frac{z^2}{z^2 - 1.272z + 0.81}$$

$$H_2(z) = \frac{z}{z^2 - 1.272z + 0.81}$$

and $H(z)$ equals

$$H(z) = (zI - A)^{-1}bz = \begin{bmatrix} z - 1.27 & 0.81 \\ -1.0 & z \end{bmatrix} \begin{pmatrix} z \\ 0 \end{pmatrix} = \frac{(z^2, z)^T}{z^2 - 1.272z + 0.81}$$

$$= \begin{pmatrix} H_1(z) \\ H_2(z) \end{pmatrix}$$

The poles of this filter are found at $z = 0.9 \exp(\pm j\pi/2)$. Based on linear algebraic considerations, it follows that

$$G(z) = d + z^{-1}cH(z) = 1 + \frac{2.272z - 0.81}{z^2 - 1.272z + 0.81}$$

$$= \frac{z(z-1)}{z^2 - 1.272z + 0.81}$$

which is in complete agreement with the previously stated result. The potential advantage of the state-space representation is in comparing architectures. Using standard linear (matrix) algebraic methods, the four-tuple (A, b, c, d) can be mapped into the alternative architecture (A', b', c', d'). This concept is developed in detail in the next section.

EXAMPLE 11.5.1 A second-order ladder network is abstracted in Fig. 11.5.2. In a state-space format, the network three-tuple (A, b, c) is given by

$$A = \begin{bmatrix} -a_1a_2 & -a_3 \\ a_1a_2a_3 & -a_2a_4 \end{bmatrix}; \quad b = \begin{bmatrix} ba_2 \\ ba_3a_4 \end{bmatrix}, \quad c = (c_1, c_2)$$

It follows that

$$(zI - A)^{-1}b = \frac{\begin{bmatrix} z + a_2a_4 & a_3 \\ -a_1a_3a_4 & z + a_1a_2 \end{bmatrix} \begin{bmatrix} ba_2 \\ ba_3a_4 \end{bmatrix}}{z^2 + (a_1a_2 + a_2a_4)z + a_1a_3a_4 + a_1a_2a_4 = \Delta(z)}$$

which simplifies to

$$= \begin{bmatrix} \dfrac{zba_2 + a_2^2a_4b + a_3^2a_4b}{\Delta(z)} \\ \dfrac{za_3a_4b}{\Delta(z)} \end{bmatrix}$$

This result can be verified through the use of Mason's gain formulas as follows:

Item	To x_1	To x_2
	$M_1 = z^{-1}a_2 b$	$M_1 = a_3 a_4 bz^{-1}$
	$M_2 = a_3^2 a_4 bz^{-2}$	
Individual loops	$\Delta_{11} = -a_1 a_2 z^{-1}$	$\Delta_{12} = -a_1 a_3^2 a_4 z^{-2}$
		$\Delta_{13} = -a_2 a_4 z^{-1}$
Dual nontouching loops	$\Delta_{21} = (-a_1 a_2 z^{-1})(-a_2 a_4 z^{-1}) = a_1 a_2^2 a_4 z^{-2}$	
Characteristic equation	$\Delta = 1 - (\Delta_{11} + \Delta_{12} + \Delta_{13}) + \Delta_{21} = \Delta(z)$	as before

Induced loops

$$M_1: \qquad \Delta_1 = 1 - (-a_2 a_4 z^{-1}) \qquad\qquad M_2: \qquad \Delta_1 = 1 - 0$$

$$M_2: \qquad \Delta_2 = 1 - 0$$

Transfer function	$H_1(z) = \dfrac{M_1 \Delta_1 + M_2 \Delta_2}{\Delta(z)}$	$H_2(z) = \dfrac{a_3 a_4 bz^{-1}}{\Delta(z)}$

$$= \frac{a_2^2 a_4 bz^{-2} + a_3^2 a_4 bz^{-2} + a_2 bz^{-1}}{\Delta(z)}$$

which agrees with the state-variable result. Finally G(z) can be seen to satisfy

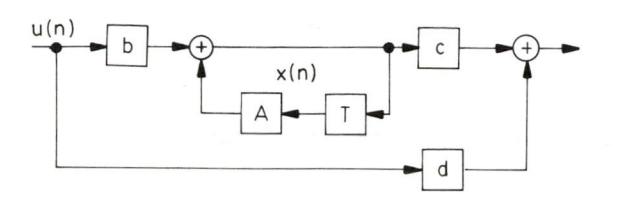

FIG. 11.5.1 Block diagram description of a state-determined system.

$$G(z) = d + cH(z) = d + (c_1, \ c_2)(H_1(z), \ H_2(z))^T$$

$$= 1 + \frac{z(a_3 a_4 c_2 + a_2 bc_1) + (c_1 a_3^2 a_4 b + a_2^2 a_4 bc_1)}{\Delta(z)}$$

If $H(z)$ is defined to be

$$H(z) = \frac{z(z - 1)}{z^2 - 1.272z + 0.81} = 1 + \frac{2.272z - 0.81}{z^2 - 1.272z + 0.81}$$

then a nonunique set of ladder coefficients $(a_1, \ a_2, \ a_3, \ a_4, \ b, \ c_1, \ c_2)$ can be computed. These coefficients can be used to architect the Gray-Markel filter found in Fig. 11.5.3.

In terms of Mason's gain formula, it follows that

$$\Delta_{11} = -\alpha_0 \alpha_1 z^{-1} \qquad \Delta_{13} = -\alpha_0 z^{-1}$$

$$\Delta_{12} = -\alpha_1 z^{-2} \qquad \Delta_{14} = -\alpha_0^2 \alpha_1 z^{-2}$$

Therefore, $\Delta(z) = 1 + (\alpha_0 \alpha_1 + \alpha_1)z^{-1} + (\alpha_1 + \alpha_1 \alpha_0^2)z^{-2}$. In tabular form, the data represents:

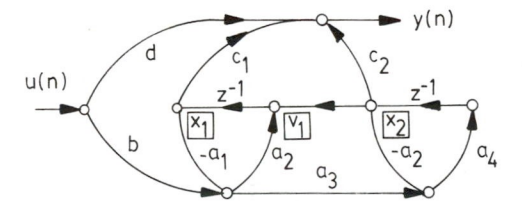

FIG. 11.5.2 Flow diagram of a second-order ladder network indicating state locations.

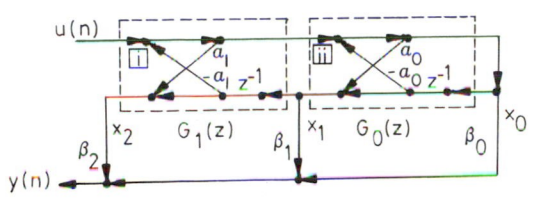

FIG. 11.5.3 Flow diagram for a second-order Gray-Markel filter section decomposed into two first-order filter sections.

To x_2	To x_1	To x_0
$M_1 = \alpha_1$	$M_1 = \alpha_1$	$M_1 = 1$
$\Delta_1 = 1 + \alpha_0 z^{-1}$	$\Delta_1 = 1$	$\Delta_1 = 1$
$M_2 = \alpha_0 z^{-1}$	$M_2 = z^{-1}$	
$\Delta_2 = 1$	$\Delta_2 = 1$	
$M_3 = z^{-2}$		$H_0(z) = \dfrac{1}{\Delta(z)}$
$\Delta_3 = 1$		

$$H_1(z) = \frac{\alpha_0 + z^{-1}}{\Delta(z)}$$

$$H_2(z) = \frac{\alpha_1 + (\alpha_1 \alpha_0 + \alpha_0)z^{-1} + z^{-2}}{\Delta(z)}$$

The connection between the coefficient set $\{a_i\}$ and $\{\alpha_i\}$ can be derived in a straightforward manner. For example, in terms of the architecture exhibited in Fig. 11.5.4, one obtains

$$x_2(n + 1) = -\alpha_0 x_2(n) + \alpha_1 x_1(n) + u(n)$$

$$x_1(n + 1) = x_2(n) - \alpha_0 \alpha_1 x_1(n) + \alpha_0 u(n)$$

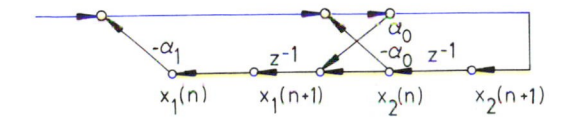

FIG. 11.5.4 Detail of a Gray-Markel second-order subsection.

or

$$x(n + 1) = \begin{bmatrix} -\alpha_0 \alpha_1 & 1 \\ -\alpha_1 & -\alpha_0 \end{bmatrix} x(n) + \begin{bmatrix} \alpha_0 \\ 1 \end{bmatrix} u(n) \triangleq \widetilde{A}x(n) + \widetilde{b}u(n)$$

The transfer function that exists between $u(n)$ and $x_1(n)$ or $x_2(n)$ can be defined in terms of

$$(zI - \widetilde{A})^{-1} \widetilde{b}z = \begin{bmatrix} z + \alpha_0 \alpha_1 & -1 \\ \alpha_1 & z + \alpha_0 \end{bmatrix}^{-1} \begin{pmatrix} \alpha_0 \\ 1 \end{pmatrix} z$$

$$= \frac{\begin{pmatrix} \dfrac{\alpha_0 z(z + \alpha_0) + z}{z^2} \end{pmatrix}}{z^2 + (\alpha_0 \alpha_1 + \alpha_0)z + \alpha_0^2 \alpha_1 + \alpha_1}$$

Continuing, the set (α_0, α_1) can be equated, if desired, to the coefficient set found in the model

$$H(z) = \frac{N(z)}{D(z)} = \frac{\displaystyle\sum_{i=0}^{n} c_i z^{-i}}{\displaystyle\sum_{i=0}^{n} a_i z^{-i}}$$

Using the ladder synthesis formulas found in Section 8.2, it follows that

1. $N_j(z) = \displaystyle\sum_{i=0}^{j} c_{j,i} z^{-i}$ and $D_j(z) = \displaystyle\sum_{i=0}^{j} a_{j,i} z^{-i}$

2. $\alpha_{j-1} = a_{j,j}$ and $\beta_j = c_{j,j}$ with $\beta_0 = c_{0,0}$

3. $D_{j-1}(z) = [D_j(z) - \alpha_{j-1} z B_j(z)]/(1 - \alpha_{j-1}^2)$

4. $N_{j-1}(z) = N_j(z) - z B_j(z)\beta_j$

5. $z B_j(z) = D_j(1/z)z^{-j}$

For example, if

$$H(z) = \frac{z^2}{z^2 - 1.272z + 0.81} = \frac{1}{1 - 1.272z^{-1} + 0.81z^{-2}}$$

Then, for $j = 2$,

1. $\alpha_1 = 0.81$ and $\beta_2 = 0$

2. $zB_2(z) = D_2(1/z)z^{-2} = (1 - 1.272z + 0.81z^2)z^{-2}$

$$= z^{-2} - 1.272z^{-1} + 0.81$$

3. $N_1(z) = N_2(z) - zB_2(z);\ \beta_2 = N_2(z)$

4. $D_1(z) = \dfrac{D_2(z) - \alpha_1 zB_2(z)}{1 - \alpha_i^2}$

$$= \frac{(1 - 1.272z^{-1} + 0.81z^{-2}) - (0.81z^{-2} - 1.0308z^{-1} + 0.6561)}{1 - 0.6561}$$

$$= 1 - 0.703z^{-1}$$

5. For $j = 1$

6. $\alpha_0 = -0.703$ and $\beta_1 = 0$

7. $\beta_0 = 1$

Upon substitution,

$$\Delta(z) = 1 + [0.81(-0.703) - 0.703]z^{-1} + [0.81 + 0.81(-0.703)^2)]z^{-2}$$

$$= 1 - 1.272z^{-1} + 0.81z^{-2}$$

$$H_2(z) = \frac{0.81 + [0.81(-0.703) - 0.703]z^{-1} + z^{-2}}{\Delta(z)}$$

$$H_1(z) = \frac{-0.703 + z^{-1}}{\Delta(z)}$$

$$H_0(z) = \frac{1}{\Delta(z)}$$

Extending this result to its final conclusion, it follows that

$$H(z) = H_2(z)\beta_2 + H_1(z)\beta_1 + H_0(z)\beta_0 = \frac{z^2}{z^2 - 1.272z + 0.81}$$

which is the desired result.

11.6 MATHEMATICS OF SCALING

Jackson and others have studied the problem of magnitude scaling from a functional analytic standpoint. Their work is based on the assumption that a 1's- or 2's-complement encoding of data is to be used. The reason for this was suggested in Section 11.4 when it was noted that these two numbering systems are tolerant of partial sum overflows. The development found in this section makes the same assumption and also assumes that a filter's input time series and internal variables are bounded by some value M. To exceed this value will result in register overflow. A scale factor, say k_0, can be used to guarantee that this case will never arise by requiring that

$$k_0 \leq \frac{1}{\displaystyle\sum_{k=0}^{\infty} |h(n)|} \tag{11.6.1}$$

A more rigorous statement can be made in terms of the theory of normed linear vector spaces. Those spaces, which influence the design and analysis of engineering systems, will be singled out for special study at this time.

For the purpose of clarity, the scaling problem will be briefly reviewed. Suppose that the dynamic range of all system variables (registers) is [-M, M]. The problem is simply to find a scale factor k_0 so that even under worst-case conditions, registers will not overflow. More specifically,

$$x_i(n) = k_0 h(n) * u(n) = k_0 \sum_{k=0}^{\infty} h(k)u(n-k) \tag{11.6.2}$$

where $x_i(n)$ is the ith-state variable, which can be bounded by

$$|x_i(n)| \leq k_0 \sum_{k=0}^{\infty} |h_i(k)| M \leq M \tag{11.6.3}$$

If $h_i(n)$ is the impulse response relating the input time series $\{u(n)\}$ to $\{x_i(n)\}$, choose

$$k_0 \le \frac{1}{\displaystyle\sum_{k=0}^{\infty} h_i(k)} \tag{11.6.4}$$

Such a choice of k_0 will guarantee that even under worst-case conditions, registers will not overflow.

The study of norms and linear vector spaces is a topic of advanced mathematics. Much of this subject is beyond the interest of the average reader. However, to gain insights into and awareness of this subject, some key topics will be singled out for special consideration. These topics have been chosen on the basis of their relationship to the problem of scaling.

Suppose that V and W are two linear vector spaces in which distance is measured in terms of some suitable norm. Consider for the moment a linear operator (or transform) ϕ, defined on V, such that $\phi : V \longrightarrow W$. If ϕ is a continuous map, then
is

$$|\phi(x_i) - \phi(x)| \longrightarrow 0 \text{ as } |x_i - x| \longrightarrow 0 \tag{11.6.5}$$

The mapping ϕ is said to be bounded if there exists an M such that

$$|\phi(x)| \le M|x| \quad \text{for all x in V} \tag{11.6.6}$$

For example, the output y(t) at t = 1 of a linear system having an impulse response h(t) is given by

$$y(1) = \phi(x) = \int_0^1 h(x, 1)x(s) \, ds \tag{11.6.7}$$

where h and $x \in C[0, 1]$ (the space of continuous functions on [0, 1]). It follows that y(t) also belongs to C[0, 1]. For $M = \max|h(r)|$, $0 \le r \le 1$, the norm of y, denoted $|y|$, may be defined as follows:

$$|y| = \max |y(s)| \le M \max |x(s)| = M|x| \quad \text{for } 0 \le s \le 1$$

Therefore, $\phi(x)$ is a bounded linear operator.

Many of the normed linear spaces found in popular use are familiar to the signal processing scientist. They are summarized in Table 11.6.1.

Some important linear vector-space concepts can be developed in the context of these familiar norms. First, the set of all convergent sequences, denoted c, is a closed subspace of l_∞. The distance metric on c is the supremum operator associated with l_∞. If the sequence limits are zero, the set is denoted c_0. Furthermore, l^n and l_∞ are special cases of C(X)

TABLE 11.6.1 Norms Induced by Common Linear Vector Spaces

Space	Norm
1. Real (complex) R or C	$\lvert x \rvert = \lvert x \rvert$
2. Real (complex) Euclidian space R^N or C^N	$\lvert x \rvert = (\Sigma_{i-1}^{N} \lvert x_i \rvert^2)^{1/2}$
3. ℓ_p, $1 \le p <$ (infinite sequence)	$\lvert x \rvert_p = [\Sigma_{i=1}^{\infty} \lvert x(n) \rvert^p]^{1/p}$
4. ℓ_p^k, $1 \le p < \infty$ (finite sequence)	$\lvert x \rvert_p = [\Sigma_{i=1}^{K} \lvert x(n) \rvert^p]^{1/p}$
5. $\ell_p[0, T]$, $1 \le p < \infty$	$\lvert x \rvert_p = [\int_T \lvert x(t) \rvert^p \, dm(t)]^{1/p}$
6. ℓ_∞ (infinite sequence)	$\lvert x \rvert_\infty = \sup[\lvert x(n) \rvert]$
7. ℓ^k (finite sequence)	$\lvert x \rvert_\infty = \max[\lvert x(n) \rvert; n \le k]$
8. $\ell_\infty [0, T]$	$\lvert x \rvert_\infty = \sup[\lvert x(t) \rvert]$
9. C(x) space of all continuous real valued functions on x	$\lvert x \rvert = \sup[f(x)]$: uniform norm $f(x) \in C(x)$
10. bv sequence of bounded variation	$\lvert x \rvert_{bv} = \lvert x_0 \rvert + \Sigma_{n=0}^{\infty} \lvert x_{n+1} - x_n \rvert$

over a discrete topology. In signal processing, a discrete topology is equivalent to the set of sample instances.

The L_p norms, generally studied by engineers, are usually parameterized by p = 1, 2, or ∞. They are graphically interpreted in Fig. 11.6.1 in a two-dimensional-space framework. Observe that the geometry associated with each of these norms is different. Therefore, one must specify a priori which norm or norms are being used to compare the performance of discrete systems and filters.

The inner product of two vectors, say x and y [denoted (x, y)], provides a user with a wealth of system information. Inner products, for example, can be used to measure angles in a generalized sense (e.g., orthogonality) as well as to define projections. A familiar example of the projection potential of inner products is found in the familiar Fourier transform. The Fourier transform represents a series of inner products where the nth coefficient, say c_i, is given by $c_i = (x, W_N^i)$ with $W_N = \exp(-j2\pi/N)$. That is, c_i is the projection of x into the W_N axis (or basis). This is suggested in Fig. 11.6.2.

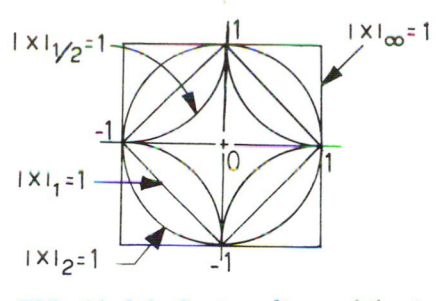

FIG. 11.6.1 Contour lines of the L_p norm of x equal to unity.

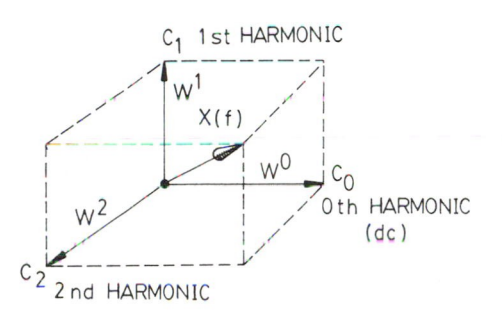

FIG. 11.6.2 Graphical interpretation of the first three coefficients of a Fourier transform in a linear vector space.

11.7 LINEAR FUNCTIONALS

We may extend this work to the concept of the <u>linear functional</u>. A linear real functional, say f(x), is a mapping of $x \in X$ onto the real line R (i.e., $f:X \longrightarrow R$). A functional is defined over a normed linear space X. The functional differs from an inner product in several ways (e.g., an inner product is a bilinear form $\phi :X \times X \longrightarrow R$ whereas a functional has the form $f:X \longrightarrow R$). A linear functional can be associated with a constant M such that $|f(x)| < M|x|$ for all $x \in X$. The smallest of these choices is called the <u>norm of f(x)</u> and is denoted f(x) . For an arbitrary f(x), all linear functionals defined on X are said to belong to the <u>conjugate space of X</u> and are denoted X^*. The study of conjugate spaces is important to signal processing and is therefore developed in greater detail.

The study of conjugate spaces makes heavy use of two important inequalities. They are:

1. Hölder inequality:

$$\sum_{i=1}^{n} |a_i b_i| \le |a|_p |b|_q; \quad \frac{1}{p} + \frac{1}{q} = 1 \tag{11.7.1}$$

2. Minkowski inequality:

$$|x + y| \le |x|_p + |y|_p \tag{11.7.2}$$

Hölder's inequality will be used to begin the study of conjugate spaces.

Linear Functionals on $\mathcal{L}_p[0, t]$

The form of a functional on $L_p[0, T]$ is

$$f(x) = \int_0^T x(t)\alpha(t)\, dt; \quad x(t) \in \mathcal{L}_p[0, T], \quad p > 1 \tag{11.7.3}$$

and is constrained by

$$\alpha(t) \in \mathcal{L}_q[0, T]; \quad \frac{1}{p} + \frac{1}{q} = 1 \tag{11.7.4}$$

The norm is formally defined to be

$$|f| = \left(\int_0^T |\alpha(t)|^q\, dt \right)^{1/q} < M \tag{11.7.5}$$

EXAMPLE 11.7.1 A stable, first-order, linear, continuous, time-invariant system can be modeled as $dy(t)/dt = ay(t) + x(t)$. The convolution integral, which defines $y(T)$, can be interpreted in terms of a functional given by

$$y(T) = f(x) = \int_0^T x(t) \exp[a(t - T)]\, dt \tag{11.7.6}$$

where the system's impulse response is given by $h(t) = \exp(-at)$. That is,

$$y(T) = f(x) = \exp(-aT) \int_0^T x(t) \exp(at)\, dt \tag{11.7.7}$$

If $x(t) \in \mathcal{L}_p[0, T]$, $p > 1$ (e.g., $x(t) = \cos(\omega_0 t) \in \mathcal{L}_p[0, T]$ since $[\int_0^T x(t)^p \, dt] <$ M for some M], then $y(t)$ is bounded since $|\exp(at)|_q$ is bounded.

EXAMPLE 11.7.2 If $p = 2$, then $q = 2$. For this reason $L_2[0, T]$ is called a <u>self-conjugate</u>. Therefore, the inner product of any two $L_2[0, T]$ functions is bounded.

Linear Functionals on ℓ_p, $p > 1$

The form of a linear functional on ℓ_p is

$$f(x) = \sum_{k=0}^{\infty} x_k c_k; \quad x_k \in l_p, \quad \{x\} = x_0, x_1, \ldots \tag{11.7.8}$$

under the constraint that

$$\{c_k\} \in \ell_q; \quad \frac{1}{p} + \frac{1}{q} = 1 \tag{11.7.9}$$

The norm is defined to be

$$|f| = \left(\sum_{k=0}^{\infty} |c_k|^q \right)^{1/q} \tag{11.7.10}$$

EXAMPLE 11.7.3 For $p = 2$, let $f(x) = \Sigma x_k c_k = c^T x$. If $\Sigma c_k^2 < M$, then $|f|$ is bounded and $|f| = (\Sigma c_k^2)^{1/2}$. We can note that ℓ_2 is also self-conjugate.

The self-conjugacy of L_2 and ℓ_2 account for much of their popularity. However, there are other equally important spaces that must be considered. They are summarized in Table 11.7.1. It is tempting to complete the list with $X = \ell_\infty$ and $X^* = \ell_1$. However, this is not the case. It has been shown that only if $X = \ell$ or c_0, is $X^* = \ell_1$. The analysis in this conjugate space requires a very technical argument, involving Stieltjes integration, and it is beyond the scope of this work to pursue this statement on a rigorous basis. Instead, only the major results will be presented.

Using the functional form $f(x) = g(0)x(0) + \Sigma (g(i + 1) - g(i))x(i)$, if $x \in X_\infty$, then the conjugate space of X_∞ is the space of all functions of bounded variation [i.e., $|f| = \text{var}(g)$].

EXAMPLE 11.7.4 Consider again the first-order shift-invariant system having an impulse response $h(i) = a^i$. The system's output is given by a convolution sum. The duality between the convolution sum and a linear functional can be developed as follows:

TABLE 11.7.1 Common Linear Vector Spaces and Their Conjugates, Induced Norm Definitions, and Linear Functional Formula

Primary space x	Conjugate space x^*	Functional description	Norm definition
ℓ_p^n; $p \leq 1$	ℓ_q^n	$f(x) = \sum_{i=0}^{n-1} \alpha_i x_i$; $x \in \ell_p$	$\|f\| = (\sum_{i=0}^{n-1} |\alpha_i|^p)^{1/p}$
ℓ_1^n	ℓ_∞^n	$f(x) = \sum_{i=0}^{n-1} \alpha_i x_i$; $x \in \ell_1^n$	$\|f\| = \sup(|\alpha_i|)$; all i
ℓ_∞^n	ℓ_1^n	$f(x) = \sum_{i=0}^{n-1} \alpha_i x_i$; $x \in \ell_\infty^n$	$\|f\| = \sum_{i=0}^{n-1} |\alpha_i|$
ℓ_p $(\mathcal{L}_p)$	ℓ_q $(\mathcal{L}_q)$	$f(x) = \sum_{i=0}^{n-1} \alpha_i x_i$; $x \in \ell_p$	$\|f\| = (\sum_{i=0}^{\infty} |\alpha_i|^p)^{1/p}$
ℓ_1 $(\mathcal{L}_1)$	ℓ_∞ $(\mathcal{L}_\infty)$	$f(x) = \sum_{i=0}^{n-1} \alpha_i x_i$; $x \in \ell_1$	$\|f\| = \inf(\sup|\alpha_i|)$

Convolution: $x(0) = \sum a^i u(n-1)$

$x(0) = u(0)$	$x(0) = g(0)u(0) = g(0) = 1$
$x(1) = u(1) + au(0)$	$x(1) = g(0)u(0) + g(1)u(1) - g(0)u(1) + g(1)$
$x(2) = u(2) + au(1) + a^2 u(0)$	$= 1 + a$
$\vdots$	$x(2) = g(0)u(0) + g(1)u(1) - g(0)u(1)$
	$\qquad + g(2)u(2) - g(1)u(2) + g(2) = 1 + a + a^2$
or	or
$h(i) = a^i$	$g(i) = \sum a^i$

It is immediately apparent that the norm of g [i.e., var (g)] is bounded by $1/(1 - |a|)$ for $|a| < 1$. As a point of intellectual reinforcement, we can interpret the previous example problem in the context of a lowpass filter. The maximal output of this class of filter occurs when the input is dc [i.e., $u(n) = 1$ for all n]. For a dc forcing function, the convolution sum produces an output also bounded by $\sum a^i 1/(1 - |a|)$.

The problem of finding a scale factor can also be managed in the frequency domain if we consider inputs to be periodic and bounded. It is well known that from the definition of a Fourier transform, that if x(n) is band-limited to B hertz, then

$$x(n) = \frac{1}{B} \int_0^B k_0 H(\omega) X(\omega) \exp(jn\omega t) \, d\omega \qquad (11.7.11)$$

Equation (11.7.11) represents a functional representation of $x(n)$ in terms of the variable $X(\ )$. If we choose to test at frequencies found on the periphery of the unit circle [i.e., $\exp(jn\omega T) = 1$], it can be shown that

$$|x(n)| \leq |k_0 H|_p |X_q| \qquad (11.7.12)$$

where

$$|X|_p = \left[\int_0^B |X(\omega)|^p \, d\omega/B \right]^{1/p} \qquad (11.7.13)$$

This result is related to Parseval's representation of a DFT. Again the study will be restricted to the case where $p = 1$, 2, and ∞. Here, if:

1. $|X|_1 \leq 1$, then $k_0 \leq |1/\, H|_\infty$.
2. $|X|_2 \leq 1$, then $k_0 \leq |1/\, H|_2$.
3. $|X|_\infty \leq 1$, then $k_0 \leq |1/\, H|_1$[†].

The first case was investigated in the introductory section of this chapter. Here it was conjectured that if a system was driven by a pure tone, the output would be bounded by the maximal gain found in the filter's magnitude response. For example, suppose that $u(n)$ is given by the complex signal

$$u(n) = \sum_{i=0}^k \cos(niT)$$

where T is the sample period. Here $|u(n)|_1 = K$, when interpreted as a one-sided spectrum. If one now considers the filter, whose magnitude response (see Fig. 11.7.1) is $H(\omega)$, such that $|H|_\infty = 1$. Then it follows that $|x(n)| \leq K$.

Case 2 represents a familiar analysis methodology. In practice, $u(n)$ is often modeled as a random process where $|U|_2^2 = \sigma_u^2$ [i.e., the variance of $u(n)$]. From the Wiener-Khinchine theorem of linear system theory, it is known that

$$\sigma_x^2 \leq |H|_2^2 \sigma_u^2$$

[†]If $H(\omega)$ is continuous over [0, B], it is also of bounded variation.

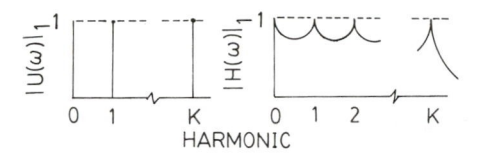

FIG. 11.7.1 Hypothetical frequency response used to interpret the L_1 norm of a system.

where $|H|_2^2 = (\int_0^B H(\omega)H * (\omega) \, d\omega)/B$. Case 3 states that if the maximal frequency component of an input process is bounded by unity (i.e., $|U|_\infty \leq 1$), the filter's output is bounded by H_1. Suppose, for example, that $u(n)$ is a N-sample periodic process with $u(n) = u(n + N)$ and

$$u(n) = \sum_{i=0}^{K} \cos(inT)$$

(i.e., sum of K harmonics). Then $U(k)$ is the DFT of $\{u(n)\}$ and is given by

$$U(k) = \sum_{i=0}^{N} u(n)W_N^{-ik} = \begin{cases} N & \text{for } k = 0, 1, \ldots, K \\ 0 & \text{otherwise} \end{cases}$$

Suppose that $u(n)$ is passed through an ideal lowpass filter characterized by (see Fig. 11.7.1) so that

$$k_0 H(k) = \begin{cases} k_0, & 0 \leq k \leq S, \; S \leq K \\ 0, & \text{otherwise} \end{cases}$$

Direct computation yields $|H|_1 = S$. The output time series can readily be computed through the use of the inverse DFT. Therefore,

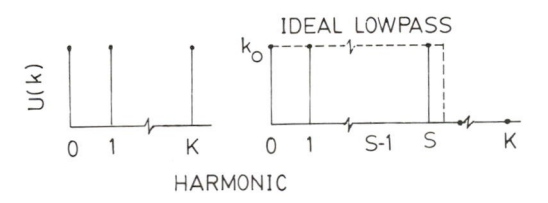

FIG. 11.7.1 Diagram showing the combined effect of L_1 and L_∞ norms on the output spectrum.

$$x(n) = k_0 \sum_{i=0}^{S} \cos(niT)$$

and $|x(n)| \leq |H|_1 |X|_\infty$.

11.8 SUMMARY

Two scaling methodologies have been developed. They were based on time and frequency domain considerations. There are advantages and disadvantages to both. However, the reader is reminded that it is simpler to specify norm bounds on a time series. This is especially true for $|x|_\infty$ since the "front end" of many signal processing systems is an A/D unit having a fixed dynamic range. The error bounds considered in this chapter are pessimistic. Many authors will claim that using worst-case and norm bounds produces too large a scaling (attenuating) policy, which reduces filter precision. They claim that in practice these bounds are never achieved. Nevertheless, sometimes worst-case inputs do exist (e.g., dc forcing functions of a lowpass filter). These decisions are difficult for the designer to make. There is always an element of risk if one wishes to ignore the computed bounds. Finally, in Chapter 12, a computer-aided ℓ_2 bounding algorithm is presented. However, before this routine can be understood and appreciated, several major topics must be considered.

BIBLIOGRAPHY

Antonious, A. (1979), Digital Filters: Analysis and Design, McGraw-Hill, New York.

Avenhaus, E. (1972), On the Design of Digital Filters with Coefficient of Limited Wordlength, IEEE Trans. Audio Electroacoust., AU-20, August, pp. 206-212.

Halmos, P. R. (1958), Finite-Dimensional Vector Spaces, 2nd ed., D. van Nostrand, Princeton, N.J.

Jackson, L. B. (1973), Roundoff Noise Analysis in Cascase Realizations of Finite Impulse Response Digital Filters, Bell Sys. Tech. J., 52, March, pp. 329-345.

Oppenheim, A. V., and R. W. Schafer (1975), Digital Signal Processing, Prentice-Hall, Englewood Cliffs, N.J.

Rabiner, L. R., and B. Gold (1975), <u>Theory and Application of Digital Signal Processing</u>, Prentice-Hall, Englewood Cliffs, N.J.

Tretter, S. A. (1976), <u>Introduction to Discrete Time Signal Processing</u>, Wiley, New York.

12
Roundoff Error Analysis

12.1 INTRODUCTION

In Chapter 10, the concept of errors in a finite-word-length digital filter was motivated. In this chapter quantization and roundoff errors are introduced together with the concept of limit cycling. Further, these concepts will be developed analytically. As a result, a digital filter architecture may be thoroughly analyzed in the context of these error sources and its performance predicted as a function of word length. It will be generally assumed that a fixed-point arithmetic structure is being used to implement a digital filter. However, floating-point architectures will also be discussed where and when applicable.

12.2 ROUNDOFF ERRORS

In Section 10.3 it was noted that roundoff errors occur in digital filtering. At that time, the roundoff error source was modeled to be a uniformly distributed, random noise process distributed over $[-Q/2, Q/2]$, where $Q = 2^{-n}$. It was assumed that the rounded product, say ax, can be adequately modeled by corrupting the real (infinite-precision) product ax with uniformly distributed noise. This model was suggested in Fig. 10.3.1. It is generally assumed that the ith system multiplication, say $a_i x_i$, is corrupted by $e_i(n)$, where

1. $\mathcal{E}(e_i(n)e_i(n+k)) = Q^2/12)\delta(k)$ (i.e., white noise source).
2. $\mathcal{E}(e_i(n)e_j(n+k)) = 0$ if $i \neq j$ (i.e., uncorrelated noise sources).
3. $\mathcal{E}(e_i(n)u(n+k)) = 0$, where u(i) is the filter's input.

The noise found at the output of a filter consists of the collective influences of all the distributed noise sources found within that system. For example, consider the system abstracted in Fig. 12.2.1. Here the roundoff noise found at the output, due to e_i alone, is given by

$$e_x(n) = \Sigma h_i(k) e_i(n - k); \quad k = 0, 1, \ldots, n \tag{12.2.1}$$

where $h_i(j)$ is the impulse response of a single input-single output shift-invariant filter having an output $y(n)$ and an input $e_i(n)$. The roundoff error variance, due to $e_i(n)$, can be directly computed to be

$$\sigma_x^2(n) = \mathcal{E}(e_{x_i}(n - m) e_{x_i}(n - k))$$

$$= \sum_{m=0}^{n} \sum_{k=0}^{n} h_i(m) h_i(k) \, \mathcal{E}\, (e_i(n - m) e_i(n - k))$$

$$= \sum_{m=0}^{n} \sum_{k=0}^{n} h_i(m) h_i(k) \sigma^2 \delta_k(k - m) = \sigma_e^2 \sum_{m=0}^{n} h_i^2(m) \tag{12.2.2}$$

That is, the noise variance found at the output of the filter, due solely to $e_i(n)$, is defined in terms of the impulse response between the output and the signal injection point [i.e., $e_i(n)$]. Other noise sources would generally be processed through different impulse responses. Therefore, computing the composite noise contributions at the output can become a nontrivial task. A state-variable approach to computing the many impulse responses required of this type of analysis is therefore attractive. Here, using linear algebraic operations, the tedious process of computing M individual impulse responses can be made more efficient. In particular, one would identify the impulse response $h_{jk}(n)$ with the propagation of an error to the jth-state variable $[x_j(n)]$ due to the rounding of the product $[a_{jk} x_k(n - 1)]$. Here $h_{jk}(n)$ can be written in terms of $Z^{-1}(H_{jk}(z))$, where a_{jk} is the jkth location of the state coefficient matrix A and $H_{jk}(z)$ is the jkth element of $(zI - A)^{-1} z$.

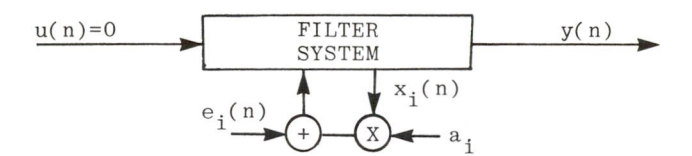

FIG. 12.2.1 Conceptual model of the roundoff error associated with a fixed-point multiplication.

EXAMPLE 12.2.1 Consider the first-order system shown in Fig. 12.2.2 It has previously been noted that $h_0(n) = a_0^n$. Therefore, at steady state, the output roundoff noise variance is given by

$$\sigma_x^2 = \frac{Q^2}{12} \sum_{k=0}^{2} a_0^{2k}$$

$$= \frac{2^{-2n}}{12} \frac{1}{1 - a_0^2}$$

where the roundoff is n bits in length. It can be seen that if $a_0 = 0$, the feedback path is broken and the output $y(n) = x(n)$. As a result, the output roundoff error variance degenerates to the familiar $Q^2/12$. Second, if $a_0 \longrightarrow 1$, the filter emulates an ideal discrete integrator (i.e., zero-order hold). As a result, all the stochastic energy injected into the filter recirculates, and never truly dissipates.

EXAMPLE 12.2.2 This analysis can be continued for the second-order filter shown in Fig. 12.2.3. The impulse response connecting $e_1(n)$ and $e_2(n)$ to the output are given by

$$h_1(n) = h_2(n) = \frac{r^n \sin[(n + 1)\phi]}{\sin(\phi)}$$

or

$$H_2(z) = \frac{1}{1 - 2r\cos(\phi)z^{-1} + r^2 z^{-2}}$$

Then

$$\sigma_x^2 = \sigma_1^2 + \sigma_2^2 = \frac{2^{-2n}}{6} \sum_{k=0}^{\infty} r^{2k} \frac{\sin^2[\phi(k + 1)]}{\sin^2(\phi)}$$

which, after some algebraic manipulation, can be expressed as

$$\sigma_x^2 = \frac{2^{-2n}}{6} \frac{1 + r^2}{(1 - r^2)[1 + r^4 - 2r^2 \cos(2\phi)]}$$

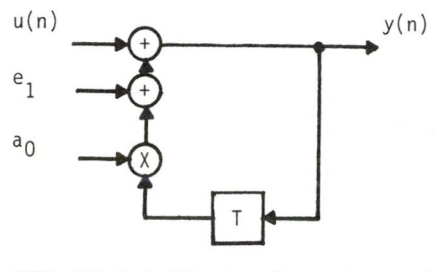

FIG. 12.2.2 First-order noise model.

The amplification of a noise source $e_i(n)$ by $h_i(n)$ can be interpreted in the frequency domain as well. Using Parseval's theorem,

$$\sum_{k=0}^{\infty} h_i^2(k) = \frac{1}{2\pi j} \oint H_i(z)H_i(z^{-1})z^{-1} \, dz$$

which can be evaluated using the residue theorem.

EXAMPLE 12.2.3 Reconsider the first-order system found in Fig. 12.2.1. Here $H_0 = 1/(1 - a_0 z^{-1})$. Then

$$\sum_{k=0}^{\infty} h_0^2(k) = \frac{1}{2\pi j} \oint H_0(z)H_0(z^{-1}) \, dz \, z^{-1}$$

$$= \frac{1}{2\pi j} \oint \frac{z^{-1} \, dz}{(1 - a_0 z^{-1})(1 - a_0 z)}$$

where $|a_0| \le 1$ for a stable filter. Therefore, upon integrating along the closed contour defined by the periphery of the unit circle, and noting there

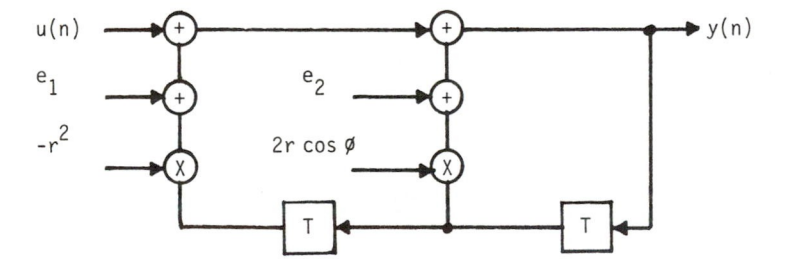

FIG. 12.2.3 Second-order noise model.

exists but one pole within this contour (i.e., $z = a_0$), it follows that σ_x^2 is given by

$$\sigma_x^2 = \frac{2^{-2n}}{12} \frac{1}{1 - a_0^2}$$

which agrees with the previous result.

12.3 FLOATING POINT

Floating-point encoding schemes are covered in detail in Chapter 15. However, for the sake of completeness, this scheme is outlined briefly here. A real number x can be approximated in terms of a mantissa-exponent representation of the form $x \simeq m_x e^y$. In particular, m_x will be considered to be a $(n + 1)$-bit signed data word and y, the exponent of the radix e, a $(b + 1)$-bit signed word. Typical values for e are 2, 8, 10, and 16. Liu and and Kaneko have modeled the floating-point operations of addition and multiplication as follows:

$$\phi(x + y) = \text{float}(x + y) = (x + y)(1 + \epsilon)$$

$$\phi(ax) = \text{float}(ax) = ax(1 + \delta)$$

(12.3.1)

where ϵ and δ represent the statistical errors associated with rounding the mantissa to its n most significant magnitude digits. Recall that in fixed-point operations, the roundoff errors were additive. In the floating-point model, besides being additive, they are scaled by $(x + y)$ or ax. Therefore, the floating-point errors, unlike their fixed-point counterparts, are amplitude sensitive. In particular, ϵ and δ can be modeled as a white, uniformly distributed, random process having a variance $Q^2/12$, where $Q = 2^{-n}$. A simple, first-order filter, satisfying the discrete equation $y(n) = ay(n - 1) + bu(n)$ would therefore be interpreted in the manner suggested by Fig. 12.3.1. It can be seen that even a first-order system possesses a complex roundoff error model. The analysis of high-order systems often is too complex to be considered seriously. In addition, to ensure that the input time series is uncorrelated with the floating-point noise sources (condition 3 of Section

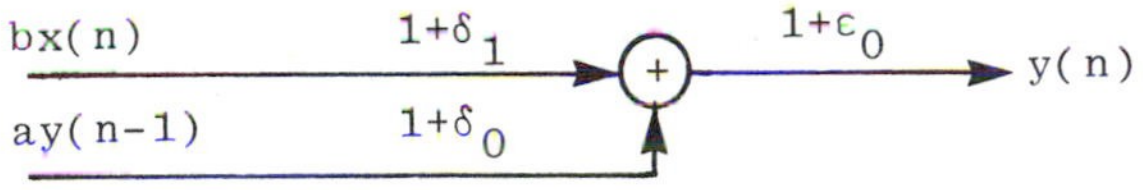

FIG. 12.3.1 Conceptual model of the roundoff error associated with a floating-point multiplication.

12.2), the input time series must be white. Since this is usually not the
case, the value of such an analysis is questionable.

12.4 ROUNDOFF ERROR PREDICTION

The roundoff noise estimate scheme, developed to this point, is predicated
on a uniform noise source being amplified by some internal impulse response or transfer function. Several other analysis methodologies are also
available to support this study. One is <u>simulation</u> in software. Here the
filter is modeled using fixed-point arithmetic having a user-supplied word
length. The impulse response of this system can be compared to that obtained using double-precision floating point (assumed to be infinite precision). The error variance would be directly computed and analyzed. This
experimental approach is often found to be rewarding and efficient. Because
of the large number of transfer or impulse response functions needed to
quantify a high-order filter, this direct approach will often save many
hours of rigorous analytic study. However, the shortcoming of simulation
is that it will not give the designer the mathematical machinery needed to
optimize a design. To this end, a very powerful state-space analysis
scheme will be presented.

In general, a linear shift-invariant filter can be characterized in
terms of the four-tuple (A, b, c, d) (see Fig. 12.4.1) where the state equations are given by

$$x(n + 1) = Ax(n) + bu(n) \tag{12.4.1}$$

and the output equation is

$$y(n) = cx(n) + d(0)u(n) \tag{12.4.2}$$

As before, it will be required that the system's impulse response must be
analyzed in the context of potential register overflow. The critical nodes
of a state-determined system are those at which the state variables $x_i(n)$ of
Eq. (12.4.1) are defined. For an n-dimensional filter, the n-dimensional
state impulse response is given by

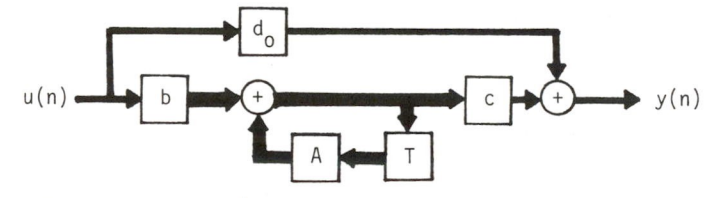

FIG. 12.4.1 State variable block diagram.

$$h(k) = \begin{cases} d(0) & \text{if } k = 0 \\ cA^{k-1}b & \text{if } k \geq 1 \end{cases} \qquad (12.4.3)$$

or in terms of a transfer function,

$$H(z) = d(0) + c(zI - A)^{-1}b \qquad (12.4.4)$$

It will again be assumed that all system variables are to be encoded into a finite-word-length format. In many cases, the data word length for a system is fixed. However, they may also be independently assigned. For example, the input may be derived from an 8-bit A/D converter and internally manipulated with 12×12-bit multipliers and a 27-bit accumulator. Following this line of reasoning, let the ith state variable be budgeted n_i bits of precision (n_i to be determined as part of the design procedure) so that $x_i(n)$ is bounded by $\pm Q2^{n_i-1}$, where Q, as before, is the quantization step size. Furthermore, it will be assumed that there exists a total of N bits of register which will be jointly occupied by all data words. That is,

$$N = \sum_{i=1}^{n} n_i \qquad (12.4.5)$$

Thus the state variables are to be considered concatenated into an N-bit data field of possibly different block sizes.

The unscaled state outputs, which may potentially overflow their allocated dynamic range, are given by

$$x(n) = \sum_{k=1}^{\infty} A^{k-1}bu(n - k) \qquad (12.4.6)$$

where the ith column of $(A^{k-1}b)$ will be denoted $f_i(n)$. This parameter can be used to provide an L_2 bound on $x_i(k)$ by observing that $|x_i|_2 = |f_i|_2$. Therefore, overflow prevention information is found in the data string $\{f_i\}$. However, it has been previously stated that this is a pessimistic overflow bound. To increase the utility of this bound, a scale factor δ, $\delta \geq 1$, is often used. In particular, it will be required that $x_i(n)$ be bounded by $\delta |f_i|^2$. The parameter δ can be loosely related to the probability of overflow. That is, if $\delta = 1$, the probability of overflow is zero, whereas for other admissible values of δ, the probability of overflow exists. Therefore, for nonunity choices of δ, it is suggested that saturating arithmetic be used. Under this hypothesis, one obtains

$$\delta^2 \|f_i\|_2^2 = \delta^2 \left[\sum_{k=1}^{\infty} f_i^2(k) \right] = (Q2^{n_i-1})^2 \qquad (12.4.7)$$

for all i. This condition was referred to as <u>scaling</u> in Chapter 11 and is required to prevent register overflow.

It may be of interest to note that there exists a possible flaw in this philosophy. Although the analysis is rigorous, the fact that an ℓ_2 norm is used does impose some limitations of the interpretability of the results. Recall that the dual space of ℓ_2 is ℓ_2 (i.e., self-adjoint). Therefore, the statement $|f_i|_2 |u_i|_2 \leq |x_i|_2 \leq 1$ is with respect to f_i and u being ℓ_2 functions. There are many important signals which do not possess an ℓ_2 interpretation. For example, u(n) = 1, or u(n) = $\sin(\omega n/T)$, <u>for all n</u>, are not ℓ_2 bounded. It would be desirable to use an ℓ_1 norm when dealing with f_i. Unfortunately, this imposes some severe analysis demands on the signal processing scientist. The self-adjoint ℓ_2 spaces offer a more simplified analysis media at the expense of interpretation of results.

The analysis must now turn to the question of roundoff errors. The usual assumptions that errors are uncorrelated uniformly distributed, random processes over [-Q/2, Q/2] provides a $Q^2/12$ error variance estimate per multiplier. If in the computation of the state variable $x_i(n)$, there are m_i required multiplications ($0 \in m_i \in n + 1$), then m_i roundoff error sources will corrupt $x_i(n)$. If the filter is properly scaled, the nested error sources will not overflow system registers. As a result, multipliers shall be considered the only internal noise source. Input error sources will be considered to be due to quantization effects. The output time series, given by y(n) = cx(n) = $cA^n x(0)$, where the input is an impulse, will also have to be analyzed with respect to roundoff errors. To accomplish this, a parameter g_i is used where g_i is equal to the ith component of cA^i. Therefore, the noise contribution in y(n) due to $x_i(n)$ is given by

$$\sigma_{x_i}^2 = \frac{m_i Q^2 |g_i|_2^2}{12} \qquad (12.4.8)$$

Collecting all the statistical quantities together, the roundoff noise found at the output is given by

$$\sigma_x^2 = \frac{1}{12} \sum_{i=1}^{n} m_i Q^2 |g_i|_2^2 \qquad (12.4.9)$$

A more compact formalization of this state has been offered by Mullis and Roberts (1976). In terms of positive-definite symmetric matrices K and W, one obtains[†]

[†]A constant-coefficient, positive-definite matrix has only positive eigenvalues.

$$K = AKA^T + bb^T = \sum_{k=0}^{\infty} (A^k b)(A^k b)^T \tag{12.4.10}$$

$$W = A^T WA + c^T c = \sum_{k=0}^{\infty} (cA^k)^T (cA^k) \tag{12.4.11}$$

with $W_{ij} = |g_i g_j|$, $K_{ij} = |f_i f_j|$. The previously derived bounds can be interpreted in terms of the elements of K and W by observing that $\delta^2 K_{ii} = (Q2^{n_i-1})^2$ (scaling condition) and $\sigma_x^2 = \Sigma m_i Q^2 W_{ii}/12$ for $i = 1, \ldots, n$. To complete this analysis, the noise contribution due to input quantization effects and the rounding of the output time series, given by $y(n) = cx(n) + h(0)u(n)$, must be added to σ_x^2 (assuming that rounding is desired, or required). These two auxiliary terms are independent of architecture and would add at most

$$\sigma_0^2 = \frac{(n+1)\left[\sum_{i=0}^{\infty} h^2(n)\right] Q^2}{12} = \frac{(n+1)|h|^2 Q^2}{12} \tag{12.4.12}$$

to the output error variance due to the rounding of $x_i(n)$ only. Here h is given by Eq. (12.4.3).

Up to this point, the word length n_i has been unspecified. Mullis and Roberts suggested an optimal choice algorithm such that the sum of all n_i's equaled N. The criterion of optimization was a minimal roundoff error variance. The budgeting policy essentially allocated word length on the basis of a statistical need. This policy attempts to ensure that all error sources found in y(n) were of equal weight. This is consistent with Rabiner and Gold's experimental and heuristic observation on how to order filter subsections. It was noted that the order rule should be such that the round-off noise does not vary significantly from section to section.

Using the so-called geometric-arithmetic mean inequality, it follows that

$$\frac{1}{n} \sum_{i=1}^{n} \frac{W_{ii} K_{ii}}{(2^{n_i})^2} \geq \left[\prod_{i=1}^{n} \frac{W_{ii} K_{ii}}{(2^{n_i})^2} \right]^{1/n} = \frac{1}{2^{N/n}} \left(\prod_{i=1}^{n} K_{ii} W_{ii} \right)^{1/n} \tag{12.4.13}$$

which is optimized when equality is achieved. This occurs when

$$n_i = N + \left(\frac{1}{2} \log_2 K_{ii} W_{ii} - \frac{1}{n} \sum_{i=1}^{n} \log_2 W_{ii} K_{ii} \right) \tag{12.4.14}$$

and it follows that:

$$\sigma_x^2 = \left(\frac{n+1}{3}\right)n\left(\frac{\delta}{2^n}\right)^2 \left(\prod_{i=1}^{n} K_{ii}W_{ii}\right)^{1/n} \tag{12.4.15}$$

A more traditional approach is to assign all words equal word lengths, say $n_0 = N/n$ (i.e., $n_i = n_0$ for all i). Under this hypothesis, the foregoing result degenerates to

$$\sigma_x^2 = \frac{(n+1)n}{3}\left(\frac{\delta}{2^{n_0}}\right)^2 \left(\frac{1}{n}\sum_{i=1}^{n} K_{ii}W_{ii}\right) \tag{12.4.16}$$

If it is assumed that $m_i = n + 1$ [the maximal number of multiplies to form $x_i(n)$], the error analysis above can provide a useful design vehicle. These formulas provide the user with a figure of merit by which filter architectures may be compared. Given a filter (A, b, c, d), alternative architectures, say $(A', b'c', d')$, can be developed using a simple linear transformation T. In particular, it will be assumed that T defines a mapping of the state of the existing system into the states of the rearchitected filter, say x' (i.e., $T: x \longrightarrow x'$, so that $x' = TX$). Under T, it follows directly that K' and W', for the new system, satisfy $K'W' = T^{-1}$ [KW]T, where K' and W' are also symmetric positive definite. The mapping T defines $(A, b, c, d) \longrightarrow$ $(T^{-1}AT, T^{-1}b, cT, d) = (A', b', c', d')$. Since K' and W' are explicit functions of T, it therefore follows that <u>scaling and roundoff performance is architecture dependent.</u> Since architecture will directly influence a filter's roundoff error performance, it would be desirable to seek those structures which exhibit low roundoff error sensitivity. One structure that was claimed to possess desirable roundoff error sensitivity was that due to Gray and Markel (Chapter 8). It was noted that, a trade-off between performance (roundoff errors) and complexity (multiplier count and throughput) existed. Therefore, if one seeks to use the ℓ_2 methods developed in this section to optimize a filter architecture, these trade-offs must be subjectively applied to the design evaluation.

12.5 COMPUTATION METHODS

The computation of the positive-definite symmetric matrices K and W, found in the preceding section, has received much attention in recent years. One of the most efficient procedures was developed by Mullis and Roberts and involves the use of a Toeplitz matrix. The algorithm can be abstracted as follows:

K-W Algorithm. Let the denominator of the filter under study be given by $D(z) = z^n + p_1 z^{n-1} + \cdots + p_{n-1} z + p_n$. Define a coefficient set $\{r_0, r_1, \ldots, r_{n-1}\}$ as follows:

$$r_i + \sum_{i=1}^{n} p_j r_{|i-j|} = \begin{cases} 1 & \text{if } i = 0 \\ 0 & \text{otherwise; } 0 \leq i \leq n-1 \end{cases} \tag{12.5.1}$$

Computation of K and W.   Define

For K	For W
$x(1) = b$; $x(k+1) = Ax(k) + p_k b$	$x(i) = c^T$; $x(k+1) = A^T x(k) + p_k c^T$
(i) $K = XRX^T$	(ii) $W = XRX^T$ (12.5.2)

for

$$X = [x(n), x(n-1), \ldots, x(1)] \tag{12.5.3}$$

and

$$R = \begin{bmatrix} r_0 & r_1 & r_2 & & r_{n-1} \\ r_1 & r_0 & r_1 & \cdots & r_{n-2} \\ r_2 & r_1 & r_0 & \cdots & r_{n-3} \\ . & . & . & \cdots & . \\ . & . & . & \cdots & . \\ . & . & . & \cdots & . \\ r_{n-1} & r_{n-2} & r_{n-3} & & r_0 \end{bmatrix} \qquad \text{Toeplitz matrix}$$

EXAMPLE 12.5.1 Let us again consider the second-order system having an impulse response $h(n) = r^n \sin(\omega(n+1)T)/\sin(\omega T)$ (see Fig. 12.2.3). Equivalently, $H(z) = z^{-1}/[1 - 2r \cos(\omega T) z^{-1} + r^2 z^{-2}]$. It was noted in Example 12.2.2 that $|h(n)|_2^2 = (1 + r^2)/(1 - r^2)[r^4 + 1 - 2r \cos(2\omega T)]$. For the purpose of demonstration, the following problem parameterization will be used:

$$r^2 = \frac{1}{2}$$

$$2r \cos(\omega T) = 1$$

$$\omega T = \phi = \frac{\pi}{4}$$

Then $|h(n)|_2^2 = (1 + 0.5)/(1 - 0.5)(0.25 + 1 - 0) = 12/5$. This can be verified using state-variable methods by noting

$$A = \begin{bmatrix} 0 & 1 \\ -0.5 & 1 \end{bmatrix}; \quad b = \begin{pmatrix} 0 \\ 1 \end{pmatrix}; \quad c = (1 \ \ 0); \quad d = 0$$

The characteristic equation is given by $\det(zI - A)$ and equals $z^2 - z + 0.5$. Therefore, the values of p_1 and p_2, found in the computing algorithm, are -1 and 0.5, respectively. Continuing, one obtains

$$\left. \begin{array}{l} r_0 + p_1 r_1 + p_1 r_2 = 1 \\ r_1 + p_1 r_0 + p_2 r_1 = 0 \\ r_2 + p_1 r_1 + p_2 r_0 = 0 \end{array} \right\} \quad Pr = s$$

or solving for $r = P_s^{-1}$

$$\begin{bmatrix} 1 & -1 & 0.5 \\ -1 & 1.5 & 0 \\ 0.5 & 1 \end{bmatrix} \begin{bmatrix} r_0 \\ r_1 \\ r_2 \end{bmatrix} = \begin{bmatrix} 1 \\ 0 \\ 0 \end{bmatrix} ; \begin{bmatrix} r_0 \\ r_1 \\ r_2 \end{bmatrix} = \begin{bmatrix} \dfrac{12}{5} \\ \dfrac{8}{5} \\ \dfrac{2}{5} \end{bmatrix}$$

Therefore, the Toeplitz matrix R is formed as follows:

$$R = \begin{pmatrix} \dfrac{12}{5} & \dfrac{8}{5} \\ \dfrac{8}{5} & \dfrac{12}{5} \end{pmatrix} = \frac{1}{5} \begin{pmatrix} 12 & 8 \\ 8 & 12 \end{pmatrix}$$

and $x(1) = (0, 1)^T$ with $x(2) = Ax(1) + p_1 b = (1, 0)^T$ or $X = I$. The direct computation of $K = XRX^T = R$ yields $k_{11} = k_{22} = 12/5$ (as previously computed). To compute W, one observes that $x(1) = c^T = (1, 0)^T$ and $x(2) = A^T x(1) + p_1 c^T = (1, 0)^T$. Thus W is given by

$$W = XRX^T = \frac{\begin{bmatrix} 12 & 8 \\ 8 & 12 \end{bmatrix}}{12}$$

The resulting unscaled error variance, due to roundoff errors, is given by $\sigma_x^2 = (m_1 W_{11} + m_2 W_{22}) Q^2/12$. Here $\sigma_x^2 = ((0 \times 12/5 + 2 \times 12/5) Q^2)/12$

[Note: There does not exist a nonunity or zero multiplication in the creation of $x_1(n)$ since $m_1 = 0$. Also, there are two multiplications required to form $x_2(n)$.] This formula is in agreement with the published error variance estimate of Rabiner and Gold (1975) for this filter. Using their results, developed using classical methods, it follows that the unscaled error variance is

$$
\sigma_x^2 = \frac{Q^2}{12} \; \frac{2(1 + r^2)}{(1 - r^2)[1 + r^4 - 2r^2 \cos(2\phi)]}
$$

$$
= \frac{Q^2}{12} \; \frac{2(1/2)}{(3/2)(5/4)} \; = \frac{Q^2}{12} \; \frac{24}{5} \quad \text{(unscaled)}
$$

If the scale factor, defined by K, is included in the analysis, Eq. (12.5.16) can be used to compute a noise variance bound on σ_x^2. This formula is based on the assumption that $m_i = n + 1$ (its maximal value). A direct analysis of the scaled system given by

$$
A = \begin{bmatrix} 0 & 1 \\ -0.5 & 1 \end{bmatrix}; \quad b = \left(\frac{0}{1/\sqrt{K_{22}}} \right); \quad c = (1 \quad 0)
$$

yields $m_1 = 0$ and $m_2 = 2$ nonzero, or nonunity, multiplications. Therefore,

$$
\sigma_x^2 = (m_1 K_{11} W_{11} + m_2 K_{22} W_{22}) \frac{Q^2}{12} = \frac{11.5 Q^2}{12}
$$

for $\delta = 1$. In terms of bits, if $Q = 2^{-(n_0-1)}$, then $\sigma_x^2 = 3.84(2^{-2n_0})$. Using Eq. (12.4.6), a value of $11.52 \, (2^{-2n_0})$ is obtained. Here the effects of scaling the input to prevent overflow is seen through a reduced output precision (i.e., added noise variance).

   Since the algorithm is exact and does not require the precomputation of eigenvalues, it can be easily implemented on a digital computer. Using linear algebraic service routines, the software system found in Fig. 12.5.1 was developed (IBM or DEC routines). A source code for this routine is shown below. The user is required to supply the following data:

1. Order of the filter, $0 \leq n \leq 11$.
2. Enter p's $\{p_1, \ldots, p_{n-1}, p_n\}$: free format.
3. Enter A matrix in row order: free format.
4. Enter b vector: free format.
5. Enter c vector: free format. The software package will output the Toeplitz matrix T, echo print (A, b, c), the K and W matrices, the

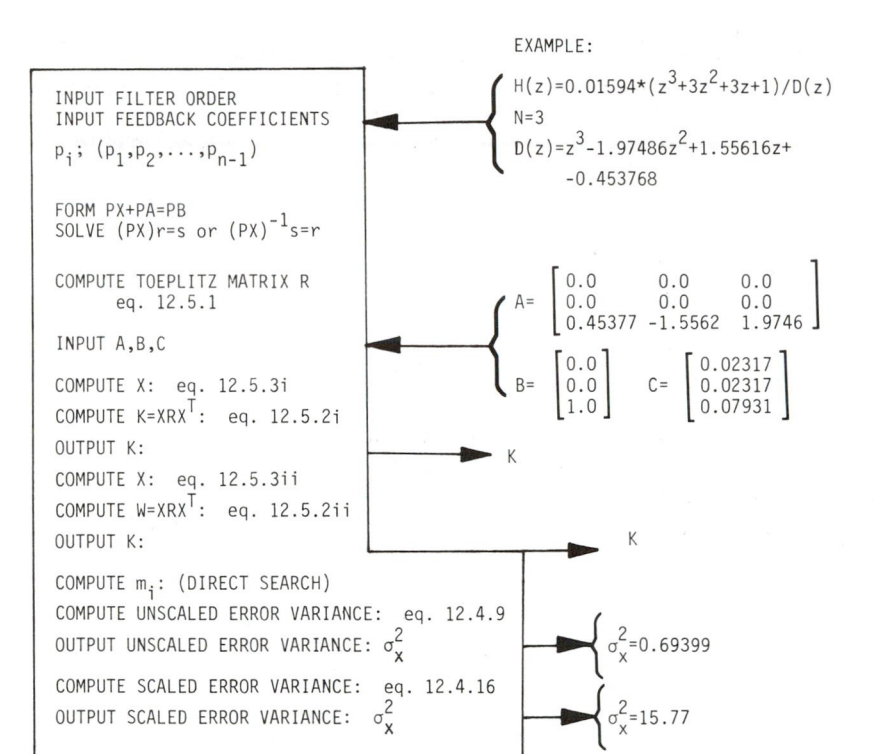

FIG. 12.5.1 Flow diagram for the computation of the ℓ_2 scale factor and roundoff error variance using a Toeplitz matrix.

number of multiplications per state variable (i.e., m_i), and the scaled and unscaled error variance.

The source code is as follows:


```
PURPOSE
         THIS PROGRAM IS TO CALCULATE THE K AND W MATRICES
         IN THE ERROR ANALYSIS OF ROUNDOFF ERRORS BY A STATE
         VARIABLE APPROACH.
REMARKS
         THIS PROGRAM CAN CALCULATE AN UP TO 10TH-ORDER
         FILTER.
DESCRIPTION OF PARAMETERS
         A      - MATRIX A IN THE STATE EQUATIONS.
         AT     - TRANSPOSE OF MATRIX A.
         B      - MATRIX B IN THE STATE EQUATIONS.
         C      - MATRIX C IN THE OUTPUT EQUATIONS.
         HERE   - WORKING VECTOR USE FOR CALCULATE THE INVERSE
                  MATRIX.
```

```
       MP    - NOISE SOURCES
       MZ    - THE LOGICAL UNIT NUMBER FOR WRITE.
       N     - NUMBER OF ORDERS OF THE FILTER.
       P     - VECTOR P IN THE DENOMINATOR POLYNOMIAL OF
               THE FILTER.
       PA    - MATRIX OF THE FIRST DECOMPOSITION OF THE
               COMPUTING ALGORITHM OF MATRIX PX.
       PB    - MATRIX OF THE SECOND DECOMPOSITION OF THE
               COMPUTING ALGORITHM OF MATRIX PX.
       R     - THE TOEPLITZ MATRIX.
       S     - TEMPERORY VECTOR TO CALCULATE P(I)*B.
       T     - VECTOR WHICH IS EQUAL TO THE FIRST COLUMN
               OF INV(PX).
       TEMP  - MATRIX USE FOR CALCULATE THE INVERSE MATRIX.
       TEMPW - MATRIX USE FOR CALCULATE THE INVERSE MATRIX.
       TEMPX - MATRIX USE FOR CALCULATE THE INVERSE MATRIX.
       TEMPY - MATRIX USE FOR CALCULATE THE INVERSE MATRIX.
       TEMPZ - MATRIX USE FOR CALCULATE THE INVERSE MATRIX.
       W     - THE W MATRIX.
       X     - THE X MATRIX.
       XA    - TEMPERORY VECTOR TO CALCULATE X(I).
       XB    - TEMPERORY VECTOR TO CALCULATE X(I).
       XK    - THE K MATRIX.
       Y     - THE RESULTANT OF TWO MATRICES PRODUCT.
       Z     - TRANSPOSE OF MATRIX X.

       DIMENSION A(15,15),AT(15,15),B(15,1),C(15,1),P(15,1),PA(15,15)
       DIMENSION PX(15,15),R(15,15),T(15,1),X(15,15),XA(15,1),Y(15,15)
       DIMENSION S(15,1),XB(15,1),XK(15,15),W(15,15)
       DIMENSION Z(15,15),PB(15,15),TEMP(15,15),TEMPW(15,15)
       DIMENSION TEMPX(15,15),TEMPY(15,15),TEMPZ(15,15)
       INTEGER HERE(15)
       WRITE(4,1)
1      FORMAT(/,'  THE ORDER OF THE FILTER IS  (0<N<11) :',/)
       READ(4,*)N
       M=N+1
       CALL ABC(A,B,C,P,PA,PB,PX,R,T,X,XA,XB,Y,Z,W,XK,M,N,AT,S,HERE,
     1 TEMP,TEMPW,TEMPX,TEMPY,TEMPZ)
       STOP
       END

       SUBROUTINE ABC(A,B,C,P,PA,PB,PX,R,T,X,XA,XB,Y,Z,W,XK,M,N,AT,
     1 S,HERE,TEMP,TEMPW,TEMPX,TEMPY,TEMPZ)
       DIMENSION A(15,15),B(15,1),C(15,1),P(15,1),PA(15,15),PB(15,15)
       DIMENSION AT(15,15),R(15,15),T(15,1),X(15,15),XA(15,1),Y(15,15)
       DIMENSION XB(15,1),XK(15,15),W(15,15),Z(15,15),PX(15,15),S(15,1)
       DIMENSION TEMP(15,15),TEMPW(15,15),TEMPX(15,15),TEMPY(15,15)
       DIMENSION TEMPZ(15,15),MP(15)
       INTEGER HERE(15)
       WRITE(4,2)
2      FORMAT(/,'  NOW TYPE IN ALL THE P(N), USE FREE FORMAT :',/)
       P(1,1)=1.0
       READ(4,*)(P(I,1),I=2,M)

       CONSTRUCT THE PX(N,N) MATRIX

       DO 3 I=1,M
       DO 3 J=1,M+1-I
3      PA(I,J)=P(I+J-1,1)
       DO 4 I=2,M
       DO 4 J=2,I
4      PB(I,J)=P(I-J+1,1)
       CALL GMADD(PA,PB,PX,M,M)
       WRITE(4,5)
5      FORMAT(/,'  READ IN THE LOGICAL UNIT NUMBER FOR WRITE.')
       WRITE(4,6)
```

```
6       FORMAT('  TI: USE 4, TTO: USE 5, TT7: USE 6',/)
        READ(4,*)MZ
        CALL MINV(PX,M,TEMP,TEMPX,TEMPY,TEMPZ,TEMPW)

        CONSTRUCT THE T(N,1) VECTOR AND TOEPLITZ MATRIX R(N,N)

        DO 20 I=1,M
20      T(I,1)=PX(I,1)
        DO 21 I=1,N
        KK=1
        DO 21 J=I,N
        R(I,J)=T(KK,1)
21      KK=KK+1
        DO 22 I=2,N
        DO 22 J=1,I-1
22      R(I,J)=R(J,I)
        WRITE(MZ,25)
25      FORMAT(/,'  THE TOEPLITZ MATRIX R(N,N) IS AS FOLLOWS:',/)
        DO 27 I=1,N
        WRITE(MZ,26)(R(I,J),J=1,N)
26      FORMAT(10(2X,E12.5,2X))
27      CONTINUE

        NOW WE CAN COMPUTE K AND W

        WRITE(MZ,28)
28      FORMAT(/,'  NOW INPUT MATRIX A , USE FREE FORMAT',/)
        READ(4,*)((A(I,J),J=1,N),I=1,N)
        WRITE(MZ,29)
29      FORMAT(/,'  INPUT MATRIX B, USE FREE FORMAT',/)
        READ(4,*)(B(I,1),I=1,N)
        WRITE(MZ,30)
30      FORMAT(/,'  INPUT MATRIX C, USE FREE FORMAT',/)
        READ(4,*)(C(I,1),I=1,N)
        WRITE(MZ,31)
31      FORMAT(/,'  THE A MATRIX IS AS FOLLOWS:',/)
        DO 33 I=1,N
        WRITE(MZ,32)(A(I,J),J=1,N)
32      FORMAT(10(2X,E12.5,2X))
33      CONTINUE
        WRITE(MZ,34)
34      FORMAT(/,'  THE B MATRIX IS AS FOLLOWS:',/)
        WRITE(MZ,35)(B(I,1),I=1,N)
35      FORMAT(2X,F8.5)
        WRITE(MZ,36)
36      FORMAT(/,'  THE TRANSPOSE OF MATRIX C IS AS FOLLOWS:',/)
        WRITE(MZ,35)(C(I,1),I=1,N)

        FINALLY WE CAN COMPUTE K

        DO 37 I=1,N
        XA(I,1)=B(I,1)
37      X(I,N)=B(I,1)
        DO 40 I=2,N
        CALL GMPRD(A,XA,XB,N,N,1)
        DO 38 J=1,N
38      S(J,1)=P(I,1)*B(J,1)
        CALL GMADD(XB,S,XA,N,1)
        II=N-I+1
        DO 39 J=1,N
39      X(J,II)=XA(J,1)
40      CONTINUE
        CALL GMPRD(X,R,Y,N,N,N)
        CALL GMTRA(X,Z,N,N)
        CALL GMPRD(Y,Z,XK,N,N,N)
```

```
          COMPUTE W

          DO 41 I=1,N
          XA(I,1)=C(I,1)
41        X(I,N)=C(I,1)
          CALL GMTRA(A,AT,N,N)
          DO 44 I=2,N
          CALL GMPRD(AT,XA,XB,N,N,1)
          DO 42 J=1,N
42        S(J,1)=P(I,1)*C(J,1)
          CALL GMADD(XB,S,XA,N,1)
          II=N-I+1
          DO 43 J=1,N
43        X(J,II)=XA(J,1)
44        CONTINUE
          CALL GMPRD(X,R,Y,N,N,N)
          CALL GMTRA(X,Z,N,N)
          CALL GMPRD(Y,Z,W,N,N,N)
          WRITE(MZ,45)
45        FORMAT(/,'  THE K MATRIX IS AS FOLLOWS:',/)
          DO 47 I=1,N
          WRITE(MZ,46)(XK(I,J),J=1,N)
46        FORMAT(10(2X,E12.5,2X))
47        CONTINUE
          WRITE(MZ,48)
48        FORMAT(/,'  THE W MATRIX IS AS FOLLOWS:',/)
          DO 50 I=1,N
          WRITE(MZ,49)(W(I,J),J=1,N)
49        FORMAT(10(2X,E12.5,2X))
50        CONTINUE
          DO 52 I=1,N
          MP(I)=1
          IF(ABS(B(I,1)).EQ.1.OR.B(I,1).EQ.0) MP(I)=0
          DO 53 J=1,N
          MP(I)=MP(I)+1
          IF(ABS(A(I,J)).EQ.1.OR.A(I,J).EQ.0) MP(I)=MP(I)-1
53        CONTINUE
          WRITE(MZ,51) I,MP(I)
51        FORMAT(/,'  NUMBER OF NOISE SOURCES FOR',I2,'TH STATE=',I3)
52        CONTINUE
          DO 55 I=1,N
          VAR=MP(I)*W(I,I)+VAR
          MP(I)=MP(I)+1
          B(I,1)=B(I,1)/SQRT(XK(I,I))
          IF(ABS(B(I,1)).EQ.1.OR.B(I,1).EQ.0) MP(I)=MP(I)-1
          SVAR=MP(I)*W(I,I)*XK(I,I)+SVAR
          WRITE(MZ,54) I,B(I,1)
54        FORMAT(/,'  SCALED VALUE OF B(',I2,')=',E10.3)
55        CONTINUE
          WRITE(MZ,56) VAR,SVAR
56        FORMAT(/,'  UNSCALED ERROR VARIANCE=',E10.4,'  SCALED=',E10.4)
          RETURN
          END
          SUBROUTINE GMADD(A,B,R,M,N)
          DIMENSION A(15,15),B(15,15),R(15,15)
          DO 1 I=1,M
          DO 1 J=1,N
1         R(I,J)=A(I,J)+B(I,J)
          RETURN
          END
          SUBROUTINE GMPRD(A,B,C,N,M,L)
          DIMENSION A(15,15),B(15,15),C(15,15)
          DO 1 I=1,N
          DO 1 J=1,L
```

```
            C(I,J)=0.0
            DO 2 K=1,M
2           C(I,J)=C(I,J)+A(I,K)*B(K,J)
1           CONTINUE
            RETURN
            END
            SUBROUTINE GMTRA(A,C,N,M)
            DIMENSION A(15,15),C(15,15)
            DO 1 I=1,N
            DO 1 J=1,M
1           C(I,J)=A(J,I)
            RETURN
            END
            SUBROUTINE MINV(A,N,TEMP,TEMPX,TEMPY,TEMPZ,TEMPW)
            DIMENSION A(15,15),TEMP(15,15),TEMPX(15,15),TEMPY(15,15)
            DIMENSION TEMPZ(15,15),TEMPW(15,15)
            INTEGER HERE(15)
            DO 1 I=1,N
            DO 1 J=1,N
            TEMPZ(I,J)=0.0
1           TEMPW(I,J)=0.0
            DO 2 I=1,N
            TEMPZ(I,I)=1.0
2           TEMPW(I,I)=1.0
            DO 8 I=1,N
            AMAX=0.0
            DO 3 J=1,N
            IF(ABS(A(I,J)).LT.AMAX) GO TO 3
            AMAX=ABS(A(I,J))
            HERE(I)=J
3           CONTINUE
            AMAX1=A(I,HERE(I))
            DO 4 II=1,N
            DO 4 JJ=1,N
            TEMPX(II,JJ)=0.0
4           TEMPY(II,JJ)=0.0
            DO 5 II=1,N
            TEMPX(II,II)=1.0
5           TEMPY(II,II)=1.0
            TEMPX(I,I)=1.0/AMAX1
            CALL GMPRD(TEMPX,A,TEMP,N,N,N)
            CALL GMPRD(TEMPX,TEMPZ,TEMPW,N,N,N)
            DO 6 IJ=1,N
            IF(IJ.EQ.I) GO TO 6
            TEMPY(IJ,I)=-TEMP(IJ,HERE(I))
6           CONTINUE
            CALL GMPRD(TEMPY,TEMPW,TEMPZ,N,N,N)
            CALL GMPRD(TEMPY,TEMP,TEMPX,N,N,N)
            DO 7 IJ=1,N
            DO 7 IL=1,N
7           A(IJ,IL)=TEMPX(IJ,IL)
8           CONTINUE
            DO 10 I=1,N
            DO 10 J=1,N
            IF(HERE(J).NE.I) GO TO 10
            DO 9 IP=1,N
9           A(I,IP)=TEMPZ(J,IP)
10          CONTINUE
            RETURN
            END
```

EXAMPLE 12.5.2 In this chapter, the architecture dependence of the third-order Chebyshev filter, given by

$$H(z) = \frac{0.01594(1 + 3z^{-1} + 3z^{-2} + z^{-3})}{1 - 1.97486z^{-1} + 1.55616z^{-2} - 0.453768z^{-3}}$$

will be studied. Huang (1977) proposed a canonical realization of the transfer function to be the three-tuple (A, b, c) where

$$A = \begin{bmatrix} 0 & 1 & 0 \\ 0 & 0 & 1 \\ 0.453768 & -1.55616 & 1.97461 \end{bmatrix}; \quad b = \begin{pmatrix} 0 \\ 0 \\ 1 \end{pmatrix};$$

$$c = (0.02317, \; 0.02317, \; 0.07931)$$

A computer analysis of these data follows:

```
  THE ORDER OF THE FILTER IS (0<N<11) :
  3
  NOW TYPE IN ALL THE P(N), USE FREE FORMAT :
  -1.97461,1.55616,-.453768

  READ IN THE LOGICAL UNIT NUMBER FOR WRITE.
  TI: USE 4, TTO: USE 5, TT7: USE 6
  4
1  THE TOEPLITZ MATRIX R(N,N) IS AS FOLLOWS:
    0.17045E+02        0.14868E+02        0.95796E+01
    0.14868E+02        0.17045E+02        0.14868E+02
    0.95796E+01        0.14868E+02        0.17045E+02

  NOW INPUT MATRIX A , USE FREE FORMAT
  0,1,0
  0,0,1
  .453768,-1.556161,1.97461
  INPUT MATRIX B, USE FREE FORMAT
  0,0,1
  INPUT MATRIX C, USE FREE FORMAT
  .02317,.02317,.07931

  THE A MATRIX IS AS FOLLOWS:
    0.00000E+00        0.10000E+01        0.00000E+00
    0.00000E+00        0.00000E+00        0.10000E+01
    0.45377E+00       -0.15562E+01        0.19746E+01

  THE B MATRIX IS AS FOLLOWS:
    0.00000
    0.00000
    1.00000

2 THE TRANSPOSE OF MATRIX C IS AS FOLLOWS:
    0.02317
    0.02317
    0.07931
```

```
THE K MATRIX IS AS FOLLOWS:
   0.17045E+02      0.14868E+02      0.95796E+01
   0.14868E+02      0.17045E+02      0.14868E+02
   0.95796E+01      0.14868E+02      0.17045E+02

THE W MATRIX IS AS FOLLOWS:
   0.48169E-01      0.66477E+00      0.95532E-01
   0.66477E+00      0.10332E+02      0.11002E+01
   0.95532E-01      0.11002E+01      0.23133E+00
```

NUMBER OF NOISE SOURCES FOR 1ST STATE= 0

NUMBER OF NOISE SOURCES FOR 2ND STATE= 0

NUMBER OF NOISE SOURCES FOR 3RD STATE= 3

3 SCALED VALUE OF B(1)= 0.000E+00

SCALED VALUE OF B(2)= 0.000E+00

SCALED VALUE OF B(3)= 0.242E+00

UNSCALED ERROR VARIANCE=0.6940E+00   SCALED=0.1577E+02

It can be noted from the preceding computer printout that K_{ii}, for $i = 2$, and 3, has a value of 17.045. An inspection of the pair (A, b) verifies that $m_1 = m_2 = 0$ and $m_3 = 3$ (scaling by 0 or ± 1 will not be considered a multiplication operation). To prevent register overflow, the gain vector b must be scaled by $1/\sqrt{K_{ii}}$. That is, a vector b preventing overflow would have the following form:

$$b = \frac{b_1}{\sqrt{K_{11}}}, \frac{b_2}{\sqrt{K_{22}}}, \frac{b_3}{\sqrt{K_{33}}}$$

The unscaled error variance is

$$\sigma_x^2 = \left(\sum_{i=1}^{3} m_i W_{ii}\right)\frac{Q^2}{12} = \frac{3(0.23133)Q^2}{12} = \frac{0.69399Q^2}{12}$$

The scaled error variance is given by

$$\sigma_x^2 = \left(\sum_{i=1}^{3} m_1' W_{ii} K_{ii}\right)\frac{Q^2}{12}$$

where m_1' is the multiplication count associated with the scaled system (A, b') where

$$A = \begin{bmatrix} 0 & 1 & 0 \\ 0 & 0 & 1 \\ 0.453768 & -1.55616 & 1.97461 \end{bmatrix}; \quad b' = \begin{pmatrix} 0 \\ 0 \\ 0.24221 \end{pmatrix}$$

Therefore, $m_1' = m_2' = 0$ and $m_3' = 4$, and

$$\sigma_x^2 = 4W_{33}K_{33} = 4(17.045)(0.23133) = 15.77Q^2/12$$

This agrees with the results of a previously published analysis of this filter by Huang (1977).

Huang (1975), Barnes (1979), and Mullis and Roberts (1976) have studied the problem of designing filters having a minimal roundoff error variance. These filters are intellectually interesting but usually require a more complex architecture than those previously studied. Rather than concentrate on these complex architectures, less complex, near-optimal (suboptimal) filters, having excellent roundoff error performance, will be developed.

12.6 NORMAL FILTERS

Jackson et al. (1977) have studied the problem of minimizing the roundoff errors associated with the system (A, b, c, d) in an ℓ_2 sense. Their work concentrated on the design of <u>normal filters</u>. A normal filter satisfies the commutative condition that $AA^T = A^TA$. From linear algebra it is known that such structures possess an orthonormal set of eigenvectors, say $\{\phi_i\}$. Barnes (1979) used this eigenvector set to develop a simple roundoff error synthesis and analysis methodology. As before, consider $f(n) = \text{col}\{f_1(n), \ldots, f_N(n)\}$ where col denotes "column of" and $f(n) = A^{n-1}b$. It was required that $|f_i(n)|_2^2 = \Sigma_1^\infty |f_i(n)|^2 \le \delta^2$ (δ as defined in Section 12.4). Furthermore, it was argued that there exists a Hermitian matrix K, such that [†]

$$K = \sum_{n=0}^{\infty} f(n)f^T(n)$$

$$= \sum_{n=0}^{\infty} A^n bb^T(A^n)^T \quad \text{(steady-state covariance)} \quad (12.6.1)$$

$$|f|_2^2 = \text{trace K}$$

[†] A Hermitian matrix H satisfies the following relation: $H = [h_{ij}]$; $h_{ij}^* = h_{ji}$.

which, for normal systems, simplifies to

$$K = \sum_{i,j=1}^{N} \frac{a_i a_j^*}{1 - \lambda_i \lambda_j} \phi_i \phi_j^T$$

$$(\phi_i, \phi_j) = \phi_i \phi_j^T = \delta_k(i - j) \tag{12.6.2}$$

$$= \begin{cases} 1 & \text{if } i = j \\ 0 & \text{otherwise} \end{cases}$$

It follows that

$$|f|_2^2 = \text{trace } K = \sum_{i=1}^{N} \frac{|a_i|^2}{1 - |\lambda_i|^2} \tag{12.6.3}$$

If $|f_i|_2^2$ is to be bounded by δ^2, then the trace of K, namely $|f|_2^2$, must be bounded by $N\delta^2$. As before, assume that all roundoff errors satisfy the prevailing assumptions. Define M (a multiplier count matrix) to be

$$M = \begin{bmatrix} m_1 & \cdots & 0 \\ \cdot & \cdot & \cdot \\ \cdot & \cdot & \cdot \\ \cdot & \cdots & \cdot \\ 0 & & m_N \end{bmatrix}; \quad m_i = \text{number of multipliers (i.e., noise sources) used in defining } x_i(n) \tag{12.6.4}$$

where m_i = number of multipliers (i.e., noise sources) used in defining $x_i(n)$. Then the total error variance is given by

$$\sigma_0^2 = \left(\sum_0^\infty \frac{c^T A^n MA^{nT} c}{\sigma_x^2} + m_0 \right) Q^2/12 \tag{12.6.5}$$

where m_0 is the number of noise sources found at the output node. A normal filter is architecturally robust and highly interconnected. It can be assumed that the number of multiplies per state variable is m (i.e., M = mI) and that m = 1 is for rounding performed on the overall extended precision sum of products $\Sigma a_{ij} x_j(n) + b_i u(n)$, $i = 1, \ldots, N$. Alternatively, for m = N + 1 each product is rounded before addition. The term σ_x^2, more specifically, the errors due to product rounding in the formation of the filter states x_i, has a suggested minimal value of

$$\sigma_x^2 = \frac{m}{N\delta^2} \left(\sum_{i=1}^{N} \frac{|\alpha_i|}{1 - |\lambda_i|^2} \right)^2 Q^2/12 \qquad (12.6.6)$$

where m is the multiplication count, with α_i and λ_i being filter parameters when expanded in a Heaviside manner. That is,

$$H(z) = d_0 + \sum_{i=1}^{N} \frac{\alpha_i}{z - \lambda_i} \qquad (12.6.7)$$

6

This is a particularly useful formula in that σ_x^2 can be computed in terms of given filter parameters. However, the reader should remember that this analysis is applicable only to the normal filter.

EXAMPLE 12.6.1   A second-order section of a normal filter can be expressed in terms of

$$\lambda_1 = r \exp(j\phi), \qquad \lambda_2 = r \exp(-j\phi); \qquad \alpha_1 = \alpha_2^*$$

$$b \Rightarrow b_1^2 + b_2^2 = 2(1 - r^2)\delta^2$$

$$A = r \begin{bmatrix} \cos\theta & -\sin\theta \\ -\sin\theta & \cos\theta \end{bmatrix}$$

$$\frac{b_2}{b_1} = \frac{r^2 \sin(2\theta) + [1 + r^4 - 2r^2 \cos(2\theta)]^{1/2}}{1 - r^2 \cos(2\theta)}$$

Note that this formalization admits a nonunique value for b_i. Continuing, the c vector is given by

$$c \Rightarrow \begin{cases} c_1 = \dfrac{b_1(\alpha_1 + \alpha_2) + b_2(j\alpha_1 - j\alpha_2)}{2(1 - r^2)\delta^2} \\[4mm] c_2 = \dfrac{b_1(j\alpha_2 - j\alpha_1) + b_2(\alpha_1 + \alpha_2)}{2(1 - r^2)\delta^2} \end{cases}$$

This concept is even more generalizable. For $\{\phi_i\}$, an orthogonal set of basis vectors induced by A, it follows that

$$A = \sum_{i=1}^{N} \lambda_i \phi_i \phi_i^{T*} ; \quad b = \sum_{i=1}^{N} \beta_i \phi_i ; \quad c^{T*} = \sum_{i=1}^{N} \gamma_i \phi_i^{T*} \tag{12.6.8}$$

where c is a linear combination of these basis vectors and $\beta_i \gamma_i = \alpha_i$. From the preceding section on state-variable representations, it was shown that the filter input must be scaled by $|f_i|_2$ in order to prevent register overflow. In this specific case, $|f_i|_2^2 = |\beta_i|^2/(1 - |\lambda_i|^2)$. Similarly, the unscaled roundoff error due to multiplier roundoff (i.e., σ_x^2) is

$$\sigma_x^2 = \frac{m}{N} \left(\frac{\alpha_i^2}{1 - \lambda_i^2} \right)^2 Q^2/12 \tag{12.6.9}$$

It is interesting to observe that σ_x^2 is defined, for a normal architecture, in terms of the given filter parameters (i.e., no eigenvector computation is required).

EXAMPLE 12.6.2 Consider the second-order filter given by

$$H(z) = \frac{1}{1 - 2r\cos\phi z^{-1} + r^2 z^{-2}} = \frac{1 + 2r\cos\phi z - r^2}{z^2 - 2r\cos\phi z + r^2}$$

$$= 1 + \frac{a_1}{z - \lambda_1} + \frac{a_2}{z - \lambda_2} ; \quad \begin{array}{l} a_1 = a_2^* \\ \lambda_1 = \lambda_2^* \end{array}$$

For $r\cos(\phi) = 1/2$ and $r^2 = 1/2$, it follows that (see Fig. 12.6.1)

$$H(z) = 1 + \frac{z - 1/2}{z^2 - z + 1/2}$$

$$= 1 + \frac{0.5}{z - r\exp(j\pi/4)} + \frac{0.5}{z - r\exp(-j\pi/4)}$$

From the computing formula for σ_x^2, one obtains

$$\sigma_x^2 = \frac{m}{2}(1+1)^2 \frac{Q^2}{12} = m\frac{4}{2}\frac{Q^2}{12} = 2m\frac{Q^2}{12}$$

where the properties of m are as discussed previously. The normal architecture of the filter under study is shown in Fig. 12.6.2.

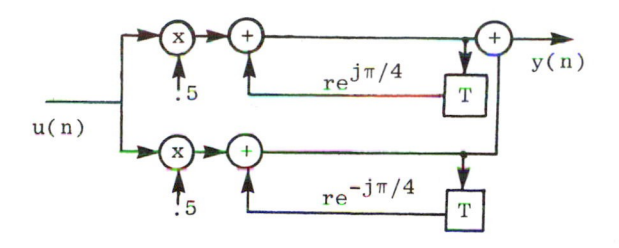

FIG. 12.6.1 Block diagram of a parallel architecture.

The overflow scale factor for this filter is found by computing the eigenvectors ϕ_1 and ϕ_2 and β_1 and β_2. Here $\lambda_1 = 0.5 + j0.5$, $\lambda_2 = 0.5 - j0.5$, and $Ae_i = \lambda_i e_i$, where e_i is an eigenvector. Therefore,

$$\begin{bmatrix} -0.5 & 0.5 \\ -0.5 & 0.5 \end{bmatrix} \begin{bmatrix} r_{11} + ji_{11} \\ r_{12} + ji_{12} \end{bmatrix} = (0.5 + j0.5) \begin{bmatrix} r_{11} + ji_{11} \\ r_{12} + ji_{12} \end{bmatrix}$$

or

$$\not{r}_{11} + r_{12} = \not{r}_{11} - i_{11}, \quad \not{i}_{11} + i_{12} = \not{i}_{11} + r_{11}$$

$$-r_{11} + \not{r}_{12} = \not{r}_{12} - i_{12}, \quad -\not{i}_{11} + i_{12} = \not{i}_{12} + r_{12}$$

which implies that

$$r_{12} = -i_{11}, \quad r_{12} = -i_{11}, \quad -r_{11} = -i_{12}, \quad -r_{11} = -i_{12}$$

If one chooses $r_{11} = 1$ and $i_{11} = 0$, then $e_1 = (1, j)^T$. Similarly, $e_2 = (1, -j)^T$. The orthonormal versions of e_1 and e_2 are $\phi_1 = (1, j)^T/\sqrt{2}$ and

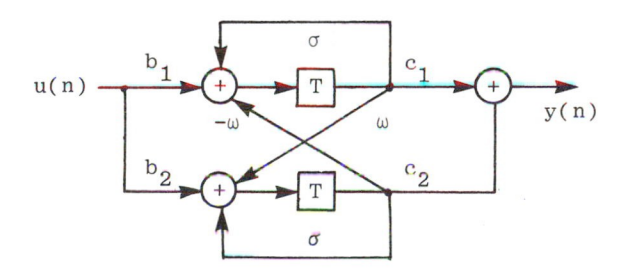

FIG. 12.6.2 Block diagram of a normal filter architecture.

$\Phi_2 = (1, -j)^T/\sqrt{2}$. From $b_1 = 0.526$ and $b_2 = 0.851$, it follows that from $b = \Sigma \beta_i \phi_i$.

$$\begin{cases} 0.526 = \dfrac{\beta_1}{\sqrt{2}} + \dfrac{\beta_2}{\sqrt{2}}; \quad \beta_1 = R_1 + jI_1 \\[4mm] 0.851 = j\dfrac{\beta_1}{\sqrt{2}} - j\dfrac{\beta_2}{\sqrt{2}}; \quad \beta_2 = R_2 + jI_2 \end{cases}$$

$$\left. \begin{array}{l} 0.745 = R_1 + R_2 \\[2mm] 0 = I_1 + I_2 \\[2mm] 1.203 = I_2 - I_1 \\[2mm] 0 = R_1 - R_2 \end{array} \right\} \begin{array}{l} R_1 = R_2 = 0.3715 \\[6mm] \\ I_2 = I_1 = 0.6015 \end{array}$$

Thus

$$\beta_1 = 0.3715 - j0.6015; \quad |\beta_1|^2 = 0.5$$
$$\beta_2 = \beta_1^*; \quad |\beta_2|^2 = 0.5$$

and

$$|f_i| = \frac{|\beta_i|^2}{(1 - |\lambda_1|^2)^2} = \frac{0.5}{0.5} = 1$$

This result can be verified by observing that

$$A = \frac{1}{\sqrt{2}} \begin{bmatrix} \cos \dfrac{\pi}{4} & \sin \dfrac{\pi}{4} \\[3mm] -\sin \dfrac{\pi}{4} & \cos \dfrac{\pi}{4} \end{bmatrix} = \begin{bmatrix} 0.5 & 0.5 \\[2mm] -0.5 & 0.5 \end{bmatrix}$$

$$b_1^2 + b_2^2 = 2(1 - r^2)^2 = 2\left(\frac{1}{2}\right) = 1$$

for $\delta = 1$. Then

$$\frac{b_2}{b_1} = \frac{r^2 \sin(\pi/2) + [1 + r^4 - 2r^2 \cos(\pi/2)]^{1/2}}{1 - r^2 \cos(\pi/2)} = \frac{1 + \sqrt{5}}{2}$$

Choosing $b_1 = 0.526$ and $b_2 = 0.851$ satisfies this constraint. Also,

$$c_1 = \frac{b_1(\alpha_1 + \alpha_2) + b_2(j\alpha_1 - j\alpha_2)}{1 - r^2 \cos(2\phi)} = b_1$$

$$c_2 = \frac{b_1(j\alpha_1 - j\alpha_2) + b_2(\alpha_1 + \alpha_2)}{1 - r^2 \cos(2\phi)} = b_2$$

Using the analysis method developed in Section 12.5, it can be shown that K and W satisfy:

For K:

$$x(1) = b = (0.526,\ 0.851)^T$$
$$x(2) = Ax(1) + p_1 x(1) = (0.163,\ -0.688)^T$$

$$X = \begin{bmatrix} 0.162 & 0.526 \\ -0.688 & 0.851 \end{bmatrix}$$

For W:

$$x(1) = b = (0.526,\ 0.851)^T$$
$$x(2) = Ax(1) + p_1(1) = (0.163,\ -0.688)^T$$

$$X = \begin{bmatrix} 0.163 & 0.526 \\ -0.688 & 0.851 \end{bmatrix}$$

Then, in conjunction with a precomputed value of R,

$$K = XRX^T = \begin{bmatrix} 1 & x \\ x & 1 \end{bmatrix}, \quad W = XRX^T = \begin{bmatrix} 1 & x \\ x & 1 \end{bmatrix}$$

where x denotes a number whose value is not significant. The fact that $K_{11} = K_{22} = 1$ is characteristic of this class of filters. Here the values of the b vector are automatically scaled so as to prevent overflow in an ℓ_2 sense. Finally, the resulting value of roundoff error variance, namely σ_x^2, satisfies

$$\sigma_x^2 = \frac{(m_1 W_{11} + m_2 W_{22})Q^2}{12} = \frac{2mQ^2}{12}$$

as previously determined. Because of the self-scaling property of K, the scaled and unscaled error variances are equal in this case.

12.7 SECTION FILTERS

Another low roundoff error variance filter, due to Jackson et al. (1977), is called the section optimal filter. This filter is based on the observation that the generic filter (A, b, c) is not scaled. That is, $|f_i|$ is, in general, not bounded by unity. In previous analysis, coefficients found in a flow diagram were scaled by $|f_i|$ to protect against overflow. If the output is defined by nonrecursively filtering state variables (i.e., linear combinations), the feedthrough paths should be scaled upward by $|f_i|$ to correct for the downward scaling at the input (i.e., replace f_i by $f_i/|f_i|$ and $g_i/|g_i|$). The desired scaling can be performed by a diagonal transform D such that $x_i' = x_i/d_{ii}$. Furthermore, in the section optimal filter, a fixed structure (A, b, c, d) is chosen to represent the second-order filter given by

$$H(z) = d + \frac{c_2 z^{-2} + c_1 z^{-1}}{a_2 z^{-2} + a_1 z^{-1} + 1} \tag{12.7.1}$$

Using a fairly detailed development, the design of an unscaled section optimal filter was shown to be (see Fig. 12.7.1)

$$\hat{A} = \begin{bmatrix} \hat{a}_{11} = \dfrac{-a_1}{2}; & \hat{a}_{12} = \dfrac{\lambda^+ (1 + c_2)}{c_1^2} \\[4ex] \hat{a}_{12} = \dfrac{\lambda^-}{1 + c_2}; & \hat{a}_{22} = \dfrac{-a_1}{2} = \hat{a}_{11} \end{bmatrix}$$

$$\lambda^{\pm} = \left(c_2 - \frac{a_1 c_1}{2} \right) + \sqrt{c_2^2 - c_1 c_2 a_1 + c_1^2 a_2}$$

$$\hat{b}_1 = \frac{1 + c_2}{2}, \qquad \hat{b} = \frac{c_1}{\sqrt{2}}$$

$$\hat{c}_1 = \frac{c_1}{1 + c_2}, \qquad \hat{c}_2 = 1$$

$$\hat{d} = d$$

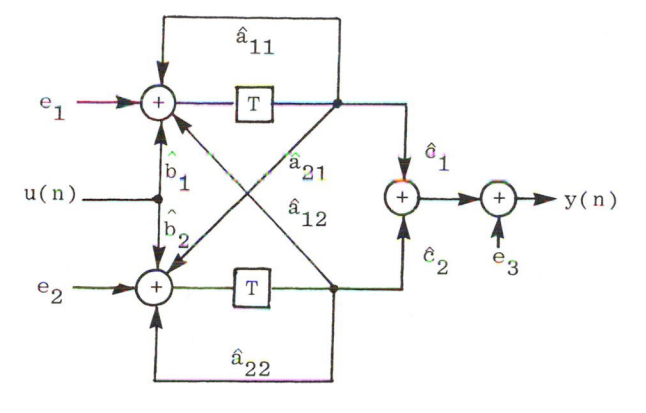

FIG. 12.7.1 Block diagram for a section optimal second-order subfilter.

where λ is a real number. The realized second-order section optimal fil-
ter, say (A', b', c', d'), can be written in terms of $(D^{-1}\hat{A}D, D^{-1}\hat{b}, \hat{c}^T D, \hat{d})$,
where D has been previously defined to be a diagonal matrix. In particu-
lar, D satisfies

$$D = \begin{bmatrix} |f_1|_2 & 0 \\ 0 & |f_2|_2 \end{bmatrix}$$

for an ℓ_2 filter, and

$$D = \begin{bmatrix} |f_1|_\infty & 0 \\ 0 & |f_2|_\infty \end{bmatrix}$$

for an ℓ_∞ filter. For the ℓ_2 case, the algorithms for K and W can be used
to quantify the norms found in the definition of D.

EXAMPLE 12.7.1 Again consider the second-order system given by

$$H(z) = \frac{1}{1 - 2r\cos\phi z^{-1} + r^2 z^{-2}} = 1 + \frac{2r\cos\phi z^{-1} - r^2 z^{-2}}{1 - 2r\cos\phi z^{-1} + r^2 z^{-2}}$$

with respect to the parameterization

$$\hat{a}_{11} = \hat{a}_{22} = \frac{-a_1}{2} = r\cos\phi; \quad \hat{b}_1 = \frac{1 - r^2}{2}, \quad \hat{c}_1 = \frac{2r\cos\phi}{1 - r^2}$$

$$\hat{a}_{12} = \frac{(1 - r^2)(2r^2 \cos^2 \phi)}{4r^2 \cos^2}; \quad \hat{b}_2 = r \cos \phi, \quad \hat{c}_2 = 1$$

$$\hat{a}_{21} = \frac{2r^2 (\cos^2 \phi - 1)}{1 - r^2}; \quad d = 1$$

$c_1 = 2r \cos \phi$	$c_2 = -r^2$
$a_1 = -2r \cos \phi$	$a_2 = r^2$

For $r \cos(\phi) = 1/2$ and $r^2 = 1/2$, it follows that $H(z) = 1 + (z^{-1} - 0.5z^{-2})/(1 - z^{-1} + 0.5z^{-2})$ and $\hat{a}_{11} = \hat{a}_{22} = 1/2$, $\hat{a}_{12} = 1/4$, $\hat{a}_{21} = -1$, $\hat{b}_1 = 1/4$, $\hat{b}_2 = 1/2$, $\hat{c}_1 = 2$, and $\hat{c}_2 = 1$. To scale the system (A, b, c, d) into a section optimal form, the scaling matrix D must be computed. Here D is given by

$$D = \begin{bmatrix} |f_1|_2 & 0 \\ 0 & |f_2|_2 \end{bmatrix}$$

where the diagonal elements of D are determined from computing K as suggested in Section 12.5. For the purpose of clarity, the analysis is repeated. From

$$A = \begin{bmatrix} 0.5 & 0.25 \\ -1 & 0.5 \end{bmatrix}; \quad b = \begin{bmatrix} 0.25 \\ 0.5 \end{bmatrix}; \quad c = (2, 1)$$

$$p(z) = z^2 - 2r \cos \phi z + r^2 = z^2 - z + 0.5$$

Using the following computer analysis of these data, it was determined that K and W satisfy

$$K = \begin{bmatrix} 0.15 & 1.0 \\ 1.0 & 0.4 \end{bmatrix}, \quad W = \begin{bmatrix} 6.4 & 1.6 \\ 1.6 & 2.4 \end{bmatrix}$$

```
THE ORDER OF THE FILTER IS (0<N<11) :

2

NOW TYPE IN ALL THE P(N), USE FREE FORMAT :

-1,0.5

READ IN THE LOGICAL UNIT NUMBER FOR WRITE.
TI: USE 4, TT0: USE 5, TT7: USE 6

4
```

1 THE TOEPLITZ MATRIX R(N,N) IS AS FOLLOWS:

   0.24000E+01   0.16000E+01
   0.16000E+01   0.24000E+01

  NOW INPUT MATRIX A , USE FREE FORMAT

 .50,.25
 -1.0,.50

  INPUT MATRIX B, USE FREE FORMAT

 .25,.50

  INPUT MATRIX C, USE FREE FORMAT

 2,1

  THE A MATRIX IS AS FOLLOWS:

   0.50000E+00   0.25000E+00
  -0.10000E+01   0.50000E+00

2 THE B MATRIX IS AS FOLLOWS:

  0.25000
  0.50000

  THE TRANSPOSE OF MATRIX C IS AS FOLLOWS:

  2.00000
  1.00000

  THE K MATRIX IS AS FOLLOWS:

   0.15000E+00   0.10000E+00
   0.10000E+00   0.40000E+00

  THE W MATRIX IS AS FOLLOWS:

   0.64000E+01   0.16000E+01
   0.16000E+01   0.24000E+01

3 NUMBER OF NOISE SOURCES FOR 1ST STATE= 3

  NUMBER OF NOISE SOURCES FOR 2ND STATE= 2

  SCALED VALUE OF B(1)= 0.645E+00

  SCALED VALUE OF B(2)= 0.791E+00

  UNSCALED ERROR VARIANCE=0.2400E+02 SCALED=0.6720E+01

The variance of the system $(\widehat{A},\ \widehat{b},\ \widehat{c},\ \widehat{d})$ is given by

$$\sigma_x^2 = \frac{[3(6.4) + 3(2.4)]Q^2}{12} = \frac{(26.4)Q^2}{12}$$

The section optimal filter can be derived by equating $d_{11} = k_{11} = \sqrt{0.15}$ and $d_{22} = k_{22} = \sqrt{0.4}$. It is obvious that D is given by $D = \text{diag}(0.3873)$, $0.6325)$ (where diag denotes "diagonal matrix"), and

$$A' = D^{-1}\hat{A}D = \begin{bmatrix} 0.5 & 0.4082 \\ -0.6134 & 0.5 \end{bmatrix}; \quad b' = D^{-1}\hat{b} = \begin{pmatrix} 0.6455 \\ 0.7096 \end{pmatrix};$$

$$c' = \hat{c}D = (0.7746, \ 0.6325)$$

This system was analyzed with a computer and the results are as follows:

```
THE ORDER OF THE FILTER IS (0<N<11) :

2

NOW TYPE IN ALL THE P(N), USE FREE FORMAT :

-1,.50

READ IN THE LOGICAL UNIT NUMBER FOR WRITE.
TI: USE 4, TTO: USE 5, TT7: USE 6

4

THE TOEPLITZ MATRIX R(N,N) IS AS FOLLOWS:

   0.24000E+01      0.16000E+01
   0.16000E+01      0.24000E+01

NOW INPUT MATRIX A , USE FREE FORMAT

.50,.4082
-.6134,.50

INPUT MATRIX B, USE FREE FORMAT

.6455
.7906

INPUT MATRIX C, USE FREE FORMAT

.7746,.63250

THE A MATRIX IS AS FOLLOWS:

   0.50000E+00      0.40820E+00
  -0.61340E+00      0.50000E+00

THE B MATRIX IS AS FOLLOWS:

   0.64550
   0.79060
```

THE TRANSPOSE OF MATRIX C IS AS FOLLOWS:

0.77460
0.63250

THE K MATRIX IS AS FOLLOWS:

0.99995E+00   0.40761E+00
0.40761E+00   0.10009E+01

THE W MATRIX IS AS FOLLOWS:

0.96085E+00   0.39130E+00
0.39130E+00   0.96002E+00

NUMBER OF NOISE SOURCES FOR 1ST STATE=   3

NUMBER OF NOISE SOURCES FOR 2ND STATE=   3

SCALED VALUE OF B(1)= 0.646E+00

SCALED VALUE OF B(2)= 0.790E+00

UNSCALED ERROR VARIANCE=0.5763E+01   SCALED=0.7687E+01

Note that the filter (A', b', c', d') is self-scaled. The error variance, due to rounding of the states x_i, is given in terms of the diagonal elements of W, where $W = XRX^T$. Here w_{11} and w_{22} are computed to have values given by $w_{11} = 0.96 \simeq =1$ and $w_{22} \simeq =1$. The scaled error variance satisfies, for $m_1 = m_2 = 3$,

$$\sigma_x^2 = [3(1) + 3(1)]Q^2/12 = 6Q^2/12$$

which is <u>less</u> than the $23.04Q^2/12$ associated with the scaled error variance for the canonical filter (Example 12.5.1) and equal to the error variance of the normal filter (Example 12.6.2 for $m = 3$).

Although the section optimal filter enjoys an improved noise figure of merit, it requires six multiplications per second-order section. This compares with only three multipliers (including input scaling) per second-order section required by a canonical realization. Therefore, the question of filter complexity should be considered when making a design choice.

12.8 COMPARISONS

In an attempt to provide the reader with additional insight into the question of roundoff errors, several filter realizations of a common transfer function will be studied. The comparative study will make use of the software techniques discussed in Section 12.5. The transfer function to be synthesized is given by

techniques discussed in Section 12.5. The transfer function to be synthesized is given by

$$H(z) = \frac{0.01591(z + 1)^2}{z^3 - 1.9786z^2 + 1.556161z - 0.453768}$$

having poles located at 0.657874 and 0.658494 ± j0.506098 (i.e., third-
order Chebyshev). This transfer function can be studied with respect to
canonical, direct II, normal, and section optimal architecture. The three-
tuple (A, b, c) for these filters is summarized below:

Filter	A			b	c
Canonical	0	1	0	0	0.02301
	0	0	1	0	0.02301
	0.453768	-1.55616	1.97461	1.0000	0.07930
Direct II	0	1	0	0	0.00000
	0	0	1	0	0.04782
	0.453768	-1.55616	1.97461	1	0.04782
Normal	0.66724	0.05888	0.58666	0.6221	0.29179
	0.09511	0.64881	0.58666	-0.1549	0.28067
	0.08939	-0.46606	0.65881	0.6111	-0.09612
Section optimal	0.65787	0.00000	0.00000	0.7531	0.37656
	0.00000	0.65849	0.50610	0.3370	-0.17724
	0.00000	-0.50610	0.65849	0.7110	-0.37183

The result of processing these filter parameters through the software analysis package resulted in the following:

Filter	Maximal ℓ_2 state			Noise gain σ^2 (scaled)	Noise gain (bits) ($\log_2 \sigma$)
	K_{11}	K_{22}	K_{33}		
Canonical	17.045	17.045	17.045	11.829	~1.8
Direct II	17.062	17.062	17.062	7.4792	~1.3
Normal	0.99759	1.0002	1.0002	2.6192	~0.7
Section	1.0000	1.0000	1.0000	2.1408	~0.5

Referring to the preceding table, one notes that the last two filters are markedly superior to the first two in terms of noise gain. However, it should also be noted that these two filters are more complex. The normal and section optimal filters require 15 multiplications per filter cycle compared to 6 for the others. This translates into potentially higher cost and lower throughput. It can also be noted that the last two filters have similar values for W_{ii} (i.e., W_{ii} W_{22} W_{33}). This is characteristic of optimal filters. As previously noted, they will have a tendency to distribute the noise error equally. Finally, one should note that the last two filters are <u>autoscaled</u> in that the diagonal values of K are unity (for all practical purposes).

The noise gain found in the summary should be interpreted as a relative figure of merit. For example, one could redesign the direct II filter to include the scale factor $1/\sqrt{K_{ii}} = 0.24209$. The resulting b vector would become b' = (0, 0, 0.24209). Presenting this new filter parameterization to the software analysis routine, one would obtain an autoscaled design having an error variance of 0.58445. This is less than the optimal filter's error variance. However, the reader is reminded that the filter realized is the prescaled filter $0.24209H(z)$ rather than the H(z). Accounting for this scale factor, the scaled error variance of the direct II filter would again be larger than that associated with the optimal filter.

If one wishes to extend this analysis to a more quantifiable level, an attempt can be made to predict the roundoff error in bits. However, since most of the analysis machinery developed in this section relies on ℓ_2, rather than ℓ_∞ concepts, the results may have a questionable interpretation. Nevertheless, suppose for the purpose of clarity that a 16-bit word is symbolically formated as follows:

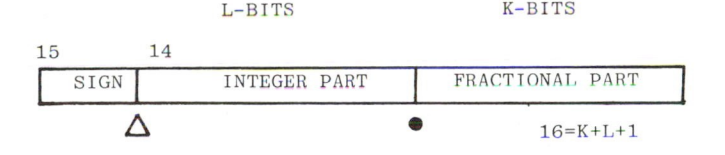

Since the canonical filter has a maximal value of $\sqrt{17.06} = 4.13 \leq 2^3$, three integer digits are required to encode the data (i.e., L = 3). However, the section optimal filter has a maximal value of one (i.e., L = 0). Therefore, the canonical filter and section optimal filters assume values of K equal to 12 and 15 bits, respectively. In terms of the quantization parameter Q, the values of $Q = 2^{-12}$ and $Q = 2^{-15}$ are realized. The ratio of estimated error variance is computed to be:

$$\frac{\text{canonical}}{\text{section}} = \frac{\text{gain(unscaled)}}{\text{gain (scaled)}} = \frac{0.69399(2^{-24}/12)}{2.4107(2^{-30}/12)}$$

$$= 18.42 - 2^{4.2} \cong (4.2 \text{ bits})$$

If one is willing to accept a modest risk of overflow, and use saturating arithmetic, the value of K_{ii} can be assumed to be equal to 4 (rather than 4.13). Under such an assumption, the previous computation reduces to

$$\frac{\text{canonical}}{\text{section}} = 4.602 \sim 2^{2.2} \quad (2.2 \text{ bits})$$

It is now up to the designer to determine how the error variance-complexity trade-offs will affect the design choice.

12.9 FREQUENCY DOMAIN METHODS

In practice, a network would be composed of several interconnected systems of subfilters. For example, $H(z)$ may be realized as a cascade filter satisfying $H(z) = \Pi H_i(z)$, $i = 1, \ldots, n$. If the impulse response of the ith subfilter is $h_i(n)$, the output process can be defined in terms of the linear combination of subfilter outputs as suggested in Fig. 12.9.1. In this figure, a system of scale factors, denoted k_j, are embedded into the architecture. The purpose of these scale factors, k_0 and k_1, satisfies

$$k_0 = \frac{1}{\max(|H_1|_q, |H_1'|_q)}; \quad k_1 = \frac{1}{\max(|H_1 H_2|_q, |H_1 H_2|_q)} \tag{12.9.1}$$

The other scale factors have a similar form. In this analysis, an ℓ_q norm has been assumed. Various choices of q will produce different roundoff error interpretations. For example, suppose that u(n) is a random process having a known covariance $R_{ui}(k)$. Furthermore, let $x_i(n)$ be a system (filter) variable given by $x_i(n) = h_i(n) * u(n)$. Then the covariance metric associated with $x_i(n)$, say $R_{x_i}(n)$, can be shown to be bounded by

$$R_{x_i}(n) \le |H_i^2|_p |G_u|_q; \quad \frac{1}{p} + \frac{1}{q} = 1 \tag{12.9.2}$$

where $\mathcal{F}(h_i(n)) = H_i(\omega)$ and $\mathcal{F}(R_u(n)) = G_u(\omega)$ (i.e., power spectrum). This second-order statistical expression can be rewritten to read

$$R_{x_i}(n) \le |H_i|_{2p}^2 |G_u|_q \tag{12.9.3}$$

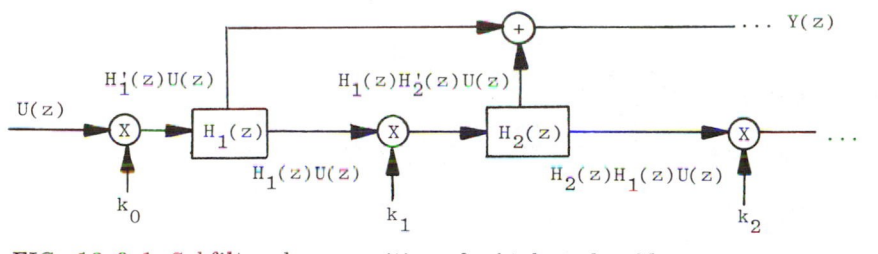

FIG. 12.9.1 Subfilter decomposition of a high-order filter

Recall $\sigma_{x_i}^2 = R_{x_i}(0)$ and note that if $p = 1$, then $q = \infty$. This result is well known in communication theory and is often represented as

$$\sigma_{x_i}^2 \leq \sigma_u^2 |H_i|_2^2 = \sigma_u^2 H_i(\omega) H_i^*(\omega) \tag{12.9.4}$$

Furthermore, the norm $|H_i|_2^2$ can be computed (if desired) using a contour integral of the form

$$|H_i|_2^2 = \frac{1}{2\pi j} \oint \frac{H_i(z) H_i(z^{-1})}{z} \, dz \tag{12.9.5}$$

It can be seen that $|G_u|_q$ can be interpreted in an ℓ_2 sense. Under the usual assumption that u is a white stochastic process which satisfies $R_u(n) = \sigma_u^2(n)$ (i.e., flat spectrum with amplitude σ_u^2), the maximal value of $G_u(\omega)$ is simply σ_u^2. Other interpretations can be made for different choices of p and q. However, once these parameters are chosen, the roundoff errors are to be interpreted in an $\ell_p - \ell_q$ sense.

EXAMPLE 12.9.1 Consider the direct II filter described in Fig. 12.9.2. With $|x_1(n)|_p \leq 1$, it follows that $|x_i(n)|_p \leq 1$ for all other i. To ensure this condition, $k_0 \geq 1/|1/B(z)|_q$. Since all the x_i's would now be bounded by unity, it follows that

$$|y| \leq \sum_{i=0}^{n} |a_i|_p \tag{12.9.6}$$

Obviously, $|y|$ may exceed unity. Two approaches to this problem can be taken. One is to increase the value of k_0 to the point where $|y| \leq 1$. The other is to increase the size of the accumulator's dynamic range (e.g., 16-bit multiplier and a 24-bit accumulator).

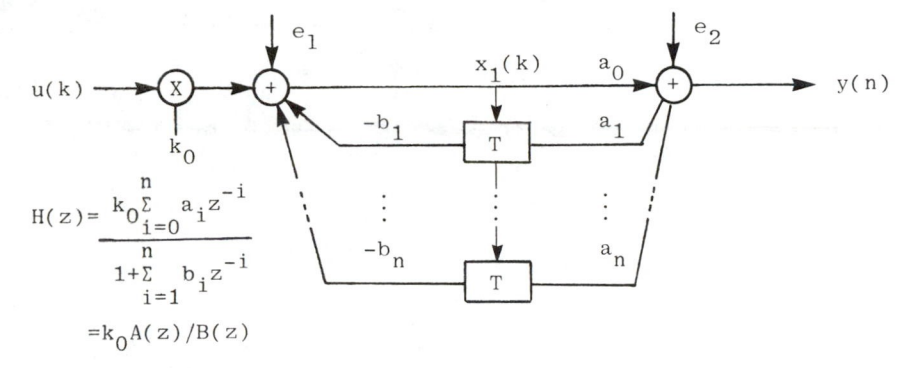

FIG. 12.9.2 General form of a direct II discrete filter.

In terms of roundoff errors, the noise power spectrum at the output, say $N_y(\omega)$, is given by

$$N_y(\omega) = \underbrace{\frac{Q^2}{12}(n+1)}_{\substack{e_2 \text{ of}\\ \text{Fig. 12.9.2}\\ n+1 \text{ noise}\\ \text{to } a_i\text{'s}}} + \underbrace{\frac{Q^2}{12}(n+1)\frac{\left|H(\exp(j\omega)\right|^2}{k_0}}_{\substack{e_1 \text{ of Fig. 12.9.2}\\ n+1 \text{ noise sources}\\ \text{from feedback paths}}} \tag{12.9.7}$$

Recall that

$$R_y(k) = \mathcal{F}^{-1}(N_y(\omega)) = \frac{1}{j2\pi}\int_{-\pi}^{\pi} N_y[\exp(j\omega)]\exp(j\omega)\,d\omega \tag{12.9.8}$$

and it follows that

$$\sigma_y^2 = R_y(0) = \frac{1}{j2\pi}\int_{-\pi}^{\pi} N_y[\exp(j\omega)]\exp(j0)\,d\omega \tag{12.9.9}$$

with

$$N_y[\exp(j\omega)] \geq 0, \qquad \sigma_y^2 = N_{y1} \tag{12.9.10}$$

12.10 SUMMARY

In this chapter the effect of arithmetic roundoff was studied. It was noted that in a fixed-point environment, roundoff noise is architecture dependent. Several architectures were presented which are known to have superior roundoff error immunity. These filters were second-order modules which would be integrated into higher-order filters. The trade-off facing the designer is that of reduced noise variance at the expense of filter complexity. The added arithmetic operations required of an optimal filter will compromise filter throughput, and in some cases, reliability and cost. The designer should be aware of these negative attributes. Since roundoff error variances are of the form $kQ^2/12$, where $Q = 2^{-n}$, the choice of word length will have a scaling effect. It may turn out that a 16-bit simple filter will provide a more acceptable overall design than a 12-bit optimal filter. If ℓ_2 norms are the mechanism by which filters are compared, the software-based analysis scheme presented in this chapter will greatly simplify this study.

BIBLIOGRAPHY

Barnes, C. W. (1979), Roundoff Noise and Overflow in Normal Digital Filters, IEEE Trans. Circuits Syst., CAS-26, March, pp. 154-159.

Fam, A. T., and C. W. Barnes (1979), Nonminimal Realization of Fixed-Point Digital Filters that are Free of All Finite Word-Length Limit Cycles, IEEE Trans. Acoust. Speech Signal Process., ASSP-27, April, pp. 149-153.

Huang, S. Y. (1975), Dynamic Range Constraint in State-Space Filtering, IEEE Trans. Acoust. Speech Signal Process., ASSP-23, December, pp. 591-593.

Huang, S. Y. (1977), Minimum Uncorrelated Unit Noise in State Digital Filtering, IEEE Trans. Acoust. Speech Signal Process., ASSP-25, August, pp. 273-281.

Kaneko, T. (1973), Limit-Cycle Oscillations in Floating Point Digital Filters, IEEE Trans. Audio Electroacoust., AU-21, April, pp. 100-106.

Liu, B. (1971), Effect of Finite Word Length on the Accuracy of Digital Filters, IEEE Trans. Circuit Theory, CT-18, November, pp. 670-677.

Mullis, C. T., and R. A. Roberts (1976), Synthesis of Minimum Roundoff Noise Fixed Point Digital Filters, IEEE Trans. Circuits Syst., CAS-23, September, pp. 551-562.

Oppenheim, A. V., and R. W. Schafer (1975), Digital Signal Processing, Prentice-Hall, Englewood Cliffs, N.J.

Oppenheim, A. V., and C. J. Weinstein (1972), Effects of Finite Register
    Length in Digital Filtering and the Fast Fourier Transform, Proc.
    IEEE, 60, August, pp. 957-976.

Oppenheim, A. V., and C. J. Weinstein (1972), Effects of Finite Register
    Length in Digital Filtering and the Fast Fourier Transform, Proc.
    IEEE, 60, August, pp. 957-976.

Rabiner, L. R., and B. Gold (1975), Theory and Application of Digital
    Signal Processing, Prentice-Hall, Englewood Cliffs, N. J.

13
Coefficient-Related Errors

13.1 INTRODUCTION

A typical filter synthesis process involves three steps. First, an analog prototype filter is designed with specific frequency domain attributes. The analog filter is then transformed into a discrete filter using a suitable transform. Finally, the discrete filter is realized as a digital filter. The coefficient space for the first two operations is the real numbers. However, a digital filter is defined in terms of filters having a finite precision (i.e., finite word length). As a result, there may exist differences between the responses of a discrete and realized digital filter.

13.2 COEFFICIENT QUANTIZATION

To this point, roundoff errors have been the major consideration. There exist other error sources as well. One such error source is due to coefficient quantization. Whereas roundoff error will typically degrade a filter's performance, coefficient quantization error can alter dramatically its response. This can best be argued in terms of pole locations of a general filter. Suppose that a filter's characteristic equation is given by $D(z)$, where $D(z) = \Sigma a_i z_i$, $i = 1, \ldots, n$. Because of finite-word-length effects, the quantized characteristic equation will be given by $D_Q(z)$, where $D_Q(x) = \Sigma [a_i]_Q z^i$, with $[x]_Q$ denoting the quantization (rounding) of x. If the pole locations of $D(z)$ are $z = z_i$, so that $D(z) = \Pi(z - z_i)$, it follows that $D_Q(z) = \Pi(z - \bar{z}_i)$, where, in general, $z_i \neq \bar{z}_i$. That is, because of quantization, the factors of $D(z)$ and $D_Q(z)$ may differ. The different factors will give rise to dissimilar pole-zero locations. Changes in the pole locations can dramatically alter the performance of a filter. For example, a steep-skirt (stonewall) filter's poles are finely tuned and reside near the periphery of the unit circle. Quantization errors, in this case, can shift the filter's center frequency or, in some cases, produce poles which are outside the unit circle (i.e., destabilizing).

EXAMPLE 13.2.1 Before developing this thesis further, a simple example will be offered. Suppose that $H_0(z) = (2z - 3/2)/[z^2 - (3/2)z - 35/64]$. This filter can be realized in canonical, cascade, or parallel form.

Canonical:

$$H_1(z) = \frac{2z - 3/2}{z^2 - (3/2)z - 35/64}$$

Cascade:

$$H_2(z) = \frac{2z - 3/2}{z - 7/8} \frac{1}{z - 5/8}$$

Parallel:

$$H_3(z) = \frac{1}{z - 7/8} + \frac{1}{z - 5/8}$$

Note that it would require only 4–bits (three fractional) words to encode the coefficients of $H_3(z)$. However, 5–bit (three fractional) and 8–bit (six fractional) words would be required for $H_2(z)$ and $H_1(z)$, respectively. They may also be compared on the basis of a fixed word length. Letting $H^\infty(z)$ denote the discrete filter, and $H^4(z)$ and $H^{4'}(z)$ denote the filter realized using the 4–bit word $\pm x_\Delta xx$ and $\pm_\Delta xxx$, respectively (rounding assumed), then

$$H_1^\infty = \frac{2z - 3/2}{z^2 - (3/2)z + 35/64} \longrightarrow H_1^4 = \frac{2z - 3/2}{z^2 - (3/2)z + 1/2}$$

$$H_2^\infty = \frac{2z - 3/2}{(z - 7/8)(z - 5/8)} \longrightarrow H_2^4 = \frac{2z - 3/2}{z - 3/4} \frac{1}{z - 1/2}$$

$$H_3^\infty = \frac{1}{z - 7/8} + \frac{1}{z - 5/8} \longrightarrow H_3^4 = \frac{1}{z - 7/8} + \frac{1}{z - 5/8}$$

Formally, one can consider a filter of the form $H(z) = N(z)/D(z)$ or an inner product of the form, $\underline{a}^T \underline{z}/(1 + \underline{b}^T \underline{z})$, where

$$\underline{a}^T = (a_1, \ldots, a_n); \quad \underline{b}^T = (b_1, \ldots, b_n) \tag{13.2.1}$$

$$\underline{z}^T = (z^0, z^{-1}, \ldots, z^{-n+1})$$

and

$$H(z) = \frac{\displaystyle\sum_{i=1}^{n} a_i z^{-i}}{1 - \displaystyle\sum_{i=1}^{b} b_i z^{-i}} \qquad (13.2.2)$$

Denote the quantized version of $H(z)$ to be $H_Q(z)$, where

$$H_Q(z) = \frac{N_Q(z)}{D_Q(z)} = \frac{a_Q^T z}{1 + b_Q^T z} \qquad (13.2.3)$$

and

$$a_Q^T = ([a_1]_Q, \ldots, [a_n]_Q)$$

$$b_Q^T = ([b_1]_Q, \ldots, [b_n]_Q)$$

Furthermore, it shall be assumed that the quantization error may be modeled in the following manner:

$$[a_i]_Q \simeq a_i + \widetilde{e}_i; \qquad [b_i]_Q \simeq b_i + \widehat{e}_i \qquad (13.2.4)$$

The difference between the unquantized and quantized system outputs produces an error process given by $e(n) = y(n) - y_Q(n)$. In terms of a transform domain analysis, retaining only the first-order terms, $E(z)$ is given by

$$Z(e(n)) = E(z) = \widetilde{E}(z)X(z) - \widehat{E}(z)Y(z) - V(z)E(z) \qquad (13.2.5)$$

with

$$\widetilde{E}(z) = \Sigma \widetilde{\epsilon}_i z^{-i}; \qquad i = 1, \ldots, n$$

$$\widehat{E}(z) = \Sigma \widehat{\epsilon}_i z^{-i} \qquad (13.2.6)$$

$$V(z) = \Sigma b_i z^{-i}$$

Alternatively, (13.2.6) may be compressed to read

$$0 = E(z)X(z) - \widehat{E}(z)Y(z) - B(z)E(z); \qquad B(z) = 1 + V(z) \qquad (13.2.7)$$

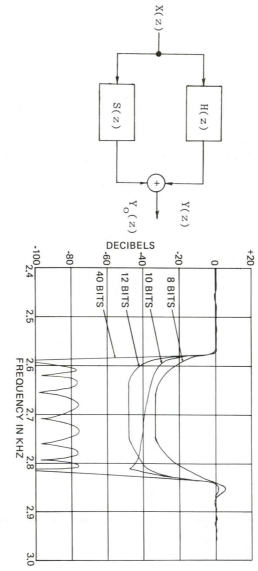

FIG. 13.2.1 Degeneration in the frequency response of a 22nd-order bandstop digital filter. (After Knowles and Olcayto, 1968; reprinted with the permission of the IEEE.)

Since $Y(z)$ and $Y_Q(z)$ are given by $H(z)X(z)$ and $H_Q(z)X(z)$, respectively, it follows that the quantization error variance, based on first-order statistical effects, satisfies

$$\sigma^2 = \mathcal{E}\left(\frac{1}{2}\int_{-\pi}^{\pi} |H(z) - H_Q(z)|^2 \, dz\right)$$

$$= \mathcal{E}\left(\frac{1}{2\pi}\int_{-\pi}^{\pi} \frac{|(\widetilde{E}(z) - \widehat{E}(z)H(z))|}{B(z)}\right) \tag{13.2.8}$$

Knowles and Olcayto (1968) modeled this error to be induced by a "stray" filter operating in parallel with the infinte-precision discrete filter. The stray filter, say $S(z)$, is modeled as $S(z) = (\widetilde{E}(z) - \widehat{E}(z)H(z))/B(z)$. If $H(z)$ were a bandstop filter, a typical frequency response, as a function of word length, is suggested in Fig. 13.2.1. In this figure, the response of a 22nd-order bandpass filter is analyzed with respect to coefficient word length.

EXAMPLE 13.2.2 The effects of coefficient quantization can be analyzed using simulation. The quantization effects on third-through sixth-order Butterworth and Chebyshev filters are reported in Fig. 13.2.2 where F.P. denotes floating point.

13.3 COEFFICIENT SENSITIVITY

The problem of coefficient quantization can be embedded into the classical study of coefficient sensitivity. Again, consider $H(z)$ and $H_Q(z)$, where the poles of $H(z)$ are found interior to the unit circle. For reasonably long word length, the rounding errors may be assumed to be small. Under this hypothesis, the poles of $H_Q(z)$ may be approximated to be given by $\widehat{p}_i = \widetilde{p}_i + \Delta p_i$, where Δp_i is given by

$$\Delta p_i = \sum_{j=1}^{M} \frac{\partial p_i}{\partial c_j} \, \Delta c_j \tag{13.3.1}$$

where c_j is a filter coefficient. From the fact that $\partial p_k/\partial a_i$ has the general form $\partial p_k/\partial a_i = z_i^{n-k}/\Pi(p_k - p_j)$, $j = 1, \ldots, N$, $k \neq j$, it can be seen that if $p_k \sim p_j$, the sensitivity to small parameter variations is large. This supports a well-known article of linear constant-coefficient control theory (servomechanisms) that pole clustering results in a coefficient-sensitive design. Frequency-selective filters, having a high Q, often exhibit a tight clustering of poles in the z domain. As a result, these filters often require larger than normal word lengths in order to achieve satisfactory performance.

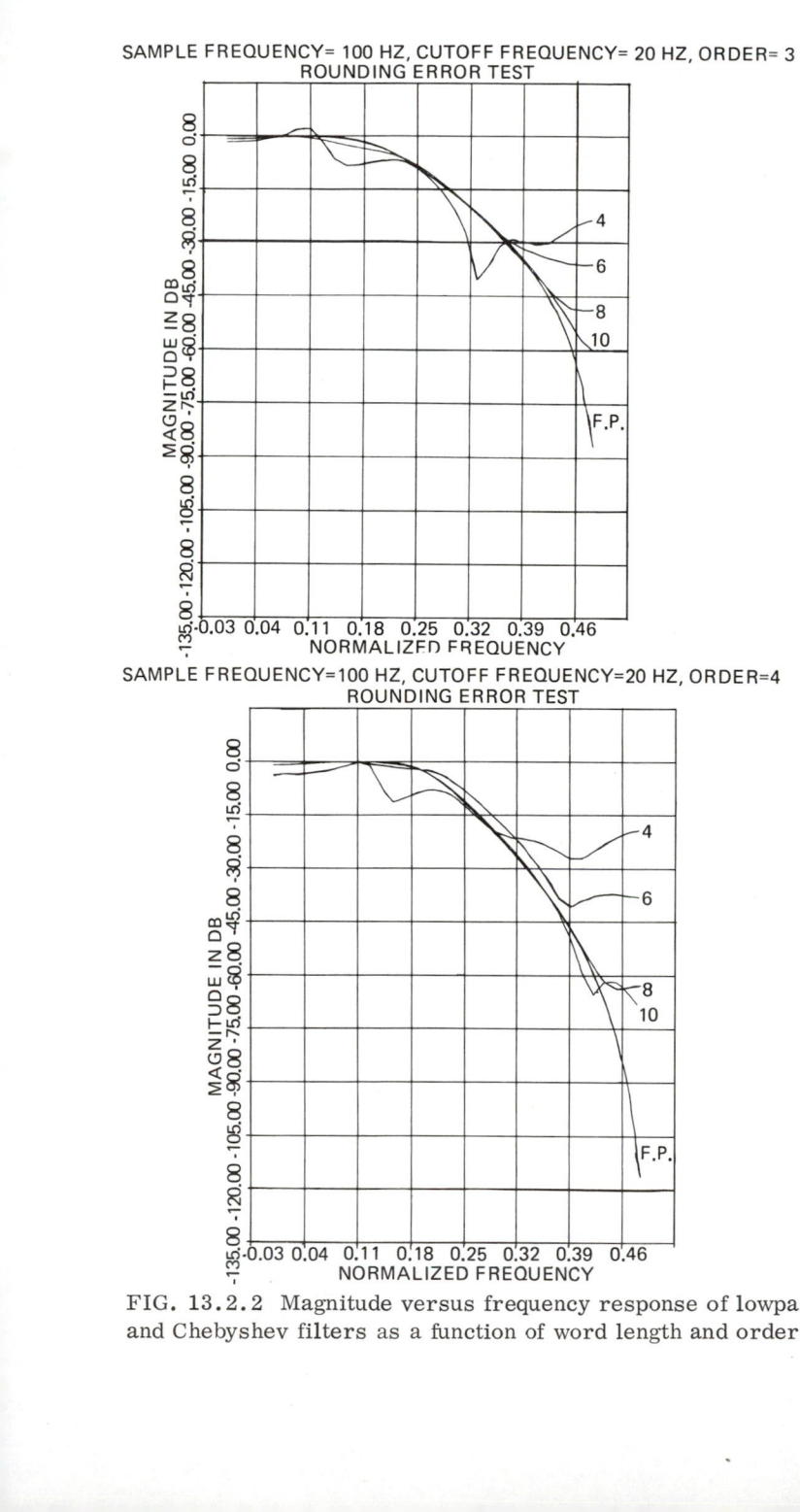

FIG. 13.2.2 Magnitude versus frequency response of lowpass Butterworth and Chebyshev filters as a function of word length and order.

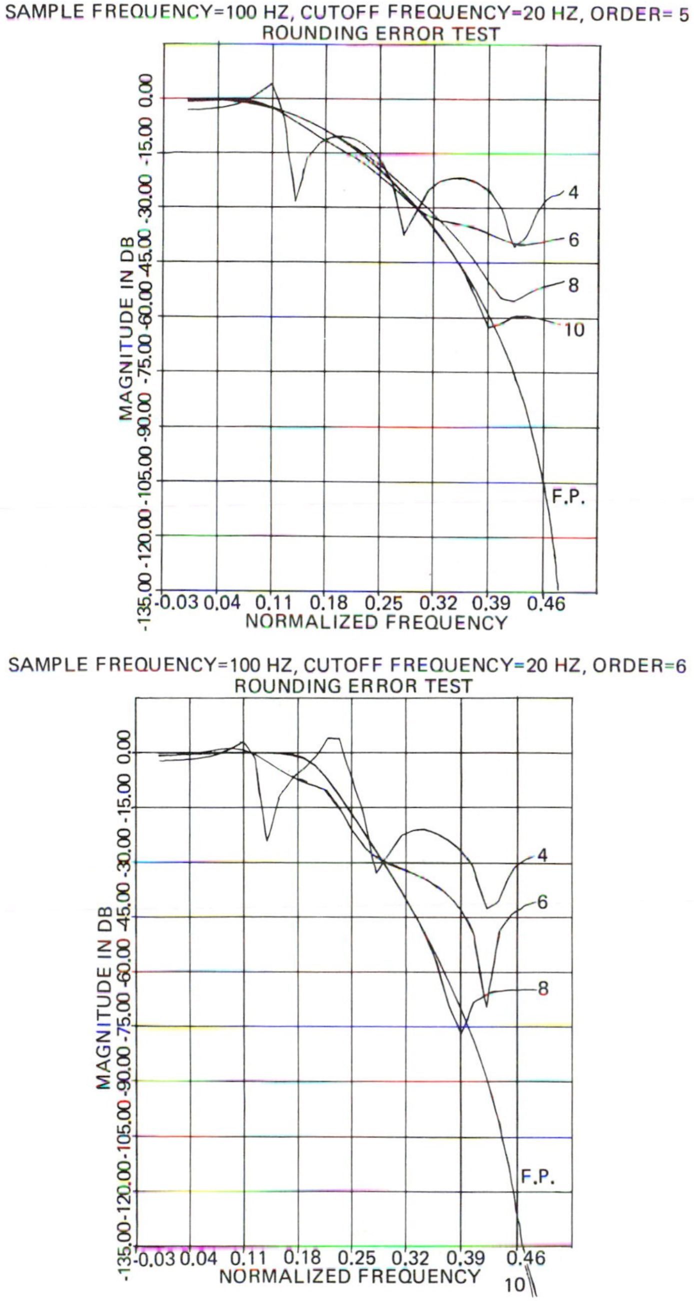

SAMPLE FREQUENCY=100 HZ, CUTOFF FREQUENCY=20 HZ, ORDER= 5
ROUNDING ERROR TEST

SAMPLE FREQUENCY=100 HZ, CUTOFF FREQUENCY=20 HZ, ORDER=6
ROUNDING ERROR TEST

FIG. 13.2.2 (Continued)

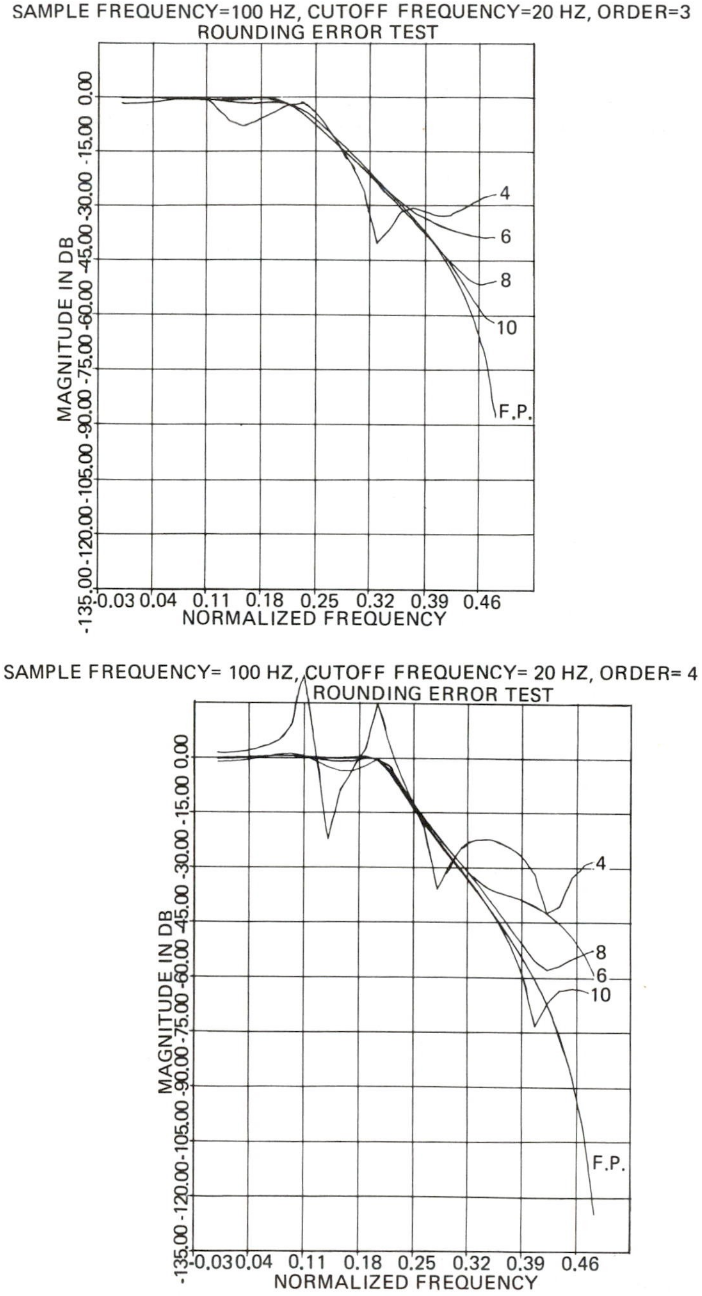

SAMPLE FREQUENCY=100 HZ, CUTOFF FREQUENCY=20 HZ, ORDER=3
ROUNDING ERROR TEST

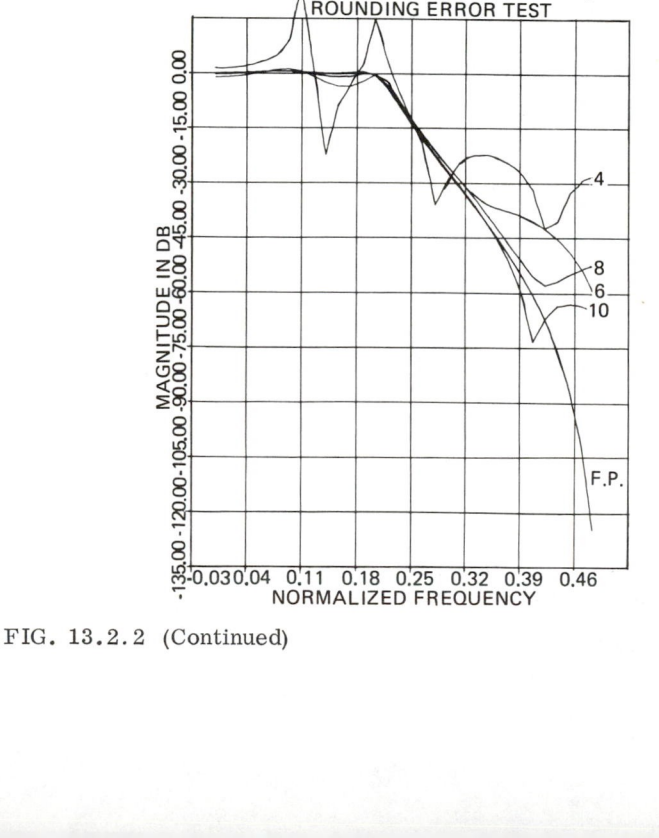

FIG. 13.2.2 (Continued)

SAMPLE FREQUENCY=100 HZ, CUTOFF FREQUENCY=20 HZ, ORDER=5
ROUNDING ERROR TEST

SAMPLE FREQUENCY=100 HZ, CUTOFF FREQUENCY=20 HZ, ORDER=6
ROUNDING ERROR TEST

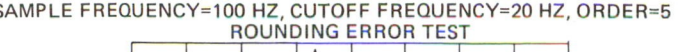

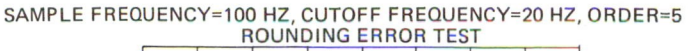

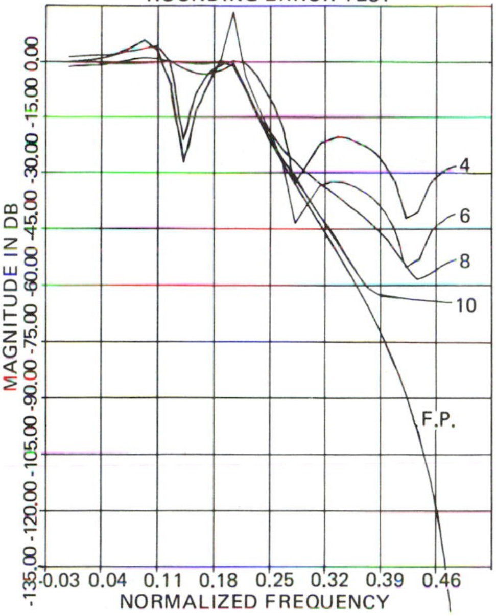

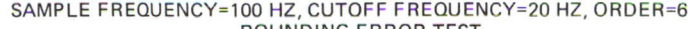

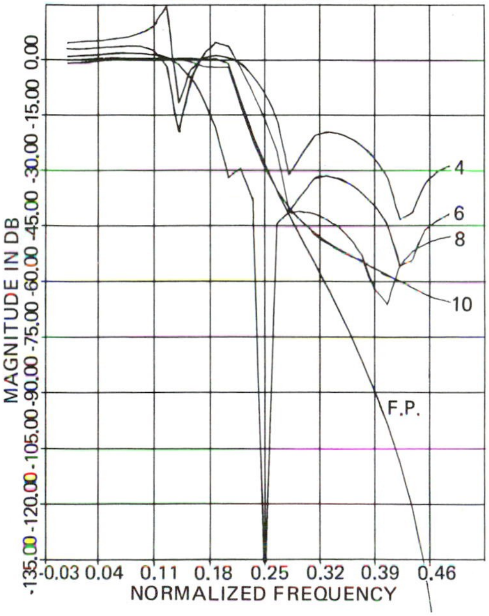

FIG. 13.2.2 (Continued)

13.4 COEFFICIENT FIELDS

The problem of coefficient quantization can be investigated in a geometric setting. Consider for the moment the canonical filter having the characteristic equation $D(z) = z^2 = 2r \cos(\phi) + r^2$. The admissible stable pole locations, for a filter having coefficients defined in terms of a signed 5-bit word in the form $\pm_\Delta$xxxx, are summarized in Fig. 13.4.1. The longer the word length, the tighter the mesh of admissible pole locations (i.e., increased precision). The poles of this filter reside on arcs having a radius-squared value of $[r^2]_Q$ with a real-axis projection given by $r \cos(\phi)_Q$. This geometry establishes a dense field of admissible pole locations near $0 + j1$ but leaves it sparse near $1 + j0$. Therefore, the effects of coefficient quantification can be seen to be a function of the pole locations of the discrete

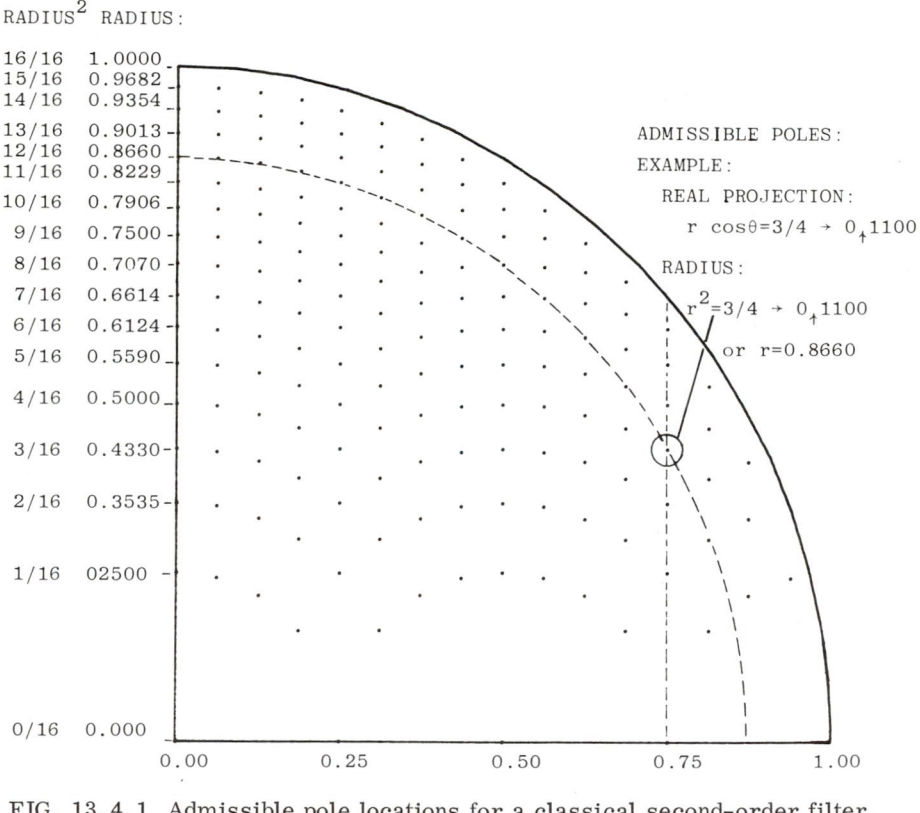

FIG. 13.4.1 Admissible pole locations for a classical second-order filter. Word length = 5 bits.

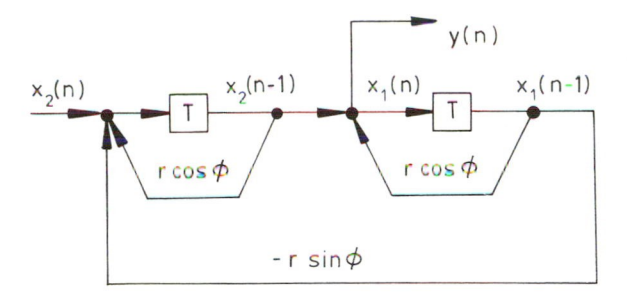

FIG. 13.4.2 Modified second-order IIR digital filter architecture.

filter. A prototype filter, for example, having poles near $0 + j1$ would suffer less than a lowpass filter whose poles are near $1 + j0$.

An alternative architecture can be used which distributes the errors more uniformly over the z plane. This architecture is compared to the canonical form in Fig. 13.4.2. In state-variable form, the filter satisfies

$$\underline{x}(n) = \begin{bmatrix} x_1(n) \\ x_2(n) \end{bmatrix}$$

$$= \begin{bmatrix} r\cos & r\sin \\ -r\sin & r\cos \end{bmatrix} [x(n-1)] + \begin{pmatrix} 0 \\ 1 \end{pmatrix} u(n) \qquad (13.4.1)$$

The characteristic equation, namely $\det(zI - A)$, is given by $z^2 + 2r\cos(\phi)z + r^2$. The coefficients found in this filter have values $r\cos(\phi)$ (real projection) and $r\sin(\phi)$ (imaginary projection). The pole geometry induced by this filter is summarized in Fig. 13.4.3. It can be noted that the realized pole locations now reside at the quantized projection values of $r\cos(\phi)$ and $r\sin(\phi)$.

The problem of minimizing coefficient quantization errors has been geometrically interpreted using integer programming. This optimization methodology can be used to minimize error, subject to a set of integer-valued side constraints. The side constraints for this problem are the restricted set of admissible pole locations. Avenhaus and Schuessler studied an eighth-order elliptic filter using integer programming. It was discovered that the tightly clustered poles of this filter were sensitive to quantization errors. By using integer programming to optimize the pole locations, a filter having 6 fractional bits was shown to perform as well as one having 36 fractional bits and straight coefficient rounding. However, integer programming methods historically require a large amount of computer time

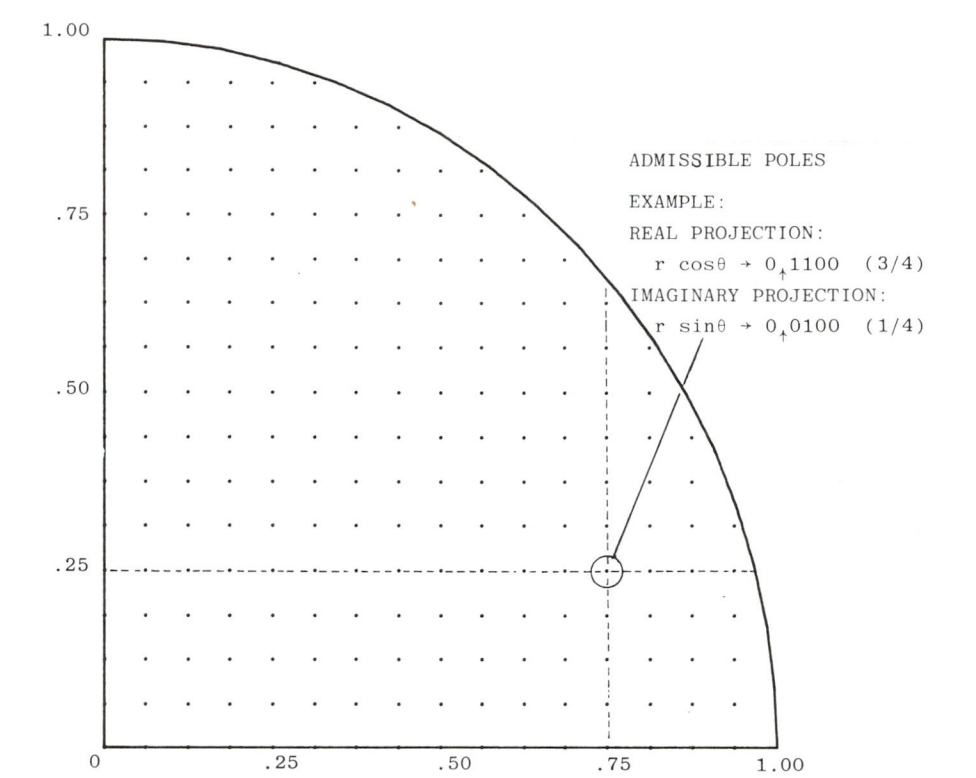

FIG. 13.4.3 Admissible pole locations for the modified IIR second-order architecture. Word length = 5 bits.

to perform. A less exotic method by which a filter's coefficient sensitivity can be studied is called the statistical word length method.

13.5 STATISTICAL WORD LENGTH

The statistical word length method is rooted in the field of classical filter analysis. In this study, filter sensitivity is measured with respect to the cause-and-effect relationships associated with small variations in coefficient values (see Section 13.3). The sensitivity measure generally takes the form $\partial|H(\omega)|^2/\partial a_i$, where a_i is a lumped filter coefficient. In classical analog filter analysis, small variations in a_i are assumed to be due to aging, drift, manufacturing tolerances, and so on. In a digital filter, variations are assumed to be attributable only to quantization effects (i.e., rounding of discrete filter coefficients). Nevertheless, one could hope that

a low-sensitivity analog filter would be transformed under a z transform
into a low-sensitivity discrete filter. As a result, the digital realization
of the discrete filter would have a certain tolerance to coefficient roundoff
errors. Crochiere, Oppenheim, Rabiner, and many others have followed
this line of reasoning. This philosophy will now be reviewed. However,
the reader is reminded that this work is valid if and only is small coefficient
roundoff errors are considered. Therefore, one is generally considering
designs which are at least 8 bits (usually much more) in size.

The concept of statistical word length is defined in terms of incremental changes in the system's transfer function due to coefficient roundoff.
A filter's magnitude response, say $H(\omega)$, can be modeled as

$$|H(\omega)| = \Delta|H(\omega)| + |H_0(\omega)| \qquad\qquad (13.5.1)$$

where $H_0(\omega)$ is assumed to be the filter's infinite precision (i.e., discrete)
frequency response, while $H(\omega)$ models the realized response. The incre-
mental analysis is premised on the assumption that the error term $\Delta|H(\omega)|$
is small. Furthermore, the discrete response $|H(\omega)|$ is often an approxi-
mation to some prespecified ideal response $H_I(\omega)$ (e.g., third-order Chebyshev approximation of an ideal lowpass filter). Typically, a designer
would specify some region (or cone) of acceptance about the ideal filter
response. Any realizable filter, whose response fell within this region
would be considered to be an acceptable design. For example, referring to
Fig. 13.5.1, the region of acceptance about the ideal $H_I(\omega)$ is denoted
$\pm\delta(\omega)$. That is, $|H(\omega) - H_I(\omega)| \le \delta(\omega)$.

The design problem now becomes one of trade-offs. Filter order and
word length will introduce approximation errors that will cause the realized
filter to deviate from the ideal. Errors due to coefficient roundoff can be
modeled as

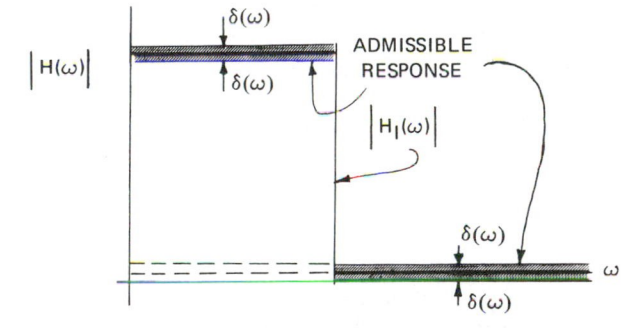

FIG. 13.5.1 Constrained optimal filter design model.

$$\Delta |H(\omega)| = \sum_{i=1}^{M} \frac{\partial |H_0(\omega)|}{\partial c_i} \Delta c \qquad (13.5.2)$$

where c_i is a filter coefficient. Here Δc_i is assumed to represent the error due to finite-word-length effects. This error has been previously argued to be uniformly distributed, mean zero, variance $Q^2/12$. Under this hypothesis, it follows that

$$\sigma^2 \simeq \frac{Q^2}{12} S^2(\omega); \quad S^2(\omega) = \sum_{i=1}^{M} \left(\frac{\partial |H_0(\omega)|}{\partial c_i} \right)^2 \qquad (13.5.3)$$

If M is sufficiently large, the law of large numbers (see Example 4.9.2) implies that $\Delta H(\omega)$ may be assumed to be normally distributed. That is, the linear combination of a large number of independent uniformly distributed errors tends to become normally distributed. Using this statistical model, the probability that $\Delta |H(\omega)|$ will exceed some prespecified value, say $s \sigma_\Delta$, $s \geq 0$, can be quantified. If σ_Δ represents the standard deviation of the normal error process, error function tables can be used to interpret the results. For example, if $s = 2$, the probability that $\Delta |H(\omega)|$ will <u>not</u> exceed two standard deviations is approximately 0.95. For a "tighter" confidence bound, s would be increased.

To satisfy the design objective, it is obvious that

$$|\Delta H(\omega)| \leq |\delta(\omega) - |H_0(\omega) - H_I(\omega)|| \triangleq G(\omega) \qquad (13.5.4)$$

For a given confidence level, one would choose Q so that $sQS(\omega)/\sqrt{12} \leq \min [G(\omega)]$. By observing that $Q \leq \sqrt{12} \min [G(\omega)]/sS(\omega)$ and by defining $Q' = \max \{ \min [\sqrt{12} G(\omega)]/sS(\omega) \}$, the statistical estimate of word length needed to satisfy the statistical design constraints can be specified. The computed word length n will be interpreted in terms of a signed, binary-valued word having length $n = 1 + n_M - n_L$, where

1. n_M denotes the weight of the most significant data bit.
2. n_L denotes the weight of the least significant data bit.

The value of n_M is established by the largest filter coefficient. The second parameter, namely n_L, can be estimated to be given by

$$n_L = \log_2 Q' \qquad (13.5.5)$$

To use these equations, the filter design three-tuple ($H_0(\omega)$, $H_I(\omega)$, $\delta(\omega)$) would have to be specified and the <u>sensitivity metric</u> $S(\omega)$ computed.

For clarity, let us temporarily ignore $S(\omega)$. In a classical filter design problem (Chapter 7), the designer often specifies the filter's magnitude frequency response in a piecewise constant form. If the realized discrete filter is to be a close approximation of the desired response, a high-order filter may be required. High-order filters can generally be expected to follow the frequency response contours of the ideal $H_I(\omega)$ very closely. As a result, the filter's response will probably have a small probability of exiting the $\delta(\omega)$ region, about the ideal, due to minor coefficient roundoff errors. However, reducing $\delta(\omega)$ by some factor would increase this probability (i.e., tighter design constraints). A lower-order filter would be expected to have a greater separation between the ideal and the synthesized discrete filter. The deviations would naturally place certain sections of the discrete response near the edge of the acceptance boundary $\pm\,\delta(\omega) + H_I(\omega)$. Therefore, the probability of leaving the region of acceptance, because of coefficient roundoff errors, can be high.

Once an analog prototype filter has been synthesized, it is converted into a discrete filter using a suitable z transform. The resulting discrete filter can be realized in one of many possible architectures. Once an architecture is chosen, the lumped filter coefficients are derived. If the chosen filter is a low-sensitivity filter, rounding these coefficients to form a digital filter would have a minor effect on performance. However, it should be stressed that if the sensitivity metric found in Eq. 13.5.1 is used, only individually small roundoff errors can be tolerated. This means that one should consider filters with reasonably long word lengths to ensure that this conjecture is satisfied.

EXAMPLE 13.5.1 Crochiere and Oppenheim (1975) have experimentally studied this problem in terms of direct II, cascaded second-order, parallel second-order, continued fraction, Gray-Markel ladder, and wave filter architectures for $\delta(\omega) = 0.05$. The target ideal filter is abstracted in Fig. 13.5.2a. It was determined that an eighth-order elliptic filter would satisfy the specified design constraints. The resulting discrete filters were synthesized and their sensitivity metrics [i.e., $S(\omega)$] computed. The resulting filter architectures are summarized in Fig. 13.5.2b-f. The result of this experiment were in general agreement with the observations reported by others. The data are given in Table 13.5.1 and provide us with an excellent opportunity to evaluate design choice trade-offs. The data found in this table are admittedly conservative (usually by a few bits). Note that the wave filter presented here exhibits the previously established wave property that stable filter coefficients are bounded by unity (i.e., $n_m = -1$). Antoniou has summarized the coefficient sensitivity trade-offs as listed in Table 13.5.2.

The bit-multiplier product found in Table 13.5.1 provides the reader with a figure of merit for studying accuracy-complexity trade-offs. Quantifying the roundoff error using this method unfortunately requires that the

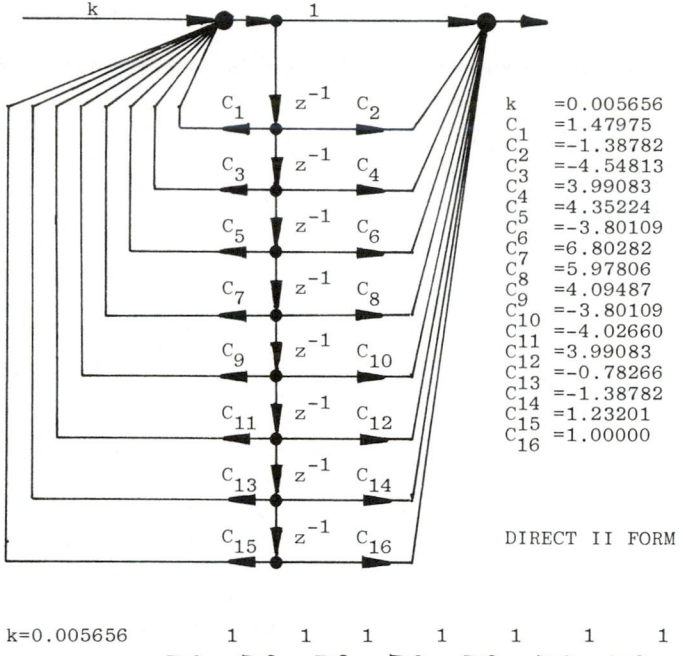

k   =0.005656
C_1 =1.47975
C_2 =-1.38782
C_3 =-4.54813
C_4 =3.99083
C_5 =4.35224
C_6 =-3.80109
C_7 =6.80282
C_8 =5.97806
C_9 =4.09487
C_{10} =-3.80109
C_{11} =-4.02660
C_{12} =3.99083
C_{13} =-0.78266
C_{14} =-1.38782
C_{15} =1.23201
C_{16} =1.00000

DIRECT II FORM

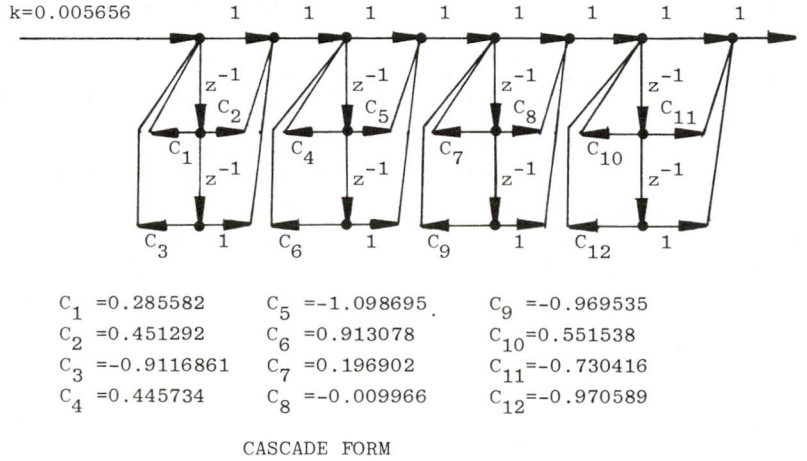

C_1 =0.285582 C_5 =-1.098695 C_9 =-0.969535
C_2 =0.451292 C_6 =0.913078 C_{10}=0.551538
C_3 =-0.9116861 C_7 =0.196902 C_{11}=-0.730416
C_4 =0.445734 C_8 =-0.009966 C_{12}=-0.970589

CASCADE FORM

FIG. 13.5.2 Comparison of common filter architectures in terms of an eighth-order bandpass design model. The passband and stopband word lengths (in bits) represent the minimum word lengths required to satisfy the a priori frequency domain constraints associated with the filter.

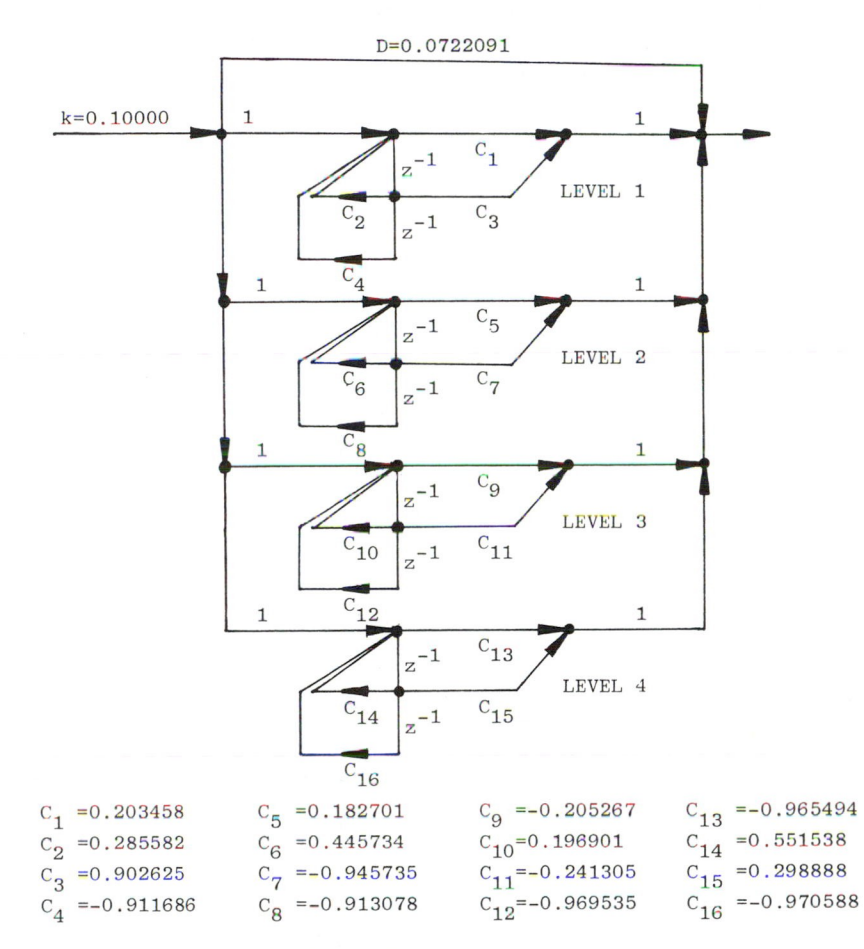

C_1 =0.203458 C_5 =0.182701 C_9 =-0.205267 C_{13} =-0.965494
C_2 =0.285582 C_6 =0.445734 C_{10}=0.196901 C_{14} =0.551538
C_3 =0.902625 C_7 =-0.945735 C_{11}=-0.241305 C_{15} =0.298888
C_4 =-0.911686 C_8 =-0.913078 C_{12}=-0.969535 C_{16} =-0.970588

PARALLEL FORM

FIG. 13.5.2 (Continued)

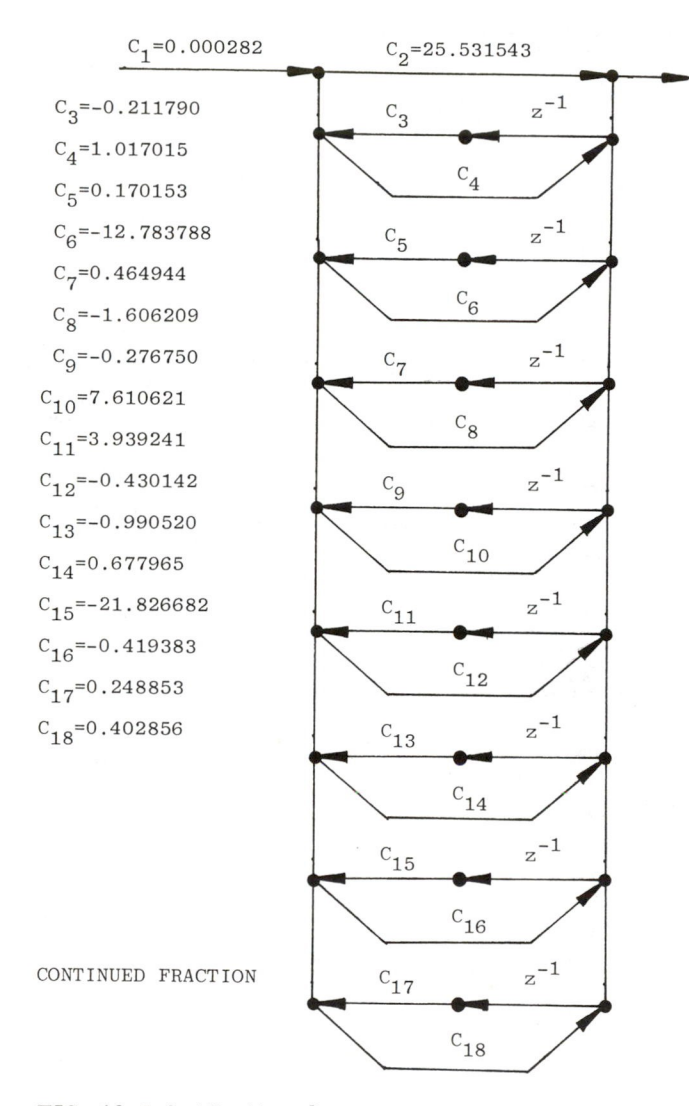

$C_1 = 0.000282$

$C_2 = 25.531543$

$C_3 = -0.211790$

$C_4 = 1.017015$

$C_5 = 0.170153$

$C_6 = -12.783788$

$C_7 = 0.464944$

$C_8 = -1.606209$

$C_9 = -0.276750$

$C_{10} = 7.610621$

$C_{11} = 3.939241$

$C_{12} = -0.430142$

$C_{13} = -0.990520$

$C_{14} = 0.677965$

$C_{15} = -21.826682$

$C_{16} = -0.419383$

$C_{17} = 0.248853$

$C_{18} = 0.402856$

CONTINUED FRACTION

FIG. 13.5.2 (Continued)

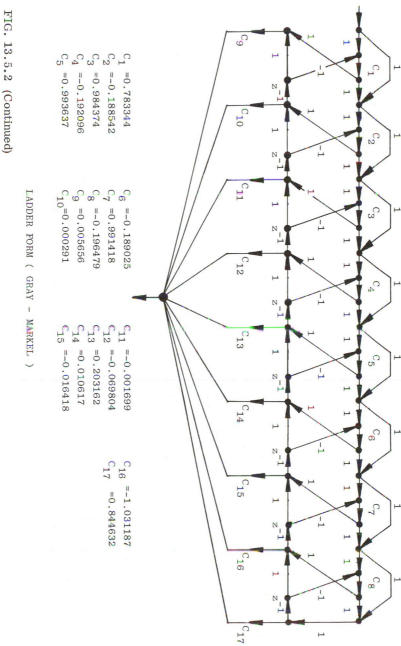

LADDER FORM (GRAY - MARKEL)

$C_1 = 0.783344$
$C_2 = -0.188542$
$C_3 = 0.984374$
$C_4 = -0.192096$
$C_5 = 0.993637$

$C_6 = -0.189025$
$C_7 = 0.991418$
$C_8 = -0.196479$
$C_9 = 0.005656$
$C_{10} = 0.000291$

$C_{11} = -0.001699$
$C_{12} = -0.069804$
$C_{13} = 0.203162$
$C_{14} = 0.010617$
$C_{15} = -0.016418$

$C_{16} = -1.031187$
$C_{17} = 0.844632$

FIG. 13.5.2 (Continued)

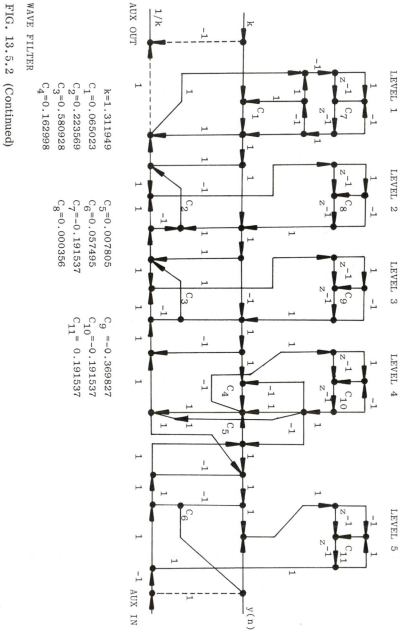

AUX OUT

LEVEL 1 LEVEL 2 LEVEL 3 LEVEL 4 LEVEL 5

k=1.311949

$C_1 = 0.065023$ $C_5 = 0.007805$ $C_9 = -0.369827$
$C_2 = 0.223569$ $C_6 = 0.057495$ $C_{10} = -0.191537$
$C_3 = 0.580928$ $C_7 = -0.191537$ $C_{11} = 0.191537$
$C_4 = 0.162998$ $C_8 = 0.000356$

WAVE FILTER

FIG. 13.5.2 (Continued)

TABLE 13.5.1 Experimental Results Detailed in Fig. 13.5.2

Example index	Filter type	Passband bits	Stopband bits	n_M	Number of		Bit–multiplier product
					Multipliers	Adders	
1	Direct II	20.86	10.85	3	16	16	334
2	Cascade	11.33	5.76	1	13	16	147
3	Parallel	10.12	7.95	0	18	16	182
4	Continued fraction	22.61	14.46	5	18	16	408
5	Gray–Markel	13.97	10.81	1	17	32	238
6	Wave	11.35	5.76	0	11	31	125

Note: n_M = number of magnitude bits to the left of the binary point (i.e., $|x| \leq 2^{n_M}$).
Source: Suggested by Crochiere and Oppenheim, 1975.

5.2 Summary of the Coefficient Sensitivity of Common Filter Sensitivity

Architecture	Coefficient sensitivity	Roundoff error	Complexity
Direct II	High	High	Low
Continued fraction	High	High	Low
Cascade	Low	Higher than parallel	Low
Parallel	Low	Lower than cascade	Low
Wave	Low	Good	Significant
Gray-Markel	Low	Moderate	Moderate

sensitivity term $S(\omega)$ be computed. Besides being a complex calculation, system transfer functions must also be explicitly determined. The reader may find these computational burdens to be too high a price to pay for studying high-order filters. If so, numerical simulation and experimentation may be considered to be a feasible alternative.

13.6 LIMIT CYCLES

Another noise source in digital filters is referred to as limit cycling. This noise source can appear in one of two forms:

1. Overflow limit cycling
2. Zero input (roundoff) limit cycling

The first source can be a serious problem. It results from a system register being driven beyond its dynamic range. This condition was outlined in Section 10.5. At that time, the modular and saturation adders (see Fig. 10.5.2) were seen to be directly involved with the creation of this noise source. However, the distortion introduced by such nonlinear operations can, in some cases, be controlled. Fam and Barnes (1979) have shown that stable filters, using 2's-complement arithmetic, are free of overflow limit cycling provided that all internal states are properly scaled. This is based on the previously established fact that the 2's-complement numbering system is tolerant of partial sum overflows as long as the final sum is within the dynamic range (i.e., does not overflow). As a result, 2's-complement encoding is commonly found in digital signal processing systems.

Zero input limit cycling is characterized by the system's output being corrupted by low-amplitude oscillations. The problem of zero input limit cycling is less severe than overflow limit cycling and generally is considered to be merely a nuisance. In Chapter 10 this class of limit cycling was motivated in terms of a simple first-order example problem. For the purpose of clarity, another simple motivational example will be presented.

EXAMPLE 13.6.1 Consider the first-order filter satisfying $H(z) = a/(z - 1)$, $|a| \leq 1$. If the system is initialized to a value of unity [i.e., $y(0) = 1$], and the input is removed (thus the notion of <u>zero input</u> limit cycling), the discrete filter response is given by $\{1, 1/a, 1/a^2, 1/a^3, \ldots\}$. At steady state, the unforced response of the filter will converge to zero. However, if the filter is to be realized as a digital filter, using a 4-bit data word having the format $\pm xxx$, the realized output is

$$y(n) = \left\{ \frac{7}{8}, \frac{1}{2}, \frac{1}{4}, \frac{1}{8}, \frac{1}{8}, \ldots, \frac{1}{8}, \ldots \right\} \quad \text{if } a = \frac{1}{2}$$

$$= \left\{ \frac{7}{8}, -\frac{1}{2}, \frac{1}{4}, -\frac{1}{8}, \frac{1}{8}, \ldots, \frac{(-1)^n}{8}, \ldots \right\} \quad \text{if } a = -\frac{1}{2}$$

It can be seen that low-amplitude oscillations of period 2 exist and can be attributed to finite-word-length effects. As a point of terminology, the amplitude range $-1/8 \leq y(n) \leq 1/8$ is called the <u>deadband</u>.

In an analysis provided by Jackson (1970), deadbands for first-order filters were found to exist whenever $|y(n)| \leq V$, where

$$y(n) = ay(n - 1) + x(n)$$

$$\left| ay(n) \right|_Q - ay(n) \leq 0.5Q \tag{13.6.1}$$

$$V = \frac{0.5Q}{1 - |a|}$$

with $[v]_Q$ denoting the quantized value of v, and Q denotes the quantization step size. That is, as long as the difference between the discrete and digital filter response does not exceed one-half the quantization step size, rounding will not induce limit cycling. Alternatively, it can be seen that

$$|y(n)| \leq \frac{0.5Q}{1 - |a|} = V; \quad Q = 2^{-b} \tag{13.6.2}$$

under the assumption that a data word has b magnitude digits [i.e., $(b + 1)$-bit word].

EXAMPLE 13.6.2 Numerical tests were performed on the second-order filter considered in the last two examples. The results of this test are summarized in Fig. 13.6.1. At steady state (assumed to exist over the sample interval 512 to 536), the results found in Table 13.6.1 and Fig. 13.6.1 were obtained.

The effects of zero input limit cycling can be visually interpreted in the context of an experiment performed by Kieburtz. A 10th-order multiple bandpass filter, architected with five second-order subfilters and a direct II structure, was numerically tested. The results are reported in Fig. 13.6.2. In Fig. 13.6.2A, the response of an interior second-order subfilter is analyzed. It can be seen that there exists zero input limit cycling which is bounded in magnitude by ± 5 (equivalently, $\pm 5Q$ in analog units). The spectrum of this noise can be seen to be a multifrequency tonal process. The zero input response for the entire 10th-order filter can be found in Fig. 13.6.2B. Here the limit cycling for this case can be seen to be of greater amplitude and possess a more complex spectrum. The frequency response of the tested filter is displayed in Fig. 13.6.2C. Here the round-off noise, which has an almost white spectrum, is found to be uniformly corruptive across the entire spectrum. The limit-cycle noise is more concentrated into narrow bands of frequencies. This is due to the fact that

TABLE 13.6.1 Limit Cycle Behavior of a Second-Order System

a_1	a_2	B-bits	Limit cycle maximum	Minimum
-1.0	-0.9	16	0.00012	-0.00015
-1.0	-0.9	12	0.0020	-0.0024
-1.0	-0.9	8	0.03	-0.04
-1.0	-0.9	4	0.50	-0.63
-1.0	-0.75	16	0.00003	-0.00003
-1.0	-0.75	12	0.0005	-0.0005
-1.0	-0.85	8	0.008	-0.008
-1.0	-0.75	4	0.12	-0.25
-1.0	-0.51	16	0.00003	-0.00003
-1.0	-0.51	12	0.0005	-0.0005
-1.0	-0.51	8	0.008	-0.008
-1.0	-0.51	4	0.12	-0.12

limit-cycle noise has a periodic pattern, whereas roundoff noise is essentially random. It can also be seen that limit-cycle noise will affect the least significant bits of the output processes [e.g., $-84.3 = \log(2^{-14})$].

The analysis of zero input limit cycling requires that the bounds on the filter state be quantified. Long and Trick considered modeling the errors associated with each multiplication operation in the usual additive error sense. That is, $v(n) = cu(n) + e(n)$, where $e(n)$ is a white, uniformly distributed random process. The output of a linear shift-invariant filter can then be represented as

$$y(n) = \sum_{i=0}^{n} a_i x(n - i) - \sum_{i=1}^{n} b_i y(n - i) \tag{13.6.3}$$

if infinite precision arithmetic is assumed. The zero-input filter response can be modeled as

$$y(n) = \sum_{i=0}^{n} [a_i x(n - i) + \hat{e}_i(n)] - \sum_{i=0}^{n} [b_i y(n - i) - e_i(n)] \tag{13.6.4}$$

if finite-word-length arithmetic is considered. It can be shown that the zero input ℓ_1 bound on $y(n)$ is given by

$$|y(n)| \le \frac{nQ}{2} \left[\sum_{p=0}^{M-1} \left| \sum_{i=0}^{M-1} h(i) \right| \right] \tag{13.6.5}$$

where the limit cycling, if it exists, is assumed to have a period of M samples. Using this relationship, deadbands for the important first- and second-order filters can be derived. They are summarized below:

First order:

$$H(z) = \frac{z}{z - b}$$

$$y(n) = x(n) + by(n - 1)$$

$$\left| \sum_{i=0}^{\infty} h(i) \right| = \frac{1 - 1}{|b|}$$

$$|y(n)| \le \frac{Q}{2(1 - b)}$$

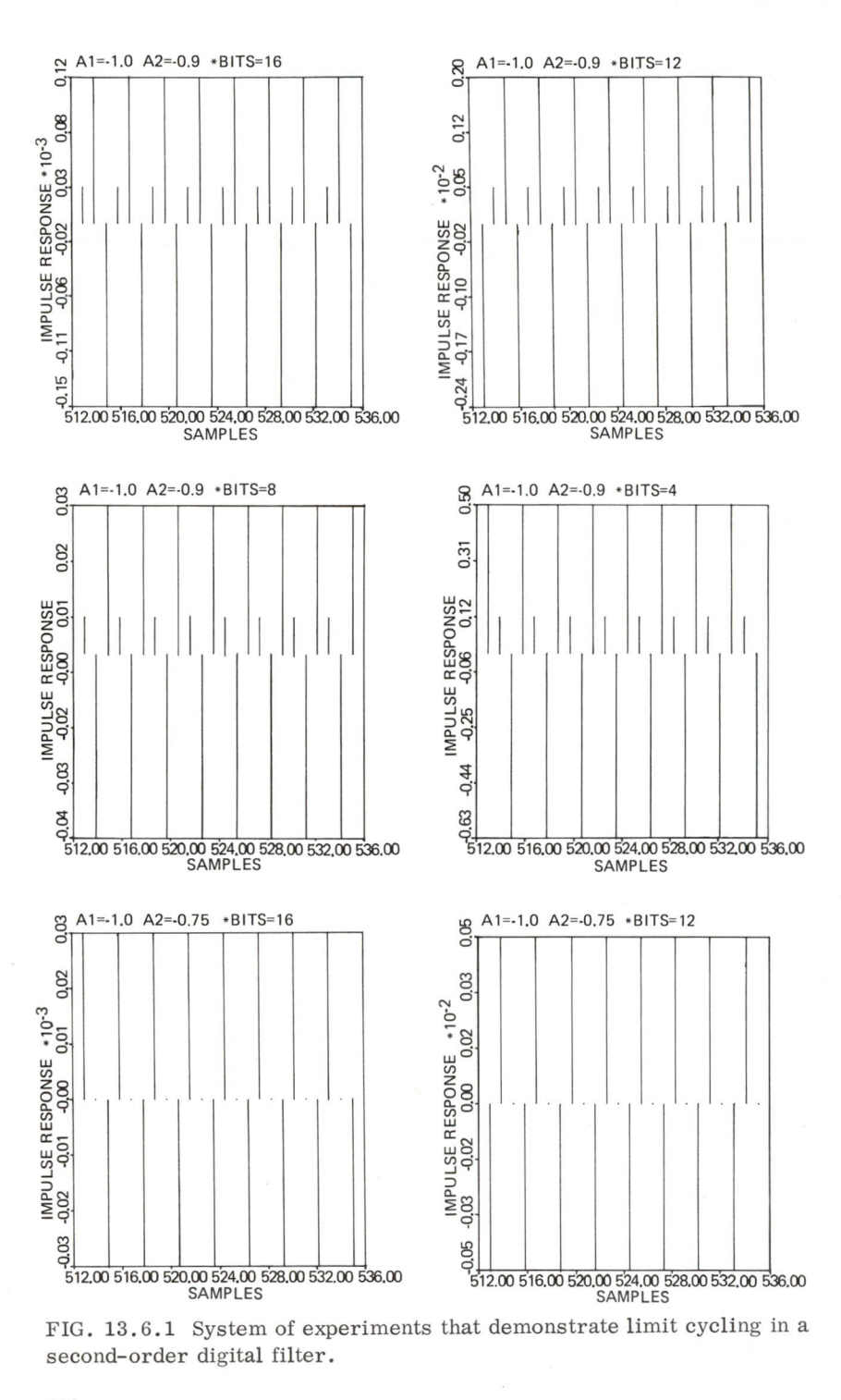

FIG. 13.6.1 System of experiments that demonstrate limit cycling in a second-order digital filter.

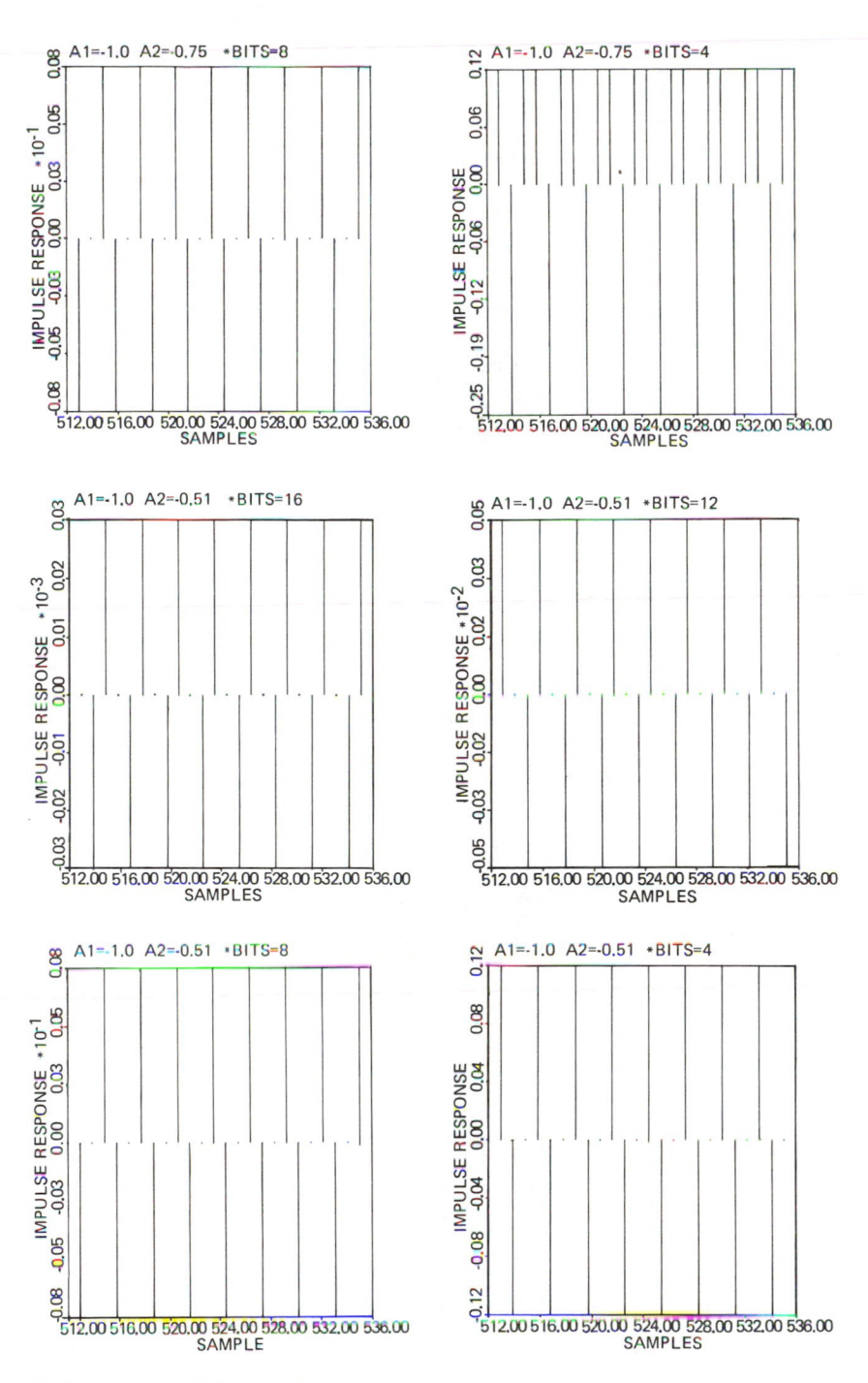

FIG. 13.6.1 (Continued)

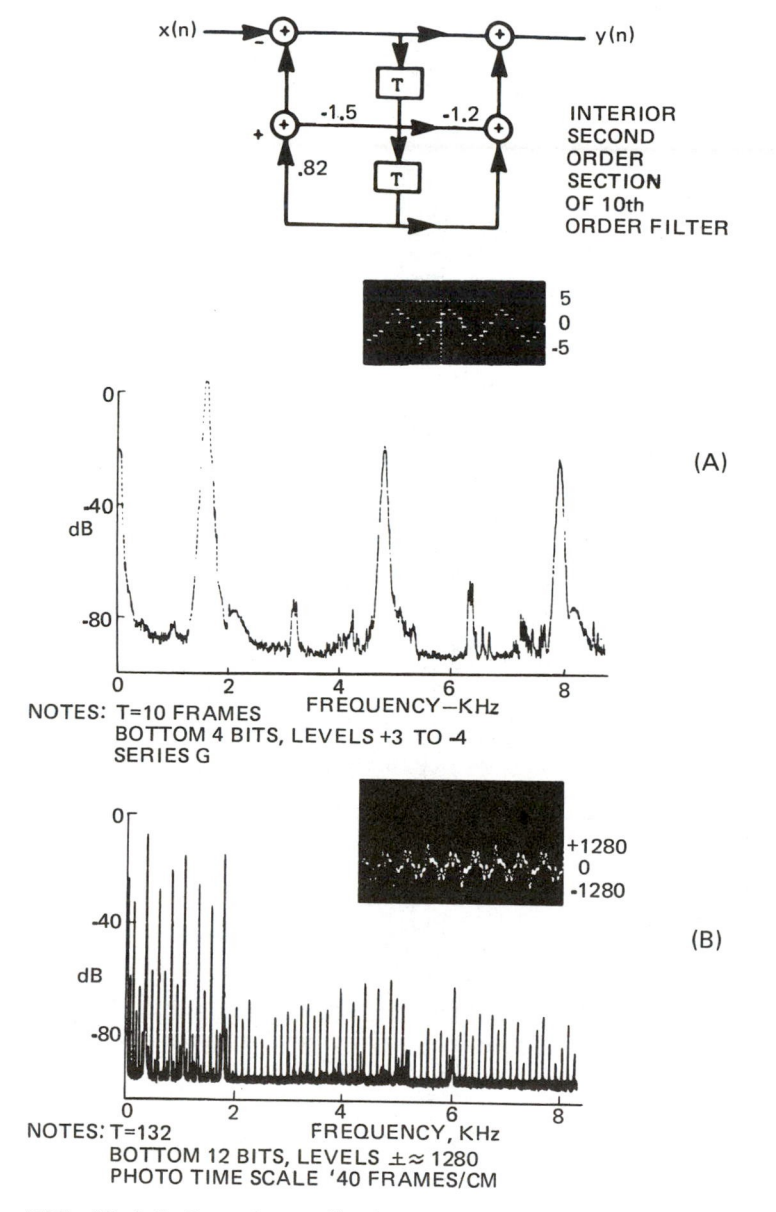

(A)

NOTES: T=10 FRAMES
BOTTOM 4 BITS, LEVELS +3 TO -4
SERIES G

(B)

NOTES: T=132
BOTTOM 12 BITS, LEVELS ±≈ 1280
PHOTO TIME SCALE '40 FRAMES/CM

FIG. 13.6.2 Experimentally derived limit-cycle behavior of a high-order digital filter. (After Kieburtz, 1975; portions of these data have been re-printed with the permission of the IEEE.)

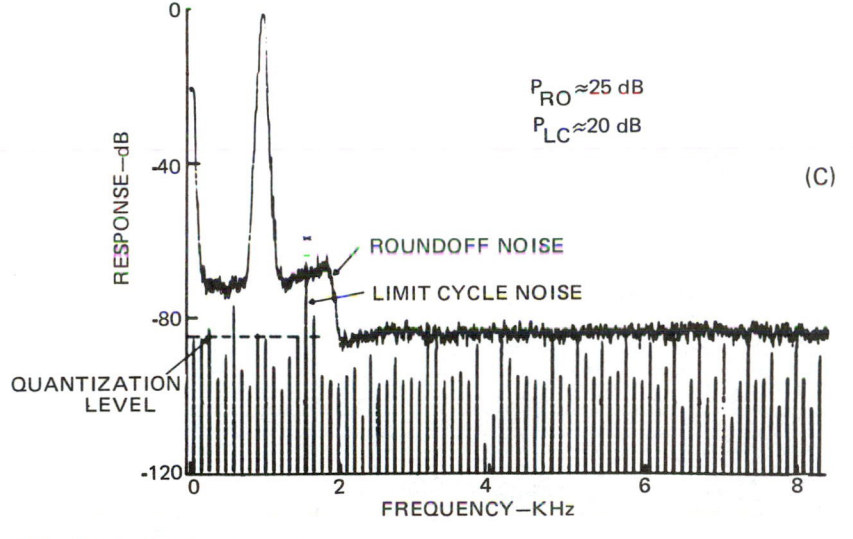

FIG. 13.6.2 (Continued)

Second order (due to Chang):

$$H(z) = \frac{z^2}{z^2 + a_1 z + a_2}$$
(13.6.6)

$$y(n) = x(n) - a_1 y(n-1) - a_2 y(n-2)$$

$$\left| \sum_{i=0}^{\infty} h(i) \right| = \sum_{i=0}^{\infty} \frac{i^n \sin[(n+1)\phi]}{\sin \phi}$$

$$\leq \frac{\sin \phi + 4R \sin(\phi/2)}{(1 - |a_1| + a_2) \sin \phi}$$

with

$$R = \frac{r^m}{1 - r^m}$$

$$m = \text{integer}\left(\frac{\pi}{\phi}\right)$$

$$\phi = \cos^{-1} \frac{|a_1|}{2\sqrt{a_2}}$$

$$r = \sqrt{a_2}$$

$$|y(n)| \leq \frac{Q \sin \phi + 4R \sin(\phi/2)}{(1 - |a_1| + a_2) \sin \phi}$$

A weaker (but less complex) bound for second-order filters has been suggested by Jackson (1970). This bound satisfies

$$|y(n)| \leq \frac{0.5Q}{1 - a_2}$$
(13.6.7)

for $0 \leq a_2 \leq 1$. Now, define a nonzero integer deadband parameter I to be given by $I = |y(n)|/Q = 2^B |y(n)| \leq 0.5/(1 - a_2)$. For $0 \leq |a_2| \leq 0.5$, the deadband parameter is $I = 0$ (i.e., no limit cycling).

EXAMPLE 13.6.3 Consider the case where $a_1 = 0$, $B = 4$, and the forcing function is an impulse. The rounded second-order filter response is zero for all odd samples [i.e., $y(2n+1) = 0$]. For the even samples, the data are summarized in Table 13.6.2. In general, if $(2k-1)/2k \leq |a_2| \leq (2k+1)/2(k+1)$, then $I \in [-k, k]$.

TABLE 13.6.2 Even Sample Limit Cycles for Second-Order Filter

Sample		$a_2 = 0.25$ Discrete $y(2i) = -0.25y(2i-2)$	$a_2 = 0.25$ Digital $Y(2i) = [-0.25Y(2i-2)]_R$	$a_2 = 0.50$ Discrete $y(2i) = -0.50y(2i-2)$	$a_2 = 0.50$ Digital $Y(2i) = [-0.50Y(2i-2)]_R$	$a_2 = 0.75$ Discrete $y(2i) = -0.75y(2i-2)$	$a_2 = 0.75$ Digital $Y(2i) = [-0.75Y(2i-2)]_R$
i	$2i$						
1	0	1	$0_\triangle 1111$ 15/16	1	$0_\triangle 1111$ 15/16	1	$0_\triangle 1111$ 15/16
2	2	-1/4	-4/16	-1/2	-8/16	-3/4	-12/16
3	4	1/16	1/16	1/4	4/16	9/16	9/16
4	6	-1/64	0	-1/8	-2/16	-27/64	-7/16
5	8	1/256	0	1/16	1/16	81/256	5/16
6	10	-1/2096	0	-1/32	-1/16	-729/2096	-4/16
7	12	1/16384	0	1/64	1/16	2187/16384	3/16
8	14	-1/65536	0	-1/128	-1/16	-6561/65536	-2/16
9	16	1/262144	0	1/256	1/16	26245/262144	2/16
			$\vdots$		$\vdots$		$\vdots$
			± 0		$\pm 1/16$		$\pm 2/16$
			$I \in [0, 0]$		$I \in [-1, 1]$		$I \in [-2, 2]$

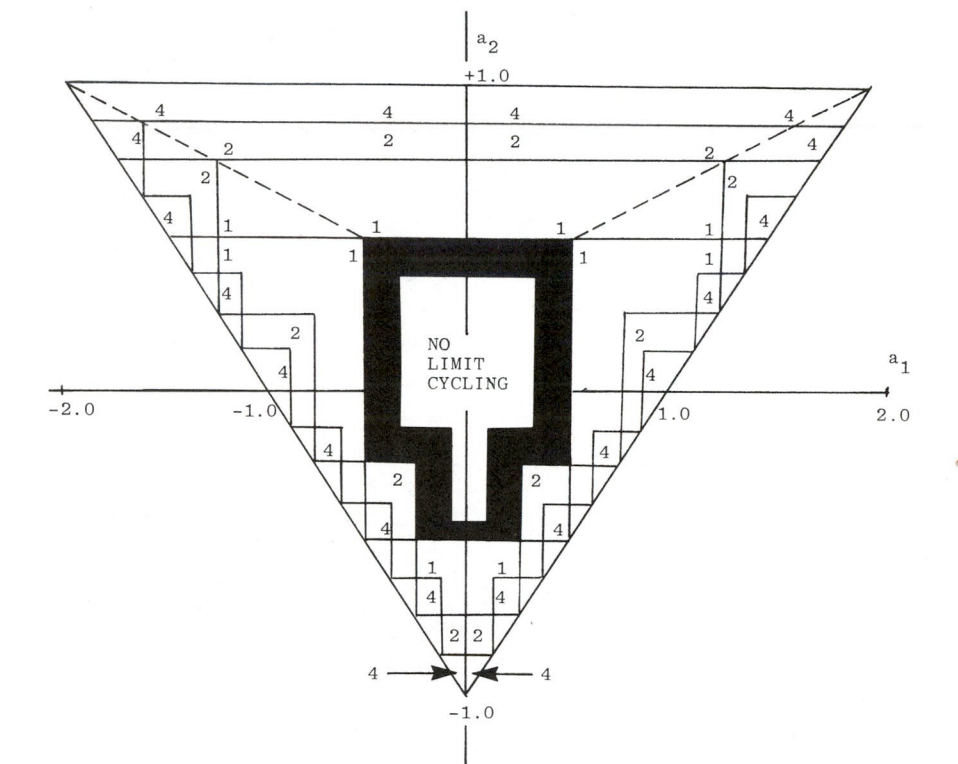

FIG. 13.6.3 Diagram of the amplitude of limit cycles in a second-order filter as a function of filter coefficients. (Portions of these data are suggested by Jackson, 1970.)

It can be observed that no zero-input limit cycling will occur when $|y(n)| < |y(n - k)|$, $k > 0$, $y(n) \neq 0$. That is, if a system is asymptotically stable, the output will continue to contract toward $y(n) = 0$ rather than oscillate. In order to oscillate, the system's poles (in conjunction with the preceding observations) must reside in a restricted coefficient domain. This domain, for a second-order filter, is graphically interpreted in Fig. 13.6.3. The integers 1, 2, and 4 represent the maximal value of the limit-cycle oscillations in multiples of $Q = 2^{-n}$. The crosshatched area represents limit cycling with amplitude zero (i.e., no limit cycles).

It is interesting to observe that in Example 13.6.3, there exists a region in which no limit cycling can occur. Therefore, if filters can be synthesized which contain eigenvalues residing in these restricted zones, a limit-cycle invariant filter can be realized. Fam and Barnes (1979) along

with Barnes (1979) have studied this problem in the context of a state-determined filter model of the form

$$x(n + 1) = Ax(n) + bu(n) \tag{13.6.8}$$

which can be rewritten to read

$$x(n + 1) = A^k x(n + 1 - k) + \sum_{i=0}^{k-1} A^i bu(n - i) \tag{13.6.9}$$

This representation suggests the <u>transversal-recursive</u> filter shown in Fig.
13.6.4. The purpose of this architecture is to control the location of the
filter's poles, and, therefore, influence its limit-cycle behavior. For the
purpose of stability, it can be argued that an eigenvalue of A, say λ_i, is
bounded by $|\lambda_i| < 1$. By forming A^k, it may be possible to force the eigen-
values of A^k to be bounded by $|\lambda_i|^k$ and thereby create a limit-cycle-free
filter.

EXAMPLE 13.6.4 Consider the first-order system given by $y(n) = 0.75y(n -$
$1) + x(n)$. From a previous analysis [Eq. (13.6.5)], it can be seen that zero-
input limit cycling can exist. However, for $A = 3/4$, $b = 1$, $c = 1$, $d = 0$,
and $k = 3$, it follows that (see Figs. 13.6.4 and 13.6.5)

$$|A| = \frac{3}{4} = 0.75; \quad |A|^2 = \frac{9}{16} = 0.5625; \quad |A|^3 = \frac{27}{64} = 0.422 < 0.5$$

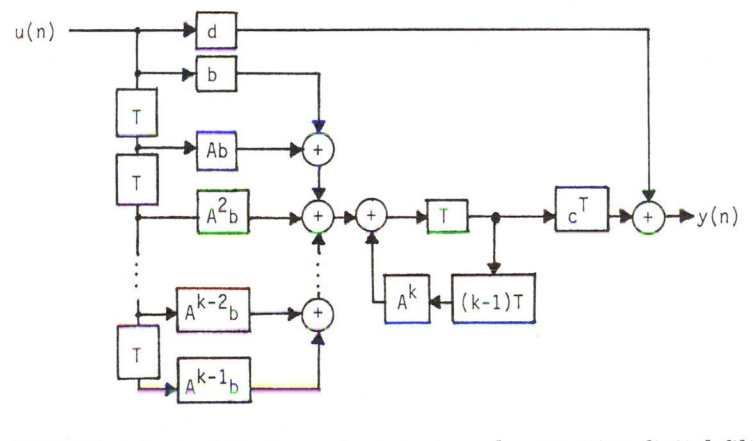

FIG. 13.6.4 Architecture of a transversal-recursive digital filter.

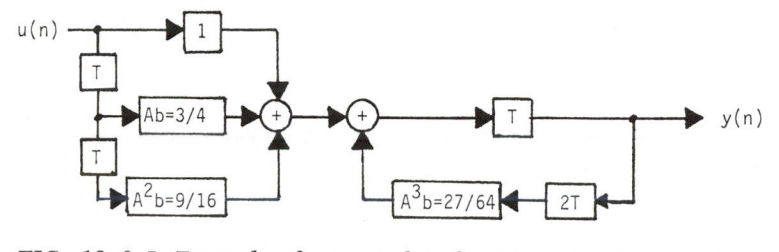

FIG. 13.6.5 Example of a second-order transversal-recursive digital architecture.

Here the autonomous solution is given by $y(n + 1) = A^3 x(n - 2)$, where $A^3 = 27/64 < 1/2$. As a result, the hybrid filter will not limit cycle.

Formally, what is desired is to place a bound on A^k so that limit cycling is eliminated. The bound is defined in terms of the <u>operator norm</u> which satisfies[†]

$$|A| = \sup \frac{|AX|_2}{|x|} = \max |\lambda_i|; \quad |x|_2 \leq 1 \tag{13.6.10}$$

or

$$|A^k| = \max\{|\lambda_i|^k\} \tag{13.6.11}$$

The autonomous solution associated with the A^k system is assumed to be defined in terms of infinite-precision arithmetic (i.e., discrete response). In a 2's-complement environment, note that $|A^k v|_2 \leq 2|v|_2$, which implies that

$$|x(n + 1)| \leq 2|A^k x(n + 1 - k)|_2$$
$$= 2|A^k| \, |x(n + 1 - k)|_2$$
$$< |x(n + 1 - k)|_2 \tag{13.6.12}$$

if $|A^k|_2 \leq 1/2$. That is, asymptotic stability can be guaranteed if $|A^k|_2 < 1/2$ or, equivalently, $(1/2)^k > \max(|\lambda_i|)$ for all i.

To ensure that an internal summing node of the filter does not over-flow, scaling is generally required. Similarly, to complete the analysis of

---

[†]A filter having a minimal value for $|A|$ is called a <u>minimum norm filter</u>.

a limit-cycle-free filter, the error variance due to rounding should be estimated. Since second-order sections are the basis for many filter designs, a comprehensive study of this filter will now be undertaken.

EXAMPLE 13.6.5 Consider the filter having poles located at $z = r \exp(\pm j\phi)$, $r < 1$ and a transfer function given by

$$H(z) = d_0 + \frac{a_1}{z - z_1} + \frac{a_2}{z - z_2} ; \quad a_1 = a_2^*$$

The design of a limit-cycle-free filter can be accomplished as follows. For

$$x(n + 1) = A^k x(n + 1 - k) + \sum_{i=0}^{k-1} A^i bu(n - i)$$

$$y(n) = c^T x(n) + d_0 u(n)$$

1. Choose k so that $r^k < 1/2$.
2. Form the minimum norm A matrix as follows:

$$A = r \begin{bmatrix} \cos\phi & \sin\phi \\ -\sin\phi & \cos\phi \end{bmatrix} \quad \text{(antisymmetric)}$$

3. Compute $b^T = (b_1, b_2)$ using

$$b_1^2 + b_2^2 = 2(1 - r^2)\delta$$

$$\frac{b_1}{b_2} = \frac{r^2 \sin(2\phi) \pm \sqrt{1 - r^4 - 2r^2 \cos(2\phi)}}{1 - r^2 \cos(2\phi)}$$

where b is not unique and δ relates to the probability of overflow.
4. Compute $c^T = (c_1, c_2)$ as follows:

$$c_1 = \frac{b_1(a_1 + a_2) + jb_2(a_1 - a_2)}{2(1 - r^2)\delta^2}$$

$$c_2 = \frac{b_2(a_1 + a_2) + jb_1(a_1 - a_2)}{2(1 - r^2)\delta^2}$$

5. Compute the roundoff error variance to be

$$\sigma_x^2 = \frac{2\,a_1^2}{\delta^2(1 - r^{2k})(1 - r^2)} \frac{Q^2}{12}$$

For $k = 1$, σ_x^2, degenerates to $(2|a_1|^2/\delta^2(1 - r^2)^2$, which agrees with a previous calculation. Now consider

$$H(z) = \frac{1}{z^2 - 2r\cos\phi z + r^2} = 1 + \frac{2r\cos\phi z - r^2}{z^2 - 2r\cos\phi z + r^2}$$

$$= 1 + \frac{a_1}{z - z_1} + \frac{a_2}{z - z_2}$$

which can be parameterized in terms

$$r = \cos\phi = \frac{1}{2}$$

$$r^2 = \frac{1}{2}$$

$$H(z) = 1 + \frac{0.5}{[z - \exp(j\pi/4)/\sqrt{2}][z - \exp(-j\pi/4)/\sqrt{2}]}$$

Now, using the recipe, it follows that:

Step 1: (note strict inequality)

$$|\lambda_1| = \frac{1}{\sqrt{2}}, \qquad k = 3 \ni |\lambda_1|^3 = \frac{1}{2\sqrt{2}} < \frac{1}{2}$$

Step 2:

$$A = \frac{1}{\sqrt{2}} \begin{bmatrix} \cos\frac{\pi}{4} & \sin\frac{\pi}{4} \\ -\sin\frac{\pi}{4} & \cos\frac{\pi}{4} \end{bmatrix} = \frac{1}{2} \begin{bmatrix} 1 & 1 \\ -1 & 1 \end{bmatrix}$$

Step 3: Let $\delta^2 = 1$; $b_1 + b_2 = 2(1 - 1/2) = 1$; then (since b is not unique) choose

$$\frac{b_1}{b_2} = \frac{1}{2}\left(1 \pm \sqrt{1 - \frac{1}{4}}\right) = \frac{1}{2}\left(1 \pm \frac{\sqrt{5}}{2}\right)$$

or $b_1 = 1.618 b_2$. Therefore,

$$b_2^2 + \left(\frac{1}{1.618}\right)^2 b_2^2 = 1; \quad b_2 = 0.851; \quad b_1 = 0.526$$

Step 4:

$$c_1 = \frac{0.526(1) + j0.851(0)}{2(1 - 1/2)} = 0.526$$

$$c_2 = \frac{0.851(1) + j0.526(0)}{2(1 - 1/2)} = 0.851$$

Step 5:

$$\sigma_x^2 = \frac{2(1/2)^2}{(1 - (1/2)^3)(1 - 1/2)} \frac{Q^2}{12} = 1.428 \frac{Q^2}{12}$$

The error variance for the limit-cycle-prone filter [i.e., poles at $\exp(j\pi/4)/\sqrt{2}$] can be computed in the manner suggested in Chapter 12. The resulting second-order transversal-recursive filter is suggested in Fig. 13.6.6.

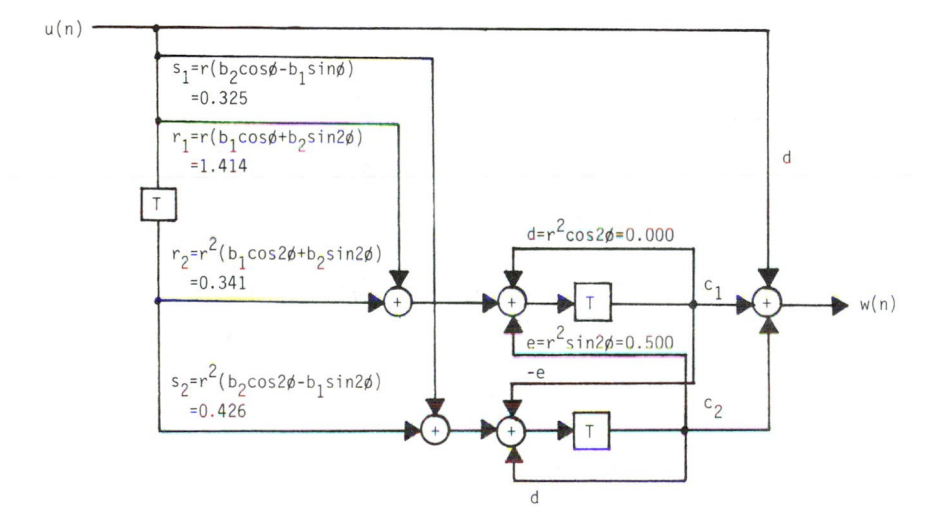

FIG. 13.6.6 Example of a second-order transversal-recursive filter.

BIBLIOGRAPHY

Antonious, A. (1979), Digital Filter Analysis and Design, McGraw-Hill, New York.

Avenhaus, E. (1972), On the Design of Digital Filters with Coefficients of Limited Wordlength, IEEE Trans. Audio Electroacoust., AU-20, August, pp. 206-212.

Avenhaus, E., and W. Schussler (1970), On the Approximation Problem in the Design of Digital Filters with Limited Wordlength, Arch. Elektron. Uebertrag, 24, pp. 571-572.

Barnes, C. W. (1979), Roundoff Noise and Overflow in Normal Digital Filters, IEEE Trans. Circuits Syst., CAS-26, March, pp. 154-159.

Chang, T. L. (1976), A Note on Upper Bounds on Limit Cycles in Digital Filters, IEEE Trans. Acoust. Speech Sig. Proc., ASSP-24, February, pp. 99-100.

Crochiere, R. E. (1973), A New Statistical Approach to the Coefficient Wordlength Problem for Digital Filters, IEEE Trans. Commun., COM-21, June, pp. 757-763.

Crochiere, R. E., and A. V. Oppenheim (1975), Analysis of Linear Digital Networks, Proc. IEEE, April, pp. 581-595.

Fam, A. T., and C. W. Barnes (1979), Nonminimal Realizations of Fixed-Point Digital Filters That Are Free of All Finite Word-length Limit Cycles, IEEE Trans. Acoust. Speech Signal Process., ASSP-27, April, pp. 149-153.

Jackson, L. B. (1970), On the Interaction of Roundoff Noise and Dynamic Range in Digital Filters, Bell Syst. Tech. J., February, pp. 159-184.

Kieburtz, R. B. (1975), An Experimental Study of Roundoff Effects in a Tenth-Order Recursive Digital Filter, IEEE Trans. Circuits Syst., March, pp. 190-196.

Knowles, J. B., and E. M. Olcayto (1968), Coefficient Accuracy and Digital Filter Response, IEEE Trans. Circuit Theory, CT-15, March, pp. 31-41.

Mullis, C. T., and R. A. Roberts (1976), Synthesis of Minimum Roundoff Noise Fixed Point Filters, IEEE Trans. Circuits Syst., CAS-23, September, pp. 551-562.

APPENDIX   State-Determined Filter
Software Package

The software reported in this section was written in ANSI standard FORT-
RAN for use on a general-purpose digital computer. The purpose of the
software package is to compute, in a state-variable format, pole-zero locations, state-determined coefficients, overflow scaling coefficients, and
roundoff error parameters.

The user need only supply the program the $n + 1$ numerator/denomina-
tor filter coefficients, and the order of the filter (i.e., n). The required
coefficients and order information, for classically defined low-, band-, and
highpass Butterworth, Chebyshev, and elliptic filters can be obtained using
the software developed in Appendix 7.A. The user-supplied parameters
and coefficients are echo-printed and displayed together with pole-zero
values. The user has a choice of six architectures to study. They are:

1. Cascade
2. Parallel
3. Direct II
4. Gray-Markel (one- and two-multiplier versions)
5. Normal
6. Wave

The appropriate state-determined A, B, C, and D matrices are computed
for the chosen architecture. The covariance matrix K and unit noise matrix
W (see Section 12.5) are also computed and displayed. These data are used
to compute the overflow scaling parameters for each second-order subfilter.
A multiplier count is maintained for the purpose of calculating the roundoff
error variance.

In the following pages, a set of examples are offered to assist the
reader in using the software and interpreting the results. The software can
be obtained at a modest cost by contacting J. Marshall, State-Variable Software, Cincinnati Electronics, Glendale-Milford Road, Cincinnati, Ohio, or
F. Taylor, Electrical Engineering, University of Florida, Gainesville, Florida.

The examples in this appendix fall into one of two categories. The
first block of examples represent an analysis of a fourth-order Chebyshev
lowpass filter configured in cascade, parallel, direct II, normal, and Gray-
Markel form. The second is a sixth-order lowpass wave filter example.
The fourth-order lowpass model is given by

$$H(z) = \frac{0.001836z^4 + 0.007344z^3 + 0.011020z^2 + 0.00734z + 0.001836}{z^4 - 3.0544z^3 + 3.8291z^2 - 2.2924z + 0.5507}$$

and possesses the pole-zero diagram, shown in Fig. A.1.

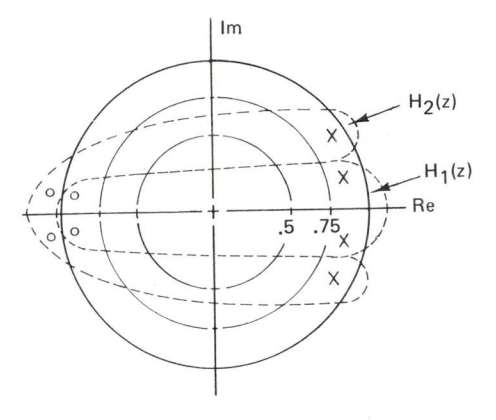

FIG. A.1 Pole-zero selection process for fourth-order example.

The embedded pole-zero selection rule, used to create the cascaded second-order sections, namely $H_1(z)$, is discussed in Section 11.4. The error properties of the fourth-order filter are summarized below:

Filter	Scaled error variance $\times \dfrac{Q^2}{12}$
1. Cascade	$14.677 + 10.754 = 25.431$
2. Parallel	$8.507 + 2.954 = 11.491$
3. Direct II	258.151
4. Gray-Markel	3.573
5. Normal	$1.641 + 1.654 = 3.295$

Computer-Aided Design of a Fourth-Order Cascade Filter

```
UNIT NOISE DIGITAL FILTER DESIGN
     USING STATE VARIABLES
     OVERALL FILTER ORDER =   4
THE TRANSFER FUNCTION COEFFICIENTS ARE:

NUMERATOR          DENOMINATOR
  0.001836           0.550700
  0.007344          -2.292400
  0.011020           3.829100
  0.007344          -3.054400
  0.001836           1.000000
```

Observe that for the normal filter, $\sigma^2_{H_1} \cong \sigma^2_{H_2}$, as suggested by theory.

```
                    ZEROS
         REAL            IMAGINARY
      -0.848120         -0.130326
      -0.848120          0.130326
      -1.151884         -0.177004
      -1.151884          0.177004
                    POLES
         REAL            IMAGINARY
       0.777197         -0.212333
       0.777197          0.212333
       0.750004         -0.534673
       0.750004          0.534673
```

CASCADE ARCHITECTURE

THE NORMALIZED SECOND ORDER SECTIONS ARE:

	NUMERATOR	DENOMINATOR
H1 (Z):		
	0.042849	1.000000
	0.072681	-1.554394
	0.031549	0.649121
H2 (Z):		
	0.042849	1.000000
	0.098713	-1.500007
	0.058195	0.848381

FOR H1 (Z):
THE A MATRIX IS:
```
   0.0000000        1.000000
  -0.6491205        1.554394
```
THE B MATRIX IS:
```
   0.0000000        1.000000
```
THE TRANSPOSE OF THE C MATRIX IS:
```
   3.7351996E-03    0.1392850
```
THE D MATRIX IS: 0.0428
THE INVERSE MATRIX IS:
```
   0.15488E+02    0.51223E+01   -0.10054E+02
   0.14598E+02    0.54344E+01   -0.94761E+01
   0.12638E+02    0.51223E+01   -0.72036E+01
```
THE T(N,1) VECTOR IS:
```
     0.15488067E+02
     0.14598425E+02
     0.12638085E+02
```
THE TOEPLITZ MATRIX IS:
```
   0.15488E+02      0.14598E+02
   0.14598E+02      0.15488E+02
```
THE K MATRIX IS AS FOLLOWS:
```
   0.15488E+02      0.14598E+02
   0.14598E+02      0.15488E+02
```
THE W MATRIX IS:
```
   0.13311E+00     -0.19295E+00
  -0.19295E+00      0.31588E+00
```
FOR THE UNSCALED SYSTEM:

STATE X(I)	PRODUCT NOISE SOURCES
STATE X(I)	PRODUCT NOISE SOURCES
1	0
2	2

SIGMA**2(Y)= 0.631759 *SIGMA**2(X)

```
FOR THE SCALED SYSTEM:
STATE X(I)        PRODUCT NOISE SOURCES
  1                      0
  2                      3
SIGMA**2(Y) (SCALED)= 14.677087 *SIGMA**2(X)
FOR H2 (Z):
THE A MATRIX IS:
  0.0000000        1.000000
 -0.8483812        1.500007
THE B MATRIX IS:
  0.0000000        1.000000
THE TRANSPOSE OF THE C MATRIX IS:
  2.1843564E-02  0.1629863
THE D MATRIX IS:     0.0428
THE INVERSE MATRIX IS:
  0.10451E+02   0.12859E+01  -0.88664E+01
  0.84812E+01   0.15846E+01  -0.71953E+01
  0.38555E+01   0.12859E+01  -0.22709E+01
THE T(N,1) VECTOR IS:
     0.10450979E+02
     0.84812298E+01
     0.38554921E+01
THE TOEPLITZ MATRIX IS:
   0.10451E+02       0.84812E+01
   0.84812E+01       0.10451E+02
THE K MATRIX IS AS FOLLOWS:
   0.10451E+02       0.84812E+01
   0.84812E+01       0.10451E+02
THE W MATRIX IS:
   0.24735E+00      -0.23422E+00
  -0.23422E+00       0.34300E+00
FOR THE UNSCALED SYSTEM:
STATE X(I)        PRODUCT NOISE SOURCES
  1                      0
  2                      2
SIGMA**2(Y)=  0.686004 *SIGMA**2(X)
FOR THE SCALED SYSTEM:
STATE= 10.754112 *SIGMA**2(X)
```

Magnitude Frequency Response of a Fourth-Order Cascade Filter

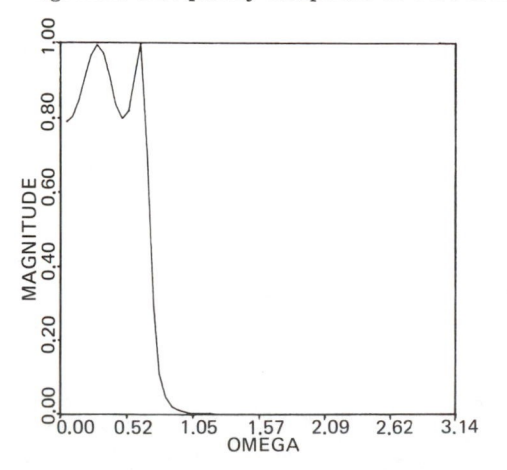

Computer-Aided Design of a Fourth-Order Parallel Filter

```
UNIT NOISE DIGITAL FILTER DESIGN
     USING STATE VARIABLES
     OVERALL FILTER ORDER =   4
THE TRANSFER FUNCTION COEFFICIENTS ARE:
NUMERATOR        DENOMINATOR

 0.001836          0.550700
 0.007344         -2.292400
 0.011020          3.829100
 0.007344         -3.054400
 0.001836          1.000000
          ZEROS
    REAL          IMAGINARY
-0.848120         -0.130326
-0.848120          0.130326
-1.151884         -0.177004
-1.151884          0.177004
          POLES
    REAL          IMAGINARY
 0.777197         -0.212333
 0.777197          0.212333
 0.750004         -0.534673
 0.750004          0.534673
    PARALLEL ARCHITECTURE

INPUT CONSTANT=  0.003334
THE REDUCED TRANSFER FUNCTION COEFFICIENTS ARE:
               NUMERATOR        DENOMINATOR

               0.000000          0.550700
               0.014987         -2.292400
              -0.001746          3.829100
               0.017527         -3.054400
              -0.001498          1.000000
THE NORMALIZED SECOND ORDER SECTIONS ARE:
               NUMERATOR        DENOMINATOR

H1 (Z):
               0.073112          1.000000
               0.038653         -1.554394
                                 0.649121
H2 (Z):
              -0.074610          1.000000
              -0.027430         -1.500007
                                 0.848381
FOR H1 (Z):

THE A MATRIX IS:
  0.0000000        1.000000
 -0.6491205        1.554394
THE B MATRIX IS:
  0.0000000        1.000000
THE TRANSPOSE OF THE C MATRIX IS:
 -4.7458354E-02   0.1522973
THE D MATRIX IS:      0.0731
THE INVERSE MATRIX IS:
  0.15488E+02      0.51223E+01      -0.10054E+02
  0.14598E+02      0.54344E+01      -0.94761E+01
  0.12638E+02      0.51223E+01      -0.72036E+01
THE T(N,1) VECTOR IS:
   0.15488067E+02
   0.14598425E+02
   0.12638085E+02
```

```
THE TOEPLITZ MATRIX IS:
   0.15488E+02      0.14598E+02
   0.14598E+02      0.15488E+02

THE K MATRIX IS AS FOLLOWS:
   0.15488E+02      0.14598E+02
   0.14598E+02      0.15488E+02

THE W MATRIX IS:
   0.79400E-01     -0.11641E+00
  -0.11641E+00      0.18309E+00
FOR THE UNSCALED SYSTEM:
STATE X(I)          PRODUCT NOISE SOURCES
   1                         0
   2                         2
SIGMA**2(Y)=  0.366186 *SIGMA**2(X)
FOR THE SCALED SYSTEM:
STATE X(I)          PRODUCT NOISE SOURCES
   1                         0
   2                         3
SIGMA**2(Y) (SCALED)=  8.507267 *SIGMA**2(X)
FOR H2 (Z):
THE A MATRIX IS:
   0.0000000        1.000000
  -0.8483812        1.500007
THE B MATRIX IS:
   0.0000000        1.000000
THE TRANSPOSE OF THE C MATRIX IS:
   6.3297480E-02 -0.1393454
   6.3297480E-02 -0.1393454

THE D MATRIX IS:    -0.0746
THE INVERSE MATRIX IS:
   0.10451E+02      0.12859E+01     -0.88664E+01
   0.84812E+01      0.15846E+01     -0.71953E+01
   0.38555E+01      0.12859E+01     -0.22709E+01
THE T(N,1) VECTOR IS:
   0.10450979E+02
   0.84812298E+01
   0.38554921E+01
THE TOEPLITZ MATRIX IS:
   0.10451E+02      0.84812E+01
   0.84812E+01      0.10451E+02

THE K MATRIX IS AS FOLLOWS:
   0.10451E+02      0.84812E+01
   0.84812E+01      0.10451E+02

THE W MATRIX IS:
   0.72518E-01     -0.70307E-01
  -0.70307E-01      0.95188E-01
FOR THE UNSCALED SYSTEM:
STATE X(I)          PRODUCT NOISE SOURCES
   1                         0
   2                         2
SIGMA**2(Y)=  0.190377 *SIGMA**2(X)
FOR THE SCALED SYSTEM:
STATE X(I)          PRODUCT NOISE SOURCES
   1                         0
   2                         3
SIGMA**2(Y) (SCALED)=  2.984433 *SIGMA**2(X)
```

Magnitude Frequency Response of a Fourth-Order Parallel Filter

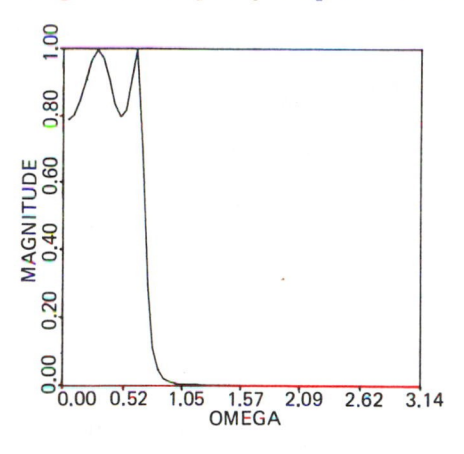

Computer-Aided Design of a Fourth-Order Direct II Filter

```
UNIT NOISE DIGITAL FILTER DESIGN
     USING STATE VARIABLES
     OVERALL FILTER ORDER =   4
THE TRANSFER FUNCTION COEFFICIENTS ARE:
  NUMERATOR       DENOMINATOR
  0.001836         0.550700
  0.007344        -2.292400
  0.011020         3.829100
  0.007344        -3.054400
  0.001836         1.000000
              ZEROS
    REAL          IMAGINARY
 -0.848120        -0.130326
 -0.848120         0.130326
 -1.151884        -0.177004
 -1.151884         0.177004
              POLES
    REAL          IMAGINARY
  0.777197        -0.212333
  0.777197         0.212333
  0.750004        -0.534673
  0.750004         0.534673
              DIRECT FORM II
              ARCHITECTURE
THE A MATRIX IS:
  0.0000000        1.000000        0.0000000        0.0000000
  0.0000000        0.0000000        1.000000        0.0000000
  0.0000000        0.0000000        0.0000000        1.000000
 -0.5507000        2.292400       -3.829100        3.054400
THE B MATRIX IS:
  0.0000000
  0.0000000
  0.0000000
  1.000000
```

```
THE TRANSPOSE OF THE C MATRIX IS:
  8.2491478E-04  1.1552846E-02  3.9897729E-03  1.2951879E-02
THE D MATRIX IS:     0.0018
THE INVERSE MATRIX IS:
1E0036660E+03    -0.32853E+03    -0.10458E+03     0.34364E+03    -0.1468
5E+031337        0.24288E+03    -0.29611E+03    -0.94020E+02     0.31131E+03

9E+029865        0.17915E+03    -0.20974E+03    -0.65294E+02     0.22485E+03

5E+025208        0.94579E+02    -0.96878E+02    -0.27389E+02     0.11208E+03

4E+016079        0.12855E+02     0.93576E+01     0.84222E+01     0.57473E+01

THE T(N,1) VECTOR IS:
    0.26659625E+03
    0.24288095E+03
    0.17915279E+03
    0.94579453E+02
    0.12855279E+02
THE TOEPLITZ MATRIX IS:
  0.26660E+03    0.24288E+03    0.17915E+03    0.94579E+02
  0.24288E+03    0.26660E+03    0.24288E+03    0.17915E+03
  0.17915E+03    0.24288E+03    0.26660E+03    0.24288E+03
  0.94579E+02    0.17915E+03    0.24288E+03    0.26660E+03
THE K MATRIX IS AS FOLLOWS:
  0.26660E+03    0.24288E+03    0.17915E+03    0.94579E+02
  0.24288E+03    0.26660E+03    0.24288E+03    0.17915E+03
  0.17915E+03    0.24288E+03    0.26660E+03    0.24288E+03
  0.94579E+02    0.17915E+03    0.24288E+03    0.26660E+03
```

Magnitude Frequency Response of a Fourth-Order Direct II Filter

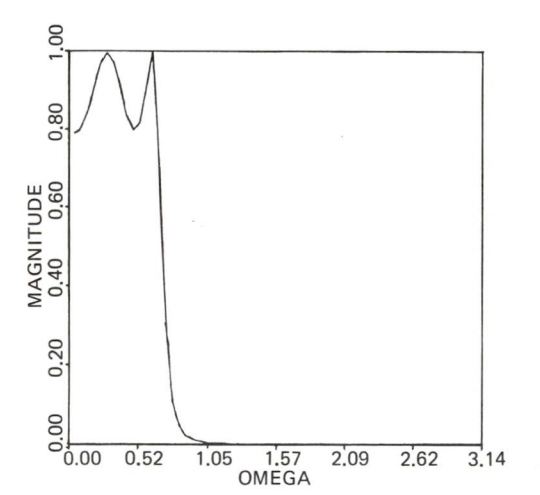

Computer–Aided Design of a Third–Order Gray–Markel Filter

```
UNIT NOISE DIGITAL FILTER DESIGN
     USING STATE VARIABLES
     OVERALL FILTER ORDER =   3
THE TRANSFER FUNCTION COEFFICIENTS ARE:
   NUMERATOR         DENOMINATOR
   0.015400          -0.458300
   0.046200           1.572000
   0.046200          -1.990000
   0.015400           1.000000
              ZEROS
     REAL          IMAGINARY
  -1.001506        -0.002716
  -1.001506         0.002716
  -0.994821         0.000000
              POLES
     REAL          IMAGINARY
   0.660508         0.000000
   0.664746        -0.501969
   0.664746         0.501969
             GRAY-MARKEL
        LADDER ARCHITECTURE
THE TWO-MULTIPLIER VALUES ARE:
   MULTIPLIERS                    TAPS
  K(2)=-0.4583000        NU(3)= 0.0154000
  K(1)= 0.8354627        NU(2)= 0.0768460
  K(0)=-0.8755870        NU(1)= 0.1454910
  ZH(0)= 1.0000000       NU(0)= 0.0856459
THE OPTIMAL SIGN CHOICE PARAMETERS ARE:
   -1.0
   -1.0
   -1.0

THE VALUES OF THE MULTIPLIERS, PI(N), ARE:
0.4500373
0.2399448
1.4583000
1.0000000
THE MODIFIED ONE MULTIPLIER TAP PARAMETERS ARE:
NUHAT(3)= 0.0154000
NUHAT(2)= 0.0526956
NUHAT(1)= 0.6063520
NUHAT(0)= 0.1903084

THE VALUES OF SIGMA(I) ARE:
SIGMA(3)=  1.0000000
SIGMA(2)=  1.2658851
SIGMA(1)=  4.1916437
SIGMA(0)= 17.9631023
THE OUTPUT ERROR VARIANCE IS:
SIGMA**2(Y)= 0.2282030 *SIGMA**2(X)

THE A MATRIX IS:
   0.3829     1.8355      0.0000
  -0.0660     0.7315      0.1244
   0.1414    -1.5670      0.8756
THE B MATRIX IS:
   1.218355
  -0.2100925
   0.4500373
```

```
THE C MATRIX IS:
 1.5400002E-02  0.2420696       0.2420696
THE D MATRIX IS:      0.0154
THE INVERSE MATRIX IS:

   0.17963E+02    -0.42900E+01    -0.13821E+02    0.82325E+01
   0.15728E+02    -0.30666E+01    -0.11786E+02    0.72083E+01
   0.10269E+02    -0.76407E+00    -0.61279E+01    0.47065E+01
   0.39440E+01     0.13341E+01    -0.15364E-02    0.28075E+01


THE T(N,1) VECTOR IS:
   0.17963076E+02
   0.15728238E+02
   0.10269492E+02
   0.39439766E+01


THE TOEPLITZ MATRIX IS:
   0.17963E+02     0.15728E+02     0.10269E+02
   0.15728E+02     0.17963E+02     0.15728E+02
   0.10269E+02     0.15728E+02     0.17963E+02


THE K MATRIX IS AS FOLLOWS:
   0.26921E+01    -0.74835E-06     0.63831E-06
  -0.19466E-06     0.24133E+00     0.82034E-06
   0.19073E-05     0.59605E-06     0.36381E+01


THE W MATRIX IS:
   0.22752E-01    -0.12100E+00     0.27023E-01
  -0.12100E+00     0.89897E+00    -0.14718E+00
   0.27023E-01    -0.14718E+00     0.17333E+00

FOR THE UNSCALED SYSTEM:
STATE X(I)        PRODUCT NOISE SOURCES
   1                      3
   2                      4
   3                      4
SIGMA**2(Y)=  4.357470 *SIGMA**2(X)

FOR THE SCALED SYSTEM:
STATE X(I)        PRODUCT NOISE SOURCES
   1                      3
   2                      4
   3                      4
SIGMA**2(Y) (SCALED)=  3.573936 *SIGMA**2(X)
```

Magnitude Frequency Response of a Third-Order Gray-Markel Filter

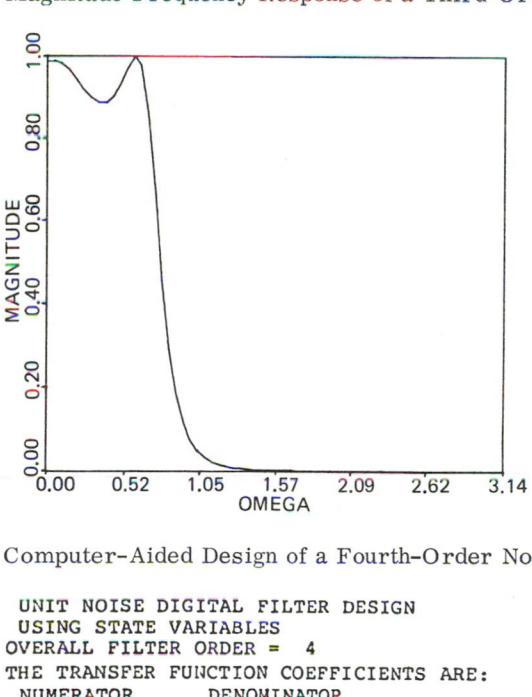

Computer-Aided Design of a Fourth-Order Normal Filter

```
UNIT NOISE DIGITAL FILTER DESIGN
USING STATE VARIABLES
OVERALL FILTER ORDER =   4
THE TRANSFER FUNCTION COEFFICIENTS ARE:
 NUMERATOR        DENOMINATOR
  0.001836         0.550700
  0.007344        -2.292400
  0.011020         3.829100
  0.007344        -3.054400
  0.001836         1.000000
              ZEROS
     REAL         IMAGINARY
 -0.848120       -0.130326
 -0.848120        0.130326
 -1.151884       -0.177004
 -1.151884        0.177004
              POLES
     REAL         IMAGINARY
  0.777197       -0.212333
  0.777197        0.212333
  0.750004       -0.534673
  0.750004        0.534673
      NORMAL ARCHITECTURE
INPUT CONSTANT=  0.001836
THE REDUCED TRANSFER FUNCTION COEFFICIENTS ARE:
 NUMERATOR        DENOMINATOR
  0.000825         0.550700
  0.011553        -2.292400
  0.003990         3.829100
  0.012952        -3.054400
  0.000000         1.000000
```

```
THE FIRST ORDER SECTIONS ARE:
      NUMERATOR                        DENOMINATOR
      0.0761      0.1670        0.7772    -0.2123
      0.0761     -0.1670        0.7772     0.2123
     -0.0697     -0.0385        0.7500    -0.5347
     -0.0697      0.0385        0.7500     0.5347
THE A MATRIX IS:
   0.7771971         0.2123325
  -0.2123325         0.7771971
THE B MATRIX IS:
   0.3750375         0.7490700
THE C MATRIX IS:
   0.4378469        -1.5901890E-02
THE INVERSE MATRIX IS:
   0.15488E+02      0.51223E+01     -0.10054E+02
   0.14598E+02      0.54344E+01     -0.94761E+01
   0.12638E+02      0.51223E+01     -0.72036E+01
THE T(N,1) VECTOR IS:
   0.15488067E+02
   0.14598425E+02
   0.12638085E+02
THE TOEPLITZ MATRIX IS:
   0.15488E+02      0.14598E+02
   0.14598E+02      0.15488E+02
THE K MATRIX IS AS FOLLOWS:
   0.10000E+01      0.63696E+00
   0.63696E+00      0.10000E+01
THE W MATRIX IS:
   0.42025E+00      0.93998E-01
   0.93998E-01      0.12684E+00
FOR THE UNSCALED SYSTEM:
STATE X(I)        PRODUCT NOISE SOURCES
   1                        3
   2                        3
SIGMA**2(Y)= 1.641270 *SIGMA**2(X)

FOR THE SCALED SYSTEM:
STATE X(I)        PRODUCT NOISE SOURCES
   1                        3
   2                        3
SIGMA**2(Y) (SCALED)= 1.641270 *SIGMA**2(X)

THE A MATRIX IS:
   0.7500037         0.5346733
  -0.5346733         0.7500037

THE B MATRIX IS:
   0.1975606         0.5140110

THE C MATRIX IS:
  -0.2214393        -0.1859840

THE INVERSE MATRIX IS:
   0.10451E+02      0.12859E+01     -0.88664E+01
   0.84812E+01      0.15846E+01     -0.71953E+01
   0.38555E+01      0.12859E+01     -0.22709E+01
THE T(N,1) VECTOR IS:
   0.10450979E+02
   0.84812298E+01
   0.38554921E+01
```

```
THE T(N,1) VECTOR IS:
    0.10450979E+02
    0.84812298E+01
    0.38554921E+01
THE TOEPLITZ MATRIX IS:
   0.10451E+02      0.84812E+01
   0.84812E+01      0.10451E+02
THE K MATRIX IS AS FOLLOWS:
   0.10000E+01      0.14038E+00
   0.14038E+00      0.10000E+01
THE W MATRIX IS:
   0.25194E+00      0.30505E-01
   0.30505E-01      0.29961E+00
FOR THE UNSCALED SYSTEM:
STATE X(I)       PRODUCT NOISE SOURCES
   1                      3
   2                      3
SIGMA**2(Y)=  1.654650 *SIGMA**2(X)
FOR THE SCALED SYSTEM:
STATE X(I)       PRODUCT NOISE SOURCES
   1                      3
   2                      3
SIGMA**2(Y) (SCALED)=  1.654649 *SIGMA**2(X)
```

Magnitude Frequency Response of a Fourth-Order Normal Filter

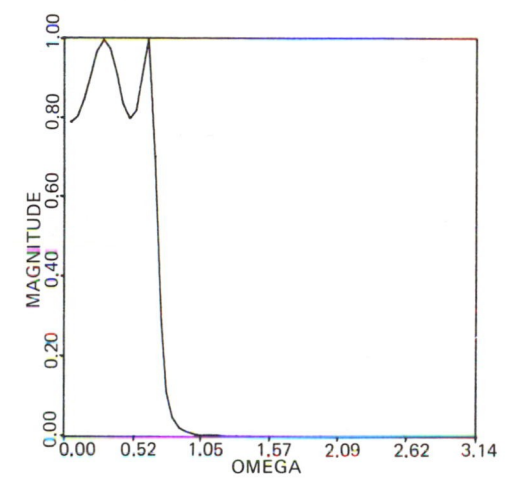

Computer-Aided Design of a Sixth-Order Wave Filter

```
UNIT NOISE DIGITAL FILTER DESIGN
     USING STATE VARIABLES
   OVERALL FILTER ORDER =  6
        WAVE
   ARCHITECTURE
FOR M1, M2:
     -0.9020
      0.6917
```

```
THE A MATRIX IS:
   0.9019570         -1.691663
   9.8043025E-02  0.6916630
THE B MATRIX IS:
   1.901957          9.8043025E-02

THE C MATRIX IS:
   9.8043025E-02    1.691663
D IS:
   9.8043025E-02
THE INVERSE MATRIX IS:
   0.12828E+02    0.24021E+01   -0.10130E+02
   0.11423E+02    0.26977E+01   -0.90206E+01
   0.80729E+01    0.24021E+01   -0.53752E+01

THE T(N,1) VECTOR IS:
     0.12828166E+02
     0.11422670E+02
     0.80729160E+01
THE TOEPLITZ MATRIX IS:
   0.12828E+02        0.11423E+02
   0.11423E+02        0.12828E+02
THE K MATRIX IS AS FOLLOWS:
   0.10189E+02        0.10000E+01
   0.10000E+01        0.46622E+00
THE W MATRIX IS:
   0.46622E+00       -0.47684E-06
  -0.74506E-06        0.80443E+01
FOR THE UNSCALED SYSTEM:
STATE X(I)         PRODUCT NOISE SOURCES
   1                         3
   2                         3
SIGMA**2(Y)= 25.531498 *SIGMA**2(X)
FOR THE SCALED SYSTEM:
STATE X(I)         PRODUCT NOISE SOURCES
   1                         3
   2                         3
SIGMA**2(Y) (SCALED)= 25.502392 *SIGMA**2(X)
FOR M1, M2:
     -0.9171
      0.4307
THE A MATRIX IS:
   0.9170850         -1.430687
   8.2915008E-02  0.4306870
THE B MATRIX IS:
   1.917085          8.2915008E-02
THE C MATRIX IS:
   8.2915008E-02    1.430687

D IS:
   8.2915008E-02
THE INVERSE MATRIX IS:
   0.65582E+01    0.28404E+01   -0.33683E+01
   0.58397E+01    0.31899E+01   -0.29993E+01
   0.45022E+01    0.28404E+01   -0.13124E+01

THE T(N,1) VECTOR IS:
   0.65581636E+01
   0.58396525E+01

STATE X(I)         PRODUCT NOISE SOURCES
   1                         3
   2                         3
```

```
SIGMA**2(Y) (SCALED)=  6.008472 *SIGMA**2(X)
FOR M1, M2:
      -0.9239
       0.3135
THE A MATRIX IS:
   0.9238720      -1.313526
   7.6128006E-02  0.3135260
THE B MATRIX IS:
   1.923872      7.6128006E-02
THE G MATRIX IS:
   7.6128006E-02  1.313526
D IS:
   7.6128006E-02
THE INVERSE MATRIX IS:
  0.56923E+01   0.30936E+01  -0.22180E+01
  0.50686E+01   0.34743E+01  -0.19750E+01
  0.4023E+01    0.50686E+01
  0.50686E+01   0.56923E+01
THE K MATRIX IS AS FOLLOWS:
  0.10169E+02   0.10000E+01
  0.10000E+01   0.12473E+00
THE W MATRIX IS:
  0.12473E+00   0.00000E+00
 -0.14901E-07   0.21521E+01
FOR THE UNSCALED SYSTEM:
STATE X(I)       PRODUCT NOISE SOURCES
   1                      3
   2                      3
SIGMA**2(Y)=  6.830492 *SIGMA**2(X)
FOR THE SCALED SYSTEM:
STATE X(I)       PRODUCT NOISE SOURCES
   1                      3
   2                      3
SIGMA**2(Y) (SCALED)=  4.610565 *SIGMA**2(X)
```

Magnitude Frequency Response of a Sixth–Order Wave Filter

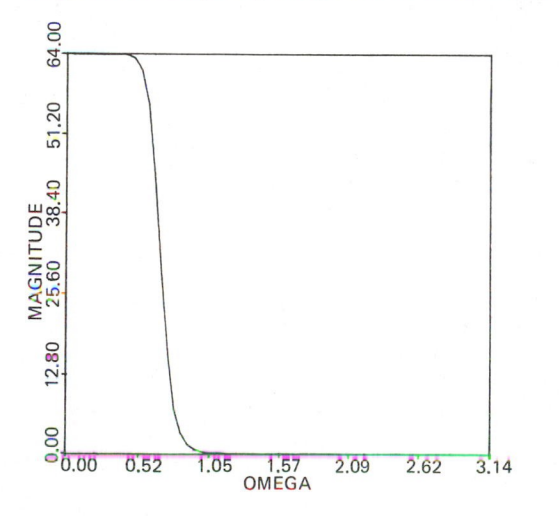

14
Introduction to Hardware

14.1 INTRODUCTION

Digital signal processing is basically a technology-driven science. The availability of low-cost high-performance hardware and systems has made this digital signal processing both useful and practical. Throughput, complexity (packaging), reliability, and economic metrics continue to improve because of technological innovations. Without these breakthroughs, the study of advanced algorithms and filter architectures would be but an academic exercise. The contemporary digital signal processing scientist must be conscious of these hardware trends. The material in the next few chapters provides the basis for understanding digital signal processing hardware. The treatment of this subject is pedagogical. It should be remembered that this is a highly dynamic field and the material discussed here will be obsolete in a few years. Therefore, the best way to maintain competence in this area is through continued involvement with questions of design and technology.

14.2 LOGIC FAMILIES

The designer of digital signal processing hardware systems must be familiar with the logic families used to design contemporary systems. Questions of interface, cost, complexity, speed, and reliability are answered in terms of electronic hardware. The traditional logic groups used to design most digital systems are summarized in Table 14.2.1.

Transistor-transistor (TTL) logic is the mainstay of the digital industry. It is a very mature technology and therefore offers certain cost and availability advantages. The historical disadvantages of this family have been the tendency of gates to go into saturation. This reduces throughput since there is a finite amount of time required for the gates to come

TABLE 14.2.1 Summary of Logic Families Currently Found in Common Use

Type	Gate delay (nsec)	Clock (MHz)	Supply (V)	Power mW/gate	Noise immunity	Cost
TTL	12	15–30	5.0	12	High	Med.
TTL Schottky	4–5	20–75	5.0	20–40	High	High
ECL	1–4	50–400	−5.2	40–60	Med.	High
I^2L	50–500	1–8	5.0	0.01–0.1[a]	Med.	Low
PMOS	200–600	0.5–2	−13, −27	0.2–2[a]	Low	Low
NMOS	100–300	1–4	8, 15, 5	0.1–1[a]	Low	Low
CMOS	20–100	1–12	5, ±15	0.01–2	High	Low

[a]Power dissipation is a function of clock rate.

out of a saturated state. To overcome this problem, diode clamps have been added. This has resulted in the Schottky TTL family. Emitter-coupled logic (ECL) uses nonsaturating bipolar configurations to allow faster than Schottky TTL performance. Unlike the TTL technology, ECL output can be wire-ORed. However, they are more noise sensitive and have a large power appetite. A newer technology, which is gaining in popularity, is called the integrated injection logic (I^2L) family. They are devoid of all resistive elements and therefore offer excellent power savings. As a result, a high degree of integration can be achieved. The price paid is one of gate speed.

On the metal-oxide semiconductor (MOS) side of the ledger is found a wide variety of devices. One of the most popular members of this family is the MOS field effect transistor (MOSFET). The MOFSET is a voltage-controlled low-power electronic device generally appearing in a n- or p-channel (NMOS, PMOS) or silicon-on-sapphire (SOS) configurations. The SOS devices, because of a more complex fabrication process, are the more expensive of the three. NMOS devices can operate at 5 V (TTL levels) and therefore offer certain design advantages. "High-threshold" MOSFET logic units are known to possess good noise rejection capabilities. Complementary MOS (CMOS) configurations exhibit a high noise immunity at low power. Therefore, the CMOS often represents a good design choice. MOSFET gates can be found in static or dynamic forms. Dynamic units consume less power than their static counterparts but need to be refreshed with injected current. The cost and packaging of the refresh circuitry should be included in any design trade-off study. Also, MOSFET devices have a low "fanout" capability.

The MOS era ushered in the phenomenon known as the microprocessor. The microprocessor is no longer exclusively a MOS device. These highly integrated programmable units are discussed in some detail in a later section. In addition, specialized integrated signal processing devices are reviewed in another section. It should be remembered that because the field of integrated circuitry is in a constant state of revolution and evolution, the material discussed in detail in the next few chapters will be replaced with superior versions in the near future. The successful signal processing technologist will utilize this resource to its fullest. Truly, it would appear that the only limit of growth in this area is the ability of the language to form new acronyms.

14.3 MEMORY TECHNOLOGY

Digital signal processing systems make heavy use of memory for:

1. Storing instructions
2. Storing filter coefficients and parameters

Memory appears in random access and read-only form. The nomenclature of the field defines access time to be the time required to select and read (or write) a unit. Memory cycle time is defined to be the time required between successive accesses. Memory types and technologies are summarized in Table 14.3.1.

In a computer environment, one generally studies the design of systems in terms of main and bulk (mass) storage requirements. For special purposes, speed may be the principal design objective; in others cost may be the main concern. Choosing the correct memory will dictate the performance level of these systems.

Memory is grouped in the following categories:

1. Nonvolatile: information remains stored in place after power-down.
2. Destructive and nondestructive readout.
3. Random access (RAM) and read-only memory (ROM).
4. High-speed, low-speed (gap) and electromechanical memories. These memory classifications will be further refined later in this chapter.

Functionally, access time will influence, not dictate, the use of the various memory types. They are summarized in Table 14.3.2.

Core

Magnetic core RAM is a nonvolatile memory but suffers a destructive readout. Therefore, data are often read back onto core immediately after a read

TABLE 14.3.1 Summary of the Various Types of Memory Class Found in Common Use

Type	Access time (nsec)	Cycle time (nsec)	Power (mW/bit)	Cost (¢/bit)
Core	300–800	600–1600	variable	0.5
Dynamic PMOS	150–500	250–800	0.200	0.1–0.2
Dynamic NMOS	100–400	200–600	0.100	0.1–0.2
Static NMOS	60–200	70–300	0.3–.700	0.2–0.5
Static CMOS	100–150	150–800	0.15–.300	0.5
Static SOS	80–200	120–500	0.03–1.00	0.7–1
Bipolar	20–100	60–150	0.1–2.00	0.2–2
ROM	35–1000	70–1500	0.1–.500	0.1–1
PROM	50–500	80–800	0.100	0.5–2
EPROM	250–500	250–500	0.03–0.008	0.1–1

operation. Therefore, a read cycle in effect occupies two memory cycles. Core units are generally more expensive, less reliable, and slower than their semiconductor counterparts.

Semiconductor Memory

Semiconductor RAM offers three basic types of data access modes. One is sequential, using a fixed number of clock cycles, such as that found in a stack

TABLE 14.3.2 Summary of the Utility of Various Types of Memory Units

Application	Time (nsec)	Type
Microprocessor	200–500	Static and dynamic RAM
Minicomputer	100–250	Dynamic MOS RAM, bipolar RAM
Cache	20–80	Bipolar RAM, static MOS RAM
Scratchpad	LT –50	Bipolar RAM (ECL or TTL)
Writable control	50–100	Bipolar RAM, static MOS RAM

operation. Another is a conventional RAM operation where any data address is accessible in one memory cycle. The last is a content-addressable (associative) operation where data are outputted whenever a portion of stored data matches the input. Such a memory policy is useful in performing direct searches. The RAM in turn can be subdivided into three more groupings based on bipolar, static MOS, and dynamic MOS technology. Bipolar memories are the fastest of the three. ECL units can be found with 10-nsec cycle times, whereas the more commonly encountered units work in the range 20 to 70 nsec. Referring to Table 14.3.1, it can be noted that bipolar devices have higher-than-average cost and power budgets. Also, the packing densities of the bipolar units are less than those found in MOS due to the higher power dissipations. Bipolar memories can be found in sizes from 1K to 4K. Typically, speeds for a 1K-unit range from 12 nsec for ECL devices to 45 nsec for TTL units. Speeds and packing densities will improve in the future.

Static MOS memory devices are essentially flip-flops. Newer units use a 5-V TTL-compatible power supply (NMOS). These devices cover a wide range of cost and speed metrics. Some semiconductor memories, such as those based on CMOS technology, have very low "standby" power dissipations. Because of their basic simplicity, static RAMs are popular for the design of small (8K or less) memory systems. However, the newer static devices should lend themselves to economically justifiable larger memory systems in the near future. For example, the field has come from 500-nsec 1K devices in the mid-1970s to memory units which are routinely available at 55 nsec with a size of 4K.

Future MOS devices should provide the system designer with higher performance. The key to this technology is "scaling." Using high-resolution lithography, a scale constant $K > 1$ will result which will yield the following attributes:

Scaled by $1/K$: voltage, current, capacitance, gate delay
Scaled by $1/K^2$: power dissipation

The use of integrated injection logic (I^2L) is providing users with packing densities normally associated with MOS devices. However, MOS units still possess cost advantages over this technology down to 50 nsec. Below this metric, bipolar technology is required. The empirical relationship used by some designers to forecast speed is NMOS = $2 \times$ TTL, TTL = $2 \times$ ECL. The state of the static MOS memory art is currently at 4K (50 to 70 nsec), which can be further classified as follows:

1. Nonclocked (address change produces output from selected chip).
2. Clocked (chip must be selected in each case).
3. Nonclocked power-down (same as class 1 except that it takes advantage of a low standby current).

Technology forecasts indicate that 16K (120 nsec) type 3 memories should be available soon.

Dynamic MOS memory is a high-density technology. Dynamic memory uses integrated capacitors to store binary-valued data. To replenish the charge leaked from these capacitors, a refresh policy must be used. The refresh procedure is to externally inject current into the device every few milliseconds. As a result, additional clock and control circuitry is required. The relative cost of this peripheral circuitry can be significant if the memory size to be served is small. For high-density memory, say 16K or greater, the cost becomes less significant. The state of MOS dynamic memory is currently 64K × 1, having access times on the order of 100 to 150 nsec with 4-msec refresh. This corresponds to a 1.3 to 1.6% refresh efficiency. Such devices can be found in a 16-pin package operating off a single 5-V supply while dissipating 200 mW. However, many existing high-density dynamic RAMs do require two power supplies. Using dc-dc converters, specifically made for this use, a single 5-V supply can be split into the needed multiple voltage levels.

Commercially available memory units have been selectively summarized in Table 14.3.3. The criterion of choice was speed within each member classification.

The MOS memories can be integrated in large memory banks in a variety of ways. Vendors' technical literature is usually the best source of information in this area. For example, memory units may be configured as stand-alone using select lines as shown in Fig. 14.3.1. The memory address is composed of N_S select bits and N_D device address bits. Within a given bank, memory chips may be "sandwiched" together as suggested in the same figure. A wide choice of decoder and controller chips can be found to support these memory configuration schemes. In many cases, these support functions are integrated into the memory chip itself.

EXAMPLE 14.3.1 Consider the Intel 2102 (1024 × 1) static RAM having the attributes detailed in Fig. 14.3.2. Using this 1K unit, larger memory sections can be synthesized. For example, a 4K × 4 memory array can be designed in the manner suggested in Fig. 14.3.3.

14.4 READ-ONLY MEMORIES

Read-only memories are becoming increasingly popular in signal processing. Some of their more common uses are in:

1. Storage of programs and operating systems
2. Storage of special-purpose library routines
3. Storage of filter parameters and coefficients

TABLE 14.3.3 Summary of Sample Memory Units Currently
Commercially Available

Group	Organization	Access time (nsec)	Power, active/standby (mW)	Pins	Vendor
TTL,	16 × 4	40	500	16	National
RAM	16 × 4	25	500	16	AMD
	256 × 1	30	500	16	TI
	256 × 4	45	675	22	Fairchild
	1024 × 1	35	600	16	Fairchild
	256 × 8	55	675	22	Signetics
	4096 × 1	75	500	18	TI
	4096 × 1	45	675	18	Signetics
ECL,	64 × 1	15	550	16	Motorola
RAM	128 × 1	15	525	16	Motorola
	256 × 1	10	450	16	National
	1024 × 1	20	775	16	Fujitsu
	1024 × 1	10	775	16	Fairchild
	4096 × 1	35	1000	16	National
	1024 × 4	25	825	22	Siemens
	16K × 1	35	925	16	Fairchild
Dynamic	16K × 1	80	140/15	16	Mostek
RAM	2K × 8	150	150/25	28	Mostek
	32K × 1	150	375/20	28	TI
	64K × 1	100	200	16	TI
	64K × 1	150	325	16	Intel
Static	4K × 1	200	275/15	20	TI
(NMOS)	4K × 1	35	900/175	18	Intel
	4K × 1	45	950	18	Hitachi
	1K × 4	55	600/150	18	Intel
	1K × 4	20	600	18	Intel
	1K × 4	200	350	22	AMD
	1K × 4	250	400	22	National
	1K × 8	120	425	24	Intel
	1K × 8	55	650	24	Mostek
	16K × 1	45	550/85	20	Intel
	2K × 8	80	200	16	Intel
	4K × 8	200	200/100	28	Zilog
	64K × 1	100	150/10	16	Intel
	64K × 1	120	300	16	Mostek
RAM	256 × 4	45	125/0.1	22	Intel
(CMOS)	256 × 4	215	2.5/0.5	22	Intersil
	2048 × 1	250	5/0.05	18	Harris
	1024 × 4	200	150/0.2	18	NEC
	2048 × 8	100	175/0.02	24	Hitachi

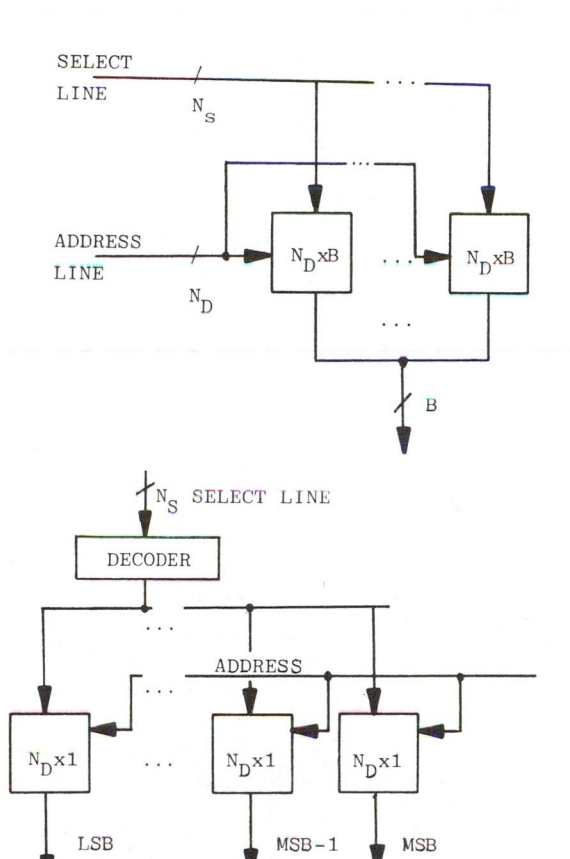

FIG. 14.3.1 Block diagram of the development of larger memory units from smaller memory modules.

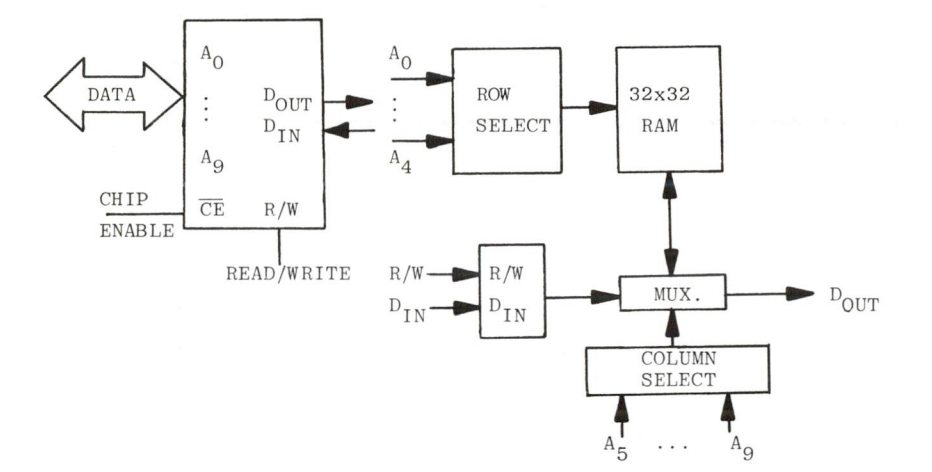

FIG. 14.3.2 Detail of a typical memory module.

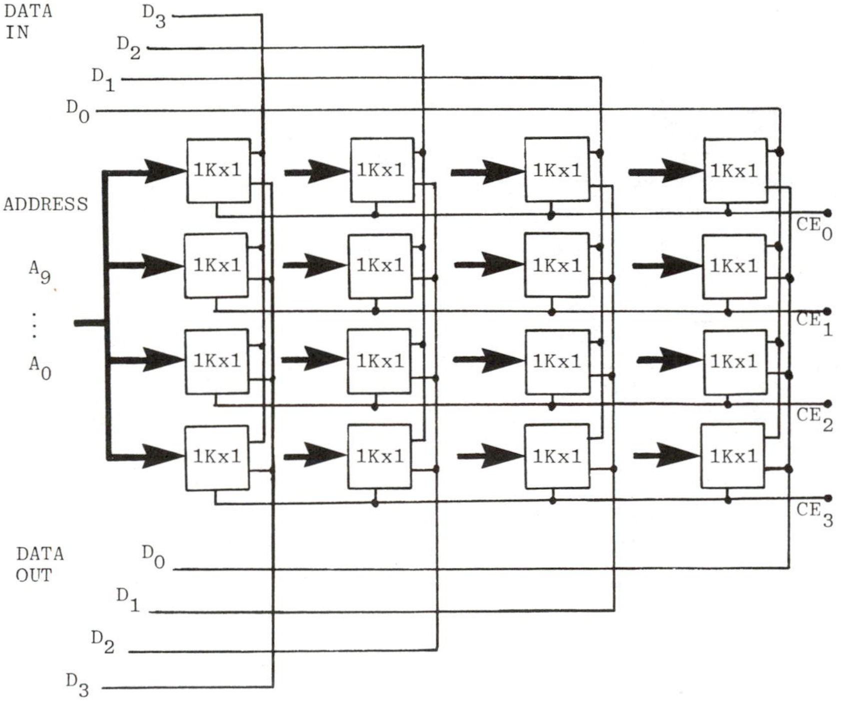

FIG. 14.3.3 Interconnection of 1K × 1 memory modules to form a 4K × 4 memory bank.

TABLE 14.4.1 Example of Commercially Available ROM Memory Units

Size[a]	Organization	Pins	Access time (nsec)	Power (mW)	Technology
1024	256 × 4	16	40	460	Schottky
2048	256 × 8 or 512 × 4	20(16)	45	525	Schottky
8K	2K × 4	18	70	500	Schottky
16K	2K × 8	24	150	500	I^2L

[a] User must specify mask information.

Semiconductor ROMs are matrix arrays which can be permanently masked by the vendor or user. Some devices are designed to be programmed in the field (PROM), while others require a laboratory setting.[†] A variation of PROM is the erasable PROM (EPROM). Fusible-link high-speed memory units are also available with access times below 100 nsec. The state of the ROM area is summarized in Table 14.4.1.

The bipolar ROM field should continue to grow with higher speeds and greater packing densities (especially in the I^2L area). In addition, the field of MOS ROMs is maturing rapidly. These memory units can be found over a wide range of performance, cost, and density parameters (e.g., 4K × 8 bits, 350 nsec, 24 pins). Improved micron and submicron fabrication methods will continue to force improvements in this area.

Currently, EPROMs are available in a variety of configurations. The state of the EPROM art is summarized in Table 14.4.2. The outputs of these devices may be chosen to open collector or tri-state.

Units of 64K or more should be routinely available in the near future. Electronically alterable ROM (EAROM) memories are becoming more available as well (e.g., N-channel, 350 nsec, 16K). These devices (4K to 8K) allow the user to reprogram memory electronically, with program retention being measured in months to years. Program retention is affected, however, by reprogramming. One EAROM is designed to survive power failures by combining 1K of RAM with 1K of EAROM. That is, the EAROM serves as a power fail buffer for the volatile RAM.

[†] The user must supply mask information to the vendor, usually in the form of Hollerith cards.

TABLE 14.4.2 Example of Commercially Available EPROMs

Size	Organization	Pins	Access time (nsec)	Cycle time (nsec)
4K	512 × 8	20	45	135
8K	1024 × 8	24	60	175
16K	2048 × 8	24	80	175
32K	4K × 8	24	450	450
64K	8K × 8	28	450	450

[a] Power active less than 500 mW; power standby, 50 mW.

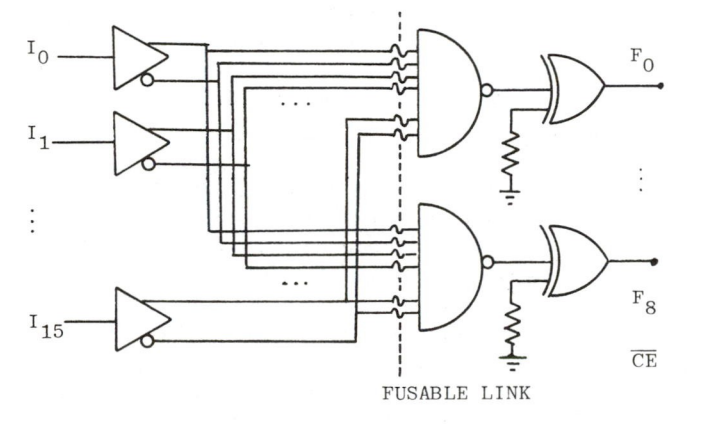

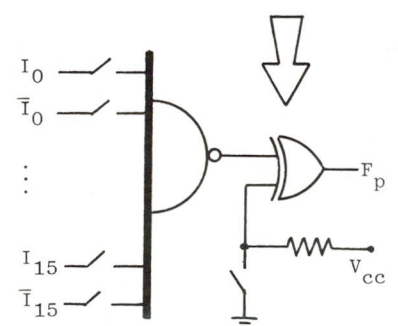

FIG. 14.5.1 Structure and diagram of a PLA.

14.5 PROGRAMMABLE LOGIC ARRAY

A specialty device is the programmable logic array (PLA). PLA is programmed by the vendor or in the field as a FPLA. PLA will map a n-bit
input into a m-bit output under a finite set (repertoire) of Boolean operations. These operations are subject to the side constraint imposed by a
finite number of admissible AND and OR terms. These devices are useful
in performing special encoding, decoding, or combinational logic operations.
A typical masked PLA may map a 9-bit input into one of 459 20-bit outputs
(note that a RAM would admit 2^9 = 512 unique outputs) consisting of at most
62 AND and 20 OR operations. Due to a different internal structure, a PLA
will typically run faster than a RAM having the same input space size. For
the purpose of completeness, the structure of a PLA, having 16 inputs and
9 outputs is suggested in Fig. 14.5.1.

　　　To this point, various programmable devices have been summarized.
The design of a system, using these items, is a study in trade-offs: cost
versus performance, device availability, and electronic compatibility.

14.6 FUTURE TECHNOLOGIES

Today we live in the silicon age. However, the supremacy of silicon may
soon be challenged by gallium arsenide (GaAs). This technology is receiv-
ing increased attention because of its potential speed (100-psec gate delays)
and lower power dissipation. For example, Lockheed has reported testing
a GaAs device with a 50-psec gate delay (12-GHz clock) configured as a 4K
static RAM having a 1- to 2-nsec access time. Although years away from
commercial availability, GaAs deserves close monitoring. The gate delays
associated with the devices of the 1970s are defined in terms of ECL, bipolar, and MOS metrics (see Fig. 14.6.1). The future may be found in the

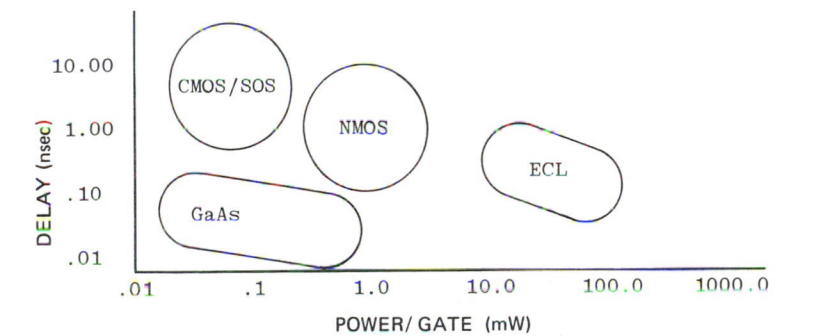

FIG. 14.6.1 Diagram indicating the relative performance of GaAs and
other commonly encountered technologies.

new ultra-high-speed technologies such as the Josephson devices. This
technology is being advanced by IBM and others. It has been under study
since 1965 and exhibits the ability to achieve high speeds on the order of
1-nsec processor cycle times, 2-nsec cache, and 7-nsec 64M-byte memory
cycle times. All of this can potentially be packaged in a 8 × 8 cm 10-W
package. For example, a hypothetical computer was configured by IBM
using Josephson technology. It was reported to have the following attributes:

1. Performance: 70 million instructions per second
2. CPU: 4 nsec
3. Cache: 32K, 4 nsec
4. Main memory: 16M bytes 20 nsec
5. Power: 7 W at 4K; 15 W at 300K

However, since this is a cryogenic technology, its use in the signal processing arena will probably be limited.

14.7 HIGH-DENSITY RAM

The high-density RAM is a very attractive design tool. In 1979, devices
such as the Motorola 64K × 1-bit RAM with integrated refresh, needing 5 V
at 250 mW, and having an access time of 150 nsec were available in 16-pin
packages (see Table 14.7.1). NMOS and similar technologies will be continually improving on these performance specifications.

TABLE 14.7.1 Sample of Commercially Available 64K Dynamic RAMs

Access time (nsec)	Refresh cycles/period	Power, active standby (mW)	Voltage (V)	Vendor
150	256/4	125/17.5	5	TI
150	128/2	275/30	5	Motorola
100-120	128/2	300/20	5	Mostek
100-200	256/4	200/20	5	National
100-200	128/2	200/20	5	AMD
60-120	256/4	300/15	5	Signetics

14.8 FIELD-PROGRAMMABLE DEVICES

Field-programmable memory units are popular in signal processing applications. They provide the user with the attributes of a ROM without the permanency of a fixed program device. Therefore, the field-programmable unit is an excellent development tool which can also be useful in producing low-volume systems which may be in need of field adjustments.

14.9 VLSI AND VHSIC

A new technology that has received much attention is known as very large scale integration (VLSI). This technology is characterized by integration up to 10,000 active gates on a 1/3-in. silicon wafer. The high-density and speed capabilities of these units have made them a feasible signal processing device technology. VLSI developments should allow megabit memory chips to be realized in the 1980s. Device houses are announcing general-purpose signal processing chips, with more to come in the near future. Some of the more commonly available chips are discussed in Sections 16.3 and 17.7.

Very high speed integrated circuits (VHSIC) may have a major impact on future military electronics. This technology is currently in a product definition phase. Some of the more promising noncommercial VHSIC applications are:

1. Parallel pipeline processors
2. Image-processing devices
3. Frequency synthesis
4. Frequency, phase, and code manipulation
5. Correlators and matched filters
6. FFT subsystems
7. Linear predictive coding and signal processing

The commercial semiconductor industry has evaluated the potential of VHSIC in replacing their standard product line. Generally, industry does not believe at this time that no presently produced nonmemory products will result from VHSIC. Although this may be true for the commercial sector, it is not true in the digital signal processing area. Many high-speed and high-performance secure communications, countermeasures, and image-processing systems may develop directly from the VHSIC program.

For example, some of the forecasted applications of the VHSIC technology are:

1. Surveillance radar (Navy) outfitted with an ultra-high-speed programmable matched filter, signal synthesizer, and detector

2. Tactical radar signal processor (Navy) using 100-MHz A/D and FFTs and 256 × 16-bit RAMs having access times on the order of 4 nsec
3. Multifunction radar signal processor (MRSP-Air Force) using 20-MHz FFTs
4. ESM signal sorter (Navy) to reduce a 1500-chip LSI design to 30 VHSIC chips and to improve content-addressable memories (64-bit at 300 nsec versus 1K-bit at 30 nsec)
5. General-purpose 32-bit computer (Navy) capable of executing 50 million instructions per second
6. Battlefield information distribution system (BIDS-Army) to support spread-spectrum modulation rates of 100 MHz or greater

14.10 MEMORY TYPES

Memory can be classified as main (previously discussed), gap, or magnetic. They are subdivided in terms of their access speeds. A summary of these memory classifications can be found in Fig. 14.10.1.

Gap memories are semiconductor memories that fill the speed gap between main and magnetic memory. Access time runs between 2 sec and 20 msec. Since many signal processing operations require large volumes of memory (e.g., FFT, ensemble averaging, etc.), gap memory may play a major future role in future system development. The idle time associated

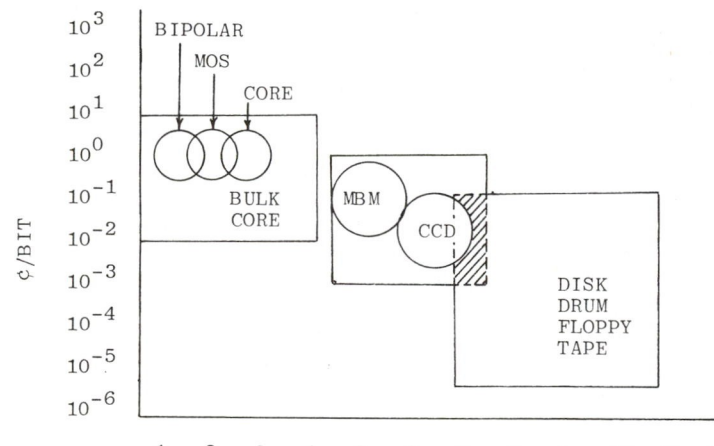

FIG. 14.10.1 Performance versus cost of contemporary storage devices.

with a processor waiting for a slow-speed electromechanical disk transaction to take place can be substantially reduced using gap memory.

Gap memories are found in charge-coupled devices (CCDs) or magnetic bubble devices (MBDs). For a CCD to complete with the simpler RAM, a 4:1 density improvement is generally assumed to be necessary. In light of improved RAM fabrication techniques, this is becoming more and more difficult to achieve. Nevertheless, 16K and 64K CCD memories are now available with 128K to 256K units soon to be released. The CCD can prove useful as a bulk storage medium, where it offers some economic and speed advantages over the competitive floppy disk. However, they are volatile and are therefore often found backed up with low-cost cartridge tapes.

The MBD is a 1967 product of Bell Laboratory. Although slower than the CCD, it is nonvolatile and offers outstanding packing densities. If the MBD can be found to cost two to three times less than higher-speed but volatile CCDs, the MBD should be considered to be a good design choice. CCD memories are basically low-cost bulk storage devices which can benefit directly from the MOS technological explosion. CCD memories are volatile and therefore are at a disadvantage when compared to some MOS RAM. The future of this field is uncertain as a memory unit. However, interesting signal processing devices should continue to evolve. Also, the MBD warrants close attention. Specifically, if capabilities exceed 64K bits per chip, magnetic bubble memories should be considered. Densities on the order of 1M bit per chip are found in dual in-line packages. Magnetic bubble memories are already cost effective in harsh environments (e.g., severe vibration, dust, etc.). Typical magnetic bubble memories are characterized by:

Capacity	Package	Manufacturer
1M	Leadless	Intel
92K	14	Texas Instruments
1M	24	Texas Instruments

For example, the Texas Instruments 92K MBD has a 50K-pps I/O rate while dissipating only 0.7 W.

14.11 MASS STORAGE

Many signal processing operations require large volumes of memory. Memory-intensive operations such as data logging and ensemble averaging can consume large amounts of memory. As a result, the system designer must often turn to mass-storage resources to support the signal processing task. The mass-storage units attached to programmable systems take many forms. Some of the most common are:

TABLE 14.11.1 Summary of Mass Storage Units Found in Common Use

Type	Bytes/sec	Access (max)	Storage (bytes)	Cost ($)
Cassette	0.5-1.5K	250 sec	5-12K	800-3K
Cartridge	5-6K	200 sec	50K-2.5M	800-3K
9-Track	6-60K	240 sec	2.5K-2.5M	5-14K
Floppy	20-50K	200-400 msec	10-50K	3-30K
Moving	40K-1M	50-200 msec	100K-8M	3-30K
Fixed	40K-1M	8-20 msec	4-75K	6-30K
Winchester	40K-1M	30-70 msec	5-30M	3-5K

1. Cassette: low-cost, low data rates
2. Cartridge: inexpensive, slow data rates, good media lifetime
3. Nine-track tape: economical, appearing with embedded controllers and formatters
4. Floppy disk: economically suitable for nonvolatile I/O data exchanges; 8 in., 5.25 in. (4 in. and 3 in. for portable use)
5. Disk: fixed or moving head, high-density storage offering moderate data rates

The various mass storage units are summarized in Table 14.11.1. Another mass storage device at a higher cost per bit is bulk core. Appearing in modular units (e.g., 256K-byte increments), they can support data trafficking at rates some 10,000 in excess of that found on some fixed-head disks.

14.12 ANALOG VERSUS DIGITAL TECHNOLOGIES

LSI and VLSI will strongly influence the course of analog, digital, and hybrid signal processing. Improved lithographic techniques will allow smaller geometries to be realized. Analog circuits require the highest component precision. As a result, analog circuits will become increasingly difficult to realize as device line widths shrink. However, smaller line widths mean faster device speeds and larger bandwidths. This attribute will be shared by both analog and digital devices. In addition, as geometries become smaller, supply voltages will reduce. As a result, signal strengths will be scaled downward as a consequence of a reduced supply voltage. Devices that are configured with on-board sampling circuits will suffer from this scaling due to an increased noise component. Device noise can be modeled, in RMS voltage units, as $v(RMS) = SQRT(kT/C)$, where k is Boltzmann's constant, T the temperature in kelvin, and C the sampling capacitor. Since small geometries result in small capacitors, a proportionally higher noise voltage will result. Therefore, as a result of reduced supply voltages and increased noise levels, the signal-to-noise ratio in small-geometry VLSI devices may be poor.

In a digital environment, poor signal-to-noise ratios can be tolerated. Consider the states of a digital gate to be $V + v_i$ and $0 + v_i$. The noise voltage is modeled to be a normally distributed random process v_i having mean zero and variance kT/C. Using classical signal detection theory, it can be shown that in an equiprobable channel, where the probability of a logical one (V volts) is equal to a logical zero (0 volts), a 20-dB signal-to-noise ratio will result in a detection error probability of 10^{-10}. This is an acceptable digital figure of merit, whereas a 20-dB signal-to-noise ratio is unacceptable in most analog applications.

BIBLIOGRAPHY

Bolan, M. (1979), Design Memory Boards for RAM/ROM/EPROM Interchange, EDN, August, pp. 100-103.

Foster, M. J., and M. T. Kung (1980), The Design of Special-Purpose VLSI Chips, IEEE Comput., January, pp. 26-40.

Hayes, J. (1980), MOS Scaling, IEEE Comput., January, pp. 8-13.

Hnateh, E. R. (1978, 1980), Semiconductor Memories, Parts 1, 2, and 3, Comput. Des., December 1978, January and February 1980.

Sumney, L. (1980), Sumney and Processing, MSN, January, pp. 28-39.

15
Numbering Systems

15.1 INTRODUCTION

The digital technologies reviewed in Chapter 14 establish the speed and cost potentials of a design. Another design consideration is, of course, word length. Long-word-length filters would require more hardware than a shorter-word-length version. Complex filters would generally trade accuracy for cost, complexity, and sometimes throughput degradation. However, all architectures of a given word length are not alike. Their precision is also a function of the data-encoding scheme chosen. In this chapter the more familiar numbering systems are discussed and developed. It will be shown that serious thought should go into this choice since it can indeed strongly influence the design.

15.2 NUMBERING SYSTEMS

In a digital computer or filter environment, all numbers must be encoded so as to reside in a register of fixed word length. The choice of code significantly influences the efficiency of the system's arithmetic section. Besides affecting throughput, the encoding choice will establish accuracy limits as well.

Numbering systems fall into one of two classes. They are <u>weighted</u> and <u>unweighted</u>. The weighted numbering system is by far the most popular of the two. The form of a weighted representation of a nonnegative integer $x \in [0, r^n - 1]$ is

$$x = \sum_{j=0}^{n-1} a_i r^i \tag{15.2.1}$$

where r is the radix (or root) of the system and a_i is the ith digit in the n-tuple representation of x given by $(a_0, a_1, \ldots, a_{n-1})$, $a_i \in [0, r-1]$. A characteristic of weighted numbers is that they possess a <u>least significant digit</u> (LSD) (namely, a_0) and a <u>most significant digit</u> (MSD) (namely, a_{n-1}). The other digits can be referenced with respect to their position relative to the MSD or LSD. The familiar decimal numbering system is defined in terms of r = 10 with binary, octal, and hexidecimal given by r = 2, 8, and 16, respectively. The digits for these four numbering systems are:

1. <u>Decimal:</u> $a_i = \{0, 1, 2, \ldots, 9\}$
2. <u>Binary:</u> $a_i = \{0, 1\}$
3. <u>Octal:</u> $a_i = \{0, 1, 2, \ldots, 7\}$
4. <u>Hexidecimal:</u> $a_i = \{0, 1, 2, \ldots, 9, A, B, C, D, E, F\}$

Fractional forms are also admissible where

$$x = \sum_{i=0}^{n-1} a_i r^{-i} \tag{15.2.2}$$

and is assumed to have a dynamic range given by $[0, 1 - r^{-n}]$. For <u>signed</u> integer representations, an additional digit (called sign digit) is needed. This can be best biewed in terms of the data found in Table 15.2.1. In this table, the more popular binary codes are presented.

The 2's-complement code has become the standard binary-valued weighted code. The popularity of this code is found in its ability to support high-speed nested addition. This thesis will be developed in more detail later in this chapter. At that time it will be shown that the so-called carry-out bit of a 2's-complement accumulator can, under certain circumstances, be discarded. In addition, it will be shown that if the final result of a nested summation is within the admissible dynamic range of the code, partial sum overflows can be tolerated.

15.3 2's COMPLEMENT

Since the 2's-complement system is the most commonly encountered en-code used in digital signal processing systems, a more detailed development of its properties will now be presented. In general, for $L \geq 0$, $K \geq 0$, and M defining the <u>message length</u>, it follows that

$$x = -x_k 2^K + \sum_{i=-K+1}^{L} x_i 2^{-i}$$

TABLE 15.2.1   Common Binary-Valued Numbering System

Type	Integer	Fractional
2's Complement	$-2^{n-1} \le x \le 2^{n-1}$ $x = -a_{n-1}2^{n-1} + \sum_{j=0}^{n-2} a_i 2^i$	$-1 \le x \le 1 - 2^{1-n}$ $x = -a_{n-1} + \sum_{j=0}^{n-2} a_i 2^{i-(n-2)}$
Sign magnitude	$1 - 2^{n-1} \le x \le 2^{n-1} - 1$ $x = (-1)a_{n-1} * \sum_{i=0}^{n-2} a_i 2^i$	$-1 + 2^{1-n} \le x \le 1 - 2^{1-n}$ $x = (-1)a_{n-1} * \sum_{i=0}^{n-2} a_i 2^{i-(n-2)}$
Offset binary	2's complement with sign reversed	
1's Complement	$1 - 2^{n-1} \le x \le 2^{n-1} - 1$ $x = (1 - 2^{n-1})a_{n-1} + \sum_{i=0}^{n-2} a_i 2^i$	$-1 + 2^{1-n} \le x \le 1 - 2^{1-n}$ $x = (2^{1-n} - 1)a_{n-1} + \sum_{i=0}^{n-2} a_i 2^{-i}$

EXAMPLE 15.3.1 Consider $M = 6$, $K = 4$, $L = 1$; then

$$x = -x_4 2^4 + x_3 2^3 + x_2 2^2 + x_1 2^2 + x_0 2^0 + x_{-1} 2^{-1}$$

$$\Longrightarrow (x_{4\triangle} x_3 x_2 x_1 x_0 \cdot x_{-1})$$

If $x = -13.5$, then $-13.5 \cong 1_\triangle 0010.1$. If $x = 13.5$, then $13.5 \cong 0_\triangle 1101.1$, where the symbol $\triangle$ is used to separate the sign digit from the magnitude digits and "." denotes the binary point.

When developing digital computational systems and algorithms, data must sometimes be truncated or rounded. The statistical properties of these operations were discussed in Section 12.2. The mechanics of truncation are summarized in Table 15.3.1. The three types of truncation studied in this table are:

1. Rear truncation of L bits
2. Front truncation by K bits
3. Front and rear truncation

Some of the basic rules of truncation are summarized below.

1. The order of truncation is irrelevant.
2.
$$x_*^* = (x_*)^* = (x^*)_*$$
(15.3.2)

2. If $x_{i*} = x_i$ for all i, then

$$\left(\sum_{i=1}^{n} x_i \right)_* = \sum_{i=1}^{n} x_i$$
(15.3.3)

e.g., $M = 6$, $K = 4$, $L = 0$, $x_1 = -14 = x_{1*}$, $x_2 = -14 = x_{2*}$, $(x_1 + x_2)_* = (-28)_* = -28$

3. If $x_{i*} = x_i$ for $i = 2, 3, \ldots, n$, then

$$\left(\sum_{i=1}^{n} x_i \right)_* = x_{1*} + \sum_{i=2}^{n} x_i$$
(15.3.4)

4. For x given, the binary complement $\overline{x}$ is given by

$$\overline{x}_* = -\overline{x}_k 2^k + \sum_{i=-K+1}^{L} \overline{x}_i 2^{-i}$$
(15.3.5)

TABLE 15.3.1 Summary of Front and Rear Truncation Forms on Integer and Fractional 2's Complement Data

Type	Form	Example ($K = 4$, $L = 1$, $M = 6$, $x = 13.5$)
Rear truncation L-bits from LSB	$$x_* = \left\{ -x_k 2^k + \sum_{i=-K+1}^{L} x_{-i} 2^{-i} \right\}_*$$	$x_{10} = -13.5 \rightarrow 1 \,_\Delta 0010.1$ $x_{*10} = -14 \rightarrow 1 \,_\Delta 0010.$ $x_{10} = 13.5 \rightarrow 0 \,_\Delta 1101.1$ $x_{*10} = 13 \rightarrow 0 \,_\Delta 1101.$
Front truncation scale by 2^{-K}, keep $M - L$ bits	 $$x^* = \left\{ -x_0 + \sum_{i=1}^{M-L} x_i 2^{-i} \right\}^*$$	$x_{10} = (-13.5/16) = -.84375$ $x_{10}^* = -0.875 \; 1 \,_\Delta 0010$ $x_{10} = (13.5/16) = 0.84375$ $x_{10}^* = \,_\Delta 8125 \; 0 \cdot 1101$

so that

$$(-x)_* = \begin{cases} x_* & \text{if } x_* \neq x \\ -x_* & \text{if } x_* = x \end{cases}$$

for example,

$M = 16, \; K = 4, \; L = 1$ if $x = -13.5$ $1_{\triangle}0010.1$

$x_* = -14 \neq x, \quad \therefore \; \overline{x}_* = 13 \rightarrow 0_{\triangle}110$

and

$(-x)_* = (13.5)_* = 13 = \overline{x}_*$ if $x = -14$ $1_{\triangle}0010.0, \; x_* = -14 = x$

Therefore,

$\overline{x}_* = 13$ $0_{\triangle}110.1$ and $(-x)_* = (14)_* = 14 = -x_*$

5. Finally, $y = (\Sigma x_1)^*$ is equivalent to a modulo 2 addition if $|y| \leq 1$.

for example,

$x_1 = x_2 = -13.5$

$y = -27, \quad y^* = \dfrac{y}{16} = -1.6875 \text{ mod } 2 = 0.3125$

$x_1 \longrightarrow 1_{\triangle}0010.1$

$x_2 \longrightarrow 1_{\triangle}0010.1$

$1:0_{\triangle}0101.0 \longrightarrow (0_{\triangle}0101.0)^* = (0_{\triangle}0101) = 0.3125 = y^*$

15.4 CANONICAL NUMBERS

The integer coding schemes considered to this point belong to a class of numbers called <u>fixed point</u>. Fixed point refers to the fact that the binary point is found in a fixed location. There exist a few other fixed-point systems which hold some interest to the digital signal processing scientist. One such is called the <u>canonical signed digit</u> code.

In the early 1950s, this code enjoyed a high degree of popularity. However, due to semiconductor advancements, other binary-valued numbering systems rapidly replaced the canonical system. This numbering system does, however, possess a unique feature which makes it attractive to certain signal processing applications.

For a given n-digit representation of a number x $[-2^{n-1}, 2^{n-1}]$, the canonical representation can be related to its 2's-complement version through

$$x = -x_n 2^{n-1} + \sum_{i=0}^{n-2} x_i 2^i \triangleq \sum_{i=0}^{n-1} c_i 2^i \tag{15.4.1}$$

where $x_i = (0, 1)$ and $c_i = (0, +1, -1)$. Whereas the 2's-complement is binary, the canonical system is ternary. The digits c_i can be chosen so that $c_i c_{i-1} = 0$, $n \geq 0$, $i \geq 0$. Since the product of two adjacent digits is zero, at least one digit must be in fact zero. Using this coding scheme, data words can be encoded which have a maximal number of zeros embedded into their structure. This can be motivated in terms of the following simple example.

EXAMPLE 15.4.1 For $n = 3$, the integers $[-4, 3]$ are encoded as follows:

2's-complement	x	Canonical
011	3	10-1
010	2	010
001	1	001
000	0	000
111	-1	00-1
110	-2	0-10
101	-3	-101
100	-4	-100

2's-complement: }12 zeros   Canonical: }15 zeros

It has been pointed out by Peled (1976) that the probability of a canonical digit c_i having a nonzero value is

$$P(|c_i| = 1) = \frac{1}{3} + \frac{1}{9n} \left[1 - \left(-\frac{1}{2} \right)^n \right] \tag{15.4.2}$$

As n becomes large, this probability tends to 1/3. This compares to a value of 1/2 for 2's-complement digits. As a result, operations such as multiplication, which involve nested weighted additions, can be accelerated. Using the canonical structure, the probability of the weight being zero is approximately 2/3. Therefore, this term will contribute nothing to the nested sum and thus may be ignored. By ignoring these operations, multiplier throughput can be accelerated. For example, suppose that one is interested in multiplying a variable x by a constant c which is known a priori. This represents a scaling operation and has a major importance in digital signal processing (e.g., butterfly coefficients, filter coefficients, etc.). It will be assumed that c possesses the following canonical representation:

$$c = \sum_{i=0}^{n-1} c_i 2^i \qquad (15.4.3)$$

with c_i having values 0, +1, or -1. Since c is known, the coefficients c_i can be precomputed. It then follows that the product $y = cx$ can be written as

$$y = \sum_{i=0}^{n-1} (c_i x) 2^i \qquad (15.4.4)$$

Since the canonical system maximizes the number of $c_i = 0$ occurrences the number of nontrivially weighted (i.e., 2^i) additions and subtractions (i.e., $c_i = +1$ or -1) can be minimized. Therefore, the product y can be computed in fewer steps than required for a 2's-complement mechanization. However, special control logic must now be used to execute the add or subtraction operation dictated by the nonzero value of c_i.

The recoding of a 2's-complement integer into a canonical form can be performed in the manner suggested by the data found in Table 15.4.1.

TABLE 15.4.1 Conversion of 2's Complement to Canonical Digits

c_i	x_{i+1}	x_i	c_{i+1}	C_i	Comment
0	0	0	0	0	x_{i+1} and x_i = 2's complement digits of x
0	0	1	0	1	$x_{n+1} = 0$ if x G.E. 0 and 1 if x L.T. 0
0	1	0	0	0	
0	1	1	1	-1	c_i = ith carry digit of the ith set
1	0	0	0	1	$c_0 = 0$
1	0	1	1	0	
1	1	0	1	-1	C_i = ith canonical signed digit
1	1	1	1	0	

EXAMPLE 15.4.2 Let $x = -3$ (decimal) $= 101$ (2's complement); then

$c_0 = 0, \; x_1 = 0, \; x_0 = 1 \quad c_1 = 0, \; C_0 = 1$

$c_1 = 0, \; x_2 = 1, \; x_1 = 0 \quad c_2 = 0, \; C_1 = 0$

$c_2 = 0, \; x_3 = 1, \; x_2 = 1 \quad c_3 = (NA), \; C_2 = -1$

$x_{(canonical)} \longrightarrow (-1, \; 0, \; 1) \;$ or $\; x = (-1)(2^2) + 0(2^1) + (1)(2^0) = -3$

Let $x = -2$ (decimal) $= 110$ (2's complement); then

$c_0 = 0, \; x_1 = 1, \; x_0 = 0 \quad c_1 = 0, \; C_0 = 0$

$c_1 = 0, \; x_2 = 1, \; x_1 = 1 \quad c_2 = 1, \; C_1 = -1$

$c_2 = 1, \; x_3 = 1, \; x_2 = 1 \quad c_3 = (NA), \; C_2 = 0$

$x_{(canonical)} \longrightarrow (0, \; -1, \; 0) \;$ or $\; x = 0(2^2) + (-1)(2^1) + 0(2^0) = -2$

The canonical number system can be extended in utility one step further by enriching the distribution of zeros within a set of admissible coded numbers. It has been suggested that those words which have a large concentration of distributed nonzero digits (i.e., ± 1) be ignored. Therefore, those integer numbers having a canonical representation containing a high incidence of nonzeros would be considered inadmissible digits. For example, there are eight ternary-valued 5-bit numbers which have the maximal allowable number of nonzero digits (namely three):

MSB			LSB		Integer	Decimal
1	0	1	0	1	21	0.65625
1	0	1	0	-1	19	0.59375
1	0	-1	0	1	13	0.40625
1	0	-1	0	-1	11	0.34375
-1	0	1	0	1	-11	-0.34375
-1	0	1	0	-1	-13	-0.40625
-1	0	-1	0	1	-19	-0.59375
-1	0	-1	0	-1	-21	-0.65625

Therefore, at most eight numbers can be deleted from a 5-bit canonical set of integers over $[-16, 15]$. All the remaining numbers would have at least three of their five digits being zero. This obviously would accelerate the arithmetic.

In general, there are $(n + 1)/2$ (n odd) or $n/2$ (n even) canonical n-bit numbers which have a maximal number of nonzero digits in their code. Since there are 2^n possible n-bit code words, this translates into the following percentages.

1. For $n = 5$, $\dfrac{8}{32} \times 100 = 25\%$

2. For $n = 8$, $\dfrac{16}{256} \times 100 = 6.25\%$

3. For $n = 16$, $\dfrac{256}{65,536} \times 100 = 0.39\%$

It can be seen that as n increases, the amount of damage done to the data set by removing those numbers having a high density of nonzero digits is small.

The reader may be concerned about modifying the number set at all. There are instances where the modified canonical data set can be used with a minimum of distortion. Consider the filter given by $H(z) = \Sigma\, a_i z^{-i} / \Sigma\, b_i z^{-i}$, where a_i and b_i are real numbers. For a given architecture (e.g., direct II), the a_i and b_i will translate into a set of rounded filter coefficients. With a large probability the digital filter coefficients will be found in the modified table of admissible coefficients. If not, one can choose to consider another architecture (e.g., cascade) or round the discrete filter coefficient to its closest admissible value. Integer programming methods, such as those discussed in Section 13.4, can also be used to minimize disorder.

15.5 OTHER CODES

Other fixed-point codes are also found in popular use. One family is known as the binary-coded decimal (BCD) codes. The BCD provides a radix 10 (decimal) representation for a string of binary-valued digits. The radix 10 representation simplifies the problem of human interpretation of a data field. The basic BCD codes are summarized in Table 15.5.1. Even though arithmetic can be performed in BCD form, it is generally implemented using a weighted binary code.

Signal processing scientists are often responsible for supporting control systems using stepping motors (i.e., digital control). The Gray code is often used in this application (see Table 15.5.2). The interesting feature of this code is that the difference between two adjacent code words is one bit location. Therefore, sequentially moving from one motor shaft location to another can be accomplished in a simplified manner.

The fixed-point codes developed to this point can also be modified to accept a degree of error detecting and correcting. To provide this service,

TABLE 15.5.1 Summary of 8, 4, 2, 1, Excess -3, and
2, 4, 2, 1 BCD Codes

Decimal	8, 4, 2, 1	Excess -3 8, 4, 2, 1 x = x + 3	2, 4, 2, 1	
0	0000	0011	0000	
1	0001	0100	0001	Lower
2	0010	0101	0010	half
3	0011	0110	0011	range
4	0100	0111	0100	
5	0101	1000	1011	
6	0110	1001	1100	Upper
7	0111	1010	1101	half
8	1000	1011	1110	range
9	1001	1100	1111	

data word lengths must be extended. In addition, data security can be purchased by using the more complex secure coding policies found in the

TABLE 15.5.2 Gray Coding of Integers Showing
that the Difference Between Any Two Adjacent
Word Locations Is One Bit

Decimal	Code
0	0 . . . 000
1	0 . . . 001
2	0 . . . 011
3	0 . . . 010
4	0 . . . 110
5	0 . . . 111
.	.
.	.

literature. All these operations will add to the cost and complexity of the design. It is assumed that the reader will take all these factors into account when configuring a system.

15.6 SCALING

One remaining and important fixed-point topic is scaling. Scaling is a special type of multiplication. Whereas multiplication is an operation performed on two variables, scaling represents the algebraic composition of a variable with a constant having a known value (i.e., yx versus cx). Scaling is important to digital signal processing. For example, input scaling is sometimes needed to prevent register overflow. Constant-coefficient IIR filtering represents a system of scale factors applied to the states of the filter. In a weighted numbering system, scaling can be performed with relative ease. By shifting right a 1's- or 2's-complement word by n bits, an effective scaling by $1/2$ can be accomplished. It is required, however, that the sign bit be copied before shifting the data. The sign bit is then returned to the scaled word afterward.

EXAMPLE 15.6.1 Let x = 7 (decimal) = $0_\triangle 11.0$ (2's complement) and x = -7 (decimal = $1_\triangle 001.0$ (2's complement); then

Step	Operation	Register	Comment
1	None	$7_{10} = 0_\triangle 111.0$	Initialize
2	Scale by $1/2$	$3.5_{10} = 0_\triangle 011.1$	Copy sign bit and shift
1	None	$-7_{10} = 1_\triangle 001.0$	Initialize
2	Scale by $1/2$	$-3.5_{10} = 1_\triangle 100.0$	Copy sign bit and shift

In a sign-magnitude system, the magnitude digits are shifted while the sign bit remains "frozen" in place.

If upward scaling is required, a left-shift policy is needed. For a 1's- or 2's-complement code, shifting n bits to the left upon copying the sign bit is equivalent to scaling by 2^n. However, caution must be taken to ensure that dynamic range overflow will not occur.

EXAMPLE 15.6.2 Let x = 3 (decimal) = $0_\triangle 011$ (2's complement) and x = -3 (decimal) = $1_\triangle 101$ (2's complement); then

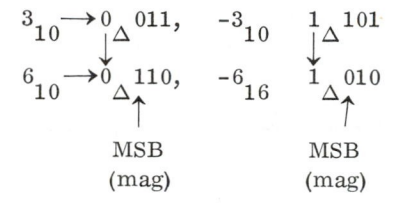

$$3_{10} \longrightarrow 0_\triangle 011, \quad -3_{10} \quad 1_\triangle 101$$

$$6_{10} \longrightarrow 0_\triangle 110, \quad -6_{16} \quad 1_\triangle 010$$

MSB   MSB
(mag)   (mag)

Note: Since the MSB magnitude digit equals 1 (0), no further scaling is allowed.

15.7 WORD FORMATS

The fixed-point systems developed in this section can be conveniently encoded for storage in a register. For the sake of completeness, the storage protocol used in the Digital Equipment PDP-11 series minicomputer is abstracted in Fig. 15.7.1.

Fixed-point systems offer the user simplicity in coding and performing arithmetic. The shortcoming of this coding scheme is its limited dynamic range. To overcome this problem, a floating-point system was developed. Floating-point numbers have the general form $x = mr^e$, where m is a fixed-point n-bit number referred to as the mantissa and e is a p-bit exponent of the radix r (r = 2 or 16 typically). The value of a floating-point number is generally found in normalized form. For r = 2, a normalized mantissa will be bounded between 1 and 1/2 This means that the MSB data digit is always equal to "1" and $|x| \leq 2^e$, $e \leq 2^{p-1}$. In a PDP-11 environment, a floating-point number is represented as shown in Fig. 15.7.2. Since the MSB data digit for a normalized radix 2 mantissa is always one ("1"), it can be considered to be redundant and is therefore omitted (hidden).

16-BIT SINGLE PRECISION WORD:

15 14	0	S=SIGN BIT
S	NUMBER	S=0 FOR POSITIVE NUMBERS
		S=1 FOR NEGATIVE NUMBERS

32-BIT DOUBLE PRECISION WORD:

15 14	0	
s	HIGH NUMBER PART	NUMBERS IN TWO'S COMPLEMENT

15	0
	LOW NUMBER PART

FIG. 15.7.1 Fixed-point data format for a 16-bit machine.

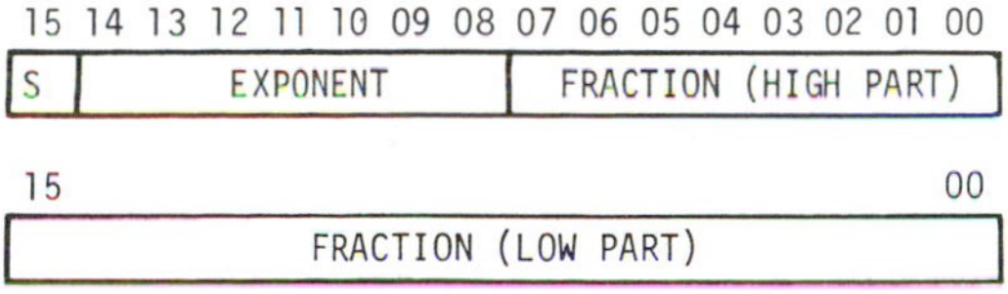

```
15 14 13 12 11 10 09 08 07 06 05 04 03 02 01 00
```

S	EXPONENT	FRACTION (HIGH PART)

```
15                                              00
```

FRACTION (LOW PART)

Exponent = 8 bits in excess (200) notation (signed)
Fraction = 23 bits plus 1 hidden bit (ie; since the MSB data digit of
a normalized fraction is always "1", it is redundant information and
can therefore be ignored)

FIG. 15.7.2 Typical floating-point format for a 16-bit machine.

EXAMPLE 15.7.1 The number +12 (decimal) = 1100 (binary) possesses a
floating-point representation given by

$$12 \text{ (decimal)} = (2^4) \times \left(\frac{3}{4}\right) \text{ (decimal)} = (2^4) \times (0.11) \text{ (fractional)}$$

or

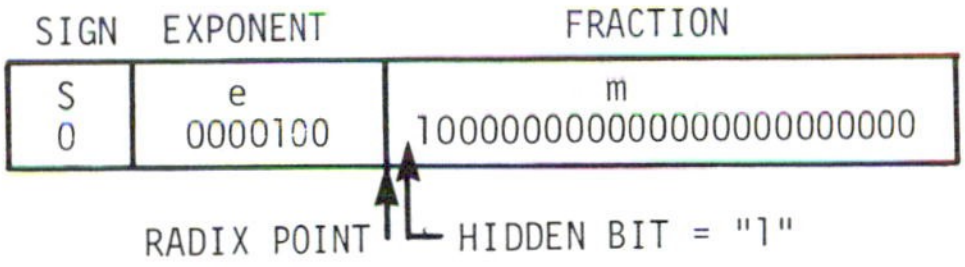

```
SIGN   EXPONENT          FRACTION
```

S	e	m
0	0000100	10000000000000000000000

RADIX POINT ┘└ HIDDEN BIT = "1"

        The floating-point system provides the user with superior dynamic
range at the expense of added system complexity and reduced throughput. A
variation of this theme, which overcomes some of these objections, is
called block floating point. In this system, an array of data is scanned and
its maximal absolute value determined. The exponent for this number is
defined in the usual floating-point way. The exponent, however, is used for
all the other words in the data list. The mantissa's sign and magnitude
digits are computed in accordance with this fixed exponent. Since all data
now have a common exponent (and generally an unnormalized mantissa),
data storage efficiencies can be obtained by ignoring exponent data. Also,
certain arithmetic operations are simplified. These improvements are
bought at a cost of reduced mantissa accuracy, however. The block floating-
point system, due to its fixed exponent, possesses a higher statistical accuracy than an equivalent-word-length fixed-point system.

15.8 UNWEIGHTED NUMBER SYSTEMS

A special branch of modular arithmetic has come to be known as residue
arithmetic. Arithmetic operations based on residue concepts have an

advantage over conventional weighted number procedures (e.g., decimal) in that no "carry information" need be passed between digits. This is not true for weighted systems where the management and propagation of carry bits introduces computational delays. The absence of such information in a residue system admits pure parallel arithmetic data processing and as a result, improved speed.

Residue numbers have been known for centuries. In more modern times, residue arithmetic was formalized by Gauss in the nineteenth century. In 1968, Szabo and Tanaka published a book on residue arithmetic in which they demonstrated the ability to do high-speed algebra. However, the technology of the period could not economically support this work. In particular, specialized table-lookup operations were required. The core-based memory systems of the late 1960s were too expensive to be used in this mode. Today, however, high-speed RAM and ROM provide affordable media for performing residue arithmetic.

Residue arithmetic concerns itself with representing and manipulating integers of the form $y = x \bmod(p)$.[†] The integers represented by a residue system reside over an interval $[0, M - 1]$. The digits of this system are defined in terms of a set of relatively prime numbers[‡] called the moduli set and denoted $P = (p_1, p_2, \ldots, p_L)$. It is the relative independence of the members of the moduli set which "decouples" the residue digits and makes the concept of carry moot. The dynamic range of the system, namely M, can be explicitly computed to be

$$M = \prod_{i=1}^{L} p_i; \quad p_i \in P \tag{15.8.1}$$

From the Euclidian algorithm, there exist integers k_i and x_i such that

$$x = k_i p_i + x_i, \quad i = 1, 2, \ldots, L$$

The quantity x_i is called the ith <u>residue of x</u> and is often denoted $|x|_{p_i}$ or $x \bmod(p_i)$.

It is apparent that the two numbers x and x + kM have identical residue representations. Only if $x \in [0, M - 1]$ can the uniqueness of the residue representation be guaranteed. This representation can be formalized in terms of the L-tuple $x \longrightarrow (x_1, x_2, \ldots, x_L)$, where x_i is the ith residue. This basic representation can be slightly modified to accommodate

---

[†] $y = x \bmod(p)$ is to be read as "y equals x modulo p," where y is the remainder of the quotient x/p. For example, $23 \bmod(10) = 3$.
[‡] Two numbers are relatively prime if their only common divisor is unity.

TABLE 15.8.1 Residue Numbers with Respect to Module 2, 3, and 5

Integer x	Signed–integer	$x \bmod(2) = x_1$	$x \bmod(3) = x_2$	$x \bmod(5) = x_3$
0	0	0	0	0
1	1	1	1	1
2	2	0	2	2
3	3	1	0	3
4	4	0	1	4
5	5	1	2	0
6	6	0	0	1
7	7	1	1	2
8	8	0	2	3
9	9	1	0	4
10	10	0	1	0
11	11	1	2	1
12	12	0	0	2
13	13	1	1	3
14	14	0	2	4
15	−15	1	0	0
16	−14	0	1	1
17	−13	1	2	2
18	−12	0	0	3
19	−11	1	1	4
20	−10	0	2	0
21	−9	1	0	1
22	−8	0	1	2
23	−7	1	2	3
24	−6	0	0	4
25	−5	1	1	0
26	−4	0	2	1
27	−3	1	0	2
28	−2	0	1	3
29	−1	1	2	4

signed numbers over the range $[-M/2, M/2]$. Here the L-type representation is $x \longrightarrow (r_1, r_2, \ldots, r_L)$, where

$$r_i = \begin{cases} x \bmod(p_i) & \text{if } x \geq 0 \\[2mm] (p_i - |x|) \bmod(p_i) & \text{if } x \leq 0 \end{cases} \tag{15.8.2}$$

The signed encoding scheme has been referred to as the p-complement system.

EXAMPLE 15.8.1 Let $P = (2, 3, 5)$; then $M = 30$. The residue codes for the integers over $[0, 29]$ and $[-15, 14]$ are reported in Table 15.8.1.

A dramatic property possessed by the residue number system is its modularity. The arithmetic operation ($\circ$ = addition, subtraction, or multiplication) is given by

$$z \bmod(M) = (x \circ y) \bmod(M) \longrightarrow (t_1, t_2, \ldots, t_L)$$

where $x \longrightarrow (r_1, r_2, \ldots, r_L)$, $y \longrightarrow (s_1, s_2, \ldots, s_L)$, and $t_i = (r_i \circ s_i) \bmod(p_i)$. At first glance, this may seem to be a rather benign result. The importance of this statement should not be underestimated, however. It maintains that the residue representation of $x \longrightarrow y$ can be computed as L separate and parallel operations! Thus residue arithmetic is fundamentally different from that encountered in weighted number systems where carry information must be managed and which are inherently sequential.

EXAMPLE 15.8.2 Let $P = (2, 3, 5)$; then for $x = 9 \longrightarrow (1, 0, 5)$ and $y = 2 \longrightarrow (0, 2, 2)$, it follows that $x + y = 1 \longrightarrow (1, 2, 1)$, or

$$\begin{array}{rccc} x \rightarrow & 1 & 0 & 4 \\ +y \rightarrow & 0 & 2 & 2 \\ \hline & 1 & 2 & 6 \end{array} = (1 \bmod 2, 2 \bmod 3, 6 \bmod 5) = (1, 2, 1)$$

while $xy = 18 \longrightarrow (0, 0, 3)$, or

$$\begin{array}{rccc} x \rightarrow & 1 & 0 & 4 \\ *y \rightarrow & 0 & 2 & 2 \\ \hline & 0 & 0 & 8 \end{array} = (0 \bmod 2, 0 \bmod 3, 8 \bmod 5) = (0, 0, 3)$$

BIBLIOGRAPHY

Korn, G. A. (1977), Microprocessors and Small Digital Computer Systems for Engineers and Scientists, McGraw-Hill, New York.

Peled, A. (1976), On the Hardware Implementation of Digital Signal Processors, IEEE Trans. Acoust. Speech Signal Process., ASSP-24, February, pp. 76-86.

Szabo, N. S., and R. I. Tanaka (1967), Residue Arithmetic and Its Application to Computer Technology, McGraw-Hill, New York.

Taylor, B. L. (1978), Digital Networks and Computer Systems, Wiley, New York.

16
Arithmetic

16.1 INTRODUCTION

Digital arithmetic, in the form of addition and multiplication, is intrinsic to signal processing. The elegant theory found in the earlier chapters would have little or no importance unless a digital computer could efficiently process data algebraically. Computers can do algebra in software, firmware, or hardware. Where cost is the overriding consideration, and throughput requirements are modest, software-based signal processing systems are often found to be the most cost effective. However, as speed requirements increase, firm and hardware realizations are generally required. In the following sections, an introduction to computer algebra is presented. Here addition and multiplication, from a theoretical and hardware viewpoint, are stressed.

16.2 ADDITION (SUBTRACTION)

The realization of an efficient adder is critical to performance of many digital signal processing systems. The concept of digital addition over a binary coefficient field is well known to many computer engineers. However, the average systems-oriented signal processing scientist's knowledge of the structure and constraints associated with high-speed addition may be limited. The basis for most hardware digital adders is the <u>half adder</u> or <u>modulo 2 adder</u>. The half adder, found in Fig. 16.2.1, can be noted to sum its binary-valued inputs and generate carryout information. However, the half adder's inability to accept input carry information limits its general use. To overcome this limitation, the <u>full adder</u> is used. The structure and performance of the full adder is summarized in Fig. 16.2.1.

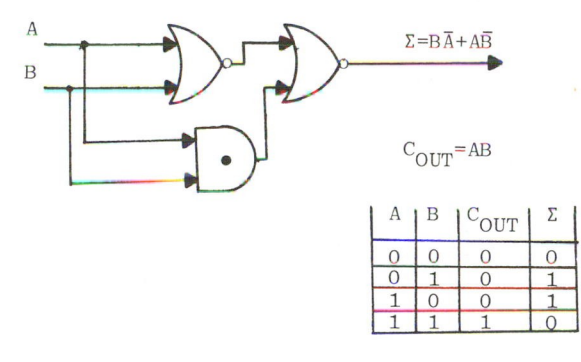

A	B	C_{OUT}	Σ
0	0	0	0
0	1	0	1
1	0	0	1
1	1	1	0

FIG. 16.2.1 Functional diagram of a half adder.

Integrated circuit technology has been able to replace the individual full or half adder with full-adder systems. Full adders can now be found in 4×4, 8×8, and so on, configurations. A typical 4×4 full adder is suggested in Fig. 16.2.3. The delay found between C_{in} and S_3 can range from 15 to 20 nsec in a modern device, while the delays found between A_i (or B_i) and C_{out} typically range from 10 to 20 nsec. The fact that the later delay is shorter will prove beneficial when lookahead adders are considered.

Even though larger-word-length adders are commercially available, there is a practical upper bound to their size. However, long-word-length adders can be configured by interconnecting small-word-length adders. This will be studied in a future section. If versatility is desired, one can turn to the device known as the <u>arithmetic logic unit</u> or ALU. ALUs usually appear in 4-, 8-, or 16-bit architectures. Their speed is related to technology and word length. The ALU offers a user a set of admissible operations which may be executed upon presenting to the device the proper code words. For example, a 4-bit phantom ALU, configured with three select lines, could allow the user to perform the following set of operations:

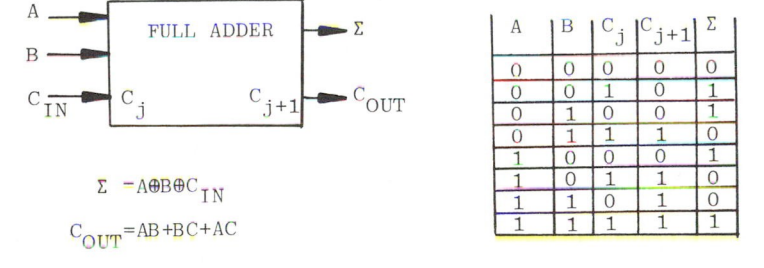

A	B	C_j	C_{j+1}	Σ
0	0	0	0	0
0	0	1	0	1
0	1	0	0	1
0	1	1	1	0
1	0	0	0	1
1	0	1	1	0
1	1	0	1	0
1	1	1	1	1

$$\Sigma = A \oplus B \oplus C_{IN}$$

$$C_{OUT} = AB + BC + AC$$

FIG. 16.2.2 Model and table for a full adder.

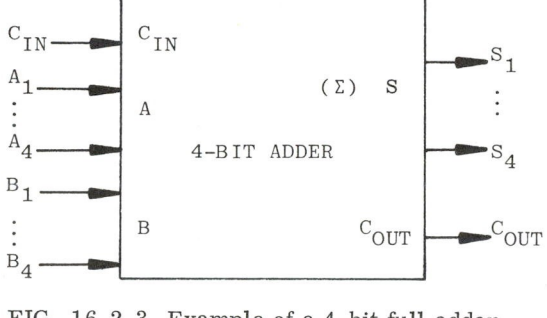

FIG. 16.2.3 Example of a 4-bit full adder.

$$S = A + B, \ \overline{A} + B, \ A + \overline{B}, \ A + B, \ 0, \ 1, \ A \oplus B, \ AB \qquad (16.2.1)$$

Most contemporary 4-bit ALUs would have an instruction repertoire measured in the fifties or hundreds. Detailed study of the ALU will be deferred to a later section.

Before leaving this brief review of adder hardware, a few additional digressions will be undertaken. In some sense, counters may be considered to be a special adder. Counters are available in binary, octal, hexidecimal, and decimal form. They can be used to increment addresses, count periods, or intervals, to generate random sequences, and so forth. Counters can add an offset to the count by presetting registers to a prespecified value. Other counters specialize in up/down counting.

16.3 HIGH-SPEED ADDERS

The signal processing designer is often faced with the problem of supporting high-speed arithmetic. Often these speed requirements exceed that obtainable using general-purpose digital hardware. In such cases, specialized digital devices and techniques are called for.

Relativistically speaking, the slowest adder configuration is the ripple adder. Here carry information is passed (rippled) from one level of the adder to another in the manner suggested in Fig. 16.3.1. In general, $S_i = A_i + B_i + C_i$ and $C_{i+1} = A_i B_i + B_i C_i + A_i C_i$. For an N-bit ripple adder, execution speed is essentially limited to the time delay associated with propagating the carry information from C_{in} to C_{n-1}. This propagation delay can be reduced through the use of algebra by observing that

$$C_{i+1} = A_i B_i + B_i C_i + A_i C_i = A_i B_i + C_i (A_i + B_i) \qquad (16.3.1)$$

which can be further simplified in terms of $G_i = A_i B_i$ (referred to as the carry generator) and $P = A + B$ (referred to as the carry propagate) to read

$$C_{i+1} = G_i + C_i P_i \tag{16.3.2}$$

The carry information presented to the $(i + 1)$st stage of the composite adder is the previous stage's generator G_i and propagation parameter P_i, which is produced with the prior carry bit C_i. If these terms are grouped together into groups having a 4-bit range, then:

1. $S_i = A_i \oplus B_i \oplus C_i \tag{16.3.3}$

2. $S_{i+1} = A_{i+1} \oplus B_{i+1} + [G_i + P_i C_i] \tag{16.3.4}$

3. $S_{i+2} = A_{i+1} \oplus B_{i+2} \oplus [G_{i+1} + P_{i+1} G_{i+1} + P_{i+1} P_i C_i] \tag{16.3.5}$

4. $S_{i+3} = A_{i+3} \oplus B_{i+3} \oplus [G_{i+2} + P_{i+2} G_{i+1}$

$$+ P_{i+2} P_{i+1} G_i + P_{i+2} P_{i+1} P_i C_i] \tag{16.3.6}$$

5. $C_{i+4} = G_{i+3} \oplus P_{i+3} G_{i+2} \oplus P_{i+3} P_{i+2} G_{i+1}$

$$+ P_{i+3} P_{i+2} P_{i+1} G_i + P_{i+3} P_{i+2} P_{i+1} P_i C_i \tag{16.3.7}$$

It is noteworthy to observe that this system of equations has a functional dependence on A, B, and C only. That is, intermediate carries (i.e., C_{i+1} through C_{i+3}) are not <u>explicitly</u> found in Eq. (16.3.7). A dedicated combinational logic circuit can be used to realize the carry parameter C_{i+4} found in this equation. The entire adder network, in this case, would be called a 4-bit <u>carry adder.</u> There are practical limitations to the size a single n-bit carry adder may have. This size is determined basically by device technology and pin-out considerations. However, like the full adder,

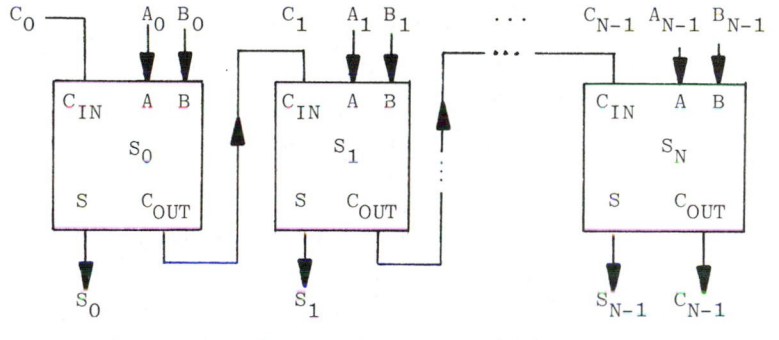

FIG. 16.3.1 System diagram for a set of full adders connected as a ripple adder.

they may be interconnected to form long-word-length adders. This will be
explored momentarily.

Four-bit carry adders can be commonly found in contemporary use.
A typical representative of this group is the AM25L2517 (Advanced Micro-
devices) adder shown in Fig. 16.3.2. This mature device allows the user
to select one of several basic algebraic operation to be performed by properly coding a 3-bit select line. The device outputs the algebraic result (S_1
through S_4), carry information C_{i+4}, and an <u>overflow flag</u> denoted OVR.
The overflow flag will indicate that the algebraic result exceeds the admissible dynamic range of the device. Overflow detection is performed by
investigating the values of the two adjacent high-order carry bits in the
manner suggested by the following example.

EXAMPLE 16.3.1   Assuming that a fractional weighted number system is
used, the sum of -1 (decimal) $\rightarrow 1.000...0$ and -1 (decimal) $\rightarrow 1.000...0$,
would produce -2 (decimal) which is out of range (i.e., overflow). In particular,

$$1.000...0$$
$$+1.000...0$$
$$\overline{10.000...0}$$

It can be observed that $C_S = 0$ and $C_{S+1} = 1$, where C_S and C_{S+1} denote the
carry-in and carry-out (from the sign bit), respectively. Formally, the
overflow state OVR is indicated by

$$OVR = C_s \oplus C_{s+1} \qquad\qquad\qquad (16.3.8)$$

In the example under study, $C_S = C_{i+3}$ and $C_{S+1} = C_{i+4}$. These two binary-
valued digits are OR'ed in the adder chip and presented to the user as the
OVR bit.

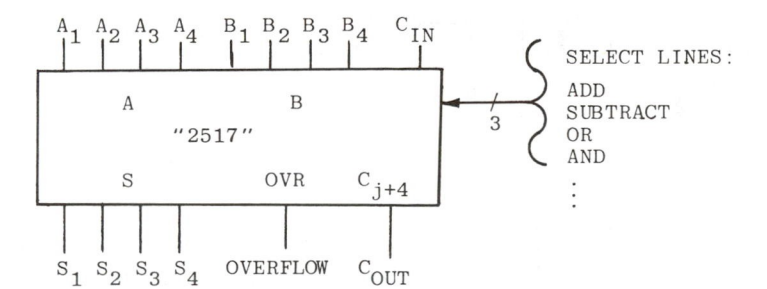

FIG. 16.3.2 Diagram for a commercial 4-bit adder.

16.4 RIPPLE ADDERS

Using adders, such as the 2517, long-word-length adders can be configured.
A 16-bit ripple adder is shown in Fig. 16.4.1. It is a logical extension of
the ripple adders based on 1-bit full-adder modules. Using the n-bit carry
adder, the ripple propagation speed can be reduced and therefore provide a
higher throughput.

Adder speed can be predicted with a reasonably high degree of accuracy by using vendor-supplied data. The execution speed of an adder can
be estimated by quantifying the longest adder propagation delay. Using
typical vendor-supplied parameters, the diagrammed adder can be analyzed
in the following manner:

Path	Output (nsec)			Comments
	S_i	C_{i+4}	OVR	
A_i or B_i to C_{i+4}	36	36	36	Delay associated with the 1st carry (C_4)
C_i to C_{i+4}	22	22	22	Delay in producing 2nd carry (C_8)
C_i to C_{i+4}	22	22	22	Delay in producing 3rd carry (C_{12})
C_i to S_i	23	—	—	Delay in producing MSBs $(S_{12}\ldots S_{15})$
C_i to C_{i+4} or OVR	—	22	22	Delay in producing carry out and OVR data
Predicted speed	103	102	102	

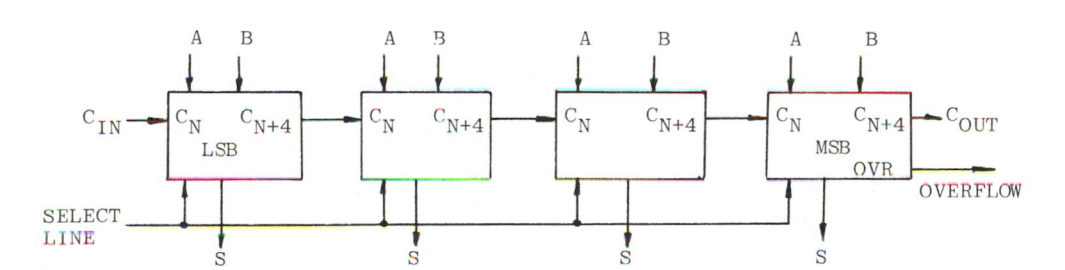

FIG. 16.4.1 Interconnection diagram for a typical ripple adder.

That is, the hypothesized 16-bit addition would require 103 nsec to complete while carry-out and overflow information would be available in 102 nsec. Worst-case predictions can be made if the vendor supplies maximal delay data in addition to that considered to be typical.

16.5 LOOKAHEAD ADDERS

Higher addition speeds can be obtained by using carry lookahead adders which accelerate the carry propagage process. These devices provide the user with an output data string (sum), a generator bit (G), and propagater bit (P). Such a device is suggested in Fig. 16.5.1. Within this device is high-speed electronic logic system which produces the carry lookahead data bits G and P. Carry lookahead information is produced in the following manner:

$$C_{i+4} + G + C_i P \text{ [function of G, P, and } C_i \text{ (carry into the adder)]}$$

$$G = G_{i+3} + P_{i+3} G_{i+3} + P_{i+3} P_{i+2} G_i + P_{i+3} P_{i+2} P_{i+1} G_i$$

$$P = P_{i+3} P_{i+2} P_{i+1} P_i \tag{16.5.1}$$

where G_i and P_i are given in Eq. (16.3.2). It can be noted that the system of equations (16.5.1) is defined in terms of internally derived data and one external variable (i.e., the carry-in C_i). As a result, a special combination logic network known as a <u>lookahead generator</u> can be configured to mechanize the carry generation and propagation process. For example, the AM381 carry adders, similar to those discussed in the preceding section, can be outfitted with an AM2902 lookahead generator. The resulting system is shown in Fig. 16.5.2. Referring to this figure, it can be noted that the last 4-bit stage of this 16-bit lookahead adder is functionally

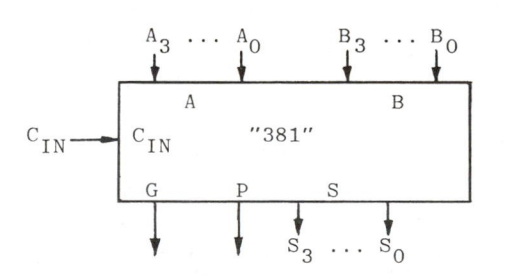

FIG. 16.5.1 Diagram for a commercial-grade carry lookahead adder module.

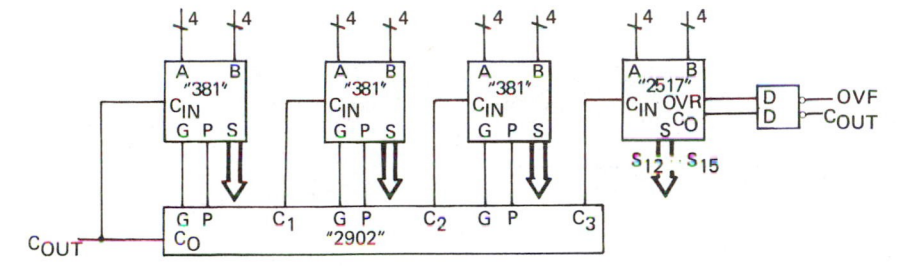

FIG. 16.5.2 System diagram for a complete 16-bit carry lookahead adder.

different from its predecessors. This unit, shown as the AM2517, was discussed in the preceding section. The other 4-bit adder sections have been
configured with a G and P output not found in the ripple adder configuration.
The G and P data are manipulated by the lookahead generator, which returns
the needed carry information back to the 4-bit adder sections with minimum
delay. Finally, overflow and carry-out information is processed from data
available at the last stage of the adder system.

The execution speed of this typical lookahead adder can be computed
by determining the system's propagation delays. Using suggested vendor-
supplied data, the following analysis would result:

Path	S_i (nsec)	C_{i+4} (nsec)	OVR (nsec)	Comments
A_i or B_i to G and P	27	27	27	Setup time for lookahead generator
G_i or P_i to C_{i+4}	10	10	10	Lookahead carry generation delay
C_i to S_i	23	—	—	*: Delay in producing S_i's
C_i to C_{i+4} or OVR	—	22	22	#: Delay in producing OVR
	60*	59#	59*	Delay

Lookahead adders can be nested to extend their word length as well.
For example, using a 260-mW Texas Instruments 74182 lookahead carry
generator, long-word-length adders can be configured. This chip was designed to support a I^2L 4-bit slice microprocessor whose ALU contains a
4-bit-wide adder. A 16-bit and 64-bit carry lookahead adder, based on this

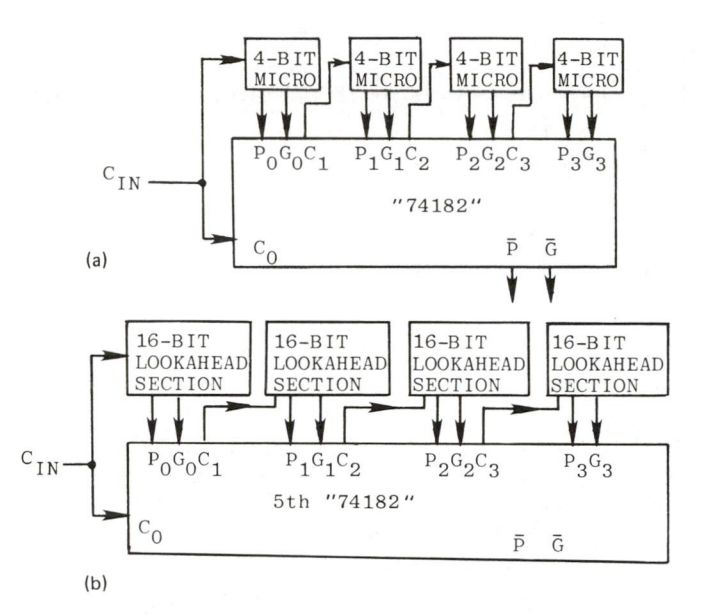

(a)

(b)

FIG. 16.5.3 Realization of (a) 16- and (b) 64-bit carry lookahead adders.
(Note: Presence of P and G carry and generator bits generated by lookahead
unit.)

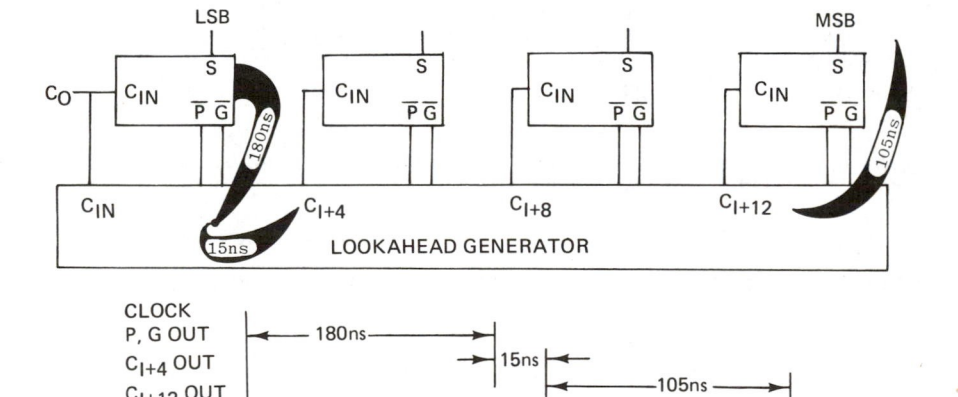

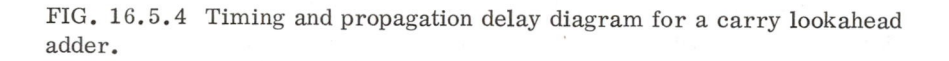

FIG. 16.5.4 Timing and propagation delay diagram for a carry lookahead
adder.

technology, are summarized in Fig. 16.5.3. It can be noted that the 16-bit version requires one lookahead generator (one level), whereas the 64-bit version requires five (two levels). The nesting of these lookahead sections can be conducted to any desired depth. However, these devices are costly and have a large power appetite. Therefore, the designer must address the usual cost-complexity-performance trade-off questions.

EXAMPLE 16.5.1 Assume that the delay between a system clock phase and the generation of P and G on the ALU chip is on the order of 180 nsec with a 15-nsec delay between the receipt of P and G and the production of C_{i+4}. The last stage of a 4n-bit lookahead adder can produce the final 4 bits of the sum once its carry-in has been received. It shall be assumed that the delay between C_{IN} and the availability of the sum bit S and overflow flag OVR is on the order of 105 nsec. Then the following analysis can be made (see Fig. 16.5.4):

Critical time		Word length (nsec)				
		4	8	16	32	64
$C_{IN} \to S$: 105 nsec	Ripple adder	240	240 +60 300	240 105 60 405	240 105 (5)60 645	240 105 13(60) 1125
CLK $\to$ S: 240 nsec CLK $\to \overline{P}, \overline{G}$, 180 nsec	One-level lookahead	N.A.	300	300	300 105 405	300 (2)60 105 525
ALU carry-in ALU carry-out: 60 nsec	Two-level lookahead	N.A.	N.A.	N.A.	300 15 315	300 15 315

16.6 MULTIPLIERS

Most digital arithmetic units possess at least rudimentary fixed-point addition, subtraction, and shift-rotate (i.e., arithmetic shifts). However, multiplication is generally not found in an ALU's lexicon. Instead, multiplication must be synthesized from these more primitive operations using soft- or firmware support. Microcoded multiplication routines will often provide the throughput needed to support many signal processing tasks. For example,

TABLE 16.6.1 Summary of Arithmetic Speeds Obtained from Common Mini- and Microprocessors

Digital Equipment Corporation model	Floating point arithmetic (μsec)		
	Add	Multiply	Divide
LSI-11 (EPA microcode)	42	74	
11/03	97	90	
11/40	25	22	51
11/55 (with core)	6.6	8.2	9.2
11/55 (with bipolar)	5.0	6.6	7.6
11/60 (FPE11-E support)[a]	2.0	1.0	6.8

[a]The FPE11-E microcode option has been optimized to do multiplication at the expense of division.

the Digital Equipment Corporation's LSI-11 system provides the user with a 16-bit fixed-point microcoded multiplication time of 24 to 35 μsec (typical) with a worst-case time given to be 64 usec.[†] Fixed-point division has a worst-case delay of 78 μsec. Floating-point operations, because of their complexity, require longer intervals of time, as the data found in Table 16.6.1 would indicate.

If higher instruction rates are required, hardware realizations are required. Over a binary field, multiplication is formally defined to be

$$(1) \ (1) = 0, \quad (0) \ (1) = 1, \quad (1) \ (0) = 1, \quad (0) \ (0) = 0 \qquad (16.6.1)$$

It is well known that multiplication can be performed as a series of nested additions and that the product of two n-bit words is 2n bits in length. Because of the last consideration, a full-precision multiplier would require its output register be twice the length of the input registers.

16.7 SHIFT-ADD MULTIPLY

The mechanics of a shift-add multiplication algorithm can best be dramatized through the following 4-bit example. It will be considered that a

_____

[†]The FP11-E microcode has been optimized for performing multiplication.

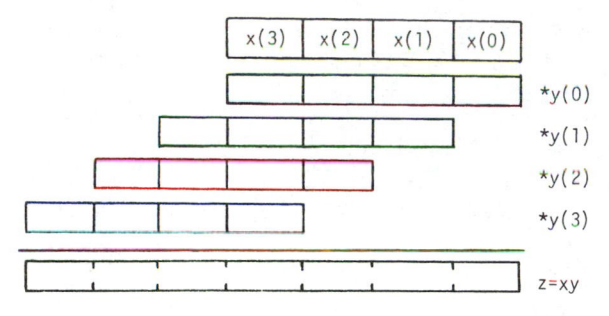

FIG. 16.7.1 Functional model of a simple shift–add multiplier.

multiplicand x and multiplier y are two positive unsigned 4-bit words having the following representations:

$$x = \sum_{i=0}^{3} x[i]2^i; \quad y = \sum_{i=0}^{3} y[i]2^i \qquad (16.7.1)$$

The product of x and y is given by z = xy, where

$$z = \sum_{i=0}^{3} x(y[i]2^i) = \sum_{i=0}^{3} (x[i]2^i)y \qquad (16.7.2)$$

is graphically interpreted in Fig. 16.7.1. Since a weighted binary-valued numbering system is used, the term $x2^k$ is formed by simply shifting x k bits to the left, as indicated in the diagram.

EXAMPLE 16.7.1 Let x = 10(decimal) ⟶ 1010 and y = 5(decimal) ⟶ 0101; then

Step	Multiplier weight y_i	Accumulator content (bits)		Comment
0	N.A.	0000	0000	Clear
1	1	+	1010	Copy y
		0000	1010	
2	0	+0	000	Add 0
		0000	1010	
3	1	+10	10	Copy 2^2y
		0011	0010	
4	0	+000	0	Add 0
	3 =	0011	0010 =	50(decimal)

The process of division can similarly be illustrated by observing that for a dividend, say $x = 55$(decimal) $\rightarrow 110111$ and a divisor $y = 5$(decimal)$\rightarrow$ 1010, long division produces

$$
\begin{array}{r}
1\ 0\ 1\ 1 \ \ (\text{quotient}) = 11(\text{decimal}) \\
1\ 0\ 1\ \overline{\big)\ 1\ 1\ 0\ 1\ 1\ 1} \\
\underline{1\ 0\ 1} \\
0\ 0\ 1\ 1 \\
\underline{0\ 0\ 0\ 0} \\
1\ 1\ 0 \\
\underline{1\ 0\ 1} \\
0\ 1\ 0\ 1 \\
\underline{0\ 1\ 0\ 1} \\
0
\end{array}
$$

The algebraic structure of the long-division model suggests that division is nested subtraction.

Shift-add multipliers can be directly realized in hardware. The general form of such a multiplier is abstracted in Figure 16.7.2. This structure is sometimes referred to as a sequential multiplier for unsigned data. The functional modules found in this figure are generally MSI or LSI units similar to the 4-bit adder units previously studied. The key features of this multiplier are the extended-size output register, a binary-valued multiplexer, and an adder which is configured as an accumulator.

EXAMPLE 16.7.2 Let $x = y = 3$(decimal) $= 0011$; then

Step		A	B	A + B	Accumulator
1	1 +	0000	0011	0.0011	00011001
2	1 +	0001	0011	0.0100	00100100
3	0 = NOP	0010	0000	0.0010	00010010
4	0 = NOP	0001	0000	0.0001	00001001 = 9(decimal)

This oversimplification may give the reader a false impression of the technical difficulties associated with hardware-based multiplication. Fortunately, as long as an n-bit multiplier is outfitted with a 2n-bit output register, overflow cannot occur. However, the question of algebraic sign has yet to be addressed. For example, suppose that $x = y = -3$ are coded in 2's-complement. At the end of the last shift-add cycle, the output register will contain $10110100 \neq 9$(decimal). Therefore, corrective measures must be taken. One obvious remedy is to recode all 2's-complement data into sign-magnitude form and derive the sign from the following rule:

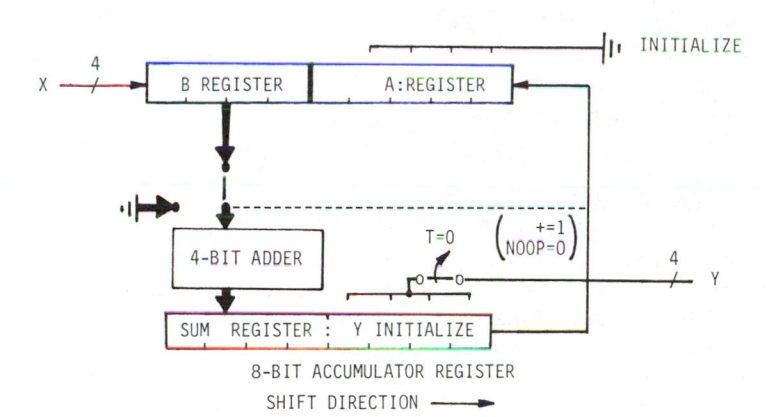

FIG. 16.7.2 Diagram of a sequential multiplier configured with a double-word-length-product register.

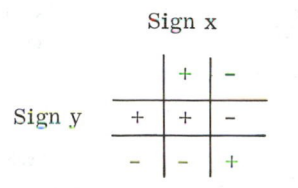

This simple sign generation rule can be easily mechanized with an EXOR gate.

EXAMPLE 16.7.3 Let $x = -y = 3$(decimal). In 2's complement $x \longrightarrow 0_\wedge 11$ and $y \longrightarrow 1_\wedge 01$, and:

Step 1: Multiply $|x||y|$ in sign-magnitude form:

$$
\begin{array}{r}
|x| = 3 \longrightarrow 0\ 1\ 1 \\
|y| = 3 \longrightarrow 0\ 1\ 1 \\
\hline
0\ 1\ 1 \\
0\ 1\ 1 \\
0\ 0\ 0 \\
\hline
{}_\wedge 0\ 1\ 0\ 0\ 1 = 9\text{(decimal)}
\end{array}
$$

Step 2: Compute sign, sign $= 0 \oplus 1 = 1$ (negative).

Step 3: Convert 9(sign-magnitude) to 9(2's complement):

$$
\begin{array}{r}
-9_{10} \longrightarrow {}_\wedge \overline{0\ 1\ 0\ 1} = {}_\wedge 1\ 0\ 1\ 1\ 0 \\
+ {}_\wedge 0\ 0\ 0\ 0\ 1 \\
\hline
{}_\wedge 1\ 0\ 1\ 1\ 1 \quad \text{in 2's complement}
\end{array}
$$

Step 4:   Add sign bit:

$$1_{\wedge}1\ 0\ 1\ 1\ 1_{2's} \longrightarrow -9(\text{decimal})$$

16.8 BOOTH'S ALGORITHM

An alternative approach to the recoding problem previously studied is to use Booth's algorithm. This method differs from the sequential multiplication scheme by virtue of the fact that it can directly accommodate signed data and manages the accumulator shifts at the end of each multiplier cycle. Booth's algorithm is premised on the observation that a string of logical o's require no formal add-to-accumulator transactions. The only operation required under such a hypothesis is the shifting of the contents of the accumulator a sufficient number of locations upon receipt of a string of 0's. Returning to the previous example, one notes that at steps 3 and 4, the select line input is 0. The net effect of this was to shift the contents of the accumulator 2 bits but leave it otherwise unaltered. If a string of 1's are encountered, starting at the rth-bit location and ending at the sth bit ($r \le n$ and $s > 0$), its contribution to the product $z = xy$ is z^*, where

$$z^* = \sum_{i=r}^{s} 2^{n-i}x \qquad (16.8.1)$$

The digital engineer is familiar with this result and often expressed it as

$$z^* = (2^s + 2^{s-1} + \cdots + 2^r)x = (2^{s+1} - 2^r)x \qquad (16.8.2)$$

This algebraic result can be simply demonstrated in terms of the following binary-valued experiment. Let $a = 2^{s+1}$ and $b = 2^r$; then $a - b$ is

$$\begin{array}{ll} & \text{s+1st location \quad rth location} \\ 2^{s+1}_{10} = & 000...0\ 1\ 0\ 0\ 0\ 0\ 0\ 0\ 0\ 0...00 \\ -2^{r}_{10} = & \underline{-000...0\ 0\ 0\ 0\ 0\ 0\ 0\ 1\ 0...00} \\ & 000...0\ 0\ 1\ 1\ 1\ 1\ 1\ 1\ 1\ 0...00 \end{array}$$

Here, in a Booth's sense, a string of 1's is treated as a subtraction of x when the first element in the string is encountered. This is followed by the addition of $2^{s+1}x$ to the accumulator after the string of 1's are extinguished. At intermediate locations, within the string of 1's, no action is needed. Therefore, a number of intermediate operations required to perform a multiplication can possibly be reduced and throughput increased.

Formally, Booth's algorithm is defined in terms of the following select table (K = accumulator):

y_{i-1}	y_i	Fraction	Partial product	
0	0	Do nothing	K	K + 0
1	0	Add x	K	K + A
0	1	Subtract X	K	K - A
1	1	Do nothing	K	K + 0

EXAMPLE 16.8.1 Consider again Example 16.7.3, where $x = y = -3$ in 2's-complement form. Using Booth's algorithm the product $z = xy$ may be computed directly. As in the case of the sequential multiplier, data in the system's shift register will be moved one biy upon the completion of each multiplication cycle. Recall that the problem of scaling is solved through the use of right and left shifts. If a 1's- or 2's-complement system is used, the sign bit must be copied into the MSB location during this operation. Mechanically, to realize a Booth's multiplier, the previously considered 4-bit rendering of y must be augmented to 5 bits through the addition of y_{-1}. The 5-bit representation of y will be modeled by $y \longrightarrow [y_{-1}, y_0, y_1, y_2, y_3]$. These data will be presented to the Booth's truth table in a pairwise sense and the resulting function (control) will be performed. This is summarized in Fig. 16.8.1. Again considering $x = y = -3$, the accumulator contents are given by ($y_{-1} = 0$)

Step	Select	Event	Accumulator	Accumulator after shift
1	01	K $\longrightarrow$ K - X	00110000	00011000
2	10	K $\longrightarrow$ K + X	00011000 +1101 —————— 11101000	Note the copying of sign bit in the 2's complement shift-scaling operation 11110100
3	01	K $\longrightarrow$ K - X	11110100 +0011 —————— 00100100	00010010
4	11	K $\longrightarrow$ K	No op	$00001001 = 9_{10}$

A variation of Booth's algorithm is the modified Booth's algorithm. The select table for this algorithm is given in Table 16.8.1.

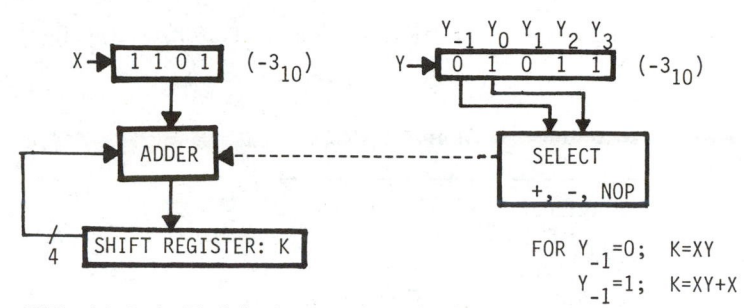

FIG. 16.8.1 Model of a Booth multiplier.

16.9 CARRY SAVE MULTIPLIER

Another method used to realize high-speed multipliers can be motivated in terms of the 4 × 4 sequential multiplication table for unsigned data given in Fig. 16.9.1. Multiplication speed is essentially limited by the ripple delay indicated in this figure. Here the delay consists of a nesting of adder ripple delays. In particular, the delay exists from p_{00} through p_{03} and then diagonally down to s_y. One could obviously improv the throughput of the multiplier if the last level was configured as a lookahead adder. Nevertheless, the carry propagation delay time to the last level still limits the multiplier's performance. To overcome this problem, a <u>carry save</u> adder may be used instead of the previously considered carry propagated adder. These adders use a flip-flop to save carry information. As a result, the long propagation delays associated with rippling can be reduced.

Multipliers configured using carry save adders are called carry save multipliers. This philosophy accelerates the carry information to the last level by using path b of Fig. 16.9.1 rather than the slower path, a. Once the carry information is presented to the last level, high-speed lookahead addition can take place.

EXAMPLE 16.9.1 Using carry-save adders, the multiplication of x = y = 11 would proceed in the manner suggested in Fig. 16.9.2.

16.10 OTHER ALGORITHMS

A third type of high-speed multiplier structure is in common use. Called the Wallace tree, it uses a device called a <u>pseudoadder</u>. These units add three numbers together and output both sum and carry information (see full adder with carry input used as the third entry). The advantage of this structure is its ability to minimize carry propagation path lengths. As a result, high-speed operation is obtainable. Data are grouped together as three-tuples and interconnected in a tree structure. The tree has come to

TABLE 16.8.1 Input–Output Mapping for a Modified Booth Multiplier

Y_{i-1}	Y_i	Y_{i+1}	Event
0	0	0	$K \longrightarrow K$
1	0	0	$K \longrightarrow K + X$ (end of string)
0	1	0	$K \longrightarrow K + X$
1	1	0	$K \longrightarrow K + 2X$ (end of string)
0	0	1	$K \longrightarrow K - 2X$ (beginning of string)
1	0	1	$K \longrightarrow K - X$
0	1	1	$K \longrightarrow K - X$ (beginning of string)
1	1	1	$K \longrightarrow K$

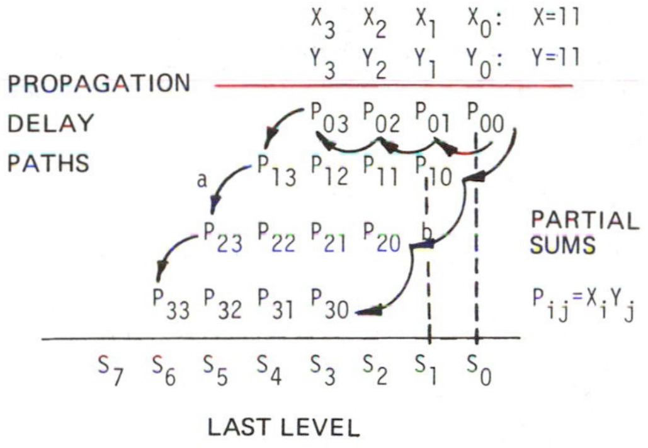

$$X_3 \quad X_2 \quad X_1 \quad X_0: \quad X=11$$
$$Y_3 \quad Y_2 \quad Y_1 \quad Y_0: \quad Y=11$$

PROPAGATION
DELAY
PATHS

$$P_{03} \quad P_{02} \quad P_{01} \quad P_{00}$$
$$a \quad P_{13} \quad P_{12} \quad P_{11} \quad P_{10}$$
$$P_{23} \quad P_{22} \quad P_{21} \quad P_{20} \quad b$$
$$P_{33} \quad P_{32} \quad P_{31} \quad P_{30}$$

PARTIAL
SUMS

$$P_{ij} = X_i Y_j$$

$$S_7 \quad S_6 \quad S_5 \quad S_4 \quad S_3 \quad S_2 \quad S_1 \quad S_0$$

LAST LEVEL

FIG. 16.9.1 Model for the sequence of operations found in a sequential multiplier. *: 1 + 1 + 0 implies x = 0 and c = 1. **: 1 + 1 + 1 implies s = 1 and c = 1.

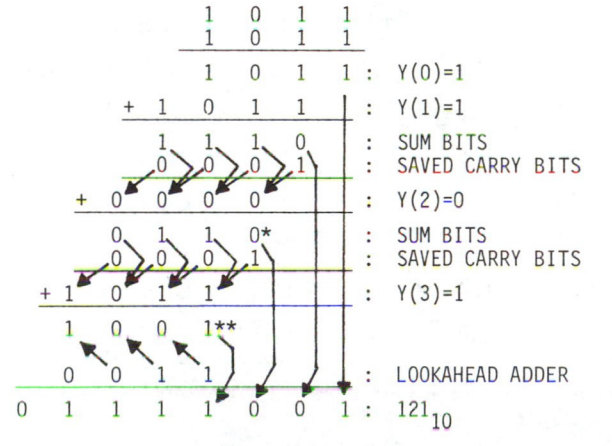

1	0	1	1	
1	0	1	1	

```
          1   0   1   1  :  Y(0)=1
    +  1   0   1   1     :  Y(1)=1
      1   1   1   0      :  SUM BITS
       0   0   0   1     :  SAVED CARRY BITS
  +  0   0   0   0       :  Y(2)=0
      0   1   1   0*     :  SUM BITS
       0   0   0   1     :  SAVED CARRY BITS
  + 1   0   1   1        :  Y(3)=1
    1   0   0   1**
     0   0   1   1       :  LOOKAHEAD ADDER
  0   1   1   1   1   0   0   1  :  121₁₀
```

$$0 \quad 1 \quad 1 \quad 1 \quad 1 \quad 0 \quad 0 \quad 1 : \quad 121_{10}$$

FIG. 16.9.2 Model for the sequence of operations found in a carry-save multiplier.

be known as the <u>Wallace tree.</u> Wallace trees will be reinvestigated in the
multiplier array section. Here long-word-length multipliers will be configured by networking smaller multiplier modules.

A fourth multiplier strategy is based on direct table lookups. Here
the product of two n-bit words are interpreted as a 2n-bit address to a high-
speed RAM or ROM. The memory unit is programmed to output the product
of the two inputs (usually rounded). For example, if $n = 4$, all possible
products can be precomputed and stored on a $2^8 \times 4 = 1024$-bit fast-access
memory unit. It can be noted that, because of table size restrictions,
longer-word-length lookup multiplies (i.e., 8-bit or greater) are physically
unrealizable with today's technology. However, the 4-bit lookup units provide excellent support for the new generation of 4-bit slice microprocessors.

16.11 COMMERCIAL DEVICES

Generally speaking, the digital designer will not be involved in the detailed
design of a multiplier module or chip. Rather, the concern of the reader is

TABLE 16.11.1 Summary of Some Commercially Available
Multiplier Units

Vendor	Size	Pins	Speed (nsec)	Power (W)	Coding	Algorithm
AMD	2×4	24	25	0.6	2's	Booth
AMD	8×1	16	50	0.5	2's	Booth
AMD	8×8	40	200	1.0	2's	Modified Booth
MMI	8×8	40	100	1.0	2's/uns	Modified Booth
MMI	16×16	24	800	1.0	2's	Modified Booth
MMI	8×8	20	400	0.7	2's	Modified Booth
Motorola	2×4	24	20	0.8	2's	Booth
TI	4×4	20	50	0.5	uns	Lookup
TRW	8×8	40	45	1.0	2's	Carry-save
TRW	12×12	64	80	2.0	2's	Carry-save
TRW	16×16	64	100	3.0	2's	Carry-save
TRW	24×24	64	200	4.0	2's	Carry-save

Note: AMD = American Micro Devices, TI = Texas Instruments, MMI =
Monolithic Memories, uns = unsigned.

how to integrate them into a design having specified speed, chip complexity, word length, power, and cost side constraints. As a guide to the reader, a brief review of commercially available multiplier modules is presented in Table 16.11.1. The entries in this list are to be interpreted as a guide. New higher-performance units continually enter the marketplace. The reader can remain current on this subject only through the study of the trade literature and liaison with the major semiconductor vendor representatives.

It may be noted that the multiplier algorithms presented to this point can be classified as being sign or unsigned routines. Most programmable or dedicated arithmetic systems are designed to work in one of these two modes. In a signal processing application, input is most conveniently interpreted in fractional form (i.e., $|x(n)| \leq 1$). The product of two unsigned fractional numbers is itself bounded between zero and unity. However, care must be taken in defining the location of the binary point when multiplying signed fractional numbers. If the binary point is placed immediately following the sign bit, the resulting product will be in error by a factor of $1/2$. The correct answer can be effected by scaling by $1/2$, which involves a simple arithmetic shift.

EXAMPLE 16.11.1

Unsigned × unsigned	Signed fraction × signed fraction

$\dfrac{3}{4} \times \dfrac{3}{4} = \dfrac{9}{16}$

$\dfrac{3}{4} \longrightarrow 0.11$

$(0.11) \times (0.11) = 11$

$\dfrac{11}{0.1001} = \dfrac{9}{16}$ (decimal)

$\dfrac{3}{4} \times \dfrac{3}{4} = \dfrac{9}{16}$

$\dfrac{3}{4} \longrightarrow 0.11$

Step 1: multiply magnitudes

$$\begin{array}{r} 0\ 1\ 1 \\ 0\ 1\ 1 \\ \hline \wedge 0\ 1\ 0\ 0\ 1 \end{array}$$

Step 2: sign bit $0 \oplus 0 = 0$(positive)

Step 3: $\overline{x}\,\overline{y} \longrightarrow 0 \underset{\wedge}{.}\ 0\ 1\ 0\ 0\ 1$

(Decimal) $\longrightarrow \left(\dfrac{1}{4} + \dfrac{1}{32}\right) = \dfrac{9}{32} = \dfrac{1}{2} xy$

Another minor problem is associated with the accuracy of a rounded product. Multiplying two n-bit fraction 2's-complement words results in a 2n-bit product. To maintain a common word length, this product is usually rounded or truncated to n bits. Truncation will produce a product having an expected error of $-Q/2$, where Q is the quantization step size. This statistic has an accumulative effect in digital filters and transforms. For example, if inner products of the form

$$y(n) = \sum_{i=0}^{N-1} a_i x(n - i) \qquad (16.11.1)$$

are individually truncated, the expected error can be shown to be equal to $\log_2(N) * (-Q/2)$. If rounding is used, the expected error is zero. However, rounding is an overhead bearing operation and therefore may affect throughput.

16.12 MONOLITHIC MULTIPLIERS

For expository reasons, a typical $n \times n$ multiplier chip will be studied. The theme chip is modeled in Fig. 16.12.1 and consists of a three-tuple input of n-bit words, namely (x, y, k), and a 2n-bit output given by $z = xy + k$. In a 2's-complement system, the input space is generally defined in terms of an a-bit x, b-bit y, and c-bit k, having the representations

$$x = X - x_s 2^{a-1}, \quad y = Y - y_s 2^{b-1}, \quad k = K - k_s 2^{c-1} \qquad (16.12.1)$$

where the subscript s denotes sign and the capitalized variable indicates magnitude digits [e.g., $x = -10$ (decimal) implies $x = 6 - (16)$ or $x = 6$, $x_s = 1$]. It immediately follows that $z = xy + k$ satisfies

$$z = x_s y_s 2^{a+b-2} - X y_s 2^{b-1} - Y x_s 2^{a-1} + XY - k_s 2^{c-1} + K \qquad (16.12.2)$$

For the special case where $a = b = c = n$, Eq. (16.12.2) degenerates to

$$z = x_s y_s 2^{2n-2} - (X y_s + Y x_s + k_s) 2^{n-1} + XY + K \qquad (16.12.3)$$

A point of potential concern is that the product $z = xy + k$ may overflow a 2n-bit output register. This is not the case with 2's complement, however, It can be noted that the maximal value of K is at most $1/2^{n-1}$ of its full-scale dynamic range. Note also that the maximal value of z for $k = 0$ is $+1$ (resulting from the multiplication of -1 and -1). However, in 2's complement, -1 is characterized by $x_{-s}(y_s) = 0$. Under this hypothesis, $z_{max} = 2^{2n-2}$ is not in an overflow state. For k nonzero, the maximal value of z is obtained if $x = y = -1$ and $k = 1$. In 2's complement, $k = 1$ is equivalent to $k = 2^{n-2} - 1$. Collecting these terms together, it follows that $z_{max} = 2^{2N-2} + 2^{N-2} - 1 \le 2^{2N} - 1$. Therefore, even under the worst conditions, the output register will not overflow.

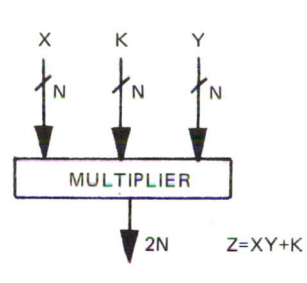

FIG. 16.12.1 Diagram representing a common (n × n)-bit multiplier and a double-precision 2n-bit product.

The utility of the offset parameter k is application determined. This parameter is normally used to support nested multiplications such as found in Section 16.16 in multiplier arrays. In this mode, the less significant digits from earlier partial products will be presented to the modular multiplier chip in terms of an offset constant k. Also, the parameter k can be used to force the rounding of a product xy. Operating in this mode, truncating the 2n-bit product x = xy + k to its n most significant bits is equivalent to rounding up.

The parameter k is sometimes used in polynomial manipulation. For example, consider the Chebyshev interpolating polynomial for $\cos(\phi)$ given by

$$\cos(\phi) = 0.99995795 - 0.49924045(\phi)^2 + 0.03962764(\phi)^4$$

which is of the form $a(x) = b + c(x) + d(x)^2$ with $(\phi)^2 = x$. The polynomial can be realized with the architecture suggested in Fig. 16.12.2. However, care must be taken to ensure that data errors do not occur because of finite-word-length effects.

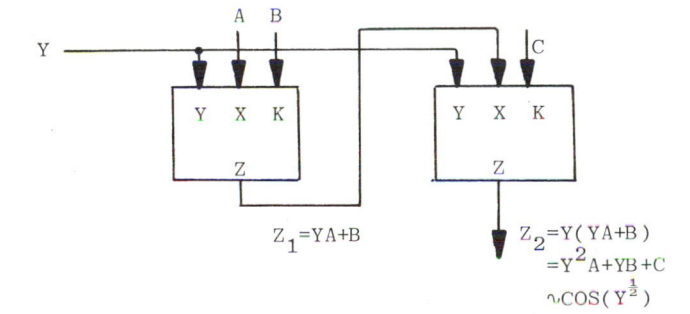

FIG. 16.2.2 Chebyshev interpolation realized using two multipliers of the form found in Fig. 16.12.1.

EXAMPLE 16.12.1 Referring to Fig. 16.12.2, consider $y = 0.5$ and $n = 5$ bits. Then

$y = 0.5$(decimal) $\longrightarrow 0.100$

$a = 0.03962644$*decimal) $\longrightarrow 0.001 \longrightarrow 0.0625$(decimal)

$b = -0.49924045$(decimal) $\longrightarrow 1.100 \longrightarrow -0.5$(decimal)

$c = 0.9995$(decimal) $\longrightarrow 0.111 \longrightarrow 0.9375$(decimal)

Step	Event
$t = 0$	$z = 0$
$t = 1$	$z_1 = (a_5 y_5 + b_5)_5 = [(0.0625)(0.5) + 2^{-4}(-0.5)]_5 = 0$
$t = 2$	$z_2 = [((a_5 y_5 - b_5)_5 (y_5 + c_5)_5$
	$\quad = [0(0.5) + 2^{-4}(0.93750]_5$
	$\quad = 0.625$(decimal)
	$\quad \neq \cos(\sqrt{y}) = 0.74$

Obviously, the error found in this example is due to the restricted word length of k. A more satisfactory architecture is suggested in Fig. 16.12.3

EXAMPLE 16.12.2 Consider again the data given in Example 16.12.1. Then, referring to Fig. 16.12.3, one notes:

Step	
$t = 0$	Clear, load x
$t = 1$	$z \longrightarrow (1C_5 + RND)_5 = (0.11100001 + RND)_5 = (0.1110) = 0.875$(decimal)
	$s \longrightarrow z$
$t = 2$	$z \longrightarrow (0.25B_5 + RND)_5 = (1.0010) = -0.125$(decimal)
	$s \longrightarrow z + 5 = (0.1100) = 0.25$(decimal)
$t = 3$	$z \longrightarrow (0.0625A_5 + RND)_5 = (0.0000) = 0$(decimal)
	$s = \ z + 5 = (0.1100) = 0.75$(decimal) $\simeq \cos(0.5) = 0.74$

The utility of k can be extended into other application areas as well. Consider the problem of designing shift-invariant digital filters. A subsection

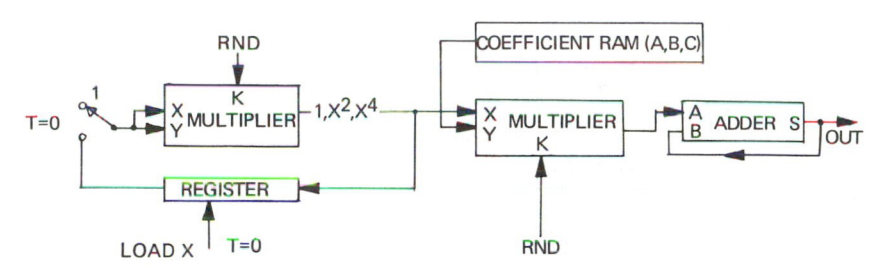

FIG. 16.12.3 Alternative realization of a Chebyshev interpolation.

of the filter may be required to perform the operation $s = xy + z$, where x and y are filter parameter and variables and z is the contents of an accumulator (i.e., sum of products). The system abstracted in Fig. 16.12.4 has been shown to be useful, under certain constraints, in performing this familiar filter task. Referring to the figure, it is assumed that s_0 is the sum of xy and the low-order bits of z. The output s_1 is formed by adding to s_0 the high-order bits of z. It should be noted that by truncating $s_0 = 2^n(xy)_{HI} + (xy)_{LO} + z_{LO}$ to its n MSBs, the effect of z_{LO} is that of rounding the product xy upward.

Another variation is found in Fig. 16.12.5. This configuration can support certain recursive filtering operations. The network can best be analyzed in terms of the following simple example.

EXAMPLE 16.12.3 Let $x = -7/8$, $y = 1/2$, and $k = 5/16$. Then

$$x = -\frac{7}{8} \text{ (decimal)} \longrightarrow 1.001$$

$$y = \frac{1}{2} \text{ (decimal)} \longrightarrow 0.100$$

$$k = -\frac{15}{16} \text{ (decimal)} \longrightarrow 1.1011000$$

$$z = xy + k = -\frac{3}{4}$$

$$x: 1.001$$

$$y: 0.100$$

$$z: 1.1001000 = -\frac{7}{16} \text{ (decimal)}$$

Note that in the example, the sign bit K_s is presented as data bit k_s, which is not the sign bit location of the data field for $k = (k_0, \ldots, k_6, k_s)$. Since

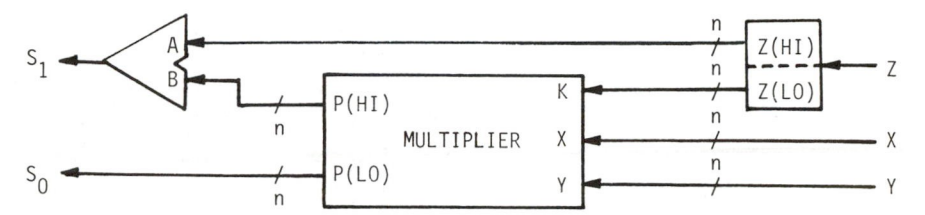

FIG. 16.12.4 Architecture for computing the multiply-add form $xy + k$.

$k_3k_2k_1k_0 = 1000$, the multiplier assumes that $K = 1.000$ or $K = -1$(decimal). However, K affects only the low-order bits of S. In this case $K = 1$ appears in S with weight $-2^{-4} = -1/16$. Therefore, $S = -7/16 - 1/16 = -8/16$. The high-order bits of k, namely $k_7k_6k_5k_4 = 1101$, equal $-3/8$(decimal). A direct sum of S and $-3/8$ yields $-14/16$, which is not the desired answer of $xy + k = -3/4$. The reason for this error is the bogus interpretation of the k_3 bit of k. This problem can be corrected by connecting the troublesome k_3 bit to the carry-in to the system adder. Recall that the multiplier interprets $k_s = -k_32^{-4}$ in S and the carry-in C_0 contributes k_32^{-3} to the output. The collective result of these two weights is $k_3(2^{-3} - 2^{-4}) = k_s2^{-4}$, which is the desired result. As a result, the product discussed previously is corrected by $1/8$ and the correct answer results.

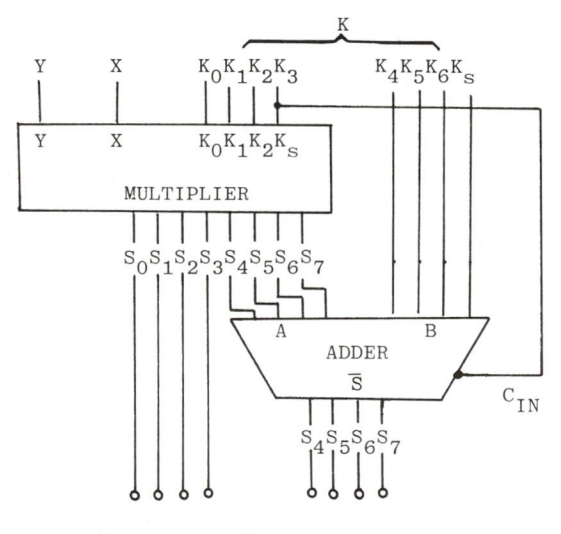

FIG. 16.12.5 Detailed description of an architecture that will compute $xy + k$.

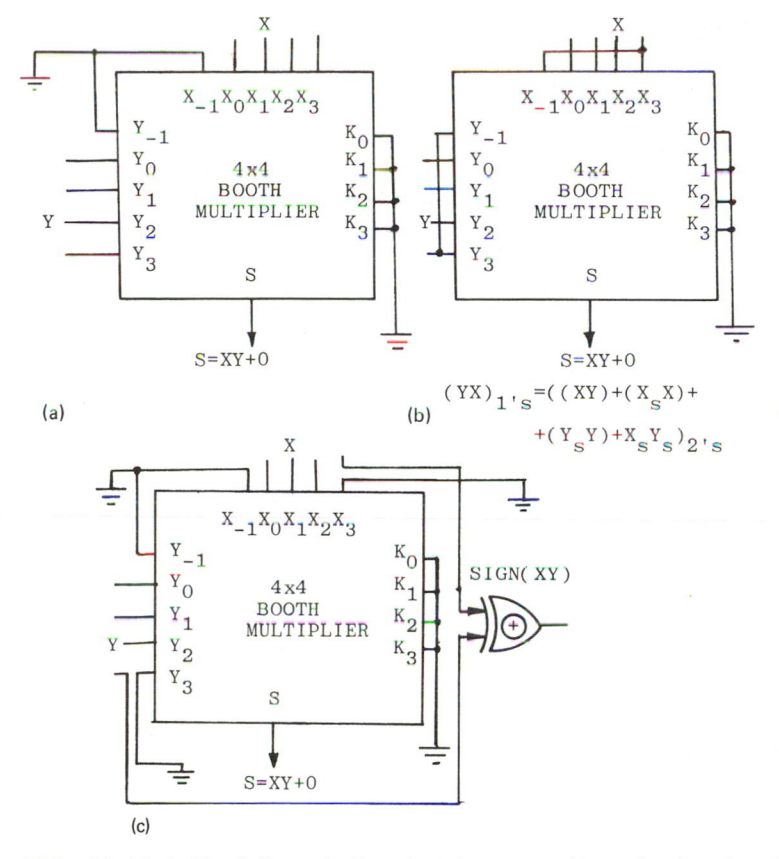

FIG. 16.12.6 Variations in the pin interconnection of a 4 × 4 multiplier to support (a) 2's-complement, (b) 1's-complement, and (c) sign-magnitude arithmetic.

It should be apparent that the chosen numbering system will affect how a monolithic multiplier will be configured in a system. Suppose that one concentrates on the Booth multiplier for a moment. Recall that this algorithm required an initialization (denote x_{-1}). Summarized in Fig. 16.12.6 are the required configurations for a 1's-complement, 2's-complement, and sign-magnitude 4 × 4 monolithic Booth multiplier.

16.13 VLSI MULTIPLIERS

Recent breakthroughs in integrated circuitry packing densities have allowed device manufacturers to improve the performance of digital multipliers.

Through the use of VLSI, high-speed long-word-length multipliers have been realized. Some of these new units have integrated into their architecture extended word length accumulators. A typical representative of this class of VLSI multiplier is shown in Fig. 16.13.1. In the TRW TDC series multipliers, the accumulator word length $m = 2n + 3$. This means that at least $2^3 = 8$ worst-case full-precision (i.e., 2n-bit) products can be summed without fear of register overflow. This can be particularly beneficial in many signal processing applications. For example, the FFT butterfly operation given by $W^{ij}x$, where W and x are complex numbers, represents four real multiply-add operations. Using the extended register accumulator, the butterfly operation $W^{ij}x$ could not overflow.

EXAMPLE 16.13.1 Using TRW performance parameters as a guide, a simple second-order recursive filter can be realized and analyzed in the manner indicated in Fig. 16.13.2. Since a second-order filter cycle requires 600 nsec, a filter throughput rate of 1.667 MHz is obtainable. The user would, of course, have to provide the control bits in real time. The control tuple (ACC, RND, SUB, IO) would normally be precomputed and stored in a circular shift register, RAM, or ROM and supplied to the multiplier/accumulator in a synchronized fashion.

A familiar problem to the signal processing scientist is complex multiplication. The multiplication of two complex numbers, say $x_1 = X_1 + jY_1$ and $x_2 = X_2 + jY_2$, is defined in terms of four real multiplies. If the multiplier has a real multiplication rate of 200 nsec, then a complex multiply would require 800 nsec. This problem can be dramatized in terms of an FFT which is dominated by complex (butterfly) multiplications.

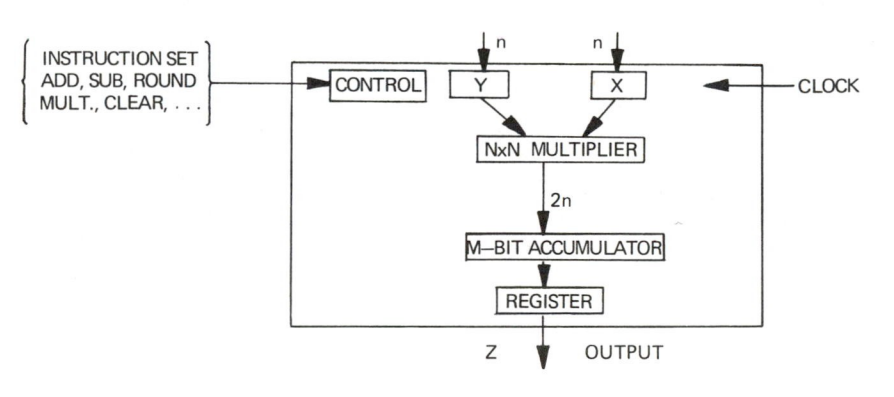

FIG. 16.13.1 Functional diagram of a VLSI multiplier with an integrated accumulator.

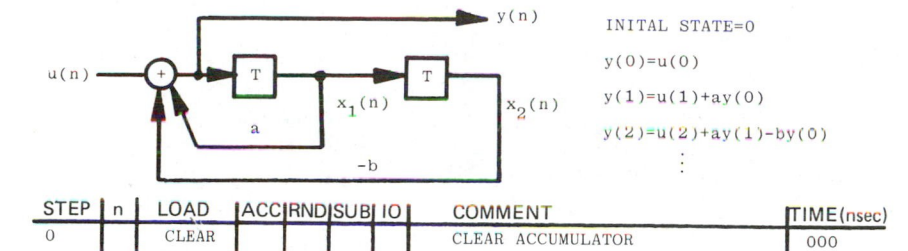

INITAL STATE=0

$y(0)=u(0)$

$y(1)=u(1)+ay(0)$

$y(2)=u(2)+ay(1)-by(0)$

$\vdots$

STEP	n	LOAD	ACC	RND	SUB	IO	COMMENT	TIME(nsec)
0		CLEAR					CLEAR ACCUMULATOR	000
1	0	u(0),1	1	1	0	1	y(0)=u(0); SEND TO DESTINATION	200
2	1	x_1,a	1	0	0	0	ACC=ay(0)	400
3	1	u(1),1	1	1	0	1	ACC=ay(0)+u(1); SEND	600
4	2	x_2,b	1	0	1	0	ACC=-by(0)	800
5	2	x_1,a	1	0	0	0	ACC=-by(0)+ay(1)	1000
6	2	u(2),1	1	1	0	1	ACC=-by(0)+ay(1)+u(2); SEND	1200

FIG. 16.13.2 Second-order filter and source code for a VLSI realization.

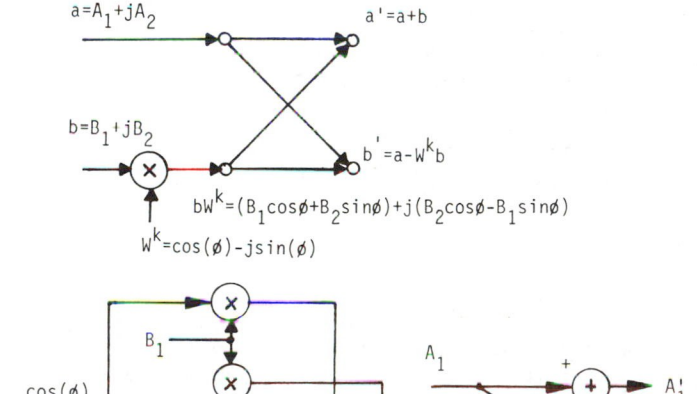

FIG. 16.13.3 Butterfly diagram for a radix-2 FFT. $W = \cos \phi + j \sin \phi$.

TABLE 16.13.1 Source Code of a VLSI Realization of a Radix-2 FFT

STEP	L=LOAD ; H=HOLD	ACC	SUB	RND	CONTENTS OF OUTPUT REG.	COMMENT	
1	(L)B_1; (L)0.5cosø*	0	0	1		0.5B_1cosø	+
2	(L)B_2; (L)0.5sinø*	1	0	0		ACC=0.5(B_1cosø+B_2sinø)=0.5Z_1	:
3	(L)A_1; (L)0.5*	1	0	0	0.5B_1cosø	ACC=0.5Z_1+0.5A_1=0.5A_1'	:
4	(H)A_1; (L)0.9999	1	1	0	0.5Z_1	ACC=0.9999A_1-0.5A_1'=0.5A_2'	:
5	(L)B_2; (L)0.5cosø*	0	0	0	0.5A_1'	0.5B_2cosø	:
6	(L)B_1; (L)0.5sinø*	1	1	0	0.5A_2'	ACC=0.5(B_1sinø-B_2cosø)=0.5Z_2	:
7	(L)A_2; (L)0.5*	1	0	0	0.5B_1sinø	ACC=0.5A_2+0.5Z_2=0.5B_1'	:
8	(H)A_2; (L)0.9999	1	1	0	0.5Z_2	ACC=0.9999A_2-0.5B_1'=0.5B_2'	+
9	NEXT INPUT	0	0	0	0.5B_1'		
10	NEXT INPUT	1	0	0	0.5B_2'	8 PROGRAM STEPS PER BUTTERFLY	
						AT 200 nsec PER STEP=1600nsec OR (625 KHz BUTTERFLY RATE)	

EXAMPLE 16.13.2 A simple radix-2 DIT FFT butterfly unit can be diagrammed in the manner suggested in Fig. 16.13.3. The control sequence
needed to realize the radix-2 DIT FFT is suggested in Table 16.13.1.

The VLSI multipliers, although slower than their Schottky counterparts, can achieve high-speed performance through pipelining. This concept is developed in the next section.

16.14 PIPELINING

A logical extension of the shift-add multiplication routine is the series-
parallel multiplier. The series-parallel multiplier accepts x as an n-bit
parallel word and y as an n-bit serial data string. Using such a device, a
radix-2 FFT butterfly (as well as many other signal processing tasks) can
be realized. If speed is of primary concern, plural multipliers may be
considered. For example, the functional structure of a high-speed 8-bit
radix-2 FFT butterfly unit is suggested in Fig. 16.14.1. It is composed of
four series multipliers and six adders. The timing diagram for the hypothesized multiplier is abstracted in Fig. 16.14.2a. Note that a complete
multiplication cycle requires 18 clock cycles, which at a 20-MHz clock rate,
translates to 1.1M multiplications per second. Based on this model, a
system timing diagram can be derived. Such a timing diagram is suggested
in Fig. 16.14.2b.
The analysis of the butterfly network remains incomplete. The problem of register overflow needs to be addressed. For example, at data

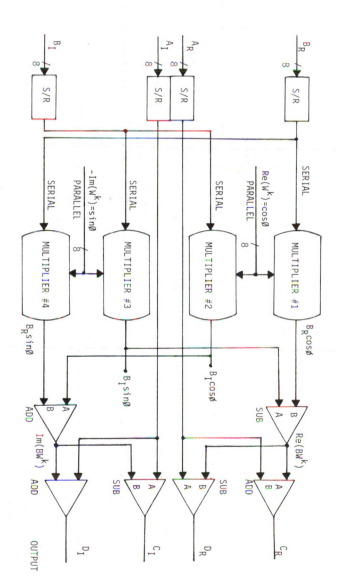

INPUT SHIFT REGISTERS

8-BIT 2'S COMPLEMENT SERIAL-
PARALLEL MULTIPLIERS

LEVEL 1
SERIAL ADDERS

LEVEL 2
SERIAL ADDERS

FIG. 16.14.1 Architecture for a radix-2 FFT Butterfly module using a series-parallel multiplier.

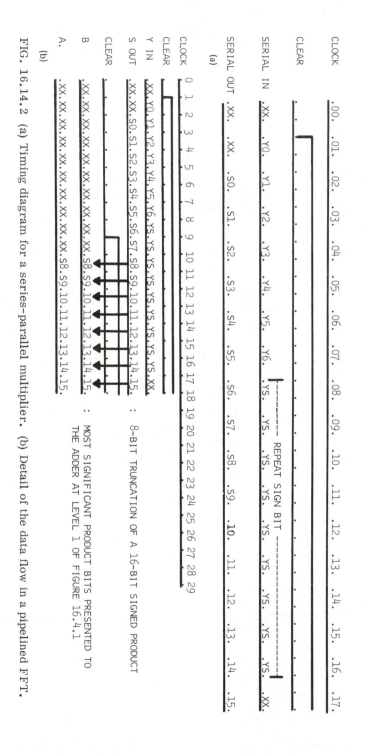

FIG. 16.14.2 (a) Timing diagram for a series-parallel multiplier. (b) Detail of the data flow in a pipelined FFT.

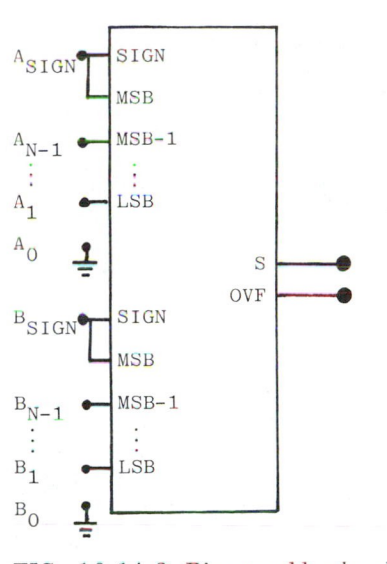

FIG. 16.14.3 Binary adder hard-wired to perform 2's-complement scaling by one-half.

location y_1 of Fig. 16.14.1, the real part of the product BW is found. If $|BW| \leq 1$, no register overflow will occur at that location. However, at the next level addition is required. Here the sum of two numbers bounded in modulus by unity can obviously overflow. To ensure that register overflow does not occur at this level, the data presented to the adder should be scaled by 1/2. This scaling can be imbedded into the hardware if desired. Consider for the moment the adder configured in Fig. 16.14.3. In this figure it can be noted that the sign bit of the incoming 2's-complement 8-bit data word is copied using a dedicated data path. In addition, the arithmetic shift of data by one bit is also accomplished using dedicated data paths. Using this architecture, a scaling by one-half can be accomplished with zero overhead.

In a pipelined architecture, computational routine is broken down into a series of independent operations. These operations are chained together as a series of tasks. Data are entered to the "pipe" and moved along, form one task to the next, in a piecewise constant manner. Each level of the pipeline will simultaneously execute its assigned task. If there are N levels to a pipeline, with the longest task execution rate being T^* seconds, the pipelined data rate will be $(1/T^*)$ words per second. It should be noted, however, that it will take NT^* seconds for a particular data word, once entered into the pipeline, to reappear at the output. In the case of the radix-2 FFT butterfly, the routine can be factored into four distinct operations. They are:

1. Input data
2. Multiplication
3. Adder level 1
4. Adder level 2

The pipeline would be architected in the manner suggested by the data given in Fig. 16.14.3. The overlapped operations would form the integral parts (tasks) of the pipeline. Nevertheless, the objective of any pipeline design is to keep the pipe maximally full of data during run time. This almost always requires that there exist no conditional branches into or from the pipe.

EXAMPLE 16.14.1 A three-stage radix-4 $N = 4^3 = 64$ point FFT is abstracted in Fig. 16.14.4. Since no conditional branching is required during an FFT operation, the data paths (internal to the radix-4 FFT) can be pipelined through the addition of registers and a pipeline schedule. The butterfly units are denoted BFU, with BFU1 being used for only 16 of the 64 FFT cycles. Using two 64-point FFT systems, and the data structure/timing diagram found in Fig. 16.14.4, a two- or fourfold   oughput improvement can be achieved. The 2:1 and 4:1 overlap data policies suggests that in a radix-r FFT application, throughput improvements of 2, 4, 8, . . ., r can be achieved at a cost of running 2, 4, 8, . . ., r concurrent radix-r FFT subsystems.

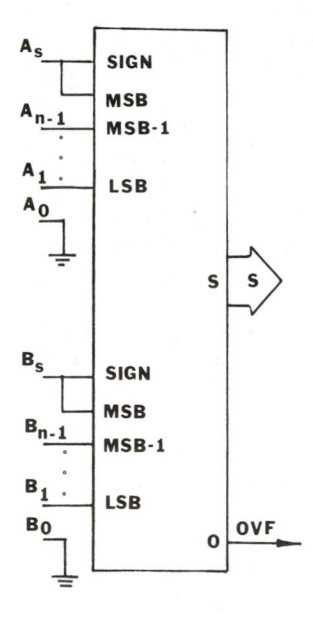

FIG. 16.14.4 Diagram of a 64-point radix-4 FFT.

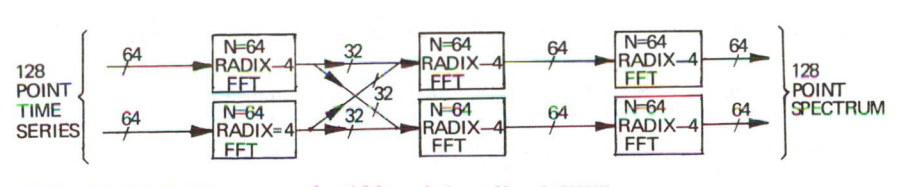

FIG. 16.14.5 Diagram of a 128-point radix-4 FFT.

EXAMPLE 16.14.2 The Signal Processing System's SPS-1000 FFT processor uses a complex radix-4 FFT module as its fundamental building block.
It contains both arithmetic and PROM butterfly coefficient tables. The FFT
module can be interconnected with others to perform long-record-length
FFTs. Each module also contains embedded memory for data buffering.
The basic unit has an 8-bit complex word length (precision) which can be
extended to 16, 24, or 32 bits through the addition of two, three, or four
more units. An 8-bit module can process data at a 3.3-MHz rate.

The modules can be pipelined if high speed is required. To double
the throughput, two modules will have to operate in parallel. Each module
will process N/2-complex data words if a N-point transform is architected.
However, there is known to be an interconnection between the data paths in
an FFT structure. At some point of the SPS-1000 architecture, two of its
four radix-4 butterfly module outputs must be routed to its parallel processing neighbor. That is, there is an exchange of data between the parallel
subsystems. Figure 16.14.5 describes a 128-point radix-4 FFT in a SPS-
1000 sense. It can be seen that the data paths initially mix (output of level
1) but remain separate thereafter. If four parallel units were configured
along similar lines, a fourfold increase would result. The performance of
the SPS-1000 is summarized as follows:

Complex FFT module	Word length (bits)		
	8	16	24
Butterfly delay (μsec)	1.2	1.9	2.6
Basic unit: data rate (MHz)	3.3	2.1	1.5
Dual-parallel (MHz)	6.6	4.2	3.0
Quadruple parallel	13.2	8.4	6.0

EXAMPLE 16.14.3 Advanced Micro Devices bipolar 29500 series consists
of a microprogrammable signal processing chip, multiplier chip, multilevel pipeline chip (see next section), and a FFT address chip. The signal
processing chip is a byte-slice (8-bit) microprogrammable device capable
of managing external memory transactions, I/O to the multiplier chip, ALU
operations, internal data manipulations, and so forth. The multiplier chip

is a high-speed 16-bit multiplier which appears in several versions. One can be made to be compatible with the TRW 16-bit VLSI multiplier while the other exhibits more efficient I/O operations. These two chips can be used to architect digital filters as well. The pipeline and the FFT address are designed to integrate the other chips into a FFT system.

16.15 FLOATING POINT

Floating-point operations can be realized using the hardware systems developed in this chapter. Such devices as the 16- or 24-bit VLSI multipliers provide a useful floating-point design tool. Recall that a floating-point word is expressed as $x = X2^k$ (assuming radix-2), where X is the mantissa and K is the exponent. If a normalized system is used $0.5 \leq |X| \leq 1$. Therefore, the product floating number, namely $z = XY2^{k_x + k_y}$, produces a mantissa bounded by $0.25 \leq |XY| \leq 1$. If this fixed-point mantissa exhibits a MSB data bit of unity, then $0.5 \leq |XY| \leq 1$ and no renormalizing is required. If the opposite is true, the mantissa would be scaled by 2 (in a 2's-complement sense) and the exponent decremented by 1. Commercial VLSI floating-point chips are available, capable of running at 100 to 200 nsec, using standard formats. The adder-multiplier pairs are typified by a 24-bit mantissa and a 4-bit exponent and include underflow/overflow flags plus a rounding control.

16.16 MULTIPLIER ARRAYS

To this point, only single monolithic multiplier units have been considered. In order to construct a long-word-length multiplier from these small- to medium-word-length modules, interconnections are required. To focus this study on something tangible, a 2×4-bit multiplier will be used to represent the small-word-length module. Using this unit, long-word-length multiplier structures will be derived.

Suppose that x and y are two N-bit words partitioned into blocks of K-bits each such that $K = N/M$, M an integer. Then the representation can be further refined as follows:

$$x \rightarrow (x_1, x_2, \ldots, x_M); \quad x_i \rightarrow (x_{i1}, x_{i2}, \ldots, x_{ik})$$

$$y \rightarrow (y_1, y_2, \ldots, y_M); \quad y_i \rightarrow (y_{i1}, y_{i2}, \ldots, y_{ik})$$

The product of x and y can therefore be computed to be

$$
\begin{array}{ccccc}
x_M & x_{M-1} & \cdots & x_2 x_1 \\
y_M & y_{M-1} & \cdots & y_2 y_1 \\
\hline
x_M y_1 & x_{M-1} y_1 & \cdots & x_2 y_1 \ x_1 y_1 \\
x_M y_2 & x_{M-1} y_2 & x_{M-2} y & \cdots & x_2 y_1 \\
& \cdots & & \\
\cdots y_3 y_{M-1} & y_2 y_{M-1} & x_1 y_{M-1} \\
x_M y_M \cdots x_2 y_M & x_1 y_M \\
\hline
s_{2M} \cdots s_{M+1} & s_M & s_{M-1} & \cdots s_2 & s_1
\end{array}
$$

$\left.\right\}$ partial products

EXAMPLE 16.16.1 Suppose that x = 21(decimal) and y = 17(decimal). Let the representations of x and y have the following decimal form:

x = 20 + 1; y = 10 + 7

Then, in terms of Eq. (16.16.2), one obtains

$$
\begin{array}{rr}
20 & 01 \\
10 & 07 \\
\hline
140 & 07 \\
200 \quad 10 & \\
\hline
\end{array}
$$

200 + 150 + 07 = 357

Upon close inspection of this example, it can be noted that a decimal overflow has occurred at several points of the algorithm. In particular, the digits 140 and 200 exceed their field limits of [0, 99]. Therefore, a multiplier system must be constructed so that overflow information is properly managed. Consider representing x and y instead as the two–tuples x = (2, 1) and y = (1, 7). The multiplication of x and y can now be symbolically represented as

$$
\begin{array}{l}
x \longrightarrow (2, 1) \\
y \longrightarrow (1, 7) \\
\hline
1, 4, 7 \\
0, 2, 1 \\
\hline
0, 3, 5, 7
\end{array}
$$

$\left.\right\}$ shift–add

0, 3, 5, 7   0(1000) + 3(100) + 5(10) + 1(7) = 357

$$z = \sum_{j=0}^{7} \sum_{i=0}^{7} x_i y_j 2^{i+j}$$

$$
\begin{array}{ll}
\quad\quad I & \quad\quad K
\end{array}
$$

-------------------------------   --------------------

$z = y_0 x_0 + 2y_0 x_1 + 4y_0 x_2 + 8y_0 x_3 + 16y_0 x_4 + \cdots$ $\cdots + 128 y_0 x_7$

$\quad\quad 2y_1 x_0 + 4y_0 x_2 + 8y_0 x_3 + 16y_0 x_3 + \cdots$ $\cdots + 128 y_1 x_6 + 256 y_1 x_7$

-------------------------------   --------------------

$\quad\quad\quad 4y_2 x_0 + 8y_2 x_1 + 16y_2 x_2 + \cdots$ $\cdots + 128 y_2 x_5 + 256 y_2 x_6 + 512 y_2 x_7$

$\quad\quad\quad\quad 8y_3 x_0 + 16y_3 x_1 + \cdots$ $\cdots$

$\quad\quad\quad\quad\quad 16y_4 x_0 + \cdots$

FIG. 16.16.1 Mapping of partial product found in an 8 × 8 multiply based on 4 × 2 multiplier modules.

Based on this model, it should be apparent that a long-word-length multiplier can be developed from shorter-word-length modules. To demonstrate this thesis, the mechanics of realizing an 8 × 8 product from 2 × 4 multiplier devices, the data in Fig. 16.16.1 are offered. Here it is assumed that both x and y are unsigned data words. The data found in this figure suggest that the synthesized multiplier has a parallelogram geometry. This geometry is further motivated in Fig. 16.16.2. Here, in terms of hardware, the conjectured multiplier is realized.

Additional speed can be obtained through pipelining. By inserting storage registers between the various levels of the parallelogram, a pipeline can be realized. In a nonpipelined configuration, multiplier execution rates are limited by the worst-case propagation delay. Referring to Fig. 16.16.3, the longest propagation delay for an 8 × 16-bit multiplier is seen to exist between y_0 and s_{15}. Using vendor-supplied data, this delay can be quantified as follows:

1. $y_0 C_{n+4} = 22$ nsec (a)

2. $^* C_n s_{0:3} = 12$ nsec (b)

3. $^* k_i$ to $C_{n+4} = 10$ nsec (c)

4. (Repeat * for paths b and c [i.e., 2(22) nsec = 44 nsec]

5. $C_n S_{15}$ (last bit) = 15 nsec (d)

Total = 103 nsec

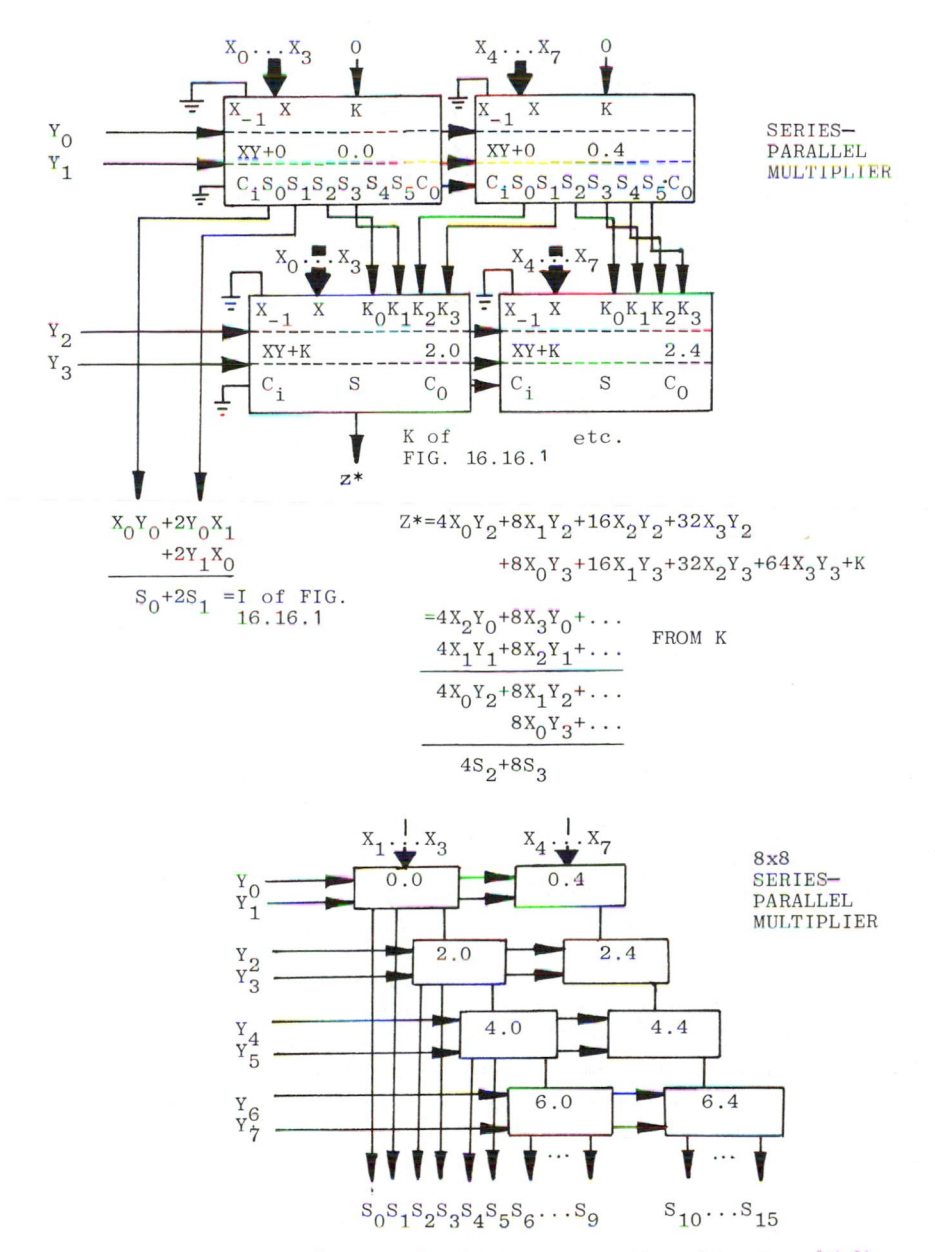

FIG. 16.16.2 Circuit diagram for the interconnection of 4 × 2 multipliers to achieve longer-word-length multiplication.

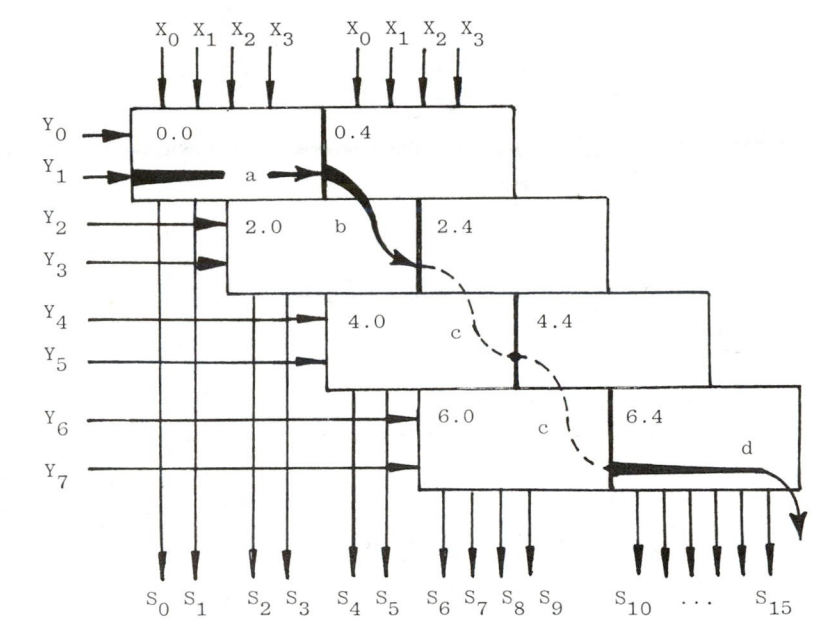

FIG. 16.16.3 Propagation delays and paths found in a typical modular multiplier (8 × 8 series-parallel multiplier).

There are many other array configurations presently in use. One is the Wallace tree (see Section 16.10). This architecture makes use of a modified carry-save adder and a trapezoidal array configuration. The Pezaris and Baugh-Wooley 2's-complement multiplier structures are used to a lesser degree than the more familiar Wallace tree.

Technical literature is generally supplied by the MSI or LSI multiplier supplier which will enable the user to configure multiplier arrays from their commercially available products. As a guide to the reader, typical array configuration are summarized in Table 16.6.1. In this table, 8 × 8 and 16 × 16 array performance metrics are compared.

Finally, in long-word-length multiplications, truncation or rounding operations are usually required. Truncation can be accomplished by paying a slight (if any) overhead price. If the user wishes to improve the precision of the result, rounding is suggested. However, rounding is an overhead-bearing operation.

16.17 DIVISION

Compared to multiplication, division plays a secondary role in digital signal processing as division is essentially a trial-and-error operation. The structure of a division operation can best be seen through the use of a

TABLE 16.16.1 Summary of Commercially Available MSI-LSI-VLSI Multipliers

Vendor	Size	Pins	8 × 8 Multiplier			16 × 16 Multiplier		
			Number of devices	Speed (nsec)	Power (W)	Number of devices	Speed (nsec)	Power (W)
AMD S05	2 × 4	24	8	75	5	32	150	19
AMD S2516	8 × 8	40	8	400	1	2	800	2
MMI 67558	8 × 8	40	1	100	2	14	140	9
MMI 67516	16 × 16	24	-	-	-	1	800	1
TI S274	4 × 4	20	12	75	5	45	120	21
TRW 8AJ	8 × 8	40	1	130	1	14	170	10
TRW 8AJ-1	8 × 8	40	1	45	1 1/2	-	-	-
TRW 16AJ	16 × 16	64	-	-	-	1	160	4
TRW 16HJ	16 × 16	64	-	-	-	1	100	4

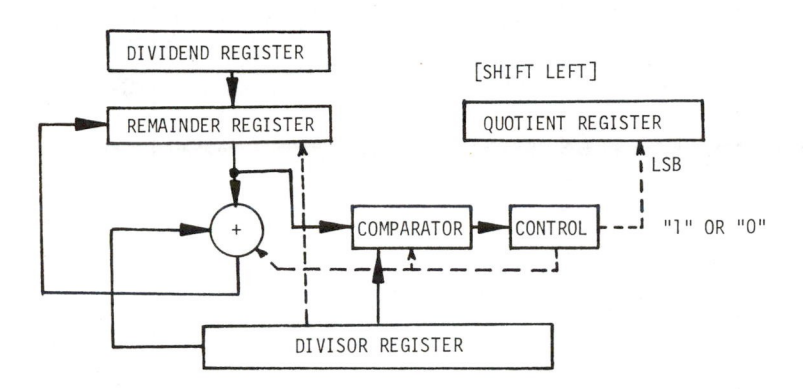

FIG. 16.17.1 Circuit diagram detailing the structure of a digital divider.

long-division model. For example, for a dividend of $x = 19$ and a divisor of $y = 5$, a long-division model would yield

Step	R	Q
1	$16 > 0$	0
2	$16 - 1(5) = 11 \geq 0$	1
3	$11 - 1(5) = 6 \geq 0$	2
4	$6 - 1(5) = 1 \geq 0$	3
5	$1 - 1(5) = -4 \not\geq 0$	terminate $Q = 3$ (quotient)
		$R = 1$ (remainder)

In block diagram form, one obtains the structure suggested in Fig. 16.17.1. If 2's-complement division is required, the sign information must be integrated into the division routine. If P is the dividend, Q_2 the quotient sign bit, and D_S the sign bit of the divisor, then:

1. If $D_S = 0$ and $-D/2 \leq P$, then set $Q_S = 0$; otherwise, set $Q_S = 1$.
2. If $D_S = 1$ and $-D/2 \leq P$, then set $Q_S = 1$; otherwise, set $Q_S = 0$.

EXAMPLE 16.17.1 Suppose that $D = 7$; then $D_S = 0$ and the rounded quotient is positive if $-3.5 \leq P$. That is, if $P = -3.5 + e$, then $(3.5 + e)/7 = 0$, whereas if $P = -3.5 - e$, then $[(-3.5 - e)/7] = -1$.
     The remaining digits of the quotient are given by:

1. If $D_S = 0$ and $T_{i-1}D + D/2 \leq P$, set $Q_i = 1$; otherwise, set $Q_i = 0$.
2. If $D_S = 1$ and $T_{i-1}D + D/2 \leq P$, set $Q_i = 0$; otherwise, set $P_i = 1$.

Here T_i is the ith partial product. Because of the presence of these overhead-bearing operations, division throughput is generally slower than multiplication. If possible, division should be avoided in a signal processing application.

BIBLIOGRAPHY

Booth, T. L. (1978), Digital Networks and Computer Systems, Wiley, New York.

Hwang, K. (1979), Computer Arithmetic, Wiley, New York.

Korn, G. A. (1977), Microprocessors and Small Digital Computer Systems for Engineers and Scientists, McGraw-Hill, New York.

Pezaris, S. D. (1971), A 40ns 17-Bit-by-Bit Array Multiplier, IEEE Trans. Comput., C-20, April, pp. 442-447.

Rabiner, L. R., and B. Gold (1975), Theory and Applications of Digital Signal Processing, Prentice-Hall, Englewood Cliffs, N.J.

Wallace, C. S. (1964), A Suggestion for Parallel Multipliers, IEEE Trans. Electromag. Compat., EC-13, February, pp. 14-17.

Specific technical information may be obtained from the following firms:

Advanced Micro Devices, Schottky and Low Power Schottky Data Book, 901 Thompson Place, Sunnyvale, CA 94086.

Fairchild Semiconductors, TTL and MSI Data Book, 313 Fairchild Drive, Mountain View, CA 94043.

National Semiconductor, Data Book, 2900 Semiconductor Drive, Santa Clara, CA 95051.

Texas Instruments, Data books, Dallas, TX.

TRW, VLSI Products Data Book, Redondo Beach, CA.

17
General-Purpose Computers

17.1 INTRODUCTION

The phase "general-purpose digital computer" means many things to many people. Because of this diversity of opinion, a comprehensive treatment of this subject here would be impossible. Instead, a brief overview of selected topics is presented.

General-purpose computers span a wide range of size, cost, and performance capabilities. Signal processing scientists are generally interested in developing systems that are found at the low end of this spectrum. That is, minimum cost and complexity is often the design objective. Furthermore, design flexibility is often required. Here, a high degree of programmability is desired. This will allow the system to be more easily maintained in the field as well as to accept modifications more readily. As a result, the digital signal processing scientist often works directly with mini- and microcomputing systems.

Broadly stated, a programmable digital machine is organized in the manner suggested by Fig. 17.1.1. The major elements of this system are:

1. Memory (main storage): stores data and instructions
2. ALU (arithmetic logic unit): temporarily stores and manipulates data in registers
3. Control: controls flow of data and instructions, fetches, executes instructions, and enables required resources.
4. I/O: interfaces processor with external world

From a slightly more microscopic viewpoint, the two-step processor of Fig. 17.1.2 may be studied. This system is composed of a fetch cycle and an execute cycle. Referring to the system of Fig. 17.1.2, the following operations can be sequentially identified:

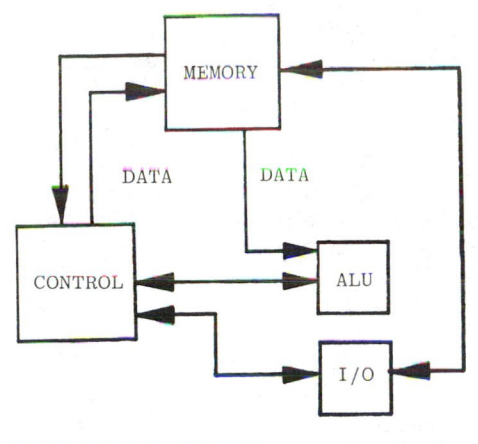

FIG. 17.1.1 Organization of a simple digital processor.

Fetch Cycle

1. The control unit fetches address n from the program counter.
2. The decoder decodes that address.
3. The control unit logic fetches the contents of the decoded address.
4. The control unit interprets the memory word as an instruction and sends it to the instruction register.

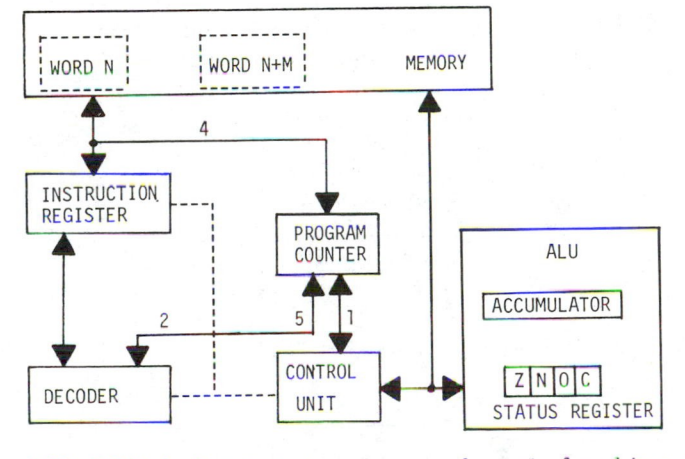

FIG. 17.1.2 Detail of the sequence of events found in a simple digital processor.

5. The control unit increments the program counter.
6. The control unit fetches the contents of the instruction register and
    decodes it.

Execution Cycle

Cycle 2, the execution cycle, mechanizes the decoded instruction. For example, a typical request may entail adding the contents of word n + m to the
accumulator. The status register will contain information regarding the
algebraic consequence of this operation (e.g., Z = ZERO, N = NEGATIVE,
O = OVERFLOW, C = CARRY).

EXAMPLE 17.1.1 A computing machine that has wide acceptance in the
industry is the PDP-11 series machine marketed by Digital Equipment
Corporation. The architecture of this machine is abstracted in Fig. 17.1.3.
A typical minicomputer in the PDP-11 series possesses eight 16-bit registers. Registers R_0 through R_5 serve as general-purpose accumulators,
pointers, or index registers. Register R_6 serves as a stack pointer, with
R_7 used as a program counter. Data buses are 16-bit bidirectional data
lines that can be used for programmed DMA transfers. An 18-bit address
memory address and device select capability (i.e., 256K bytes of addressable memory) is used plus 22 control lines.

17.2 MICROPROCESSORS

Ted Hoff of Intel conceived of the integrated programmable system, now
referred to as the microprocessor, in 1969. Since its commercial introduction in 1971, it has been widely and enthusiastically accepted as a basic
digital design tool. The power and utility of these integrated devices are
well known. In fact, the microprocessor has developed a "cult" of followers

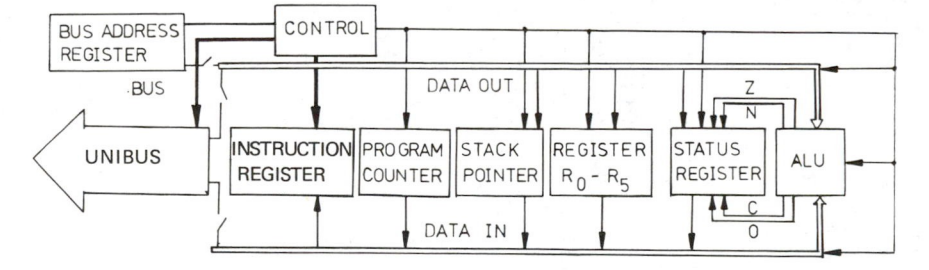

FIG. 17.1.3 Example of the type of architecture found in a Digital Computer Corporation PDP-11 minicomputer.

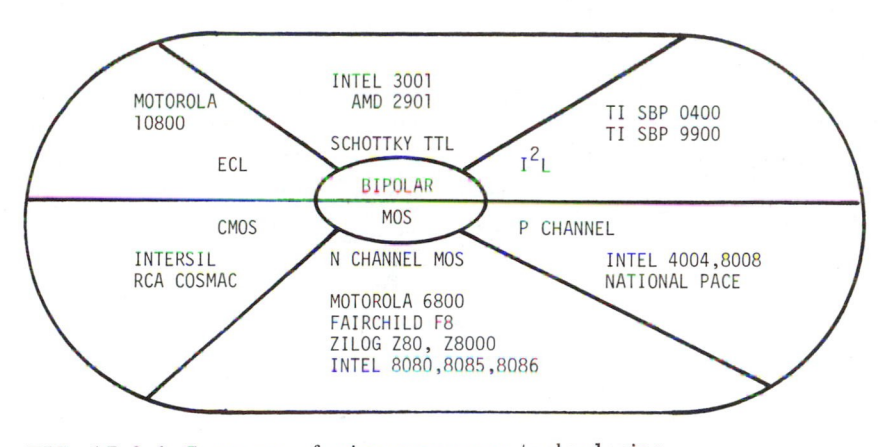

FIG. 17.2.1 Summary of microprocessor technologies.

who treat these devices as a separate science rather than a digital design medium. The fact is that these electronic widgets are tools and are there-fore subject to the strengths and weaknesses of any general class of instruments. For a given task, a given tool is required. The microprocessor makes sense when cost and packing constraints are to be met (assuming, of course, that execution and throughput performance metrics are satisfied). However, there are many cases where the microprocessor is not capable of achieving the needed speed and capacity demands of a signal processing application.

At one time in the history of microprocessors, these units were exclusively MOS devices. Recently, other technologies have invaded the microprocessor field. A partial summary of selected microprocessors and their supporting technology is given in Fig. 17.2.1.

Microprocessors originally appeared on the market as 4-bit devices (e.g., Intel 4004). Today, microprocessors can be found in 4-, 8-, and 16-bit configurations with performance capabilities that often approach their minicomputer cousins. The evolution of the microprocessor can be traced in terms of the data given in Table 17.2.1.

Concentrating on 16-bit microprocessors, the following can be ob-served (see Table 17.2.2). The choice of which microprocessor to use is based on many variables. Some are beyond the control of the designer. For example, in addition to basic performance specifications, the question of availability, second sources, up and down compatibility with other products, software support, and the manufacturer's reputation for after-the-sale service influence the design decision. Considering the 16-bit microprocessor data in Table 17.2.2, for example, based on memory expansion and power supply considerations, the 8086 and 8000 are good design choices. If long-word-length transforms (e.g., two-dimensional

TABLE 17.2.1 Survey of Popular Microprocessors

Item	8008	8080	Z80	990	8086	Z8000
Word length	8	8	6	16	16	16
Memory addressing	16K	64K	64K	32K	1024K	8172K
Power	5, -9	12, ±5	5	12, ±3	5	5
Relative throughput	1	10	20	66	100	100
Year introduced	1972	1973	1976	1975	1978	1979

FFTs) are needed, large memory capacity is a prerequisite. If, however, a digital recursive filter is considered, memory requirements are often secondary to arithmetic speed. Here the TI9900 is superior. When speed is the primary design consideration, another version of the basic micro-processor theme, the bit-slice processor, should be considered.

17.3 BIT-SLICE PROCESSORS

In many applications, microprocessors are embedded into signal processing designs. Often the relatively slow speed of these processors limit their use in high-speed applications. In recent years, high-performance MOS, I^2L, and CMOS 16-bit 5-MHz microprocessors have, in some cases, filled this void (e.g., TI SBP 9900, Intel 8086, Motorola 9600). At a performance

TABLE 17.2.2 Survey of Popular 16-Bit Microprocessors

Item	TI9900	Z8000	8086
Maximum clock rate	4.17M	8M	5M
Instruction word length	1-3W	2-6B	2-6B
Instruction lexicon	69	110+	97
Longest instruction	Divide (31 μsec)	Divide (90 μsec)	Divide
Memory-to-memory ADD	9 μsec	NA	NA
Multiply	15.6 μsec	17.5 μsec	26 μsec
Register-to-register ADD	4.2 μsec	2.25 μsec	600 nsec

rate of three to ten times this figure can be found a new family of units
known as bit-slice microprocessors (e.g., AMD 2901, TI 481, Intel 3000,
Fairchild 10K ECL and ECL8). The two high-performance families derive
their attributes and limitations from differing solid-state technologies.
Bit-slice processors use high-speed-device technologies to achieve high
throughput rates. Since speed and power dissipation parameters are generally inversely related, the word lengths of a high-speed bit-slice processor must be small (e.g., 2 or 4 bits). These high-power-dissipating
small-word-length devices can be interconnected to form longer-word-
length processors. Therefore, the bit-slice register-ALU units (sometimes
referred to as RALUs) offer a great deal of design flexibility. Since they
have a loose architecture, extensive software support is often difficult to
find. In those cases where the manufacturer has structured the architecture
(as chip set or complete microcomputers), high-level programming language support may be found. In Fig. 17.3.1, a typical 16-bit bit-slice processor is configured. Using carry lookahead units, a 16×16 multiply can
be completed in several microseconds (e.g., TI 54S48 = 3 μsec). Since
bit-slice processors are microprogrammed, their performance is often
related to the concept of microcycle. A National IDM2901A has microcycle
times as low as 60 nsec. The newer bit-slice processors are making use
of pipelining techniques to decrease the effective microcycle times associated with standard arithmetic operations.

The future of these high-technology processors is extremely bright.
Bit-slice (typically 4 bits wide) microprocessors are giving way to byte-
slice microprocessors. These high-speed flexible processors have many
applications in digital signal processing. The semiconductor industry is
now designing integrated bit- and byte-slice chip sets for signal processing
applications. For example, Advanced Micro Devices is developing a bipolar
29500 series system consisting of a microprogrammable signal processing
chip, multilevel pipeline chip, and an FFT address chip. This ensemble of
chips can be configured to do fast Fourier transforms. The byte-slice
microprogrammable chip is capable of managing external memory transactions,

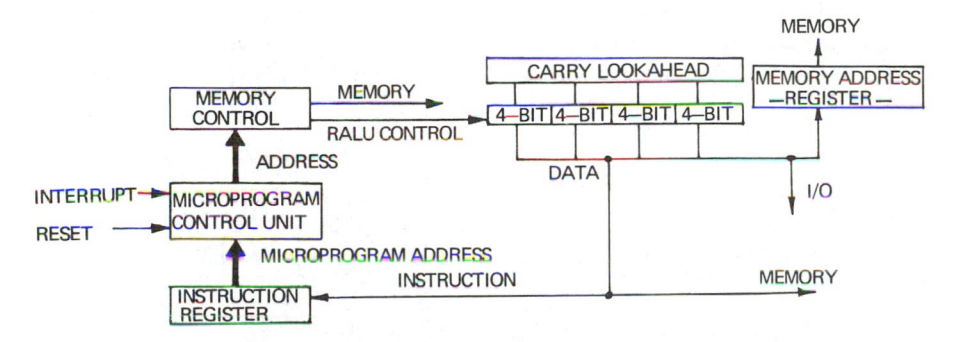

FIG. 17.3.1 Architecture of a bit-slice microprocessor.

I/O operations to a peripheral multiplier chip, ALU operations, internal
data manipulations, and others. The multiplier chip is a high-speed 16-bit
multiplier that appears in several versions, one of which is comparable to
the TRW 16-bit VLSI multiplier. The other multiplier version exhibits a
more efficient set of I/O operations. The pipeline and FFT address chips
are designed to integrate the other devices into a FFT system.

17.4 PROGRAMMING CONSIDERATIONS

Small-word-length processors require more skill than theory to program.
Here, experience is the best teacher. Therefore, this topic will only be
surveyed. Only topics having major signal processing importance will be
given special attention.

Microprocessors use flags and test bits to support branching. Historically, branching is classified as conditional or unconditional. Conditional branching allows the processor to pursue a logical course of action
based on a computed event. The data may be arithmetic or logical. In a
signal processing application, logical branchings are rare due to its algo-
rithmatic nature. Branching can also be caused by the information found in
the status register (i.e., Z, N, O, C). Care must be exercised in interpreting these bits. For example, the overflow in a 2's-complement sense
is indicated by N and O collectively (i.e., $N \oplus O$). When working with
unsigned data, negating requires the carry bit to be manipulated. Servicing
overflow is an overhead-bearing operation. Therefore, if overflows are
allowed to occur, throughput will suffer. If a real-time system is to be
designed, the worst-case condition must be budgeted for each epoch. The
random occurrences of overflows and the like will force longer-than-desired
epochs to be used.

One of the more important microprocessor operations is the cyclic
rotate/shift. These operations are shown in Fig. 17.4.1. By clearing or
setting the carry bit (i.e., copy sign), rotations suggested in this figure
can be used as an arithmetic shift (scaling by 2 or 1/2). Many machines
possess a self-contained arithmetic shift operation for unsigned data. Here
the carry flag is used to indicate dynamic range overflow. In a 2's-complement system, multiplication by 2 can be treated as an unsigned left shift
with overflow indicated by the operation [sign bit] $\oplus$ [carry bit]. Scaling by

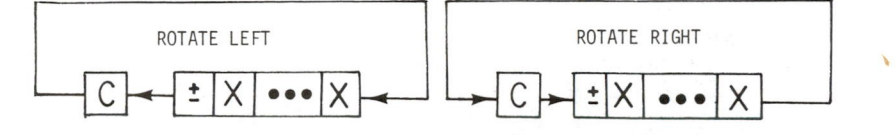

FIG. 17.4.1 Elementary left- and right-shift operations.

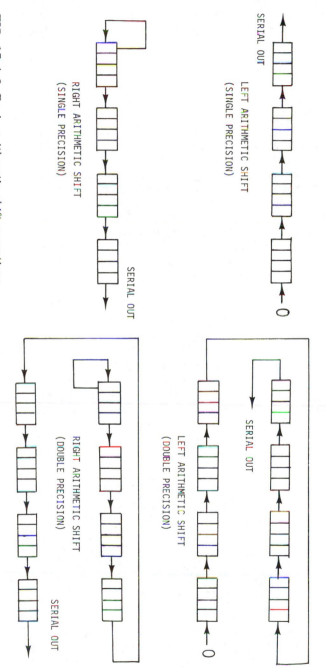

FIG. 17.4.2 Basic arithmetic shift operations.

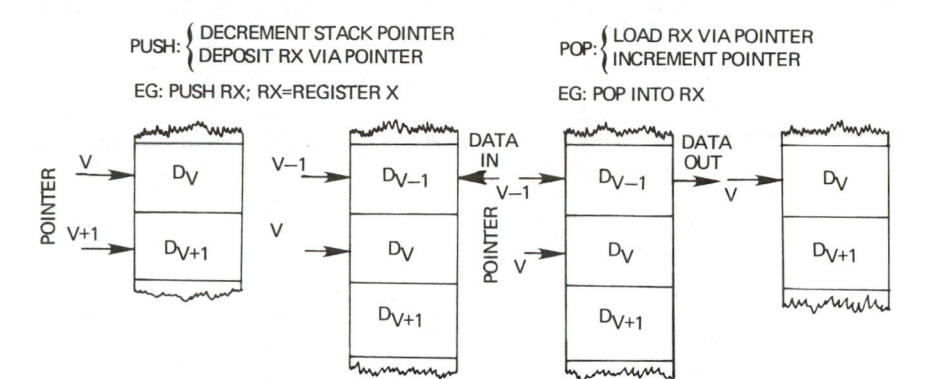

FIG. 17.4.3 PUSH and POP operations interpreted as stack manipulations.

1/2 is performed by copying the sign bit into the carry location and then right-shifting the data. These operations are suggested in Fig. 17.4.2, using as a model the TI SBP0400A processor.

Stack operations are also of major importance to the signal processing system designer. Data, representing the states of a digital filter, are constantly changing in time. These data are often stored in chronological order using a last-in first-out (LIFO) stack. This data list can be manipulated through the operations commonly known as PUSH and POP. Here a stack pointer is used to monitor the movement of data within a LIFO. Suppose at some instant of time that the stack pointer is directed toward word V. Then the PUSH operation "pushes" new data onto the top of the stack. This is represented symbolically in Fig. 17.4.3. POP literally "pops" data to the list in the manner suggested in Fig. 17.4.3. The user of these schemes must reserve a block of memory for use as a stack which will not be exhausted during run time. Obviously, the data from another part of the program must be denied accidental access to the stack. A data list of length N can be managed through the use of a modulo(N) counter if desired. In some of the more sophisticated operations, POP can be supported with arithmetic (e.g., POP and ADD, POP and MULT). Stacks can also be interpreted as a singly dimensioned array which can be randomly accessed by an index register. However, unlike large mainframe computers, microprocessors generally do not have sufficient index register capacity to support multidimensional data access.

17.5 DIRECT MEMORY ACCESS

The problem of moving data from one part of a signal processing system to another often arises. For example, in a high-speed DSP application, large volumes of digitized data may have to be stored rapidly into prespecified

sections of memory. After performing the "burst" transfer, the data block
may have to be moved to another location for processing. Using registers,
the data could technically be transferred by servicing interrupts. The time
required to support this task would be the sum of:

1. Latency time: the time needed to complete the instruction currently
   being executed
2. Delay in reaching the first interrupt service statement
3. Overhead associated with stacking processor states and registers

This would be repeated for each piece of data transferred. A more efficient approach is the use of direct memory access (DMA). DMA operations
generally involve a cycle stealing operation. In small machines, the processor is "frozen" in place by inhibiting the processor's clock. This causes
the processor to go into a comatose state with all its registers and pointer
data fixed in place. The DMA transfer is completed synchronously at a
high data rate. A block of data having a prespecified size is transferred
from a given device, starting at a given location, to a specified unit starting
at a directed location. The DMA transfer is block oriented and does not
pass data through the processor's registers. Once the DMA has been ser-
viced, the processor's clock is returned. The processor then continues to
work on its data as if nothing had interrupted it.

In a DMA operation, the destination address and number of words
to be transferred are prespecified. In addition, the DMA controller will
arbitrate all conflicts in priority. Under DMA, data rates from 100K to
1M words per second can be achieved. A typical DMA network is shown in
Fig. 17.5.1.

DMA transfers can be device or program initiated. Device-initiated
transfers could be used to empty a buffer that is periodically being filled.
Program-initiated transfers often occur in signal processing applications
as well. For example, the memory address register can be directed to
sequence a 2^N-word data array in a bit-reversed sense. Thus a DIT FFT

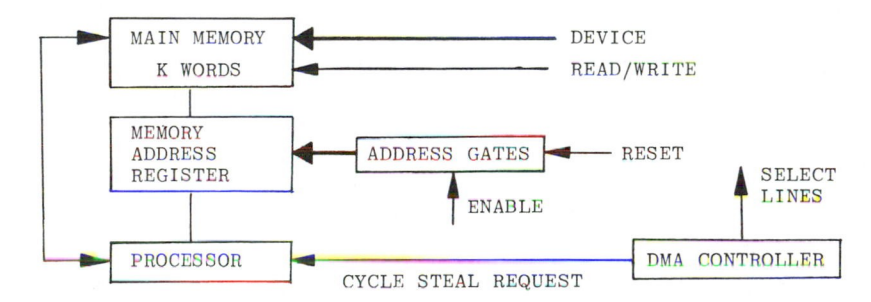

FIG. 17.5.1 Block diagram interpretation of a DMA transfer and hardware.

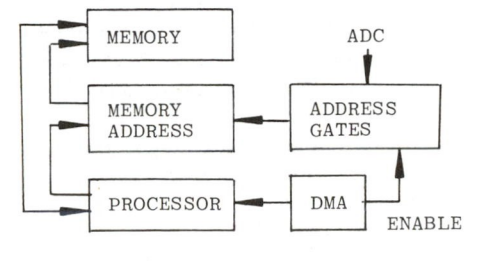

FIG. 17.5.2 Method of generating histograms under DMA control.

data base can be efficiently created in memory. Korn (1977) has suggested several other novel uses of the DMA. An often encountered
signal processing problem is one of developing histograms of a data
field. For example, there are 2^{12} possible outcomes of a 12-bit
analog-to-digital conversion. In 2^{12} contiguous memory locations a
count will be kept of the number of times a particular word has been
encountered (i.e., histogram). Each time a particular word occurs,
its assigned memory location will be incremented by 1, as shown in
Fig. 17.5.2. Another useful DMA function is ensemble averaging.
Using an add-to-memory (if it exists) and the strategy suggested in
Fig. 17.5.3, ensembling can be performed.

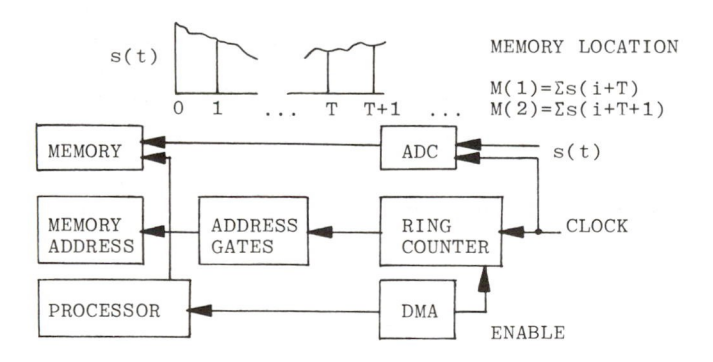

FIG. 17.5.3 Method of ensemble averaging under DMA control.

17.6 ARITHMETIC LOGIC UNITS

A general-purpose mini- or microcomputer consists of the following items:

1. Memory (RAM, ROM, etc.)
2. Memory buffer (memory to I/O transfers)
3. Memory address registers (current locations)
4. Program counter (location of current instruction)
5. Instruction register (instruction code word)
6. General-purpose registers
7. Flags (overflow, carry, etc.)
8. Control logic
9. Arithmetic logic unit (ALU)

The ALU is a collection of electronic subsystems which support arithmetic. They can be 1-bit units or as wide as 4, 8, or 16 bits in a mini- or microprocessor environment. The instruction sets vary from unit to unit. Therefore, comparisons can only be made on a case-by-case basis. Execution rates and architectural philosophies also vary widely. The basic ALU operations are:

1. ADD or SUBtract $(A \pm B \longrightarrow C)$ (e.g., a Motorola M6800 will interpret ADDA, M to mean add the contents of register A to memory location M and return the results to A.)
2. ADD or SUBtract with carry $[A \pm B + (carry) \longrightarrow C]$
3. ROtate Left (ROL)
4. Rotate Right (ROR)
5. Arithmetic Shift Left (ASL)
6. Arithmetic Shift Right (ASR)

With added complexity, the basic ALU can be expanded to do more complex fixed- and floating-point arithmetic.

In a n-bit fixed-point system, data are interpreted as a signed integer over $[-2^{n-1}, 2^{n-2} - 1]$ or in fractional form over $[-1, 1 - 2^{n-1}]$. When a finite-precision numbering system is used, care must be taken to ensure that register overflow will not occur. If overflow does occur, it must be managed through scaling. This operation can be initiated with the receipt of an overflow flag from the ALU. Scaling of data was covered in Section 15.6.

17.7 SPECIAL SIGNAL PROCESSING DEVICES

The digital revolution, combined with the maturing of digital signal processing, has resulted in a new class of electronic integrated hardware. These new chips are designed to support computation and signal processing.

They represent the "tip of the iceberg." Other specialized units are under
development and consideration. In this section some of the early entries
into this field are surveyed.

A multiply-divide unit (MDU) provides a user with a low-cost, low-
power arithmetic module (e.g., RCA CDP1856 in CMOS) having the architecture suggested in Fig. 17.7.1. Typical MDU execution rates are on the
order of 2500 nsec. This is obviously slower that bipolar systems, but
assuming that high speeds are not required, the MDU offers a very cost
effective signal processing design tool. These units are easily interfaced
with 2- to 5-MHz microprocessors as well.

The signal processing scientist is often faced with the problem of
designing a low-cost system having only modest throughput requirements.
In certain circumstances, the arithmetic processor chips used in multifunction hand-held calculators can be used. These chips, sometimes called
"number crunchers," are relatively slow but offer outstanding packaging
and cost savings. They are typically externally clocked and controlled
using a programmable device (e.g., microprocessor using microcode),
dedicated sequential logic, or programmed control stored on ROM or RAM.
An inte esting example of this class of device is the Advanced Micro Devices
Am 9511 arithmetic processor. This unit is capable of being clocked at
4 MHz and therefore may serve as a peripheral processor to a large class
of microprocessors. The 9511 instructions are summarized in Table
17.7.1. All transfers (including operands, data transactions, etc.) are
communicated over an 8-bit bidirectional bus and stored on an integrated
8×16-bit stack. The transfers can be performed I/O on DMA control.
Transfers have a 10- to 20-nsec clock cycle overhead penalty due to required handshaking. The Am 9512 version of this family conforms to the
IEEE standard floating-point format. Other manufacturers are developing
chips and chip sets in this area as well. Some of these units are targeting
the floating-point FFT, while others seem to be interested in FIR or IIR
filtering. Nevertheless, such units can inexpensively offer the user a 5- to
50-fold increase in arithmetic throughput over their software-based counterpart. They therefore represent a feasible design alternative.

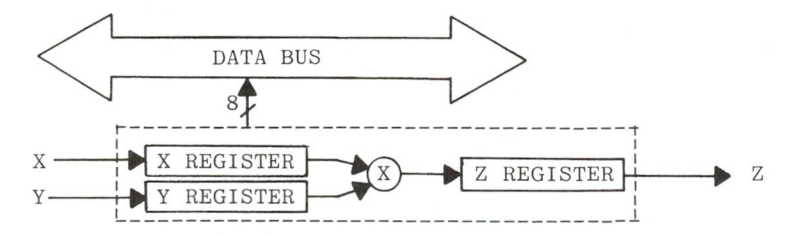

FIG. 17.7.1 Architecture for a multiply-divide unit.

TABLE 17.7.1 Performance Capabilities of a Typical
Number-Cruncher Chip

Instruction	Time (μsec)	
Fixed point (16–bit single precision at a 4 MHz clock)		
ADDition	17	
SUBtraction	30	
MULtiply	92	
DIVide	92	
Fixed point (32–bit double precision)		
ADDition	21	(x. 81 single precision)
SUBtraction	38	(x. 79)
MULtiplication	208	(x. 44)
DIVide	208	(x. 44)
Floating point (25–bit mantissa, 7–bit exponent at 4 MHz)		
ADDition	56–350	
SUBtraction	58–352	
MULtiply	168	
DIVide	171	
Derived floating point functions		
SQuare root	800	
SINe	4464	
COSine	4118	
TANgent	5754	
ArcSINe	7668	
ArcCOSine	7734	
ArcTANgent	6006	
LOGarithm base 10	4490	
Logarithm base e	4478	
EXPonent e^X	4616	
EXponentiate y^X	9292	

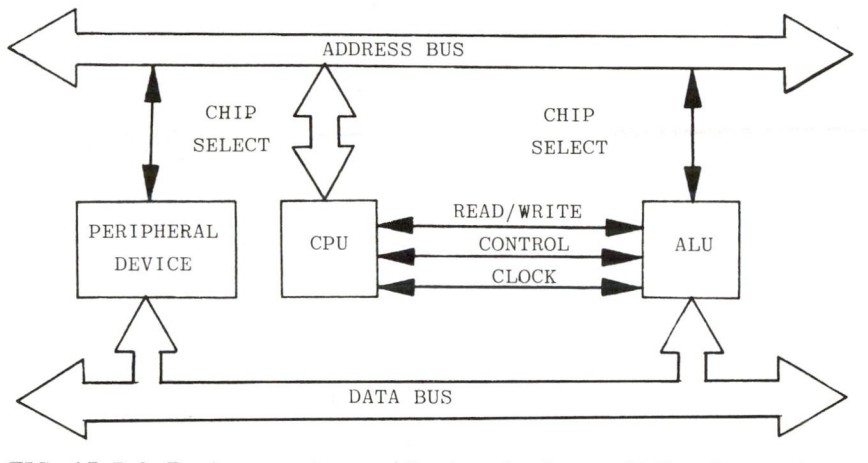

FIG. 17.7.2 Basic computer architecture having an ALU and a CPU.

The system designer is responsible for certain critical decisions if the number-cruncher chip is to be integrated into a system. They are

1. Whether to store non-data-dependent parameters [e.g., $\cos(-j2\Pi k/N)$ in the case of a FFT] in memory or compute the same during run time
2. Optimized program sequences (i.e., optimize program nesting and the use of their limited stack memory)
3. System integration and interface

The simplest architecture, suggested in Fig. 17.7.2, uses the ALU as a peripheral processor.

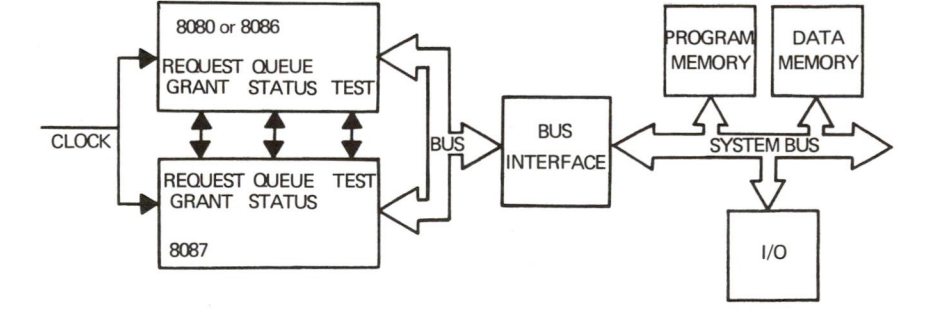

FIG. 17.7.3 Interconnection of a microprocessor and a companion numeric processor.

Another entry into the number-cruncher market is the HMOS Intel 8087 numeric processor. It is designed to be interfaced to a general-purpose 8086 16-bit microprocessor. The 8087 can provide a 100-fold throughput improvement over 8086 arithmetic done in software. The numeric processor has been designed around a 64-bit internal data path, 80-bit registers, and 68-bit ALU.

An interesting feature of the 8086-8087 system is that both chips monitor the same instructions and execute those applicable to itself. The principal responsibility of the 8086 is managing memory, while the 8087 concentrates on arithmetic and logic. The architecture of the system is summarized in Fig. 17.7.3.

Programming the 8086-8087 system is equivalent to programming a stand-alone 8086. The 8086-8087 can be thought of as a new chip (set) with 68 instructions and data sets beyond that provided by the 8086. Programming support is provided at the assembly level (ASM-86) or in a high-level language (PL/M-86). As a result, the software designer can view the 8086-8087 pair as a single-system unit. The experienced programmer can use eight 80-bit registers as a stack of addressable locations and, as a result, manipulate data rapidly and efficiently.

Data formats used by the 8086-8087 are summarized as follows:

Word type	Precision
Word	16 bits 2's complement
Short word	32 bits 2's complement
Long word	64 bits 2's complement
Binary-coded decimal	80 bits (18 digits) including sign
Short real	8-bit exponent, 22-bit mantissa and sign
Long real	11-bit exponent, 51-bit mantissa and sign
Temporary real	15-bit exponent, 64-bit mantissa and sign

Based on a 5-MHz clock rate, the 8086-8087 system will perform as follows:

Operation	8086-8087 (μsec)	8086 (μsec)	Improvement factor
ADD or SUBTRACT	14/18	1600	100
MULTIPLY (single precision)	18	1600	89
MULTIPLY (double precision)	27	2100	78
DIVIDE	39	3200	82

Finally, the 8087 will support the following operations:

1. Data transfers
2. Arithmetic (+, -, *, /, SQRT, RND, ABS, INT$\longrightarrow$FLOAT, etc.)
3. Logical (compare, examine, test)
4. Transcendental (TAN, ARCTAN, $2^X - 1$, $Y * LOG_2 X$, etc.)
5. Constants (0, 1, Π, $LOG_{10}2$, $LOG_e 2$, $LOG_2 10$, $LOG_2 e$)
6. Control (LOAD, STORE, INTERRUPT, CLEAR, etc.)

One of the weaknesses of these prototype signal processing chips and chip sets is software support. The designer must often invest a nontrivial amount of time into program development using low-level languages. A novel approach to the problem of software-device integration is the Intel 2920 chip. This device, referred to as a real-time signal processing chip, was developed with unique analytic properties of digital signal processing in mind. In particular, it can be recalled that most signal processing in arithmetic operations can be categorized as scaling rather than multiplication. For example, in linear shift-invariant filtering and fast transforming, bilinear forms of the type $z = cx$, with c known a priori, are more common than $z = yx$, where y and x are variables. The 2920 exploits this feature.

The 2920 integrates onto a single chip a EPROM, RAM, 28-bit ALU, 8-bit analog-to-digital (A/D) and digital-to-analog (D/A) converters, plus needed control logic. Besides offering packaging elegance, this chip exhibits some interesting real-time programming attributes as well. First, system throughput can be controlled with the choice of clock rate. Second, all instructions require the <u>same</u> interval of time to execute independent of their complexity. This simple real-time capability can be guaranteed only if branching is disallowed. The curse of all real-time programming is the occurrence of conditional branching (e.g., IF, ELSE, etc.) and random

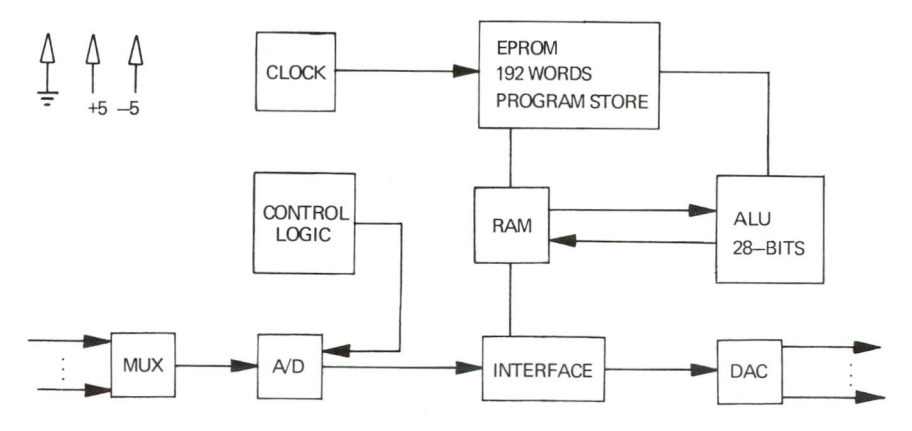

FIG. 17.7.4 Architecture for a contemporary signal processing chip.

interrupts. However, in a signal processing application, such as realizing $y(n) = \Sigma\, a_i x(i + n)$, no interrupt servicing or branching is required within a filter cycle. The architecture of this class of real-time processor is shown in Fig. 17.7.4. The ALU is designed to support scaling rather than pure multiplication. Furthermore, the 2920 uses a canonical-type encode (see Sec. 15.4), which can be explored in terms of the following example.

EXAMPLE 17.7.1 Consider the diagrammed second-order section using a canonical data structure. The unit studied has a throughput rate of 400 nsec per instruction. Using the maximal number of instructions that may be stored in EPROM as a guide, the maximal throughput rate would be

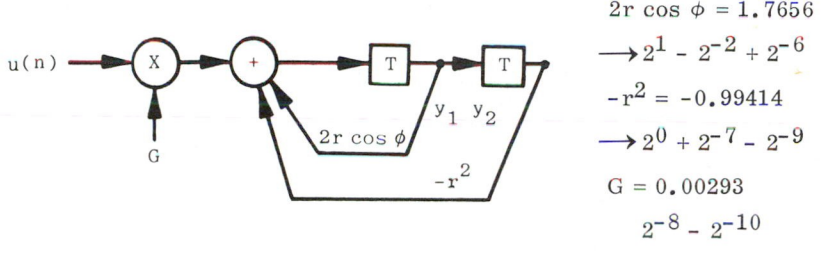

$$2r \cos \phi = 1.7656$$
$$\longrightarrow 2^1 - 2^{-2} + 2^{-6}$$
$$-r^2 = -0.99414$$
$$\longrightarrow 2^0 + 2^{-7} - 2^{-9}$$
$$G = 0.00293$$
$$2^{-8} - 2^{-10}$$

Operator	Source	Destination	Shift	Comment
LDA	Y_1	Y_2		$Y_2 \leftarrow Y_1$
LDA	Y	Y_1		$Y_1 \leftarrow Y_1$
LDA	Y_1	Y	Left 1	$Y \leftarrow (2^1 - 2^{-2} + 2^{-6})Y_1$
SUB	Y_1	Y	Right 2	
ADD	Y_1	Y	Right 6	$2r \cos \phi$
SUB	Y_2	Y	Right 0	$Y \leftarrow 2r \cos \phi\ Y_1$
ADD	Y_2	Y	Right 7	
SUB	Y_2	Y	Right 9	
ADD	U	Y	Right 8	$Y \leftarrow 2r \cos \phi\ Y_1 - r^2 Y_2$
SUB	U	Y	Right 10	$+ (2^{-8} - 2^{-10})U$

(Plus I/O instructions: A/D and D/A) G

$(192)400 \times 10^{-9}$ or 13 kHz. This translates to a Nyquist rate of 6.5 kHz.
Of course, shorter programs will result in higher throughput.

Finally, the 2920 is supported by Intel under a comprehensive software package known as SPP20. This software will run on the manufacturer's development systems. Another signal processing device that is gaining attention is referred to as the signal processing peripheral (SPP) device. For example, the American Microsystems S2811 device is a high-speed VMOS chip containing embedded multipliers. This programmable chip can be used in voice and audio application. The S2811 can fetch two operands, multiply them, accumulate the sum of products, and store data at a 300-nsec rate while dissipating 1 W (at 5 V). The processor chip contains an on-board ROM, RAM, 12-bit multiplier-accumulator, I/O, 20-MHz clock, and pipelined data paths. Reported benchmark experiments are summarized in Table 17.7.2.

The S2811 can communicate through a serial or parallel I/O link. The separate microprocessor port is directly interfaceable with the 6800 and 9900 class microprocessor, which supplies program control.

Signal processing generally requires that a signal be uniformly sampled and analyzed. Sampling strobes can be generated by the S2811 by adjusting the program length. The program length may be altered by packing the code with "wait loops" so as to achieve a desired sampling rate. Synchronization can also be achieved through external clocking.

User's code is stored on a 256×17 bit ROM having an effective storage capacity of 250 17-bit instruction words. SPP assembly-level language programming is scheduled to be supported with an in-circuit emulator and software. The manufacturer will program the ROM to the user's specifications if desired. Preprogrammed SPP operations, such as the FFT, will probably be marketed in the future.

EXAMPLE 17.7.2 Internal to the SPP is a 12×12 modified Booth multiplier having a 300-nsec multiply time. In addition, there is a system of registers and counters which can be used to support digital filtering. Using

TABLE 17.7.2 Summary of Experimental Results

Operation	Number of instructions	Speed (μsec)
Biquad filter	7	2.1
u-Law conversion (Sec. 17.8)	15	6.6
V.27 (4800 b/s) modem	248	498
32-point complex FFT	190	1.5 m/32 pt

x(n)	OP1	OP2	Operand	Comments
	NOP	CLAC		Clear ALU and overflow
		LLT1	L(XXX)	Load base, loop counter set up
	AVZ	LIBL	D(XX.X)	AVZ = load RAM-V contents to ALU
				LIBL = load input register content into base and loop counter (e.g., XX.X base = 1, loop = LC = 31)
	APZ	REPT	D(00.0)	APZ = load product in ALU
				REPT = repeat next instruction until LC = 0 (e.g., set up 1st multiplier, repeat 31 times)
	APA	TVIB	UV(X,X)	APA = add product to ALU contents
				TVIB = transfer VP contents to RAM-V and increment base (i.e., sum of products)
	ADA	TACV	D(XX.X)	TACV = transfer ALU contents to RAM-U

Accumulate last term and store in XX.X—restack

FIG. 17.7.5 Design of an FIR using a signal processing chip.

this hardware, a 31st-order FIR can be configured in the manner suggested in Fig. 17.7.5.

VLSI will have a major impact of digital signal processing systems and system design in the 1980s. High-density random access memory devices are now commercially available. Also, the VLSI multiplier accumulator units, described in Section 16.3, represent a major breakthrough in signal processing hardware. Many other dedicated signal processing tools are being envisioned in a VLSI sense. For example, Bell Laboratories is developing a VLSI echo canceller. Using NMOS, a 128-tap FIR filter has been reported.

EXAMPLE 17.7.3 A fundamental digital signal processing subsystem is the second-order direct II filter section. This filter, sometimes referred to as a biquad filter is shown in Fig. 17.7.6.

An overflow correction unit has been included in Fig. 17.7.6 to emphasize a point. Analytic scaling policies such as those described in Chapter 12 can be used to derive an overflow-free fixed-point filter architecture. The scaling bounds derived are often considered to be pessimistic. As a result, the scale factor is often relaxed in the hope of trading dynamic range off against the remote possibility that a worst-case overflow condition will occur. At other times, the rigorous derivation of overflow scale factors are replaced by an "educated guess." Under either set of circumstances, the possibility of a register overflow exists and must be managed. The most effective overflow management policy relies on the use of saturating arithmetic. NEC is developing a multibus signal processing chip which possesses many useful attributes, including saturating arithmetic. Its microinstruction time is defined by the 250-nsec delay of its 16×16 multiplier (versus 400 nsec for the Intel 2920). The device is capable of executing 500 microinstruction steps. Combined with its speed, it has the ability to mechanize a high-performance version of the filter described in Fig. 17.7.6. The program sequence used to create this filter is summarized as follows:

Step	Instruction	Comment
1	$Y_{i-1} \longrightarrow K$; $(-B_1 + 1) \longrightarrow L$; $AC - Y_{i-1} \longrightarrow C$	$AC = X_i - Y_i$
2	$Y_{i-2} \longrightarrow K$; $-B_2 \longrightarrow L$; $AC - M \longrightarrow AC$	$AC = X_i - B_1 Y_{i-1}$
3	$Y_{i-3} \longrightarrow K$; $A_2 \longrightarrow L$; $AC - M \longrightarrow AC$	$AC = X_i = B_1 Y_{i-1} - B_2 Y_{i-2} \triangleq Y_i$
4	JP to 6 if $\overline{CA0}$ (not overflow)	Saturation arithmetic
5	$C \longrightarrow AC$	Saturation arithmetic
6	$AC \longrightarrow T$; $AC + M \longrightarrow AC$	$AC = Y_i + A_2 Y_{i-2}$
7	$Y_{i-1} \longrightarrow K$; $(A_1 - 1) \longrightarrow L$; $AC + Y_{i-1} \longrightarrow AC$	$AC = (Y_i + Y_{i-1} + A_2 Y_{i-2})$
8	$T \longrightarrow Y_{i-1}$; $AC + M \longrightarrow AC$	$AC = (Y_i + A_1 Y_{i-1} + A_2 Y_{i-2}) = Y_i$
9	$K \longrightarrow Y_{i-2}$	Restack

Registers are T and C

Multiplier$\Longrightarrow K \times L = 2^{\Delta} M + N$; AC = accumulator

which equates to a $9(250 \text{ nsec}) = 2.25\text{-}\mu\text{sec}$ (444-kHz) filter cycle. Also, using this chip, a 128-point real FFT can be computed on the order of 2 msec (500 Hz).

The fast Fourier transform (FFT) is a likely candidate to benefit from the microelectronics revolution. Presently, there are no low-cost high-throughput chips or chip sets on the market. There is, however, a movement

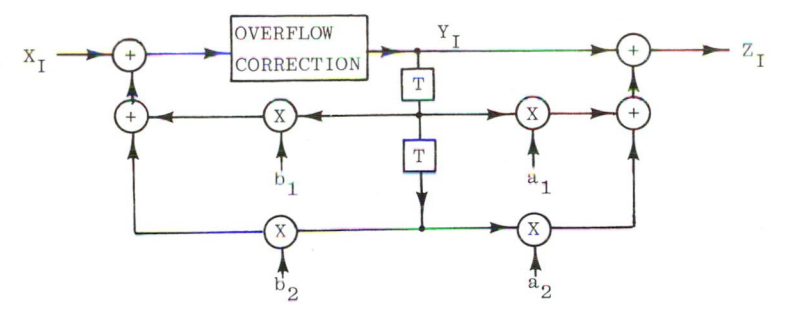

FIG. 17.7.6 Architecture of a biquad discrete filter.

to fill this void in the semiconductor industry. Some of the initial entries
into this market are:

1. American Microsystems (AMI) 2811 (in a fixed-program form it is
   marketed as the 2814)
2. NEC 7720
3. TRW CMOS 50-MHz chip set

The availability of these products is unsure, but they do indicate what will
be available in the near future. The AM2814, for example, was designed
with the audio market in mind. It is a radix-32 FFT subsystem capable of
being integrated into a 512-point configuration by simply nesting subsystems.
TRW's product is a radix-16 FFT capable of performing a transform within
a 35- to 50-μsec window. The slower 2814 has a 512-point conversion time
of 270 msec if multiplexed. If data are processed along a 2M-words per
second DMA link, the 512-point transform time can be lowered to 50 msec
in a multiplexed configuration. If 16 subsystems are parallel, and can run
concurrently, a 3-msec (333 Hz real time) conversion time can be achieved.

17.8 DATA ACQUISITION

Most digital filters require some sort of data acquisition support. The
field of analog-to-digital (A/D) and digital-to-analog (D/A) conversion is
quantified in terms of the following items:

1. Accuracy: absolute accuracy of conversion relative to a standard
   voltage or current.
2. Aperture uncertainty: uncertainty in sampling rate.
3. Bandwidth: maximum small-signal 3 dB frequency.
4. Codes: 2's complement, 1's complement, sign-magnitude, etc.
5. Common mode rejection: measure of ability to suppress unwanted noise
   at input.

TABLE 17.8.1 Classification of Analog-to-Digital and
Digital-to-Analog Converters

Analog to digital		Digital to analog	
Type	Conversion rate	Type	Settling time
Ultra-high speed	> 3 MHz	Ultra-high speed	< 100 nsec
Very high speed	300K$\leftarrow$3 MHz	Very high speed	100$\rightarrow$1 μ
High speed	30K$\leftarrow$300K	High speed	1$\rightarrow$10 μ
Intermediate	3K$\leftarrow$30K	Intermediate	10$\rightarrow$100 μ
Low speed	< 3 KHz	Low speed	> 100 μ

6. Conversion time: maximum conversion rate. If separate sample-and-hold amplifier is used, the conversion rate is the maximum conversion rate of the composite system.
7. Full scale: dynamic range in volts.
8. Glitch: undesirable noise spikes found in A/D output.
9. Linearity: linearity across dynamic rate.
10. Offset drift: worst-case variation due to parametric changes.
11. Precision: repeatability of measurement.
12. Quantization error: $\pm(1/2)$ LSB.
13. Resolution: 2^{-n}.

Analog-to-digital and digital-to-analog converters are generally classified in terms of word length and speed. The speed ranges are summarized in Table 17.8.1.

Generally speaking, the digital signal processing system designer is required to select the best available conversion unit that is available commercially. Due to rapid changes in technology, this field is constantly changing. Therefore, a rigorous comparison cannot be justified. Broadly interpreted, the A/D conversion choices are given in Table 17.8.2. An auto-zeroing system will automatically compensate for any dc offset in an A/D unit. In applications where slow conversion rates can be tolerated, auto zeroing is encouraged.

Successive approximation converters make use of integrated ladder networks. New ladder trimming techniques have driven the cost of these units down to the point where they represent a small fraction of the system cost. A typical dual-slope MOS converter (e.g., Intersil 7109 in CMOS), is a 12-bit, 30-conversion per second, 20-mW unit. Interfacing these units to a programmable processor is generally accomplished through the use of a parallel peripheral interface (PPI). There are chips and chip sets commercially available which will allow the addressable devices to be attached to a system's address-control-data bus.

Some converters are outfitted with programmable gain front ends. The gain of these amplifiers can be adjusted to a discrete set of values under program control (e.g., 1, 2, 4, 10). By adjusting the gain of these units, converter saturation and dynamic range problems can be dynamically controlled. Also, long-word-length converters can be synthesized from smaller-word-length converters in the manner suggested in Fig. 17.8.1. Here a 16-bit converter is realized by interconnecting an 8- and 9-bit converter.

A useful variation of the standard A/D converter theme is the data compression or robust converter. Using such a device, large dynamic data ranges (e.g., voice processing) can be compressed into specific dynamic range limits. A commonly encountered data compression routine is the mu-law compander, given by

$$y(x) = V \frac{\log(1 + \mu x/V)}{\log(1 + \mu)} \tag{17.8.1}$$

This relationship is graphically interpreted in Fig. 17.8.2. For $\mu = 0$, the converter is linear. The value of $\mu = 255$ has been found useful in telephone applications. As a result, commercial $\mu = 255$ companders are now found in abundance. Other companders fall into one or more of the following broad categories:

TABLE 17.8.2 Speed Ranges for Data Acquisition Systems

Type	Advantages	Disadvantages
Dual and triple slope converter	Low cost, good linearity, auto zero, low sensitivity to parameter change, low sensitivity to rate change	Slow data rates
Successive approximation	Fast, good accuracy and resolution, low cost,	Requires sample and hold, accuracy requires resistive ladder trimming
Flash	Fastest	High power, high cost

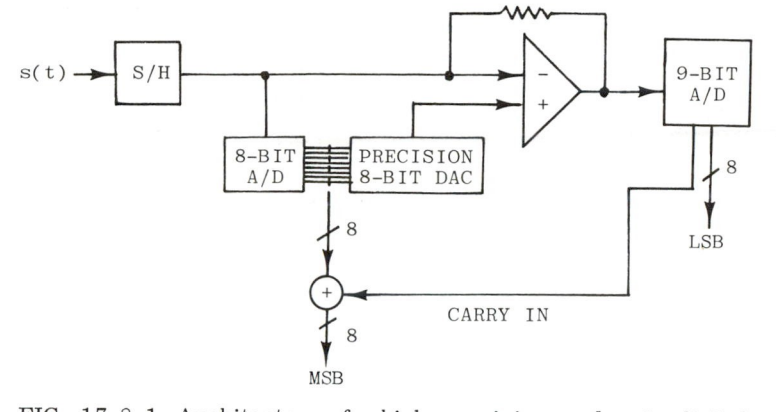

FIG. 17.8.1 Architecture of a high-precision analog-to-digital converter.

1. Differential pulse code modulation (DPCM)
2. Oversampled encoder
3. Linear predictors
4. A-law companders

Digital-to-analog converters (DACs) are of major importance as well. Some DAC units are found in a three-wire configuration. Here the difference between the ground potential of the DAC and the system it is driving is sensed. The difference is used to correct the DAC by removing this dc offset. DACs can be extremely useful in supporting test and calibration tasks. The monolithic DAC units are basically low in cost and simple to interface. Some of the standard applications of a DAC are suggested in Fig. 17.8.3.

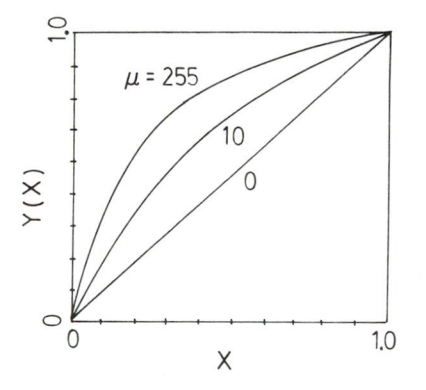

FIG. 17.8.2 Operational curve for a mu-law compander.

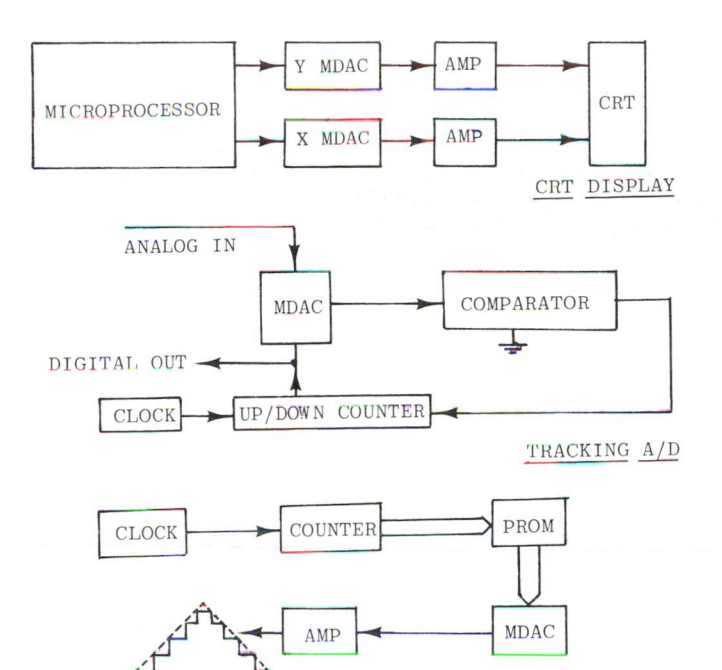

FIG. 17.8.3 Typical digital-to-analog converter applications.

Multiplying digital-to-analog converters (MDACs) are popular signal processing devices. They appear with a variety of word lengths and may have conversion rates up to the middle megahertz range. Some of the principal uses of the MDAC are:

1. Microprocessor data acquisition
2. Servomotor control
3. Pen and/or tool position control
4. Programmable attenuators
5. Programmable power supplies
6. CRT drives
7. High-speed modems
8. Analog-to-digital conversion
9. Waveform generation

17.9 GENERATORS

Signal generation is an important part of signal processing. The synthesis of standard signals, such as sinusoids and ramps can be accomplished in

software. Another time series that is often found in signal processing use
is the psuedo-random noise sequence. Uniformly distributed pseudo-random
sequences index pseudo-random noise can be generated in software, as
suggested in Example 4.9.3. Here a subroutine labeled RANDU is used to
synthesize an approximately white uniformly distributed (over [0, 1]) random sequence using a general-purpose 16-bit digital computer. Using
RANDU and the law of large numbers (see Section 4.9), a normally distributed (Gaussian) random sequence can be approximated. In Example 4.9.3,
subprogram GAUSS is used to create a normally distributed sequence having
mean AM and standard deviation S. The uniform number generator can also
be used to approximate Rayleigh distributed noise. Consider again x(n) to
be uniformly distributed over [0, 1]; then

$$\underline{\text{Rayleigh}}: \quad P(y) = \frac{y}{\sigma^2} \exp\left(\frac{-y^2}{2\sigma^2}\right)$$

$$\underline{\text{Generator}}: \quad y(n) = \text{SQRT}\left(2\sigma^2 \ln\left[\frac{1}{x(n)}\right]\right)$$

(17.9.1)

Using the Rayleigh noise model, normal noise can be approximated using

$$\underline{\text{Normal}}: \quad P(w) = \frac{1}{\sqrt{2\Pi}\sigma} \exp\left(\frac{-w}{2\sigma^2}\right)$$

$$\underline{\text{Generator}}: \quad w_1(n) = y(n) \cos[2\Pi x(n+1)]$$

$$w_2(n) = y(n) \sin[2\Pi x(n+1)]$$

(17.9.2)

where y(n) is given by Eq. (17.9.1). Furthermore, $w_1(n)$ and $w_2(n)$ are independent (quadrature). This can prove to be beneficial in some applications
(e.g., quadrature phase shift, correlation, coherence, etc.).

Uniformly distributed random noise can be synthesized in hardware
as well. Using binary-valued pseudo-random sequence generators as a
guide, uniform noise generators can be architected. It is assumed that
x(n) is an unsigned fractional number (i.e., $0 \le x(n) \le 1$), and

$$x(n) = x(n - 1) - x(n - K) \quad \text{modulo 1}$$

where K is large (e.g., $K \ge 50$). Rader suggested the network shown
in Fig. 17.9.1. Here the modulo operator can be realized by ignoring
the adder's overflow bit.

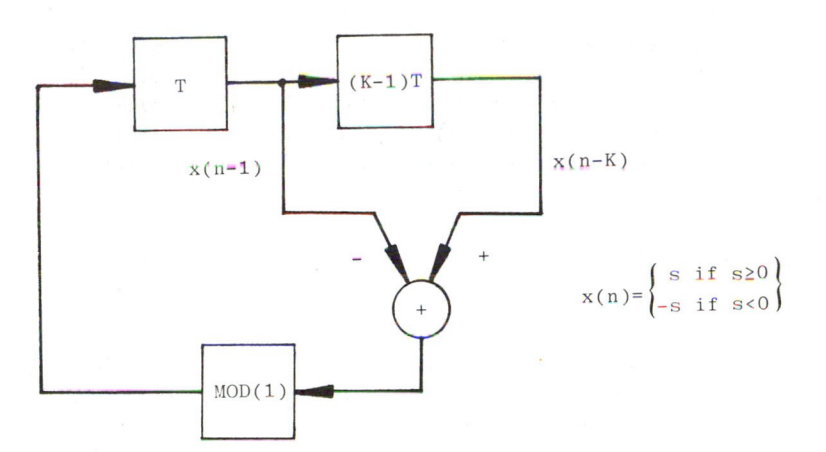

FIG. 17.9.1 Pseudo-random uniform noise generator based on a pseudo-random sequence circuit.

17.10 SPECIAL TECHNOLOGIES

Three nondigital technologies have made an impact on signal processing in recent times:

1. Charge transfer devices (CTD)
2. Surface acoustic wave devices (SAW)
3. Switched capacitance filters (SCF)

The SAW device is useful in very high frequency applications (e.g., radar). Since most of the theory and practice found in this reference is directed toward non-video-frequency applications, a detailed study of this technology will not be pursued.

The CTD family has the charge-coupled device (CCD) and bucket brigade device (BBD) as members. The devices differ slightly in dynamic range, frequency response, and fidelity. They function as analog shift registers capable of storing and transferring information in the form of "charge packages." The CCD and BBD can be used as low-cost (in volume only) fixed-coefficient FIR filters. Since these filters are basically analog (i.e., sample data) filters, they are not subject to finite-word-length effects. The error source in this class of filter is due to transfer inefficiency (de-noted ϵ and CTE). CTE is a measure of the amount of charge that remains behind after completion of a transfer operation. The CTE values for a quality commercial-grade CCD range from 10^{-3} to 10^{-4}. The effect of the CTE is that of inserting a weak feedback structure within the device itself. The transfer of charge has in fact been modeled as a Bernoulli process. Here the transfer of charge from one cell to its neighbor is given by $\delta = 1 - \epsilon$. The distribution of charge after n successive shifts can be approximated by

$$p(i, \ n) = \binom{n}{i} \delta^{n-i} \epsilon^{i} \tag{17.10.1}$$

Studies have shown that the CTE effect will produce errors of several percent in high-order filters (e.g., $N \geq 1024$). However, the practical size for most reported CCD recursive filters is 32 and transversal filters is 500.

The CCD generally possesses a maximum clock rate of 1 MHz for surface channel devices and 5 MHz for buried channel units. Minimum clock rates vary widely. Commercial CCD units are generally configured as either serial-in serial-out or serial-in parallel-out devices. For example, the Reticon SAM-128LR is a serial-in serial-out device consisting of 128 delay cells having a reported dynamic range of 70 dB. This device can be used in various time compression, bandwidth compression, linear prediction, and comb filtering applications.

Another Reticon device is the TAD-32. This unit is a serial-in parallel-out 5-MHz, 70-dB, 32-element tapped delay line. This unit can be used to synthesize recursive (IIR) filters by supplying coefficient weights to the signal found at the output taps (ports). Weighting, or coefficient scaling, can be achieved in one of the following ways:

1. Precision amplifiers: Accurate, expensive, require periodic "tuning."

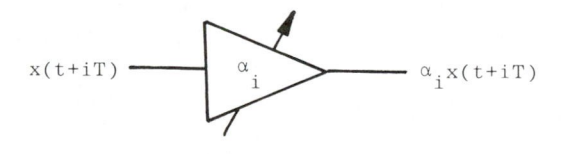

$$x(t+iT) \quad\quad\quad \alpha_i \quad\quad\quad \alpha_i x(t+iT)$$

2. Resistive voltage dividers: Less accurate, less expensive, require a more elaborate adjustment procedure.

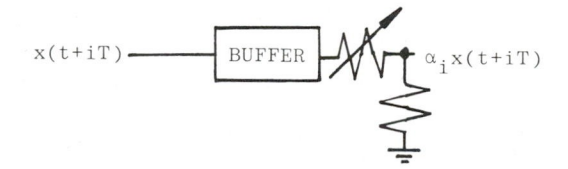

$$x(t+iT) \quad\quad \boxed{\text{BUFFER}} \quad\quad \alpha_i x(t+iT)$$

3. <u>Split electrodes</u>: This is an internally realized weighting policy based on geometry. The weighting policy must be implemented at the time of device fabrication and is therefore not alterable in the field. In those cases where high-volume production is guaranteed (e.g., chirp-z transforms, lowpass filters, etc.), the split electrode concept is feasible.

4. <u>Pulse duration modulation</u>: This policy allows the weighting coefficient to be realized using simple electronic hardware that can be programmed in real time. In the filter shown in Fig. 17.10.1, the feedback coefficient is adjusted by a FET switch. The value of this coefficient is restricted to reside in the interval (-1, 1] and has a value equal to the switching duty cycle. The duty cycle defines what percentage of charge, found at the output port, will be feedback to the input. Experimental results are reported in Fig. 7.10.2.

The disadvantages of the CTD filter are:

1. Limited dynamic range
2. Coefficient realization and tuning
3. Limited frequency response
4. Insertion loss (20 dB typical)

Monolithic active low-frequency filters are known to be extremely sensitive to RC and LC parameter variations. The limitations of the CTD filter has been discussed. An alternative technology which is exhibiting some

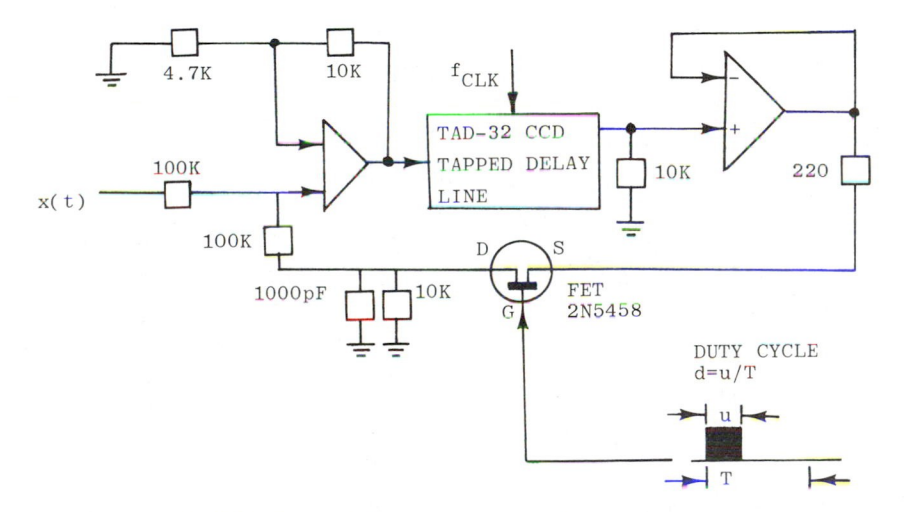

FIG. 17.10.1 CCD tap-weight scaling network.

promise in this problem area is the SCF. The physics which defines their
operation can be argued in terms of the following one-pole filter. Consider
the simple switched capacitor network shown in Fig. 17.10.3. If $1/T$ is
much larger than the Nyquist sample rate, $C_2/C_1 T$ plays the role of a
resistor in a classic RC sense. The importance of this result can be appreciated in the context of a MOS structure. In the MOS technology, high-
quality linear and stable capacitors can be realized. However, the same is
not true for resistors. Therefore, being able to replace resistors by
capacitor ratios (which can be accurately defined and fabricated) is a distinct advantage.

The SCFs have utility when high-volume fixed-coefficient filters having modest Nyquist frequencies are needed. For example, the Reticon
R5600 series SCFs are 1/3-, 1/2-, and full-octave bandpass filters which
are suited to various audio and harmonic analysis applications. These
sixth-order Chebyshev filters have a tunable center frequency (with respect

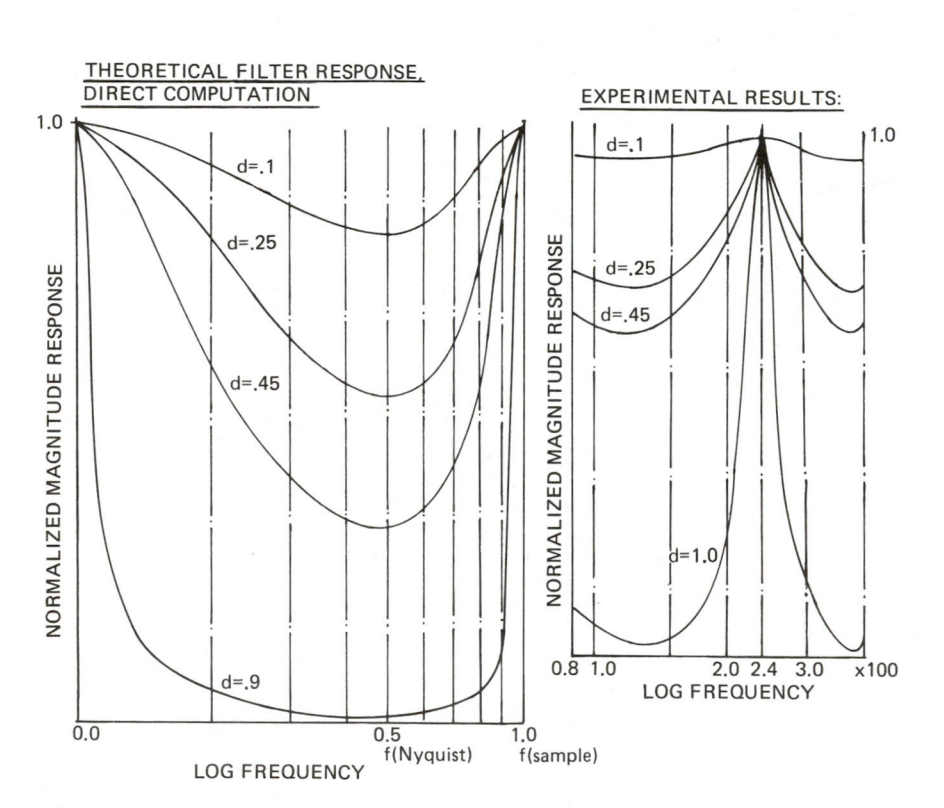

FIG. 17.10.2 Experimental behavior of a CCD filter using time-duration
modulation scaling.

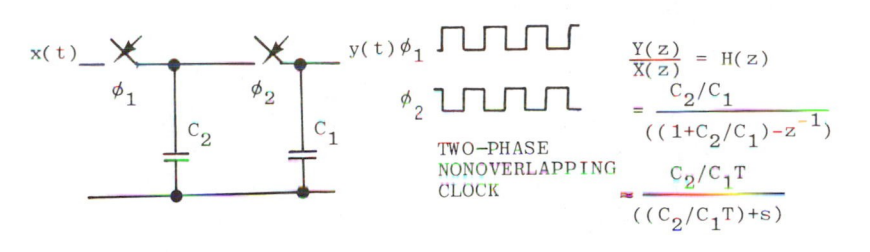

FIG. 17.10.3 Switched capacitance filter model.

to clock rate), 80 dB dynamic range, and an insertion loss on the order of ± 0.2 dB. Filter center frequencies may range from 0.5 to 10,000 Hz and passband ripple is typically 0.2 dB (see Fig. 17.10.4). The SCFs are now popular in telephone and modem applications and should have an impact on other areas in the near future.

EXAMPLE 17.10.1 The Motorola MC14413 and 14 are SC filters designed to support PCM Codec voice communication systems. The transmit and receive filters are five-pole elliptic lowpass filters with a sample rate of 128 kHz. In the transmit section the filter serves an antialiasing purpose and smoothes the output when operating as a receiver. In addition, the transmit filter includes a three-pole Chebyshev highpass filter, having a critical frequency near 100 Hz. This section rejects 50/60-Hz line noise.

17.11 RELIABILITY

Systems fail when they are stressed beyond their rated capacity. If stress is random, the failure rate can be interpreted in a statistical sense. In reliability computations, the event of interest is defined to be the satisfactory operation of the device or system. The failure rate can be parameterized in terms of its mean time before failure (MTBF). The MTBF is equal to the total operating time divided by the total number of failures over that period. A failure rate $r = 1/\text{MTBF}$ can sometimes be expressed as

$$r = \sum_i r_i; \quad r_i = \text{failure rate of the ith subsystem} \qquad (17.11.1)$$

Fortunately, modern signal processing hardware subsustems and components are extremely reliable. For example, the mean time between failure of a Motorola MC 6800 microprocessor in a ceramic package at 70° F has been determined to be 78K hr in 1974 and 16.6M hr in 1980. Unfortunately, this makes it difficult to obtain meaningful reliability data in the laboratory. Unless the sample size is large, reliability data confidence is low. Accelerated testing is sometimes used to create a large data base. However,

889

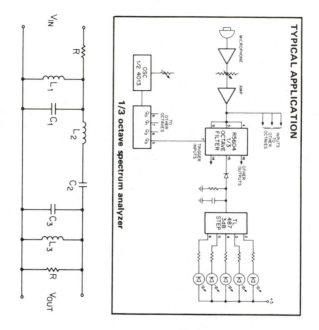

TYPICAL APPLICATION

1/3 octave spectrum analyzer

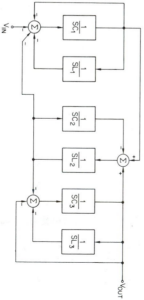

L-C circuit equivalent to the switched capacitor filters

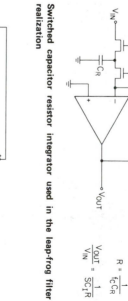

Switched capacitor resistor integrator used in the leap-frog filter realization

$$R = \frac{1}{f_C C_R}$$

$$\frac{V_{OUT}}{V_{IN}} = \frac{1}{SC_1 R}$$

Filter realization — Active leap-frog technique

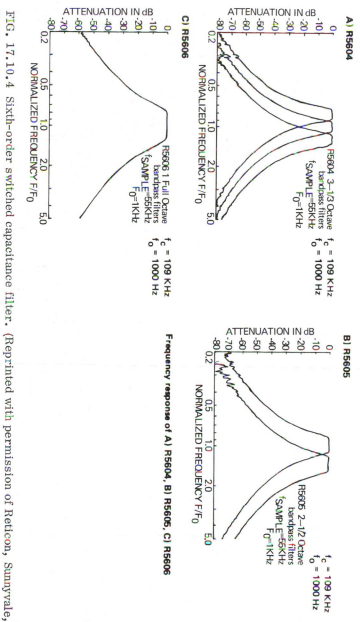

A) R5604

ATTENUATION IN dB

R5604 3–1/3 Octave
bandpass filters
f_{SAMPLE}=55KHz
F_0=1KHz

f_c = 109 KHz
f_o = 1000 Hz

NORMALIZED FREQUENCY F/F_0

B) R5605

ATTENUATION IN dB

R5605 2–1/2 Octave
bandpass filters
f_{SAMPLE}=55KHz
F_0=1KHz

f_c = 109 KHz
f_o = 1000 Hz

NORMALIZED FREQUENCY F/F_0

C) R5606

ATTENUATION IN dB

R5606 1 Full Octave
bandpass filters
f_{SAMPLE}=55KHz
F_0=1KHz

f_c = 109 KHz
f_o = 1000 Hz

NORMALIZED FREQUENCY F/F_0

Frequency response of A) R5604, B) R5605, C) R5606

FIG. 17.10.4 Sixth-order switched capacitance filter. (Reprinted with permission of Reticon, Sunnyvale, California.)

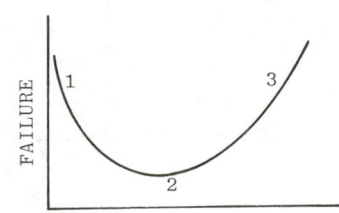

FIG. 17.11.1 Classic device failure-time history. 1, Early failure (infant mortality; 2, normal use period; 3, wearout period (old age).

sometimes such tests do not reflect the operational behavior of the component tested.

Reliability data typically has the form suggested in Fig. 17.11.1. When a large number of components are involved in a design, a Poisson failure model is often used. The probability that n failures will occur in T seconds can be approximated by

$$P_n = \frac{m^n \exp(-m)}{n!} \; ; \quad \frac{T}{MTBF} = m \qquad (17.11.2)$$

A special case is $P_0 = \exp(-T/MTBF)$ (sometimes called the <u>reliability factor</u> and denoted R) and $-\ln(R) = T/MTBF$.

EXAMPLE 17.11.1 For a 90% reliability (R = 0.9), a system will operate without failure for approximately 0.1 of the MTBF [i.e., $-\ln(0.9) = 0.1$]. Therefore, high MTBF rates are required if even modest "up-times" are to be achieved.

Using the Poisson approach in a different manner, test intervals can be estimated. For the purpose of discussion, define the acceptance quality level (usually provided by the manufacturer) to be denoted AQL and the percent of defective components in a sample to be PD. It can be shown that

$$P_n = \frac{m^n \exp(-m)}{n!} = 1 - C_n; \quad m = \frac{PD}{AQL} \qquad (17.11.3)$$

where C_n represents the <u>confidence level</u>. For commercial-grade integrated circuits, an AQL of 10% is common. For a 90% confidence level (i.e., $C_n = 0.9$), one obtains

$$P_n = 0.1 = \frac{m^n \exp(-m)}{n!} ; \quad m = \frac{PD}{AQL}$$

For $n = 0$, one obtains $\exp(-m) = 0.1$ or $m = 2.3$. Since we are dealing with percentages, this suggests that 230 experiments would have to be performed to achieve the indicated confidence in the experiment. If the components are prestressed (i.e., "burn-in"), the AQL may drop to 0.1%. Here 2300 experiments would be needed to test the 90% confidence conjecture. For $n = 0, 1, 2, 3, \ldots$, the minimum number of samples required to test the 90% conjecture, with a 10% AQL, are:

n	0	1	2	3	$\cdots$
Minimum number of tests	230	390	530	670	$\cdots$

This means that if 230 consecutive units pass the test successfully, accept with 90% confidence that the manufacturer's estimate of AQL = 10% is valid. If a single unit fails during the first 230 trials, continue the test over 390 samples. If there is but one failure in this period, accept the hypothesis, and so on.

17.12 SPECTRUM ANALYZERS

Modern spectrum analyzers appear in two basic forms: the wave analyzer and digital Fourier analyzer. Wave analyzers are essentially heterodyne devices consisting of a linear sweep oscillator and a narrowband IF filter. They appear in time compression and noncompression forms. Time compression analyzers can be found operating at rates faster than their FFT counterparts. However, these units do not possess the precision, noise immunity, or dynamic range of the FFT units.

Digital Fourier analyzers are generally equipped with a minicomputer, data acquisition subsystem, display, keyboard, and complete software support. These units are valuable in:

1. Waveform averaging (linear, peak exponential, differential, etc.)
2. Data windowing (uniform, Hann, Hamming, etc.)
3. FFT and IFFT (sometimes ZOOM as well)
4. Power spectrum analysis
5. Transfer function analysis
6. Coherence analysis
7. Correlation studies

Dedicated commercial FFT systems are available starting at about 7000 to 25,000 (and beyond). At the low end, one finds the single-channel analyzer based on microprocessor technology. At the high end is the multichannel real-time processor. The frequency ranges found in the industry range from 100 kHz to 1 Hz (e.g., HP3582A from 25 kHz to 1 Hz, Rockland 512/S from 100 kHz to 20 Hz). These are sample rates and should not be confused with the system real-time bandwidth. The real-time bandwidth (RTB) satisfies

RTB = N/T spectral lines per second or 1/T transforms per second   -

where N is the number of spectral lines in the transform and T is the processing time. For example, the Princeton Model 4512 has a 33-msec processing delay for a 512-point FFT. Therefore, the RTB = 30 transforms per second.

Some systems support ZOOM operations having zoom magnification factors of 16, 32, 64, 128, and others. These systems are sometimes supported with powerful graphics. Many also allow the user to perform data analysis (e.g., coherence, power spectrum) through a set a simple user-oriented keyboard commands. In a closely allied area, there exist software packages for general-purpose machines which are capable of supporting digital signal analysis. SPECTRUM IV is a low-cost ($20 from the Computer of the University of Texas at El Paso, El Paso, Texas) set of integrated FORTRAN programs which support signal analysis. Signal Technologies Interactive Laboratory System (ILS) is a package designed to run on Digital Equipments PDP-11 series minicomputers ($4500). It contains 85 main programs and 230 support routines which drive DEC-compatible data acquisition and display equipment.

17.13 ARRAY PROCESSORS

Array processors are high-speed computational units which are attached to a host computer. They are capable of accelerating floating-point arithmetic by a factor of 10 to 100. Array processors are expensive but become economically justifiable in those cases where a large number of floating-point instructions must be serially processed (it is typically assumed that at least 60 consecutive floating-point instructions must be executed to justify array processing).

Array processors are designed to overcome the problem known as the "von Neumann bottleneck." To execute a routine in a von Neumann sense (i.e., stored program machine), the following operations are required:

1. Locate the instruction to be executed in memory.
2. Interpret the instruction.

3. Locate the required data in memory.
4. Perform the specified operation.
5. Return data to memory.

The bottleneck is formed by requiring all of these operations to share a single common bus. Therefore, data must be interleaved and sequenced in order to provide a continuous stream of data to and from the processor. Throughput can, however, be accelerated if multiple data paths are provided. This can be achieved at the expense of system cost and complexity.

Speed enhancements can be achieved through pipelining and parallelism. Long word lengths are generally required to provide the needed control for these loosely architected processors. Often, operations are decomposed into a set of simultaneously executed small subprograms. By being able to perform more operations per unit time, throughput improvements can be realized.

Array processors belong to one of the following groupings:

1. Large array processors used to support supermaxi-computers.
2. Array processors possessing a separate mainframe and memory. These units support general-purpose minicomputers and cost from $20,000 up.
3. Collection of logic and hardware on a single card which fits into a system slot of a mini- or microcomputer. These units contain limited pipelining and cost $5000 up.

Commercially available array processors are often supported with a high-level language (e.g., FORTRAN). However, in order to use an array processor to its fullest, the processor must be kept active. This often necessitates writing software in assembly-level language or microcode.

Typical floating-point processors separate addition and multiplication into two distinct operations. The interface between the processor and its host is a system of registers that are devoted to programmed I/O control plus others used in DMA transfers. The architecture found in a typical system is suggested in Fig. 17.13.1. The DMA registers work on data stored in the host and array processor. In some circumstances, differences may exist between the data format required from one machine to another. This format problem can often be solved by managing (code conversion) the DMA registers during data transfer.

Pipelined operations require that an event be partitioned into a set of smaller units. For example, a floating-point multiply can be conveniently factored into:

1. Multiply mantissas.
2. Add exponents.
3. Normalize and round results.

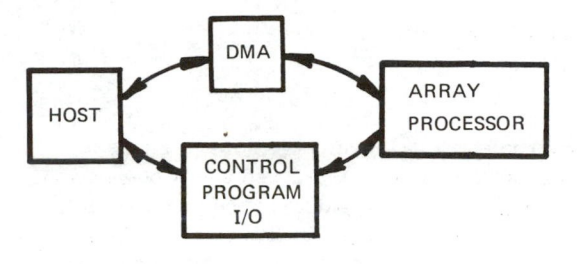

FIG. 17.13.1 Diagram showing the interface between an array processor and a host computer.

In the Floating Point Systems' AP-120B, these operations require 167 nsec each. The total multiplication time, as indicated in Fig. 17.13.2, is 500 nsec per floating-point multiply. Addition can be broken down similarly. At a 6-MHz clock rate, 12 million floating-point instructions (i.e., 12MFLOP) can be completed.

EXAMPLE 17.13.1 Floating Point Systems markets a 64-bit attached array processor known as a FPS-164. It supports a DEC VAX-11/780 of IBM 3033 or 4341 host. The array processor can handle up to 12M floating-point instruction per second plus a concurrent 6M integer/address operations per second. The FPS-164 main memory can be expanded to 1.5M 64-bit words of 12M bytes. Larger systems are also available in the form of supercomputers. For example, Control Data's Cyber 205 is a segmented computer having a main memory size of 4M 64-bit words. Each scalar operation can be executed in 20 nsec (50M instructions per second).

To gain further insight into array processors, their architecture and capabilities should be explored in greater depth. A model for this study may be the AP-400 of Analogic. The AP-400 is a relatively inexpensive

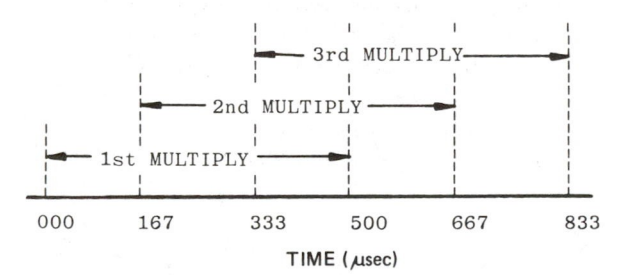

FIG. 17.13.2 Interleaved timing diagram for a commercial floating-point arithmetic unit.

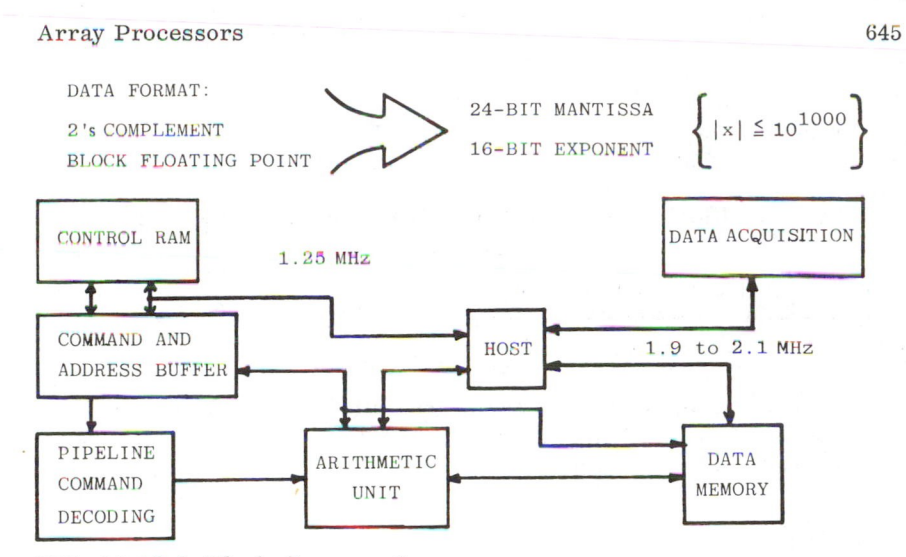

FIG. 17.13.3 Block diagram of an array processor.

array processor which is marketed to interface with general-purpose mini-computers supported with FORTRAN. This is a four-card unit selling for something in the neighborhood of $7500 if used with the host. If the unit is configured as a stand-alone device (i.e., vendor-supplied mechanical hardware, power supplies, etc.), the cost may be extended to $12,500. The four-card unit draws 20 A at 5 V dc as a support module and 103 W in the stand-alone configuration. The unit uses bipolar technology (e.g., 2901 bit-slice processor) to achieve low gate delays. The unit contains the following functional modules (see Fig. 17.13.3.):

1. Arithmetic unit
2. I/O structure to host and an external data acquisition subsystem (instructions, addressing, etc.)
3. Resident memory for storing data, coefficients (e.g., FFT butterfly coefficients), and the like
4. Controller

Along a similar line, a MAP array processor is commercially available as a multiboard unit using 300- to 170-nsec HMOS memory ($4000 to $7500 for a 16K-byte configuration).

EXAMPLE 17.13.2 The SPS-1000 of Signal Processing Systems is a highly pipelined FFT stand-alone processor. This high-speed array processor can be configured as an 8-, 16-, 24-, or 32-bit unit. Within a given main-frame, multiple processors can reside to achieve different word lengths and parallelism. The system architecture is based on the system integration of:

1. Butterfly module
2. Control module

3. Memory module
4. Multiplier module
5. Interface module

Suggested execution rates, based on a 1024-point complex FFT, are:

8-bit	16-bit	24-bit	32-bit
296	478	660	842 μsec

Software support for these units is generally found at three levels. At the highest level is a language such as FORTRAN. Here FORTRAN-callable commands will perform prespecified floating-point operations.

TABLE 17.13.1

Operation	Speed	Unit	Comment
512-point FFT (real)	1.5ms	AP-400	
1024-point FFT (real)	3.6ms	AP-400	
	3.0ms	MAP-100	
	7.0ms	MAP-200	
	2.8ms	MAP-300	
	2.7ms	AP-120B	160ns memory
	4.2ms	AP-120B	333ns memory
1024-point FFT (complex)	7.4ms	AP-400	
	60.0ms	MAP-100	
	12.0ms	MAP-200	
	4.5ms	MAP-300	
	4.8ms	AP-120B	160ns memory
	7.0ms	AP-120B	333ns memory
Convolution (512 real points)	7.3ms	AP-400	

AP = Floating Point Systems, MAP = CSP Inc.

Note: Mantissa and exponent lengths vary from system to system

EXAMPLE 17.13.3 A typical array processor will support the following operations:

1. Real and complex vector operations (single and double precision)
2. Constant with a vector operations
3. FFT
4. Data retrieval and conversion
5. Data acquisition control

EXAMPLE 17.13.4 Typical performance capabilities for some commeri-cially available array processors are given in Table 17.13.1. Using nested FORTRAN calls, complicated analysis can be performed at high data rates. At the next-lower level of programming, one finds the host's assembly-level language. Using this language, I/O drivers and interrupt calls can be realized. Finally, the array processor has an assembly-level language of its own. Here specialized programming can occur.

BIBLIOGRAPHY

Alexander, P. (1978), Array Processors, Digital Design, December, pp. 104-107.

Blascp, R. W. (1979), V-MOS Chip Joins Microprocessor to Handle Signals in Real Time, Electronics, August 30, pp. 131-134.

Curran, L. (1976), Smart Memories Speed Processing, Electronics, p. 117.

Dunweiler, D. L., and Y. S. Chen (1980), A Single-Chip VLSI Echo Can-celler, Bell Syst. Tech. J., February, pp. 149-160.

Gold, B., and C. M. Rader (1969), Digital Processing of Signals, McGraw-Hill, New York.

Hoff, M. E., and W. Li (1980), Software Makes a Big Talker Out of the 2920 Microcomputer, Electronics, January 31, pp. 102-107.

Korn, G. A. (1977), Microprocessors and Small Digital Computer Systems for Engineers and Scientists, McGraw-Hill, New York.

Nishitani, T., Y. Kavakami, and A. Awai (1980), LSI Signal Processing Development of Communication Equipment, Proc. ICASSP 80, Denver, Colo., pp. 386-389

Reticon (1979), R5604, 5605, and 5606 Switched Capacitance Filter, 345 Potrero Ave., Sunnyvale, CA 94086.

Taylor, F. J. (1978), Programmable CTD Filters Using Pulse Duration Modulation, Proc. IEEE, September.

Townsend, M., and M. E. Hoff (1980), An NMOS Microprocessor for Analog Signal Processing, IEEE Trans. Comput., C-29, February, pp. 97-101.

Wittmayer, W. R. (1978), Array Processors Provides High Throughput Rates, Computer Design, March, pp. 93-100.

General references for specific systems are available from:

American Microsystems, 3800 Homestead, Santa Clara, CA 95051

B&K Instruments, Inc., 5111 W. 164th St., Cleveland, OH 44142

Data General Corporation, 4400 Computer Drive, Westboro, MA 01581

Digital Equipment Corporation, Maynard, MA 01754

Fairchild Semiconductor, 464 Ellis, Mountain View, CA 94042

Floating Point Systems, Inc., P.O. Box 23489, Portland, OR 97223

Hewlett-Packard, 3000 Hanover, Cupertino, CA 94304

Intel Corporation, 3065 Bowers, Santa Clara, CA 45051

Kay, 13 Maple Ave., Pine Brook, NJ 07058

Motorola Semiconductor, Box 20912, Phoenix, AZ 85036

National Semiconductor Corporation, 2900 Semiconductor Drive, Santa Clara, CA 95051

Nicolet Scientific Corporation, 245 Livingston St., Northvale, NJ 07647

Tracor Northern, 2551 Beltline, Middleton, WI 53562

Plessey Microsystems, 1674 McGraw Ave., Irvine, CA 92714

Rockland Systems Corporation, 230 W. Nyack Rd., West Nyack, NY 10994

Signal Processing Systems, 223 Crescent Street, Waltham, MA 02154

Signal Technology, Inc., 5951 Encina Road, Goleta, CA 93117

Spectra Data, Inc., 18758 Bryant Street, Northridge, CA 91324

Time/Data, Division of GenRad, 2855 Bowers, Santa Clara, CA 95051

Unigon Industries, Inc., 1 Park Ave., Box 9999, Mt. Vernon, NY 10550

ZONIC Technical Laboratories, 2000 Ford Circle, Milford, OH 45150

18
Software Considerations

18.1 INTRODUCTION

General-purpose digital computers are often used to support digital signal processing tasks. The advantages of the general-purpose approach are programming simplicity, program maintenance, and the fact that no new resources are needed beyond the basic machine. The major limitation is speed and in some cases, cost. The speed metric can be improved, at a cost, through the use of arithmetic accelerators (e.g., array processors; see Section 17.13). Another speed-up option often overlooked is improved program design. This topic will now be studied.

Many signal processing algorithms are characterized by a few multiplications, many scalings [e.g., $c_k x(n - k)$, c_k a known filter coefficient], signal generation, and decisions (e.g., counting loop indexing, read/write decisions, etc.). The problem of coefficient synthesis and storage is important to linear shift-invariant filtering and constant-coefficient transforms [e.g., $W_N = \exp(j2\pi k/N)$]. These calculations are data invariant and in general are precomputed and stored for run-time retrieval.

Ideally, one would desire that the software design process be automated. This is historically called autogeneration or AUTOGEN.

18.2 AUTOGEN

The development of the autogeneration philosophy is best argued in the context of examples. One useful signal processing code worthy of study is the Pfeifer and Blankenship (PB) autocorrelation formula. Recall that the formal creation of a discrete autocorrelation metric, say R(k), is given by

$$R(k) = \Sigma \, x(n)x(n + k) \tag{18.2.1}$$

Suppose that the DFT of x(n) (i.e., FFT) is X(f). Then the autocorrelation may be alternatively expressed as IDFT(X(f)X(f)*). It has been argued that (N/2) $\log_2$(N) complex multiplies are required to form X(k), N more to create X(k)X(k)*, and (N/2) $\log_2$(N) more for the DFT inversion. This represents a total multiplication budget of N $\log_2$N + N. A direct computation of Eq. (18.2.1) would require N^2 real multiplies. The PB version of R(k) is given by

$$R(k) = \sum_{j=0}^{q-1} \sum_{i=1}^{k} x(2jk + i + k)[x(2jk + i) + x(2jk + i + 2k)] \qquad (18.2.2)$$

where k = 1, ..., p, p < N, and p represents the maximal lag index. Furthermore, q = INTEGER[n/2k] and x(n) = 0 for n < N. For p $\ll$ N, the BP algorithm essentially halves the number of multiplications normally associated with direct methods. Therefore, the multiplication count can be considerably less than that associated with FFT-based realizations if p satisfies the given constraint. Notice that the constraint requires p to be significantly less than N. However, this is not an objectionable quality in many signal processing applications. For example, if a process is white or nearly white, it is known that the autocorrelation times are short (i.e., p small). Finally, it can be noted that the arguments found in Eq. (18.2.2) can be classified into three basic operations. They are:

1. Algebra (i.e., x(a) [x(b) + x(c)])
2. Loop indexing (i.e., i versus k and j versus q - 1)
3. Argument generation (i.e., 2jk + i + k, 2jk + i, 2jk + i + 2k)

The last two operations are data invariant. In addition, their run-time computation can prove to be particularly time consuming. To overcome this overhead problem, the following software protocols will be developed:

1. Inline code
2. Threaded code
3. Knotted code

18.3 INLINE CODE

Referring to Fig. 18.3.1, it can be noted that some of the required PB indices are computed interior to the indicated DO loop as parameters I1, I2, and I3. This overhead bearing run-time computation can be eliminated through the use of an inline code. In an inline code, as the name implies, the code is arranged in a top-down fashion. Once the routine is initiated, the program flows continuously from the top to the bottom without interruption.

The automatic generation of the index set $\{I1, I2, I3\}$ can be realized without much difficulty using techniques suggested by Eckhouse and Morris (1979).

A high-level language can be used to replace the non-data-dependent (i.e., data invariant) operations with other high-level language statements (e.g., FORTRAN-to-FORTRAN compiler). For example, consider the data-invariant parameter TERM = $x(N1) * (x(N2) + x(N3))$. For specific values of I, J, and K (see Fig. 18.3.1), N1, N2, and N3 will have well-defined values. These numerical values of the three-tuple (N1, N2, N3) will evolve in the following manner:

	N1	N2	N3
S = 0			
S = S + x(2) * (x(1) + x(3))	2	1	3
S = S + x(4) * (x(3) + x(5))	4	3	5

These values can be obtained by running the original program suggested in Fig. 18.3.1 with a print statement inserted into the program location labeled "DUMMY PRINT OUT." That is, if the following statements are added to the code:

PRINT (I1, I2, I3)

the following output will result:

N1	N2	N3
2	1	3
4	3	5

$I_1 = 2jk + ik = k(2j+1) + i$

$I_2 = 2jk + i$

$I_3 = 2jk + 2k = 2k(j+1) + i$

```
                    DO a  K=1,N₁            (continue)
                        ⋮                 DO c  I=1,N₃
                    DO b  J=1,N₂           I1=T1+I
                    T1=K*(2*J+1)          I2=T2+I
                    T2=K*(2*J)            I3=T3+I
                    T3=K*2*(J+1)          (DUMMY PRINT OUT)
                        ⋮                     ⋮
                                       c    CONTINUE
                    (continue)         b    CONTINUE
                                       a    CONTINUE
```

FIG. 18.3.1 Pfeifer and Blankenship algorithm.

What is more important is that the three-tuple (N1, N2, N3) can be derived "off-line" and stored as a series of write statements. Using this policy, an AUTOGEN PB program consisting of statements of the form

CALL CORR(2, 1, 3)

CALL CORR(4, 3, 5)

      etc.

can be realized by replacing I1, I2, and I3 of Fig. 18.3.1 with the statement

WRITE ('CALL CORR(I1, I2, I3)')

As a result, the index set (N1, N2, N3) is read, rather than computed, during run time. Subroutine CORR would be used to compute

$x(N1) * (x(N2) + x(N3))$

Since the inline code has removed much of the overhead associated with the PB code, it should run faster than the original. However, the top-down structure of an inline code trades speed for program length. For each three-tuple (N1, N2, N3), an additional line of code is required.

18.4 THREADED CODE

A threaded code replaces a standard program with a series of modules that are stand-alone programs which do not require a return to the main program. A thread is a precomputed data array in which all prerequisite data-invariant information resides. Furthermore, order is faithfully preserved within a thread. The array is serially scanned at run time. As a result, data invariant information is read rather than computed during run time. Therefore, execution speed is enhanced. In addition, the data found in a long (multistatement) inline code can now be compressed into a data array for more efficient storage.

    Returning to the PB example, one notes that the following threaded program can be created:

```
    DATA L/C1, N1, N2, N3;   /1, 2, 1, 3
             C1, N1, N2, N3;   /1, 4, 3, 5
                    .
                    .
                    .
    GO TO (1, 2, · · ·), L(I)
                    .
                    .
                    .
 1     S = S + x(N1) * (x(N2) + x(N3))
                    .
                    .
                    .
```

where the parameter C1 is used to control branching. This branching control is sometimes required to handle special cases.

An AUTOGEN code can be realized by inserting a WRITE $(C_2, N_1, N_2, N3)$ statement into the original program at the location discussed for inline code. The resulting threaded code would be shorter and occupy less memory than the inline program. This savings can be accomplished without seriously reducing throughput.

18.5 KNOTTED CODE

As the name implies, knotted codes contain knots which are tied into the thread found in the threaded code. The program will move down the thread, "run around" inside a knot, and continue down the next thread. The knots in the code represent looping over a simple incremented or decremented single index. This is in contrast with the loops defined over complex index sets considered previously. It was the real-time computational burdens associated with the run-time generation of these indices that motivated the inline and threaded code. The overhead associated with looping over a simple index is small. It can be further reduced by precomputing the length of the loop and thereby eliminating the arithmetic test (e.g., IF or ELSE). A knotted code can be discussed in the context of the following operations:

1. Explicit computation: One or more CPU instruction on a piece of register or memory data (i.e., data dependent)
2. Implicit computation: Computation that occurs within a single instruction during execution (e.g., address manipulation)

Implicit computations greatly affect the time/space efficiency of a program. In the PB autocorrelation example, implicit computations are used to define N1, N2, and N3. This data can be precomputed and stored. At run time, only a small temporal penalty will be paid for "fetching" the threaded indices from memory and installing them into the program.

In more concrete terms, consider the PB autocorrelation terms:

$$R(k) = R(k) + x(m) [x(m - k) + x(m + k)]; \quad m = g(i, j, k) = 2jk + i + k$$

$$(18.5.1)$$

The production of m at run time would prove time consuming since two multiplications and additions are required per operation. In á knotted code sense, the following reduced overhead program can be created. Let

DATA L/ 1 , 2
        4
        6, 0
        .
        .
        .

where the circled parameters are used for control. Duting run time, parameter L is manipulated to ensure that the needed indices and control values are properly sequenced into the program. Tracing the flow of information in the program would result in the following identifiable events:

$J = 2$, $ND = 1$, $MD = -1$

GO TO (1, 2, 3), $L(I - D)$; TRANSFER CONTROL, FIRST $L(1) = 1$

$$\vdots$$

$S = S + x\underbrace{(L(I))}_{2} * (x\underbrace{(L(I) + ND)}_{3} + x\underbrace{(L(I) + MD)}_{1}))$; FROM DATA LIST

$I = I + 1$

IF $(L(I), NE, O)$ GO TO 1   Subsequent passes will produce N1,
$$\vdots$$ N2, N3 arguments of 3, 5, 4 $(I = 3)$;
                                    5, 7, 6 $(I = 4)$, until $L(I = 5) = 0$
                                    (i.e., true test against zero)

Memory requirements for the knotted PB example can be reduced by 75% over the threaded version, but this memory efficiency is obtained at the expense of speed. Nevertheless, the knotted code will run faster than one built using the standard software design practice of computing data-invariant parameters (using DO loops) during run time.

EXAMPLE 18.5.1 The fast Fourier transform (FFT) is rich in data-invariant operations. For example, generating the "butterfly coefficients" $W_N^k = \exp(-j2\pi k/N)$ can be time-consuming during run time. Using as a model a version of the standard FFT radix-2 routine found in Fig. 18.5.1, the problem of automatically generating inline, threaded, and knotted codes will be explored. The basic FFT routine is functionally composed of three parts:

1. Bit reversing (lines 200 to 340)
2. FFT and IFFT butterfly coefficients (lines 350 to 450)
3. Butterfly operations (lines 460 to 570)

To create an AUTOGEN inline code, the PRINT statements found in Fig. 18.5.2 are inserted into the text. Upon executing this program, the data base found in Fig. 18.5.3 was generated. First, the bit reversal information is outputted. Next comes the memory location and butterfly coefficient data on an FFT level-by-level basis. For example, it can be noted that the data forms the following groupings:

```
10    C        STANDARD RADIX-2 FFT CODE FOR 256=2**8 POINTS
20             DIMENSION A(256)
30             COMMON A
40             PI=3.141593
50             KK=-1
60             M=8
70             XKK=KK
80             N=2**M
90             NV2=N/2
100            J=1
110            NM1=N-1
120            DO 40 IN=1,2*N-1,2
130            A(IN)=SIN(4*PI*(IN-1)/(2*XN))
140   40       A(IN+1)=0.0
150   C        IF (KK=-1) DIVIDE SAMPLES VALUE BY N
160            IF(KK.NE.-1) GO TO 2
170            DO 1 JK=1,N
180            A(2*JK-1)=A(2*JK-1)/(XN/2)
190   1        A(2*JK)=A(2*JK)/(XN/2)
200   C        BIT REVERSAL ROUTINE
210   2        DO 7 I=1,NM1
220            IF(I.GT.J) GO TO 5
230            TR=A(2*J-1)
240            TI=A(2*J)
250            A(2*J-1)=A(2*I-1)
260            A(2*J)=A(2*I)
270            A(2*I-1)=TR
280            A(2*I)=TI
290   5        K=NV2
300   6        IF(K.GE.J) GO TO 7
310            J=J-K
320            K=K/2
330            GO TO 6
340   7        J=J+K
350   C        FFT (KK=-1); IFT (KK=+1)
360            DO 20 L=1,M
370            LE=2**L
380            LE1=LE/2
390            XLE1=LE1
400            UR=1.0
410            UI=0.0
420            WR=COS(PI/XLE1)
430            WI=XKK*SIN(PI/XLE1)
440            DO 20 J=1,LE1
450            DO 10 I=J,N,LE
460   C        BUTTERFLY
470            IP=I+LE1
480            TR=A(2*IP-1)*UR-A(2*IP)*UI
490            TI=A(2*IP)*UR+A(2*IP-1)*UI
500            A(2*IP-1)=A(2*I-1)-TR
510            A(2*IP)=A(2*I)-TI
520            A(2*I-1)=A(2*I-1)+TR
530   10       A(2*I)=A(2*I)+TI
540   C        NEW COS & SIN VALUES
550            X=UR
560            UR=X*WR-UI*WI
570   20       UI=UI*WR+X*WI
580            PRINT:"  REAL         IMAGINARY         MAGNITUDE"
590            PRINT:"========= ================ ================"
600   100      DO 60 IO=1,2*N-1,2
610            XMAG=SQRT(A(IO)*A(IO)+A(IO+1)*A(IO+1))
620   60       PRINT: A(IO),A(IO+1),XMAG
630            STOP
640            END
```

FIG. 18.5.1 Radix-2 FFT source code.

Level	Coefficients	Generator
1	$1 + j0$	W_1^i; $i = 0$
2	$1 + j0$, $0 + j1$	W_2^i; $i = 0, 1$
3	$1 + j0$, $0.707 - j0.707$, $0 - j1$, $-0.707 - j0.707$	W_4^i; $i = 0, 1, 2, 3$
4	$1 + j0$, $0.92 - j0.38$, $0.707 - j0.707$, $0.38 - j0.92$, $-0.38 - j0.92$, $-0.707 - j0.707$, $-0.92 - j0.38$	W_8^i; $j = 0, 1, \ldots, 7$

and so forth. These data have been used to create the inline code in Fig.
18.5.4. This code is over 10 times the length of the original code presented
in Fig. 18.5.1. Approximately a five-fold reduction in program size can
be obtained by using a threaded code (see Fig. 18.5.5). Note the use of
data arrays on a level-by-level basis. A knotted code is shown in Fig.
18.5.6. Here looping over simple index sets is allowed.

The in-place performance of the developed software is summarized
below:

Type	Lines	Time (msec)	Length ratio	Speed ratio
Standard	62	7.978	$62/62 = 1$	$7.9/7.9 = 0$
Inline	698	3.365	$698/62 = 11.25$	$3.4/7.9 = 0.43$
Threaded	121	4.445	$121/62 = 1.95$	$4.4/7.9 = 0.56$
Knotted	142	5.043	$142/62 = 2.3$	$5.0/7.9 = 0.63$

18.6 CACHE MEMORY

Recent breakthroughs in high-speed high-density semiconductor memory
technology have made cache memory a popular and affordable option on
some general-purpose machines. Cache memory is "fast" memory which
is attached to the system CPU through a short high-speed bus. Since this
shorter bus is capable of supporting higher-speed data traffic than that of
main memory bus structures, cache memory can operate at high speeds.
Due to bus length and power dissipation restrictions, cache memory normally
is limited in size to 2K or 8K bytes. In such a system, cache memory

```
2000 BIT REVERSAL ROUTINE            350C FFT (KK = -1); IFT (KK = 1)     460C BUTTERFLY
210 2 DO 7 I = 1,NM1                 360 DO 20 L = 1,M                    470 IP = 1 + LE1
220 IF(1 .GT. J) GO TO 5             365 PRINT:"                          480 TR = A(2*IP -1)*UR - A(2*IP)*UI
230 TR = A(2*J -1)                   370 LE = 2**L                        490 TI = A(2*IP)*UR + A(2*IP -1)*UI
240 TI = A(2*J)                      380 LE1 = LE/2                       500 A(2*IP -1) = A(2*I -1) - TR
244 PRINT:                           390 XLE1 = LE1                       510 A(2*IP) = A(2*I) - TI
245 PRINT: (2*J-1),(2*I-1)           400 UR = 1.0                         520 A(2*I -1) = A(2*I -1) + TR
250 A(2*J -1) = A(2*I -1)            410 UI = 0.0                         524 A(2*I) = A(2*I) + TI
260 A(2*J) = A(2*I)                  420 WR = COS(PI/XLE1)                525 PRINT: (2*I-1),(2*IP-1)
270 A(2*I -1) = TR                   430 WI =                             530 10 A(2*I) = A(2*I) + TI
280 A(2*I) = TI                      446 PRINT: "UI =",UI                 540C NEW SIN & COS VALUES
                                     450 DO 10 I = J,N,LE
```

INDEX VALUES FOR BIT REVERSING AT FFT LEVEL 0:

I	K	I	K	I	K
1	1	57	15	61	31
2	2	58	16	62	32
33	3	37	19	35	35
34	4	38	20	36	36
17	5	21	21	51	39
18	6	22	22	52	40
49	7	53	23	43	43
50	8	54	24	44	44
9	9	45	27	59	47
10	10	46	28	60	48
41	11	29	29	55	55
42	12	30	80	56	56
25	13				
26	14				

FIG. 18.5.2 Listing of bit-reverse transforms.

LEVEL 1

UR = 0.10000000E 01 (REAL.)
UI = 0. (IMAG.) } :W $^0_2$

I:	IP:
1	3
2	4
5	7
6	8
9	11
10	12
13	15
14	16
17	19
18	20
21	23
22	24
25	27
26	28
29	31
30	32
33	35
34	36
37	39
38	40
41	43
42	44
45	47
46	48
49	51
50	52
53	55
54	56
57	59
58	60
61	63
62	64

LEVEL 2

UR = 0.10000000E 01 (REAL.)
UI = 0. (IMAG.) } :W $^0_4$

I:	IP:
1	5
2	6
9	13
10	14
17	21
18	22
25	29
26	30
33	37
34	38
41	45
42	46
49	53
50	54
57	61
58	62

UR = 0.32881764E-06 (REAL.)
UI = -0.10000000E 01 (IMAG.) } :W $^3_4$

I:	IP:
3	7
4	8
11	15
12	16
19	23
20	24
27	31
28	32
35	39
36	40
43	47
44	48
51	55
52	56
59	63
60	64

LEVEL 3

UR = 0.10000000E 01 (REAL.)
UI = 0. (IMAG.) } :W $^0_8$

I:	IP:
1	9
2	10
17	25
18	26
33	41
34	42
49	57
50	58

UR = 0.70710689E 00 (REAL.)
UI = -0.70710666E 00 (IMAG.) } :W $^7_8$

I:	IP:
3	11
4	12
19	27
20	28
35	43
36	44
51	59
52	60

UR = 0.32663808E-06 (REAL.)
UI = -0.10000000E 01 (IMAG.) } :W $^6_8$

I:	IP:
5	13
6	14
21	29
22	30
37	45
38	46
53	61
54	62

UR = -0.70710643E 00 (REAL.)
UI = -0.70710712E 00 (IMAG.) } :W $^5_8$

I:	IP:
7	15
8	16
23	31
24	32
39	47
40	48
55	63
56	64

```
UR =  0.10000000E 01(REAL.)                              UR =  0.10000000E 01(REAL.)
UI =  0.            (IMAG.)                              UI =  0.            (IMAG.)
   W0                                                       W0
    16          I: 1    2        :W0                         32          I: 1        :W0
                              16                                                     32
   W0                          IP: 17  18                   W0                       IP: 33  34
    16                             49  50                     32                          1
                                   33  34                                    W0
UR =  0.92387956E                   3   4                UR =  0.98078529E            32
UI = -0.38268336E 00               35  36                UI = -0.19509028E 00           3   35
   W15                             19  20                   W21                          2   34
    16                             51  52                    32          W31
                                                                          32
UR =  0.70710689E                   5   6                UR =  0.92387956E            W20
UI = -0.70710666E 00               37  38                UI = -0.38268336E 00           32    5   37
   W14                             21  22                   W22                          4   36
    16                             53  54                    32          W30
                                                                          32
UR =  0.38268366E                   7   8                UR =  0.83146968E            W19
UI = -0.92387943E 00               39  40                UI = -0.55557013E 00           32    7   39
   W13                             23  24                   W23                          6   38
    16                             55  56                    32          W29
                                                                          32
UR =  0.33069163E-06                9  10                UR =  0.70710689E            W18
UI = -0.10000000E 01               41  42                UI = -0.70710666E 00           32    9   41
   W12                             25  26                   W24                          8   40
    16        LEVEL 4:             57  58                    32          W28
                                                                          32
UR = -0.92387969E                  11  12                UR =  0.55557040E            W17
UI = -0.38268305E 00               43  44                UI = -0.83146949E 00           32   11   43
   W11                             27  28                   W25                         10   42
    16                             59  60                    32          W27
                                                                          32      LEVEL 5:
UR = -0.70710643E 00               13  14                UR =  0.38268366E            W32
UI = -0.70710713E 00               45  46                UI = -0.92387943E 00           32   13   45
   W10                             29  30                   W26                         12   44
    16                             61  62                    32
                                                                       LEVEL 5:
UR = -0.92387932E 00               15  16                UR =  0.19509061E
UI = -0.38268397E 00               47  48                UI = -0.98078521E 00                15   47
   W9                              31  32                                                14   46
    16                            63  64
                                                        UR =  0.34243497E-06
                                                        UI = -0.99999999E 00                17   49
                                                                                        16   48

                                                        UR = -0.19509093E 00
                                                        UI = -0.98078533E 00                19   51
                                                                                        18   50

                                                        UR = -0.98078515E                   21   53
                                                        UI = -0.19509093E 00 W17/32          20   52
                                                        UR = -0.92387929E                   23   55
                                                        UI = -0.38268397E 00 W18/32          22   54
                                                        UR = -0.70710712E                   25   57
                                                        UI = -0.70710642E 00 W19/32          24   56
                                                        UR = -0.55556948E                   27   59
                                                        UI = -0.83146930E 00 W20/32          26   58
                                                        UR = -0.55556984E                   29   61
                                                        UI = -0.83146986E 00 W21/32          28   60
                                                        UR = -0.38263304E                   31   63
                                                        UI = -0.92387969E 00 W22/32          30   62
                                                        UR = -0.38263304E                        32
                                                        UI = -0.92387969E 00 W23/32          31   64
```

FIG. 18.5.3 Listing of butterfly coefficients and indices.

```
LIST                           **********INLINE CODE**********

10    DIMENSION A(256)                    620   A(46) = A(46)/XN
20    COMMON A                            630   A(47) = A(47)/XN
30    PI = 3.141592                       640   A(48) = A(48)/XN
40    KK = -1                             650   A(49) = A(49)/XN
50    M = 5                               660   A(50) = A(50)/XN
60    XKK = KK                            670   A(51) = A(51)/XN
70    N = 2**M                            680   A(52) = A(52)/XN
80    XN = N                              690   A(53) = A(53)/XN
90    NV2 = N/2                           700   A(54) = A(54)/XN
100   NM1 = N-1                           710   A(55) = A(55)/XN
110   J = 1                               720   A(56) = A(56)/XN
120   DO 40 IN= 1,2*N-1,2                 730   A(57) = A(57)/XN
130   A(IN) = SIN(4*PI*(IN-1)/(2*XN))     740   A(58) = A(58)/XN
140 40    A(IN+1) = 0.0                   750   A(59) = A(59)/XN
150 C       IF KK=-1 DIVIDE SAMPLE BY N   760   A(60) = A(60)/XN
160   XN = XN/2                           770   A(61) = A(61)/XN
170 A(1)  = A(1)/XN                       780   A(62) = A(62)/XN
180 A(2)  = A(2)/XN                       790   A(63) = A(63)/XN
190 A(3)  = A(3)/XN                       800   A(64) = A(64)/XN
200 A(4)  = A(4)/XN                       810 C     BIT REVERSAL
210 A(5)  = A(5)/XN                       820   TR = A(3)
220 A(6)  = A(6)/XN                       830   TI = A(4)
230 A(7)  = A(7)/XN                       840   A(3) = A(33)
240 A(8)  = A(8)/XN                       850   A(4) = A(34)
250 A(9)  = A(9)/XN                       860   A(33) = TR
260 A(10) = A(10)/XN                      870   A(34) = TI
270 A(11) = A(11)/XN                      880   TR = A(5)
280 A(12) = A(12)/XN                      890   TI = A(6)
290 A(13) = A(13)/XN                      900   A(5) = A(17)
300 A(14) = A(14)/XN                      910   A(6) = A(18)
310 A(15) = A(15)/XN                      920   A(17) = TR
320 A(16) = A(16)/XN                      930   A(18) = TI
330 A(17) = A(17)/XN                      940   TR = A(7)
340 A(18) = A(18)/XN                      950   TI = A(8)
350 A(19) = A(19)/XN                      960   A(7) = A(49)
360 A(20) = A(20)/XN                      970   A(8) = A(50)
370 A(21) = A(21)/XN                      980   A(49) = TR
380 A(22) = A(22)/XN                      990   A(50) = TI
390 A(23) = A(23)/XN                      1000  TR = A(11)
400 A(24) = A(24)/XN                      1010  TI = A(12)
410 A(25) = A(25)/XN                      1020  A(11) = A(41)
420 A(26) = A(26)/XN                      1030  A(12) = A(42)
430 A(27) = A(27)/XN                      1040  A(41) = TR
440 A(28) = A(28)/XN                      1050  A(42) = TI
450 A(29) = A(29)/XN                      1060  TR = A(13)
460 A(30) = A(30)/XN                      1070  TI = A(14)
470 A(31) = A(31)/XN                      1080  A(13) = A(25)
480 A(32) = A(32)/XN                      1090  A(14) = A(26)


                                          *** CONTINUE ***PAIR (15,57)
*** CONTINUE ****                         (16,58),(19,37),(20,38)
```

FIG. 18.5.4 Inline source code for a FFT.

```
1240 TR = A(23)                        1880 A(23) = A(21) - TR
1250 TI = A(24)                        1890 A(24) = A(22) - TI
1260 A(23) = A(53)                     1900 A(21) = A(21) + TR
1270 A(24) = A(54)                     1910 A(27) = A(27) + TI
1280 A(53) = TR                        1920 TR = A(27) * UR
1290 A(54) = TI                        1930 TI = A(28) * UR
1300 TR = A(27)                        1940 A(27) = A(25) - TR
1310 TI = A(28)                        1950 A(28) = A(26) - TI
1320 A(27) = A(45)                     1960 A(25) = A(25) + TR
1330 A(28) = A(46)                     1970 A(26) = A(26) + TI
1340 A(45) = TR                        1980 TR = A(31) * UR
1350 A(46) = TI                        1990 TI = A(32) * UR
1360 TR = A(31)                        2000 A(31) = A(29) - TR
1370 TI = A(32)                        2010 A(32) = A(30) - TI
1380 A(31) = A(61)                     2020 A(29) = A(29) + TR
1390 A(32) = A(62)                     2030 A(30) = A(30) + TI
1400 A(61) = TR                        2040 TR = A(35) * UR
1410 A(62) = TI                        2050 TI = A(36) * UR
1420 TR = A(39)                        2060 A(35) = A(33) - TR
1430 TI = A(40)                        2070 A(36) = A(34) - TI
1440 A(39) = A(51)                     2080 A(33) = A(33) + TR
1450 A(40) = A(52)                     2090 A(34) = A(34) + TI
1460 A(51) = TR                        2100 TR = A(39) * UR
1470 A(52) = TI                        2110 TI = A(40) * UR
1480 TR = A(47)                        2120 A(39) = A(37) - TR
1490 TI = A(48)                        2130 A(40) = A(38) - TI
1500 A(47) = A(59)                     2140 A(37) = A(37) + TR
1510 A(48) = A(60)                     2150 A(38) = A(38) + TI
1520 A(59) = TR                        2160 TR = A(43) * UR
1530 A(60) = TI                        2170 TI = A(44) * UR
1540 UR = 0.10000000E+01               2180 A(43) = A(41) - TR
1550 UI = 0.0                          2190 A(44) = A(42) - TI
1560 TR = A(3) * UR                    2200 A(41) = A(41) + TR
1570 TI = A(4) * UR                    2210 A(42) = A(42) + TI
1580 A(3) = A(1) - TR                  2220 TR = A(47) * UR
1590 A(4) = A(2) - TI                  2230 TI = A(48) * UR
1600 A(1) = A(1) + TR                  2240 A(47) = A(45) - TR
1610 A(2) = A(2) + TI                  2250 A(48) = A(46) - TI
1620 TR = A(7) * UR                    2260 A(45) = A(45) + TR
1630 TI = A(8) * UR                    2270 A(46) = A(46) + TI
1640 A(7) = A(5) - TR                  2280 TR = A(51) * UR
1650 A(8) = A(6) - TI                  2290 TI = A(52) * UR
1660 A(5) = A(5) + TR                  2300 A(51) = A(49) - TR
1670 A(6) = A(6) + TI                  2310 A(52) = A(50) - TI
1680 TR = A(11) * UR                   2320 A(49) = A(49) + TR
1690 TI = A(12) * UR                   2330 A(50) = A(50) + TO
1700 A(11) = A(9) - TR                 2340 TR = A(55) * UR
1710 A(12) = A(10) - TI                2350 TI = A(56) * UR
```

*** CONTINUE *** PAIR (15,13),   * CONTINUE * PAIR (55,53),
(16,14),(19,17),(20,18)   (56,54),(59,57),(60,58)

FIG. 18.5.4 (continued)

```
2480 A(63) = A(61) - TR          3140 A(11) = A(11) + TR
2490 A(64) = A(62) - TI          3150 A(12) = A(12) + TI
2500 A(61) = A(61) + TR          3160 TR = A(23)*UR-A(24)*UI
2510 A(62) = A(62) + TI          3170 TI = A(24)*UR+A(23)*UI
2520 UR = 0.10000000E+01         3180 A(23) = A(19) - TR
2530 UI = 0.0                    3190 A(24) = A(20) - TI
2540 TR = A(5) * UR              3200 A(19) = A(19) + TR
2550 TI = A(6) * UR              3210 A(20) = A(20) + TI
2560 A(5) = A(1) - TR            3220 TR = A(31)*UR-A(32)*UI
2570 A(6) = A(2) - TI            3230 TI = A(32)*UR+A(31)*UI
2580 A(1) = A(1) + TR            3240 A(31) = A(27) - TR
2590 A(2) = A(2) + TI            3250 A(32) = A(28) - TI
2600 TR = A(13)*UR               3260 A(27) = A(27) + TR
2610 TI = A(14)*UR               3270 A(28) = A(28) + TI


*** CONTINUE *** PAIR (13,9),    * CONTINUE * PAIR (39,35),
(14,10),(21,17),(22,18),         (40,36),(47,43),(48,44),
(29,25),(30,26),                 (55,51),(56,52),
(37,33),(38,34)                  (63,59),(64,60)


****************************     ****************************

2860 A(45) = A(41) - TR          3510 A(60) = A(60) + TI
2870 A(46) = A(42) - TI          3520 UR = 0.10000000E+01
2880 A(41) = A(41) + TR          3530 UI = 0.0
2890 A(42) = A(42) + TI          3540 TR = A(9)*UR
2900 TR = A(45)*UR               3550 TI = A(10)*UR
2910 TI = A(54)*UR               3560 A(9) = A(1) - TR
2920 A(53) = A(49) - TR          3570 A(10) = A(2) - TI
2930 A(54) = A(50) - TI          3580 A(1) = A(1) + TR
2940 A(49) = A(49) + TR          3590 A(2) = A(2) + TI
2950 A(50) = A(50) + TI          3600 TR = A(25)*UR
2960 TR = A(61)*UR               3610 TI = A(26)*UR
2970 TI = A(62)*UR               3620 A(25) = A(17) - TR
2980 A(61) = A(57) - TR          3630 A(26) = A(18) - TI
2990 A(62) = A(58) - TI          3640 A(17) = A(17) + TR
3000 A(57) = A(57) + TR          3650 A(18) = A(18) + TI
3010 A(58) = A(58) + TI          3660 TR = A(41)*UR
3020 UR = 0.32881764E-06         3670 TI = A(42)*UR
3030 UI = -0.10000000E+01        3680 A(41) = A(33) - TR
3040 TR = A(7)*UR - A(8)*UI      3690 A(42) = A(34) - TI
3050 TI = A(8)*UR + A(7)*UI      3700 A(33) = A(33) + TR
3060 A(7) = A(3) - TR            3710 A(34) = A(34) + TI
3070 A(8) = A(4) - TI            3720 TR = A(57)*UR
3080 A(3) = A(3) + TR            3730 TI = A(58)*UR
3090 A(4) = A(4) + TI            3740 A(57) = A(49) - TR
3100 TR = A(15)*UR - A(16)*UI    3750 A(58) = A(50) - TI
3110 TI = A(16)*UR + A(15)*UI    3760 A(49) = A(49) + TR
2950 A(50) = A(50) + TI          3600 TR = A(25)*UR
```

FIG. 18.5.4 (continued)

```
3770 A(50) = A(50) + TI                4560 UR = 0.100000000E-01
3780 UR =  0.70710689E+00              4570 UI = 0.0
3790 UI = -0.70710689E+00              4580 TR = A(17)*UR
3800 TR = A(11)*UR - A(12)*UI          4590 TI = A(18)*UI
3810 TI = A(12)*UR + A(11)*UI          4600 A(17) = A(1) - TR
3820 A(11) = A(3) - TR                 4610 A(18) = A(2) - TI
3830 A(12) = A(4) - TI                 4620 A(1) = A(1) + TR
3840 A(3) = A(3) + TR                  4630 A(2) = A(2) + TI
3850 A(4) = A(4) + TI                  4640 TR = A(49)*UR
3860 TR = A(43)*UR - A(44)*UI          4650 TI = A(50)*UI
3870 TI = A(44)*UR + A(43)*UI          4660 A(49) = A(33) - TR
3880 A(27) = A(19) - TR                4670 A(50) = A(34) - TI
3890 A(28) = A(20) - TI                4680 A(33) = A(33) + TR
3900 A(19) = A(19) + TR                4690 A(34) = A(34) + TI
3910 A(20) = A(20) + TI                4700 UR =  0.92387596E+00
3920 TR = A(43)*UR - A(44)*UI          4710 UI = -0.38268336E+00
3930 TI = A(44)*UR + A(43)*UI          4720 TR = A(19)*UR-A(20)*UI
3940 A(43) = A(35) - TR                4730 TI = A(20)*UR+A(19)*UI
3950 A(44) = A(36) - TI                4740 A(19) = A(3) - TR
3960 A(35) = A(35) + TR                4750 A(20) = A(4) - TI
3970 A(36) = A(36) + TI                4760 A(3) = A(3) + TR
3980 TR = A(59)*UR - A(60)*UI          4770 A(4) = A(4) + TI
3990 TI = A(60)*YR + A(59)*UI          4780 TR = A(51)*UR-A(52)*UI
4000 A(59) = A(51) - TR                4790 TI = A(52)*UR+A(51)*UI
4010 A(60) = A(52) - TI                4800 A(51) = A(35) - TR
4020 A(51) = A(51) + TR                4810 A(52) = A(36) - TI
4030 A(52) = A(52) + TI                4820 A(35) = A(35) + TR
4040 UR =  0.32663808E-06              4830 A(36) = A(36) + TI
4050 UI = -0.10000000E+01              4840 UR =  0.70710689E+00
4060 TR = A(13)*UR - A(14)*UI          4850 UI = -0.70710689E+00
4070 TI = A(14)*UR + A(13)*UI          4860 TR = A(21)*UR-A(22)*UI
4080 A(13) = A(5) - TR                 4870 TI = A(22)*UR+A(21)*UI
4090 A(14) = A(6) - TI                 4880 A(21) = A(5) - TR
4100 A(5) = A(5) + TR                  4890 A(22) = A(6) - TI
4110 A(6) = A(6) + TI                  4900 A(5) = A(5) + TR
4120 TR = A(29)*UR - A(30)*UI
4130 TI = A(30)*UR + A(29)*UI          *** CONTINUE ***
4140 A(29) = A(21) - TR
4150 A(30) = A(22) - TI                6900 A(63) = A(31) - TR
4160 A(21) = A(21) + TR                6910 A(64) = A(32) - TI
4170 A(22) = A(22) + TI                6920 A(31) = A(31) + TR
4180 TR = A(45)*UR - A(46)*UI          6930 A(32) = A(32) + TI
4190 TI = A(46)*UR + A(45)*UI          6935 PRINT:"REAL IMAG MAG"
4200 A(45) = A(37) - TR                6936 PRINT:"==== ==== ==="
4210 A(46) = A(38) - TI                6940 100 DO 60 IO=1,2*N-1,2
4220 A(37) = A(37) + TR                6950 60 PRINT:A(IO),A(IO+1)
                                                  XMAG
*** CONTINUE ***                       6960 STOP
                                       6980 END
```

FIG. 18.5.4 (continued)

```
10 DIMENSION A(256)
20 COMMON A
30 PI = 3.141592
40 KK = -1
50 M = 5
60 XKK = KK
70 N = 2**M
80 XN = N
90 NV2 = N/2
100 NM1 = N-1
110 J = 1
120 DO 4 IN = 1,2*N-1,2
130 A(IN) = SIN(8*PI*(IN-1)/(2*XN))
140 4 A(IN + 1) = 0.0
150C IF (KK = -1) DIVIDE SAMPLED VAL BY N
160 IF(KK .NE. -1) GO TO 25
170 DO 1 JK = 1,N
180 A(2*JK-1) = A(2*JK-1)/(XN/2)
190 1 A(2*JK) = A(2*JK)/(XN/2)
200C BIT REVERSAL ROUTINE
210 DIMENSION IA(24),IB(24)
220 DATA IA/3,4,5,6,7,8,11,12,13,14,15,16,19,20,23,24,27,20,
230& 31,32,39,40,47,48/
240 DATA IB/33,34,17,18,49,50,41,42,25,26,57,58,37,38,53,54,
250& 45,46,61,62,51,52,59,60/
260 25 DO 10 I = 1,24
270 T = A(IA(I))
280 A(IA(I)) = A(IB(I))
290 A(IB(I)) = T
300 10 CONTINUE
310 DIMENSION IA1(16), IB1(16), IC1(16), ID1(16)
320 DATA IA1/3,7,11,15,19,23,27,31,35,39,43,47,51,55,59,63/
330 DATA IB1/4,8,12,16,20,24,28,32,36,40,44,48,52,56,60,64/
340 DATA IC1/1,5,9,13,17,21,25,29,33,37,41,45,49,53,57,61/
350 DATA ID1/2,6,10,14,18,22,26,30,34,38,42,46,50,54,58,62/
360 UR = 0.10000000E 01
370 DO 20 I = 1,16
380 TR = A(IA1(I))*UR
390 TI = A(IB1(I))*UR
400 A(IA1(I)) = A(IC1(I)) - TR
410 A(IB1(I)) = A(ID1(I)) - TI
420 A(IC1(I)) = A(IC1(I)) + TR
430 A(ID1(I)) = A(ID1(I)) + TI
440 20 CONTINUE
450 DIMENSION IA2(16), IB2(16), IC2(16), ID2(16), UR2(16), UI2(16)
460 DATA IA2/5,13,21,29,37,45,53,61,7,15,23,31,39,47,55,63/
470 DATA ID2/6,14,22,30,38,46,54,62,8,16,24,32,40,48,56,64/
480 DATA IC2/1,9,17,25,33,41,49,57,3,11,19,27,35,43,51,59/
490 DATA ID2/2,10,18,26,34,42,50,58,4,12,20,28,36,44,52,60/
500 DATA UR2/8*0.10000000E 01,8*0.32881764E-06/
510 DATA UI2/8*0.,8*-0.10000000E 01/
520 DO 30 I = 1,16
530 TR = A(IA2(I))*UR2(I) - A(IB2(I))*UI2(I)
540 TI = A(ID2(I))*UR2(I) + A(IA2(I))*UI2(I)
550 A(IA2(I)) = A(IC2(I)) - TR
560 A(IB2(I)) = A(ID2(I)) - TI
570 A(IC2(I)) = A(IC2(I)) + TR
580 A(ID2(I)) = A(ID2(I)) + TI
590 30 CONTINUE
600 DIMENSION IA3(16), IB3(16), IC3(16), ID3(16), UR3(16), UI3(16)
```

FIG. 18.5.5 Threaded source code for a FFT.

```
610 DATA IA3/9,25,41,57,11,27,43,59,13,29,45,61,15,31,47,63/
620 DATA IB3/10,26,42,58,12,28,44,60,14,30,46,62,16,32,40,64/
630 DATA IC3/1,17,33,49,3,19,35,51,5,21,37,53,7,23,39,55/
640 DATA ID3/2,18,34,50,4,20,36,52,6,22,38,54,8,24,40,56/
650 DATA UR3/4*0.10000000E 01,4*0.70710689E 00,4*0.32663808E-06,
660&   4*-0.70710643E 00/
670 DATA UI3/4*0.,4*-0.70710666E 00,4*-0.10000000E 01,4*-0.70710712E 00/
680 DO 40 I = 1,16
690 TR = A(IA3(I))*UR3(I) - A(IB3(I))*UI3(I)
700 TI = A(IB3(I))*UR3(I) + A(IA3(I))*UI3(I)
710 A(IA3(I)) = A(IC3(I)) - TR
720 A(IB3(I)) = A(ID3(I)) - TI
730 A(IC3(I)) = A(IC3(I)) + TR
740 A(ID3(I)) = A(ID3(I)) + TI
750 40 CONTINUE
760 DIMENSION IA4(16), IB4(16), IC4(16), ID4(16), UR4(16), UI4(16)
770 DATA IA4/17,49,19,51,21,53,23,55,25,57,27,59,29,61,31,63/
780 DATA IB4/18,50,20,52,22,54,24,56,26,58,28,60,30,62,32,64/
790 DATA IC4/1,33,3,35,5,37,7,39,9,41,11,43,13,45,15,47/
800 DATA ID4/2,34,4,36,6,38,8,40,10,42,12,44,14,46,16,48/
810 DATA UR4/2*0.10000000E 01,2*0.92387956E 00,2*0.70710689E 00,
820&   2*0.38268366E 00,2*0.33069163E-06,2*-0.38268305E 00,
830&   2*-0.70710643E 00,2*-0.92387932E 00/
840 DATA UI4/2*0.,2*-0.38268336E 00,2*-0.70710666E 00,2*-0.92387943E 00,
850&   2*-0.10000000E 01,2*-0.92387969E 00,2* 0.70710713E 00,
860&   2*-0.38268397E 00/
870 DO 50 I = 1,16
880 TR = A(IA4(I))*UR4(I) - A(IB4(I))*UI4(I)
890 TI = A(IB4(I))*UR4(I) + A(IA4(I))*UI4(I)
900 A(IA4(I)) = A(IC4(I)) - TR
910 A(IB4(I)) = A(IB4(I)) - TI
920 A(IC4(I)) = A(IC4(I)) + TR
930 A(ID4(I)) = A(ID4(I)) + TI
940 50 CONTINUE
950 DIMENSION IA5(16), IB5(16), IC5(16), ID5(16), UR5(16), UI5(16)
960 DATA IA5/33,35,37,39,41,43,45,47,49,51,53,55,57,59,61,63/
970 DATA IB5/34,36,38,40,42,44,46,48,50,52,54,56,58,60,62,64/
980 DATA IC5/1,3,5,7,9,11,13,15,17,19,21,23,25,27,29,31/
990 DATA ID5/2,4,6,8,10,12,14,16,18,20,22,24,26,28,30,32/
0001000 DATA UR5/0.10000000E 01,0.98078529E 00,0.92387956E 00,
0001010&   0.83146968E 00,0.70710689E 00,0.55557040E 00,0.38268366E 00,
0001020&   0.19509061E 00,0.33423497E-06,-0.19508995E 00,
0001030&   -0.38268304E 00,-0.55556984E 00,-0.70710642E 00,
0001040&   -0.83146930E 00,-0.92387929E 00,-0.98078515E 00/
0001050 DATA UI5/0.,-0.19509028E 00,-0.38268336E 00,-0.55557013E 00,
0001060&   -0.70710666E 00,-0.83146949E 00,-0.92387943E 00,-0.98078521E 00,
0001070&   -0.99999999E 00,-0.98078535E 00,-0.92387969E 00,-0.83146986E 00,
0001080&   -0.70710712E 00,-0.55557068E 00,-0.38268397E 00, 0.19509093E 00/
0001090 DO 60 I = 1,16
0001100 TR = A(IA5(I))*UR5(I) - A(IB5(I))*UI5(I)
0001110 TI = A(IB5(I))*UR5(I) + A(IA5(I))*UI5(I)
0001120 A(IA5(I)) = A(IC5(I)) - TR
0001130 A(IB5(I)) = A(ID5(I)) - TI
0001140 A(IC5(I)) = A(IC5(I)) + TR
0001150 A(ID5(I)) = A(ID5(I)) + TI
0001160 60 CONTINUE
0001165 PRINT:" REAL              IMMAGINARY        MAGNITUDE"
0001166 PRINT:" =============== ============== ==============="
0001170 100 DO 70 IO = 1,2*N-1,2
0001180 XMAG = ((A(IO))**2 + (A(IO + 1))**2)**.5
0001190 70 PRINT:A(IO),A(IO + 1), XMAG
0001200 STOP
0001210 END
```

FIG. 18.5.5 (continued)

```
10 DIMENSION A(256)
20 COMMON A
30 PI = 3.141592
40 KK = -1
50 M = 5
60 XKK = KK
70 N = 2**M
80 XN = N
90 NV2 = N/2
100 NM1 = N-1
110 J = 1
120 DO 4 IN = 1,2*N-1,2
130 A(IN) = COS(16*PI*(IN-1)/(2*XN))
140 4 A(IN + 1) = 0.0
150C IF (KK = -1) DIVIDE SAMPLED VAL BY N
160 IF(KK .NE. -1) GO TO 25
170 DO 1 JK = 1,N
180 A(2*JK-1) = A(2*JK-1)/(XN/2)
190 1 A(2*JK) = A(2*JK)/(XN/2)
200C BIT REVERSAL ROUTINE
210 DIMENSION IAO(24),IBO(24)
220 DATA IAO/3,4,5,6,7,8,11,12,13,14,15,16,19,20,23,24,27,28,
230& 31,32,39,40,47,48/
240 DATA IBO/33,34,17,18,49,50,41,42,25,26,57,58,37,38,53,54,
250& 45,46,61,62,51,52,59,60/
260 25 DO 10 I = 1,24
270 T = A(IAO(I))
280 A(IAO(I)) = A(IBO(I))
290 A(IBO(I)) = T
300 10 CONTINUE
310 DATA IA1,IB1,IC1,ID1,UR1/3,4,1,2,0.10000000E 01/
320 DO 15 I = 1,16
330 TR = A(IA1)*UR1
340 TI = A(IB1)*UR1
350 A(IA1) = A(IC1) - TR
360 A(IB1) = A(ID1) - TI
370 A(IC1) = A(IC1) + TR
380 A(ID1) = A(ID1) + TI
390 IA1 = IA1 + 4
400 IB1 = IB1 + 4
410 IC1 = IC1 + 4
420 ID1 = ID1 + 4
430 15 CONTINUE
440 DIMENSION IA2(2), IB2(2), IC2(2), ID2(2), UR2(2), UI2(2)
450 DATA IA2,IB2/5,7,6,8/
460 DATA IC2,ID2/1,3,2,4/
470 DATA UR2,UI2/0.10000000E 01,0.32881764E-06,0., 0.10000000E 01/
480 DO 20 K = 1,2
490 IA = IA2(K)
500 IB = IB2(K)
510 IC = IC2(K)
520 ID = ID2(K)
530 DO 22 I = 1,8
540 TR = A(IA)*UR2(K) - A(IB)*UI2(K)
550 TI = A(IB)*UR2(K) + A(IA)*UI2(K)
560 A(IA) = A(IC) - TR
570 A(IB) = A(ID) - TI
580 A(IC) = A(IC) + TR
590 A(ID) = A(IB) + TI
600 IA = IA + 8
610 IB = IB + 8
620 IC = IC + 8
630 ID = IB + 8
640 22 CONTINUE
650 20 CONTINUE
660 DIMENSION IA3(4), IB3(4), IC3(4), ID3(4), UR3(4), UI3(4)
670 DATA IA3,IB3/9,11,13,15,10,12,14,16/
680 DATA IC3,ID3/1,3,5,7,2,4,6,8/
690 DATA UR3/0.10000000E 01,0.70710689E 00,0.3266380BE 00, 0.70710689L 00/
700 DATA UI3/0.,-0.70710666E 00,-0.0000000E 01, 0.70710712E 00/
710 DO 30 K = 1,4
720 IA = IA3(K)
```

FIG. 18.5.6 Knotted source code for a FFT.

```
 730 IB = IB3(K)
 740 IC = IC3(K)
 750 ID = ID3(K)
 760 DO 33 I = 1,4
 770 TR = A(IA)*UR3(K) - A(IB)*UI3(K)
 780 TI = A(IB)*UR3(K) + A(IA)*UI3(K)
 790 A(IA) = A(IC) - TR
 800 A(IB) = A(ID) - TI
 810 A(IC) = A(IC) + TR
 820 A(ID) = A(ID) + TI
 830 IA = IA + 16
 840 IB = ID + 16
 850 IC = IC + 16
 860 ID = ID + 16
 870 33 CONTINUE
 880 30 CONTINUE
 890 DIMENSION IA4(8), IB4(8), IC4(8), IB4(8), UR4(8), UI4(8)
 900 DATA IA4/17,19,21,23,25,27,29,31/
 910 DATA IB4/18,20,22,24,26,28,30,32/
 920 DATA IC4/1,3,5,7,9,11,13,15/
 930 DATA ID4/2,4,6,8,10,12,14,16/
 940 DATA UR4/0.10000000E 01,0.92387956E 00,0.70710689E 00,0.38268366E 00,
 950& 0.33069163E-06,-0.38268305E 00,-0.70710643E 00, 0.92387932E 00/
 960 DATA UI4/0.,-0.38268336E 00,-0.70710666E 00,-0.92387943E 00,
 970& -0.10000000E 01, 0.92387969E 00,-0.70710713E 00,-0.38268397E 00/
 980 DO 40 K = 1,3
 990 IA = IA4(K)
1000 IB = IB4(K)
1010 IC = IC4(K)
1020 ID = ID4(K)
1030 DO 44 I = 1,2
1040 TR = A(IA)*UR4(K) - A(IB)*UI4(K)
1050 TI = A(IB)*UR4(K) + A(IA)*UI4(K)
1060 A(IA) = A(IC) - TR
1070 A(IB) = A(ID) - TI
1080 A(IC) = A(IC) + TR
1090 A(ID) = A(IB) + TI
1100 IA = IA + 32
1110 IB = IB + 32
1120 IC = IC + 32
1130 ID = ID + 32
1140 44 CONTINUE
1150 40 CONTINUE
1160 DIMENSION IA5(16), IB5(16), IC5(16), ID5(16), UR5(16), UI5(16)
1170 DATA IA5/33,35,37,39,41,43,45,47,49,51,53,55,57,59,61,63/
1180 DATA ID5/34,36,38,40,42,44,46,48,50,52,54,56,58,60,62,64/
1190 DATA IC5/1,3,5,7,9,11,13,15,17,19,21,23,25,27,29,31/
1200 DATA ID5/2,4,6,8,10,12,14,16,18,20,22,24,26,28,30,32/
1210 DATA UR5/0.10000000E 01,0.98078529E 00,0.92387956E 00,
1220& 0.83146968E 00,0.70710689E 00,0.55557040E 00,0.38268366E 00,
1230& 0.19509061E 00,0.3342349E/E-06,-0.19508995E 00,
1240& -0.38268304E 00,-0.55556984E 00, 0.70710642E 00,
1250& -0.83146930E 00,-0.92387929E 00,-0.98078515E 00/
1260 DATA UI5/0., 0.19509028E 00,-0.38268336E 00, 0.55557013E 00,
1270& -0.70710666E 00,-0.83146949E 00, 0.92387943E 00, 0.98078521E 00,
1280& -0.99999999E 00,-0.98078535E 00, 0.92387969E 00, 0.83146986E 00,
1290& -0.70710712E 00,-0.55557068E 00,-0.38268397E 00,-0.19509093E 00/
1300 DO 60 I = 1,16
1310 TR = A(IA5(I))*UR5(I) - A(IB5(I))*UI5(I)
1320 TI = A(IB5(I))*UR5(I) + A(IA5(I))*UI5(I)
1330 A(IA5(I)) = A(IC5(I)) - TR
1340 A(IB5(I)) = A(ID5(I)) - TI
1350 A(IC5(I)) = A(IC5(I)) + TR
1360 A(IB5(I)) = A(ID5(I)) + TI
1370 60 CONTINUE
0001375 PRINT:"  REAL              IMMAGINARY      MAGNITUDE"
0001376 PRINT:"  ==============    ==============  =============="
1380 100 DO 70 IO = 1,2*N 1,2
1390 XMAG = ((A(IO))**2 + (A(IO + 1))**2)**.5
1400 70 PRINT:A(IO),A(IO : 1), XMAG
1410 STOP
1420 END
```

FIG. 18.5.6 (continued)

stores both addresses and data words (e.g., cache location 0 may reside in main memory location 0, 2048, 4096, etc., depending on the last memory write operation). The memory cache philosophy is premised on the assumption that the main program memory requirements are such that it is the most current data that is generally needed during run time. That is, a program generally does not have to look too far back into a data and parameter list to find the information needed. Therefore, the purpose of a cache memory unit is to provide the user high-speed access to current data. Data that are old will typically not be found in the cache. Instead, mature data would have to be retrieved from main memory along slower buses.

A classic use of cache is to support FFT computations. Here operations of the form $z = a + bW_N^k$ are nested. Using, for example, a PDP 11/60 (2K bytes cache), the 256 complex data words found in a 256-point FFT would normally be found in cache. Using a PDP 11/44, having 4K bytes of cache, this metric would be doubled. Experiments performed by Morris and Mudge (1977) using the PDP-11/60 indicate that a 16-bit 256-point radix-2 FFT would complete a conventional FFT in 122 msec and an AUTOGEN inline transform in 50 msec. A further speed enhancement can be obtained if some of the instructions are microcoded. Microcoding is a systematic method of implementing special control functions that are not found in the machine's lexicon. Microprogramming requires a special programming talent and skill which often precludes its use by the uninitiated. In addition, only a few machines will support user microprogramming (e.g., Hewlett-Packard HP2100, Varian V70, Data General DG Eclipse, Digital Equipment PDP-11/60, etc.).

When writing software, with or without microprogramming support, in a cache memory supported system, it is desirable to make program loops short and tight. By designing the software in such a manner, the user increases the probability that the data needed within a loop are current to the program (i.e., high "hit" rate).[†] Hit rates can be altered through the use of inline, threaded, and knotted coding. For example, it was reported that a PB correlation computation for N = 128 and p = 12 would process data at the following rates, which were discovered by Eckhouse and Morris (1979):

Program	Cache miss ratio	Program size (words)	PDP 11/55 300-nsec memory noncache (msec)	PDP 11/70 1-μsec memory cache (msec)
Conventional (twice PB mult. count)	0.016	80	9.69	9.72
Inline (PB)	0.38	7552	5.25	9.42
Knotted (PB)	0.055	768	6.70	7.05

[†]The ratio of the total number of randomly accessed words found in cache to the total number of words required during run time is defined as the hit ratio.

Observe that the high "miss ratio," which characterizes the inline AUTOGEN
PB algorithm, penalizes its performance in a cache environment. However,
the knotted code strikes an acceptable trade-off in this area.

In terms of a 1024-point radix-4 real FFT, the following performance
figures can be found in the literature:

Peripheral processor	General-purpose computer	AUTOGEN FFT
AP 120B (PDP host) 2.9-4.1 msec	PDP 11/45 (300-nsec memory)	89 msec
CSP MAD 100 (PDP host) 30 msec	PDP 11/70 (1-μsec memory)	109 msec

(see Section 17.13 for details).

Other comparisons reported by Eckhouse and Morris are summarized
below:

Machine	Memory	Cache	Threaded 1024-pt FFT	Performance index
PDP 11/45	300 nsec	No	110 msec	1.0
PDP 11/55	300 nsec	No	110 msec	1.0
PDP 11/45	1 μsec	No	160 msec	1.5
PDP 11/145	Core	No	270 msec	2.8
PDP 11/134	1 μsec	No	330 msec	3.3
PDP LSI-11	1 μsec	No	910 msec	9.1
PDP 11/70	1 μsec	Yes	110 msec	1.0

The throughput of a basic general-purpose computer can be accelerated through the use of array processors or vendor-supplied arithmetic
accelerators. Typically, vendor-supplied floating-point accelerators
are configured on a single card that fits into the computer's backplane
or a ROM chip set which contains optimized microcode. Again referring
to data published by Eckhouse and Morris, the following performance
figures were obtained:

Machine	Arithmetic unit	Standard FFT code	Performance index
PDP 11/55	FD 11-C	165 msec	1.0
PDP 11/60	FD 11-E	220 msec	1.3
PDP 11/60	Firmware	880 msec	5.3
PDP 11/34	FP 11-A	640 msec	3.9

18.7 STACKING

The foregoing software policies addressed the problem of speed and program size from an algorithmic viewpoint. Another commonly encountered signal processing software task is data stacking. For example, consider the following FIR filter:

$$y(n) = \sum_{m=0}^{N-1} h(n)x(n - m) \qquad (18.7.1)$$

A standard software realization of this FIR would exhibit the following attributes:

```
                                    ┌── data
                                    ↓
        Dimension H(N),  x(N)
            ⋮        ↑
            ⋮        └────────── coefficients

    c   READ DATA INTO STACK
        DO xx J = 1, N - 1

   xx   x(N - J + 1) = x(N - J)
        x(1) = xDATA

            ⋮
            ⋮
        DO xxx J = 1, N

  xxx   S = S + H(J) * x(J)
```

It is required that the last m values of x be randomly accessed during run time. The efficient stacking of these data can be achieved with the proper choice of software design.

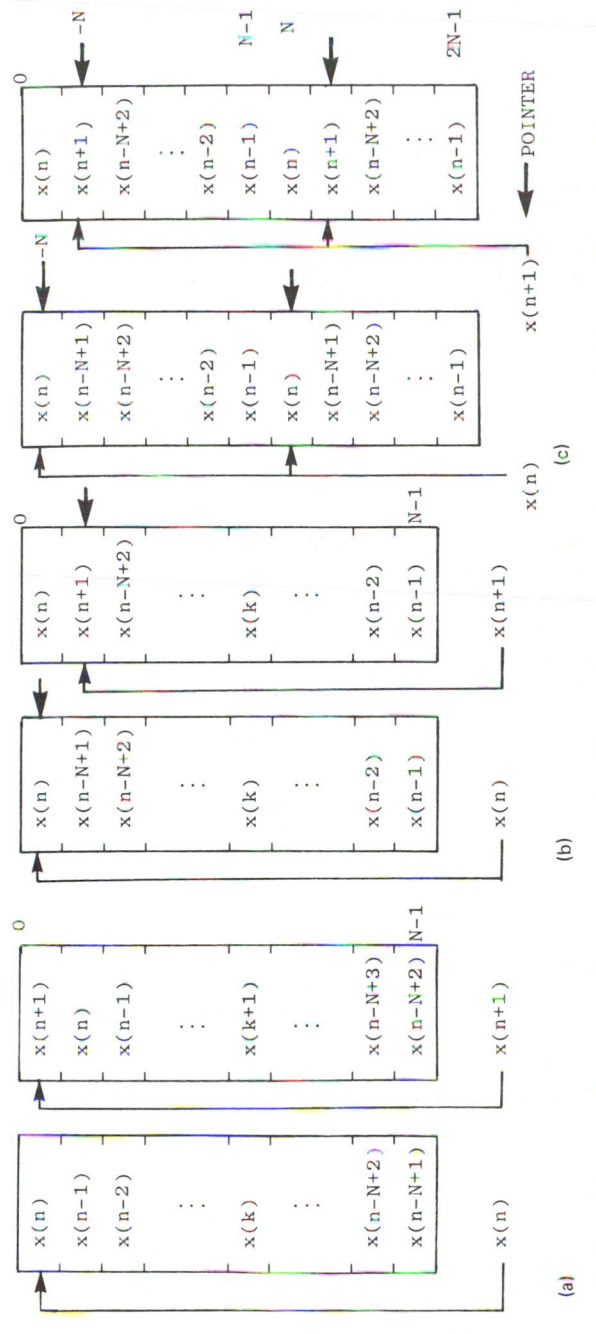

FIG. 18.7.1 (a) Simple stack arrangement. (b) Stack management using a single pointer. (c) Stack management using a dual-pointer architecture.

671

In terms of a data array, a stack would appear as the memory map found in Fig. 18.7.1a. A superior stack management policy is suggested in Fig. 18.7.1b. Here a pointer is used to indicate the entry location of the most current data element. The disadvantage of this scheme is the requirement that the pointer's location be tested against N, the stack length, during each access cycle. A dual-pointer modification is shown in Fig. 18.7.1c. Here N extra storage words are budgeted to stack operations. This scheme eliminates the time-consuming operation of testing the pointer location against N. With respect to contemporary memory cost, this is often a viable trade-off.

18.8 HIGH-LEVEL LANGUAGES

Prior to 1957, the programming of digital computing machines required a basic understanding of machine- or assembly-level languages. As a result, many potential users were functionally denied access to the digital computer. In 1957, an 18-person-year effort lead to the announcement of FORTRAN (FORmula TRANslation). FORTRAN greatly simplified scientific programming by allowing the user to state, in relatively simple algebraic-like statements, the procedural steps of a program. The advent of FORTRAN increased the user population enormously and as a result had a great deal to do with popularizing the digital computer.

In the early 1960s, BASIC (Beginner's All-purpose Symbolic Instruction Code) was introduced. This language was easier to use than FORTRAN but was more restrictive. It has been developed in many versions and can be found running at high and low (microcomputer) levels.

In 1962, Iverson of Harvard announced APL (A Programming Language) for processing information and manipulating data strings. APL is a concise, consistent, and powerful language having as its forte the handling of data arrays.

IBM introduced PL/I (Programming Language I) in the mid-1960s. Designed to support the IBM 360 systems, it was more broadly based than other popular IBM high-level languages at the time (e.g., FORTRAN or ALGOL). It could handle both arithmetic and structured information and managed to remove some of the syntax problems that plagued earlier languages. Its versatility caused it to grow in an unstandardized manner prior to 1969, when standardization was attempted.

A relatively new language is Pascal. Developed by N. Wirth of Switzerland in 1968, augmented by Urs Ammann in the early 1970s, its initial purpose was instructional. The language apparently owes its name to the author's esteem for Blaise Pascal (1623-1662). Kenneth Bowles, of the Institute for Information Systems of the University of California, San Diego, was a major force in popularizing Pascal for noninstructional purposes. It is a versatile and potentially powerful high-level language capable of working well with most processors. For example, Texas Instruments has adopted Pascal as a corporate-wide language for internal use.

A language that is sure to have a major impact on large-scale computing systems is ADA. The language is named after Augusta Ada Byron, Countess of Lovelace and the daughter of George Lord Byron, who is credited with being the world's first computer programmer based on her association with Charles Babbage's mechanical computer. ADA is a language developed in response to the U.S. defense community's desire to have a standardized language. Standardization is important from the standpoint of software transportability and maintenance. Pascal, for example, exists in many different forms and versions. ADA inherited much of its structure from Pascal, but also carries some ALGOL and PL/I structure. A significant amount of an ADA compiler is allocated to a real-time concurrent programming mission. ADA is generally considered to be a language for a large system. However, work on an ADA compiler for Western Digital's microengine, the 68000, Z8000, and VAX11/780 is reported.

18.9 CHARACTER SETS

The character sets of some major languages are summarized in Table 18.9.1. For example, in FORTRAN, the 26 uppercase letters, the 10 digits, and 11 other characters [blank, +, -, *, /, =, (,), comma, period, and $] are found. APL has many special characters and requires a special terminal.

TABLE 18.9.1 Character Set for Common Scientific Codes

Language	Alphabet set	Character set	Numerals	Others
APL	89	26	10	42
BASIC	54	26	10	18
FORTRAN	47	26	10	11
PL/I	60	26	10	24

TABLE 18.10.1 Data Types Found in Common High-Level Languages

BASIC: Integer, real, character	FORTRAN: Integer, real, logical complex, double precision (real
Pascal: Integer, real, Boolean character, scalar, subrange, pointer, user defined	PL/I: Number [attributes] i. Mode: Real or complex ii. Scale: Fixed or floating iii. Base: Decimal or binary iv. Precision: Number of digits, character string bit string, pointer, area, offset, label, entry, format, task, event

TABLE 18.10.2 Operations and Precedence Relationships
Associated with Common Scientific Languages

APL:	FORTRAN:
+ - x - = ≠	** EXPONENTIATION
* EXPONENTIATION	* /
⊛ LOGARITHM	+ -
⌈ LARGER OF	.LT. .EQ. .LE.
⌊ SMALLER OF	.NE. .GE. .GT.
! COMBINATION OF	.NOT.
∨ LOGICAL OR	.AND.
∧ LOGICAL AND	.OR.
⍱ LOGICAL NOR	
⍲ LOGICAL NAND	PL/I:
BASIC:	- ** UNARY + UNARY -
↑ EXPONENTIATION	* /
* /	= >< > ¬=
+ -	< <= ¬< ¬>
= >< > <> <	& LOGICAL AND
		LOGICAL OR
	|| CONCATENATION

18.10 TOKENS AND VARIABLES

APL has up to 77 characters in a variable name, with some keywords indicated by special characters. In BASIC, variables are limited to a single letter followed by an optional digit (e.g., A, B, C2). Literal strings are allowed if enclosed by quotation marks. In FORTRAN, variable names of one to six characters in length are permitted (IBM supports eight-character names). PL/1 has a limit of 31 characters per variable name, with the first one being alphabetic.

A knowledge of keywords is important so that they will not accidentally be used as program variables. APL has no clear-cut keyword list but does support a long list of functional operations such as CREATE, RUN, and STATEMENT. BASIC has a small keyword set which includes READ, PRINT, IF, GO TO, NEXT, DATA, REM, END, and RUN. FORTRAN uses keywords such as GO TO, IF, READ, PRINT, WRITE, FORMAT, DIMENSION, CALL, CONTINUE, DO, COMMON, STOP, and END. Both Pascal and PL/I have many recognizable keywords which depend on their intended use.

Data types are summarized in Table 18.10.1 with operations and precedences listed in Table 18.10.2.

18.11 SOFTWARE STRUCTURE

An APL program is made up of a sequence of simple statements that do not contain embedded statements of the form READ, GO TO (or assignment). It has a rich, powerful set of instructions that work on single items called scalars or, upon extension, simple arrays. The limited amount of looping makes this language concise. Data declarations are rarely required and procedure definitions are always independent of other definitions. If variables or parameters are declared, they must refer to the variable name most recently used (i.e., local definitions). The name of an identifier can be dynamically bound at run time. APL will also allow dynamic labeling. Every line of an APL program is numbered.

BASIC programs are formed by simple statements having the form

statement number;   statement (identifier);   (statement body)

where the parenthetical terms are optional. A program is executed sequentially in a step-by-step manner. All variables have a global definition with an initial value of zero unless otherwise specified.

FORTRAN programs are nonsimple in structure. (They consist of a main program, which is always present, and possibly subroutines and subprograms.) The binding of identifier names and the allocation of storage space is performed at the beginning of the program. Data or a name can be either global (COMMON) or local. The use of EQUIVALENCE statements allows several names to be associated with the same area of memory. FORTRAN has hard structures, program syntax, and I/O.

Pascal is a block-structured language consisting of two parts. They are a heading (naming the program and specifying the variables to be used in the body) and the body itself. A block is further divided into six subsections. The first four declare the labels, constants, data types, and variables. The fifth names and precedes an actual procedure or function. The last section is called the statement section and contains the executable code for the named function or procedure. Data types are numerous (arrays, sets, records, files) and procedures can be nested within procedures if references to the keywords BEGIN and END. Pascal uses a top-down structure, which makes it an efficient programming language.

In PL/I, as in FORTRAN or Pascal, a large program can be created from smaller ones. A program consists of a set of external procedures, one of which is designated to be the main program. There exists the possibility of simultaneous execution of two or more procedures through a type of subprogram called a task. Structures that are alike can be declared using the keyword LIKE. Only scalars (not arrays) can be returned from function routines. There is no "garbage collection" capability for recovering unused storage locations. The programmer must allocate and deallocate storage. Formatting input or output is optional, thereby establishing a "soft" I/O policy.

BIBLIOGRAPHY

Barron, D. W. (1977), An Introduction to Programming Languages, Cambridge University Press, New York.

Eckhouse, R. H., Jr., and L. R. Morros (1979), Minicomputer Systems, Prentice-Hall, Englewood Cliffs, N.J.

Morris, L. R. (1977), Time/Space Efficiency of Program Structures for Automatically Generated Digital Signal Processing Software, 1977 Int. Conf. ASSP.

Morris, L. R., and J. C. Mudge (1977), Speed Enhancement of Digital Signal Processing Software via Microprogramming a General Purpose Minicomputer, 1977 Int. Conf. ASSP.

Posa, J. G. (1978), "Pascal" Becomes Software Superstar, Electronics, October 12, pp. 81-84.

Rabiner, L. R. (1977), A Simplified Computational Algorithm for Implementing FIR Digital Filters, IEEE Trans. Acoust. Speech, Signal Process., June, pp. 259-261.

19
Special Topics

19.1 INTRODUCTION

To this point, standard filter design theory, analysis, and technology have been developed. There are also several special cases that need to be introduced and explored. They represent recent advances in the field of digital signal processing. The presentation of these methods has been delayed to this point for two basic reasons:

1. There exists a technology dependence that was not developed until the later chapters.
2. The error analysis is specialized.

In this chapter the filter architecture known as distributed arithmetic is explored. In addition, specialized modulation and signal conditioning policies are developed.

19.2 DISTRIBUTED ARITHMETIC

Distributed arithmetic is a data manipulation methodology. Anderseen (1971) and Zohar (1973) are generally credited with pioneering this study. However, this early work did not suggest a satisfactory hardware realization. The breakthrough came in the form of the high-speed read-only memory. The availability of these bipolar high-density devices at relatively low cost is making this architecture increasingly popular.

The study of distributed arithmetic can be motivated in terms of a simple inner-product example. Consider the inner product given by

$$y = a^T x = \Sigma \ a_i x_i; \quad i = 0, 1, \ldots, L - 1 \tag{19.2.1}$$

where x is a variable and a is a <u>known constant</u>. Both x and a are assumed to be integers. Furthermore, assume that a_j and x_j have n-bit binary representations of the form

$$a_j = \sum_i a[i, j]2^i; \quad x_j = \sum_i x[i, j]2^i; \quad j = 0, 1, \ldots, n \qquad (19.2.2)$$

where a[i, j] and x[i, j] are the jth binary coefficient of a_j and x_j, respectively. In a conventional inner-product definition, one obtains

$$y = \sum_{i=0}^{L-1} \sum_{j=0}^{n-1} \sum_{k=0}^{n-1} x[i, j]a[i, j]2^{j+k} \qquad (19.2.3)$$

In practice, y is computed using the organization found in Fig. 19.2.1. Here the s most significant bits of the full-precision products are sent to the system accumulator. As a result, there is an approximation error due to finite-word-length effects.

Alternative realizations exist. For the purpose of discussion, consider writing y as a function of a single variable in the form

$$y = \sum_{i=0}^{n-1} g(x[j])2^j; \quad j = 0, 1, \ldots, n - 1 \qquad (19.2.4)$$

where the function g(x[j]) is given by

$$g(x[j]) = \sum_{i=0}^{L=1} a_i x(i, j]; \quad i = 0, 1, \ldots, L - 1 \qquad (19.2.5)$$

That is, if x[j] is an L-typle listing of n binary-valued coefficients x[0, j] through x[L - 1, j], then g(x[j]) is uniquely specified in terms of an n-bit

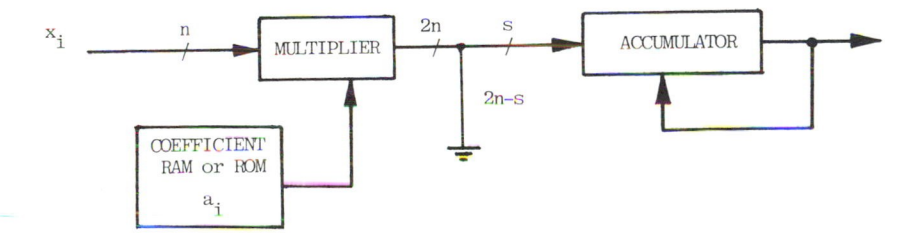

FIG. 19.2.1 Conventional digital filter hardware architecture.

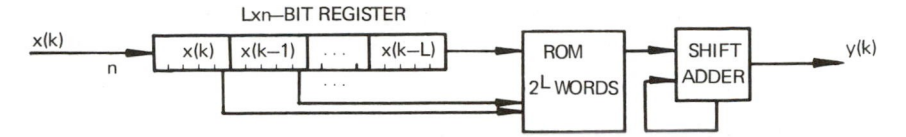

FIG. 19.2.2 Finite impulse response distributed arithmetic filter architecture.

array. There are at most 2^L possible values of $g(x[j])$ which need to be precomputed (off-line) and stored on a 2^L-word memory unit. During run time, the n-tuple $\{x[j]\}$ will be presented to the table and the precomputed value of $g(x[j])$ outputted. With the advent of high-speed high-density memory in affordable packages, this method has the potential to support practical high-performance arithmetic. Finally, the weight 2^i, found in Eq. (19.2.4), can be realized through the use of an elementary shift operation.

The concept of distributed arithmetic can be traced to the fact that the algebra is redistributed. That is, the inner product found in Eq. (19.2.1) is distributed over $i = 0, 1, \ldots, L - 1$, whereas the alternative method distributes the arithmetic over $j = 0, 1, \ldots, n - 1$ (i.e., data word length). This concept has also come to be known as the <u>bit-slice filter</u>. However, this should not be confused with the bit-slice processor. Both FIR and IIR filters can be realized using this concept. In particular, the filters will have the form

TABLE 19.2.1 Lookup Table of Contents for the Considered Example Problem

x[3:i]	x[2:i]	x[1:i]	g(x[i])		COMMENT
0	0	0	0_{10}:	0 0 0 0	(iv)
0	0	1	4_{10}:	0 1 0 0	
0	0	0	5_{10}:	0 1 0 1	
0	0	1	9_{10}:	1 0 0 1	
0	1	0	1_{10}:	0 0 0 1	
0	1	1	5_{10}:	0 1 0 1	(i)
0	1	0	6_{10}:	0 1 1 0	(ii)
0	1	1	10_{10}:	1 0 1 0	
1	0	0	5_{10}:	0 1 0 1	(iii)
1	0	1	9_{10}:	1 0 0 1	
1	0	0	10_{10}:	1 0 1 0	
1	0	1	14_{10}:	1 1 1 0	
1	1	0	6_{10}:	0 1 1 0	
1	1	1	10_{10}:	1 0 1 0	
1	1	0	11_{10}:	1 0 1 1	
1	1	1	15_{10}:	1 1 1 1	

$$g(x[i]) = 4x[0:i] + 5x[1:i] + 1x[2:i] + 5x[3:i]$$

$$\text{FIR: } y(n) = \sum_{j=0}^{L-1} a_i x(n - i)$$

(19.2.6)

$$\text{IIR: } y(n) = \sum_{i=0}^{L=1} a_i x(n - i) + \sum_{j=0}^{M-1} b_j y(n - j)$$

The architecture for the FIR case is suggested in Fig. 19.2.2.

EXAMPLE 9.2.1 Let $y = 4x_0 + 5x_1 + x_2 + 5x_3$ and $n = 4$; then there $2^4 = 16$ possible values of $g(x[j])$ which must be tabulated as shown in Table 19.2.1. Suppose at some instant of time that $x_0 = 1 \longrightarrow 0001$, $x_1 = 2 \longrightarrow 0010$, $x_3 = 3 \longrightarrow 0011$, and $x_4 = 4 \longrightarrow 0100$. Then y has the value $4 + 2 \times 5 + 3 + 4 \times 5 = 37$. Using the distributed concept, y would be computed as follows:

$$y = 2^0 g(x[0]) + 2^1 g(x[1]) + 2^2 g(x[2]) + 2^3 g(x[3])$$

or

							Table 19.2.1
2^6	2^5	2^4	2^3	2^2	2^1	2^0	comment
			0	1	0	1	$2^0 g(x[0]$	(i)
		0	1	1	0		$2^1 g(x[1])$	(ii)
	0	1	0	1			$2^2 g(x[2])$	(iii)
0	0	0	0				$2^3 g(x[3])$	(iv)
0	1	0	0	1	0	1	$= 37_{10}$

The salient features of this class of filter can be argued in the context of this example. Observe that instead of performing $L = 4$ full-precision multiplications [as in Eq. (19.2.1)], only $n = 4$ table calls are left-shifted as required. The table-lookup outputs are sent to an accumulator (i.e., shift-add). The concept is also valid for 2's-complement data. If

$$x_j = -x[i, 0] + \sum_{j=1}^{n-1} x[i, j] 2^{-j}; \quad -1 \le x_j \le 1, \quad x[i, j] = \{0, 1\} \quad (19.2.7)$$

$$a_j = -a[i, 0] + \sum_{j=1}^{n-1} a[i, j]2^{-j}; \quad -1 \le a_j \le 1, \quad a[i, j] = \{0, 1\} \quad (19.2.7)$$
$$(cont.)$$

it follows that

$$y = -g(x[0]) + \sum_{j=1}^{n-1} g(x[j])2^{-j} \qquad (19.2.8)$$

where $g(x[j])$ is defined in Eq. (19.2.5).

19.3 ERROR ANALYSIS

Distributed systems differ from their lumped-parameter counterparts in error analysis also. In a shift-invariant filter, the error due to quantization effects is given by

$$y = \sum_{i=1}^{L-1} a_i x_i - ([a_i]_R [x_i]_R) = y - y_L \qquad (19.3.1)$$

where $[\cdot]_R$ denotes rounding. Rounding errors are usually modeled to be uniformly distributed over $[-Q/2, Q/2]$, where Q is the quantization level. In a distributed system, errors, denoted $\Delta \tilde{y}$, are modeled as

$$\tilde{y} = a^T x + (a^T x)_R = y - y_D \qquad (19.3.2)$$

If full-precision arithmetic is used, $\Delta \tilde{y}$ can be represented as

$$y = -g(x[0]) + \sum_{j=1}^{n-1} g(x[i])2^{-j} \qquad (19.3.3)$$

Peled and Lui (1974) have suggested an error bound for this model in terms of its statistical error variance. If one assumes that all $g(x[j])$ and $x(i)$ are statistically independent and that the input is uniformly distributed over $[-1, 1]$, the quantization error is given by σ_D^2, where

$$\sigma_D^2 = \frac{1}{n} \sum_{j=0}^{n-1} \frac{Q^2}{12} (2^{-i})^2 \qquad (19.3.4)$$

The weighting sequence $\{1,\ 1/4,\ 1/16,\ \ldots\}$ can be bounded by

$$1 + \frac{1}{4} + \frac{1}{16} + \cdots + \frac{1}{2^{2i}} + \cdots \leq \sum_{i=0}^{\infty} 2^{-2i} = \frac{1}{1 - (1/4)} = \frac{4}{3} \qquad (19.3.5)$$

It then follows that

$$\sigma_D^2 \leq \left(\frac{Q^2}{12}\right)\left(\frac{4}{3}\right) = \frac{Q^2}{9} \qquad (19.3.6)$$

It can be noted that the lumped-parameter filter error variance of $Q^2/12$ and distributed error variance of $Q^2/9$ are comparable. However, it should be remembered that the lumped-parameter filter was synthesized using full-precision arithmetic. Only the final sum of products was rounded. If the individual products are rounded and then summed, the error variance will increase from $Q^2/12$ by a factor L (the filter order).

Kammeyer (1977) has conjectured that the distributed error variance of $Q^2/9$ may be too pessimistic when dealing with word lengths of less than 8 bits. This is based on the fact that the previously cited statistical independence assumption is bogus. This claim can be argued in the context of the definition of $g(x[j])$, namely

$$g(x[i]) = \sum_{j=1}^{L} a_j x[j,\ i] \qquad (19.3.7)$$

There exist 2^L possible combinations of L-tuples (i.e., $x[0,\ j]$, ..., $x[L-1,\ j]$) which form $x[j]$. Suppose that the input time series $\{x(i)\}$ is a white, zero-mean process over $[-1,\ 1]$; then $E(x(i)) = 0$ and $E(x(i)x(j)) = (1/3)_k(j - i)$. The statistical interrelationship between the distributed filter coefficients is given by $E(g(x[i])g(x[m]) = E(\Sigma \Sigma (a_j a_k x[j:i]x[k:m])$, where a_j and a_k are filter coefficients. If $j = k$ and $i = m$, then for all possible values of $x[j,\ i]$ and $x[k,\ m]$, one obtains

$$\left\{\begin{array}{cc} x[j,\ i] & x[k,\ m] \\ 1 & 1 \\ 0 & 0 \end{array}\right\} \Rightarrow \mathcal{E}\ (x[j,\ i]x[k, \quad = \frac{1}{2}\ (1 + 0) = \frac{1}{2}$$

If $j \neq k$ for all i amd m, or $i \neq m$ for all j and k, then for all possible values of $x[j,\ i]$ and $x[k,\ m]$, one obtains

$$
\left\{
\begin{array}{cc}
x[j,\ i] & x[k,\ m] \\
1 & 1 \\
0 & 1 \\
1 & 0 \\
0 & 0
\end{array}
\right\}
\Rightarrow \mathcal{E}(x[j,\ i] * x[k,\ m]) = \frac{1}{4}(1 + 0 + 0 + 0) = \frac{1}{4} \neq 0
$$

Therefore, the distributed filter coefficients are not statistically independent. Kammeyer used this observation to derive a "tighter" error variance
bound which was less than $Q^2/9$.

19.4 FILTER STRUCTURES

High-speed linear convolution can be performed using distribution arithmetic. Since this philosophy replaces general multiplication with table-
lookup operations, it is required that the filter coefficients be known a
priori. That is, only shift-invariant filters can be realized as distributed
filters. It will be assumed that the filters discussed in this section use a
common n-bit word.

The convolution of a binary-valued (digital) time series $\{x(n)\}$, with
an Lth-order FIR impulse response [i.e., $a(0)$, $a(1)$, ..., $a(L-1)$] is
given by $y(n) = x(n) * h(n)$. In terms of a distributed filter, $y(n)$ is given by

$$
y(n) = \sum_{j=0}^{n-1} g(x[n,\ j]),\ \dots,\ x[n - (1 - 1),\ j])2^{j} \tag{19.4.1}
$$

where

$$
g(*) = \sum_{k=0}^{L-1} a(k)x[n - k,\ j] = g(x[j]) \tag{19.4.2}
$$

and $x[r,\ s]$ is the sth bit of $x(r)$ and $a(k)$ is the kth real filter coefficient.
To implement the distributed convolution, a ROM having an L-bit input
addressing space would be chosen.

There are several variations of this basic theme. They define a
trade-off between memory size and network complexity. For example,
suppose that 1K ROMs are available for design purposes. Using this 10-bit
input address device, a 10th-order FIR can be realized. Suppose that the
design objective is a 14th-order filter. In order to realize this filter,
based on a 1K memory unit, some algebraic partitioning is required. In
general, consider that a 2^{Q}xm-bit ROM is to be considered the standard

memory-unit size. Furthermore, let the target filter be of length L where
$L = QN$, N an integer; then

$$y(n) = \sum_{j=0}^{n-1} \sum_{h=0}^{N-1} f_h 2^j \tag{19.4.3}$$

where

$$f_h = \sum_{i=0}^{Q-1} a(Nh + i)x[n - Qh - i, \ j] \tag{19.4.4}$$

By choosing the smaller size 2^Qxm-bit memory units over the larger
2Qxm-bit devices, N additional adders are required. As a result, execution speed will be reduced to the adder overhead. However, this may be
somewhat offset by being able to use smaller, higher-speed bipolar memory
devices (i.e., size and speed are generally inversely related).

   Alternatively, the filter could be partitioned with respect to word
size instead of filter length. Suppose for an integer N and word length n
that $n = n_0 N$ and

$$y(n) = \sum_{k=0}^{N-1} f_k 2^{kn_0} \tag{19.4.5}$$

where

$$f_k = \sum_{i=0}^{n-1} \sum_{h=0}^{L-1} a(h)x[n - h, \ i + kn_0]2^i \tag{19.4.6}$$

The advantage of this realization is speed. Instead of summing over $j = 0$,
$1, \ldots, n - 1$ [as found in Eq. (19.4.3)], the summation now occurs over
$k = 0, 1, \ldots, N - 1$. Therefore, there is an n_0 decrease in table calls.
Fewer calls will increase throughput. However, this may be offset by an
increase in the memory cycle time associated with larger memory units
(now of size $2^{n_0 L}$ words). A general design is, therefore, parameterized
in terms of the five-tuple (L, N, n_0, n, m). To dramatize these structural
differences, the various filter options have been summarized in Fig. 19.4.1.

EXAMPLE 19.4.1 Consider the case where $n = 2$, $x(0) = x(1) = a_0 = 3$,
and $a_1 = 1$. Examples of the various filter options are given in Fig. 19.4.2.
In these examples, full-precision multiplication has been assumed. However, the ROM implementation found in mechanization 1 (i.e., $2^2 \times 4$) can

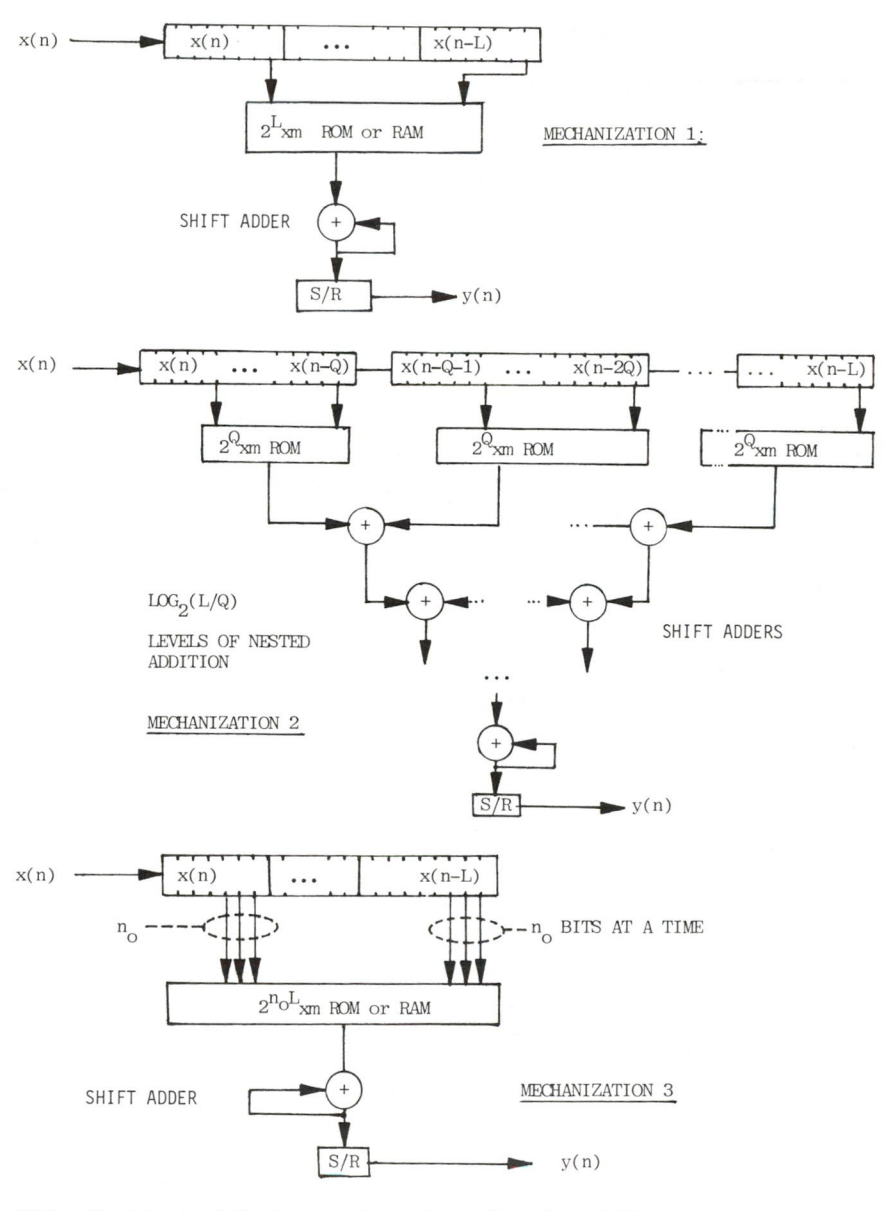

FIG. 19.4.1 Architecture options for a distributed filter.

be reduced to $2^2 \times 3$ bits if the most significant bit, supplied to the shift adder, is hard-wired to logical 0 (see Fig. 19.4.2). Similarly, the ROM found in mechanization 2 could be reduced in size to $2^1 \times 2$ bits. The rule that governs the minimal ROM size is based on the maximal bound of $f(\cdot)$. That is, for $x[i, j]$ (binary valued), suppose that

$$\max f(x[i, j], \ldots, f(x[L + i, j]) \leq 2^m$$

Then a ROM of size $2^m \times K$ would be required.

EXAMPLE 19.4.2 Consider again Example 19.4.1 using $n = 2$. Filters 1 and 2, detailed in Table 19.4.1, represent full-precision (i.e., 4-bit) and minimal-word-length (i.e., 3-bit) table structures. In both cases, the table outputs were rounded to their two most significant digits. Simulated filter behavior is abstracted below, and the error quantified. It can be seen that the minimal-word-length filter has a lower error variance. This is due to the fact that on a bit-per-bit basis, the minimal-word-length tables are of higher precision.

Recall that for sufficiently long word lengths, the error variance of a distributed filter is on the order of $Q^2/12$. This assumes a standard realization (i.e., mechanization 1 of Fig. 19.4.1). However, it would appear that mechanization 3 (Fig. 19.4.1) should have a smaller error variance since fewer rounded, table-lookup, partial products are used.

19.5 PERFORMANCE

Jenkins and Leon (1977) have tested the performance of the distributed concept using a 64th-order FIR model. The distributed filter performance metrics were compared to those obtained using conventional filter architectures. The fundamental building blocks for the conventional filter architectures. The fundamental building blocks for the conventional filter were a 25-chip 250-nsec IC multiplier or an 85-nsec MSI multiplier. System adders were budgeted 48 nsec per add. As a result, the throughput of the two conventionally architected filters can be estimated to be $64(250 + 48) = 19,072$ nsec and $64(85 + 48) = 8512$ nsec, respectively. If economy and packaging elegance is stressed, a VLSI multiplier-accumulator unit could be used at a rate of $64 \times 180 - 11,520$ nsec per filter cycle. The 64th-order distributed filter was designed using 30-nsec RAM and 30-nsec ROM. The two test filters differ basically in part count. The execution rate of the 8-bit filter was estimated to be $8(30 + 48 + 48) + 7(48) = 1334$ nsec. These timing estimates can be argued in the context of the data in Fig. 19.5.1.

Doubling the input word length (from 8 bits to 16 bits) would approximately halve the throughput of the distributed filter. However, increasing

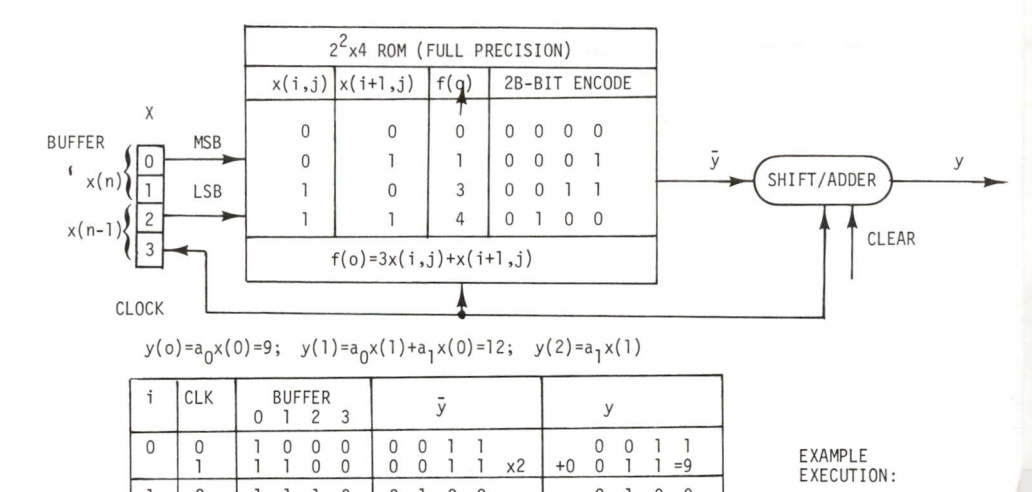

2²x4 ROM (FULL PRECISION)

x(i,j)	x(i+1,j)	f(q)	2B-BIT ENCODE
0	0	0	0 0 0 0
0	1	1	0 0 0 1
1	0	3	0 0 1 1
1	1	4	0 1 0 0

$$f(o)=3x(i,j)+x(i+1,j)$$

$y(o)=a_0x(0)=9; \quad y(1)=a_0x(1)+a_1x(0)=12; \quad y(2)=a_1x(1)$

i	CLK	BUFFER 0 1 2 3	$\bar{y}$	y	
0	0	1 0 0 0	0 0 1 1		0 0 1 1
	1	1 1 0 0	0 0 1 1 x2	+0 0 1 1	=9
1	2	1 1 1 0	0 1 0 0		0 1 0 0
	3	1 1 1 1	0 1 0 0 x2	+0 1 0 0	=12
2	4	0 1 1 1	0 0 0 1		0 0 0 1
	5	0 0 1 1	0 0 0 1 x2	+0 0 0 1	

EXAMPLE EXECUTION:

MECHANIZATION 1:

MECHANIZATION 2:

2¹x4 ROM
$f_1(o)=3x(i,j)$

x(i,j)	f(o)	2N-BIT CODE
0	0	0 0 0 0
1	3	0 0 1 1

2¹x4 ROM
$f_2(o)=1x(i,j)$

x(i,j)	f(o)	2N-BIT CODE
0	0	0 0 0 0
1	1	0 0 0 1

n=2 Q=1
L=2 N=2

i	CLK	BUFFER 0 1 2 3	$\bar{y}_1$	$\bar{y}_2$	$\bar{y}_1+\bar{y}_2$	y	
0	0	1 0 0 0	0 0 1 1	0 0 0 0	0 0 1 1		0 0 1 1
	1	1 1 0 0	0 0 1 1	0 0 0 0	0 0 1 1 x2	+0 0 1 1	=9
1	2	1 1 1 0	0 0 1 1	0 0 0 1	0 1 0 0		0 1 0 0
	3	1 1 1 1	0 0 1 1	0 0 0 1	0 1 0 0 x2	+0 1 0 0	=12
2	4	0 1 1 1	0 0 0 0	0 0 0 1	0 0 0 1		0 0 0 1
	5	0 0 1 1	0 0 0 0	0 0 0 1	0 0 0 1 x2	+0 0 0 1	=3

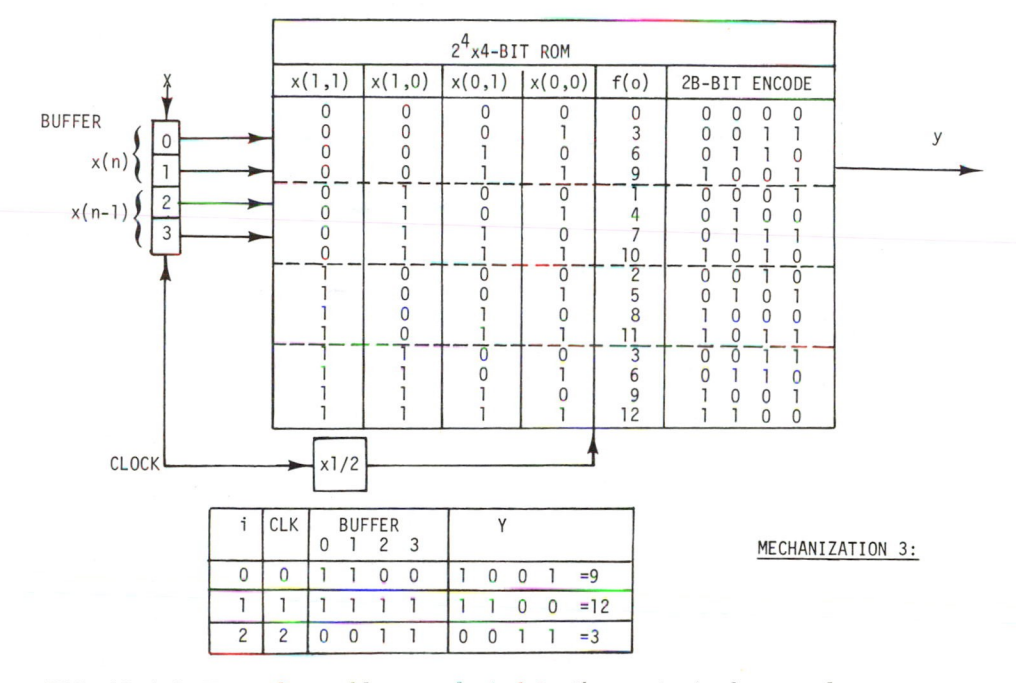

FIG. 19.4.2 Example problem evaluated in the context of several distributed architectures.

TABLE 19.4.1 Detailed Analysis of a Distributed Filter Operation

MECHANIZATION (ORIGNAL)
A: 4-BIT FULL PRECISION

	ADD 1	ADD 0	f(o)	OUTPUT	OUTPUT (ROUNDED TO TWO BITS)
	0	0	0	0000	00
	0	1	1	0001	00
	1	0	3	0011	01
	1	1	4	0100	01

TWO DISTINCT VALUES (0,4)
WEIGHT OF MSB = 8

MECHANIZATION (MODIFIED)
B: 3-BIT FULL PRECISION

	ADD 1	ADD 0	f(o)	OUTPUT	OUTPUT (ROUNDED TO TWO BITS)
	0	0	0	000	00
	0	1	1	001	01
	1	0	3	011	10
	1	1	4	100	10

THREE DISTINCT VALUES (0,2,4)
WEIGHT OF MSB = 4

CLK	BUFFER CONTENTS	FULL PRECISION Y(I)	REALIZATION A Ȳ:	YA:	|YA|2	ERROR	REALIZATION B Ȳ:	YB:	|YB|2	ERROR
0	1000	9_{10}	01 / 01x2	$01 +01\ \overline{011}=12_{10}$	$10\ =16_{10}$	7_{10}	10 / 10x2	$10 +10\ \overline{110}=12_{10}$	$11\ =12_{10}$	3_{10}
1	1100									
2	1110	12_{10}	01 / 01x2	$01 +01\ \overline{011}=12_{10}$	$10\ =16_{10}$	4_{10}	10 / 10x2	$10 +10\ \overline{110}=12_{10}$	$11\ =12_{10}$	0_{10}
3	1111									
4	0111	3_{10}	00 / 00x2	$00 +00\ \overline{000}=\ 0_{10}$	$00\ =\ 0_{10}$	3_{10}	01 / 01x2	$01 +01\ \overline{110}=\ 8_{10}$	$10\ =\ 8_{10}$	5_{10}
5	0011									

$E(|ERROR|)+14/3=4.667$

$E(|ERROR|)=8/3=2.27$

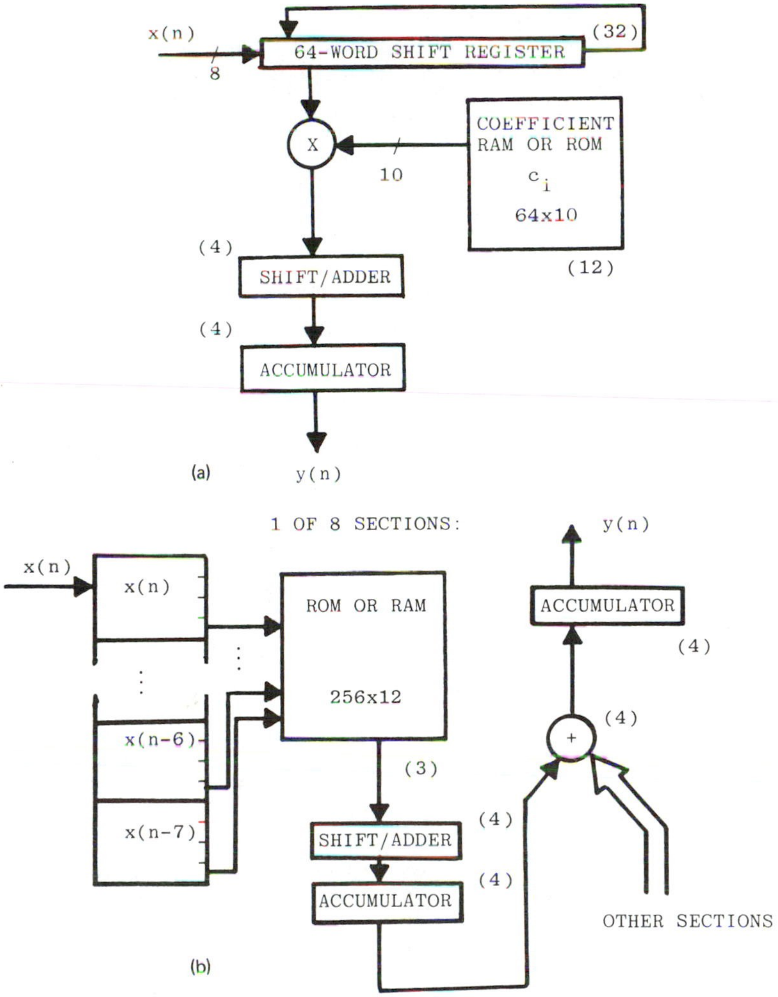

x(n)

(32)

64-WORD SHIFT REGISTER

8

X

10

COEFFICIENT
RAM OR ROM

c_i

64x10

(12)

(4)

SHIFT/ADDER

(4)

ACCUMULATOR

(a)   y(n)

1 OF 8 SECTIONS: y(n)

x(n)

x(n)

ROM OR RAM

256x12

x(n-6)

x(n-7)

ACCUMULATOR

(4)

(4)

+

(3)

SHIFT/ADDER   (4)

ACCUMULATOR   (4)

OTHER SECTIONS

(b)

FIG. 19.5.1 Comparison of two memory-intensive filter architectures.
(a) Conventional filter. (b) Distributed filter.

word length will also have a temporal retarding effect on the conventionally
architected filter.

EXAMPLE 19.5.1 The results of the Jenkins and Leon experiment are
summarized in Table 19.5.1. The reader should be aware that these
technology-intensive parameters are changing. Therefore, the data in
this table should be reinterpreted with respect to current cost, packaging,
and performance specifications.

EXAMPLE 19.5.2 A fourth-order Chebyshev filter having a transfer function given by

$$H(z) = \frac{0.00183(1 + z^{-1})^4}{(1 - 1.4999z^{-1} + 0.8482z^{-2})(1 - 1.5548z^{-1} + 0.6493z^{-2})}$$

$$= \frac{a_0 + a_1 z^{-1} + a_2 z^{-2} + a_3 z^{-3} + a_4 z^{-4}}{1 + b_1 z^{-1} + b_2 z^{-2} + b_3 z^{-3} + b_4 z^{-4}}$$

where

$$a_0 = 0.001836 = a_4, \ a_1 = 0.007344 = a_3, \ a_2 = 0.01106$$
$$b_1 = -3.0538, \ b_2 = 3.8281452, \ b_3 = -2.2920813, \ b_4 = 0.55078$$

TABLE 19.5.1 Summary of Experimental Results

Architecture	Data rate (KHz)	IC count	Percent $memory	Percent $S/R	Percent $ALU	KHz/
Conventional: 250 nsec multiplier	54.4	77	15.58	41.56	42.85	680.9
Conventional: 85 nsec multiplier	117.5	137	8.76	23.36	67.88	857.5
Conventional: VLSI ALU	86.8	49	9.92	26.41	63.60	690.5
Distributed ROM	744.0	128	18.75	25.00	56.25	5912.0
Distributed RAM	744.0	200	16.00	36.00	36.00	3720.0

Portions of the data suggested by Jenkins and Leon.

can be expressed as

$$y(n) = \sum_{i=0}^{4} a_i x(n - i) + \sum_{i=1}^{4} b_i y(n - i)$$

The system transfer function has been scaled so that $|y(n)| \leq 1$ The values of the partial products can be written as

$$y[j, k] = \sum_{i=0}^{4} a_i x[j, k - i] + \sum_{i=1}^{4} b_i y[j, k - i]$$

for $k = 0, \ldots, n - 1$, where n is the output word length in bits. The absolute value of $y[j, k]$ is bounded by the maximum of $a_0 + a_1 + \cdots + a_4 -$ $(b_1 + b_2 + \cdots + b_4) = 5.37622 < 2^3$. Therefore, an n-bit data format of the form $y[j, k] \pm xxx_\triangle x \ldots x$ would be adopted. Since there are nine binary-valued variables on the right-hand side of the example filter model, a $2^9 x n$-bit memory unit would be needed. The system is summarized in Fig. 19.5.2, with a portion of the lookup table repeated in Fig. 19.5.3. Here a 9-bit input address is interpreted as a 16-bit distributed output.

Numerical experiments were performed using this filter and the magnitude-squared frequency response computed (see Fig. 19.5.4). The data presented compare the response of a 16- and an 8-bit filter. It can be seen that the longer-word-length filter exhibits an improved magnitude response. However, since this filter will require twice as many table calls as its shorter-word-length counterpart, it will run twice as slow.

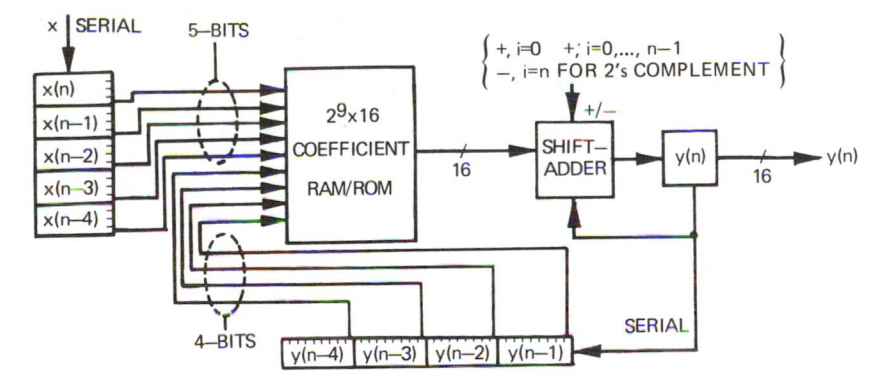

FIG. 19.5.2 Design of an IIR filter using a distributed filter architecture.

INPUT

b_4	b_3	b_2	b_1	a_4	a_3	a_2	a_1	a_0	$f(o)$	MSB — OUTPUT 16 BITS — LSM
0	1	0	1	0	0	0	0	0	5.3466790000	0101010110001100
0	1	0	1	0	0	0	1	0	5.3483880000	0101010110101001
0	1	0	1	0	0	0	1	1	5.3540030000	0101010110101011
0	1	0	1	0	0	1	0	0	5.3557120000	0101010111011100
0	1	0	1	0	0	1	0	1	5.3596190000	0101010111110000
0	1	0	1	0	0	1	1	0	5.3649900000	0101010111101011
0	1	0	1	0	0	1	1	1	5.3669430000	0101010111110111
0	1	0	1	0	1	0	0	0	5.3540030000	0101010111011110
0	1	0	1	0	1	0	0	1	5.3557120000	0101010111011101
0	1	0	1	0	1	0	1	0	5.3613280000	0101010111001010
0	1	0	1	0	1	0	1	1	5.3630370000	0101010111000110
0	1	0	1	0	1	1	0	0	5.3649900000	0101010111101010
0	1	0	1	0	1	1	0	1	5.3669430000	0101010111101111
0	1	0	1	0	1	1	1	0	5.3723140000	0101010111110111
0	1	0	1	0	1	1	1	1	5.3742670000	0101010111111101
0	1	0	1	1	0	0	0	0	5.3483880000	0101010111000111
0	1	0	1	1	0	0	0	1	5.3503410000	0101010111001001
0	1	0	1	1	0	0	1	0	5.3557120000	0101010111011000
0	1	0	1	1	0	0	1	1	5.3576660000	0101010111010001
0	1	0	1	1	0	1	0	0	5.3596190000	0101010111000010
0	1	0	1	1	0	1	0	1	5.3613280000	0101010111001010
0	1	0	1	1	0	1	1	0	5.3669430000	0101010111101110
0	1	0	1	1	0	1	1	1	5.3686520000	0101010111101110
0	1	0	1	1	1	0	0	0	5.3557120000	0101010111011000
0	1	0	1	1	1	0	0	1	5.3576660000	0101010111010001
0	1	0	1	1	1	0	1	0	5.3630370000	0101010111001010
0	1	0	1	1	1	0	1	1	5.3649900000	0101010111101010
0	1	0	1	1	1	1	0	0	5.3669430000	0101010111101111
0	1	0	1	1	1	1	0	1	5.3686520000	0101010111101110
0	1	0	1	1	1	1	1	0	5.3742670000	0101010111111101
0	1	0	1	1	1	1	1	1	5.3762200000	0101010111110001
0	1	1	0	0	0	0	0	0	-1.5363760000	1110011101011010
0	1	1	0	0	0	0	0	1	-1.5346670000	1110011101011110
0	1	1	0	0	0	0	1	0	-1.5290520000	1110011101001001
0	1	1	0	0	0	0	1	1	-1.5273430000	1110011101001000
0	1	1	0	0	0	1	0	0	-1.5253900000	1110011101000000
0	1	1	0	0	0	1	0	1	-1.5234370000	1110011101010010
0	1	1	0	0	0	1	1	0	-1.5180660000	1110011101011100
0	1	1	0	0	0	1	1	1	-1.5161130000	1110011101011001
0	1	1	0	0	1	0	0	0	-1.5290520000	1110011101000001
0	1	1	0	0	1	0	0	1	-1.5273430000	1110011101001000
0	1	1	0	0	1	0	1	1	-1.5217280000	1110011101010011
0	1	1	0	0	1	0	1	1	-1.5200190000	1110011101011100
0	1	1	0	0	1	1	0	0	-1.5180660000	1110011101011110
0	1	1	0	0	1	1	1	0	-1.5161130000	1110011101011001
0	1	1	0	0	1	1	1	1	-1.5104980000	1110011111011001
0	1	1	0	1	0	0	0	0	-1.5087890000	1110011111011000
0	1	1	0	1	0	0	0	1	-1.5346670000	1110011111011100
0	1	1	0	1	0	0	1	0	-1.5327140000	1110011111001100
0	1	1	0	1	0	0	1	1	-1.5273430000	1110011111001000
0	1	1	0	1	0	1	0	0	-1.5253900000	1110011111000000
0	1	1	0	1	0	1	0	1	-1.5234370000	1110011111010010
0	1	1	0	1	0	1	1	0	-1.5217280000	1110011111010011
0	1	1	0	1	0	1	1	0	-1.5161130000	1110011111011110

FIG. 19.5.3 Input-output distributed arithmetic mappings.

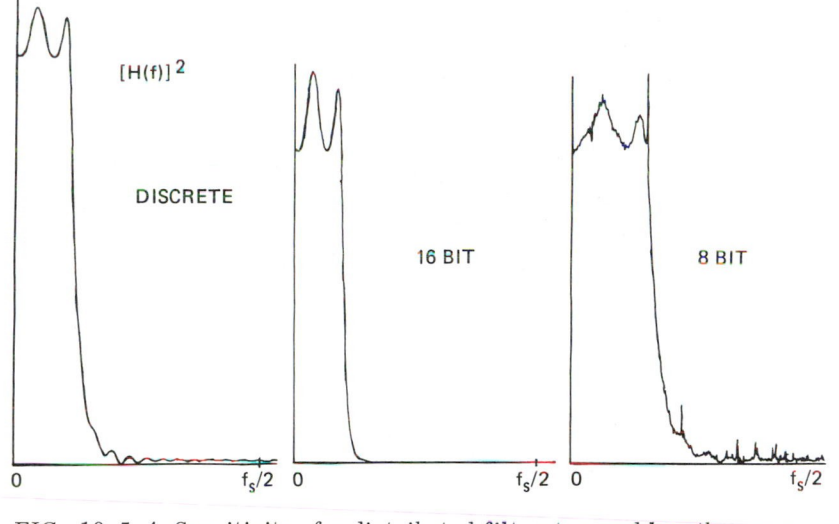

FIG. 19.5.4 Sensitivity of a distributed filter to word length.

EXAMPLE 19.5.3 The second-order filter section has been previously argued to be one of the fundamental digital filter building blocks. In Fig. 19.5.5 an architecture for a general 16-bit second-order distributed filter is suggested. The data are presented to the lookup unit at a rate of 2 bits per filter variable. The ten-tuple address is presented to four 1K × 4 HMOS memory units. Using the suggested figure of 25 to 35 nsec access and cycle times, respectively (0.5 mW per bit), throughputs on the order of 8 × (25 to 35) = 200 to 280 nsec (5 to 3.6 MHz) are obtainable.

EXAMPLE 19.5.4 Consider the fourth-order Chebyshev lowpass filter derived in Example 7.4.3. Specifically,

$$H(z) = \frac{A_0 + A_1 z^{-1} + A_2 z^{-2} + A_3 z^{-3} + A_4 z^{-4}}{B_0 + B_1 z^{-1} + B_2 z^{-2} + B_3 z^{-3} + B_4 z^{-4}}$$

$A_0 = 0.001835550371925,$ $B_0 = 1.0$

$A_1 = 0.007342201487,$ $B_1 = -3.054339676419$

$A_2 = 0.01101330223155,$ $B_2 = 3.828999227497$

$A_3 = 0.0073422014877,$ $B_3 = -2.292451729417$

$A_4 = 0.001835550371925,$ $B_4 = 0.5507445205826$

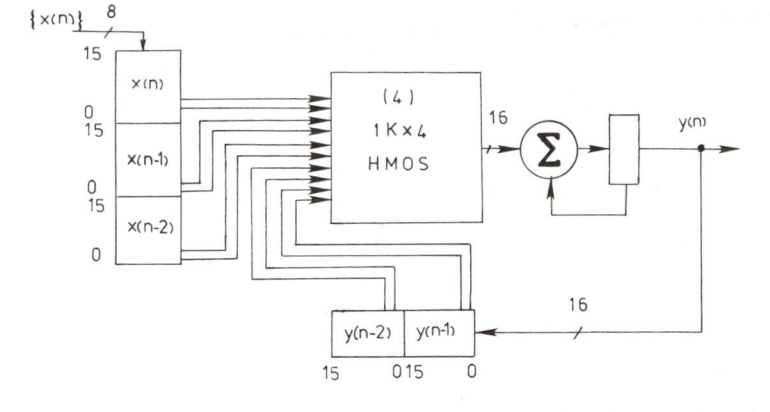

FIG. 19.5.5 Second-order digital filter realized in a distributed filter sense.

```
C
      DOUBLE PRECISION   ARRAY1, ARRAY2, SUM
C
      DIMENSION   ARRAY1(9),   ARRAY2(512)
C
C
      READ (4, 103)   ARRAY1
 103  FORMAT (E20.12)
C
      DO 163   L = 1, 512
        SUM = 0.0
        NUM = L - 1
        DO 157   I = 1, 9
          ITEMP = NUM/2
          J = NUM - 2*ITEMP
          NUM = ITEMP
          IF (J .EQ. 0)  GO TO 157
          SUM = SUM + ARRAY1(I)
 157    CONTINUE
        ARRAY2(L) = SUM
 163  CONTINUE
C
      WRITE (5, 303)  ARRAY2
 303  FORMAT (3(E27.13))
C
      END
```

FIG. 19.5.6 Source program for generating tabled data.

In the time domain one obtains

$$y(n) = \sum_{i=0}^{4} A_i x(n - i) + \sum_{i=1}^{4} B_i y(n - i)$$

Observe that there are nine system variables. There, a $2^9 = 512$-word lookup table is required to support distributed arithmetic. Using the simple FORTRAN program given in Fig. 19.5.6, the distributed arithmetic coefficient table can be computed. A section of the derived data table is given in Fig. 19.5.7. For example, octal address 113 represents

1	2	3
0 0 1	0 1 1	0 1 1
B_4 B_3 B_2	B_1 A_4 A_3	A_2 A_1 A_0 $= -2.279602876$

EXAMPLE 19.5.5 Using four 300-nsec 2920 bit-slice processors and a 67110 microprogram control sequencer, a 16-bit distributed arithmetic 12th-order elliptic IIR filter has been configured by Zeman and Nagle (1980). Using a cross assembler, the required microcode was generated. The 12th-order filter was realized as a system of three fourth-order subfilters having the form

$$H_i(z) = K_i \frac{1 + a_{i1}z^{-1} + a_{i2}z^{-2} + a_{i3}z^{-3} + a_{i4}z^{-4}}{1 + b_{i1}z^{-1} + b_{i2}z^{-2} + b_{i3}z^{-3} + b_{i4}z^{-4}}$$

where K_i is chosen to provide overflow prevention scaling. Based on this partition, three 256×16 distributed multiplication tables were derived. The filter was reported to consume 48 W of power and possessed a Nyquist frequency of 16.8 kHz.

EXAMPLE 19.5.6 Distributed arithmetic filters have been experimentally studied by Tam and Hawkins (1981). Popular microprocessors were coded to mechanize a 32nd-order FIR based on 256-word lookup tables (i.e., $L = 32$, $Q = 8$, $N = 4$). The results of their experiments are summarized in Fig. 19.5.8.

19.6 OTHER LOOKUP FILTERS

Other table-lookup methods have been developed as well. One filter struc-structure, attributable to Monkewich and Steenart, stores all possible

$$y(n) = \sum_{i=0}^{4} A_i x(n-i) + \sum_{i=1}^{4} B_i y_i(n-i)$$

HEX	y(n)	HEX	y(n)	HEX	y(n)
0	.0000000000000E+01	1	.1835550371925E-02	2	.7342201487700E-02
3	.9177751859625E-02	4	.1101330223155E-01	5	.1284885260347E-01
6	.1835550371925E-01	7	.2019105409117E-01	8	.7342201487700E-02
9	.9177751859625E-02	10	.1468440297540E-01	11	.1651995334732E-01
12	.1835550371925E-01	13	.2019105409117E-01	14	.2569770520695E-01
15	.2753325557887E-01	16	.1835550371925E-02	17	.3671100743850E-02
18	.9177751859625E-02	19	.1101330223155E-01	20	.1284885260347E-01
21	.1468440297540E-01	22	.2019105409117E-01	23	.2202660446310E-01
24	.9177751859625E-02	25	.1101330223155E-01	26	.1651995334732E-01
27	.1835550371925E-01	28	.2019105409117E-01	29	.2202660446310E-01
30	.2753325557887E-01	31	.2936880595080E-01	32	-.3053396764190E+01
33	-.3051561213818E+01	34	-.3046054562702E+01	35	-.3044219012330E+01
36	-.3042383461958E+01	37	-.3040547911587E+01	38	-.3035041260471E+01
39	-.3033205710099E+01	40	-.3046054562702E+01	41	-.3044219012330E+01
42	-.3038712361215E+01	43	-.3036876810843E+01	44	-.3035041260471E+01
45	-.3033205710099E+01	46	-.3027699059983E+01	47	-.3025863508611E+01
48	-.3051561213818E+01	49	-.3049725663446E+01	50	-.3044219012330E+01
51	-.3042383461958E+01	52	-.3040547911587E+01	53	-.3038712361215E+01
54	-.3033205710099E+01	55	-.3031370159727E+01	56	-.3044219012330E+01
57	-.3042383461958E+01	58	-.3036876810843E+01	59	-.3035041260471E+01
60	-.3033205710099E+01	61	-.3031370159727E+01	62	-.3025863508611E+01
63	-.3024027958239E+01	64	.3828999227497E+01	65	.3830834777869E+01
66	.3836341428985E+01	67	.3838176979357E+01	68	.3840012529729E+01
69	.3841848080100E+01	70	.3847354731216E+01	71	.3849190281588E+01
72	.3836341428985E+01	73	.3838176979357E+01	74	.3843683630472E+01
75	.3845519180844E+01	76	.3847354731216E+01	77	.3849190281588E+01
78	.3854696932704E+01	79	.3856532483076E+01	80	.3830834777869E+01
81	.3832670328241E+01	82	.3838176979357E+01	83	.3840012529729E+01
84	.3841848080100E+01	85	.3843683630472E+01	86	.3849190281588E+01
87	.3851025831960E+01	88	.3838176979357E+01	89	.3840012529729E+01
90	.3845519180844E+01	91	.3847354731216E+01	92	.3849190281588E+01
93	.3851025831960E+01	94	.3856532483076E+01	95	.3858368033448E+01
96	.7756024633070E+00	97	.7774380136789E+00	98	.7829446647947E+00
99	.7847802151666E+00	100	.7866157655385E+00	101	.7884513159105E+00
102	.7939579670262E+00	103	.7957935173982E+00	104	.7829446647947E+00
105	.7847802151666E+00	106	.7902868662824E+00	107	.7921224166543E+00
108	.7939579670262E+00	109	.7957935173982E+00	110	.8013001685139E+00
111	.8031357188859E+00	112	.7774380136789E+00	113	.7792735640508E+00
114	.7847802151666E+00	115	.7866157655385E+00	116	.7884513159105E+00
117	.7902868662824E+00	118	.7957935173982E+00	119	.7976290677701E+00
120	.7847802151666E+00	121	.7866157655385E+00	122	.7921224166543E+00
123	.7939579670262E+00	124	.7957935173982E+00	125	.7976290677701E+00
126	.8031357188859E+00	127	.8049712692578E+00	128	-.2292451729417E+01
129	-.2290616179045E+01	130	-.2285109527929E+01	131	-.2283273977557E+01
132	-.2281438427185E+01	133	-.2279602876814E+01	134	-.2274096225698E+01
135	-.2272260675326E+01	136	-.2285109527929E+01	137	-.2283273977557E+01
138	-.2277767326442E+01	139	-.2275931776070E+01	140	-.2274096225698E+01
141	-.2272260675326E+01	142	-.2266754024210E+01	143	-.2264918473838E+01
144	-.2290616179045E+01	145	-.2288780628673E+01	146	-.2283273977557E+01
147	-.2281438427185E+01	148	-.2279602876814E+01	149	-.2277767326442E+01
150	-.2272260675326E+01	151	-.2270425124954E+01	152	-.2283273977557E+01
153	-.2281438427185E+01	154	-.2275931776070E+01	155	-.2274096225698E+01
156	-.2272260675326E+01	157	-.2270425124954E+01	158	-.2264918473838E+01

FIG. 19.5.7 Sample of table data for a fourth-order distributed arithmetic filter.

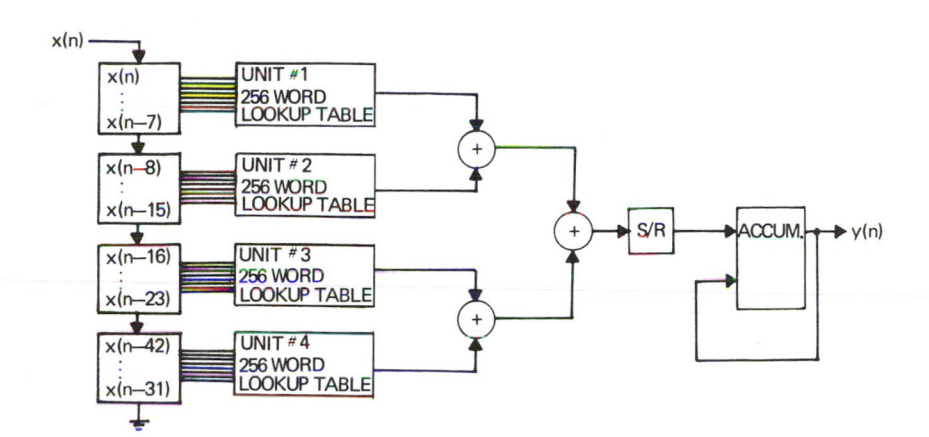

FIG. 19.5.8 Experimental distributed filter performance.

PROCESSOR	WORD LENGTH	CLOCK	NONSYMMETRICAL COEFFICIENTS PRECISION		SYMMETRICAL COEFFICIENTS PRECISION	
			SINGLE	DOUBLE	SINGLE	DOUBLE
1. Z–80A	8–BITS	4 MHz	14.8KHz	5.4KHz	25.5KHz	10.7KHz
2. MC 6800B	8	2	23.5	12.0	37.7	19.4
3. MCS6502	8	2	23.5	5.7	37.7	9.1
4. SC–MP/II	8	1	2.2	487Hz	3.9	797Hz
5. 8080A	8	2	7.5	2.4KHz	12.7	3.7KHz
6. CP1600A	16–BITS	5	7.8	N/A	13.0	N/A
7. TMS9900	16	3	5.3	N/A	8.7	N/A
8. 8086	16	5	20.0	N/A	32.8	N/A

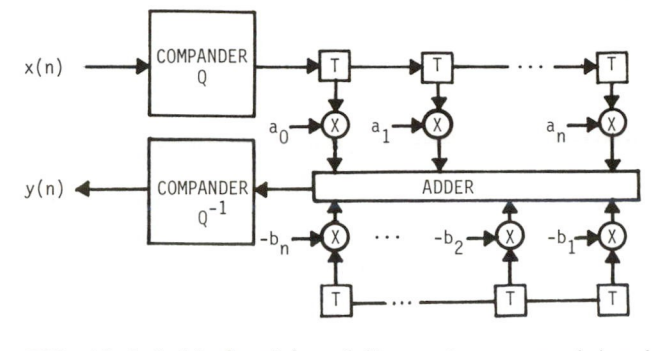

FIG. 19.6.1 Monkewich and Steenart memory-intensive filter.

signal-coefficient products. An example of this filter is suggested in Fig. 19.6.1. The filter has been outfitted with a logarithmic compander to reduce the dynamic range (and thereby the word length) of the filter input space. The major disadvantage of this architecture is the cost of the hardware. This filter, like its distributed counterpart, can only be used to realize shift-invariant filters having a prespecified coefficient set.

19.7 MODULATION METHODS

Most digital filters employ some form of pulse-coded modulation (PCM). A. H. Reeves first proposed the PCM format in 1938. Digital PCM began its commercial life in 1955 with Bell's T1 carrier system. The PCM concept has had an impact on the fields of electronics, communications, control, data processing, and signal processing because:

1. PCM allows for the transmission of information over long distances.
2. PCM supports time-division multiplexing.
3. PCM lends itself to digital signal processing.

In a PCM system, a real variable x, $-M \leq x \leq M$, is converted into an n-bit word so that $|x| \leq 2^{n-1}Q$, where Q is the quantization step size. The principal disadvantage of the PCM is its dynamic range inefficiency. Dynamic range inefficiencies result from one or more of the following:

1. To achieve high-amplitude precision, Q must be made small. To cover the range [-M, M], for Q small, n must be large. This increases the cost of the system and reduces its throughput.
2. To achieve high speed, word lengths must normally be small. Small word lengths result in increased errors due to finite-word-length effects.

An alternative encode/decode scheme, often used in communications, is known as <u>differential pulse code modulation</u> (DPCM). It has, in some cases, been applied to the design of digital filters. The design of an n-bit DPCM filter satisfying

$$y(n) = [x(n)]_R - [x(n-1)]_R \qquad (19.7.1)$$

where $[\cdot]_R$ denotes the quantization, is shown in Fig. 19.7.1.

For the special case where $n = 1$, the DPCM policy is known as <u>delta modulation</u> (DM). Intuitively, the DPCM is similar to an analog differentiator modeled as a difference equation:

$$\frac{dy(t)}{dt} \simeq \frac{y(n\,\Delta T) - y((n-1)\,\Delta T)}{\Delta T} \qquad (19.7.2)$$

The potential advantages of this form of modulation is amplitude compression. However, amplitude compression in a DPCM system is frequency dependent. For example, using the differential model and an assumed $x(t) = \cos(\omega_0 t)$, it follows that $y(t) = dx(t)/dt = \omega_0 \sin(\omega_0 t)$. As $_0$ becomes large, the dynamic range requirements of y(t) also become large. In terms of a localized Taylor series, the difference equation (19.7.2) suggests $y(n) = \cos(2\Pi f_0 n\,\Delta T) - \cos[2\Pi f_0(n+1)\,\Delta T]$ $2\Pi f_0\,\Delta T = 2\Pi\,\Delta T/T_0$, where ΔT is the sample interval and $T_0 = 1/2\Pi\omega_0$. Amplitude compression is achieved through the proper choice of sample rate (i.e., ΔT). If ΔT is made small with respect to T_0, the corresponding range of y(n) is compact. The elevated sample rate generally exceeds the Nyquist sample rate and is referred to as <u>oversampling</u>. Even when oversampling is used, the DPCM encoder may occasionally saturate (e.g., noise burst). This phenomenon is called <u>slope overload</u>. Slope overload is suggested symbolically in Fig. 19.7.2. Such errors will sometimes produce intolerable error conditions. As a result, in practice, sample rates that are two to three times the Nyquist rate are often found. Unfortunately, the need to oversample will negate some of the throughput gains obtained by using shorter word length. For example, a 2-byte 16-bit PCM filter may have a maximum clock rate of 1 MHz. An 8-bit DPCM version of the same filter, under the assumption

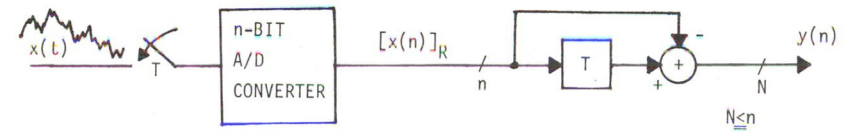

FIG. 19.7.1 Block diagram of a differential pulse-code-modulated system.

FIG. 19.7.2 Graphical interpretation of the slope overhead problem.

that speed is directly related to word length, would have to run at a clock
rate of 2 MHz or less to achieve a net improvement in throughput. The 2-
MHz sample rate may not be high enough, however, to overcome slope
overload.

    Reconstruction of a DPCM signal is accomplished using an inverse
DPCM operation, denoted $(DPCM)^{-1}$. If the DPCM is modeled as a differential process, reconstruction would require discrete integration of the
form

$$x(n) = y(n) + y(n - 1)$$

The hardware realization of a DPCM system is suggested in Fig. 19.7.3.
Using such an architecture, filters of the form $y(n) = x(n) * h(n)$ can be
realized, where $x(n)$ is an n-bit word. Explicitly, the convolution sum can
be expressed as

$$y(n) = \Sigma \ a_i x(n - i)$$

$$y(n) - y(n - 1) = \Sigma \ a_i [x(n - i) - x(n - i - 1)]$$

(19.7.3)

which can be simplified to read

$$\Delta y(n) = \Sigma \ a_i \ \Delta x(n - i)$$

(19.7.4)

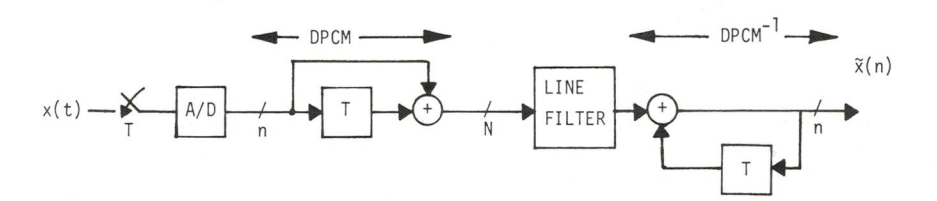

FIG. 19.7.3 Differential pulse code encoder and decoder.

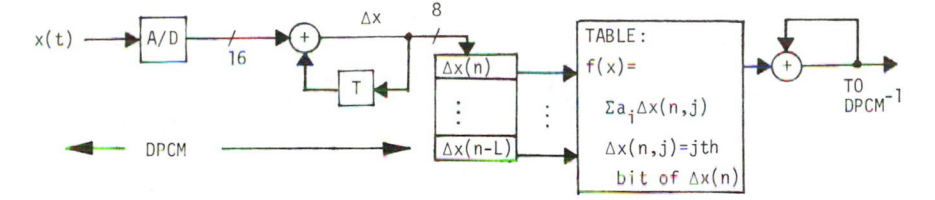

FIG. 19.7.4 Eight-bit differential pulse code encoding filter.

The incremental quantities $x(j)$ and $y(j)$, under the proper conditions, can be amplitude-compressed to b bits ($b \leq n$). For example, if $n = 16$ and $b = 8$, the conventionally architected filter would require 16×16 multipliers versus 8×8 for the DPCM system. The faster, shorter-word-length arithmetic, combined with its lower cost, may more than offset a needed oversample rate of three to five times the Nyquist frequency. For example, an audio system having a Nyquist frequency of 25 kHz, using a 16-bit PCM architecture, may prove to be an inferior design to a filter based on an 8-bit microprocessor being clocked at 100 kHz. Here it is reasonable to assume that low-cost 8-bit 100-kHz data acquisition systems can be interfaced easily to a 1- to 5-MHz microprocessor or a bit-slice processor. An 8-bit DPCM system (interfaced to a distributed arithmetic filter) is suggested in Fig. 19.7.4.

19.8 DELTA MODULATION

A very popular form of DPCM is delta modulation (DM). DM can be traced back to a 1946 French patent. DM systems are generally inexpensive but may suffer from slope overflow errors. As a result, high oversample rates must be used. DM systems also suffer from a phenomenon known as idle noise. Idle noise is caused by asymmetry found in the two current source encoders (i.e., +1 and -1) usually found in DM systems. This asymmetry often results in an audible disturbance. This problem can be controlled by using a single floating ground current source.

19.9 ADAPTIVE DPCM

An adaptive DPCM (ADPCM) policy is sometimes used to counter the slope overload problem that plagues DPCM systems. In an ADPCM system, the quantization step size Q is adjusted in real time so as to reduce the occurrence of slope overload. As a result, the ADPCM system oversample rates are typically lower than those found in DPCM configurations. However, the price paid for these attributes is system complexity and higher cost.

The heart of an ADPCM system is its quantizer. The adaptively alterable quantization levels must be integrated into the filter design. Many architectures have been proposed in this area, with that due to Goldstein and Lui (1976) holding particular interest to the FIR designer.

An FIR filter can be modeled as

$$\Delta y(n) = \Sigma \, a_i x(n - i) \tag{19.9.1}$$

or in differential terms

$$y(n) = y(n) - y(n - 1) = \Sigma \, a_i [x_i(n - i) - x(n - i + 1)]$$
$$= \Sigma \, a_i \, \Delta x_i \tag{19.9.2}$$

Here $x(n)$ and $\Delta x(n)$ are assumed to be n- and n'-bit words, respectively $(n > n')$. To overcome the historic DPCM problem of slope overload and resolution, the quantization level Q is dynamically altered. When the differential quantity $x(n)$ is small, Q should be also small so as to improve amplitude resolution. If $x(n)$ is tending to become large, Q should be increased in order to retard a possible slope overload. An ADPCM encoder capable of achieving these operational attributes is shown in Fig. 19.9.1. In this design, the lower portion of the encoder reconstructs an estimate of $x(n)$, say $\tilde{x}(n)$, from the last estimate of $x(n)$ and correction terms $\Delta \tilde{x}(n)$. That is, the estimate of $x(n)$ is obtained by using $(DPCM)^{-1}$. The adaptively adjusted quantization level, denoted Q_n, can be conveniently altered in a binary-weighted manner. In particular, consider the adaptive rule to be given by

$$Q_{n+1} = \begin{cases} 2Q_n & \text{if } |e(n)| > (2^{n'} - 1)Q_n c \\[2mm] \dfrac{Q_n}{2} & \text{if } |e(n)| \le (2^{n'} - 1)Q_n c \end{cases} \tag{19.9.3}$$

where Q_n is the previous quantization level and c is a threshold parameter satisfying $0 < c < 1$. A typical value of c is $1/2$. Using this scheme, the quantization level Q_n can be expressed as $Q_n = 2^{L_n} Q_0$, $L_n = 0$ or ± 1, and so on, and Q_n is the given initial quantization level. It can also be observed that the quantization level is forced to change by a factor of 2 each clock cycle.

The adaptive quantization levels pose a system control problem. One solution to this problem is abstracted in Fig. 19.9.2. Here the shift control circuitry contains shift registers which remember the value of the scale factor (i.e., $Q_n = 2^{\pm k} Q_0$) associated with each sample $x(j)$. The

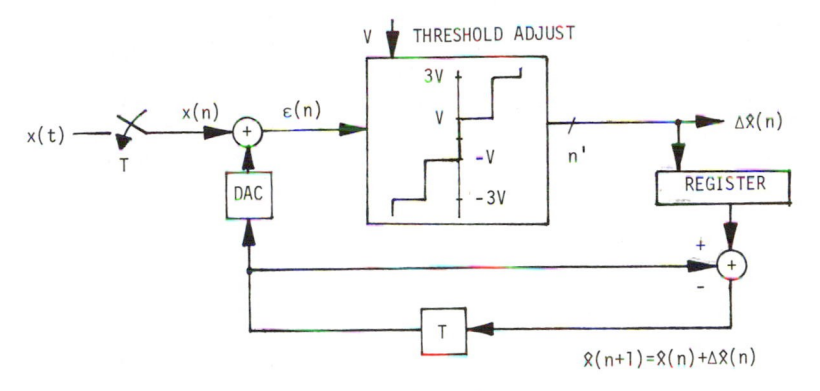

FIG. 19.9.1 ADPCM system and architecture.

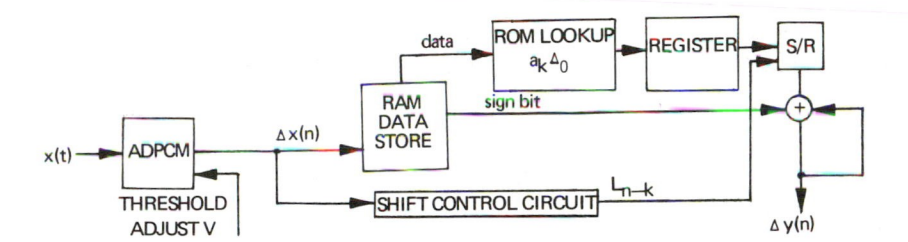

FIG. 19.9.2 Radix-2 ADPCM system and control.

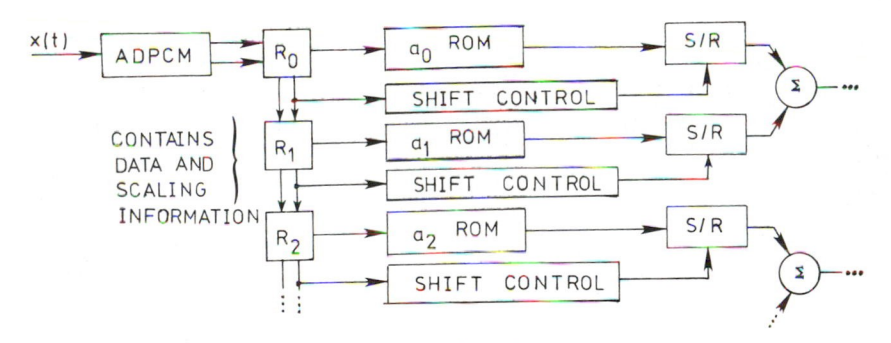

FIG. 19.9.3 Monkewich and Steenart scaling filter.

output of the control circuitry will be presented to a right-left shift register. The shift register data will control the radix-2 weight applied to data sent to the accumulator. Since short word lengths are now permissible (e.g., 10-bit ADPCM versus 16-PCM), low-density, high-speed tables (i.e., ROM and RAM) can be used in the design.

Using a Monkewich and Steenart model (see Section 19.6), the filter displayed in Fig. 19.9.3 can be realized. Here the control circuits serve the same purpose as those discussed previously. The shift control is updated in real time so as to provide current information about the value of L^k.

19.10 OVERSAMPLING

Oversampling is required in DPCM. Often, experimentation is required to determine these elevated sample rates. As a point of reference, Goldstein and Liu performed an experiment on a 32nd-order ADPCM and PCM filter. The results are summarized in Table 19.10.1. These results, based on 1976 technology, use P/f_{max} as a figure of merit. Observe that for a serial architecture, the break-even point is an oversample rate of $49/13.6 = 3.6$ and $26/4.65 = 6.5$ for serial and parallel filters, respectively.

TABLE 19.10.1 Experimentally Derived Data Demonstrating the Effect of Oversampling

Filter realization	f_{max}(word rate)	P, power (W)	IC, count	P/f_{max}
Serial ADPCM	1.25 MHz	17	8 RAM-8 ROM-771 ICs	13.6
Parallel ADPCM	40.00 MHz	185	32ROM-320 ICs	4.65
Serial PCM	450.00 KHz	22	16 RAM-1 ROM-29 ICs	49.0
Parallel PCM	14.00 MHz	360	850 ICs	26.0

BIBLIOGRAPHY

Andersson, E. (1971), A Digital Filter Implemented in Parallel Form, read at 1971 Symposium on Digital Filtering, Imperial College, London, August.

Burrus, C. S. (1977), Digital Filter Structures Described by Distributed Arithmetic Filters, IEEE Trans. Circuits Syst., CAS-24, December, pp. 674-680.

Goldstein, L. H., and B. Liu (1976), An ADPCM Realization of Non-recursive Digital Filters, IEEE Trans. Acoust. Speech Signal Process., ASSP-24, August, pp. 312-320.

Jenkins, W. K., and B. J. Leon (1977), The Use of Residue Number System in the Design of Finite Impulse Response Filters, IEEE Trans. Circuits Syst., CAS-24, April.

Kammeyer, K. D. (1977), Quantization Error of the Distributed Arith-metic, IEEE Trans. Circuits Syst., CAS-24, December, pp. 681-689.

Monkewich, O., and W. Steenart (1976), Stored Product Digital Filtering with Nonlinear Quantization, Proc. 1976 IEEE Int. Symp. Circuits Syst., pp. 157-160.

Peled, A., and B. Liu (1974), A New Hardware Realization of Digital Filters, IEEE Trans. Acoust. Speech Signal Process., ASSP-27, December, pp. 456-462.

Schindler, H. R. (1970), Delta Modulation, IEEE Spectrum, October, pp. 69-78.

Tam, B. S., and G. J. Hawkins (1981), Speed-Optimized Microprocessor Implementation of a Digital Filter, IEEE Proc., 128, No. 3, May, pp. 85-93.

Zeman, J., and H. T. Nagle, Jr. (1980), A High-Speed Microprogram-mable Digital Signal Processor Employing Distributed Arithmetic, IEEE Trans. Comput., C-29, February, pp. 134-144.

Zohar, S. (1973), New Hardware Realization of Nonrecursive Digital Filters, IEEE Trans. Comput., C-22, April, pp. 328-347.

20
Selected Applications

20.1 INTRODUCTION

The application of digital signal processing is too wide and varied to give adequate treatment in a single chapter or even a series of books. Applications to problems in radar, sonar, and geophysics are well represented in the literature. Many other technical fields have benefited directly from the products of the digital signal processing industry. In this chapter a few of these applications are discussed. No attempt is made to provide an in-depth comprehensive treatise on the subjects chosen. Instead, the reader is given an overview of the application problem with the hopes that a greater appreciation of the power and potential of the methods developed in this work will result.

20.2 SPEECH PROCESSING

Speech, and the intelligence found therein, is a field that has enjoyed a long history of active research. With new hardware and technological innovations available today, such as the Texas Instruments Speak and Spell chip set, interest in this field has accelerated. Contrasted to other types of communications, for example a Teletype, the human voice is inefficient, due to the redundancy found in human oral communication. It is therefore desirable to remove such redundancy, if possible, without reducing the intelligibility of the speaker. This would have as its by-product a reduction in the bandwidth requirement associated with the voice communication channel.

Pulse code modulation (PCM) is often used to encode and trasmit speech. The intelligence can be reconstructed into an analog message at the receiver using a variety of interpolation forms. To preserve information in this bandlimited environment, the original signal must be sampled at a rate at least twice the highest-frequency component found in the message. The higher the sample rate above the limiting Nyquist rate, the better will be the reconstructed approximation of the original. However, the higher the data rate, the larger the transmission bandwidth requirements become. For example, 8000 samples per second is generally assumed to be sufficient to produce telephone-quality speech provided that the PCM message has sufficient amplitude resolution, say 12 bits. Under this assumption, a data rate of 96,000 bits per second would result, which in turn would define the channel bandwidth requirement to be at least 48 kHz. However, the original analog signal occupied a bandwidth of only 3.3 kHz. Although there are advantages to digital voice communications, it would appear as though bandwidth reduction is not one of them. As a result, the question of bandwidth compression has received much attention.

Rather than use a linear analog-to-digital converter to produce a PCM message, one may consider using a log analog-to-digital converter, which is also known as a compander (compressor-expander; see Section 17.20). Because the response sensitivity of the human ear is logarithmic, the encoder will emulate the human voice preception process. There are two common types of compander in general use. They are the 255-mu law compander (United States; Bell Systems) and the A-law compander (standard in Europe). The differences are minor.

There are numerous hardware subsystems currently being marketed which support data companding. For example, the Motorola MC 14406 is an 8-bit PCM codec designed to operate at an 8-kHz data rate. The data transmit and receive rates are independently selectzble up to 3.088 MHz. These CMOS 28-pin devices can be pin-selected to perform mu-law or A-law companding.

Using a sign plus 7-bit companding converter, reconstructed speech can exceed that obtainable with a 12-bit linear PCM system. At a sampling rate of 8000 samples per second, a data rate of 64,000 bits per second is established. Therefore, the channel bandwidth has been reduced from 48 kHz to 32 kHz. However, this is still large compared to the 3.3-kHz requirement of the analog channel.

Delta modulation (see Section 19.7) can also be used in a speech processing application. The delta modulation encoder uses a magnitude comparator to output a sting of 0's and 1's which can be used to reconstruct an approximation to the original analog signal. The reconstruction mechanism is a simple Riemann integrator over a sequence of 0's and 1's which correspond to -1 (subtract) and +1 (add). A delta modulator is generally defined to have a fixed weight or step size. However, a delta modulator can be designed which assigns different weights to the bits as a function of the data sequence history. For example, after integrating up for three consecutive intervals [i.e., (. . ., 1, 1, 1)], if the fourth interval defines an upward integration, the designer should do so with an increased step size. This is in effect will change the slope of the integration routine.

A delta modulator that performs this task is called a continuously variable
delta modulator-demodulator (CVSD). The CVSD has the ability to follow
large excursions at the input by adjusting the step size. Experiments performed by R. Thwaits of Cincinnati Electronics indicates that telephone-
quality speech can be achieved with a CVSD operating at a 16,000-bps data
rate. At this rate, a bandwidth of only 8 kHz is required. CVSD chips are
commercially available from Harris and Motorola at a nominal cost ($8 to
$27).

Speech is often defined in terms of formants (the natural frequencies
corresponding to resonance in the vocal tract). The collection of spectra
found in Fig. 20.2.1 suggests the importance of this information. An early
method of analyzing the voice spectrum used vocoders which were a collection of bandpass filters (typically 12 to 50) covering the interval from 200
to 4000 Hz. The study of the outputs of such systems produced qualitative
information about speech. In addition, the filtered output could be used to
determine the location of formants and the pitch period of voiced and un-
voiced signals. Digital signal processing technology has replaced the cum-
bersome analog filters. Individual pitch periods can be determined digitally
and the FFT of each period can be computed with little or no leakage distortion. The envelope of the calculated spectrum is then fitted to a synthetic spectrum using least-squares successive approximation methods. A
pole-zero model for the vocal tract and the glottal source, which produces
an approximation to the synthetic spectrum, can be derived and used to fab-
ricate a speech synthesizer.

Flannagan (1972) used the cepstrum to characterize speech. To compute the cepstrum, the Fourier transform of a time series is computed.
The logarithm is taken of the resulting magnitude spectrum. The inverse
DFT of this hybrid signal is called the cepstrum. In speech applications it
has been noted that for voiced signals, the cepstrum has two peaks. The
lower is attributable to the slowly varying vocal-tract transmission function and the higher is due to vocal-cord excitation. The pitch period of the
glottal excitation can be determined from the location of the higher cepstral
peaks. Experimental results have shown that if the pitch period is quantized to at least 6-bit accuracy, the formant information can be assigned so
that intelligible speech can be synthesized at data rates on the order of 1000
bps (500-Hz bandwidth).

A hardware synthesizer reported by Waser and Peterson (1979) is
based on the MMI 67516 (16 × 16) multiplier (800 nsec), four 2901A bit-
slice microprocessors, and a 2910 sequencer, 2K × 16 main memory,
eleven 1K × 4 PROMs for microprogramming memory, and a few TTL
devices used for bus control. The hardware realized a 10th-order linear
predictive filter of the form

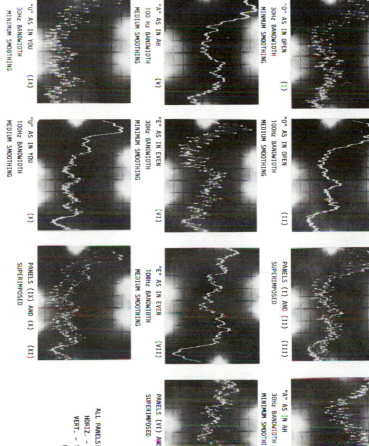

FIG. 20.2.1 Spectral properties of speech.

"O" AS IN OPEN
30Hz BANDWIDTH
MINIMUM SMOOTHING
(I)

"O" AS IN OPEN
100Hz BANDWIDTH
MEDIUM SMOOTHING
(II)

PANELS (I) AND (II)
SUPERIMPOSED
(III)

"A" AS IN AH
30Hz BANDWIDTH
MINIMUM SMOOTHING
(IV)

"A" AS IN AH
100 Hz BANDWIDTH
MEDIUM SMOOTHING
(V)

"E" AS IN EVEN
30Hz BANDWIDTH
MINIMUM SMOOTHING
(VI)

"E" AS IN EVEN
100Hz BANDWIDTH
MEDIUM SMOOTHING
(VII)

PANELS (VI) AND (VII)
SUPERIMPOSED
(VIII)

"U" AS IN YOU
30Hz BANDWIDTH
MINIMUM SMOOTHING
(IX)

"U" AS IN YOU
100Hz BANDWIDTH
MEDIUM SMOOTHING
(X)

PANELS (IX) AND (X)
SUPERIMPOSED
(XI)

ALL PANELS:
HORIZ. - 0 TO 5KHz
VERT. - 10dB PER/DIV.
(LOG UNITS)

$$H(z) = \frac{G(1)}{1 - \sum_{k=1}^{10} a_k z^{-k}} \qquad\qquad (20.2.1)$$

where G is a gain, a_k are filter coefficients, and z is the scalar z operator.
The pitch period, gain, and filter coefficients are derived using an autocorrelation analysis. The 10 coefficients of H(z) are, in fact, computed in
terms of a system of the simultaneous solution to 10 linear equations. The
times required to perform these functions are:

1. Autocorrelation = 13.0 msec
2. Pitch period = 0.1 msec
3. Coefficient a_k = 0.35 msec
4. Gain G = 0.02 msec

which sums to 13.47 msec. Since each second is divided into 44.5 time intervals (frames) of 22.5-msec duration, the 13.47-msec computation delay
will not interfere with real-time data processing. Using a 2400-bps channel
as a design model, each frame holds 2400/44.5 = 54 bits. The 54 bits are
budgeted as follows:

1. Filter coefficient a_1 through a_4 = 5 bits
2. Filter coefficient a_5 through a_8 = 4 bits
3. Filter coefficient a_9 = 3 bits
4. Filter coefficient a_{10} = 2 bits
5. Gain = 5 bits
6. Pitch period and other data = 8 bits

This 54-bit message would be transmitted and used to configure a reconstruction filter at the receiver.

20.3 VLSI APPLICATIONS—THE WIDROW FILTER

In Chapter 14, VLSI technology was introduced, and in Chapter 16 a class
of VLSI multipliers were discussed. One of the first applications of VLSI
in the digital signal processing area was an echo canceler developed at Bell
Laboratories. Echos degrade the intelligibility of speech over long-distance
voice channels. On terrestrial circuits, echo suppressors are used.
These are voice-operated switches which attempt to open the transmission
path whenever applicable. However, if both parties in a two-party net
attempt to voice information simultaneously, poor system performance may
result. For geostationary satellite operations, the echo delay paths are
extremely long (e.g., 540 msec). In an attempt to reduce the effect of echo

distortion, half-hop circuits are often configured. A half-hop circuit is a unidirectional communication path in lieu of the dual-direction (full-hop) path normally associated with terrestrial links.

More sophisticated echo cancelers attempt to create an echo replica and subtract the same from the return signal. They use an FIR digital filter typically possessing up to 250 words of storage plus a multiplier-adder unit. As a result, they can add significantly to the cost per channel if realized using standard IC-MSI-LSI products.

Echo cancelers can be modeled as a linear shift-invariant filter satisfying

$$y(k) = h(k) * x(k) + v(k) \tag{20.3.1}$$

where $y(k)$ is the return signal, $x(k)$ is the far-end speech, and $v(k)$ is the near-end speech plus noise; $\{h(k)\}$ represents the impulse response of the echo path. The echo canceler attempts to estimate optimally a facsimile of the returned message by deriving an echo path impulse response which satisfies

$$\hat{y}(k) = \hat{h}(k) * x(k) = \sum_{n=0}^{N-1} \hat{h}_n(k)x(k - n) \tag{20.3.2}$$

If the echo-path impulse response decays to zero after N samples, and if

$$\hat{h}_n(k) = h(n); \quad n = 0, 1, \ldots, N - 1 \quad \text{(stationary)} \tag{20.3.3}$$

then perfect echo cancellation (subtraction) can result. The problem formulated has been successfully stated and studied by Widrow using a class of filters now referred to as the <u>Widrow filter</u>. The echo path impulse response estimation algorithm used in the Bell Laboratory system satisfied

$$\hat{h}_n(k + 1) = \hat{h}_n(k) = Kx(k - n)e(k)$$

$$e(k) = y(k) - \hat{y}(k) \tag{20.3.4}$$

$$K = \frac{K'}{\hat{\sigma}^2_{(k)}}$$

where K' is the normalized loop gain and $\sigma^2(k)$ is an estimate of the variance (power) of $\{x(k)\}$. The basic system has been integrated onto a 24-pin VLSI chip in NMOS. The canceler is configured to be a 128-tap (16-msec delay) FIR and has a tested white-noise convergence rate of 70 dB/sec.

Rockwell International used a Widrow filter to remove unwanted noise and interference from voice records. At $25,000 per copy, the real-time adaptive speech enhancement system (developed for the FBI) uses an adaptive predictive deconvolver and Widrow filter to remove echos and reverberations from speech. The filter was based on a 150-th order FIR design. When adjusted for slow adaptation rates, the deconvolving filter has the capability of removing echos and other long-delay signal processes. Adjusting for faster adaptation speeds, the system will track and subtract background music or secondary speaker interference. Adaptation times can range from 200 msec to 5 sec and are user selectable.

In the classical Widrow filter model, two separate signals (called primary and secondary) are presented to the filter. The primary channel contains both the principal speaker and background information. The secondary channel monitors only background disturbances. The system FIR uses a cross-correlation metric to match the amplitude and phase of the background records found concurrently in both channels. The matched signals are then subtracted so that the background component in the primary channel is minimized (see Fig. 20.3.1).

Widrow filters can be synthesized in software as well. Using standard A/D conversion techniques and a 16-bit computer, a Widrow filter can be designed as suggested in Widrow (1966). In Fig. 20.3.2 the Widrow filter is shown to remove adaptively 60-Hz noise from a multitone input signal process. The removal of 60-Hz noise from a broadband input process is displayed in Fig. 20.3.3. Finally, the removal of a broadband subrange of frequencies from a broadband signal process is summarized in Fig. 20.3.4.

20.4 ROTATING MACHINES

Vibration can cause accelerated machine wear and possible failure. Therefore, it is desirable to have advanced warning of this condition so that corrective measures can be taken. This is especially true in the power and

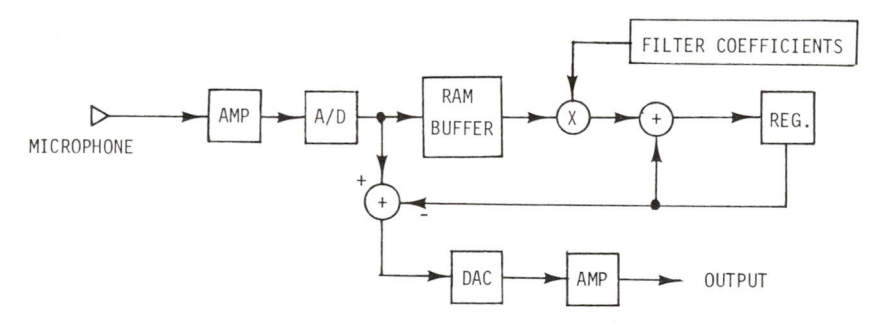

FIG. 20.3.1 Block diagram of an adaptive discrete filter.

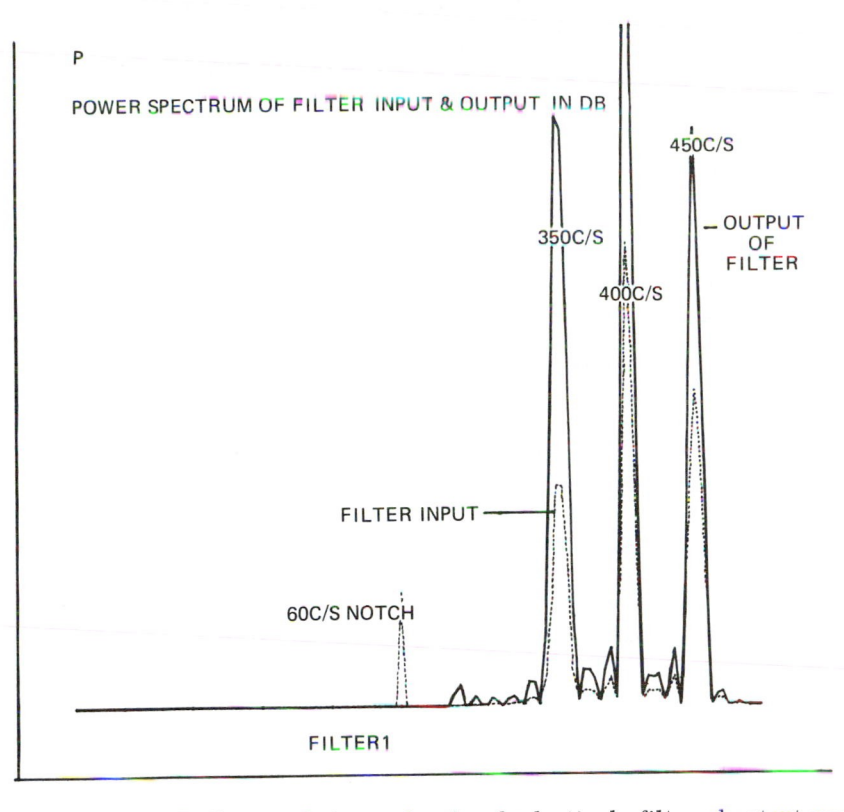

FIG. 20.3.2 Difference between input and adaptively filtered output power spectrum for a 60-Hz corrupted signal.

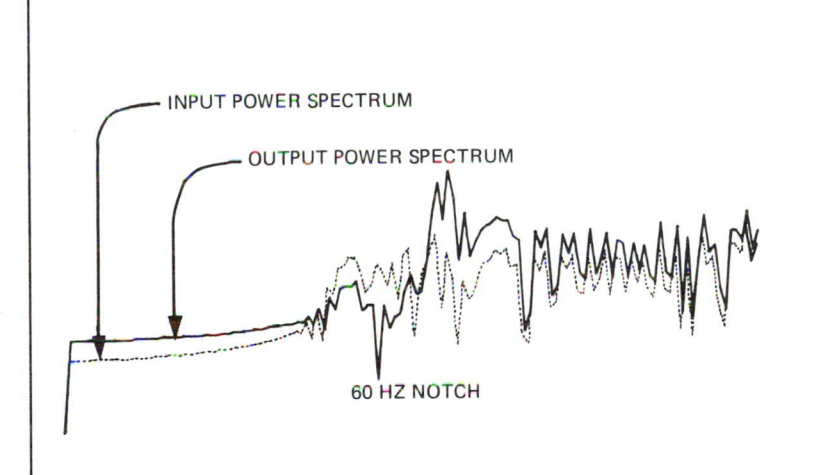

FIG. 20.3.3 Difference between input and adaptively filtered output power spectrum for a 60-Hz corrupted signal.

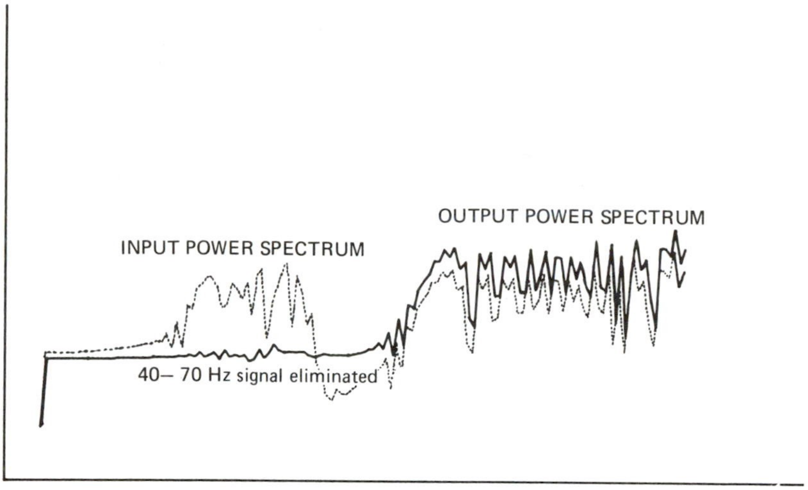

FIG. 20.3.4 Difference between input and adaptively filtered output power spectrum for a wideband additive noise condition.

chemical industries, where machine failure can range from costly to disastrous. Recently, digital signal processing has been applied to the problem of rotating machine vibration analysis.

Machine vibration is defined in terms of:

1. Relative motion: displacement between rotating and stationary components
2. Absolute vibration: displacement, velocity, or acceleration of stationary parts

Relative motion should be monitored whenever the closure of a designed gap would be catastrophic (e.g., loss of seals around the hydraulic turbine runners). When larger unbalancing forces are brought into play (e.g., bearing failure), absolute vibration measurements are required. Obviously, there is some overlap in these guidelines. Generally, it is assumed that relative displacement is useful as a monitoring policy if the current state of the machine is of principal interest. The absolute policy is applicable to the study of long-term system behavior.

A root-mean-square (rms) vibration velocity metric is often used to study the rotating dynamics of a machine. Table 20.4.1 summarizes the tolerable rms velocity as a function of machine size. These data, which should be interpreted only as a guideline, represent acceptable limits on the spatial frequencies (velocity) of a machine. Therefore, the data can be interpreted in the frequency domain using the FFT. The FFT can also be

used to perform a referenced comparison of a machine against itself
in the manner suggested in Fig. 20.4.1. In this case, changes in the
spectral signature of a machine will indicate wear or damage.

At times, a wideband spectra is needed to analyze potential de-
fects in roller bearings or gears (e.g., 20 Hz to 20 kHz). In Fig.
20.4.2 the wideband spectra of a gearbox is displayed. The labels
on the narrowband spectral components are used to indicate their harmonic relationship to a fundamental rotational speed (i.e., shaft speed).
If a rotating (or ball) bearing defect is present, energy is released to
spectral lines which are not harmonically related to the fundamental
speed of rotation (see also Fig. 20.4.2).

20.5 BIOMEDICAL APPLICATIONS

The analysis of medical signals and data is a rich and fertile area of research. Digital signal processing methods are being applied to the problem

TABLE 20.4.1 Properties of the Vibrational Effects of Rotating Machines

Source: After Thrane, 1979; reprinted with the permission of B & K
Instruments.

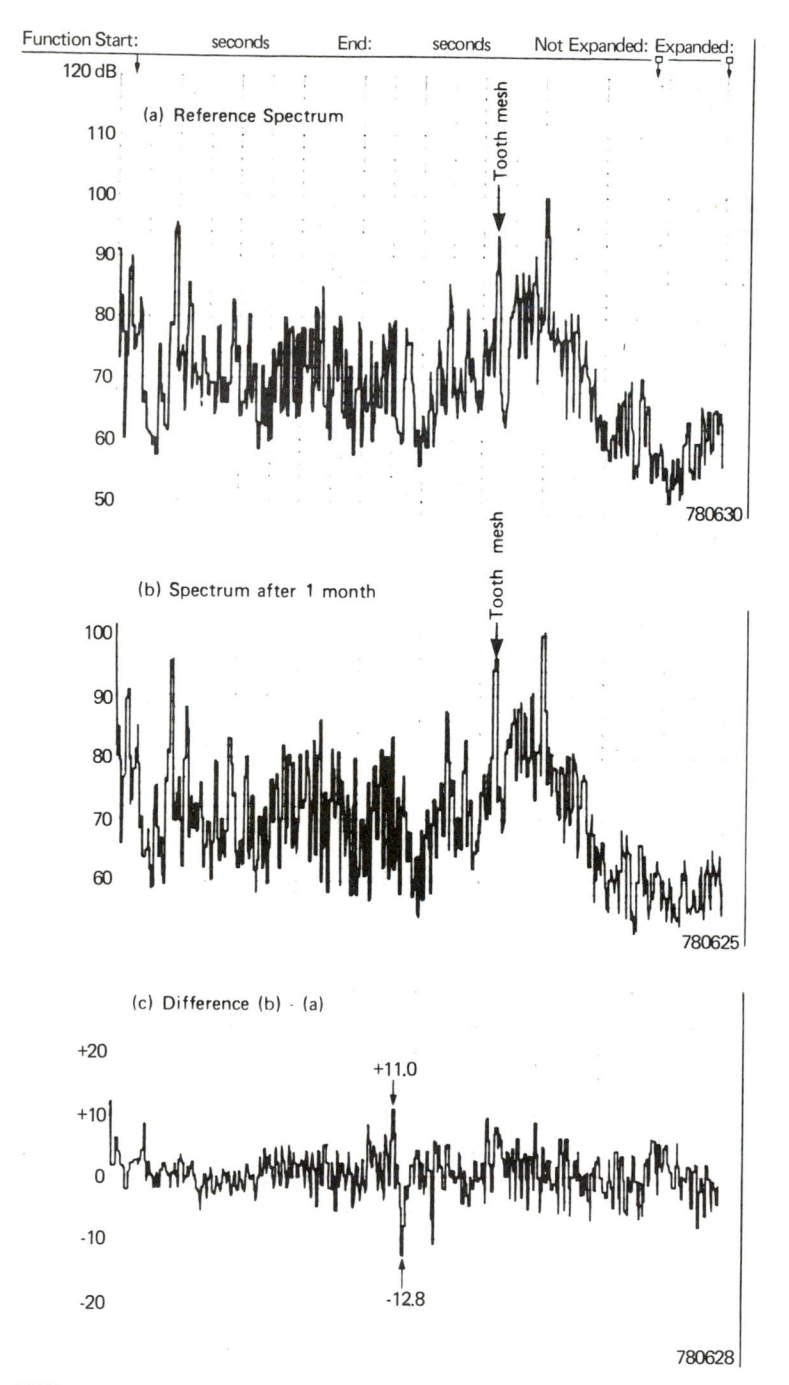

FIG. 20.4.1 Determination of wear of a rotating machine using spectral analysis. (After Thrane, 1979; reprinted with the permission of B & K Instruments.)

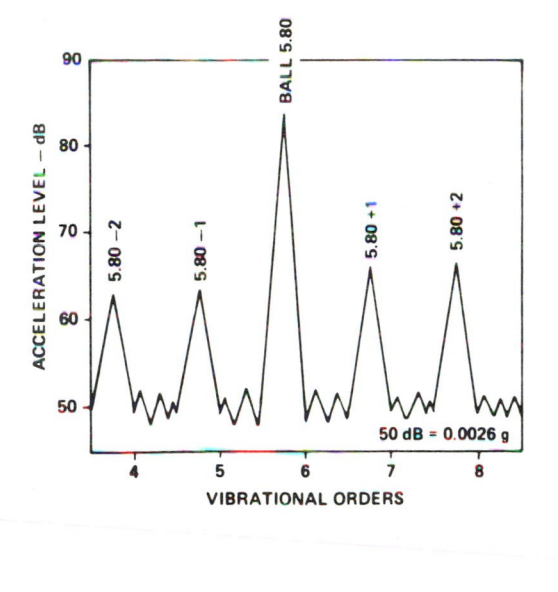

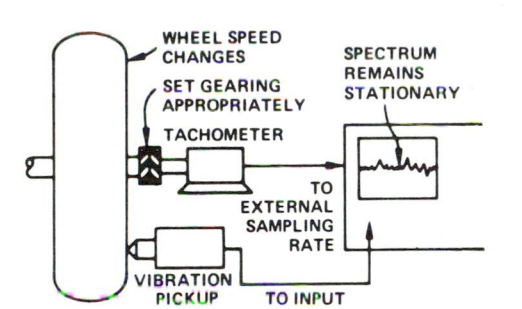

FIG. 20.4.2 Analysis of rolling bearings using spectral signature analysis.
(After Thrane, 1979; reprinted with the permission of B & K Instruments.)

of computer-aided analysis of electrocardiograms (ECG), electroencephalo-
grams (EEG), and so forth. One of the most sophisticated applications of
digital signal processing in medicine is the celebrated computerized axial
tomography (CAT) body scanner.

In many medical applications, the FFT and its derivatives are used.
One such application is the noninvasive detection of aneurysms using bio-
acoustic spectral signature analysis. C. G. Drake has written: "The
natural life history of a once ruptured aneurysm is nothing short of
bleak. ... If five patients with primary subarachnoid hemorrhage ... are
left untreated, at the end of a year only one will be alive, another will be

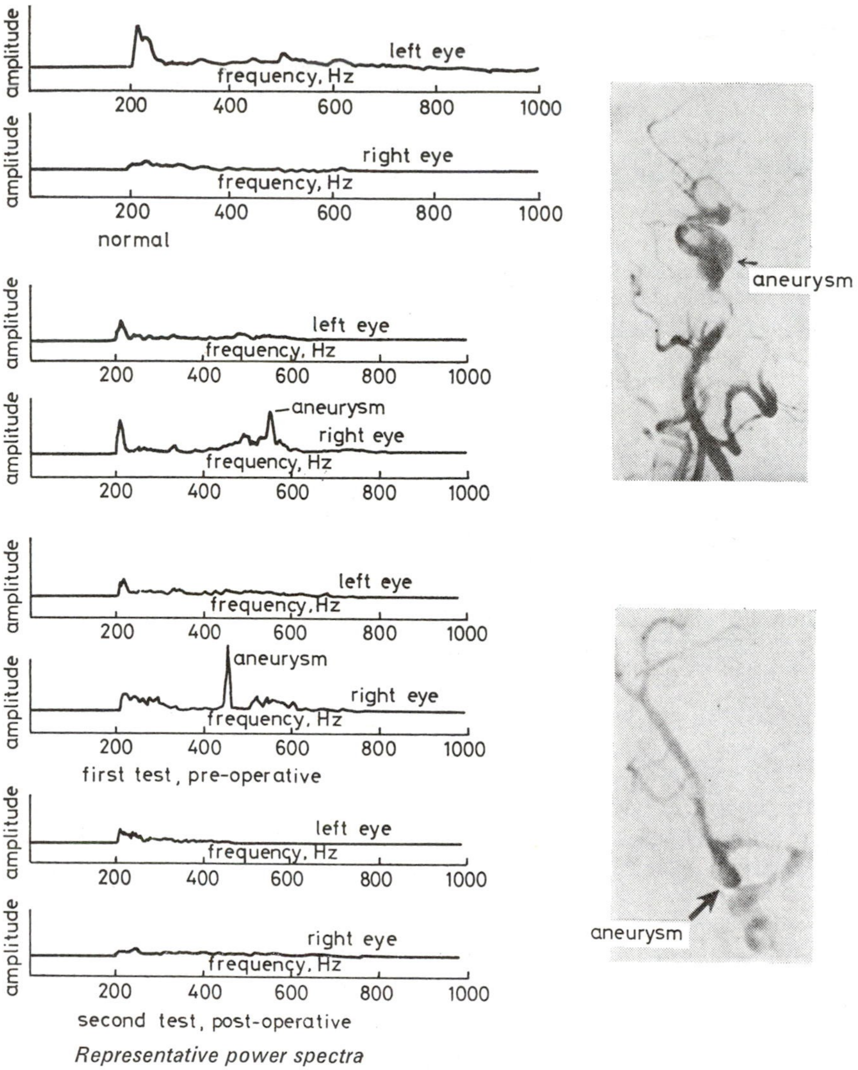

FIG. 20.5.1 Spectral properties of aneurysms.

disabled, and three will be dead." Reported statistics indicate that this class of hemorrhage is the cause of 2% of all sudden deaths, is 68% mortal, and leaves 21% mentally deficient. Surgical survival after hemorrhages is rated at 77 to 90% if detected beforehand. Statistics indicate that a patient's chances for a productive life are good if the aneurysm can be detected and treated surgically.

One possible method of detecting the presence of an aneurysm is through the acoustic signal produced by an unruptured aneurysm. Acoustic microphones have been developed which monitor the acoustic arterial sounds emitted from the left and right eyes. Heartbeat information is also recorded for synchronization purposes. The data can be processed in an FFT sense using a commercial digital spectral analyzer. Data are entered to the system on receipt of a heartbeat trigger. Normal and abnormal power spectra are presented in Fig. 20.5.1. It can be readily seen that there exist marked differences in these cases. The narrowband response found in the diseased-patient spectra are produced by the geometry on an aneurysm (see Fig. 20.5.2). It has been suggested that the aneurysm may be modeled as a flexible Helmholtz mechanical resonator. Narrowband tonal activity has been found generally to characterize the presence of an aneurysm.

The spectrum under study is similar to many found in the biosphere. Biological systems are inherently lowpass filters. As a result, the aneurysm information, generally residing between 400 and 800 Hz, can be overwhelmed by the strong low-frequency signals produced by the heart fundamental and its harmonics. To suppress the low-frequency components

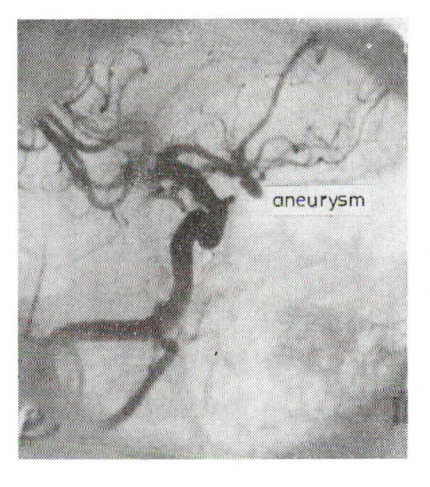

FIG. 20.5.2 Arteriogram of a human aneurysm.

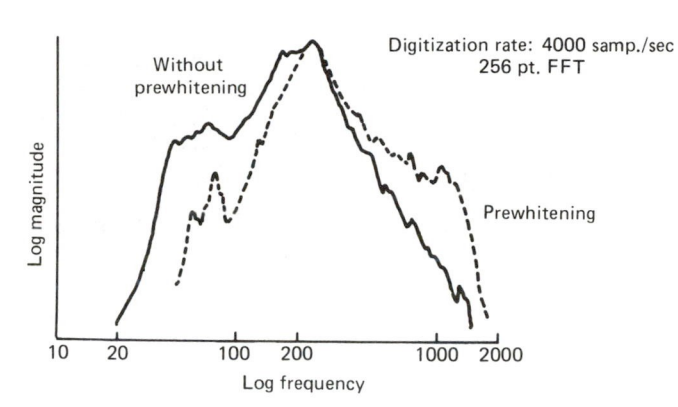

FIG. 20.5.3 Spectral analysis of an aneurysm using a chirp-z transform.
(Data after T. S. Klitner and C. F. Dewey, NIH Grant P17-HL-14209-61.)

and accentuate the high end of the spectrum, a technique called "prewhitening" can be used. Essentially, a prewhitening filter is a highpass filter
which has a magnitude response given by $|H(j\omega)| = j\omega$ (i.e., a differentiator).
For example, the spectrum of a partially occluded (closed) common carotid
artery bruit is shown in Fig. 20.5.3. It can be seen that the high end of
the spectrum possesses more resolution in the prewhitened case. Unfortunately, building such a highpass filter in analog hardware, over the frequency range needed, can be a challenging problem. The problem is less
formidable in digital hardware. In fact, using differential pulse code modulation (DPCM), a digital prewhitening filter can be realized by simply
ignoring the inverse DPCM interpolator (i.e., integrator) normally found
at the output stage. Compared to a pure differentiator having a magnitude-
squared response given by $|H(j\omega)|^2 = j\omega^2$, the prewhitening DPCM system
exhibits the response displayed in Fig. 20.5.4.

An FFT was configured using a Z-80 microprocessor system including
a floppy disk. The data acquisition system was an 8-bit A/D running at 4.5
kHz. Twelve 1024-point real sample blocks were acquired (approximately
3.75 sec of biodata) and DMAed to memory. Data were then read out of
memory and presented to a software-based FFT routine for off-line processing.

The biodata were also processed using a Reticon CZT chip and evaluation system. This device does not have the precision of a digital FFT but
offers the convenience of a small, compact hardware system and an analog-
like output. The analog output made the data interpretation problem a simple
process. However, since the CZT system contains a considerable amount
of analog hardware, it also requires constant tuning and adjustment. Results from the CZT experiments are reported in Fig. 20.5.5.

The biomedical spectra and time series derived are often averaged
to enhance their interpretability. Simple linear ensemble and temporal

averaging methods can be used. However, when more sophisticated methods are required, the familiar method of <u>least squares</u> is often employed.
One of the disadvantages of the least-squares method is that the current
estimate of a parameter is a function of the entire time history of the observed process. Unfortunately, data consisting of 10, 100, or 1000 or so
samples in the past may no longer be relevant to the phenomenon under
study. If the system or noise statistics are changing as a function of time,
it is often more advisable to use only a finite glimpse, or "snapshot" of the
past. Sometimes called a <u>sliding window</u> or <u>finite window</u> policy, this technique captures only the M most current sample in a FIFO memory unit.
For example, the blood pressure within an artery will vary (i.e., sytole
versus distole). Therefore, the amplitude of a given spectral component
will also vary. A least-squares estimate of the strength of the power spectrum, at a harmonic, can be estimated, in a sliding window sense, along
the lines suggested by the following example.

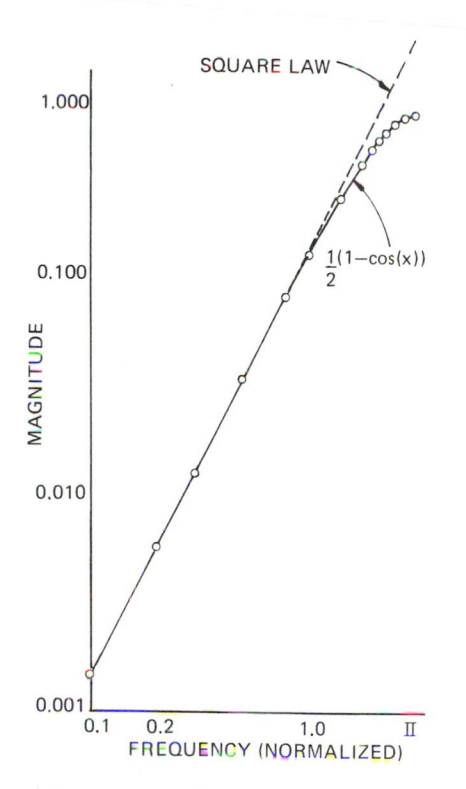

FIG. 20.5.4 Effects of prewhitening a power spectrum.

TIME DOMAIN RESPONSE   CZT DOMAIN

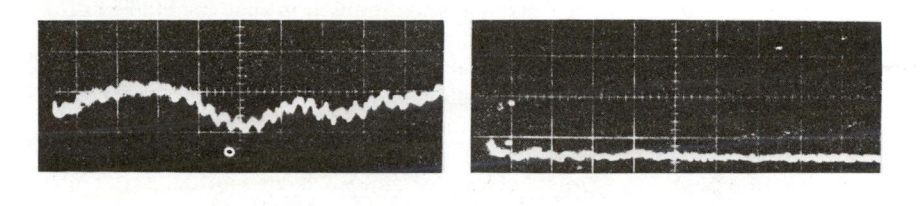

PRE-OPERATIVE CASE: LEFT EYE SOURCE

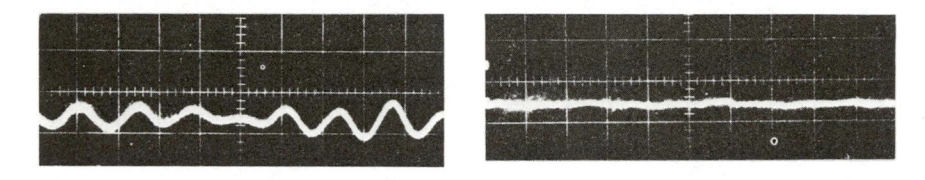

PRE-OPERATIVE CASE: RIGHT EYE SOURCE

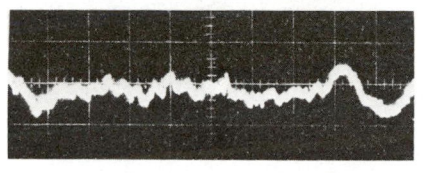 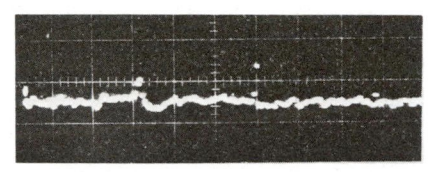

POST OPERATIVE CASE: RIGHT EYE SOURCE

FIG. 20.5.5 Spectral properties of a delta modulator.

EXAMPLE 20.5.1 Suppose that the magnitude-squared value of a particular harmonic is modeled to be given by

$$x(k + 1) = \phi(k)x(k); \quad x(0) = 1.0$$
$$y(k) = 2x(k) + v(k)$$

(20.5.1)

where $\phi(k) = 6/5$ if $k = 50$, $7/6$ if $k = 75$, and 1 otherwise. In the absence of any noise corruption, the value of $x(k)$ is 1.0 if $k \leq 50$, 1/2 if $51 \leq k \leq 75$, and 1.4 if $76 \leq k \leq 100$. If it is assumed that the additive noise process $v(k)$ is normally distributed with mean zero and variance σ^2 [i.e., $N(0, \sigma^2)$], the optimal estimate of $x(k)$, say $\hat{x}(k)$, satisfies

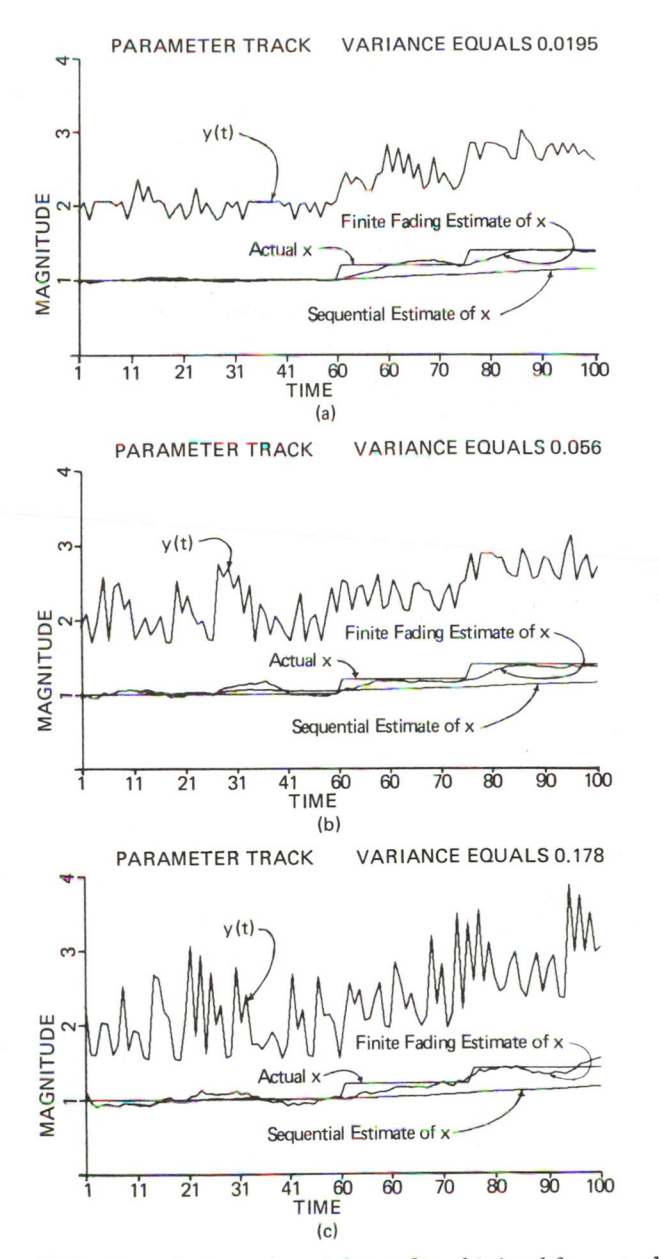

FIG. 20.5.6 Experimental results obtained from a sliding window filter.

$$\hat{x}(k + 1) = \hat{x}(k) + \frac{1}{k + 1} * \frac{y(k + 1) - 2\hat{x}(k)}{2}$$

with an error covariance metric p, at the jth sample, given by

$$p(j + 1) = \frac{1}{4 + (j + 1)}$$

Using simulation, the classic sequential estimate of x, for three choices of variances, are diagrammed in Fig. 20.5.6. If the least-squares estimate is based on only the 10 most current samples (i.e., sliding window of length 10), the results reported in Fig. 20.5.6 result. It can be seen that the sliding window filter possesses a higher degree of adaptiveness in this application. In general, the algorithm designer will be making trade-offs between statistical convergence (precision) and window length.

BIBLIOGRAPHY

Duttweiler, D. L., and Y. S. Chen (1980), A Single Chip VLSI Echo Canceler, Bell Syst. Tech. J., 59, No. 2, February, pp. 149-160.

Drake, C. G. (1971), Intracranial aneurysms, Proc. Roy. Soc. Med., 64, pp. 477-481.

Electronic Design Staff (1977), Real Time Digital Audio Processor Picks Out Distorted Conversations, Electron. Des., No. 17, August 16, pp. 34-36.

Flanagan, J. L. (1972), Speech Analysis, Synthesis, and Perception, Springer-Verlag, New York.

Flanagan, J. L., C. H. Coker, L. R. Rabiner, R. W. Schafer, and N. Umesa (1970), Synthetic Voices for Computers, IEEE Spectrum, October, pp. 22-45.

Taylor, F. J., and V. P. Shenoy (1979), On the Fourier Transform of Sequency Limited Signals with Applications to Delta Modulation and Medicine, Int. J. Syst. Sci., 10, No. 9, pp. 945-960.

Taylor, F. J., et al. (1979), Aneurysm Detection Using One Bit Correlation, Med. Bio. Eng. Comput., No. 17, July, pp. 443-488.

Taylor, F. J., et al. (1983), Non-Invasive Aneurysm Detection Using Signal Processing, J. Biomed. Eng., 5, July, pp. 201-210.

Thrane, N. (1979), An Introduction to Frequency Analysis, B & K Instruments monograph, Cleveland, Ohio, February

Waser, S., and A. Peterson (1979), Medium-Speed Multipliers Trim Cost, Shrink Bandwidth in Speech Transmission, Electron. Des., February 1, pp. 58–65.

Widrow, B. (1966), Adaptive Filters, Stanford Electron. Lab. Rep. 6264-6, December.

Index

DE